THEY ALL TOLD THE TRUTH:
THE ANTIGRAVITY PAPERS

I have, being a physicist, worked out, over a period of several years, a new theory of matter and energy (which I refer to as "The Antigravity Papers"). It does not conflict with either Relativity Theory or Quantum Theory. IN FACT, IT DERIVES AND PREDICTS THESE TWO THEORIES FROM SIMPLER PRINCIPLES. Also, I've been able to use my new theory to calculate what the mass should be for all the elementary particles of physics, within reasonable error. This is a feat which has NEVER been done before.

After I realized just what it was that I had, and its full implications, I decided to keep it secret. For several important reasons now, I no longer wish to keep it a secret. Rather than go for secrecy and monetary gain, I would rather help people. Especially the patriotic citizens of the US (and Allies) and those who believe in the Constitution of the US, and those who believe in the one, Creator GOD. And also, this book is for people who want to help save lives. That is my audience.

1. Based upon the new physics theory, I've been able to identify several people who must be telling the truth about what they are (or were) doing with antigravity. Years ago, I thought these people were quacks who didn't have the foggiest notion of what they were doing. My physics training told me that their devices would not work. Now, I'm having to change, and I've discovered that their devices very well should work, and probably DID, just as they have claimed.

2. After I looked at the drawings that Mr. David Hamel made, from memory, after he allegedly saw the these diagrams presented to him by aliens, I now know that **this is undeniable proof that these aliens exist.** There's no possible way that David, simply a carpenter, could have come up with these correct physics drawings (the new physics, not the existing, standard physics), unless he either got them from the government, or from aliens. And it wasn't the government!!

3. Release of all this information, at once, will end the years of massive government suppression of these technologies once and for all.

4. Anyone who has carpentry skills can build a working antigravity device. It's so much simpler than anyone had ever imagined! It runs on a form of acoustically induced nuclear energy release! It works like a very, very slow-release atomic bomb! And, it is not dangerous, as there's no possibility of an actual nuclear explosion or of radioactive release.

So what's this about saving lives? I've created a theory for explaining and designing devices which would be extremely useful to whoever survives the upcoming string of serious calamities of the Earth. Yes, we're facing more than one disaster, and in fact, there may be (actually, *will be*) 4 or more BIG disasters, worldwide, in the near-future. Much of the electrical power generation infrastructure (also, natural gas lines and etc…) will be knocked out and it probably won't be possible to re-build these for a *very* long time. This book explains how to build and test devices which are the basic-beginnings-of powerful, portable electrical generation devices as well as several other devices which will prove extremely useful for those who are left behind to survive, after each calamity.

Order this book online at www.trafford.com
or email orders@trafford.com

Most Trafford titles are also available at major online book retailers.

Print information available on the last page.

ISBN: 978-1-5539-5723-2 (sc)

Trafford rev. 03/29/2018

Trafford PUBLISHING® www.trafford.com
North America & international
toll-free: 1 888 232 4444 (USA & Canada)
fax: 812 355 4082

DEDICATION:

March, 2003

This book is hereby dedicated to the following people:

God (The Creator of ALL) &
Jesus, &
My Guardian Angels (777),

&

Doris C. (Mom) & Richard C. C. (Dad)
Brian C.
Paul Hicks, Estelle & Dennis, Cathi,

&...

Mike

Mike & Emily

Mike

Pooplicant

Nicodemus Skolnick

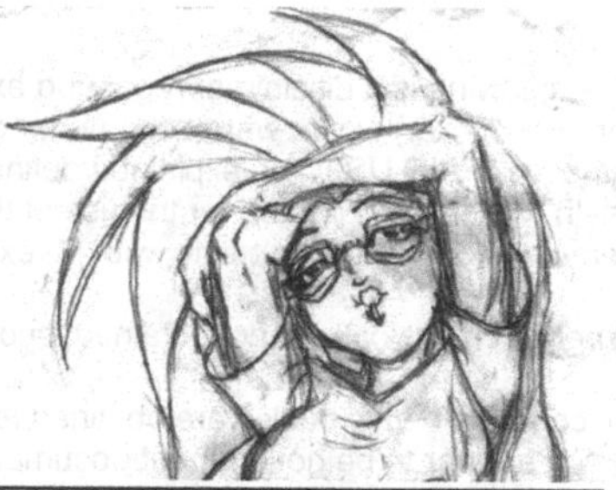

Nicodemus' Girlfriend

&
Adam Snider, Nikita Perry, Amberlee, Rich Spiekhout

Alex K., Art Bell, George Noory,
Louise Daniels (Olsen Park), Rick Devoe (CJH), Mr. Jones (THS), Mr. Clark (THS),

Kurt Leniger, Gary Brackett, Ron Stawicki, Jim Burdick,
Donna P. Hardendorf, Vic & L.White, Archer Elliott.

March, 2003

DISCLAIMERS:

The following is a disclaimer regarding excerpted works and similar-to-the-original works, including images. All excerpted (or very similar) works were used under existing copyright laws, and each excerpted work was able to be used due to either 1) "FAIR USE", as explained/defined in copyright law, OR 2) PERMISSION (either actual, or implied by the work itself), which was granted for the use of the work. Either method may be augmented by (but not replaced by) the fact that some works existed, in public, with no explicit "Copyright xxxx" notice.

Where permission was not obtained, such works may be shown herein, but are either blacked-out, masked, or-the-like.

All contents of this book were obtained legally in all respects, with regards to all state, local, and federal laws. Documents which appear to be government documents were copied from their widely-available public sources, usually the Internet.

Nothing written about or shown in this book constitutes a violation of any state, local, or federal laws. The ultimate purpose & goal for this book is to save lives: as many as possible. No ill or harmful uses are intended in any way, from the information in this book. All uses are expected to be positive and uplifting, and industrially profitable as well.

The subject matter herein includes information for building powerful devices. All readers of this book assume THEIR OWN RESPONSIBILITY & LIABILITY for ANY USES OF THE MATERIAL HEREIN, WHATSOEVER. The author and the publisher of this book are NOT RESPONSIBLE in ANY WAY WHATSOEVER for any losses, deaths, or accidents or injuries, or legal problems ("hot water") due to the use of the material in this book. The reader & user takes FULL responsibility for their actions, upon having read any part of this book.

And, remember the Golden Rule: Do unto others as <u>you would have them do</u> unto you.

THEY ALL TOLD THE TRUTH:
THE ANTIGRAVITY PAPERS

PART A: The "Big 3" & Their Forensic Analysis

PART B: The Antigravity Papers

PART C: Early Pioneers

PART D: Government & Suppression

PART G: Conclusions & Epilog

PART H: Appendices (H1 -- H4):

APPENDIX H1: The L & N.S.S. Pages For The UFT

APPENDIX H2: Operational Princ. (For The IDD)

APPENDIX H3: The I.D.D., & Explan. Supp.

APPENDIX H4: The Old A, G, & H Papers

Part A

The "Big 3" & Their Forensic Analysis

[ ]

This section of the book presents, in fact, an incredibly easy-to-understand forensic scientific analysis of the testimony of three very significant people. Since their testimonies are critical for understanding all of the different (yet similar) ways to create and use antigravity, I call them the "Big 3". Once the reader understands what these men have done, and how they did it, then he will be well-equipped to be able to analyze any and all claims of antigravity by any other people, using any of the many possible methods of creating antigravity.

An advanced model of the physics of subatomic particles is presented later in the book (in Part B). Amazingly, all of the advanced mathematics and physics "boils down" to just ONE, VERY SIMPLE PRINCIPLE:

Any device which generates or produces antigravity MUST contain, somewhere in the device, a CONE-SHAPED or a TRIANGLE-SHAPED part, which MUST be made typically of either copper or ALUMINUM.

Yes, that's all there is to it! Simple! If you have that, then you can produce antigravity. So, in this section of the book, I was able to verify & authenticate the testimony of the three people, purely by looking at their claims IN LIGHT OF the new model of physics. **You don't need any mathematics or formulas to understand the principles!**

Also, in the Prolog, the **massive government cover-up** of antigravity technology is discussed.

And, imminent danger from an orbiting asteroid is covered later in the book.

<u>Prolog & Introduction:</u>

Welcome to this Book!

This book is for EVERYONE: Physicists & Scientists, Electrical Engineers, **<u>but mainly for ANYONE</u>**, regardless of his/her background & training, who simply wishes to find out the shocking truth regarding antigravity propulsion systems and the government cover-up which has spent billions of dollars every year, for many years, to keep the knowledge and capabilities "Above Top Secret".

Here in this revolutionary book, this cover-up is now proven beyond doubt, with evidence good enough for a courtroom, through both pure scientific principles and also through some additional detective work.

This book will appeal to a wide audience, as it has something important in it for people who

- like a good mystery or "detective story", such as in a "whodunit" book,
- want to know **what our government is really up to,**
- want to know **how to build a working antigravity device,**
- want to see a simplified, easy to understand (for anyone) description of the new principles of generating and using "antigravity" levitation,
- are interested in the <u>possible</u> existence of beings not of this world,
- want to know if any ancient civilizations had antigravity devices,
- want to see the new physics model of matter and energy which I have independently developed over the last several years. Using this unique model I've developed, I was able to verify testimony of various persons, purely by looking at their claims in light of this new mathematical model of physics.
- want to learn about "42", the secret to life, the universe, and everything.

>>>Basically, this book vindicates and proves truth of, beyond doubt, the testimonials of 3 people in particular. These three are the "smoking gun" and the "motherlode" of information regarding secret antigravity systems built both by the government and by <u>individuals</u>. The stories of these three, once understood, can then be used by anyone as <u>a measure</u> to verify the claims of anyone else who has claimed involvement with antigravity.<<<

Physicists, especially, TAKE NOTE:
(all others may skip these 2 short paragraphs that follow) :

The best possible mathematical model of physical matter (and energy) in the universe that we have, which is accepted and verified, is called the "Standard Model of QCD" (Quantum ChromoDynamics). Yes, there are theories out there that try to go beyond this, such as "string theory", but these are on very shaky ground, at best, and are not yet "officially" accepted. So, we really know little or nothing about the <u>internal structure</u>, if any, of **quarks**, which are the particles which comprise the nucleons (i.e. protons and neutrons) in all atoms of matter. Until now. **My newly formulated model of protons and neutrons actually takes us 2 levels of detail below that of the quarks, and it shows the internal dynamic substructure of the proton. This is 2 levels of detail below any accepted current models**

that we have. You can see that the proton is actually made up of a total of 168 vacuons ("vacuon" is a term I created to name the "protoparticles" I've discovered). I have discovered that the vacuon, NOT the quark, is the smallest possible constituent of matter (i.e. of particles). In fact, I've found that everything in the universe, including photons as well as material particles, are composed of different numbers of vacuons, where there is only one known type of vacuon. Vacuons are all identical, and thus all have the same size and mass. **So everything truly is made up of just one kind of little particle, and all of these little particles are identical.** How much simpler could it get!! [Actually, to be completely correct, there are two types of vacuons, but they are both the same particle. The difference is that one of these types comes in clumps of 7 vacuons, and the other is just a single free vacuon; they are the same particle, but the masses are very different, because one is bound in a clump with others, and the other is free. The difference in mass is called "binding energy".] It has often been said that the simpler and more elegant that a theory of physics is, the more chance it has of being correct, because nature, in principle, should be very simple at the lowest level. For years, physics got harder and more complicated. Now, it just got simpler!

So for you readers who are physicists & scientists, I was able to show that **a quark is actually a ring of vacuons,** where I simply balanced the centripetal forces with the centrifugal forces to find the equilibrium radius of a quark. These two forces are "gravitomagnetic" in nature (explained in detail, in my theory). Then, the relativistic **Dirac Quantum Theory of the Electron** pops right out of the equations, as a natural result of the vacuon model. Also, the mass of the quarks can be computed, and you will see later (Part B) that with this new model, **I have been able to predict/calculate the masses of all of the known particles,** within only a few percent error. Surely you can recognize the significance of this achievement, as such feat has never been done before. Similarly, and also never done before, I have shown **a complete and simple explanation of fractional electric charge (+/-1/3).** So this new theory demands a great deal of study and experimentation, obviously, as it is very worthy of such expenditure of time and materials. The physical "pieces of the puzzle" of matter and energy now make much, much more sense than they ever did previously with all other quantum theories.

Here's a direct quote from later on in the book:
So, all the skeptics can be totally satisfied, as we see that Energy, Momentum, and Angular momentum are ALL conserved, just as conventional physics has always required! And, the enormous energy needed to power a device actually does come from an identifiable "somewhere", and there is an infinite amount of free energy available from that "somewhere", and it can be effortlessly "pumped out" via the use of the energy cycle.

Now, back to non-physicists (i.e. everyone else!):

Now, I present some more introductory information regarding the subjects of this book.

My background, as the author, is one of physics and electrical engineering. I graduated from Texas Tech University in 1984, with a degree in electrical engineering. My gradepoint average there was 3.6. Over the many years since, I have acquired tremendous knowledge in other areas, and now I have the equivalent of a master's degree in each of these areas (physics and E.E.). Since 1984, I have, for reasons I don't really know, been extremely interested in antigravity or electromagnetic propulsion systems. A strange interest indeed.

As far as work experience goes, I have worked for two different government contractors, at different nuclear plant sites. During my employment there, I had to maintain a "Q" clearance, which is a security clearance allowing access to both CONFIDENTIAL and SECRET (but not Top Secret) nuclear or national security information. My experience in those places taught me many things regarding how classified (i.e. government secret) information and documents must be handled. The experience has been helpful (although not tremendously so) in understanding how antigravity devices could be kept so secret (in fact, they are several levels above Top Secret-- so we never had anything like this at the two sites I worked at). But, the handling of Secret does help me in understanding what the handling of Top Secret must be like. One of the companies I worked for was E.G.&G., and so I know a number of things about this company, which was helpful in verifying the testimony of one of the 3 people which I mentioned earlier. And I also know that a company called S.A.I.C. exists (also helpful in understanding testimony), because we used to work with them on some major projects.

As a final point regarding my background, I have presented and explained my new theories of physics to no less than Dr. Hal Puthoff of EarthTech International (www.earthtech.org), and Al Mouton, ex-VP of Motorola. Both parties have expressed a great interest in my material, and the presentations were very successful. Hal's eyes really lit up when he started looking at my physics theory, and then he canceled all of his afternoon appointments, just so he and Al could learn more. Dr. Puthoff showed me several ways in which he could tell that my theory was sound, and he concluded that it was a sound theory, which could be used to build devices. But I can't and won't go into more detail about possible future plans at this time.

SECRECY:

For many people, the idea that the Government could keep antigravity and antigravity-based flying devices such a secret for all these years is preposterous & laughable. But they are wrong. Dead wrong.

From the book "Disclosure", by Dr. Steven M. Greer:

What is a USAP (this is an acronym for "Unacknowledged Special Access Project")? It is a top secret, compartmentalized project requiring special access <u>even for those with a top secret clearance</u>, AND it is <u>unacknowledged</u>. This means that if someone (anyone) including your superiors, including the commander and chief, the president, asks you about it, <u>you reply that no such project exists</u>. You lie. People in these USAPS are dead serious about keeping their project secret, and will do nearly anything to keep the story covered, and to keep both other officials and the public disinformed.

The US government builds almost nothing (thank goodness...). The B2 Stealth bomber is not built by the US government, but FOR the US government by private industry. And private industry keeps secrets even better than USAPS. It makes sense: After all these years no body knows the formula for Coca Cola. Not even the President of the United States can get it. The formula is secret, and private. Now if you will, combine the proprietary power of private secrets with a combined liaison with USAPS and you build a covert fortress which is virtually impenetrable. Because if you try to get at it through the private sector, it is protected by proprietary privilege. And if you try to get to it through the public sector-government it is hidden in USAPS, and the "government" as you and I ordinarily think of it is clueless.

So, what is the essential profile of this covert operation?

Description: This group is a quasi-governmental, USAPS related, quasi-private entity operating internationally/transnationally. The majority of operations are centered in private industrial "work for others" contract projects related to the understanding and application of advanced ... technologies. Related compartmentalized units, which are also USAPS, are involved in disinformation, public deception, active disinformation, ..., space-based weapons systems and specialized liaison groups (for example to media, political leaders, the scientific community, the corporate world, etc.). Think of this entity as a hybrid between government, USAPS, and private industry.

The group consists primarily of mid-level USAPS-related military and intelligence operatives, USAPS or black units within certain high-tech corporate entities, and select liaisons within the international policy analysis community, certain religious groups, the scientific community and the media, among others. The identities of some of these entities and individuals are known to us, though most remain unidentified.

The vast majority of political leaders, including White House officials, military leaders, congressional leaders, UN leaders and other world leaders are not routinely briefed on this matter. When and if inquiries are made, they are told nothing about the operations, nor is the existence of any operation confirmed to them. In general the nature of this covert entity ensures that such leaders do not even know to whom such inquiries should be addressed...

The majority of personnel as well as the leadership of most if not all of these agencies and private groups are uninvolved and unaware of these compartmentalized, unacknowledged operations. For this reason, sweeping accusations related to any particular agency or corporate entity are wholly unwarranted. "Plausible deniability" exists at many levels. Moreover, specialization and compartmentalization allows a number of operations to exist without those involved knowing that their task is related to the ... subject.

Both positive inducements to cooperate and penalties for violating secrecy are extraordinary. A senior military source has related to us that at least 10,000 people have received $1 million or more each to ensure their cooperation, over the past few decades. Regarding penalties, we know of more than one credible case where individuals have had their families threatened should they break the code of silence, and we have learned of two recent alleged 'suicides' at a private contract industrial firm which occurred after the victims began to violate secrecy on a ... technology.

Funding: A senior congressional investigator has privately related to us that "black budget" funds apparently are used for this and similar operations which are USAPS. This 'black budget' involves conservatively $10 billion, and may exceed $80 billion per year.

...Additionally, significant funds are derived from overseas sources and private and institutional sources. Amounts deriving from these activities are also unknown by us.

Testimony From the "Disclosure Project":

Dr. Tom Bearden is a retired Lieutenant Colonel of the U.S. Army and holds a Ph.D. and an M.S. in Nuclear Engineering from the Georgia Institute of Technology. He is currently CEO of CTEC, Inc., Director of the Association of Distinguished American Scientists, and Fellow Emeritus of the Alpha Foundation's Institute for Advanced Study. In Colonel Bearden's testimony he speaks extensively about how it is possible to derive useable energy from the vacuum without violating any currently known laws of physics. He and others have built electromechanical devices which actually demonstrate this technology.

In "Disclosure", Bearden testifies that:

".... Lethal force is used. I worked with an inventor, for example, Sparky Sweet, who is quite well known. He was shot at once by a sniper rifle from about 300 yards. The only thing that saved his life was, he was an old guy and very feeble, there towards that part of his life, and he stumbled as he was coming up the steps and he fell down. As his head went forward, the bullet went right by where his head was. And of course, the assassin was never found..."

And, also:

"...They take a deep psychological profile of an individual that they wish to suppress. They really wish to get him entangled in all kinds of difficulties he can't get out of. Now, a good trait, from a human being standpoint, may be a very valuable trait to somebody who wishes to manipulate you..."

"...So, any way that you do not have good knowledge or you have a vulnerability, then they arrange scenarios, just like you would write a movie. In fact, they do it with computers. It's all computerized. And in this scenario, we have, we write a play where this particular vulnerability is going to be exploited, and the target neutralized."

"Now, they keep big psychological profiles on lots of useful fellows. These are people who basically have knee jerk reactions or something. Or they are radical. Or they have some kind of way they interact which, if it could be connected with you, would be in the area we wish the interaction to occur to get you off into something else, totally different from what you are doing. All it takes to set that up may be a phone call, and stimulate the action to occur. And then the controllers sit back and watch the game go. It is gaming. But it is just like watching a movie scenario."

"I can tell you they are very effective. You can get so many different games from so many different walks of life by so many charming folks who are really oily characters that you would not believe it. And those come at you en mass. And usually, they bury you. They bury you off in the courts, they bury you off, or they get you tricked into doing an illegal act...."

"Machiavellian is not dead. He still lives. And these are the games that are played."

Brigadier General Stephen Lovekin: Army National Guard Reserves:

In "Disclosure", he testifies as follows:

"...One older officer discussed with me what possibly could happen if there was a revelation. He was talking about being erased and I said, 'Man, what do you mean erased?' And, he said, 'Yes, you will be erased — disappear.' And I said, 'How do you know all this?' And he said, 'I know. Those threats have been made and carried out. Those threats started way back in 1947. The Army Air Force was given absolute control over how to handle this. This being the biggest security situation that this country has ever dealt with and there have been some erasures...' "

"...I don't care what kind of a person you are. I don't care how strong or courageous you are. It would be a very fearful situation because from what Matt [this older officer] said, 'They will go after not only you. They will go after your family.' Those were his words. And, so I can only say that the reason that they have managed to keep it under wraps for so long is through fear. They are very selective about how they pull someone out to make an example of. And I know that that has been done."

Lance Corporal Jonathan Weygandt: US Marine Corps

In "Disclosure", he testifies as follows:

"'You weren't supposed to be there.' 'You are not supposed to see this.' 'You are going to be dangerous if we let you go.' I thought that they were going to kill me, really...
"They had a Lieutenant Colonel from the Air Force and he did not identify himself. And he told me, 'If we just took you out in the jungle, they would never find you out there.' I didn't want to test him to see if he would really do that so I just said, 'Yeah.' And, he said, 'You have got to sign these papers. You never saw this.' I "don't exist" and "this situation never happened." And if you tell anybody, you will just come up missing...'
"They are yelling at me and hollering and cursing. 'You didn't see anything. We will do you and your whole goddamn family.'
"It was basically that for about eight or nine hours... 'We are going to take you off in a helicopter and we are going to kick your ass out in the jungle and we are going to end you...'
"These different agencies are on their own. They don't obey the law. They are rogue. Do I think that this is a project that goes up through the government and everyone has a piece in it? No. I think these guys operate on their own and no one knows what they do. It is so easy to do today. And there is no oversight, no control. They just do whatever they want...
"Lethal, deadly force has been used. For those of you who don't know, I know marine snipers and I have heard other guys talk about it and I've heard that these guys go on the streets and they stalk people and they kill them. I know that the Army Airborne snipers do the same thing. They use Delta Force to go grab these people and silence them by killing them."

Well, we see that all of the above definitely appears, MOST DISTURBINGLY, to have been going on. I highly recommend that everyone read the book "Disclosure" by Steven Greer M. D. It is a very thorough book which investigates all of the things the government has covered up. What little I've excerpted here is NOTHING in comparison to the content of that book, which bears the quote on the front cover: "MILITARY AND GOVERNMENT WITNESSES REVEAL THE GREATEST SECRETS IN MODERN HISTORY". The book is full of testimonies of insiders, and it is these kinds of testimonies that can be (and maybe someday will be) used in a courtroom to prove things beyond a doubt.

"That classified, above top-secret projects possess fully operational antigravity propulsion devices and new energy generation systems that, if declassified and put into peaceful uses, would empower a new human civilization without want, poverty, or environmental damage."

That's the <u>good</u> news. But... The book "Disclosure" also contains information that everyone should know, and should have the right to know, because very bad things are coming down the pike, and we must read this book in order to be able to prevent these things. GET THIS BOOK NOW and read it! Your and our future depends on it!

Now, a closely related subject will be discussed, which has to do with filing of U. S. patents. If someone just happens to create a device which just so happens to closely resemble a device already in possession of the military, and if it just happens to be a classified device (secret, etc...), then the government can issue to the inventor what's called a SECRECY ORDER (Title 35, United States Code (1952), sections 181-188). This is very nasty, as you will see later below.

It is important to know that U.S. Patents undergo a lengthy review process. Part of that process is a detailed review by a committee to see if any of the technology within the patent is applicable to "groups" in support of the national security. One panel, called the ASPAB, is set up each year, composed of members of the four armed services to perform this function. If ANY member of the panel feels that the patent information could be used "in the interest of

national security," then the patent is removed from the patent process, and the inventor is sent the "Secrecy Order" reprinted below. The inventor is ordered: "in nowise to publish or disclose the invention or any material information with respect thereto, ... ". In other words, they say: "thank you for the information, now SEND IT IN AND FORGET IT!" Over 3,000 patents have been classified this way (MacNeill 1983). Each of these is supposed to be reviewed on a yearly basis for possible release back to the inventor.

SECRECY ORDER
(Title 35, United States Code (1952), sections 181-188)
NOTICE: To the applicant above named, his heirs, and any and all of his assignees, attorneys and agents, hereinafter designated principals:
You are hereby notified that your application as above identified has been found to contain subject matter, the unauthorized disclosure of which might be detrimental to the national security, and you are ordered in nowise to publish or disclose the invention or any material information with respect thereto, including hitherto unpublished details of the subject matter of said application, in any way to any person not cognizant of the invention prior to the date of the order, including any employee of the principals, but to keep the same secret except by written consent first obtained of the Commissioner of Patents, under the penalties of 35 U.S.C. (1952) 182, 186.
Any other application already filed or hereafter filed which contains any significant part of the subject matter of the above identified application falls within the scope of this order. If such other application does not stand under a security order, it and the common subject matter should be brought to the attention of the Security Group, Licensing and Review, Patent Office.
If, prior to the issuance of the secrecy order, any significant part of the subject matter has been revealed to any person, the principals shall promptly inform such person of the secrecy order and the penalties for improper disclosure. However, if such part of the subject matter was disclosed to any person in a foreign country or foreign national in the U.S., the principals shall not inform such person of the secrecy order, but instead shall promptly furnish to the Commissioner of Patents the following information to the extent not already furnished: date of disclosure; name and address of the disclosee; identification of such part; and any authorization by a U.S. government agency to export such part. If the subject matter is included in any foreign patent application, or patent, this should be identified. The principals shall comply with any related instructions of the Commissioner.
This order should not be construed in any way to mean that the Government has adopted or contemplates adoption of the alleged invention disclosed in this application; nor is it any indication of the value of such invention.

READ THE FOLLOWING CAREFULLY!

Recent Documented Examples of Suppression :

From Paul Brown's written Open Letter (Brown 1991): "I have been involved with alternative energy research since 1978, while still a college student. Over the years I have heard many nightmare stories about people who developed something significant only to be persecuted, harassed, prosecuted, and even killed. ... As time went on, and in about 1982, I became involved in work of some significance and received some minor criticism and skepticism that I found to be beneficial as well as practical ... However, things began to change, slowly and alarmingly. The more success I had in my endeavors, the more I began to attract dishonest and greedy people. In 1987 we decided it was time to let the world know what we were working on and the results we were getting. ... But this was the real beginning of the worst. Since that February 1987, I or my company have been persecuted by the State Dept. of Health; then the Idaho Dept. of Finance filed a civil complaint; ... I began to receive threats; securities fraud charges were then filed against my company and my self; then the tax man; then the SEC; then my wife was assaulted; my house was robbed three times; twice I was accused of drug manufacturing; I lost my home; and most recently my mother's car was pipe bombed. ... I am here to tell you it is not coincidence. I now understand why some inventors drop out from society. My advice to you is keep a low profile until you have

completed your endeavor; be selective in choosing your business partners; protect yourself and your family; know that the nightmare stories are true." "God speed, Good Luck in your endeavors, and Never lose The Faith."

From the letter for "Urgent Appeal" by inventor and patent recent holder William Hyde (Hyde 1992): "The United States Department of Energy and various large corporations have and are attempting to suppress this new emerging energy technology. They use harassment, death threats, blackmail, extortion and employ people to file frivolous lawsuits to drain off resources to stop development and manufacture. I have been in court for five years over every conceivable frivolous matter one could think of. Unless I receive help from the American people, I will not be able to work in this important energy technology area, as the costs of fighting a cabinet level federal bureaucracy and billion dollar corporations is prohibitive. My legal costs over the last five years has been over $35,000. I have lost over 30 foreign patents because of the government action." "Again, I ask for you help in this important matter. I personally thank all of you in advance for your help. I look forward to hearing from you all."

There are other examples of evident suppression of advanced research and of newsworthy stories in the U.S. media. "Project Censored" at Sonoma State University in California publishes a newsletter and a yearly compilation of what journalism students at that university think are the most actively suppressed stories by the U.S. media. The are many theories as to the reasons for such suppression and the groups that could be performing it. A good review of such groups is given by John Coleman in his book Conspirators' Hierarchy: The Story of the Committee of 300, (Coleman 1992). Possible reasons for such control are the main subject of another book, Bankruptcy 1995, (Figgie Jr. 1992). These and other valuable books are also available from the Liberty Library - get their free catalogue using their 800 number.
AND...
Cold Fusion Over-Unity Research :
The verified and repeatable over-unity effects and phenomena of 'Cold Fusion' also will just not go away. Edmund Storms of Los Alamos National Laboratory has reviewed much of the work done internationally in the so-called area of cold fusion and has documented the results of the now world-wide research in this area (Storms 1993). Carol White has documented a summary of research and repeated results presented at the 3rd International Cold Fusion Conference (White 1993). Within the U.S., Bob Shaubach at Thermacore has presented a paper entitled "Electrolysis of Light Water Produces Excess Energy - 78 Watts Output with 28 Watts Input." Also, the Electric Power Research Institute (EPRI) research at the Stanford Research Institute (SRI) will be published this August that verifies the existence of the cold fusion phenomena, including nuclear product formation (McKubre 1994). How such cold fusion systems could be commercialized in the near future is summarized in a recent paper by Hal Fox (Fox 1993).

The above was transcribed from:
PROCEEDINGS; The Second International Symposium on Non-Conventional Energy Technology, September 9-11, 1983, Atlanta, GA, pp 125-126.
Contact Ken MacNeill at Cadake Industries, P.O. Box 1866, Clayton, GA 30525.

I'm sure the reader is horrified at reading the above. Sounds a lot like a good corroboration of Tom Bearden's testimony, doesn't it? These people play for keeps. And now, here's some more:

The following is from the website
http://www.amasci.com :

From: "cliff"
To: "William Beaty" <billb@eskimo.com>
Subject: RE: funny thing
Date: Thu, 25 May 2000 22:05:49 -0700

I hope you do well in making this funny thing. I hope to continue when I have the time. I have also shown a local science teacher, he has a Ph.D. and really knows his stuff. I also have a few friend of mine at Offit AFB north of Lancaster CA. where I use to work for NASA. My friend there still works there even after Pres. Clinton. of which I lost my job during his so called cut backs. Never liked him since. However, my friend and I will be meeting each other sometime next month on this device. Should get some attention or at lease a eyebrow.

AND....

Date: Fri, 26 May 2000 20:08:28 -0700 (PDT)
From: William Beaty <billb@eskimo.com>
Subject: RE: funny thing

On Thu, 25 May 2000, cliff wrote:
> I hope you do well in making this funny thing. I hope to continue when I
> have the time.

I started cutting out hundreds of tin squares. If it works, it should be easy to make several capacitors, since the print shop can punch them out quickly.

> I have also shown a local science teacher, he has a Ph.D. and
> really knows his stuff.

Watch out. If he gets "dollar signs in the eyes" disease, that means trouble.

People in the fringe sciences typically have their discoveries destroyed by money, and by paranoia/secrecy it brings.

It's like discovering a deposit of gold, and the snakes and vultures come running. I personally watched this happen to two separate people recently (Greg Watson and his magnetic ramp, and Norm Wootan and the MRA device.) It keeps happening over and over, yet EVERY SINGLE TIME the victim says "it can't happen to me." And then it does.

The only cure I know of it the way of science: publish openly, keep no secrets.

> I also have a few
> friend of mine at Offit AFB north of Lancaster CA. where I use to work for
> NASA.

The military can actually stop you from working on an invention. This is not just paranoia, it's a common occurrence. Adam Trombly tried to patent a homopolar generator design in 1992 or so, and the military immediately slapped a secrecy order on it. Around 1959, one of the original inventors of the laser took his ideas to the military. They made them classified, but the guy had no security clearance, so he was excluded from working on it.

Before you show this stuff to professionals and the military, I recommend that you do some research and find out how other inventors were screwed over in the past. Your greatest enemy is the idea that "it can't happen to me." If you don't know how to protect yourself, then "it" probably will happen to you. If you want to talk to Greg Watson about his experiences, I can look up his email address.

> Should get some attention or at lease a eyebrow.

Raised eyebrows are safe. It's when outsiders start taking you seriously that the trouble starts.

The above was from:
(((((((((((((((((((((((((O)   )))))))))))))))))))))))))

William J. Beaty	SCIENCE HOBBYIST website
billb@eskimo.com	http://www.amasci.com
EE/programmer/sci-exhibits	science projects, tesla, weird science
Seattle, WA 206-781-3320	freenrg-L taoshum-L vortex-L webhead-L

AND NOW, the TABOO subject:

I really, really, really did NOT want to bring up the subject of **UFO's** and **ALIENS** in this entire book. These subjects have caused many a person discreditation and slander, and these subjects often bring about laughs, snickers, and serious disbelief. The reader may notice that up until this point, I have purposely avoided mention of such things, as they are quite controversial.

However, I have found it advantageous to go ahead now and at least acknowledge the existence of aliens, if nothing else. Why, you ask?
Can't we just do without that crap? **Here's why:**

One of the "big 3" witnesses that I will discuss at length in following chapters, Mr. David Hamel, is an individual who is not at all related to the government or to any secret projects whatsoever. He is a person who is teaching himself how to build and refine devices which exhibit antigravity. He has been doing this over a period of many years, and with success. When you read the section of this book which discusses this person and his amazing achievements, you will see that my new theory of physics and antigravity does an excellent job of explaining how his devices work. He, of course, has some ideas as to how the devices work, but he is not a physicist or engineer, but a carpenter. You will see that there are several ways my physics theory confirms his claims to be true and the theory shows clearly how his devices can function in the desired manner. **HOWEVER,** I have several MORE WAYS to prove he's telling the truth, in the form of his spectacular drawings which he says he drew from memory, after having seen these drawings that were made by, yes, **ALIENS.**
Or so he says.

I decided I just can't ignore these additional items of proof, because there are a significant number of things which Mr. Hamel claims he got from aliens, and when I look at these drawings, I can understand many of them rather well based upon my new theory of physics. By existing theories of physics, the drawings look like nonsense. An also, consider this: If Mr. Hamel is building sophisticated, very specifically-designed devices which in fact work and rely on advanced theories of physics, then where in the heck do we figure that he got this information from? He would have either found it out from the secret government (not likely), or he would have found it out from advanced visitors from elsewhere, namely aliens (likely).

SO, WE WILL JUST BITE OUR TONGUE, GRIT OUR TEETH, AND PROCEED ON WITH THE UNDERSTANDING THAT ALIENS FROM OTHER WORLDS DO INDEED PROBABLY EXIST AND CAN COMMUNICATE WITH US. Sorry, but this issue just won't go away, and if you were now to take a look at the book **"Disclosure"** or at the web site **www.majesticdocuments.com,** you would see that they are very UFO/alien slanted, in addition to government conspiracy and secrecy.

As an aside, below is an interesting document which is a reply to Timothy Good's request, via Freedom Of Information Act. Here's what the government replied with:

DEPARTMENT OF THE ARMY
OFFICE OF THE ASSISTANT CHIEF OF STAFF FOR INTELLIGENCE
WASHINGTON, DC 20310-1001

REPLY TO
ATTENTION OF

March 12, 1987

Directorate of Counterintelligence

Mr. Timothy Good

Dear Mr. Good:

This is in response to your Freedom of Information Act request of October 7, 1986, for information on the Interplanetary Phenomenon Unit of the Scientific and Technical Branch, Counterintelligence Directorate, Department of the Army. Your request was received in this office on 15 October 1986.

Please excuse the delay in responding to your request, but due to an extraordinary administrative and operational burden placed on this office by a shortage of personnel, a considerable backlog of requests has accumulated.

Please be advised that the aforementioned Army unit was disestablished during the late 1950's and never reactivated. All records pertaining to this unit were turned over to the US Air Force, Office of Special Investigations in conjunction with operation "BLUEBOOK."

Therefore, we suggest that you initiate a Freedom of Information Act request with the Air Force for the answers to your questions. You should address your request to:

United States Air Force
HQ AFOSI/DADF
Attention: Freedom of Information
Bolling Air Force Base
Washington, D.C. 20332

Again my apologies for taking so long to answer your request.

Sincerely,

The received document proves that the "Interplanetary Phenomenon Unit" of the Army really did exist. The question is, WHY would we need such a department? "Interplanetary" means "between-planets". This department existed in the 1940's or early 1950's, at a time when we humans on Earth had no way of ourselves going out into space. So again, why did we need an INTER-planetary department? If it was just a "planetary" department, it could be explained as something to do with astronomy. But a between-planet phenomenon department?? Many "leaked-out" government documents, which can be seen on the www.majesticdocuments.com web site, have references to an "Interplanetary Phenomenon Unit" of the Army, or "IPU". This is just another way to verify authenticity of those leaked documents.

Also, as another quick aside:

Here's a press release written and released by the Army, so that it could be printed in the "Roswell Daily Record". Most believe that it was an unauthorized accident, because the next day, the Army, et. al., completely retracted these statements, and vehemently insisted that the object found was just a "weather balloon". And never mind that a highly-reliable & honest member of the community, a hardware store owner, and his wife, reported seeing a flying saucer-like thing fly by at high speeds right before the crash. (his name is Mr. Wilmot)

July 8, 1947:
Press Release by Public Information Officer 1st Lt. Walter Haut on the direct orders of base commander Col. William Blanchard.

The many rumors regarding the flying disk became a reality yesterday when the Intelligence office of the 509th Bomb Group of the Eighth Air Force, Roswell Army Air Field, was fortunate enough to gain possession of a disc through the cooperation of one of the local ranchers and the sheriff's office of Chaves County.

The flying object landed on a ranch near Roswell sometime last week. Not having phone facilities, the rancher stored the disc until such time as he was able to contact the sheriff's office, who in turn notified Maj. Jesse A. Marcel of the 509th Bomb Group Intelligence Office.

Action was immediately taken and the disc was picked up at the rancher's home. It was inspected at the Roswell Army Air Field and subsequently loaned by Major Marcel to higher head-Quarters.

Anyway, and moving on...

If you want to look at the mountains of evidence for aliens and UFO's, here are some additional good sources as well:

- **"Crash At Corona" and "Top Secret / MAJIC", by Stanton Friedman.**
- **"Above Top Secret : The worldwide UFO COVER-UP", by Timothy Good.**
- **The IUFOMRC web site: www.iufomrc.com, which is the UFO museum in Roswell, New Mexico.**

Disclaimer:
I, as the author of this book, am NOT trying to either convince or un-convince the reader of the possible existence of Aliens. It is totally the reader's responsibility to research carefully these things, which are outside the scope of this book.

Again, my position is that I neither advocate nor dis-advocate the possible existence of UFO's or aliens.

Clarifications regarding my new theory:

I want to say at this point that my new theory of physics does not in any way suggest any production of or even possibility of "free-energy". It is strictly a theory about how antigravity effects are coupled to known electromagnetic effects. There are no places in the theory where energy and momentum are NOT CONSERVED. Everything, including energy, comes from "somewhere", even if that somewhere looks a bit mysterious at first glance. There is no "something-for-nothing". Conservation of energy prevails.

You may have noticed some references in this book which allude to free-energy. Note that all such references are NOT due to my theory, but come from other people's sources.

THERE ARE NO PERPETUAL MOTION MACHINES, AND THERE IS NO ACTUAL "FREE-ENERGY".

But, there are ways to release so-called "Zero-Point" energy, because it exists "somewhere", and it can be tapped from that "somewhere", just like a windmill spinning in the wind does tap energy from somewhere. There's an unlimited supply of ZPE, too! It is a form of gravitomagnetic energy, as can be seen later.

Interesting history of my science & detective work:

The reader must realize that, prior to discovery of the new physics theory, I had the same education that every other physicist and engineer gets at a university. So, I can really, really understand the skepticism that physicists have toward antigravity-producing devices and similar controversial devices. Current theories just won't allow for devices which can produce really powerful waves, sufficient to achieve enough lift to defeat gravity. And so years ago, upon seeing the web sites and claims of several people who were allegedly doing antigravity, I just laughed out loud, and said to others, "Hey, take a look at this quack!" Their claims were hilarious and absurd.

But now, I owe all these people an apology. I have now "put my tail between my legs" and can back down and admit that these guys really have something. Something awesome. So for certain devices which I read about, where they claim the device levitated and caused their TV set to explode, I now believe their story.

This is precisely why, if you are reading this and you are a scientist or engineer, please do yourself the favor of completely studying my new theories of matter and energy. Once you understand it, you'll be glad you did. It will not have been a waste of your time, like you probably think it will be right now as you read this.

And this sets the stage for the next section, so read on below.

Now, interesting facts regarding science & its history:

FAMOUS QUOTES:
These really put into perspective how the "establishment" of the day usually will reject and even try to slander and debunk new ideas.

It's all about reputation (what others think of you) and fear of losing one's job or high-ranking position.

From:
New Scientist book review, May 13, 2002, page 48: Margaret Wertheim reviews Robert Marc Friedman's "The Politics of Excellence" (Time Books):

"Seen as a purveyor of metaphysical nonsense that would corrupt the vigorous strain of experimental physics admired by conservative Nobel committee members, Einstein's nomination provoked an extraordinary depth of hostility."

And...
Listen to Millikan on the subject of photoelectric theory... He is probably best known for his "oil drop" experiment, but he also made a vital contribution to photoelectric theory. His experiments confirming that Nature really does seem to obey the law that Einstein had predicted in 1905 are still taken as definitive. In his main paper on the subject, (Millikan, R A, "A Direct Photoelectric Determination of Planck's 'h'", Physical Review 7, 355-388, 1916) he says in the introduction:
It was in 1905 that Einstein made the first coupling of photo effects and with any form of quantum theory by bringing forward the ***bold, not to say reckless, hypothesis of an electro-magnetic light corpuscle of energy h□****, which energy was transferred upon absorption to an electron. This hypothesis may well be called reckless, first because an electromagnetic disturbance which remains localized in space seems a violation of the very conception of an electromagnetic disturbance, and second because it* ***flies in the face of the thoroughly established facts of interference****.*

Other great minds in the history of physics are as follows:

James L. Baird (television camera) :
When the first television system was demonstrated to the Royal Society (British scientists,) they scoffed and ridiculed it.

Arrhenius (ion chemistry) :
His idea that electrolytes are full of charged atoms was considered crazy. The atomic theory was new at the time, and everyone "knew" that atoms were indivisible (and hence they could not "lose" or "gain" any electric charge.) Because of his heretical idea, he only received his university degree by a very narrow margin.

C.J. Doppler (Doppler effect) :
He proposed a theory of the optical Doppler Effect in 1842, but was bitterly opposed for two decades because it did not fit with the accepted physics of the time (the Ether theory). He was finally proven right in 1868 when W. Huggins observed red shifts and blue shifts in stellar spectra. Unfortunately this was fifteen years after he had died.

Karl F. Gauss (non-Euclidean geometry) :
Kept secret his discovery of non-Euclidean geometry for thirty years because of fear of ridicule. Lobachevsky later published similar work and WAS ridiculed. After Gauss' death his work was finally published, but even then it took decades for non-Euclidean Geometry to win acceptance among the professionals.

Binning/Roher/Gimzewski (scanning-tunneling microscope) :
Invented in 1982, surface scientists refused to believe that atom-scale resolution was possible, and demonstrations of the STM in 1985 were still met by hostility, shouts, and laughter from the specialists in the microscopy field. It's discoverers won the Nobel prize in 1986, which went far in forcing an unusually rapid change in the attitude of colleagues.

Julius R. Mayer (The Law of Conservation of Energy) :
Mayer's original paper was contemptuously rejected by the leading physics journals of the time.

George S. Ohm (Ohm's Law) :
Ohm's initial publication was met with ridicule and dismissal. His work was called "a tissue of naked fantasy." Approx. ten years passed before scientists began to recognize its great importance.

N. Tesla (brushless AC motor) :
An AC motor which lacks brushes was thought to be an instance of a Perpetual Motion Machine.

Wright bros (flying machines) :
After their Kitty Hawk success, The Wrights flew their machine in open fields next to a busy rail line in Dayton Ohio for almost an entire year. American authorities refused to come to the demos, and Scientific American Magazine published stories about "The Lying Brothers." Even the local Dayton newspapers never sent a reporter (but they did complain about all the letters they were receiving from local "crazies" who reported the many flights.) Finally the Wrights packed up and moved to Europe, where they caused an overnight sensation and sold aircraft contracts to France, Germany, Britain, etc.

George Zweig (quark theory) :
Zweig published quark theory at CERN in 1964 (calling them 'aces'), but everyone knows that no particle can have 1/3 electric charge. Rather than receiving recognition, he encountered stiff barriers and was accused of being a charlatan.

"*The study of history is a powerful antidote to contemporary arrogance. It is humbling to discover how many of our glib assumptions, which seem to us novel and plausible, have been tested before, not once but many times and in innumerable guises; and discovered to be, at great human cost, wholly false.*"
-Paul Johnson

"*Men show their character in nothing more clearly than by what they think laughable.*"
-J. W. Goethe

"Concepts which have proved useful for ordering things easily assume so great an authority over us, that we forget their terrestrial origin and accept them as unalterable facts. They then become labeled as 'conceptual necessities,' etc. The road of scientific progress is frequently blocked for long periods by such errors."
- Einstein

"When a true genius appears in this world, you may know him by this sign, that the dunces are all in confederacy against him."
- Jonathan Swift

So,.... ANTIGRAVITY?..... Preposterous!!

[ ]

[THIS PAGE IS INTENTIONALLY BLANK]

Bob Lazar:

Who is Bob Lazar? Bob Lazar is a now well-known scientist who has worked on one of our government's most highly classified projects, Galileo, which involves the technology of large scale gravito-magnetic flying disks (also more commonly referred to as "flying saucers"). His employment as a scientist required a clearance 38 levels above "Q" (A "Q" level clearance is what government employees need to access "Secret" and "Confidential" nuclear weapons information). Since going public and telling of his work as a senior staff physicist at Area S4 in the Nellis Air Force Range (this is part of the so-called "Area 51" in Nevada), he has had his life threatened and he has been shot at. Government operatives have also erased his hospital birth records, college transcripts, and employment records, including those of his employment with Los Alamos National Laboratories and through EG&G. The shocking fact that the government can and does use unorthodox and sometimes severe & illegal means to cover up it's antigravity research secrets will become very apparent throughout this book, as the various stories unfold.

Evidence supporting Bob's claims is now much more than sufficient. In addition to his claiming Naval Intelligence work at S4 from late 1988 to early 1989, Lazar claimed to have worked at the Meson Physics lab, a part of the Los Alamos National Laboratories. The FBI is still "dragging its feet" in investigating his employment there, even though former Nevada Congressman James Bilbray asked it to investigate several years ago. Evidently, FBI agents are still scratching their heads, wondering how to both deny his employment at Los Alamos and explain why his name is in an old telephone directory of Los Alamos scientists. An article by staff writer Terry England in the June 27, 1982 edition of the *Los Alamos Monitor*, which shows a picture of Lazar standing next to a jet car and refers to his employment as a scientist with Los Alamos, is also hard to explain. Two-dozen or-so Los Alamos employees told former KLAS-TV anchor George Knapp that they remembered Lazar. Some of them said that they had been warned not to talk about Lazar and that they were afraid to talk about him. Four of them, though, confirmed for Knapp that Lazar had been working on classified projects there. After denying Lazar's employment there since 1989, Los Alamos in April 1994 finally changed its story and said that he had been employed there. Knapp also talked to former employees of the super-secret Groom Lake base (Area 51), who corroborated Lazar's description of such details as how one gets to the base dining room, what the dining room looks like, and how one pays for meals there. It's extremely unlikely that an outsider would know such information. A respected, no-nonsense reporter, Knapp has a master's degree in communications and has won AP and UPI awards for his quality UFO journalism. He accepts Lazar's story because too much of it checks out. In 1989, Lazar passed a lie-detector test arranged by Knapp. At MUFON's 1992 Midwest Conference in Springfield, Missouri, Knapp presented further strong evidence of Lazar's credibility. Lazar had mentioned that a man by the name of Mike Thigpen had visited his house and interviewed him in connection with his S4 employment. Kristen Merck and Mrs. Wayne Higdon, two witnesses who happened to be at Lazar's house, confirmed Thigpen's visit. Knapp rhetorically asks, "How did Bob Lazar know the name Mike Thigpen?" The Department of Energy confirmed for Knapp that the Office of Federal Investigations (whose phone number is not even listed in Las Vegas phone books) performs background checks on people who get clearances to work at the Nevada Test Site or at Nellis AFB. An employee of OFI called Knapp and confirmed that Thigpen worked for OFI. How did Lazar know that Thigpen did background checks? It took Knapp phone calls, friendly insider governmental contacts, and all his award-winning investigative skills before he found out who Thigpen was. The W-2 form Naval Intelligence mailed Lazar is hard to explain away as well. Knapp has examined this W-2 form, and copies of it have been seen on TV. Further boosting Lazar's credibility, John Andrews, plastic kit division manager of the Testor Corporation, found out that the U.S. Postal Service sends mail with the zip code NIC-01, the code on Lazar's W-2 form, to Naval Intelligence command in Maryland. Also hard to explain away is the unusual response the State of Nevada received when it requested documents about Lazar from the federal government. The reply said that information on Bob Lazar was on a need-to-know basis, and you don't need to know!

Specifically where Lazar worked & first time on TV news:
Between December 1988 and April 1989, the area known as Groom Lake, on the Nellis Air Force Range in central Nevada became unusually popular. The now infamous Area 51 and especially the Groom and Papoose dry lake beds were relatively unknown terms to the mainstream community from the mid 1970's to1989. The scientific circles knew it as "Dreamland" or as "The Skunkworks" or simply as Groom Lake. One night in May 1989, a reporter broadcast a story from a satellite link in Las Vegas Nevada. A young physicist spoke under hidden identity, and told us of nine antigravity discs held near Groom Lake by a small, autonomous group of the American government. He used the pseudonym "Dennis" which turned out to be the name of his superior at the base. A few weeks later he went on camera using his real name, Bob Lazar, and he has been the subject of world-wide curiosity, speculation and controversy ever since. Lazar was flown from McCarran Airport in Las Vegas to Area 51, which is a highly secure government base on the Nevada Test Site. Area 51 is located about 125 miles north of Las Vegas near the Groom mountains and the Groom dry lake bed. "From Area 51, I was bussed to an even more highly secure facility located about 15 miles from Area 51, called S4 ." Carved into the base of the Papoose mountain is a laboratory including nine hangars, each one housing an antigravity disc. It was at this base where Lazar worked.

The S4 installation is built into the mountain with the hangar doors built on an angle matching the slope of the mountain. These doors are covered with a sand textured coating to blend in with the side of the mountain and the desert floor. This was designed to hide the base from soviet spy satellites. There was a bus with blacked out windows that drove Lazar and the other scientists from Groom Lake (Area 51) to S4. The bus drove past all nine hangar doors, which are numbered one through nine, respectively, before turning to the left at the end of the ninth hangar and parking at an indentation in the mountain. This is where the entrance to S4 is. Groom Lake lies on the other side of the mountain range.

For typical Wednesday night flight preparations, the bus with the blacked out windows drove south in the early evening. The sun would usually be setting to Lazar's right side as he rode the 15 mile shuttle to S4. Once approaching S4, the bus would drive the length of the nine hangar doors and make a sharp left turn to park. The scientists were let out of the bus and directed to the entrance. This was simply a steel door in the side of the mountain. They would go past guards holding barking snarling dogs. Except these dogs were different. For all the fangs and snarling and barking, there was no sound coming out. The dogs had their vocal chords cut out to avoid being detected. Once inside the door, there was a single guard sitting at a desk in a barren room. Behind him was a door. The entrants were checked and led to another small room with a hand reader and another door. This door led into the long hallway which ran the length of the hangar doors and also led to the briefing rooms, restroom and nurse's station. The white areas are the only areas Lazar was allowed to cover. And even then under guard.

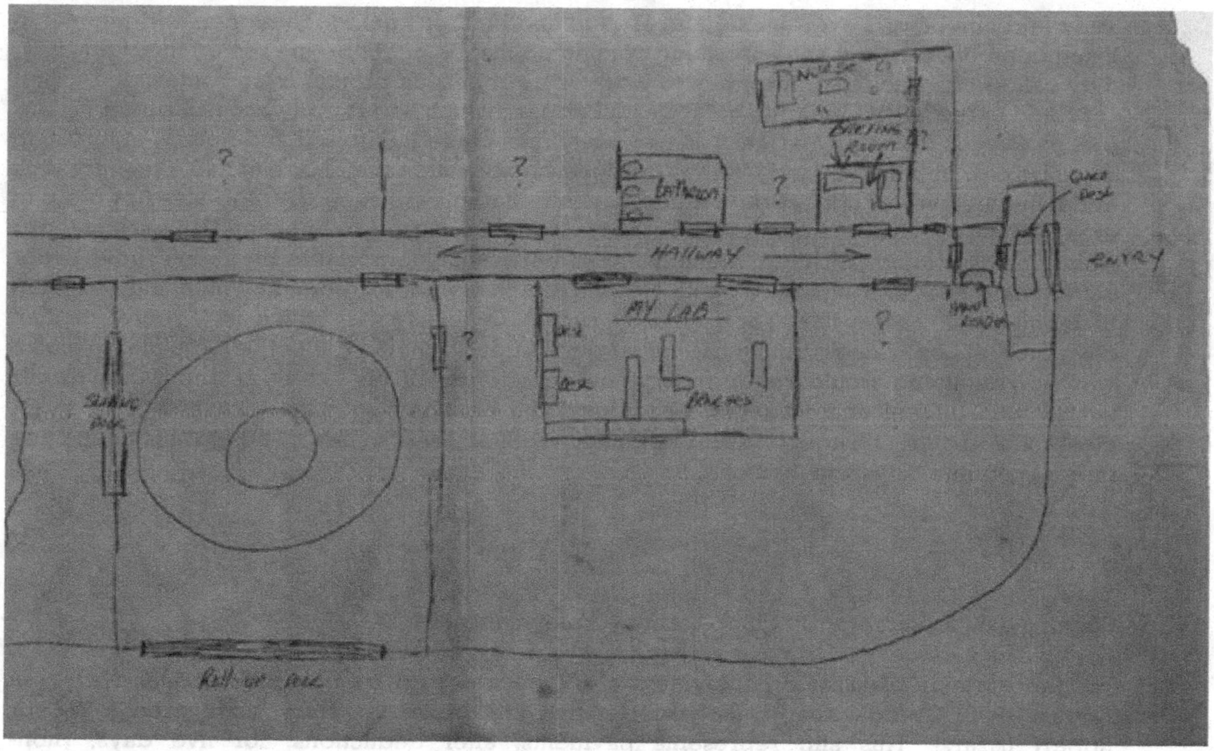

How Lazar got his job at S4:

In 1988, Bob was bored with his current photo lab job and sent out resumes to various government labs to try and get back into another scientific job. Included on that list was Dr. Edward Teller, the father of the hydrogen bomb and scientific consultant to the last 5 or 6 presidents. Bob had met Teller in 1982 in Los Alamos when Teller was there lecturing. Coincidentally, at this same time, Bob's picture was on the front page of the Los Alamos Monitor newspaper. A staff writer named Terry England had done an article on Bob and a Honda CRX in which he had installed a jet engine. This article ran in the June 27, 1982 edition of that paper and in it they refer to Bob as "Lazar, a physicist at the Los Alamos Meson Physics Facility." It's difficult to believe that in a town of 10,000 scientists, who all have egos about their credentials, the only local newspaper could refer to Bob as a scientist if he was not one, at least not without any public outrage. Anyway, when Bob arrived early to hear Teller speak, there Teller sat reading the newspaper article about Bob. Bob used this as an avenue to introduce himself and they had a short chat about jets, etc. Bob later heard Teller's speech and that was that.

So in 1988 when Bob sent out all of these resumes, he thought, why not take a shot? Teller later responded by phone and said that he was no longer active, but currently worked in a chief consultant capacity. However, Teller did give Bob the name of a gentleman to call in Las Vegas. Bob called this guy and talked to him and a short time later, someone from EG&G called Bob to set up an interview. EG&G (Edgerton, Germeshausen and Grier) is a company that has all kinds of high tech interests at the Nevada Test Site. So Bob went to EG&G for the interview.

They told him he was overqualified for the job and he didn't get it. They then told him not to lose hope because they had something coming up that he might be interested in. Ultimately, they called him back and hired him to work on a propulsion project in an "outer area." Bob reported to the EG&G building at McCarran airport, which had runway access, where he was met by Dennis Mariani. Mariani was a security man of medium build and height, about 35 to 40 years old, blonde hair and a tightly cropped blonde mustache. Mariani had a military look and manner, but he didn't wear a uniform. With Mariani as his escort, Bob was flown out to area 51 at Groom Lake. At area 51, Bob had to sign a secrecy agreement and an agreement to waive his constitutional rights, which is illegal but was made possible by an executive order with Ronald Reagan's signature on it. He also had to sign an agreement which allowed them to monitor his phone line. Bob already had a "Q" clearance, which is a secret civilian clearance, at Los Alamos, but he had never gone through anything like this. The clearance he was now attaining would require perpetual monitoring of his activities and would never simply be attained and forgotten about until the next review date (like it is with a "Q" clearance). Later, OFI agents would randomly visit Bob Lazar's house, and also his phone was tapped continuously from then on.

Specific data found regarding Bob Lazar's employment at S4:

For the duration of Lazar's employment at S4, he was paid by the United States Navy. In August 1990, George Knapp, the reporter that first broke the story, uncovered a W2 for Robert Lazar. This slip represents payments, after deductions, for five days, (non-consecutive) working at the S4 base. Lazar started at the end of December 1988. And he had only been at the base five times before the new year. Note at the top of the slip is a field reserved for the O.M.B. (Office of Management and Budget). Note Lazar's Employee number: E-6722MAJ. Compare the following badge, issued to Lazar by the security department at S4. Lazar noted that he had "Majestic" clearance. This clearance level was 38 levels above "Q" clearance. At the bottom of the badge were the different sectors that this clearance would admit entrance to. Notice the S4 indication in the lower left corner. The other three sectors are unfamiliar to Lazar: DS - ETL - WX.

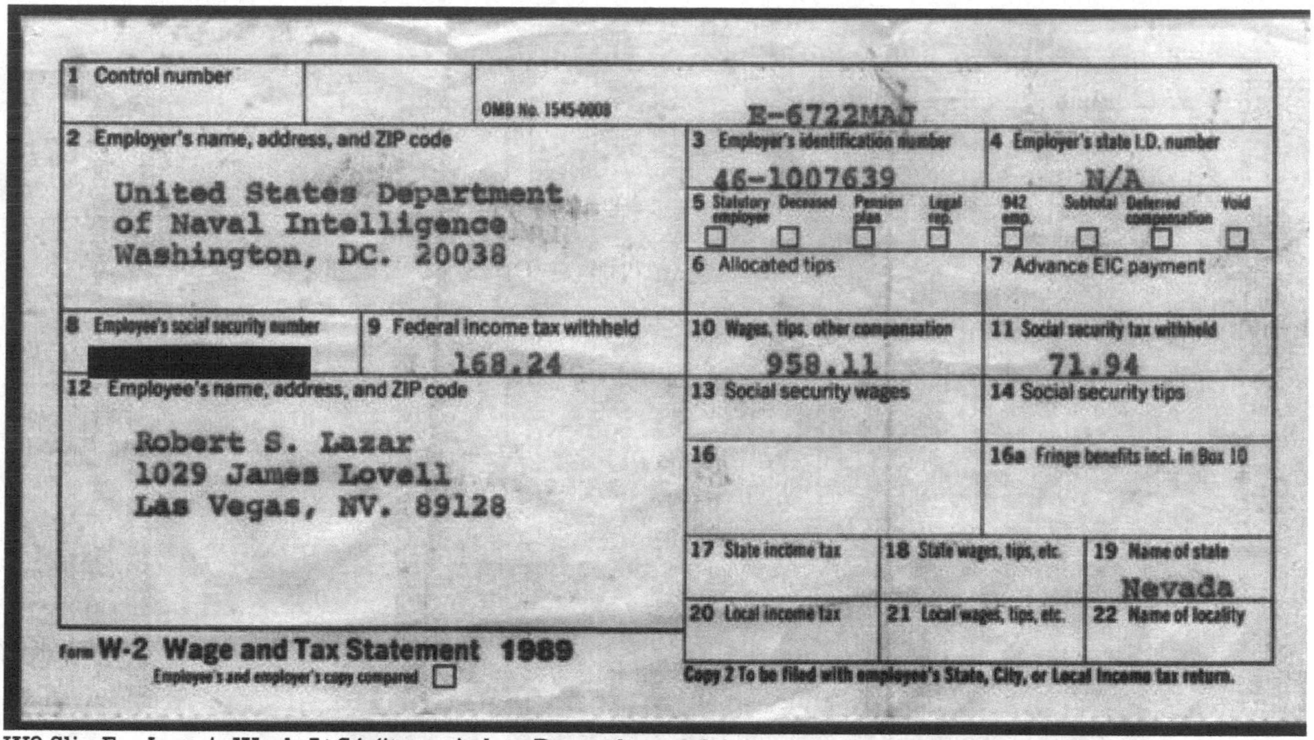

1 Control number

OMB No. 1545-0008

E-6722MAJ

2 Employer's name, address, and ZIP code

United States Department
of Naval Intelligence
Washington, DC. 20038

3 Employer's identification number: 46-1007639

4 Employer's state I.D. number: N/A

5 Statutory employee / Deceased / Pension plan / Legal rep. / 942 emp. / Subtotal / Deferred compensation / Void

6 Allocated tips

7 Advance EIC payment

8 Employee's social security number: [redacted]

9 Federal income tax withheld: 168.24

10 Wages, tips, other compensation: 958.11

11 Social security tax withheld: 71.94

12 Employee's name, address, and ZIP code

Robert S. Lazar
1029 James Lovell
Las Vegas, NV. 89128

13 Social security wages

14 Social security tips

16

16a Fringe benefits incl. in Box 10

17 State income tax

18 State wages, tips, etc.

19 Name of state: Nevada

20 Local income tax

21 Local wages, tips, etc.

22 Name of locality

Form W-2 Wage and Tax Statement 1989

Employee's and employer's copy compared

Copy 2 To be filed with employee's State, City, or Local income tax return.

W2 Slip For Lazar's Work At S4 (it was in late December, only).

Security Clearance Badge,
Worn During Work at S4.

Bob Lazar.

Bob Lazar, On The Right.

The craft that Lazar worked on:

Bob Lazar saw several different types of flying disks at the facility, which he described as "looking like different cake pans and Jell-O molds one would typically find in a kitchen" -- rather stupid looking, he thought! There was one particular craft there which Bob thought was rather nice looking, and, in fact, he was assigned solely to this particular disk. Since it was decent looking, he nicknamed it the "Sport Model". The dimensions of the "Sport Model" are 16 feet tall and 52 feet, nine inches in diameter. The exterior skin of the disc is metal and coloring similar to unpolished stainless steel. The craft sits on its belly when it is not energized. The entry hatch is located on the upper half of the disc, with just the bottom portion of the door wrapping around the center lip of the disc. The fuel that the disk uses is Element 115 (also called "ununpentium").

AutoCAD rendering of The Sport Model.

Test Flight.

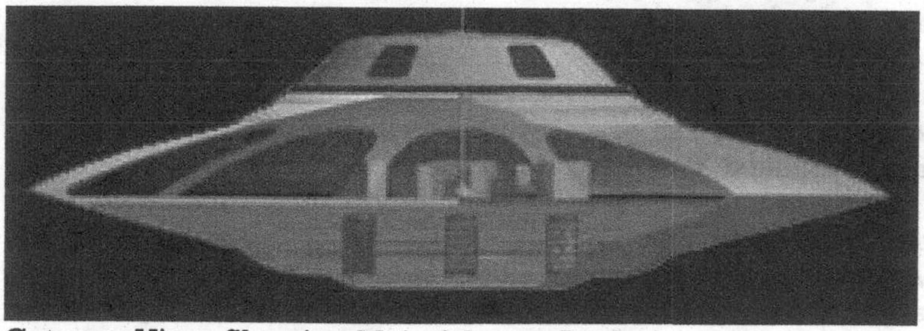

Cutaway View, Showing Main & Lower Deck.
The Three Lower Tubes are Gravity Wave Amplifiers.

The Reactor.

The interior of the disc is divided into three levels. The lower level is where the three gravity amplifiers and their wave guides are located. These are the integral components of the propulsion system that are used to amplify and focus the gravity "A" wave. The Reactor is located directly above the three gravity amplifiers on the center level and is in fact centered between them. The reactor is a closed system which uses the Element 115 as its fuel. The element is also the source of the gravity-A wave which is amplified for space/time distortion and travel. The Center level of the disc also houses the control consoles and seats, both of which were small and low to the floor. The walls of the center level are all divided into archways. Lazar was never given access to the upper level of the disc, so what the porthole-like areas are can't be illustrated, but they're definitely not portholes.

The Reactor's physics, per Bob: The power source is a reactor which uses Element 115 as its fuel. In this reactor, a **wedge-shaped** piece of Element 115 is used as a target and is bombarded with protons in a small, highly sophisticated particle accelerator. When a proton fuses into the nucleus of an atom of Element 115, it is transmuted and becomes an atom of Element 116. As soon as each atom of 115 is transmuted into 116, it immediately decays and produces a radiation unlike that which we normally observe in nuclear decay. Each atom of Element 116 decays and releases two antiprotons (anti-hydrogen), a form of antimatter. Antimatter can be produced in some particle accelerators around the world, but only in minute quantities and only stored for short periods of time. The flux of antimatter particles produced in the reactor are channeled down an evacuated, tuned tube (which keeps it from contacting with the matter that surrounds it) and reacted with a gaseous matter target. This Total Annihilation reaction is the most efficient and energetic nuclear reaction there is. The more familiar nuclear reactions are Fission, producing energy from the splitting of atoms as used in nuclear reactors & atomic bombs, and Fusion, the fusing or combining of atoms (typically hydrogen nuclei) to release even more energy. Fusion is the reaction that powers the sun and other stars and is what gives hydrogen bombs their "punch." These two more common nuclear reactions are dwarfed by the power and efficiency of the annihilation reaction used in the disk's reactor. The reaction between the gaseous matter target and the antimatter particles produces a continuous release of tremendous amounts of heat. This heat is converted directly into electricity by the use of a thermionic generator. The thermionic generator used in this reactor is so efficient, that there is no detectable waste heat produced. This is an apparent violation of one of the basic laws of thermodynamics. Similar, but not nearly as efficient or powerful, thermionic generators are used as power sources in our satellites and space probes. As amazing and efficient as all this seems, it is only secondary to the primary function of the reactor. The antiparticle flux emitted from the transmuting Element 115 is not the only energy radiated during operation. This is the point at which the gravity "A" wave is first produced. Lazar thinks the gravity wave emitted by the 115 reaction then appears on the hemispherical cover of the reactor, propagating up the tuned waveguide pipe to the top of the craft in a fashion very similar to the way microwaves behave. He says the reactor provides an enormous amount of power which is then used to amplify the gravity "A" wave so it can cause the requisite space/time distortion for space travel. In an interview of Bob Lazar on the Art Bell Show, Bob said that they could throw golf balls at the outer cover of the reactor (which contains the wedge-shaped piece of element 115), and these golf balls would just bounce off of an invisible field, rather than impacting on the reactor.

Bob says that this invisible, but very real, gravity field would then normally pass through the 3 amplifier heads, where the gravity wave field becomes much stronger and also focused. In fact, these amplifier tubes (made mostly of copper plating and tubing) were adjustable, and could be set up to focus the gravity wave into a small spherical volume about the size of a baseball. Out of curiosity, the scientists would light a wax candle, and place it so that the flame would be inside of this baseball-sized spherical "black-hole". Perhaps not surprisingly, the candle would continue to burn, but one very weird effect did occur. The candle flame, which normally flickers as candle flames tend to do, would "freeze" its motion, looking like it was "frozen in time." They concluded that not just space was warped in this region, but time itself, as predicted by Einstein's theories. Time itself stopped, or nearly so, inside the volume. Lazar also did some other interesting things, such as using a laser beam to pass close by, to show how the beam's path would curve significantly when in the vicinity of the gravity distortion wave.

Geometric details of the fuel element , also as per Bob:

This is where things get really interesting (for me, at least). The wedge of Element 115 has to be created using a complicated, strange-looking, and extremely wasteful machining process. Bob apparently had no clue why the particular shape, created by the special machining process, was critically important for the design. Here are the steps Bob said must be used to create the wedge-shaped piece of Element 115:

THE MACHINING PROCESS:

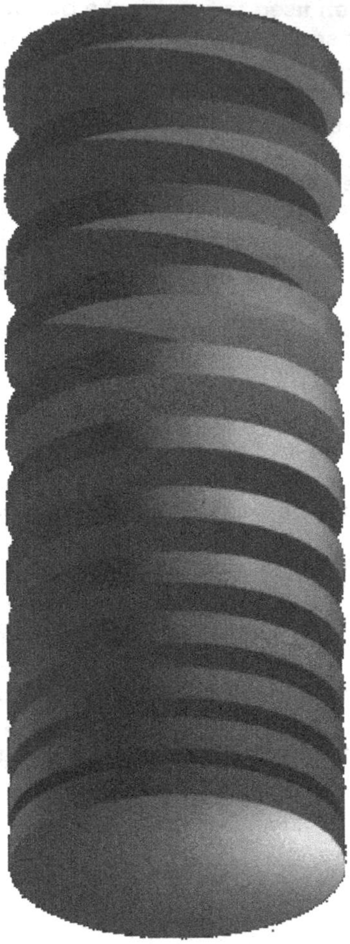

First Step: Fuse together a stack of Element 115 disks. Each disk is an orange colored, solid piece of pure Element 115.

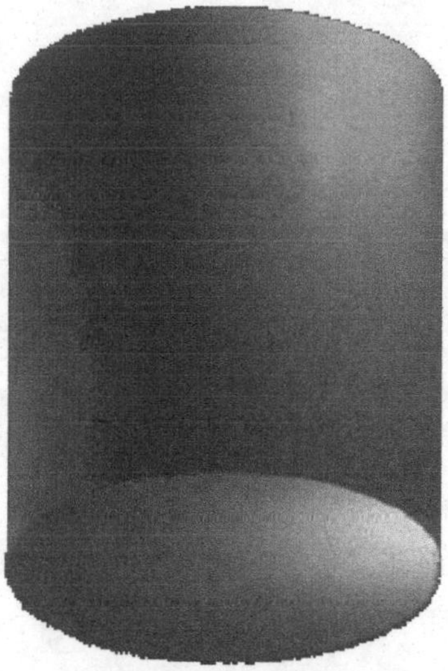

The completed cylinder of Element 115.

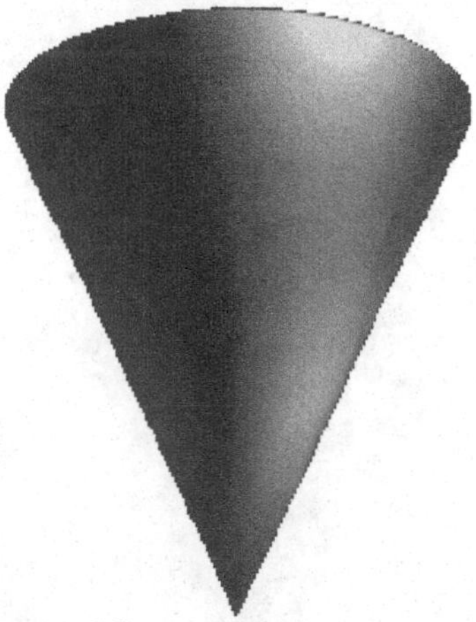

Second Step: Machine the piece so as to create a cone shape.

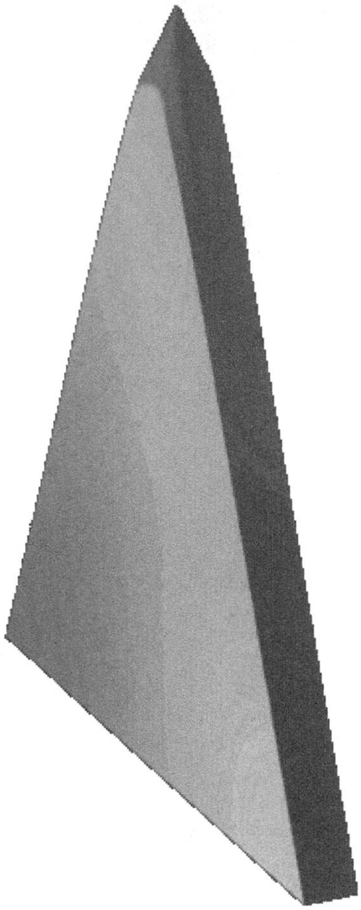

Final Step: Cut out a narrow slice from right down the cone's center. This is the final piece (of Element 115). It is now ready to be placed into the Reactor.

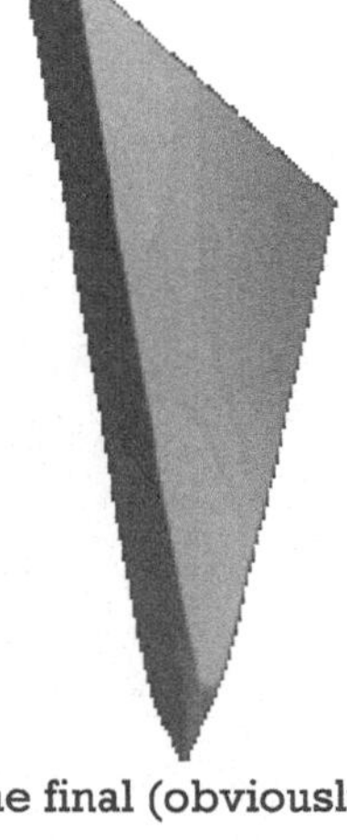

The final (obviously wedge-shaped)

piece (again). This piece of material (Element 115) is placed just inside the top end of the short vertical tube (this tube is situated at the <u>bottom</u> of the hemispherical Reactor dome). It must be placed in the tube so that the apex of the wedge points downward, as shown here, per Bob.

FORENSIC ANALYSIS [Bob Lazar]:

My website, <u>www.antigravitypapers.com</u>, once contained a page about Bob Lazar. I wrote this about Bob **AFTER** I had discovered four linearized gravity field types, "A", "B", "C", and "D", and that a proton or neutron can be modeled as several small "counter-rotating rings" of positive and effectively "negative" mass, but **BEFORE** I had mathematically discovered the detailed substructure of quarks, which is my new theory of physics. My web page stated the following:

Bob Lazar's involvement with antigravity can best be seen by following the link below: www.boblazar.com

I rank Bob Lazar as one big question mark. This is not necessarily good or bad, but rather undecided.

According to Bob's website, he was part of a government team which worked with antigravity flying craft. I must applaud Bob as having hands-down the BEST and MOST BEAUTIFULLY DETAILED web site I have ever seen!! I'm sure you will like it too, and there are enough clickable links to keep you occupied for hours on end.

So here's my take on Bob. He seems to be a rather non-credible person, at least according to Stanton Friedman, the UFO researcher and physicist. Stan's take on Bob Lazar can be seen at Stan's website: Stan Friedman's Page. [Comment for this book: According to Stan, there were NO employment records, and his college records showed he had only attended some mediocre college, and with unspectacular grades! **Can you spell G-o-v-e-r-n-m-e-n-t S-u-p-r-e-s-s-i-o-n & T-a-m-p-e-r-i-n-g !**]

What really amazes me though, is Bob's description of the gravity-based propulsion systems onboard the craft. Bob almost got it right, as far as creating an antigravity propulsion system which would actually work correctly. In fact, he was a little bit TOO RIGHT. My designs don't require "ununpentium", and are much more straightforward to build than are Bob's described designs. But, there are a few things that Bob talks about which are right in line with my discoveries and designs. I don't think it's a coincidence. Bob comes across to me as a person who has a rather confused knowledge of physics (but he is apparently very interested in the subject). This is right in line with what Stan Friedman says regarding Bob's educational background. So to me, it's AMAZING to see all that's right about what Bob is claiming, and I don't think he could have made up these descriptions himself.

As an example, Bob says that there are actually 2 forms of gravity, "A" and "B". This is essentially correct, but he got them exactly backwards. His "A" gravity sounds like it could be the "gravitomagnetic" field, which is now known about. Interestingly, though, I have discovered that there are actually two other kinds of gravity, in addition to the "A" and the "B". However, I cannot disclose what they are exactly and how they can be used. [they are "C" and "D", as I call them]

So, here are some possible explanations for Bob's claims:

.....

.

.

Like I said earlier, Bob is one big question mark...
[END]

Bob, the Federal Government tried to nullify & discredit you, but NOW (see below) we will all know the truth!

Many months later, I finally realized that Bob Lazar's story was unquestionably true, as it could then be proven by my new understanding of matter, the physics of matter and gravity, and of energy (magnetic and gravitomagnetic, electric and gravitoelectric). I had developed this new theory of physics only recently, and it is explained in **"The Antigravity Papers"** section **(Part B of this book).** The more detailed parts of the theory are in the Appendix.

Physics doesn't lie. It is solidly mathematical and ponderable and measurable. It is also testable by experiment and repeatable. If physics says that a device which is desired to do "so-and-so" MUST be built with exactly "such-and-such" a thing in order to work, then we would expect all devices claimed to do "so-and-so" must contain at least one or more "such-and-such".

So if Bob Lazar was working with real devices, based upon gravitomagnetism, for creating antigravity lift, **then his devices MUST have contained certain special parts (to be explained shortly), as per the requirements of this new physics theory.** The existing physics theories before this new one are not incorrect, nor is this new theory. The new theory supports and explains the old existing theories, but adds new levels of detail and additional structure. It also allows for some new effects related to gravity. This new theory is more correct in details than are the existing ones.

As is discussed later on in **"The Antigravity Papers"** section of this book in Chapter 7 "Next step of the Journey--the simplest model for a solid", we see that in order to produce significantly powerful antigravity effects, based upon our new understanding of gravitomagnetism, we **MUST build such devices based upon CONES and / or TRIANGLES** (rather than, say, a cylinder). A cylinder shaped part will not work since it just can't cause a separation of positive and (effectively) "negative" mass to occur, but a cone (or triangle) shaped part can do this (as per my new physics theory, shown later in Part B of this book).

Antigravity devices are all about "cones" & "triangles"!
Please refer to (in Part B) Figures 9A&B, 11A&B, 12, and 13 in order to understand this fact.

All of Lazar's claims are based upon having a **triangular** wedge of special material (Element 115) in each spacecraft, with the **apex pointing downward.** This wedge-shaped piece is located just inside of the top end of the vertical circular pipe (waveguide), which is inside of the hemispherical cover of the Reactor (at the bottom). Note that if Lazar's wedge piece were to be spinning, it would then **PERFECTLY** re-create the outline of a **CONE.** And, for all antigravity devices, the new physics theory requires parts made of either cones or triangles!

Now, look at (Part B) Figure 32, one of my inventions. This is a type of 1-cone device, as the "cone" is created by the spinning of the wedge (an **inverted triangle**) shape. It is effectively a **cone,** when it rotates rapidly. Also, a similar wedge-shaped device I invented is shown in Figure 26, and it is similar to my device in Figure 32. Figure 26 looks amazingly similar to Lazar's alleged design for the "Reactor", containing a circular pipe or "waveguide", which is thicker towards the bottom.

So Bob Lazar must have somehow been exposed to specific information about what's needed to build and/or test an antigravity device. **Before the publishing of this book, I have never seen this antigravity physics information and theory published anywhere, in any form.** This includes books, scientific journals, the news media, and the Internet. I have certainly kept my information secret, to myself, and furthermore, I have never met Lazar at all, and I have no ties with him whatsoever at the time of this writing.

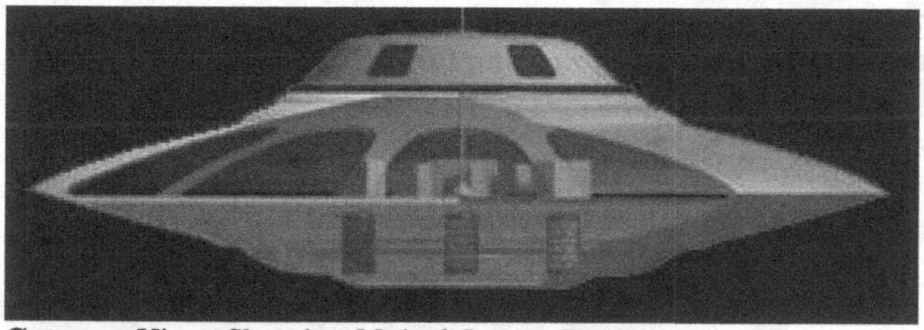

Cutaway View, Showing Main & Lower Deck.
The Three Lower Tubes are Gravity Wave Amplifiers.
(Repeated)

The Reactor.
(Repeated)

Please refer to the above two drawings (repeated). To prove that Lazar's claims are true, start by looking again at what's inside the hemisphere in the second drawing. What you see is a waveguide pipe (a short, vertical pipe inside of the hemispherical shell), which is like the design I invented in (Part B) Figure 26. We can see that the lower part of each waveguide is flared out somewhat. Bob said that inside of the top of the waveguide (he called it a "tuned pipe"), there is a triangular slab of Element 115 ("ununpentium") material. This corresponds to my design in Figure 26, except I know that other materials such as aluminum or copper will also work. And Bob never did mention anywhere whether he knew or not that the triangle has to be spinning. I know that this is required in order to make the amplifiers work in a "pulsed" manner. Bob did at least mention that the amplifiers are "pulsed", and not on constantly. Also, Bob probably wasn't allowed to view the R.F. source, and so he did not know that it has to spin or move around, as well.

And now regarding amplifiers:
Referring to (Part B) Figure 23, we see what I have designed as the basic "building" block, out of which all GW amplifiers must be made. From the side, it looks like an inverted copper triangle. Now back to Bob's drawings. I have seen a large poster version of the saucer-like craft shown. I looked closely at the cutaway view of the 3 amplifiers, and lo-and-behold, I could see **several inverted copper triangle shapes,** and basically the whole thing is made of copper plating. So there you have it. **Bob Lazar is completely telling the truth,** as proven by PHYSICS ALONE. Remember, physics doesn't lie.

Check out his website at www.boblazar.com. This website is really worth looking at, and it is the most beautifully done website I've ever seen. You could literally spend hours looking at all its different pages. It is now, however, a pay-only web site.

Bob, we deeply thank you for coming forth with this information, for the betterment of us all. You must have put yourself at great risk by coming forth.

Thomas Townsend Brown:

The following (formerly SECRET, or above) report is presented before any specific information about Thomas Townsend Brown is given, in order to set the stage for what all was going on at the time of Brown's involvement with antigravity. I found the following on an unnamed website several years ago. The report is presented as-is, except I have changed all important wording to **boldface,** so the reader need only skim over this report and look at just the relevant information only. How "Dr. StrangeQ" got this official document is not known by me, but I have several reasons to believe it to be totally authentic. These I will discuss after the end of the document.

ELECTROGRAVITICS SYSTEMS
The **Declassified Report:** Transcribed and Hereby Given to the World.

[My comments are in square brackets]
[And only a few have been inserted for clarification.]
[Spelling and punctuation have been retained as in the original report.]
[I give this to You and the World this day with deep Love.]
[Now Get Out of Your Head and Do Something About It!]
[If YOU Do Not Do Anything, Who Actually Will? And - For What?]
[Dr. StrangeQ. Christmas 1992.]

[Title Page]
ELECTROGRAVITICS SYSTEMS

An examination of electrostatic motion,
dynamic counterbary and barycentric control.

Prepared by:
Gravity Research Group
Aviation Studies (International) Limited
Special Weapons Study Unit
29-31 Cheval Place, Knightsbridge
London, S.W.7. England

Report GRG-013/56 **February 1956.**

AF Wright Aeronautical Laboratories
Wright-Patterson Air Force Base
Technical Library
Dayton, Ohio 45433

CONTENTS

* * *

PROPERTY OF USAF

[Page 1]
ELECTROGRAVITICS SYSTEMS

An examination of electrostatic motion,
dynamic counterbary and barycentric control.

It has been accepted as axiomatic that the way to offset the effects of gravity is to use a lifting surface and considerable molecular energy to produce a continuously applied force that, for a limited period of time, can remain greater than the effects of gravitational attraction. The original invention of the glider and evolution of the briefly self-sustaining glider, at the turn of the century led to progressive advances in power and knowledge.
This has been directed to refining the classic Wright Brothers' approach. Aircraft design is still fundamentally as the Wrights adumbrated it, with wings, body, tails, moving or flapping controls, landing gear and so forth. The Wright biplane was a powered glider, and all subsequent aircraft, including the supersonic jets of the nineteen-fifties are also powered gliders. Only one fundamentally different flying principle has so far been adopted with varying degrees of success. It is the rotating wing aircraft that has led to the jet lifters and vertical pushers, coleopters, ducted fans and lift induction turbine propulsion systems.

But during these decades there was always the possibility of making efforts to discover the nature of gravity from cosmic or quantum theory, investigation and observation, with a view to discerning the physical properties of aviation's enemy.

[Page 2]
It has seemed to Aviation Studies that for some time insufficient attention has been directed to this kind of research. If it were successful such developments would change the concept of sustentation, and confer upon a vehicle qualities that would now be regarded as the ultimate in aviation.

This report summarizes in simple form the work that has been done and is being done in the **new field of electrogravitics.** It also outlines the various possible lines of research into the nature and constituent matter of gravity, and how it has changed from **Newton** to **Einstein** to the modern **Hlavaty** concept **of gravity as an electromagnetic force that may be controlled like a light wave.**

The report also contains an outline of opinions on the feasibility of different electrogravitics systems and there is reference to some of the barycentric control and electrostatic rigs in operation.

Also included is a list of references to electrogravitics in successive Aviation Reports since a drive was started by Aviation Studies (International) Limited to suggest to aviation business eighteen months ago that **the rewards of success are too far-reaching to be overlooked,** especially in view of the hopeful judgment of the most authoritative voices in microphysics. Also listed are some relevant patents on electrostatics and electrostatic generators in the United States, United Kingdom and France.

[Signed]

Gravity Research Group

25 February 1956

[Page 3]
DISCUSSION

Electrogravitics might be described as a synthesis of electrostatic energy used for propulsion - either vertical propulsion or horizontal or both - and gravitics, or dynamic counterbary, in which energy is also **used to set up a local gravitational force independent of the earth's.**

Electrostatic energy for propulsion has been predicted as a possible means of propulsion in space when the thrust from a neutron motor or ion motor would be sufficient in a dragless environment to produce astronomical velocities. But the ion motor is not strictly a part of the science of electrogravitics, since barycentric control in an electrogravitics system is envisaged for a vehicle operating within the earth's environment and it is not seen initially for space application. Probably large scale space operations would have to await the full development of electrogravitics to enable large pieces of equipment to be moved out of the region of the earth's strongest gravity effects. So, though electrostatic motors were thought of in **1925,** electrogravitics had its birth after the War, when **Townsend Brown** sought to improve on the various proposals that then existed for electrostatic motors sufficiently to produce some visible manifestation of sustained motion. Whereas earlier electrostatic tests were essentially pure research, Brown's rigs were aimed from the outset at producing a flying article. As a private venture he produced evidence of motion using condensers in a couple of saucers suspended by arms rotating round a central tower with input running down the arms. The **massive-k** situation was summarized subsequently in a report, **Project Winterhaven,** in **1952.** Using the data some conclusions were arrived at that might be expected from ten or more years of

[Page 4]
intensive development - similar to that, for instance, applied to the turbine engine. Using a number of assumptions as to the nature of gravity, **the report postulated a saucer as the basis of a possible interceptor with Mach 3 capability.** Creation of a **local gravitational system** would confer upon the fighter the sharp-edged changes of direction typical of motion in space.

The essence of electrogravitics thrust is the use of a very strong positive charge on one side of the vehicle and a negative on the other. The core of the motor is a condenser and the ability of the condenser to hold its charge **(the k-number)** is the yardstick of performance. **With air as 1, current dielectrical materials can yield 6 and use of barium aluminate can raise this considerably, barium titanium oxide (a baked ceramic) can offer 6,000 and there is promise of 30,000, which would be sufficient for supersonic speed.**

The **original Brown rig** produced **30 fps** on a voltage of around 50,000 and a small amount of current in the milliamp range. There was no detailed explanation of gravity in Project Winterhaven, but it was assumed that particle dualism in the subatomic structure of gravity would coincide in its effect with the issuing stream of electrons from the electrostatic energy source to produce counterbary. The Brown work probably remains a realistic approach to the practical realization of electrostatic propulsion and sustentation. Whatever may be discovered by the Gravity Research Foundation of New Boston a complete understanding and synthetic reproduction of gravity is not essential for limited success. **The electrogravitics saucer can perform the function of a classic lifting surface - it produces a pushing effect on the under surface and a suction effect on the upper, but, unlike the airfoil, it does not require a flow of air to produce the effect.**

[Page 5]
First attempts at electrogravitics are unlikely to produce counterbary, but may lead to development of an electrostatic VTOL vehicle. Even in its developed form this might be an advance on the molecular heat engine in its capabilities. But hopes in the new science depend on an understanding of the close identity of electrostatic motivating forces with the source and matter of gravity. It is fortuitous that lift can be produced in the traditional fashion and if an understanding of gravity remains beyond full practical control, electrostatic lift might be an adjunct of some significance to modern thrust producers. Research into electrostatics could prove beneficial to turbine development, and heat engines in general, in view of **the usable electron potential round the periphery of any flame.** Materials for electrogravitics and especially the development of commercial quantities of **high-k material** is another dividend to be obtained from electrostatic research even if it produces no counterbary. This is a line of development that Aviation Studies, Gravity Research Group is following.

One of the interesting aspects of electrogravitics is that a breakthrough in almost any part of the broad front of general research on the intranuclear processes may be translated into a meaningful advance towards the feasibility of electrogravitics systems. This demands constant monitoring in the most likely areas of the physics of high energy sub-nuclear particles. It is difficult to be overoptimistic about the prospects of gaining so complete a grasp of gravity while the world's physicists are still engaged in a study of fundamental particles - that is to say those that cannot be broken down any more. Fundamental particles are still being discovered - the most recent was the Segre-Chamberlain-Wiegand attachment to the bevatron, which was used to isolate the missing anti-proton, which must - or should be presumed to - exist according to **Dirac's theory of the electron.**

[Page 6]
Much of the accepted mathematics of particles would be wrong if the anti-proton was proved to be non-existent. Earlier Eddington has listed the fundamental particles as:-

e. The charge of an electron

m. The mass of an electron.

M. The mass of a proton.

h. Planck's constant

c. The velocity of light.

G. The constant of gravitation, and

[Greek letter, small lambda]. The cosmical constant.

It is generally held that no one of these can be inferred from the others. But electrons may well disappear from among the fundamental particles, though, as Russell says, it is likely that e and m will survive. The constants are much more established than the interpretation of them and are among the most solid of achievements in modern physics.

* * *

Gravity may be defined as a small scale departure from Euclidean space in the general theory of relativity. The gravitational constant is one of four dimensionless constants: first, the mass relation of the nucleon and electron.
Second is (e*e)/(h*c) [equation form], third, the Compton wavelength of the proton, and fourth is the gravitational constant, which is the ratio of the electrostatic

[Page 7]
to the gravitational attraction between the electron and the proton.

One of the stumbling blocks in electrogravitics is the absence of any satisfactory theory linking these four dimensionless quantities. Of the four, moreover, gravity is decidedly the most complex, since any explanation would have to satisfy both cosmic and quantum relations more acceptably and intelligibly even than in the unified field theory. A gravitational constant of around 10.E-39 [equation form] has emerged from quantum research and this has been used as a tool for finding theorics that could link the two relations. This work is now in full progress, and developments have to be watched for the aviation angle. Hitherto Dirac, Eddington, Jordan and others have produced differences in theory that are too wide to be accepted as consistent. It means therefore that (i) without a cosmical basis, and (ii) with an imprecise quantum basis and (iii) a vague hypothesis on the interaction, much remains still to be discovered. Indeed some say that a single interacting theory to link up the dimensionless constants is one of three major unresolved basic problems of physics. The other two main problems are the extension of quantum theory and a more detailed knowledge of the fundamental particles.

All this is some distance from Newton, who saw gravity as a force acting on a body from a distance, leading to the tendency of bodies to accelerate towards each other. He allied this assumption with Euclidean geometry, and time was assumed as uniform and acted independently of space. Bodies and particles in space normally moved uniformly in straight lines according to Newton, and to account for the way they sometimes do not do so, he used the idea of a force of gravity acting at a distance, in which particles of matter cause in others an acceleration proportional to their mass, and inversely proportional to the

[Page 8]
square of the distance between them.

But Einstein showed how the principle of least action, or the so-called cosmic laziness means that particles, on the contrary, follow the easiest path along geodesic lines and as a result they get readily absorbed into space-time. So was born non-linear physics. The classic example of non-linear physics is the experiment in bombarding a screen with two slits. When both slits are open particles going through are not the sum of the two individually but follows a non-linear equation. This leads on to wave-particle dualism and that in turn to the Heisenberg uncertainty principle in which an increase in accuracy in measurement of one physical quantity means decreasing accuracy in measuring the other. If time is measured accurately energy calculations will be in error; the more accurate the position of a particle is established the less certain the velocity will be; and so on. This basic principle of the acausality of microphysics affects the study of gravity in the special and general theories of relativity. Lack of pictorial image in the quantum physics of this interrelationship is a difficulty at the outset for those whose minds remain obstinately Euclidean.

In the special theory of relativity, space-time is seen only as an undefined interval which can be defined in any way that is convenient and the Newtonian idea of persistent particles in motion to explain gravity cannot be accepted.
It must be seen rather as a synthesis of forces in a four dimensional continuum, three to establish the position and one the time. The general theory of relativity that followed a decade later was a geometrical explanation of gravitation in which bodies take the geodesic path through space-time. In turn this means that instead of the idea of force acting at a distance it is assumed that space, time, radiation and particles are linked and variations in them from gravity are due rather to the nature of space.

[Page 9]
Thus gravity of a body such as the earth instead of pulling objects towards it as Newton postulated, is adjusting the characteristics of space and, it may be inferred, the quantum mechanics of space in the vicinity of the gravitational force. Electrogravitics aims at correcting this adjustment to put matter, so to speak, 'at rest'.

* * *

One of the difficulties in **1954** and **1955** was to get aviation to take **electrogravitics** seriously. **The name alone was enough to put people off.** However, in the trade much progress has been made **and now most major companies in the United States are interested in counterbary.** Groups are being organised to study electrostatic and electromagnetic phenomena. **Most of industry's leaders have made some reference to it. Douglas** has now stated that it has counterbary on its work agenda but does not expect results yet awhile. **Hiller** has referred to new forms of flying platform, Glenn **Martin** say gravity control could be achieved in **six years,** but they add that it would entail a Manhattan District type of effort to bring it about. **Sikorsky,** one of the pioneers, more or less agrees with the **Douglas** verdict and says that gravity is tangible and formidable, but there must be a physical carrier for this immense trans-spatial force. This implies that where a physical manifestation exists, a physical device can be developed for creating a similar force moving in the opposite direction to cancel it. **Clarke Electronics** state **they have a rig,** and add that **in their view the source of gravity's force will be understood sooner than some people think. General Electric** is working on the **use of electronic rigs** designed to make adjustments to gravity - this line of attack has the advantage of using **rigs already in existence for other defence work. Bell** also has an **experimental rig** intended, as

[Page 10]
the company puts it, to cancel out gravity, and Lawrence Bell has said **he is convinced that practical hardware will emerge from current programs.** Grover Leoning is certain that what he referred to as an electro-magnetic contra-gravity mechanism **will be developed for practical use. Convair** is **extensively committed** to the work **with several rigs. Lear** Inc., autopilot and electronic engineers have **a division of the company working on gravity research** and so also has the Sperry division of **Sperry-Rand.** This list embraces **most of the U.S. aircraft industry.** The remainder, Curtiss-Wright, **Lockheed, Boeing** and **North American** have not yet declared themselves, but **all these four are known to be in various stages of study with and without rigs.**

In addition, the **Massachusetts Institute of Technology** is working on gravity, **the Gravity Research Foundation** of New Boston, the **Institute for Advanced Study at Princeton,** the **CalTech Radiation Laboratory, Princeton University** and the **University of North Carolina** are **all active in gravity.** Glenn L. **Martin** is setting up a Research Institute for Advanced Study which has a small staff **working on gravity research** with **the unified field theory** and this group is committed to extensive programs of applied research. Many others are also known to be studying gravity, some are known also to be planning a general expansion in this field, such as in the proposed Institute for Pure Physics at the University of North Carolina.

A certain amount of work is also going on in **Europe.** One of the French nationalized constructors and one company outside the nationalized elements have been making preliminary studies, and a little company money has in one case actually been committed. Some work is also going on in **Britain** where **rigs are now in existence.** Most of it is private venture work, such

[Page 11]
as that being done by **Ed Hull** a colleague of **Townsend Brown** who, as much as anybody, introduced Europe to electrogravitics. Aviation Studies' Gravity Research Group is doing some work, **mainly on k studies,** and is sponsoring dielectric investigations.

One **Swedish** company and two **Canadian** companies have been making studies, and quite recently the **Germans** have **woken up to the possibilities.** Several of the companies have started digging out some of the early German papers on wave physics. They are almost certain to plan a gravitics program. Curiously enough the Germans during the war paid no attention to electrogravitics. This is one line of advance that they did not pioneer in any way and it was basically a U.S. creation. **Townsend Brown** in **electrogravitics** is the equivalent of **Frank Whittle** in **gas turbines.** This German overlooking of electrostatics is even more surprising when it is remembered how astonishingly advanced and prescient the Germans were in nuclear research. (The modern theory of making thermonuclear weapons without plutonium fission initiators returns to the original German idea that was dismissed, even ridiculed.) The Germans never went very far with fission, indeed they doubted that this chain would ever be made to work. The **German** air industry, still in the embryo stage, has included **electrogravitics** among the subjects it intends to examine when establishing the policy that the individual companies will adopt after the present early stage of foreign licence has enabled industry to get abreast of the other countries in aircraft development.

* * *

It is impossible to read through this summary of the widening efforts being made to understand the nature of matter of gravity without sharing the hope that many groups now have, of major theoretical breakthroughs occurring before very long.

[Page 12]
Experience in nucleonics has shown that when attempts to win knowledge on this scale are made, advances are soon seen. There are a number of elements in industry, and some managements, who see gravity as a problem for later generations. Many see nothing in it all and they may be right. But as said earlier, if Dr. Vaclav Hlavaty thinks gravity is potentially controllable that surely should be justification enough, and indeed inspiration, for physicists to apply their minds and for management to take a risk. Hlavaty is the only man who thinks he can see a way of doing the mathematics to demonstrate Einstein's **unified field theory** - something that Einstein himself said was beyond him. Relativity and the unified field theory go to the root of electrogravitics and the shifts in thinking, the hopes and fears, and a measure of progress is to be obtained only in the last resort from men of this stature.

Major theoretical breakthroughs to discover the sources of gravity will be made by the most advanced intellects using the most advanced research tools.
Aviation's role is therefore to impress upon physicists of this calibre with the urgency of the matter and to aid them with statistical and peripheral investigations that will help to clarify the background to the central mathematical and physical puzzles. Aviation could also assist by recruiting some of these men as advisers. **Convair** has taken the initiative with its recently established panel of advisers on nuclear projects, which include **Dr. Edward Teller** of the **University of California.** At the same time **much can be done in development of laboratory rigs, condenser research and dielectric development,** which **do not require anything like the same cerebral capacity** to **get results** and make a **practical contribution.**

As gravity is likely to be linked with the new particles, only the highest powered particle accelerators are likely to be of use

[Page 13]
in further fundamental knowledge. The country with the biggest tools of this kind is in the best position to examine the characteristics of the particles and from those countries the greatest advances seem most likely.

Though the United States has the biggest of the bevatron - the Berkeley bevatron is 6.2 bev - the Russians have a 10 bev accelerator in construction which, when it is completed, will be the world's largest. At Brookhaven a 25 bev instrument is in development which, in turn, will be the biggest.

Other countries without comparable facilities are of course at a great disadvantage from the outset in the contest to discover the explanation of gravity. Electrogravitics, moreover, unfortunately competes with nuclear studies for its facilities. The clearest thinking brains are bound to be attracted to localities where the most extensive laboratory equipment exists. So, one way and another, results are most likely to come from the major countries with the biggest undertakings. Thus the nuclear facilities have a direct bearing on the scope for electrogravitics work.

The OEEC report in January made the following points:- The U.S. has six to eight entirely different types of reactor in operation and many more under construction. Europe has now two different types in service.

The U.S. has about 30 research reactors plus four in Britain, two in France.

The U.S. has two nuclear-powered marine engines. Europe has none, but the U.K. is building one.

[Page 14]
Isotope separation plants for the enrichment of uranium in the U.S. are roughly 11 times larger than the European plant in Britain.

Europe's only heavy water plant (in Norway) produces somewhat less than one-twentieth of American output.

In 1955 the number of technicians employed in nuclear energy work in the U.S. was about 15,000; there are about 5,000 in Britain, 1,800 in France, and about 19,000 in the rest of Europe. But the working party says that pessimistic conclusions should not be drawn from these comparisons. European nuclear energy effort is unevenly divided at the moment, but some countries have notable achievements to their credit and important developments in prospect.
The main reason for optimism is that, taken as a whole, "Europe's present nuclear effort falls very far short of its industrial potential".

Though **gravity research,** such as there has been of it, **has been unclassified,** new principles and information gained from the nuclear research facilities that have a vehicle **application is expected to be withheld.**

The heart of the problem to understanding gravity is likely to prove to be the way in which the very high energy sub-nuclear **particles convert something, whatever it is, continuously and automatically into the tremendous nuclear and electromagnetic forces.** Once this key is understood, attention can later be directed to finding laboratory means of duplicating the process and reversing its force lines in some local environment and returning the energy to itself to produce counterbary. Looking beyond it seems possible that gravitation will be shown to be a part of the universal electro-magnetic processes and controlled

[Page 15]
in the same way as a light wave or radio wave. This is a synthesis of the Einstein and Hlavaty concepts. Hence it follows that though in its initial form the **mechanical processes for countering gravity** may initially be massive to deal with the massive forces involved, eventually this **could be expected to form some central power generation unit.** Barycentric control in some required quantity could be passed over a distance by a form of radio wave. **The prime energy source to energise the waves would of course be nuclear in its origins.**

It is difficult to say which lines of detailed development being processed in the immediate future is more likely to yield significant results. Perhaps the three most promising are: first, the new attempt by the team of men led by Chamberlain working with the Berkeley bevatron to find the anti-neutron, and to identify more of the characteristics of the anti-proton* and each of the string of high energy particles that have been discovered during recent operation at 6. 2 bev.

A second line of approach is the United States National Bureau of Standards program to pin down with greater accuracy the acceleration values of gravity. The presently accepted figure

----

* The reaction is as follows: protons are accelerated to 6.2 bev, and directed at a target of copper. When the proton projectile hits a neutron in one of the copper atoms the following emerge: the two original particles (the projectile and the struck neutron) and a new pair of particles, a proton and anti-proton. The anti-proton continues briefly until it hits another proton, then both disappear and decay into mesons.

[Page 16]
of 32.174 feet per second per second is known to be not comprehensive, though it has been sufficiently accurate for the limited needs of industry hitherto.
The NBS program aims at re-determining the strength of gravity to within one part of a million. The present method has been to hold a ball 16 feet up and chart the elapsed time of descent with electronic measuring equipment. The new program is based on the old, but with this exceptional degree of accuracy it is naturally immensely more difficult and is expected to take 3 years.

A third promising line is the new technique of measuring high energy particles in motion that was started by the University of California last year. This involves passing cosmic rays through a chamber containing a mixture of gas, alcohol and water vapour. This creates charged atoms, or positive ions, by knocking electrons off the gas molecules. A sudden expansion of the chamber results in a condensation of water droplets along the track which can be plotted on a photographic plate. This method makes it easier to assess the energy of particles and to distinguish one from the other. It also helps to establish the characteristics of the different types of particle. The relationship between these high energy particles and their origin, and characteristics have a bearing on electrogravitics in general.

So much of what has to be discovered as a necessary preliminary to gravity is of no practical use by itself. There is no conceivable use, for instance, for the anti-proton, yet its discovery even at a cost of $9-million is essential to check the mathematics of the fundamental components of matter. Similarly it is necessary to check that all the nuclear ghosts that have been postulated theoretically do in fact exist. It is not, moreover, sufficient, as in the past, only to observe the particles by

[Page 17]
radiation counters. In each instance a mechanical maze has to be devised and attached to a particle accelerator to trap only the particle concerned. Each discovery becomes a wedge for a deeper probe of the nucleus. Many of the particles of very high energy have only a fleeting existence and collisions that give rise to them from bevatron bombardment is a necessary prerequisite to an understanding of gravity. There are no shortcuts to this process.

Most of the major programs for extending human knowledge on gravity are being conducted with instruments already in use for nuclear research and to this extent the cost of work exclusively on gravitational examinations is still not of major proportions. This has made it difficult for aviation to gauge the extent of the work in progress on gravity research.

* * *

[Page 18]
CONCLUSIONS

1. No attempts to control the magnitude or direction of the earth's gravitational force have yet been successful. But if the explanation of gravity is to be found in the as yet undetermined characteristics of the very high energy particles it is becoming increasingly possible with the bevatron to work with the

constituent matter of gravity. It is therefore reasonable to expect that the new bevatron may, before long, be used to demonstrate limited gravitational control.

2. An understanding and identification of these particles is on the frontiers of human knowledge, and a full assessment of them is one of the major unresolved puzzles of the nucleus. An associated problem is to discover a theory to account for the cosmic and quantum relations of gravity, and a theory to link the gravitational constant with the other three dimensionless constants.

3. Though the obstacles to an adequate grasp of microphysics still seem formidable, the transportation rewards that could follow from electrogravitics are as high as can be envisaged. In a weightless environment, movement with sharp-edged changes of direction could offer unique maneuverability.

4. Determination of the environment of the anti-proton, discovery

[Page 19]
of the anti-neutron and closer examination of the other high energy particles are preliminaries to the hypothesis that gravity is one aspect of electromagnetism that may eventually be controlled like a wave. When the structure of the nucleus becomes clearer, the influence of the gravitational force upon the nucleus and the nature of its behaviour in space will be more readily understood. This is a great advance on the Newtonian concept of gravity acting at a distance.

5. Aviation's role appears to be to establish facilities to handle many of the peripheral and statistical investigations to help fill in the background on electrostatics.

6. A distinction has to be made between electrostatic energy for propulsion and counterbary. **Counterbary is the manipulation of gravitational force lines;** barycentric control is the adjustment to such manipulative capability to produce a stable type of motion suitable for transportation.

7. Electrostatic energy **sufficient to produce low speeds** (a few thousand dynes, **has already been demonstrated.** Generation of a region of **positive electrostatic energy** on one side of a plate and **negative on the other** sets up the same lift or propulsion effect as the **pressure and suction below and above a wing,** except that in the case of electrostatic application **no airflow is necessary.**

8. Electrostatic energy sufficient to produce a **Mach 3 fighter** is possible with megavolt energies and a **k of over 10,000.**

[Page 20]
9. **k figures** of **6,000** have been obtained from some ceramic materials and **there are prospects of 30,000.**

10. Apart from electrogravitics there are other rewards from investment in electrostatic equipment. Automation, autonetics and even turbine development use similar laboratory facilities.

11. Progress in electrogravitics probably awaits a new genius in physics who can find a single equation to tie up all the conflicting observations and theory on the structure and arrangement of forces and the part the high energy particles play in the nucleus. This can occur any time, and the chances are improved now that bev. energies are being obtained in controlled laboratory conditions.

* * *

[New Page]
APPENDIX I

EXTRACTS FROM AVIATION REPORT

[Page 21]
ANTI-GRAVITATION RESEARCH
The basic research and technology behind electro-anti-gravitation is so much in its infancy that this is perhaps one field of development where **not only the methods but the ideas are secret. Nothing therefore can be discussed freely at the moment. Very few papers on the subject have been prepared so far,** and the only schemes that have seen the light of day are for pure research into rigs designed to make objects float around freely in a box. There are various radio applications, and aviation medicine departments have been looking for something that will enable them to study the physiological effects on the digestion and organs of an environment without gravity. **There are however long term aims of a more revolutionary nature that envisage equipment that can defeat gravity.**
Aviation Report 20 August 1954

MANAGERIAL POLICY FOR ANTI-GRAVITICS
The prospect of engineers devising gravity-defeating equipment - or perhaps it should be described as the creation of pockets of weightless environments - does suggest that as a long term policy aircraft constructors will be required to place even more emphasis on electro-mechanical industrial plant, than is now required for the transition from manned to unmanned weapons. Anti-gravitics work is therefore likely to go to companies with the biggest electrical laboratories and facilities. It is also apparent that anti-gravitics, like other advanced sciences, **will be initially sponsored for its weapon capabilities.** There are perhaps two broad ways of using the science - one is to postulate the design of advanced type projectiles on their best inherent capabilities, and the more critical parameters (that now constitutes the design limitation) can be eliminated by anti-gravitics. The other, which is a longer term plan, is to create an entirely new environment with devices operating entirely under an anti-gravity envelope.
Aviation Report 24 August 1954

[PAGE 22]
THE GREATER THE EASIER
Propulsion and atomic energy Trends are similar in one respect: the more incredible the long term capabilities are, the easier it is to attain them. It is strange that the greatest of nature's secrets can be harnessed with decreasing industrial effort, but greatly increasing mental effort. The Americans went through the industrial torture to produce tritium for the first thermonuclear experiment, but later both they and the Russians **were able to achieve much greater results** with the help of **lithium 6 hydride.** The same thing is happening in aviation propulsion: the nuclear fuels are promising to be tremendously powerful in their effect, but excessively complicated in their application, unless there can be some means of direct conversion as in the strontium 90 cell. But lying behind and beyond the nuclear fuels is the linking of electricity to gravity, which is an incomparably more powerful way of harnessing energy than the only method known to human intellect at present - electricity and magnetism. Perhaps the magic of **barium aluminium oxide** will perform the miracle in propulsion **that lithium 6 hydride has done in the fusion weapon.** Certainly it is a well-known material in dielectrics, but when one talks of **massive-k,** one means of course five figures. At this early stage **it is difficult to relate k to Mach numbers** with any certainty, but **realizable k can, with some kinds of arithmetic, produce astounding velocities.** They are achievable, moreover, **with decreasing complexity,** indeed **the ultimate becomes the easiest in terms of engineering, but the most hideous in terms of theory.** Einstein's general theory of relativity is, naturally, and important factor, but some of the postulates appear to depend on the unified field theory, which cannot yet be physically checked because nobody knows how to do it. Einstein hopes to find a way of doing so before he dies.
Aviation Report 31 August 1954

GRAVITICS FORMULATIONS

All indications are that there has still been little cognizance of the potentialities of electrostatic propulsion and it will be a major

[Page 23]
undertaking to re-arrange aircraft plants to conduct large-scale research and development into novel forms of dielectric and to improve condenser efficiencies and to develop the novel type of materials used for fabrication of the primary structure. Some extremely ambitious theoretical programs have been submitted **and work towards realization of a manned vehicle has begun. On the evidence, there are far more definite indications that the incredible claims are realizable than there was, for instance, in supposing that uranium fission would result in a bomb. At least it is known, proof positive, that motion, using surprisingly low k, is possible.** The fantastic control that again is feasible, has not yet been demonstrated, but there is no reason to suppose the arithmetic is faulty, **especially as it has already led to a quite brisk** example of **actual propulsion. That first movement was indeed an historic occasion, reminiscent of the momentous day at Chicago when the first pile went critical, and the phenomenon was scarcely less weird.** It is difficult to imagine just where a well-organised examination into long term gravitics prospects would end. Though a circular platform is electrostatically convenient, it does not necessarily follow that the requirements of control by differential changes would be the same. Perhaps the strangest part of this whole chapter is how the public managed to foresee the concept though not of course the theoretical principles that gave rise to it, before physical tests confirmed that the mathematics was right. It is interesting also that there is no point of contact between the conventional science of aviation and the New: it is a radical offshoot with no common principles. Aerodynamics, structures, heat engines, flapping controls, and all the rest of aviation is part of what might be called the Wright Brothers era - even the Mach 2.5 thermal barrier piercers are still Wright Brothers concepts, in the sense that they fly and they stall, and they run out of fuel after a short while, and they defy the earth's pull for a short while. Thus this century will be divided into two parts - almost to the day. The first half belonged to the Wright Brothers who foresaw nearly all the basic issues in which gravity was the bitter foe. In part of the second half gravity will be the great provider. Electrical energy, rather irrelevant for propulsion in

[Page 24]
the first half becomes a kind of catalyst to motion, in the second half of the century.
Aviation Report 7 September 1954

ELECTRO-GRAVITICS PARADOX
Realization of electro-static propulsion seems to depend on two theoretical twists and two practical ones. The two theoretical puzzles are: first, how to make a condenser the centre of a propulsion system, and second is how to link the condenser system with the gravitational field. There is a third problem, but it is some way off yet, which is how to manipulate kva for control in all three axes as well as for propulsion and lift. The two practical tricks are first how, **with say a Mach 3 weapon in mind,** to handle 50,000 kva **within the envelope of a thin pancake of 35 feet in diameter** and second how to generate such power from within so small a space. The electrical power in a small aircraft is more than is a fair sized community the analogy being that a single rocket jet can provide as much power as can be obtained from the Hoover Dam. It will naturally take as long to develop electro-static propulsion as it has taken to coax the enormous power outputs from heat engines. True there **might be a flame** in the electro-gravitic propulsion system, but it would not be a heat engine - the temperature of the flame would be incidental to the function of the chemical burning process.
The curious thing is that though electro-static propulsion is the antithesis of magnetism,* Einstein's unified field theory is an attempt to link gravitation with electro-magnetism. This all-embracing theory goes on logically from the general theory of relativity, that gives an ingenious geometrical interpretation of the concept of force which is mathematically consistent with gravitation but fails in the case of electro-magnetism, while the special theory of relativity is concerned with the relationship between mass and energy. The general theory of relativity fails to account for electro-magnetism because the forces are proportional to the charge and not to the mass. The unified field theory is one of a number of attempts that have been made to bridge this gap, but it is baffling to imagine how it could ever be observed. Einstein himself thinks it is virtually impossible. However Hlavaty claims now

to have solved the equations by assuming that gravitation is a manifestation of electro-magnetism. This being so it is all the more incredible that electro-static

---

* Though in a sense this is true, it is better expressed in the body of this report than it was here in 1954.

[Page 25]

propulsion (with kva for convenience fed into the system and not self-generated) has actually been demonstrated. It may be that to apply all this very abstruse physics to aviation it will be necessary to accept that the theory is more important than this or that interpretation of it. This is how the physical constants, which are now regarded as among the most solid of achievements in modern physics, have become workable, and accepted. Certainly all normal instincts would support the Einstein series of postulations, and if this is so it is a matter of conjecture where it will lead in the long term future of the electro-gravitic science.

Aviation Report 10 September 1954

ELECTRO-GRAVITIC PROPULSION SITUATION

Under the terms of **Project Winterhaven** the **proposals to develop electro-gravitics to the point of realizing a Mach 3 combat type disc were not far short of the extensive effort that was planned for the Manhattan District.*** Indeed the drive to develop the new prime mover is in some respects rather similar to the experiments that led to the release of nuclear energy in the sense that both involve fantastic mathematical capacity and both are sciences so new that other allied sciences cannot be of very much guide. In the past two years since the principle of motion by means of massive-k was first demonstrated on a test rig, progress has been slow. But the indications are now that the **Pentagon** is ready to **sponsor a range of devices help further knowledge.** In effect the new family of TVs would be on the same tremendous scope that was envisaged by the X-1, 2, 3, 4 and 5 and D.558s that were all created for the purpose of destroying the sound barrier - which they effectively did, but it is a process that is taking ten solid years of hard work to complete. (Now after 7 years the X-2 has yet to start its tests and the X-3 is still in performance testing stage). Tentative targets now being set anticipate that the first disc should be complete before 1960 and it would take the whole of the 'sixties to develop it properly, even though some combat things might be available ten years from now. One thing seems certain at this stage, that the companies likely to dominate the science **will be those with the biggest**

---

* The proposals, it should be added, were not accepted.

[Page 26]

computors to work out the ramifications of the basic theory. **Douglas** is easily the world's leader in computor capacity, **followed by Lockheed and Convair.** The frame incidentally is indivisible from the engine". If there is to be any division of responsibility it would be that the engine industry might become responsible for providing the electrostatic energy (by, it is thought, **a kind of flame**) and the frame maker for the condenser assembly which is the core of the main structure.

Aviation Report 12 October 1954

GRAVITICS STUDY WIDENING

The **French** are now understood to be pondering the most effective way of entering the field of electro-gravitic propulsion systems. But not least of the difficulties is to know just where to begin. There are practically no patents so far that throw very much light on the mathematics of the relation between electricity and gravity. There is, of course, a large number of patents on the general subject of motion and force, and some of these may prove to have some application. There is, however, a series of working postulations embodied in the original **Project Winterhaven,** but no real attempt has been made in the working papers to go into the detailed engineering. All that had actually been achieved up to just under a year ago was a series of fairly accurate extrapolations from the sketchy data that has so

far been actually observed. The extrapolation of 50 mph to 1,800 mph, however, (which is what the present hopes and aspirations amount to) is bound to be a rather vague exercise. This explains American private views that nothing can be reasonably expected from the science yet awhile. Meanwhile, **the NACA** is active, and nearly all the Universities are doing work that borders close to what is involved here, and something fruitful is likely to turn up before very long.
Aviation Report 19 October 1954

GRAVITICS STEPS
Specification writers seem to be still rather stumped to know what to ask for in the very hazy science of electro-gravitic

[Page 27]
propelled vehicles. They are at present faced with having to plan the first family of things - first of these is the most realistic type of operational test rig, and second the first type of test vehicle. In turn **this would lead to sponsoring of a combat disc.** The preliminary test rigs which gave only feeble propulsion have been somewhat improved, but of course the speeds reached so far are only those more associated with what is attained on the roads rather than in the air. **But propulsion is now known to be possible, so it is a matter of feeding enough KVA into condensers with better k figures. 50,000 is a magic figure for the combat saucer - it is this amount of KVA and this amount of k that can be translated into Mach 3 speeds.** Meanwhile Glenn **Martin** now feels ready **to say in public that they are examining the unified field theory to see what can be done.** It would probably be truer to say that Martin and other companies are now looking for men who can make some kind of sense out of Einstein's equations. There's nobody in the air industry at present with the faintest idea of what it is all about. Also, just as necessary, companies have somehow to find administrators who know enough of the mathematics to be able to guess what kind of industrial investment is likely to be necessary for the company to secure the most rewarding prime contracts in the new science. This again is not so easy since much of the mathematics just cannot be translated into words. You either understand the figures, or you cannot ever have it explained to you. This is rather new because even things like indeterminacy in quantum mechanics can be more or less put into words. Perhaps the main thing for management to bear in mind in recruiting men is that essentially **electro-gravitics is a branch of wave technology** and much of it starts with Planck's dimensions of action, energy and time, and **some of this is among the most firm and least controversial sections of modern atomic physics.**
Aviation Report 19 November 1954.

ELECTROGRAVITICS PUZZLE
Back in **1948 and 49,** the public in the U.S. had a surprisingly clear idea of what a flying saucer should, or could, do. There has never at any time been any realistic explanation of

[Page 28]
what propulsion agency could make it do those things, but its ability to move within its own gravitation field was presupposed from its maneuverability. Yet all this was at least two years before electro-static energy was shown to produce propulsion. It is curious that the public were so ahead of the empiricists on this occasion, and there are two possible explanations. One is that optical illusions or atmospheric phenomena offered a preconceived idea of how the ultimate aviation device ought to work. The other explanation might be that this was a recrudescence of Jung's theory of the Universal Mind which moves up and down in relation to the capabilities of the highest intellects and this may be a case of it reaching a very high peak of perception. But for the air industries to realize an electro-gravitic aircraft means a return to basic principles in nuclear physics, and a re-examination of much in wave technology that has hitherto been taken for granted. Anything that goes any way towards proving the unified field theory will have as great a bearing on electro-gravitics efforts as on the furtherance of nuclear power generally.
But the aircraft industry might as well face up to the fact that priorities will in the end be competing with the existing nuclear science commitments. The fact that electro-gravitics has important applications other than for a weapon will however strengthen the case for governments to get in on the work going on.

Aviation Report 28 January 1955

MANAGEMENT NOTE FOR ELECTRO-GRAVITICS

The gas turbine engine produced two new companies in the U.S. engine field and they have, between them, at various times offered the traditional primes rather formidable competition. **Indeed GE** at this moment has, in the view of some, taken the Number Two position. In Britain no new firms managed to get a footing but one, Metro-Vick, might have done if it had put its whole energies into the business. It is on the whole unfortunate for Britain that no bright newcomer has been able to screw up competition in the engine field as English Electric have done in the airframe business.

[Page 29]

Unlike the turbine engine, electro-gravitics is not just a new propulsion system, it is a new mode of thought in aviation **and communications,** and it is something that may become all-embracing. Theoretical studies of the science unfortunately have to extend right down to the mathematics of the meson and there is no escape from that. But the relevant facts wrung from the nature of the nuclear structure will have their impact on the propulsion system, the airframe and also its guidance. The airframe, as such, would not exist, and what is now a complicated stressed structure becomes some convenient form of hard envelope. New companies therefore who **would like to see themselves as major defence prime contractors** in ten or fifteen years time **are the ones most likely to stimulate development.** Several typical companies in Britain and the U.S. come to mind - outfits like AiResearch, **Raytheon, Plessey** in England, Rotax and others. But the companies have to face a decade of costly research into theoretical physics and it means a great deal of trust. Companies are mostly overloaded already and they cannot afford it, **but when they sit down and think about the matter they can scarcely avoid the conclusion that they cannot afford not to be in at the beginning.**

Aviation Report 8 February 1955

ELECTRO-GRAVITICS BREAKTHROUGHS

Lawrence Bell said last week he thought that the tempo of development leading to the use of nuclear fuels and anti-gravitational vehicles (he meant presumably ones that create their own gravitational field independently of the earth's) would accelerate. He added that the breakthroughs now feasible will advance their introduction ahead of the time it has taken to develop the turbojet to its present pitch. Beyond the thermal barrier was a radiation barrier, and he might have added ozone poisoning and meteorite hazards, and beyond that again a time barrier. Time however is not a single calculable entity and Einstein has taught that an absolute barrier to aviation is the environmental barrier in which there are physical limits to any kind of movement from one point in space-time continuum to another. Bell (the company not the man) have a reputation as

[Page 30]

experimentalists and are not so earthy as some of the other U.S. companies; so while this first judgment on progress with electrogravitics is interesting, further word is awaited from the other major elements of the air business. Most of the companies are now studying several forms of propulsion without heat engines though it is early days yet to determine which method will see the light of day first. Procurement will open out because the capabilities of such aircraft are immeasurably greater than those envisaged with any known form of engine.

Aviation Report 15 July 1955

THERMONUCLEAR-ELECTROGRAVITICS INTERACTION

The point has been made that the most likely way of achieving the comparatively low fusion heat needed - 1,000,000 degrees provided it can be sustained (which it cannot be in fission for more than a microsecond or two of time) - is by use of a linear accelerator. The concentration of energy that may be obtained when accelerators are rigged in certain ways make the production of very high temperatures feasible but whether they could be concentrated enough to avoid a thermal heat problem remains to be seen. It has also been suggested that linear accelerators would be the way to develop the

high electrical energies needed for creation of local gravitational systems. It is possible therefore to imagine that the central core of a future air vehicle might be a linear accelerator which would create a local weightless state by use of electrostatic energy and turn heat into energy without chemical processes for propulsion. Eventually - towards the end of this century - the linear accelerator itself would not be required and a ground generating plant would transmit the necessary energy for both purposes by wave propagation.
Aviation Report 30 August 1955

POINT ABOUT THERMONUCLEAR REACTION REACTORS
The 20 year estimate by the AEC last week that lies between present research frontiers and the fusion reactor

[Page 31]
probably refers to the time it will take to tap fusion heat. But it may be thought that rather than use the molecular and chemical processes of twisting heat into thrust it would be more appropriate to use the now heat source in conjunction with some form of nuclear thrust producer which would be in the form of electrostatic energy. The first two Boeing nuclear jet prototypes now under way are being designed to take either molecular jets or nuclear jets in case the latter are held up for one reason or another. But the change from molecular to direct nuclear thrust production in conjunction with the thermonuclear reactor is likely to make the aircraft designed around the latter a totally different breed of cat. It is also expected to take longer than two decades, though younger executives in trade might expect to live to see a prototype.
Aviation Report 14 October 1955

ELECTROGRAVITICS FEASIBILITY
Opinion on the prospects of using electrostatic energy for propulsion, and eventually for creation of a local gravitational field isolated from the earth's has naturally polarized into the two opposite extremes. There are those who say it is nonsense from start to finish, and those who are satisfied from performance already physically manifest that it is possible and will produce air vehicles with absolute capabilities and no moving parts. The feasibility of a Mach 3 fighter (the present aim in studies) is dependent on a rather large k extrapolation, considering **the pair of saucers that have physically demonstrated the principle** only a achieved a speed of **some 30 fps.** But, and this is important, they have attained a working velocity using **very inefficient** (even by to-day's knowledge) form of **condenser** complex. These humble beginnings are surely as hopeful as Whittle's early postulations. It was, by the way, largely due to the early references in Aviation Report that work is gathering momentum in the U.S. Similar studies are beginning in France, and in England some men are on the job full time.
Aviation Report 15 November 1955

[Page 32]
ELECTRO-GRAVITICS EFFORT WIDENING
Companies studying the implications of gravitics are said in a new statement, to include Glenn **Martin, Convair, Sperry-Rand, Sikorsky, Bell, Lear Inc. and Clark Electronics.** Other companies who have previously evinced interest include **Lockheed Douglas and Hiller.** The remainder are not disinterested, but have not given public support to the new science - which is widening all the time. **The approach in the U.S. is in a sense more ambitious than might have been expected.** The logical approach, which has been suggested by Aviation Studies, is to concentrate on **improving the output of electrostatic rigs in existence that are known to be able to provide thrust.** The aim would be to concentrate on electrostatics for propulsion first and widen the practical engineering to include establishment of local gravity force-lines, independent of those of the earth's, to provide unfettered vertical movement as and when the mathematics develops.

However, the U.S. approach is rather to put money into fundamental theoretical physics of gravitation in an effort first to create the local gravitation field. Working rigs would follow in the wake of the basic discoveries. Probably the correct course would be to sponsor both approaches, and **it is**

now time that the military stepped in with big funds. The trouble about the idealistic approach to gravity is that the aircraft companies do not have the men to conduct such work. There is every expectation in any case that the companies likely to find the answers lie outside the aviation field. These would emerge as the masters of aviation in its broadest sense. The feeling is therefore that a company like **A. T. & T.** is most likely to be first in this field. **This giant company** (unknown in the air and weapons field) has already revolutionized modern warfare with the development of the junction transistor and **is expected to find the final answers to absolute vehicle levitation. This therefore is where the bulk of the sponsoring money should go.**
Aviation Report 9 December 1955

* * *

APPENDIX II

ELECTROSTATIC PATENTS

[Page 33]
[This table has been retyped to fit this 60 column file format]
[The report has the dates before the titles on the same line.]
[All other formatting and spelling is exact.]

ELECTROSTATIC MOTORS

(a) American Patents still in force.

2,413,391	**Radio Corp. America**	Power Supply System 20-6-42/31-12-46
2,417,452	**Raytheon Mfg. Co.**	Electrical System 17-1-44/18- 3-47
2,506,472	W.B. Smits	Electrical Ignition Apparatus 3-7-46Holl/ 2- 5-50
2,545,354	G.E.C.	Generator 16-3-50/13- 3-51
(-Engl.P.676,953)		
2,567,373	**Radio Corp. America**	El'static Generator 10-6-49/11- 9-51
2,577,446	Chatham Electronics	El'static Voltage Generator 5-8-50/ 4-12-51
2,578,908	**US-Atomic Energy C.**	El'static Voltage Generator 26-5-47/18-12-51
2,588,513	**Radio Corp. America**	El'static High-Voltage Generator 10-6-49/11- 3-52
2,610,994	Chatham Electronics	El'static Voltage Generator 1-9-50/16- 9-52
2,662,191	P. Okey	El'static Machine 31-7-52/8 -12-53
2,667,615	R.G. Brown	El'static Generator 30-1-52/26- 1-54
2,671,177	Consolidated Eng. Corp	El'static Charging App's 4-9-51/2 - 3-54
2,701,844	H.R. Wasson	El'static Generator of Electricity

		8-1250/ 8- 2-55
2,702,353	US-Navy	Miniature Printed Circuit Electrostatic Generator 17-7-52/15- 2-55

(b) British patents still in force.

651,153	Metr.-Vickers Electr.Co.	Voltage Transformation of electrical energy 20-5-48/14- 3-51
651,295	Ch.F.Warthen sr. (U.S.A.)	Electrostatic A.C. Generator 6-8-48/14- 3-51
731,774	"Licentia"	El'static High-Voltage Generator 19-9-52 & 21-11-52Gy/15- 6-55

(c) French patents still in force.

753,363	H. Chaumat	Moteur electrostatique utilisant l'energie cinetique d'ions gazeux 19-7-32/13-10-33
749,832	H. Chaumat	Machine electrostatique a excitation independante 24-1-32/29- 7-33

[Page 34]
[This table has been retyped to fit this 60 column file format]
[The report has the titles at the end of each line, w/o ().]
[All other formatting and spelling is exact.]

The following patents derive from P. Jolivet (Algiers), marked "A" and from N.J. Felici, E. Gartner (Centre National des Recherches Scientifique - CRNS) later also by R. Morel, M. Point etc. (S.A. des Machines Electrostatiques -SAMES-) and of Societe' d'Appareils de Controle et d'Equipment des Moteurs -SACEM-), marked "G " (because the development was centred at the University of Grenoble).

Mark of Applicant	Application Date	England	America	France	Germany	(Title)
G	8-11-44)			993,017		
	14- 8-45)	637,434	2,486,140	56,027	860,649	
						(Electrostatic Influence Machine
G	17-11-44	639,653	2,523,688	993,052	815,667	
						(Electrostatic Influence Machine
A	28- 2-45			912,444		
						(Inducteurs de Machines el'static)
G	3- 3-45	643,660	2,519,554	995,442	882,586	
						(El'static Machines
A	8- 6-45			915,929		
						(Machines electrocstatiques a flasques)
A	16- 8-45			918,547		
						(Generatrice el'statique)
G	20- 9-45)			998,397		
	21- 9-45)	643,664	2,523,689		837,267	

		(Electrostatic Machines)			
A	4- 2-46			923,593	
		(Generatrice el'statique)			
G	17- 7-46	643,579	2,530,193	1002,031	811,595
		(Generating Machines)			
G	20- 2-47	671,033	2,590,168		
		(Ignition device)			
G	21-3 -47		2,542,494		
		655,474	Re-23,560	944,574	860,650
		(El'static Machines)			
G	6- 6-47	645,916	2,522,106	948,409	810,042
		(El'static Machines)			
A	16- 6-47			947,921	
		(Generatrice el'statique)			
G	16- 1-48	669,645	2,540,327	961,210	810,043
		(El'static Machines)			
G	21- 1-49	669,454	2,617,976	997,991	815,666
		(El'static Machines)			
G	7- 2-49	675,649	2,649,566	1010,924	870,575
		(El'static Machines)			
G	15- 4-49	693,914	2,604,502	1011,902	832,634
		(Commutators for electrical machine)			
G	9-11-49	680,178	2,656,502	1004,950	850,485
		(El'static Generate)			
G	9-10-50)	702,494	2,675,516	1030,623	
	20- 2-51)	(El'static Generate)			
G	29-11-50)	702,421		1028,596	
	20- 2-51)	(El'static Generate)			
G	21-11-51	719,687		1051,430	F10421
		(El'static Machines)			
G	20- 8-52	731,773	2,702,869		938,198
		(El'static Machines)			
G	6-11-52	745,489			
		(El'static Generator)			
G	12- 2-53	745,783			
		(Rotating El'static Machines)			
G	8- 1-52	715,010	2,685,654	1047,591	
		(Rotating El'static Machines producing a periodical discharge.)			
		Appl'n.No.			

G	27- 2-54	5726/55			
		(El'static Machines)			
G	8- 3-54	6790/55			
		(El'static Machines)			
G	28- 1-55	2748/56			
		(El'static Machines)			

NOTE:- ALL THE LISTED PATENTS ARE STILL IN FORCE

[End of the Report]

Obviously, both the government and big private businesses were, at the time, intensely interested in antigravity (they called it "counterbary"). And, apparently, they had working models that could fly by actually altering or defeating gravity.

I believe the above document to be authentic. It was probably obtained by a FOIA (Freedom Of Information Act) request by Dr. StrangeQ. Also, older secret documents are routinely de-classified and released, per normal government procedure. Or, perhaps, the document was copied and "leaked" out by someone on the "inside".

At any rate, I believe the report to be authentic for several reasons. One is that I once worked for a nuclear facility, where a "Q" Secret clearance (which I had at the time) was required for access, and I am familiar with "Lithium 6 Hydride" which is mentioned in the report. This substance was for years considered secret, but information about it has somehow been released in the last few years. The nuclear industry now uses something even better than Lithium Hydride, in fact. Another reason for authenticity is the mention of "NACA", which is, in fact, what NASA was called back in those days. Also, I recognize all the big company names mentioned in the document. Many of these are still in existence today, but with a few name changes. All government documents are assigned a unique number (which this document has), and are thus retrievable and checkable at the national document archives.

Now, here are excerpts from the writings of W. L. Moore, from copies of which can be found on numerous web pages on the Internet at the time of this writing. They give historical and biographical information on **Thomas Townsend Brown.**

Originally From: "The Wizard of Electro-gravity"
by William L. Moore
May 1978

...Thomas Townsend Brown was born into a prominent Zanesville, Ohio family in 1905. Brown displayed early in life an interest in space travel - a subject considered sheer fantasy in the days when there were still those who looked askance at the Wright Brothers' flying machines. Nonetheless, young Brown was not so easily dissuaded, and enjoyed dabbling with what then regarded as "modern" electronics. It was his youthful toying with the then infant ideas of radio and electromagnetism that provided a background which was to be invaluable to him in later years; and it was during the course of this experimenting that Brown somehow acquired a Coolidge X-ray tube - an item that was to lead him to make a startling discovery. X-rays,(or Roentgen Rays) were indeed mysterious forces in those days (in fact, American physical chemist William D. Coolidge had only just invented the "Coolidge tube" itself in 1913), and even legitimate science was only beginning to learn anything about them. Brown wasn't interested in the X-rays, per se, however. Somewhere in his head rested the idea that maybe a key to space flight might be found here; and toward that end, he set up an experiment to determine whether there might be a useful force of some sort exerted by the X-rays emanating from his Coolidge tube. Trying something that no other scientist of his day had thought of, Brown mounted his tube in extremely delicate balance and began "testing" for results. To his disappointment, he was unable to detect any measurable force exerted by the rays regardless of which way he turned his apparatus; but to his amazement, he did note a very strange quality of the Coolidge tube itself. Every time it was turned on, the tube seemed to exhibit a motion of its own - a "thrust" of some sort, as if the apparatus was trying to move! Investigating further, Brown had to spend considerable time and effort before the truth finally dawned. The X-rays had nothing whatsoever to do with this new-found phenomenon - it was the high voltage used to produce the rays which was behind it! Brown now began a series of experiments designed to determine the nature of the "force" he had discovered, and after much effort finally succeeded in developing a device which he optimistically called a "Gravitor." His invention looked like nothing more than a Bakelite case some twelve inches long and four inches square, but when placed on a scale and connected to a one hundred kilovolt power source, the apparatus proceeded to gain or lose about one percent of its weight (depending on polarity). Brown was sure he had discovered a new electrical principle, but he remained unsure of just what to do with it. And in spite of the fact that there were a few newspaper accounts of his work, no scientist of any stature expressed an interest in his discovery - a not entirely surprising reaction when one considers that Brown was only then about to graduate from high school!

Readily recognizing his youth as a handicap, Brown elected to "proceed with caution," and in 1922 he entered the California Institute of Technology (Caltech) at Pasadena, CA. as a "promising young freshman," and spent his first year courting the favor of his professors - among them the late physicist and Nobel laureate, Dr. Robert A. Millikan. His success in being able to convince his instructors of his excellence as a lab man was offset by his complete inability to gain even the slightest measure of recognition for his ideas about electro-gravity. **His teachers, steeped to the last in the rigors of 19th century scientific discipline, steadfastly refused to admit that such a thing could exist, and hence, "weren't interested."** Undaunted, Brown transferred nearer to home to Kenyon College (Gambier, Ohio) in 1923, remaining there only a year and then transferring to Denison University at Granville, Ohio, where he studied as an electronics resident in the Department of Physics under Dr. Paul Alfred Biefeld, professor of physics and astronomy and former classmate, in Switzerland, of Dr. Albert Einstein. Unlike Dr. Millikan at Caltech, Biefeld proved to be interested in Brown's discovery, and together the two of them, professor and student, experimenting with charged electrical capacitors, developed a principle of physics which came to be tentatively known as the "Biefeld-Brown Effect." Basically, the "effect" concerned the observed tendency of a highly charged electrical condenser to exhibit motion toward its positive pole - the same motion observed earlier by Brown with his Coolidge tube.

...In 1939, Brown, now a lieutenant in the naval reserve went to Maryland as a material engineer for the Glenn L. Martin Company of Baltimore (later Martin Aerospace), but was there only a matter of months when he was called upon by the Navy to become officer in charge of magnetic and acoustic minesweeping research and development under the Bureau of Ships. He served faithfully, presiding over the expenditure of nearly $50 million for research (there were some fifteen Ph.D.'s responsible to Brown at one point), and even consulting with Einstein himself on occasion (the common bond, remember, was Dr. Biefeld), until after Pearl Harbor when he was transferred, with the rank of lieutenant commander, to Norfolk to continue his research while heading up the Navy's Atlantic Fleet Radar School there. The early years of the war saw Lieutenant Commander Brown deeply involved as a physicist with projects conducted under the National Defense Research Committee (NDRC), and later under its successor, the Office of Scientific Research headed by **Dr. Vannevar Bush.** Among other things, Brown performed some very valuable high-vacuum work as well as experiments centered on perfecting methods of ship degaussing. However, the combined effects of his having worked "too long and too hard," and of his personal disappointment in the failure of his projects to gain proper recognition resulted in a nervous collapse in December of 1943. Retirement from the service quickly followed and Brown was sent home to rest. Six months later, the spring of 1944 found him working as a radar consultant for the advanced design section of Lockheed-Vega Aircraft Corporation in California. Colleagues referred to him as a "quiet, modest, retiring man...a brilliant solver of engineering problems" and "exactly the sort (of man) one expects to find in important research installations."

More importantly, he was still working on his Gravitator, although, interestingly, Brown would not speak in terms of gravity when describing it - preferring rather to use the more scientific but decidedly less sensational term "stress in dielectrics."

Things began to look up just a bit in the post-war years. After leaving Lockheed, Brown went to Hawaii to take up private residence and to continue his research. It was during this time, partly through the efforts of an old friend (A.L. Kitselman) who was then teaching calculus at Pearl Harbor, that Brown's Gravitor device, somewhat improved over earlier editions, came to the interest of none other than Admiral Arthur W. Radford, Commander in Chief of the U.S. Pacific Fleet (later to become Chairman of the Joint Chiefs of Staff under President Eisenhower, 1953-57). As a result of Admiral Radford's interest, Brown was temporarily accorded consultant status to the Pearl harbor Navy Yard, but in spite of the fact that the former lieutenant commander was well treated by his navy friends, it appears from the evidence that they considered his invention as rather more of an interesting curiosity rather than any sort of key to space travel. Perhaps, to engage in a bit of speculation, had Brown been more of a salesman than a scientist, things might have been different. In the meantime, the appearance of UFO's on the American scene at the turn of the decade had succeeded in capturing Brown's personal interest. Eagerly following the controversy as it raged among the military and scientific community in the late forties and early fifties, Brown postulated that perhaps with the proper worldwide scientific approach, the question of how UFO's are powered might be solved. In those days, his belief in the abilities of modern science was such that he even dared to speculate on the possibility of a quick solution, given the proper resources and manpower, and, of course, he remained constantly aware of the possibility that he had, through his own efforts at research into electrogravity, hit upon one of the keys to the mystery. Moving to Cleveland in **1952,** Brown conceived of a project he called **"Winterhaven."** An idea which he hoped with proper refinements **could be offered for sale to the military establishment.** Through patient research, he succeeded in improving the lift force of his Gravitator apparatus until it was such that it could lift significantly in excess of one hundred per-cent of its own weight - a success that should have raised the eyebrows of any respectable scientist or pentagon official - but apparently didn't, even though the apparatus involved was quite sophisticated and, as we shall see, the demonstrations most impressive.

According to modern science, everything in the known universe owes its existence to three basic energies or forces: electromagnetism, nuclear forces and gravity. Whether these three are separate forces, or whether they are manifestations of some more basic unifying force is still a matter of conjecture. Indeed, **Albert Einstein's** life work was largely devoted to trying to perfect a theory of Unified Field, and in the process of trying to derive the field equations involved, **he came to speculate that what we call "matter" is, in reality, only a local phenomenon exhibited by areas of extreme field-energy concentration.** Even establishment science does not question the patently obvious relationship between electricity and magnetism, but the relationship of these two fields to the "gravity field" constitutes an area physics, which, more than twenty years after Einstein's death, is **still largely incomprehensible to modern science.** In general, most of **orthodox science** in the seventies **does tend to recognize** a loose linking or "coupling" effect of some sort **between electrical and gravitational forces,** but **precious few scientists** have seen fit to speculate that this coupling effect might be at all applicable. **At least, such is the case officially, although there exists sufficient reason to suspect that there may have been significant advancements in this area which are still well hidden under that proverbial "brass lid" emblazoned with the phrase "Top Secret."** In any event, Townsend Brown's departure from orthodoxy rests on the above point. Brown firmly believes there is a linking force between gravity and electricity. Whether there may be a further connection between magnetism and gravity, and hence a "unifying" field relationship between all three is yet another question. But to get back to basics; Townsend Brown believes - and his experiments seem to bear him out - that the Biefeld-Brown Effect manifests a proven link between electricity and gravity. A "dielectric" is defined as a material which has the unique ability of absorbing electrical energy or "charge" without ordinarily passing this energy on to neighboring materials. Some dielectrics are able to absorb enormous quantities of electrical energy (also referred to as "elastic stress") without discharging, providing that the energy is fed into the dielectric slowly and at low potential. Still others can be charged at extremely high potential at a rate equal to several thousand times each second. Townsend Brown concerned himself principally with this latter type. Using just such a dielectric, Brown constructed disc (or saucer) shaped condensers, and, by applying various amounts of high voltage direct current, witnessed the Biefeld-Brown effect in action. With the proper construction and electrical potential (in the kilovolt range) the disc-shaped "airfoils" were made to fly under their own power, emitting a slight hum and a bluish electrical glow as they did so. More scientifically, perhaps, this process of "flight" might best be described as "motion under the influence of interaction between electrical and gravitational fields in the direction of the positive electrode." In **1953,** Brown succeeded in demonstrating, in his laboratories, the flight of disc-shaped air foils two feet in diameter around a circular course twenty feet in diameter. The process involved tethering these saucer-shaped craft to a central pole by means of a wire through which the necessary D.C. electrical potential was supplied at a rate of fifty thousand volts with a continuous input of fifty watts. The tests produced an observable top speed of an **amazing seventeen feet per-second** (11.5 miles per hour).

Working with almost superhuman determination and at great cost to his personal finances, Brown soon succeeded in surpassing even this accomplishment. **At his next display, he exhibited a set of disks three feet across flying a fifty foot diameter course with results so spectacular** that they were **immediately classified.** Even so, most scientists who witnessed the demonstrations remained skeptical and generally attributed Brown's motive force to what they called an "electrical wind," in spite of the fact that a veritable "electrical hurricane" would have to be involved to produce the lift-potential observed. Pitiful few gave any credence whatsoever to the idea that the Biefeld-Brown Effect might represent anything new in the world of physics. Government funds were sought to enable the work to continue, but in **1955,** seeing that the money was not forthcoming, a disgruntled Brown went to Europe in hopes that perhaps he might be able to generate a little more entusiasm on the continent. Demonstrations were given first in England, but it was on the mainland under the auspices of a French corporation, La Socie'te' National de Construction Aeronautique Sud Ouest (SNCASO), that things really began to look promising. During a set of tests performed confidentially within the company's research laboratory, **Brown succeeded in flying some of his discs in a high vacuum with amazing results, thereby proving that, in fact, his discs flew more efficiently without air.** Also proven during this series of experiments was that the speed and efficiency of the "craft" could be increased by providing greater voltage to the dielectric plates. **Contemporary accounts easily visualized speeds of several hundred miles-per-hour using voltages in the range of one to two hundred thousand electron volts,** and at least one writer spoke of a "flame jet generator" then in the planning stage, which supposedly would be able to provide power potential up to 15 million volts! In fact, plans had been laid for the immediate construction of a large vacuum chamber and a one-half million volt power supply when disaster struck the project in the form of a corporate merger. SNCASO had agreed to combine with a larger company to form what was termed a "Super Douglas of France": Sud Est. The president of the emerging company proceeded to demonstrate an appalling lack of interest in "these far-out propulsion research efforts" and favored instead an increased interest in "air frame manufacture." All facilities designated and created by the former president to carry on the work on electro-gravity were summarily canceled and a thoroughly disappointed Brown was forced to return home to the U.S. in **1956.** The summer of that year found him living in the Washington, D.C. area still interested in UFO research and hoping fervently that if scientific evidence could be uncovered suggesting their possible method of propulsion, his own work would be greatly enhanced - an idea which leads us down yet another avenue of Townsend Brown's life. ...

... Within a year, he was busily engaged as chief research and development consultant for the Whitehall Rand Project, a new **anti-gravity venture being conducted under the personal auspices of Agnew Bahnson, president of the Bahnson Company of Winston-Salem, North Carolina.**

In **1958,** believing he had finally generated enough momentum to "go it alone," Townsend Brown organized **his own corporation under the name of Rand International Limited,** and set himself up as president. Although numerous patents were applied for and granted both in the U.S. and abroad, and in spite of many patiently given demonstrations to interested governmental and corporate groups, success again eluded him. In the early sixties, Brown did a brief stint as physicist for **Electrokinetics Inc.,** of Bala Cynwyd, PA. and upon terminating his employment there, went into semi-retirement. Since then, he has lived on in California, quietly pursuing his research in hopes that perhaps someday, with a little luck, the world will notice. His most recent involvement is with a project housed largely at Stanford Research Institute with additional assistance being provided by the University of

California and the **Ames Research Center of NASA.** The object of the research, details of which are still largely under wraps, is to try to determine what connection there is, if any, between the earth's gravitational field and rock electricity (petroelectricity).

Which, of course, leads us to the prime question of this article: Why indeed has Townsend Brown's impressive life's work gone so seemingly unnoticed for the past three decades? Even today, Brown is still of the opinion that further research into the Biefeld-Brown Effect could lead to a sensational breakthrough in space propulsion methods, not to mention the more domestic variety - if appropriate funding could be made available. Granted, research is expensive, but - is money the real reason for the apparent lack of interest? Perhaps. Or maybe, as Brown himself suggests, the human race is not yet ready to accept a scientific breakthrough that could place man within reach of the stars.

ALSO:

Originally From: "The Wizard of Electro-gravity <u>Revisited</u>"
by William L. Moore

During the **1930's and '40's,** working with principles developed in association with Dr. Paul Alfred Biefeld, erstwhile friend and associate of Dr. Einstein, Brown succeeded in demonstrating that certain properly constructed devices, when "energized" with a strong D.C. potential (up to 300 kilovolts), could be made to exhibit a substantial loss of weight without an accompanying loss of mass (see The Wizard of Electro-gravity, UFO Report, May 1978). Determined efforts to refine techniques and perfect methods finally bore fruit when, in the **middle 1950's,** Brown successfully flew, both in the air and later in vacuum, electro-gravitic disc-shaped "airfoils" powered only by high voltage direct current. **These were demonstrated flying in 50 feet circles at speeds so incredible they were immediately classified.** The principle involved was one discovered by Brown himself - namely that certain high-K (capacitance) dielectrics, when subjected to high-voltage charges in the 50 to 300 kilovolt range with constant input, will exhibit motion toward the positive pole. **Brown's theory is that the <u>capacitor</u> is a useful tool in demonstrating the link between electricity and gravity in the same way that the <u>coil</u> is capable of representing the link between electricity and magnetism.** He felt that given time as well as adequate funding and laboratory facilities, the problem of supplying the necessary electrical potential to the dielectrics from an internally transported power source while still enabling the craft to lift and maneuver efficiently could finally be overcome.

With respect to the saucers however, certainly it is conceivable that any technology significantly more advanced than our own could have long ago overcome the difficulties involved and perfected a propulsion system utilizing principles and techniques the existence of which is only hinted at in Townsend Brown's work.

What follows is one of the more significant patents (<u>if not THE most!</u>) created by Townsend Brown. The drawings are shown first:

June 1, 1965

T. T. BROWN

ELECTROKINETIC APPARATUS

3,187,206

Filed May 9, 1958

2 Sheets-Sheet 1

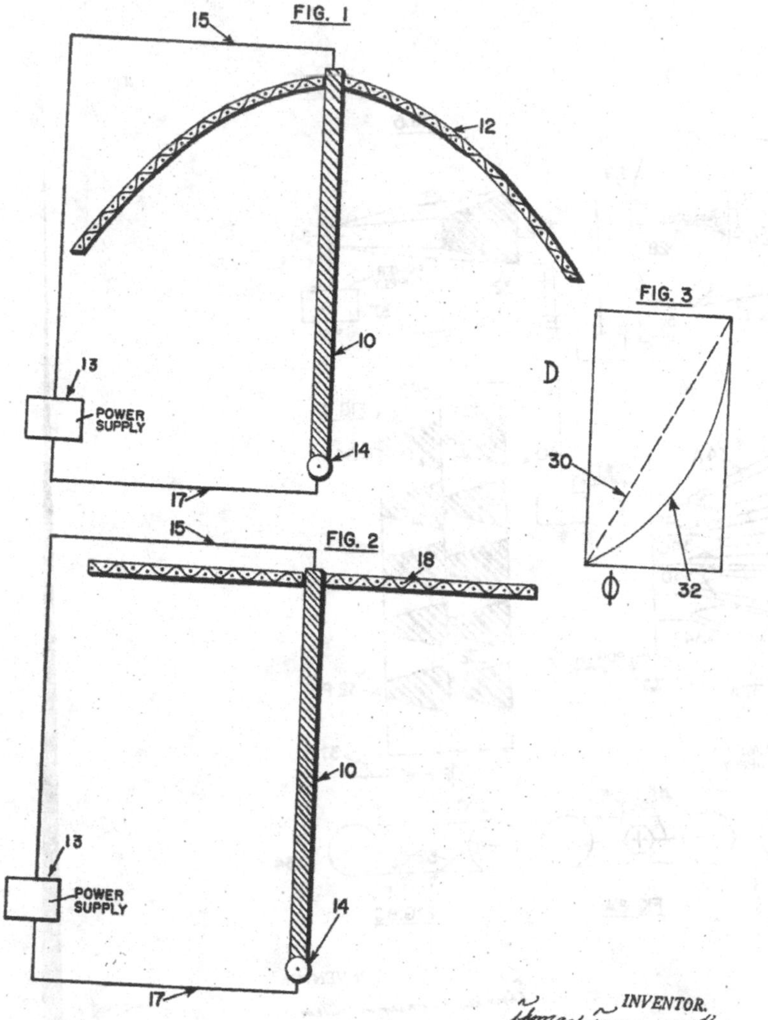

INVENTOR.
Thomas Townsend Brown

BY

Watson, Cole, Grindle & Watson
Attorneys

June 1, 1965 | T. T. BROWN | 3,187,206

ELECTROKINETIC APPARATUS

Filed May 9, 1958

2 Sheets-Sheet 2

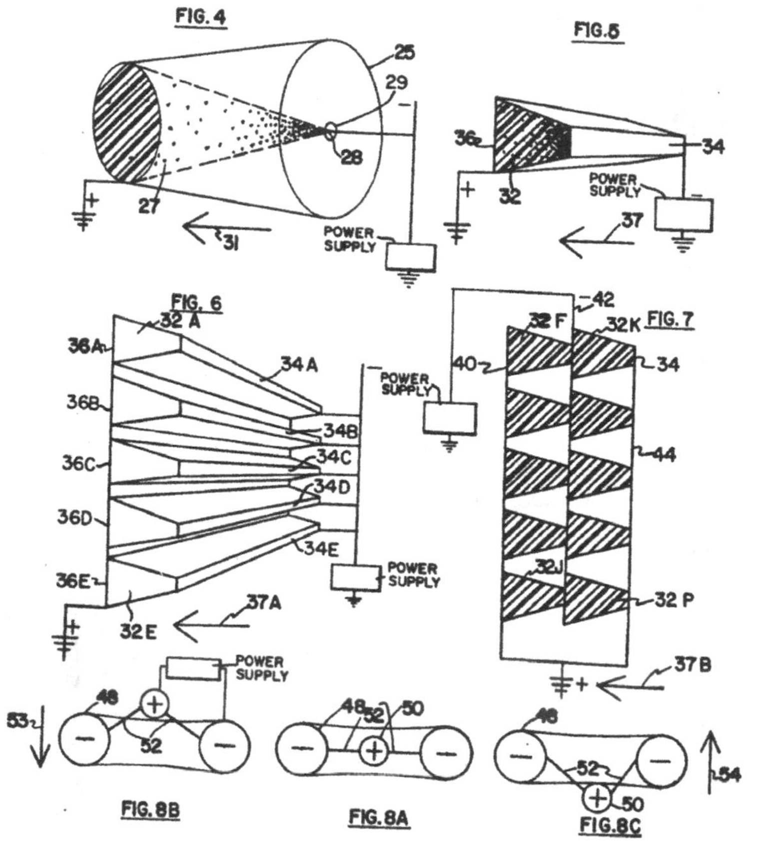

INVENTOR
Thomas Townsend Brown

BY
Watson, Cole, Grindle & Watson
ATTORNEYS

United States Patent Office
3,187,206
Patented June 1, 1965

1

3,187,206
ELECTROKINETIC APPARATUS
Thomas Townsend Brown, Walkertown, N.C., assignor, by mesne assignments, to Electrokinetics, Inc., a corporation of Pennsylvania
Filed May 9, 1958, Ser. No. 734,342
23 Claims (Cl. 310-5)

This invention relates to an electrical device for producing a thrust by the direct operation of electrical fields.

I have discovered that a **shaped electrical field** may be employed to propel a device relative to its surroundings in a manner which is both novel and useful. Mechanical forces are created which move the device continuously in one direction **while the masses making up the environment move in the opposite direction.**

When the device is operated in a dielectric fluid medium, such as air, the forces of reaction appear to be present in that medium as well as on all solid material bodies making up the physical environment.

In a **vacuum,** the **reaction forces** appear on the solid environmental bodies, **such as the walls of the vacuum chamber.** The propelling force however is **not reduced to zero** when **all environmental bodies are removed** beyond the apparent effective range of the electrical field.

By attaching a pair of electrodes to opposite end of a dielectric member and connecting a source of high electrostatic potential to these electrodes, a **force** is produced **in the direction of one electrode** provided that electrode **is of such configuration to cause the lines-of-force to converge steeply upon the other electrode.** The **force,** therefore, **is in a direction from the region of high flux density toward the region of low flux density,** generally in the direction through the axis of the electrodes.

The thrust produced by such a device is present if the Electrostatic field gradient between the two electrodes is non-linear. This **non-linearity of gradient** may result from a difference in the configuration of the electrodes, from the electrical potential and/or polarity of adjacent bodies, from the shape of the dielectric member, from a gradient in the density, electric conductivity, electric permittivity and magnetic permeability of the dielectric member or a combination of these factors.

A basic device for producing force by means of electrodes attached to a dielectric member is disclosed in my Patent 1,974,483. In one embodiment disclosed in my patent, an electrostatic motor comprises devices having a number of radially directed fins extended from one end of the dielectric body and a point electrode on the opposite end of the dielectric body. When this device is supported in a fluid medium, such as air, and a high electrostatic potential is applied between the two electrodes, a thrust is produced in the direction of the end to which the fins are attached.

Other electrostatic devices for producing thrust are disclosed and described in detail in my British Patent 300,-311, issued **August 15, 1927.**

Recent investigations in electrostatic propulsion have led to the discovery of improved devices for producing thrust by the use of electrical vectorial forces.

Accordingly, it is the primary object of this invention to provide an improved electrical device for producing thrust.

It is another object of this invention to provide a device for producing modulated thrust in response to varying electrical signals, which device produces a greater effect than the prior type devices mentioned above.

It is another object of this invention to provide **a device which shapes or concentrates electrostatic flux** to produce an improved thrust.

Broadly, the invention relates to shaping an electrical field to produce a force upon the device that shapes the field. **The electrical field is shaped by the use of an elec-**

trode of special configuration whereby the electric lines-of-force are made to converge at a distance from the electrode. One illustrative embodiment of this invention which satisfies the above requirement is an **arcuate surface,** or alternatively, a system of wires, tubes or plates embedded in a dielectric surface and forming a directive array. One such highly-charged electrode acting within and upon an ambient of differential electrical potential will move in response to the forces created by shaping of the electrostatic field.

If a smaller electrode is added at or near the focus of the field-shaping electrode, and mechchanically attached to that electrode, both electrodes as a system will move in a direction of the larger or field-shaping electrode.

As is mentioned above, the field-shaping electrode alone, when charged with respect to its electric ambient, will move or possess a force in the direction of its apex. **If another electrode carrying a different charge is added at or near the focal point of the field-shaping electrode,** then the field becomes more concentrated, i.e. shaped to a greater degree and the resulting thrust is **greater that that which exists when the field-shaping electrode alone is employed.**

Briefly in accordance with aspects of this invention, ***an electrode is connected on each end of a dielectric member*** *and one of the electrodes defines a* ***large area flat*** *or preferably* ***arcuate surface*** *which is curved in such a direction to produce, usually in co-operation with the other electrode, a shaped electrostatic field.*

Advantageously, if the arcuate electrode is in the form of a parabola or hyperbola, the length of the dielectric member may be such that the other electrode is located in the region of the focus of the parabola or hyperbola, as the case may be. If the arcuate electrode is hemispherical, the other electrode is located near the center of the hemisphere.

In accordance with other aspects of this invention the dielectric member supporting the two electrodes may have electrical conductivity and/or dielectric constant which varies progressively between its ends so that the dielectric member contributes to the non linearity of the field gradient and causes a greater thrust to be developed.

In accordance with still other aspects of this invention, an annular electrode member is secured to an electrode mounted in the region of the axis of the annular electrode. If the second electrode is located at the center of the annular electrode and the two electrodes are energized, such force is not detected. However, if the second or innermost electrode is displaced from the center of the annular electrode in the region of the axis of the annular electrode and the electrodes are energized, then thrust will be produced by the two electrodes. The annular electrode may either be a flat ring, a toroid, or a section of a cylinder.

In accordance with still other aspects of this invention, tapered dielectric members having electrodes secured to

opposite edges thereof may be employed to produce a thrust in response to the application of potentials to these electrodes. The thrust produced by these tapered dielectric members maybe further augmented by embedding massive particles, such as lead oxide, in the wedges, which particles are usually more concentrated near the points of the wedges.

Accordingly, it is a feature of this invention to provide an electrical device for producing thrust which includes a dielectric member and electrodes supported at each end of the dielectric member, one of which electrodes is located in the region of the focal point of the arc of the arcuate surface electrode.

It is another feature of this invention to provide a device for producing thrust having a dielectric member and a pair of electrodes secured to opposite ends of the dielectric rod or member, one of which electrodes de-

3

fines a parabolic or hyperbolic surface, the other electrode being located in the region of the focus of said surface.

It is another feature of this invention to employ an insulating rod or member between the two electrodes, which rod or member has a varying dielectric constant, said dielectric constant progressively increasing or decreasing along the length of the dielectric member.

It is still another feature of this invention to employ a rod or member connected between the two electrodes across which an electrostatic potential is applied, which rod or member has a varying electrical conductivity, said conductivity progressively increasing or decreasing along the length of the dielectric member.

It is another feature of this invention to employ a single electrode having an arcuate surface and to connect a source of potential to the arcuate surface which is opposite in polarity to the potential of the masses comprising the environment of the arcuate surface.

It is still a further feature of this invention to employ an arcuate electrode as a device for producing thrust and to apply a varying electrical signal to the arcuate electrode.

It is still another feature of this invention to employ a wedge of dielectric material having electrodes on opposite ends thereof to produce a thrust in response to

the application of electrical potentials.

It is still a further feature of this invention to employ a tapered dielectric material having massive particles embedded therein to produce a thrust in response to the application of potentials to the electrodes secured to the dielectric member.

It is still a further feature of this invention to employ an annular ring electrode and a second electrode secured to the annular electrode in the region of the axis of the annular electrode in the region of the axis of the annular electrode to produce a thrust in response to the application of electrical potentials thereto.

These and other various objects and features of this invention will be apparent from consideration of the following description when read in connection with the accompanying drawing wherein:

FIGURE 1 is a view in elevation of one illustrative embodiment of this invention;

FIGURE 2 is a view in elevation, partly in section, of another illustrative embodiment of this invention;

FIGURE 3 is a graphical representation of the field gradient between the electrodes of one illustrative example of this invention in which distance from one electrode is plotted as the abscissa whereas flux density is plotted as the ordinate;

FIGURE 4 is a perspective view of another illustrative embodiment of this invention;

FIGURE 5 and 6 are perspective views of still another illustrative embodiment of this invention;

FIGURE 7 is an end view of another illustrative embodiment of this invention employing a pair of devices of the type disclosed in FIGURE 6, which devices are mounted and serially connected in a single array;

FIGURES 8A, 8B and 8C are views in elevation, partly in section, of still other illustrative embodiments of this invention.

Referring to FIGURE 1, there is depicted an insulating member 10 having an arcuate electrode 12 mounted on one end thereof and a second electrode 14 mounted on the opposite end thereof. A source of direct current voltage 13 is connected to electrodes 12 and 14 through conductors 15 and 17, respectively. I have discovered that **if two electrodes are mounted on opposite end of a dielectric member,** and a field emanates from these electrodes which **produces a linear gradient** through the dielectric member as shown by the dotted line 30 of FIGURE 3, **then no thrust is produced** by the dielectric member. **However,** if the **field is distorted** to produce a **non-linear gradient** such as graphically represented by

4

line 32 in FIGURE 3, **then a thrust will be produced,** which thrust will be related to the degree of non-linearity of the field gradient. One way to produce a gradient which varies non linearly is to shape one of the electrodes in a form **of an arcuate surface such as 12.** However, **numerous other ways to influence the field gradient will be disclosed below.** Electrode 14 represents a substantial mass and it has been found that best results are obtained if the surface area of electrode 14 is greater than the surface area of the end of rod 10. In one particular example, a spherical electrode having a diameter greater than the diameter of rod 10, produced very satisfactory results. Advantageously, the dielectric member 10, may be employed to increase the non-linearity of the field gradient. For example, the dielectric member may be of material having a uniform relative dielectric constant and be tapered in the direction of electrode 14 such that the member 10 in the region of electrode 12 has a much greater cross-sectional area than the end of member 10 which is connected to electrode 14. An equivalent result may be obtained if the member 10 is of uniform diameter but has a dielectric of graduated density or which comprises a material having a progressively different electrical conductivity or dielectric constant, or alternatively the electrical conductivity, varies from a low value in the region of electrode 14 to a high value in the region of electrode 12.

The arcuate electrode 12 may be either a stitched wire surface or a solid conducting surface. In the case of stitched wire surface, the wires are very close together so that when an electrical potential is applied to these wires, they act substantially in the same manner as a conductive surface. Arcuate electrode 12 will produce a thrust when a potential is applied to the electrode 12 which is opposite in polarity to the potential of the bodies in the region of electrode 12. Such a thrust will be produced even though the dielectric member 10 and the electrode 14 are eliminated from the structure. However, the thrust produced bye the charged arcuate electrode 12 when actuating alone is less than the thrust produced by the combined device, that is, employing the dielectric member 10 and the oppositely charged electrode 14.

Referring now to **FIGURE 2,** there is depicted another illustrative embodiment of this invention in which field-shaping is accomplished. In the embodiment **of FIGURE 2,** the **planar electrode 18 is connected to a** hemispherical **electrode 14** by **means of a dielectric rod 10.**

When a source of electrical potential (not shown) is connected through wires 15 and 17 to electrodes 18 and 14, respectively, a field gradient will be produced between electrodes 18 and 14, which field gradient varies in accordance with the graph represented by the solid line 32 of FIGURE 3. In this particular embodiment, as well as in the embodiment of FIGURE 1, the non-linearity of the field gradient is further augmented by the use of a connecting rod 10 which is a dielectric with progressively different dielectric constant between electrodes 18 and 14. A similar result may be produced by the use of a rod 10 having electrical conductivity which varies progressively between electrodes 18 and 14.

Referring now to FIGURE 4 there is depicted still another illustrative embodiment of this invention in which a thrust is produced in response to the application of electrical potentials.

A frusto-conical surface 25 comprising a metal or having a metal surface to be used on an electrode is connected to a tapered member 27. The tapered member 27 is frusto-conical and is primarily of non-conductive material but contains granules of semi-conducting material, which granules are concentrated neat the tip 28. Mounted on tip 28 is a half-wave radiator 29 which may be in the form of a disk. It is noted that the axis

5

of member 27 coincides with the axis of member 25. When a source of potential is connected to electrodes 25 and 29, **a thrust is produced** in the direction of the arrow 31 **regardless of the polarity of the applied voltage.** However, **a greater thrust** is produced if the electrode 25 is **positive** with respect to electrode 29. Alternating current voltages may also be applied to electrodes 25 and 29 and the potential may be either superimposed upon or substituted for the direct current voltages. Preferably, the frequency of the applied A.C. voltage is such that the diameter of the disk 29 constitutes a half-wave length of the applied voltage.

Referring now to **FIGURE 5** there is disclosed a tapered member 32 which is non-conductive material and may contain particles of semi-conducting material in a manner similar to member 27. The semi-conducting material contained in member 32 and in member 27 may be any convenient form of massive particles such as lead oxide. Along one surface of member 32 is an electrode 34 while along the opposite surface is another electrode 36. When a potential is applied to these electrodes, **preferably** of a polarity such that electrode 36 is positive with respect to electrode 34, **a thrust is produced in the direction of the arrow 37.** In the devices disclosed in both of FIGURES 4 and 5, the thrust produced by the electrodes is augmented by the varying cross-sectional area of the non-conductive member connecting the electrodes and is further augmented by the voltage gradient produced by the embedded particles, which voltage gradient is greater than that which would be introduced by a tapered non-conductive member without embedded particles.

Referring now to **FIGURE 6** there is depicted a **bank of members** 32 such as disclosed in FIGURE 5 in which like electrodes 36A through 36E are secured together by a connector in any convenient form, such as plate 38. Each of these members 32A through 32E **produces a thrust in the direction of the arrow 37A** and the resultant force is equal to the sum of the thrust produced by the individual members 32 in response to the application of potentials to the electrodes 34A-34E and 36.

In **FIGURE 7** there is depicted a pair of banks of members, such as depicted in FIGURE 6, in which the electrodes are serially connected. In this particular instance, a plate or other member 40 comprises an electrode on which are mounted an array of members 32F through 32J. A second electrode 42 is secured between electrodes 32F through 32J and electrodes 32K through 32P. A third electrode 44 is connected to the electrode 34 on each of members 32K through 32P. It is to be noted that **electrodes 40 and 44 are connected to a source of one potential while electrode 42 is connected to a source of the opposite potential.** The **thrust produced** by this array **is in the direction of arrow 37B** and the manner in which this thrust is produced is similar to that explained in connection with FIGURES 5 and 6, although it would appear that electrode 42 will experience a natural attraction for electrodes 40 and 44. **A non-linear field gradient is produced** between these electrodes **by the varying cross-sectional area of members 32** and by the presence of semi-conducting particles in members 32. This **non-linear** field gradient gives rise to the thrust, as mentioned above.

Referring now to FIGURES 8A, 8B and 8C there is depicted other illustrative embodiments of this invention. In FIGURE 8A a toroid member 43 has an electrode 50 supported at its center by means of insulating rod 52. If the electrode 50 and the toroid member 43 are both conducting surfaces defining electrodes and these electrodes are connected to sources of opposite potential, no thrust will be developed by the device. If, however, as depicted in FIGURE 8B electrode 50 is translated along the axis of generation of toroid or annular member 48 and again supported by non-conductive members 52, this device will experience a downward thrust, as indicated by arrow 53, in response to the application of potentials of either

6

polarity. It is believed that this force is produced by the annular configuration of electrode 48 and the off central location of electrode 50. In the instance of FIGURE 8C, electrode 50 is positioned beneath the center of electrode 48 and positioned on the axis of generation of electrode 48. When potentials are applied to electrodes 48 and 50 in FIGURE 8C, a thrust is produced in an upward direction, as indicated by arrow 54. Here again the field gradient is produced by the configuration of electrode 48 and the location of electrode 50 with respect to electrode 48.

From the foregoing discussion, it is also apparent that a combination of a curved electrode, a supporting member of varying cross-sectional area, and a second electrode supported by the connecting member will produce a thrust along the axis of the curved electrode when potentials are applied to the electrodes. Similarly, a thrust may be developed between plane electrodes of unequal areas which are **connected by a member of varying cross-sectional area.** The thrust developed by this last mentioned device is further increased by the introduction of semi-conductive particles in the non-conducting member, which particles are more concentrated in the region of the smaller electrode than in the region of the larger electrode. **Further, these tapered members having planar electrodes connected to opposite surfaces may be stacked in vertical arrays and connected in parallel, or they may be stacked in vertical arrays connected in series with similar vertical arrays.**

In applying potentials to these various embodiments, it has been found **that the rate at which the potential is applied often influences the thrust.** This is especially true where the dielectric members of high dielectric constant are used and the charging time is a factor. In such cases, the field gradient changes as the charge is built up. In such cases where initial charging currents are high, **dielectric materials of high magnetic permeability likewise exhibit thrust with time.**

One advantageous manner of applying potential is that of employing potentials which vary cyclically.

It is thus apparent that one embodiment of this invention embodies a pair of electrodes mounted on an insulating member, one of which electrodes defines an arcuate surface to produce an improved thrust in response to the application of direct current potentials. It is also apparent that this thrust is augmented by increasing the non-linearity of the field gradient by a progressively-changing characteristic of the dielectric member connecting these electrodes. This non-linearity of field may be produced by a gradient in electric conductivity, electric permittivity and/or magnetic permeability along the length of the \ member, or it may result from a change in the cross-sectional area of the rod which rod has otherwise uniform characteristics.

While I have shown and described various embodiments of my invention, it is understood that the principles thereof my be extended to many and varied types of machines and apparatus. The invention therefore is not to be limited to the details illustrated and described herein.

I claim:

1. A device for producing thrust comprising a field shaping surface formed of stitched, closely spaced conductors and having a dielectric material there between to define a smooth surface, a dielectric member connected to said field shaping surface and an electrode on the end of said dielectric member remote from said field shaping surface and means for applying electrical potential between said electrode and said closely spaced conductors.

2. A device for producing thrust in accordance with claim 1 wherein said dielectric member has a dielectric constant which varies progressively between said electrode and said surface means.

3. A device for producing thrust comprising **an electrode having a relatively large surface area,** an **electrode**

7

positioned in the region of the axis of generation of said surface **and having a relatively small surface area,** dielectric means connecting said electrodes and means for applying a varying electrical potential to said electrodes.

4. A device in accordance with claim 3 wherein said dielectric means exhibits a dielectric constant which varies progressively from a relatively high value in the region of the large electrode to a relatively low value in the region of said small electrode.

5. A device in accordance with claim 3 wherein said dielectric means has an electrical conductivity which varies progressively between said electrodes.

6. A device for producing thrust comprising a planar electrode, a second electrode positioned in the region of

the axis of generation of said planar electrode and having a surface area smaller that the surface area of said planar electrode, a dielectric member connecting said electrodes and means for applying a high electrostatic potential to said electrodes.

7. A device in accordance with claim 6 wherein said dielectric member is tapered from the planar electrode towards the smaller electrode.

8. A device in accordance with claim 6 wherein said dielectric member has a conductivity which varies progressively from a relatively high value near the planar electrode to a relatively low value near the smaller electrode.

9. A device for producing thrust in response to the application of electrical potentials to the electrodes thereof comprising a first electrode, a second electrode having a relatively large planar surface area with respect to said first electrode and means including a connecting member supporting said electrodes in spaced relationship for producing a varying field gradient between said electrodes.

10. A device in accordance with claim 9 wherein said connecting member has a varying cross-section.

11. A device in accordance with claim 9 wherein said connecting member tapers between said electrodes.

12. A device in accordance with claim 9 wherein said first and second electrodes are flat electrodes of unequal area.

13. A device according to claim 9 including means for applying a varying electrical potential to said electrodes.

14. A device in accordance with claim 9 wherein said connecting member has a dielectric constant which varies between electrodes.

15. A device in accordance with claim 14 wherein said first electrode is a frusto-conical surface and wherein said

8

connecting member extends along the axis of generation of said first electrode.

16. A device in accordance with claim 14 wherein said first electrode defines a frusto-conical surface.

17. A device in accordance with claim 9 wherein said connecting member comprises semi-conducting particles whereby said connecting member is given a conductivity

gradient.

18. A device in accordance with claim 15 wherein said second electrode is a disk-shaped radiator and wherein the potentials applied to said electrodes are alternating current potentials, the diameter of said disk-shaped electrode being equal to a half-wave length of the alternating current potential.

19. A device in accordance with claim 15 wherein said connecting member contains semi-conducting particles which are more concentrated in the region of the disk radiator than in the region adjacent said first electrode.

20. A device for producing thrust in response to the application of electrical potentials to the electrodes thereof comprising an annular electrode, a second electrode, and insulating means connecting said electrodes whereby thrust is produced along the axis of generation of said annular electrode in response to the application of electrical potentials thereto.

21. A device in accordance with claim 20 wherein said annular electrode comprises a toroidal surface.

22. A device in accordance with claim 20 wherein said second electrode is mounted on the axis of generation of said annular electrode.

23. A device in accordance with claim 22 wherein said second electrode is displaced from the center of said annular electrode whereby a thrust is developed along said axis in a direction from said second electrode towards that annular electrode in response to the application of electrical potentials thereto.

References Cited by the Examiner
UNITED STATES PATENTS
1,974,483 9/34 Brown ____________________ 310-5

FOREIGN PATENTS

1,003,484 11/51 France

MILTON O. HIRSHFIELD, Primary Examiner.

ORIS L. RADER, DAVID X. SLINEY, Examiners.

And now, something interesting from William J. Beaty's website:
(http://www.amasci.com - a Science Hobbyist website)

A GRAVITY-WARP CAPACITOR http://www.amasci.com/caps/capworks.txt

----

| Overall specs (written by bill b.):
|
| 18 cm long, 0.96KG
|
| 390 tin foil layers, 781 wax paper layers
|
| .025mm tin (0.001"), 97.8% purity
|
| wax paper est. thickness: 180mm/781 - .025mm = .2mm (VERY thick!)
|
| Powered by VandeGraff electrostatic generator
|
| 5.48KG measured lift above weight-cancel point (54 Newtons),
| calculated thrust 6.44KG (63 Nt) (it lifts ~ 7grams per foil
| plate??
|
| Can two plates lift themselves?!!)
|
|
| Divided into 13 groups of 30 foils (each w/15 foil-wax-foil-wax
| layers) alternating positive and negative plates. Pos. plates
| connected together, negative plates connected together. Each group
| individually switched to high voltage in order to vary the thrust.
|
| Wax paper dielectric disks, radius 8cm, hole radius 2.7cm
|
| Tin foil circular plates, radius 6.30cm, hole radius 3.3cm, with
| nine radial gaps in foil each 0.6cm wide, gaps ending on a circle
| of 5.55cm radius circle (outer edge of foil disk is continuous.)
| See GIF diagrams linked on http://www.amasci.com/caps/capwarp.html
|
| Metal layers carefully adjusted so the gaps line up.
|

From: S
Date: Thu, 16 Mar 2000 14:22:31 EET
To: billb@eskimo.com
Subject: **it works!!**

I am a regular reader of your site, I am an electronic engineer and I am working on antigravity projects secretly in my home in my country.

I just want to tell you that the gravity capacitor works as the writer says. I complete and test it. The voltage it depends from dielectric material.

I didn't roll all the positives and negatives strips, I made groups of 30 layers for better control and testing.

Weighs only 0.962 Kg, and in full charge **its negative weight is 5.481 Kg.**

VDG "energy" is the best and safest for dielectric.

From: S
Date: Sun, 19 Mar 2000 23:39:20 EET
To: billb@eskimo.com
Subject: **Re: it works!!**

I want to tell you what I did exactly so you can or any other person can duplicate the capacitor.

Forget the capacitors you know. That thing is NOT a capacitor. Is only similar in the construction. But that device USE the capacitance to keep the charge in large quantities. Nothing more to do with the regular capacitors. Now I think that the geometrical shape of the plates and the position of them is the key, But I can not explain to you how.??

The capacitor is **a real powerful thing,** I mean that if I let it free after I charge it **I will have accident for sure.** It keeps thrusting to its positive side. I cannot measure the voltage that it builds because I don't have instruments. I use a big glass box like the one you put fish inside but bigger, with the capacitor inside with the positive up, and a weight measure machine on top of the capacitor. Closing the box I had all those stable, and the weight machine set to 0. Instant ~5,4 Kg on the scale when I hit it with working van de graff I made for testing it. **Don't use electronic weight machine, it will show instant funny characters and it will stop!!??. In fact, any electrical device above the hole and 10cm higher will not function properly, and under the hole, (and the affected area getting bigger as you go away from it, as if the force going something like a "conical thrust" of the capacitor, will not function at all. Which means that the electrons in circuits stop running.**

Now if I do not use the glass box and I lock the device with four short, equal strings to the negative end with the table or floor (I use a table) and I charge it, then the capacitor kicks itself up, now when I pressed with the stick on one of the four strings, the capacitor was turning from the side I pressed. So all I want to say to you is that the force is directional (thrust is appearing always on the negative side). In few words, it acts as it was a rocket without the fire. No noise. Careful, do not touch the glass if is not grounded after the discharge and do not touch or be near the charged capacitor. It is not a game. If you were seeing how it reacts, you could say that for your self. I am afraid to be near it when it is charged. Also, I use a barrel full of oil for discharging. And all that, I do with extreme caution, using long sticks etc. My opinion is to not use it inside the same room. Probably is emits something bad. I don't know. I am not psychotic person I am always cool (at least I try).

I have used the device 4 times. Is not easy if you do not have the correct instruments and equipment. I use continuous charge from VDG, no special circuits. The capacitor reacts instant and constant.

The discharge is very dangerous and I have never left it to discharge by itself yet to learn the time. But if you use my plate ring as it is shown in the scanned image (in scale) you will finish much faster.

Also the electrostatic leakage is very limited because of the curved angles. Also the capacitance will be increased because there will be no air trapped between the small segments, using the ring I suggest.

About the construction

I followed all instructions except that I had the waxed paper printed as the plates to be on top so I can place the plates exactly where they should be, (I think that is important) and the tin was not in small pieces but rings. There are many reasons why I made this changes. I have include a drawing how I did the rings look like.

From: S
Date: Tue, 21 Mar 2000 14:19:45 EET
To: billb@eskimo.com
Subject: Re: it works!!

billb wrote:
>I wonder, are the small slots necessary? Or will the capacitor still
>produce gravity thrust if the foil ring is made solid, with no slots?
>Or, if fewer slots are used, is the gravity thrust very small?

I think that is very important. I think the shape of the ring is the heart of the module.

I can't tell for sure for all these combinations, I have to try them first.

>Is the glass box filled with oil? Or is the capacitor dry, in air?

No, no oil inside the glass box (I haven't try that). It is dry. I only used the glass box for some kind of protection. And I do not know yet if is really protecting me or my mind just think that its protecting me..??

>Does this "field" prevent a flashlight from operating? Or a 100-watt AC
>light bulb? Perhaps it only harms transistors, but not heated wires, not
>batteries?

I think you are correct, electronics have problem around 10 cm away from cap, and not live wires (with 220 volts) (light in the room is not affected but is 4-5 meters away), I don't know about batteries but I guess you are correct. I have to experiment. I will do that.

>If you turn off the VDG, does the thrust slowly decrease? Or does the
>thrust vanish instantly when the VDG stops? (Is the thrust caused by the
>CHARGE, or by the electric current (microamperes of leakage current in the
>paper dielectric?)
>
>Also, how many metal layers in your capacitor?

781 wax paper layers and 390 metal layers, I made 13 groups of 15 positive and 15 negatives, using a primitive switching circuit having 13 switches for trust control (Those 13 switches add trust and cannot reduce of course).

Look how I do operate the module:

I switch the VDG on and after 10 sec I connect the + of VDG wire to the + (up as was) of the capacitor.(- is connected to Vdg) With 13 switches on It takes less than 1 sec for the instant thrust to its maximum (maximum for my VDG) Now (capacitor uncharged) If I switch on 1 by 1 the 13 switches the thrust do the same. Gets greater and greater. I disconnect the Vdg and still the thrust is there. When you have to discharge, better discharge step (even step is dangerous) Well I think and I feel that thrust is caused by the charge. My VDG and any Vdg I think can only give a very, very small amount of current.

NOTES FROM BILL B:
Maximum voltage will be determined by leakage, and if the VDG only outputs 10uA or so (my estimate of a typical VDG), and if the capacitance is about .01uF (my very crude estimate), then a 1-second charging time gives about 1000V on the cap. Very low! [Better estimate of capacitance: 0.5uF] And, if turning on the switched segments one by one gives easily observed thrust increases, then even a few layers (such as one of the 30-foil segments used above) should give an easily-measured thrust. I wonder if the thrust is exactly proportional to the number of layers? If so, then even a single layer-pair should be able to lift its own weight. But if the thrust is not proportional to the number of layers, then many layers would be needed before the thrust becomes (disproportionally) strong and can levitate the layers. How many foil layers are needed before they can lift their own weight? And is tin really necessary, or will other metals work too?

A GRAVITY-WARP CAPACITOR http://www.amasci.com/caps/capworks.txt

---

and,

GRAVITY-CAPACITOR, REPORT OF SUCCESS

From: "cliff"

To: <billb@eskimo.com>

Subject: **funny thing**
Date: Tue, 23 May 2000 13:09:16 -0700

I would not believed it, but this thing does have thrust. I have built it not to size but of to a smaller scale. The second attempt was built to scale. The third try was to build it larger but not to scale. Odd thing is, it still produces thrust that seems to be related to static potential. I have used wax paper with tin foil (hard to find), wax paper with house hold foil (easy to find) and lastly, paper dipped into a dielectric type bath to improve dielectric properties with aluminum foil. This last one did not work as well. Another try used mylar film with aluminum and copper foils. This worked the best although I did use just aluminum only with mylar film and found that copper produced another affect but now working on that issue. I have made a small one with x and x plates then with a mylar film between them thus making a stack of 150 plates total. Now this produces something more than just thrust **and I do not need such high voltages to get the thing to push 22lbs.**

Date: Tue, 23 May 2000 11:28:52 -0700 (PDT)
From: William Beaty <billb@eskimo.com>
Subject: **Re: funny thing**

On Tue, 23 May 2000, cliff wrote:

> I would not believed it, but this thing does have thrust. I have built
Wow! What level of thrust? At what (approx.) voltage? Could they lift their own weight?

And aluminum DOES work? A couple of people on freenrg-L built small versions with aluminum, but they did nothing. Maybe they were too small.

I found some tin, but haven't had time to get the shapes stamped out. If aluminum does work, then printshops can make these capacitors. Some printshops specialize in foil coatings (like on the titles of cheap romance novels.)

(((((((((((((((((((((((((O)))))))))))))))))))))))))))

William J. Beaty	SCIENCE HOBBYIST website
billb@eskimo.com	http://www.amasci.com
EE/programmer/sci-exhibits	science projects, tesla, weird science
Seattle, WA 206-781-3320	freenrg-L taoshum-L vortex-L webhead-L

From: "cliff"
To: "William Beaty" <billb@eskimo.com>
Subject: **RE: funny thing**
Date: Tue, 23 May 2000 14:11:34 -0700

I am using a spring type weight scale. The later device with mylar film between the plates weights in at an approximate 8lbs. The 150 plates are grouped in thirty (being a test before adding more layers). I have found that the metal foils must be smooth as possible. But so far this later device does have potential as an electric jet engine. I still am a bit puzzled on this thing. I am just a hobbyist in electronics and have made many electronic devices including pc board layout. Spent many thousands of dollars on software to aid in designing board and lots of money in silly projects. This one is as dumb as it gets, but I thought it might work as a vortex ion display if it was given a vacuum and a magnetic field.

The aluminum foil with wax paper has a tiny thrust compared to the one I am working on now. At first after making 20 or so plates and then applying 100KV to them, it did not work. Almost disappointed, I continued to add more to the stack. By the time I got to 60 plates and applying only 25KV, there was a small pull. When I went to 75KV the pull stopped, this was due to the dialectic and was being broke down. I disassembled the whole thing and switched to plastic, at that time, the device would work up to my 75KV but now melting the plastic. Switching to mylar did the trick to 100KV, but really no big changes. Tin foil works better than aluminum and copper. I would like to try lead but unable to find a lead foil that is thin as tin foil. A friend of mine has made me foil from bismuth and magnesium powder of approx. 2 mills thick. (found this formula on the Internet for discovered crashed UFO, although I do not believe in UFO).

William J. Beaty
billb@eskimo.com
EE/programmer/sci-exhibits
Seattle, WA 206-781-3320

SCIENCE HOBBYIST website
http://www.amasci.com
science projects, tesla, weird science
freenrg-L taoshum-L vortex-L webhead-L

FORENSIC ANALYSIS [T. T. Brown]:

Thomas Townsend Brown's documented works, experiments, and patented claims are unquestionably true, as can now be proven by my new understanding of matter, the physics of matter and gravity, and of energy (magnetic and gravitomagnetic, **electric and gravitoelectric**). I have developed this new theory of physics only recently, and it is explained in **"The Antigravity Papers"** section **(Part B of this book).** The more detailed parts of the theory are in the Appendix.

Physics doesn't lie. It is solidly mathematical and ponderable and measurable. It is also testable by experiment and repeatable. If physics says that a device which is desired to do "so-and-so" MUST be built using "such-and-such" concepts in order to work, then we would expect all devices claimed to do "so-and-so" must actually be built in accordance with and in regards to the "such-and-such" requirements.

THE READER MAY BE VERY CONFUSED AT THIS POINT, because Brown's devices do not appear, at first glance, to have **any cone-shaped or triangular-shaped parts in them.** His devices are NOT AT ALL like the devices described by the other two witnesses (Bob Lazar - last chapter, and David Hamel - next chapter). Also, none of Brown's claimed devices require any spinning motion, as do the devices related to those other two witnesses. Another big difference is that Brown's devices use very high voltage static electricity, which is used to charge up what is essentially a capacitor (also called a "condenser", as it is referred to in the report "Electrogravitics Systems"). The devices of both of the other two witnesses use various forms of magnetism and low voltage, instead of the very HIGH VOLTAGE static charge and NO magnetism such as Brown's devices use. So why are Brown's devices completely different from the devices claimed by the others?

Brown's devices are different, because they rely on electric fields (high voltage) and gravitoelectric fields, whereas the other devices use, and are based-upon, magnetic fields and gravitomagnetic fields. Such is possible because there has always been a very noticeable DUALITY present in physics, namely electricity vs. magnetism. For every electric concept, there exists a corresponding magnetically-based concept. And vice-versa. A good example is the fact that there are electrostatic plate motors (not very common, but DO exist), and there are the (far more common) magnetic field-based motors (with the usual wire windings).

It is stated in the "Forensic Analysis" section of the other two witnesses that:
"So if ... was working with real devices, based upon **gravitomagnetism,** for creating antigravity lift, **then his devices MUST have contained certain special parts, as per the requirements of this new physics theory."** And it was stated that such devices **MUST THEREFORE** be based typically upon **CONE and / or TRIANGLE shaped parts.**

Brown's devices are "Electrogravitic" devices (i.e. based upon **gravitoelectricity**), rather than "Gravitomagnetic" devices (i.e. based upon **gravitomagnetism**) such as described by the other two people. **HOWEVER,** the reader will shortly see that Brown's devices are ALSO based upon CONES and / or TRIANGLES. However, these descriptive words do not denote the shape of a physically observable special part in the device, but rather they describe the **SHAPE OF ELECTRICAL FIELDS** (which can't be seen (directly) by the eye) which are present (and MUST be present) in each and every device.

Thus, antigravity devices are (still) all about "cones" & "triangles"!
For review, please refer to (in Part B) Figures 9A&B, 11A&B, 12, and 13 in order to understand this fact.

All of Brown's patent claims are, in fact, based upon the creation and use of either cone-shaped or triangle-shaped electric fields, as the reader will shortly see.

We begin by looking at FIGURE 34 in Part B, The Antigravity Papers. At the top, we see the usual conical-shaped (or wedge-shaped) parts, made typically of a material which will work, such as copper or aluminum, which are capable of causing a separation of "positive mass" and "negative mass". These type parts are seen in both Bob Lazar's devices and David Hamel's devices (next chapter).

Now at the bottom of that same figure, we see a totally different device, which could be called an "**asymmetrical** capacitor". It consists of two flat (and very thin) plates of metal which are parallel to each other and which are of DIFFERENT size (surface area). We are looking at them edge-on, and we can see that the lower plate is much smaller in area than is the upper plate. Also, the two plates are separated by what's called a dielectric material (which does NOT conduct electricity, like metal does). The whole device can be seen as being like a **cone** with the apex cut off. The resulting shape is called a "**frustrum**" in the geometry community.

Normal, everyday capacitors, which are used in electronic devices, ranging from VCR's to computers and so on, are usually made of parallel metal plates of EQUAL SIZE and AREA, which are separated, again, by a dielectric. These capacitors have an overall cylindrical shape, and so are called "**symmetrical**" capacitors. No one has ever yet needed a **asymmetrical** capacitor, until now (for antigravity purposes).

In the figure, we see that the two metal plates of this **asymmetrical** capacitor are connected to a high voltage source (such as a battery, for instance). After this voltage is connected to the device, it becomes electrically charged with a typically large amount of static electricity (i.e.

many more electrons on one plate than there are on the other plate). Then, some interesting effects start to occur, as a result of the high voltage.

First, we see that the lines of force (or "flux") of the electrical field inside the device are NOT parallel like they would be in an everyday capacitor, but are at an angle to each other, such that they diverge towards the top of the device. If we paint a mental picture of the lines of flux and if we mentally extend them below the device, we se that they all would come together at a point below the device.

SO WHAT WE HAVE HERE IS A DEVICE WHICH CREATES A CONE-SHAPED (or more specifically, frustrum-shaped) ELECTRIC FIELD.

Now what else is going on here? As shown (somewhat exaggerated for clarity) in the diagram, the circulating electron cloud around the nucleus of every atom (in the dielectric material) is PULLED BACK away from the nucleus. This happens due to the fact that the electrons are negative, and so are pulled toward the upper plate, whereas the nuclei are positive, and are repelled from that same plate. The result is a distended atom.

So, since the circulating electrons are now ABOVE the nucleus of each atom, instead of AROUND each nucleus of each atom, a dipole-like magnetic field is formed around every nucleus, due to the nearby circulating electron current. In turn, this dipole-like magnetic field causes each nucleus to ALIGN with that field, because each nucleus is already like a little magnet with a North and a South end.

The effect then is that all the nuclei of all the atoms align with the electric field lines, and these electric field lines, as we've seen before, form a cone-like (or frustrum) shape.

Now, since each nucleus also produces an extremely powerful gravitomagnetic dipolar field (proven in part B), and since all nuclei are aligned exactly with the diverging electric field lines, we see that the final result is this: THE DEVICE CONTAINS **DIVERGING GRAVITOMAGNETIC FIELD LINES,** as well as the diverging electric field lines. In an everyday, symmetrical capacitor, we would have PARALLEL gravitomagnetic lines, rather than diverging or converging ones.

In order for the device to work, either the two metal plates must be aluminum, copper, or one of the materials known to work, OR the dielectric must be "doped" with suspended particles of aluminum, copper, lead, or the like. Such materials have a net overall magnetic spin of their nuclei. Iron or steel WILL NOT WORK, as there is no net spin to the nuclei!

Therefore, just like in the two drawings at the top of the figure, we get the desired effect in this electrical device for creating an antigravity, which is namely the separation of the rings of "positive mass" and "negative mass". A set of parallel field lines cannot accomplish this!

In summary, we see that an asymmetrical capacitor (bottom of Figure 34), when charged with high enough voltage, will result in the same converging/diverging gravitomagnetic field lines, and thus the same antigravity effect, as does result for the aluminum cone devices shown at the top of Figure 34.

In fact, Figure 34 contains a summary of the 4 major ways to generate antigravity. Mechanical spinning of aluminum cones, or external magnetic field applied to a cone, or external gravito-magnetic field applied to a cone, or high voltage applied across an asymmetrical capacitor. And there is listed a "fifth" method, which is using supercooling or cryogenic cooling of all the parts of a device. This isn't actually a separate method, but is a way to increase the efficiency of a device dramatically (perhaps a factor of thousands, or so). THIS ONE FIGURE (34) SAYS IT ALL.

So now, we are in a position to show how Brown's devices can produce antigravity.

After reading the report "Electrogravitics Systems" and also Brown's patent # 3,187,206, we see that there are a number of technical words (or "tech words") and phrases that are quite commonly used, such as "high k-number (means high quality of the dielectric)", "counterbary (means antigravity)", "shaped condensers (capacitors)", "arcuate electrodes", "steeply-converging Lines of Force", "shaped electrical fields", "tapered dielectrics with embedded Lead Oxide particles", "converging electric field lines", "areas of high flux density and low flux density", "concentrates electrostatic flux","electro-gravitics", and so on. Just notice these when you see them. Incidentally, also note that none of these devices needs to spin or otherwise move, as do Lazar's and Hamel's devices!

Now, please turn your attention to Brown's patent, and it's figures numbered 1 through 8C. The title of the patent is "Electrokinetic Apparatus".

In patent figures 1 and 2, we see Brown's basic and most simple device. Let us look for the required (by my new physics theory) Electric Field Lines which are non-parallel (i.e. converging, diverging, or cone shaped, frustrum shaped, or wedge-shaped), as I have just previously discussed. Remember, in any device, it must be connected to a very high voltage battery or similar power source in order to function, as can be seen in the figures.

The device in either figure 1 or 2 consists of an upper, large surface area metal "electrode", and a small surface area metal ball at the bottom. In both cases, the metal ball is held separate from the large area top electrode by a stiff dielectric rod in-between (it's part # 10 on the diagrams). This is (although odd looking) in fact an ASYMMETRICAL CAPACITOR, due to the difference in surface areas of the two electrodes. IF WE COULD ACTUALLY SEE THE ELECTRIC FIELD LINES, we would see them all emanate from the lower ball-electrode, and then spread out in a conical fashion until they reach the upper electrode, where they terminate. Thus, the shape of the electric field lines in both devices is a **cone!** So we see that this is a legitimate, capable of working, actual antigravity device! And it was patented way back in 1965!

AMAZINGLY, ACTUALLY WORKING DEVICES LIKE THESE (they look like umbrellas) CAN BE SEEN IN THE VIDEO CLIPS ON THE "JLN LABS" WEBSITE ON THE INTERNET!
They just hook up the high voltage, and up the device goes! Also to be found on the Internet (where I don't recall) are downloadable video clips (.AVI) of some of the experiments being done in Brown's Lab, long ago. Actually flying devices are shown there as well, and they look like umbrellas also!

At this point, I would like to point out that there is another type of high voltage flying device being developed on the JLN website, and it is called a "lifter". From what I have seen, I believe that by far the main force which propels a lifter is due to ionized air flow rushing through the device in a downward direction, and the T. T. Brown effect is just very small or negligible. I truly believe that NASA patented the lifter (which they DID in fact do) in order to distract experimenters and scientists from the true path to actual antigravity. Yes, NASA knows about real antigravity, and is trying to mislead us all. If one were to test a lifter alongside with a T.T. Brown "umbrella", both devices being in a large tub of water or oil, I am sure we would see that the lifter no longer functions, but the umbrella would still function.

Now, let's turn to **figure 5** in Brown's patent. **This is the most important of all of Brown's devices,** as it forms the building block needed to build much larger devices, consisting of arrays (or banks) of these simple devices. We see that this device is just an asymmetrical capacitor, very similar to the one shown in my figure 34 in Part B of this book, which we recently analyzed. Thus, we can see that upon applying the high voltage, across the large-area metal plate on the left and the small-area metal plate on the right, we will create diverging (frustrum-shaped or wedge-shaped) electric field lines within the dielectric material (item #32) which is sandwiched between the outer metal plates. Thus, per my new physics theory, this is a legitimate, actually working antigravity device! Note that Brown shows the direction of the thrust (or force) produced upon the device as item #37, which shows the direction of force is to the left, being FROM the small electrode TOWARDS the large electrode. This fits my theory as well.

I will skip figures 4 and 6 in the patent, as they may just cause confusion, but rest assured, they obey my physics theory as well and are legitimate devices.

Now, look at figure 7 in the patent. This shows a cleverly designed parallel array of the single building block device which was seen in figure 5. Note the direction of the produced force (item #37B) is correct. NOTE THAT THE POLARITY OF THE APPLIED VOLTAGE IS DIFFERENT FOR THE LEFT BANK FROM WHAT IT IS FOR THE RIGHT BANK. Thus, the diverging gravitomagnetic fields produced in the dielectric wedges are in the opposite direction in the left bank when compared with the right bank. So, wouldn't the force effect of the left bank cancel out the force effect of the right bank? The answer is no. To see this, take a look at Figure 10 and 11B in Part B of this book. They show that, regardless of the direction of the gravitomagnetic field lines, the negative mass always accumulates at the location of the more dense field lines, and the positive mass always accumulates at the wide end of the device! Therefore, the thrust force produced is always FROM the narrow end of the device TOWARDS the wide end of the device, regardless of electrical or magnetic polarity!

Finally, look at figure 8 in the patent. We can see that this is a legitimate device as well. We see that the electric field lines converge at the central point, and they diverge (spread out) at the outer ring. This device, like the others is also an asymmetrical capacitor. If all of the field lines were parallel and equally spaced, then such device would not work! Note the thrust arrows shown in the figure. They show that the thrust is from the concentrated field area (the metal ball in the middle) towards the less concentrated field area, as we would expect.

Now, I would like to clear up a confusing issue which has plagued people for years who have studied Brown's devices. Many people have mistakenly said that the thrust force is always "from the negative electrode towards the positive electrode". This statement is just plain wrong, as can easily be seen from figures 7, 8B, and 8C. The correct statement regarding the force produced is "the force of thrust produced is always from the concentrated electrical field area towards the sparse (or less concentrated) electrical field area." In fact, Brown states, right on the first page of his patent, the following:

"By attaching a pair of electrodes to opposite ends of a dielectric member and connecting a source of high electrostatic potential (voltage) to these electrodes, a **force** is **produced in the direction of one electrode** provided that the electrode **is of such configuration to cause the lines-of-force to converge steeply upon the other electrode.** The **force,** therefore, **is in a direction from the region of high flux density toward the region of low flux density,** generally in the direction through the axis of the electrodes."

So there you have it. All of Brown's claims in his patent have been forensically analyzed by my new theory of physics, and all of them check out as authentic, true antigravity-based devices. We can't say that Brown, as a "witness" to antigravity affairs, was telling a true story, because he kept quiet and didn't ever tell ANY STORY to anyone. The other two witness to antigravity actually told their story to news people and other people. Brown, on the other hand kept silent, probably under some sort of government-imposed security blanket. However, the devices Brown claimed in his patents (which is effectively his "story"-of-record) are found to be true, correct, and AUTHENTIC. So Brown, like the others, "told-the-truth".

DISCUSSION: MORE ADVANCED ELECTROGRAVITIC DEVICES:

A much more advanced electrogravitic system can be built in the following way. Imagine starting with a one-half inch thick sheet of copper. Then, use a blunt-ended hardened-steel object, which is as thin as a pencil or an ice-pick, or less, along with a hammer to pound out many, many hemispherical-shaped little dents into the copper sheet, all over it and on ONE SIDE ONLY. There should at least be thousands of these. Perhaps another way to accomplish this would be to fire thousands of hardened "B.B.'s" at the copper sheet, out of a very high pressure gun.

Then, imagine cutting the super-dented (talk about hail storm damage!) on one side copper sheet into a narrow "pie-wedge" shape (whose arc angle is one-thirty-second of a full circle, or narrower). Then, make many more such copper wedge-shapes.

Now, imagine stacking between 6 to 8 of these wedges, so that their faces are parallel, and SO THAT THEY ALL HAVE THE SUPER-DENTED SIDE FACING DOWN. Then imagine using some means to temporarily hold these copper plates with a half-inch gap between each one, and pour some kind of liquid plastic "potting" material in-between and around all the copper plates and letting said plastic material harden into a solid. Or, perhaps, one could use molten glass, poured in, and let it harden. Once the insulating material has hardened, we are left with a stack of about 8 copper plates, all separated by equal gaps (hopefully). These plates nowhere touch each other, and remember that ALL plates have the dented side face-down.

Now, to complete the device, imagine taking 32 such "stacks", and arranging them side-by-side to form a complete circle (which should be easy to do, as each piece is a pie-wedge shape). Each wedge should be designed to NOT come into electrical (i.e. copper to copper) contact with adjacent wedges. We will call this completed, circular thing, simply the "device". So now we ask, how does it work and create antigravity lift?

Let us take a look at just one pair of stacked copper plates. The upper plate will, of course, have its dents on its bottom side, and the lower plate will be just smooth copper on its top side, which is directly facing the bottom side of the upper plate. Now, imagine connecting a high voltage source between these two copper sheets. The plates will behave like any capacitor and will thus have many more electrons moved onto one of the plate faces, and will have many less electrons on the other plate face. Thus, a strong **electric field** is created between the two copper plates.

Now if both of these adjacent plate faces were just smooth copper, then all the electric field lines would be nicely parallel, and would all be exactly perpendicular to the faces.

However, this is NOT the case for our device, due to the multitude of hemispherical dents present on the bottom side of the upper plate. **Therefore, in the close vicinity of each dent,**

the lines of electric field do the exact same thing which they do in T. T. Brown's device shown in figure 1 of his patent. Namely, **a conical-shaped** electrical field is created in the region of each dent. Therefore, a true antigravity lifting force is produced upon each dent on the underside of the upper plate, as the reader now understands.

Now, imagine taking a (still wedge-shaped) stack of 8 of plates (ALL dented sides face-down), which have been embedded into a (now solid) plastic or glass dielectric material, and applying a high positive voltage to every other plate, and have all the rest of the plates electrically connected together and connected to an electrical "ground". With this arrangement, we will have half of the dented sides exposed to a positive-polarity electric field, and the other half of the dented sides will be exposed to a negative-polarity electric field. However, this is no problem, as we have seen earlier that the direction of the overall force does NOT depend upon the polarity of the electric field. It depends ONLY on where the dense part of the field is, relative to the sparse part of the field. So since all of the dents in all of the plates are facing down, we see that the force on each and every plate will be strictly UPWARD. Thus, this wedge-shaped stack of special plates can be used nicely as an antigravity device, if enough high voltage is applied.

Now, consider making a "type B" stack of wedge plates. The "type A" plates all have the dents on the down-facing side. The "type B" plates will all have the dents on the UP-facing side. Therefore, the force on a "type B" section (stack of 8 plates) will ALWAYS be DOWNWARD. The force on a "type A" section (stack of 8 plates) will always be UPWARD.

Now imagine taking 16 "type A" sections and 16 "type B" sections and placing the total 32 of them in a perfect circle. Arrange these so they alternate "A", "B", "A", "B", and so on. Then, with a little imagination, we see that we now have a large circular device which can, depending on the way it is dynamically energized with high voltage applied to different sections over time, be made to rise straight upward, or upward and tilt to the side, or downward, or in any direction conceivable. In short, we have built a completely controllable & steer-able antigravity platform, capable of just about anything! It is a legitimate antigravity device which would work and could be flown.

A graphic of (approximately) a one-eighth section of such a circular antigravity platform is shown in the diagram below.

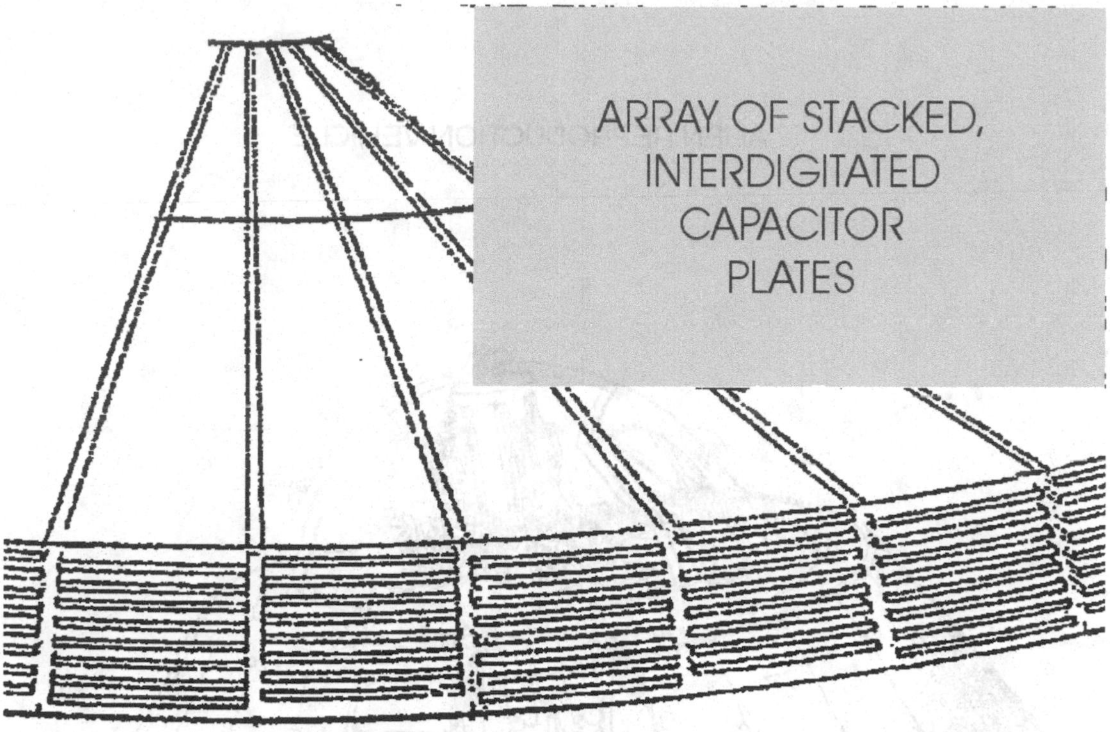

A Section of a Workable Circular Antigravity Platform Device. Every Other Section Has Its Vast Multitude of Small Dents All Facing Downward, and All The Other Sections Have Their Dented Sides Facing Upward.

One new thing to note from this diagram is that each section has its plates interstitially (i.e. interdigitated) placed with respect to its two adjacent sections. This is simply done to minimize any interaction between the adjacent wedge-shaped sections of the device. So there you have it. We have a design for a controllable antigravity platform.

Recall that the polarity of electric fields has nothing at all to do with the direction of force upon the sections. Thus, a type "A" section will always experience an upward force when energized with either plus or minus polarity of voltage, and a type "B" section will always experience a downward force when energized with either plus or minus polarity of voltage. Thus, and for various other reasons, I believe the best way to operate the platform device would be to apply a high voltage, DC PULSE to each "A" section in succession, going around quickly in a circular fashion. If the "B" sections are to be used, we need the same thing, but perhaps going around in the opposite direction.

Now that we understand that such a circular platform antigravity device would actually function if built, I can now show the reader proof that such devices have, in fact, been built, many years ago by a company in California named "S.A.I.C", which stands for "Scientific Applications International Corp.". I know this company exists because we did business with them at one place I have worked.

The completed "spaceship" they built (for the Air Force) looks much like the following crudely-done drawing:

ALIEN REPRODUCTION VEHICLE

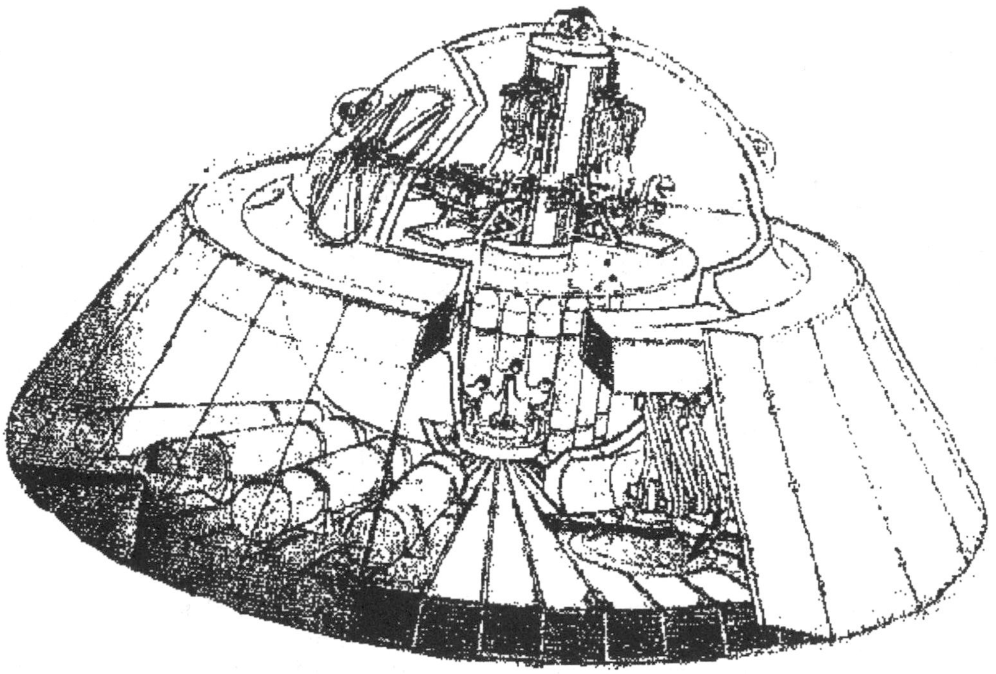

Radial Capacitor Plate Antigravity Craft. SAIC Called it the "Alien Reproduction Vehicle".

The bottom of the craft shown above is comprised of basically the same radial capacitor plate array as was described in detail earlier. If the reader were to buy and read a copy of the book "Disclosure" by Steven Greer, they would find a chapter describing the testimony of a person named Mark McCandlish, which also includes testimony of a person named Brad Sorensen. In that chapter, the reader will find a much, much better quality, intricately detailed schematic drawing of basically the same device in the crude drawing shown above. According to the testimony of those two people, Brad Sorensen (under VERY special circumstances) actually was able to see a real, working piece of hardware just like the above diagrammed device. It was "parked" in an Air Force hangar in California, and it was literally floating several feet above the floor of the hangar, with no visible signs of support! This was at Norton Air Force Base, which is just east of San Bernardino, California. Sorensen claims he saw the device on Nov. 12, 1988 (it was a Saturday). The Air Force General who showed the device to Brad, and to the V.I.P. person he was with, said that the device had been built by "SAIC", and that it could travel at "light speed or BETTER"!

I did note, that rather than making lots of dents in the sides of the copper plates, the device shown in the McCandlish drawing had its copper plates somehow constructed of billions of deformed, 15 micron copper spheres. Perhaps this is a better alternative to making thousands of "dents" as I described earlier. His drawing also said that the sections are pulsed (in a circular manner) at a rate of 30 kHz or faster.

I would highly recommend that the reader get the book "Disclosure", to see this device, as it is described in incredibly complete detail in there.

So "hats off" to Mr. McCandlish and Mr. Sorensen for their great testimony. Obviously, they were telling the complete truth, as we have just seen how to prove that their described spacecraft is indeed a device which would ACTUALLY WORK, according to the principles of my new theory of physics! So, so far, we have seen that Bob Lazar told the truth, Townsend Brown detailed the truth in his patents, and Mr. McCandlish and Mr. Sorensen told the truth, and, as we shall shortly see, Mr. David Hamel told the truth as well.

THEY ALL TOLD THE TRUTH.

At this point, I would like to digress a little into the related subject of Electrogravitic Materials. Take a look at the following pictures, taken by an electron microscope of the surfaces of two samples of material which were given to Art Bell back in 1996. The two samples allegedly came from the Roswell, NM UFO crash site. These were on Art's web site, but I believe it OK to show them here, as his web site has now disappeared forever, according to him.

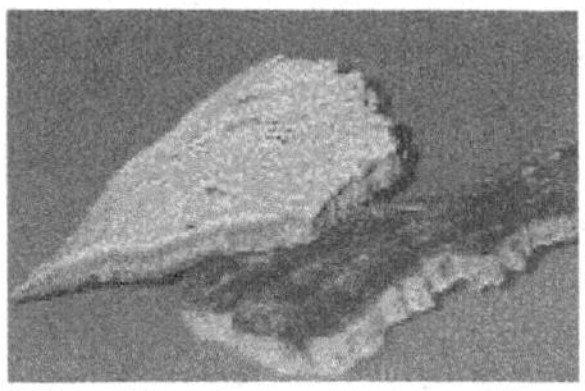

The Two Samples, as Brought To Art Bell.
(not using the electron microscope).
One is Shiny Side Up, and One is Black
Side Up.

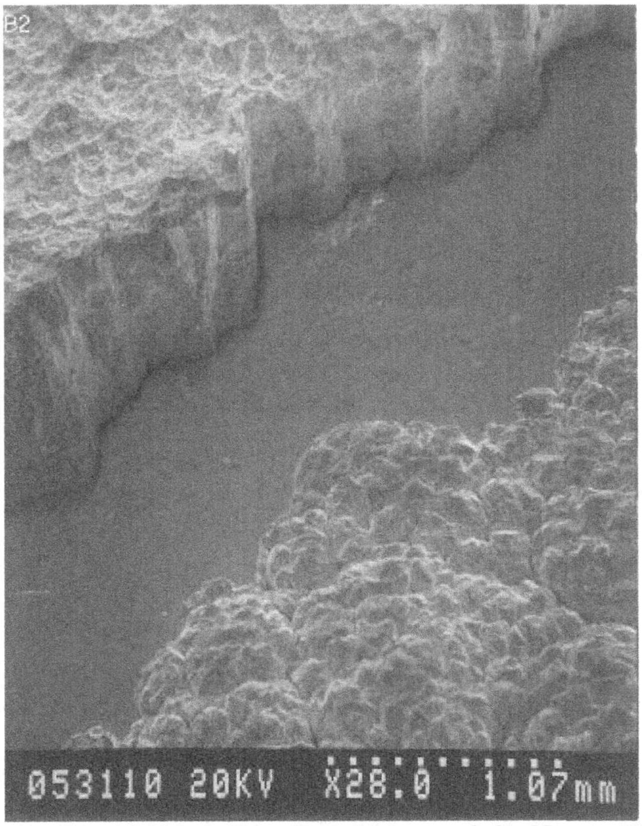

The "Black" Side of One Sample (upper left), Next to The "Shiny" Side of The Other Sample (lower right).

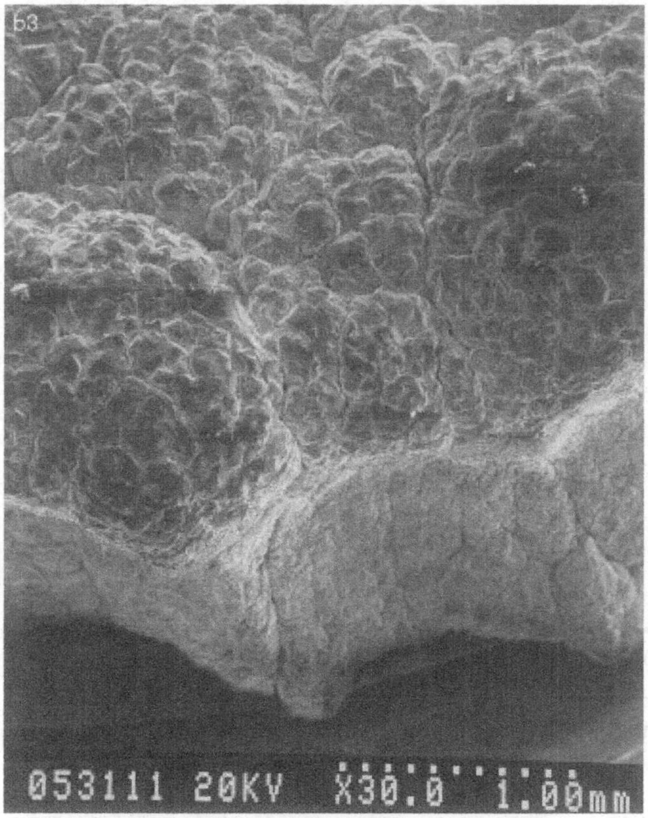

Again, The Shiny Side of One of The Samples.

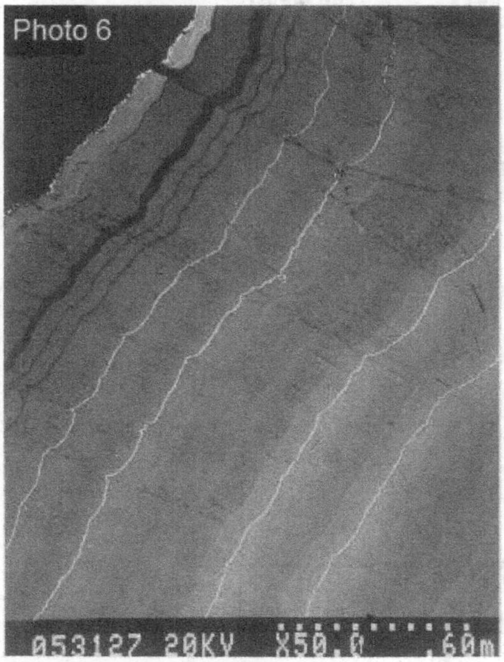

The Edge View of a Sample.
Note The Layers of Material.

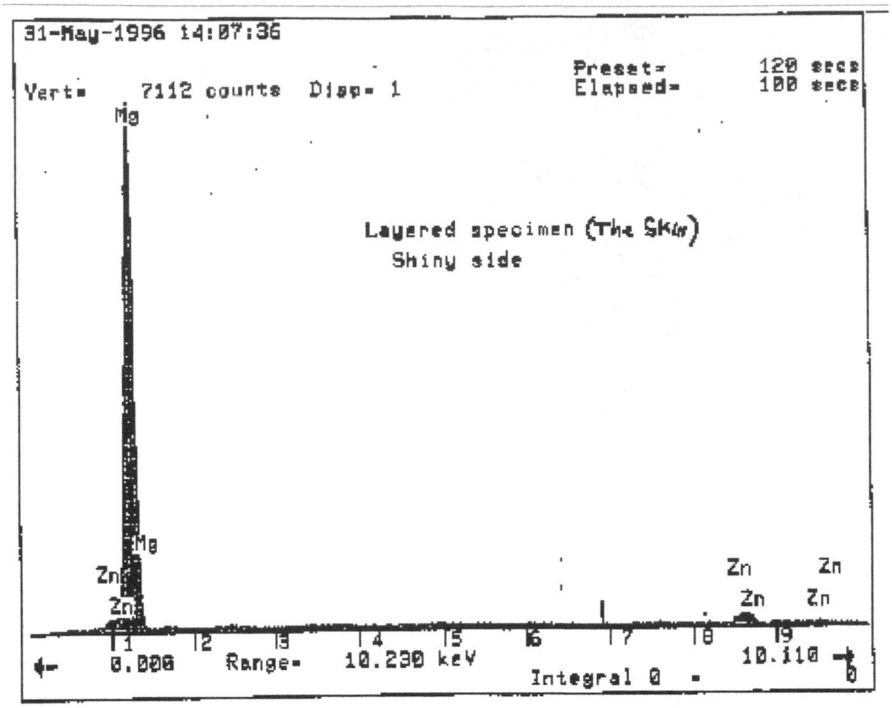

Chemical Makeup of a Shiny Layer (contains NO Bismuth, only Magnesium).

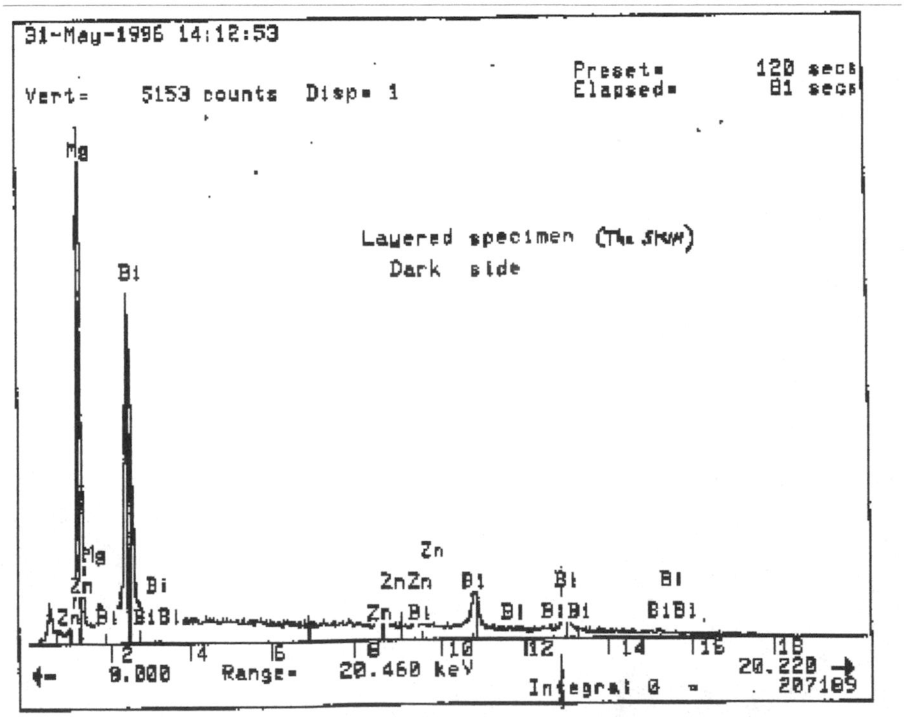

Chemical Makeup of a Dark Layer (contains **significant** Bismuth).

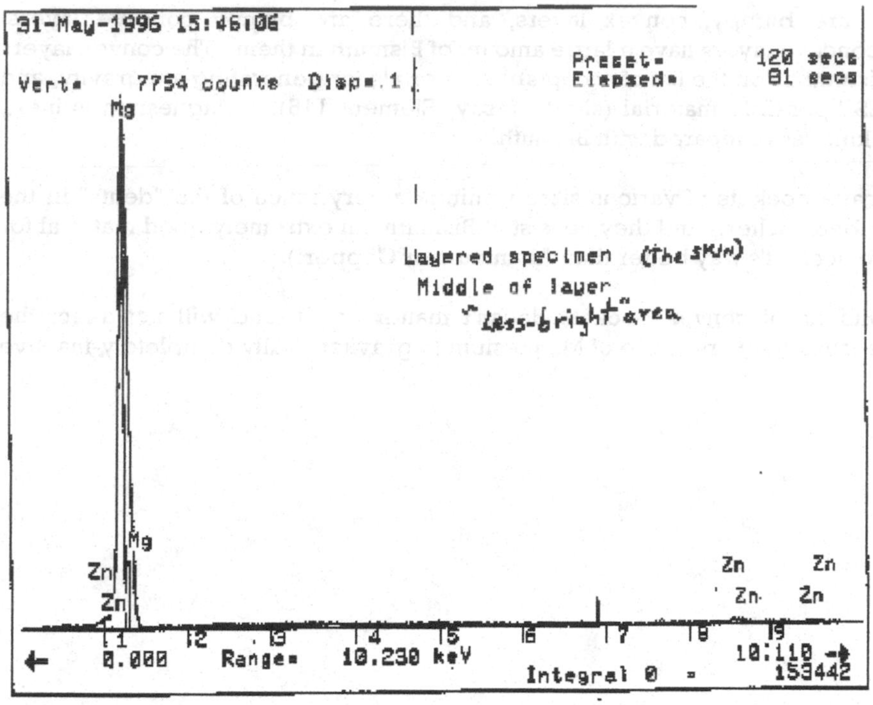

Chemical Makeup in the Middle of a Layer (contains NO Bismuth, only Magnesium).

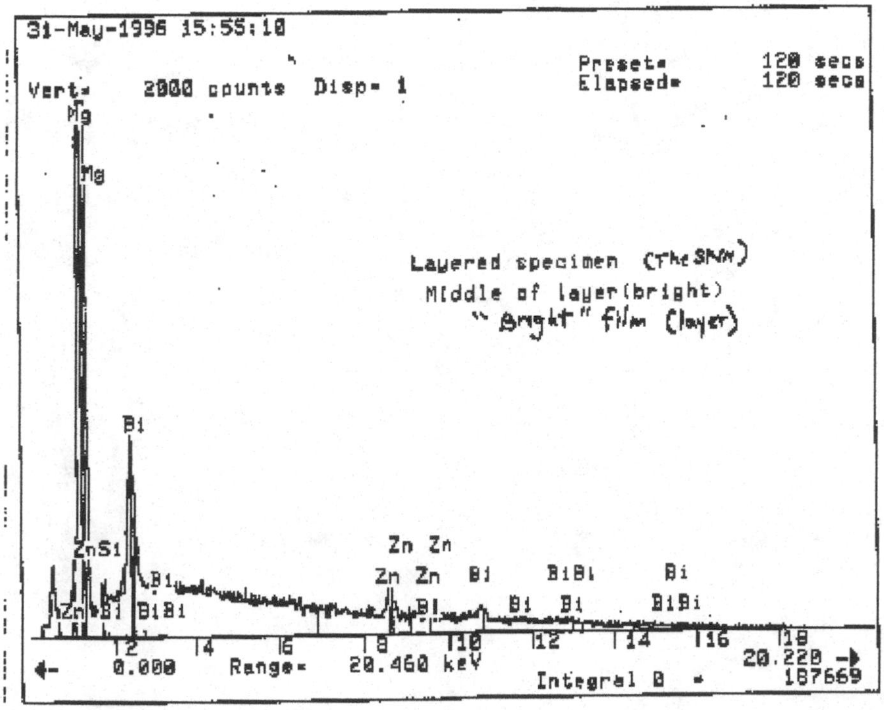

Another Location Somewhere in The Middle of a Layer.

We see that there are bumpy, convex layers, and there are bumpy concave layers. Apparently, all the concave layers have a large amount of Bismuth in them. The convex layers have no Bismuth. Bismuth is on the list of acceptable materials for generating antigravity, and in fact, it is THE BEST possible material (short of, say, Element 115). Magnesium is inert, gravitationally speaking, as compared with Bismuth.

The billions of <u>concave</u> pockets of various sizes remind us very much of the "dents" in the copper plates described earlier. And they consist of Bismuth, an extremely good material for antigravity-based devices (it's way better than Aluminum or Copper!).

The existence of billions of <u>convex</u> pockets doesn't matter at all, and will not affect the operation of this, because <u>they</u> are made of Magnesium (a gravitationally completely-inactive material).

David Hamel:

I start with an e-mail message I found on virtuallystrange.net:

TAKEN FROM:
"www.virtuallystrange.net/ufo/updates/1996/dec"
(the "Dec 31" archive).

[Start e-mail message].

From: "Steve Wingate" <swingate@crl.com>
Date: Mon., 30 Dec 1996 22:39:39 -0800
Fwd Date: Tue, 31 Dec 1996 03:10:42 -0500
Subject: [Antigravity] Hello and Working devices...!

SearchNet's IUFO Mailing List

------- Forwarded Message Follows -------

From: "Robert G. Halsey" <halseyb@govonca3.gov.on.ca>
To: <antigravity@primenet.com>
Subject: [Antigravity] Hello and Working devices...!
Date: Mon., 30 Dec 1996 01:15:46 -0500

Hello to everyone!

My name is Robert Halsey (please refer to me as Bob) from Toronto Ontario. I work in the Computer field with VAX systems mainly. I have my B.Sc. in Computer Science, but did an unofficial minor in Physics.

I joined the Antigravity mail list/conference a couple of months ago, and have been getting the volumes of E-mails every day. Most are quite interesting, although I do see a little overquoting sometimes and tangents, but overall, I quite enjoy this conference.

One of the things I expected to see when I joined was discussion of real research and real working machines that people are currently working on. While I've seen a little of that in the last few days, overall the discussion seems to be more theoretical or historical (that **ancient East-Indian Rama civilization** memo was Great!).

In pursuit of real working devices today, I would like to relay some information on one person I have met and what he is working on. The gentleman's name is **Pierre Sinclaire** in Fort Langley, B.C. (near **Vancouver**). He has been working on a device original built by **David Hamel,** now in **Ontario.** I'll make the overall story brief, but will give references on where to get more details.

David Hamel attests to **having an encounter with a UFO in 1975.** Supposedly it was a psychic experience as opposed to an actual abduction. During this encounter **he was given a "tour" of their ship and its principles** of opposing magnetic fields were explained to him. Upon release, **he was**

determined to duplicate this functionality. Through the years, he has pursued this **with significant success.**

The devices he has built have similar electromagnetic/gravitational effects as do "classic" UFO encounters. Whenever he starts his machine in motion, all electric devices within several hundred feet go completely on the fritz. His two most successful tests to date have had a hole blown through the roof of his garage, and a few years later, he built a larger device which when he inadvertently started it one night, started glowing different colours and as it speed up, it took off never to be seen again!

His main problem has been funding, and since he isn't a scientific person himself, he has also had **a problem understanding exactly what was happening.** That's where **Pierre Sinclaire** comes into the picture. Pierre has done significant research into antigravity effects on his own and joined up with David Hamel to approach the testing in a more scientific way.

I have had an involved discussion with Pierre to understand how these devices worked. The device that David/Pierre is working on is based on opposing Magnetic fields interacting at specific changing angles. The devices themselves involve a "wobbling" motion of compressed opposing fields which when contained within a metallic rim, generate strong electrogravitational effects. In the atmosphere, these devices also employ contained air flow for additional control. **The surrounding apparent electromagnetic interference effects are supposedly actual electrogravitational effects that are impeding circuits at a quantum atomic level as opposed to an electrical interference.**

Pierre expects to have his most recent creation completed sometime in the January time frame, and I am anxiously waiting word on what happens!

Various other researchers have been working on similar magnetic devices as well as tapping into the Overunity energy aspects of these magnetic effects. One person, **John Searl,** an electrical engineer, did some research on opposed wound coils and supposedly had one device he built take off on him as well!

There is one fellow **John Hutchinson,** also in **Vancouver,** who has done some testing with two positioned Tesla coils and a Van Degraf generator and has supposedly (I would like to see this one in person), **caused complete gravity negation on some testing objects,** as well, he has observed matter state effects.

Anyway, to summarize. I hope this generates some additional interest in real, happening today, types of research. **Pierre Sinclaire** is offering a book written by **Jean Manning** on the history of **David Hamel** and his research, called **The Granite Man and the Butterfly.** As well, he has a videotape available of a conference where he explained a lot of the research/history of the project. All proceeds from the sale of these items go towards the research. It's an interesting book, with some additional chapters on other research in the area. Get it and help the project!

For more information on this, take a look at "www.cascadia-net.com/magnet". I found out about this project through another source of a lot of information on several of the topics of discussion within this conference,

"www.keelynet.com"

I can't emphasize enough how fascinating the Keelynet site is!!! Moderator Jerry Decker has accumulated a massive volume of current and historical material on all aspects of what some would term "weird science". He evaluates material he receives before posting it and generally adds a "critique" of the material he posts. He has a dial-up database in Texas much larger than his Website can contain, but there are mirror sites that do contain most of his original collection.

Daniel Woolman asked me to pass on the information about Pierre when I joined, so there it is! I hope you find it interesting. Sorry for it being so long...!

Robert (Bob) Halsey
halseyb@gov.on.ca

[End e-mail message].

Now, since Hutchinson was mentioned in the above e-mail, I will take a quick side track to show some of Hutchinson's work, as it is amazing and relates to antigravity. Seeing is believing...

Here are images from John Hutchinson (Hutchison?), showing results from levitation experiments.

Ref: http://www.artbell.com/hutch.html

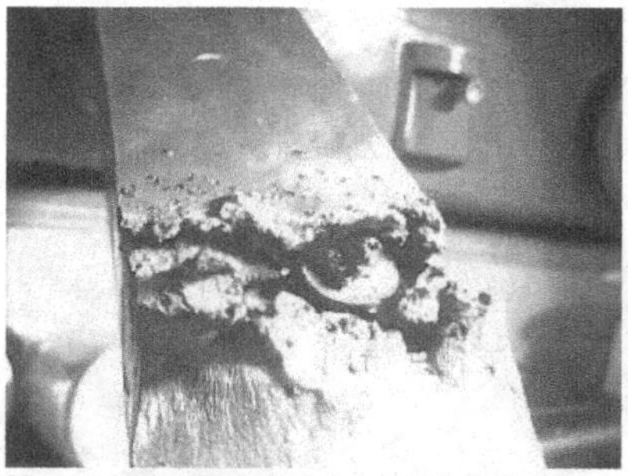

Here is a chunk of metal that is
completely distorted by the levitation process.

A metal slab with holes in it
has a knife impaled inside.

A chunk of wood impaled in it.

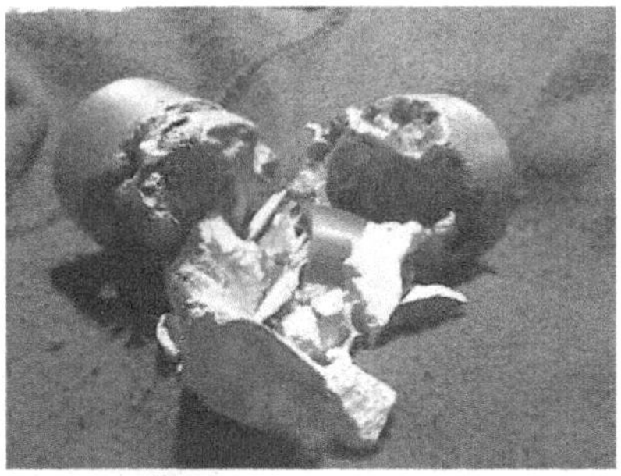
Steel cylinder is metal turned to jelly.

Ditto.

And, here are three VERY eye-opening video clips of objects affected by an **actual antigravity field** (created using two positioned Tesla coils and a Van Degraf generator): (still Hutchinson's work)

"http://www.artbell.com/mediafiles/hutchvideo.rm"

"http://www.artbell.com/mediafiles/hutchvideo2.rm"

"http://www.artbell.com/mediafiles/hutchroom.rm"

If you don't have "Real Player" viewing software, download it at the following web site: "http://www.real.com/player/8/"

Also, see:
John Hutchison 1993: Canadian Researcher, George Hathaway, VHS Tape, Toronto, Ontario, Canada.

[end of Hutchinson material].

Now, here are some excerpts from **Pierre Sinclaire's** website, www.projectmagnet.com (which disappeared earlier this year -- 2002. It's been replaced by another organization's website):

[PIERRE SINCLAIRE ; BOX 839 ; 9037 ROYAL STREET ; FORT LANGLEY, B.C. CANADA V1M 2S2]

The following was presented at the IANSPR in Denver, CO.:

TITLE:
CONSTRUCTION OF THE GRAVITO MAGNETIC DEVICE (GMD)

BY:
PIERRE SINCLAIRE

ACKNOWLEDGEMENT

This paper is dedicated to a man that I respect very much, **David Pierre Hamel.** A man who helped me tremendously in my quest into the understanding of antigravity and pushed me to look at science in a new way.

ABSTRACT

In this short paper I would like to convey enough information to help anyone who would like to undertake any research on the Gravito Magnetic Device (GMD). I will give a brief description of the hardware and the theory behind the GMD. I will also talk about various researchers who have come up with very similar theories and results, plus I will give a brief conclusion on the tremendous possibilities and potential for this technology.

INTRODUCTION

At the end of 1989 I was fortunate to meet a man named **David Pierre Hamel.** Mr. Hamel had been working for the last fourteen years on an energy system that he claimed would produce antigravity (and an abundant amount of energy). Also, to support his claim he said that he had built three prototypes that demonstrated these effects. The first one was built **in a forty-five gallon drum;** the second one was built **on a trailer,** and the third one was build **on a plywood base raised ten feet in the air** on a four-pod support structure. His work paralleled the research that I was doing and so I have **devoted the past six years** of my research in collaboration **with David Hamel** to understand the **unusual phenomena** and to build a prototype that would demonstrate and conclude that such technology is obtainable today.

[... block omitted ...]

Finally, I would like to emphasize that the power generated, surrounding the GMD, is **extremely powerful and radiates,** causing interruptions of electron flow in normal electrical systems**, i.e. lights, cars, road transformers and interferes with electromagnetic transmissions.** So it is without saying that this machine has to be taken **extremely seriously** when completed and activated. But don't be alarmed, because if a control system is in place, it is easy to stop its effect. One of the main purposes of the GMD is to understand the effects of enclosed opposing magnetic fields that have varying vector angles; like theories set forth by **Tom Bearden.**

As far as I am concerned, David Hamel has brought about an incredibly powerful and useful technology which also generates antigravity. **The GMD has a powerful upward thrust causing it to rise in the air.** Similar work has been done by **John Searl** of England with his Levi Disk experiments. When we do understand the inner workings of what causes the secondary electrogravitational fields, the possibility to create various devices enabling us to have plentiful energy of different types is mind boggling.

[... rest omitted ...]

[end of Sinclaire EXCERPTS].

And now, a picture of David Hamel, explaining a complex device:

David Hamel, inventor, will talk to anyone, freely, about anything.
The purpose why David felt he has been "chosen", was to bring this information into the public, to help mankind, and to put away the old and traditional way of acquiring energy and to stop polluting and destroying the Earth.

David is not in it for the money. He has nothing to sell, nothing to benefit from [his website was not created by him]. All his machines and equipment are built and paid for by him.

David mortgaged his house years ago to pay for the expensive magnets and red granite built exactly to his specs that are needed for his devices (a typical Granite Pinion, custom designed to his specs runs about $3,000.00 each). The plans and information he will freely give to anyone who is sincere and would like to carry it further.

Many people think that he is just an old nut, but many people have thought that the world was flat, and that the sun moved around the earth, and that alternating current had no purpose, and that the relationship between time and space was absolute. It's only after great people such as Columbus, Galileo, Tesla, and Einstein, have gone the extra "step", that we realize that things required more study, that we should have asked more questions and listened more carefully to the answers. As for David Hamel, he wants to leave behind information that the Human Race can benefit from, whether good or bad.

David is 72 years old [actually, he was about 8 years ago!] and continues to work every day to realize his mission of giving earth the technology to produce free energy. He is constantly saving whatever money he can from his pension and putting it towards the next step of his project. It is, unfortunately, a long, slow way to get it completed, but he continues to put all of his efforts and finances towards it.

He has suffered many setbacks over the twenty plus years since his first alien contact. His **plans have been stolen,** his **prototypes confiscated,** his **home ransacked** and he's **been monitored,** had **bombs mailed to him,** his **patent attempts mysteriously denied ......** but he continues to work every day on a mission that could save the world from its ultimate destruction.

The following is
from "http://www.world-famous.com/DavidHamel.html" :
[Hamel's website, created by Rudy Langen]

[A person named Paul Fulcher and someone named "Greg" were in Canada, and got directions to Hamel's house from a lady at the psychic fair, Toronto Ontario.]

Paul's story: (this is what conversations transpired)

"So who is this guy that I should meet, and why should I meet him?" I asked with enthusiasm. She responded by telling me that she was a reporter for the local newspaper and that she had written an article on him [Hamel] a few weeks earlier "...he's an extremely interesting man, **building an anti-gravity/free energy, spaceship**." This was it. This was the contact that I had been waiting for. I was on the trail and getting closer. When I asked her how to get a hold of him, she opened her purse and flipped through a note pad that she had scribbled some short hand notes and found the page with his information. She gave me his phone number and off she went.
I immediately went to the phone and dialed the number, not knowing what to really say, but knowing that it was the thing to do. The phone rang several times before it was picked up and I was addressed with a strong French-Canadian accent. "Allo" he said in his loud, course voice. I hesitated.....
"Hello Mr. Hamel, you don't know me but I am in town for the weekend with the psychic fair and I am very interested in anti-gravity and free energy......"

I hardly got it out when he interrupted with a robust enthusiasm "**Well, you'd better get over here and see what the hell I've got here in my workshop!** If you want to know anything about anti-gravity you'd better get over here and see it for yourself! Do you want to come over tonight? **Cause you won't believe** what you will see with your own eyes!

As we [Paul & Greg] slowed to a park, we looked towards each other, wondering what we had gotten ourselves into. If first impressions meant anything, we would be on our way in no time. David approached us with a warm and hearty hand shake and quickly led us to his kitchen, where he began to tell us the incredible story of his abduction and the information that he has come to understand.
Greg said, "He's been there" he added, "He really knows!"
Almost as soon as we sat down and he began to get into the details, the hair on the back of my neck stood up, I got goose bumps and I knew that this man was sincere. It was almost too much to accept, yet as I listened, I could actually feel it to be true.

We (Paul & Greg) were so inspired, that on our way home we knew we wanted to record this historical event on video. (and they did)

Mr. Hamel finally says to Paul:
"Do what you can Paul, to educate people, to let them know the simplicity and beauty of this technology. We must clean up our act, or we must get off this planet" ... David Hamel

[END OF Paul's Story]

Interesting and Memorable Quote from David Hamel:

The hell with Einstein, I will explain it better !!!

...David Hamel

I STRONGLY ENCOURAGE <u>everyone</u> to purchase the Hamel Video, as well as the Hamel book, both obtainable from
http://www.world-famous.com/DavidHamel.html. There is information in these which can affect everyone in the near future. You owe it to your future life and happiness to get these and study them!!
ELSE, you could someday soon just be left "holding the bag", so-to-speak....

The following is From:
"http://www.keelynet.com/gravity/hamel.htm"

Start of the "Keelynet" file excerpt, HAMEL.ASC.

The Hamel Flying Disc
a.k.a. : The Poor Mans' Searl Disc

File Name : HAMEL.ASC - Online Date : 07/06/96
Contributed by : Jerry Decker - Dir Category : GRAVITY
From : KeelyNet BBS - DataLine : (214) 324-3501
KeelyNet * PO BOX 870716 * Mesquite, Texas * USA * 75187
A FREE Alternative Sciences BBS sponsored by Vanguard Sciences
InterNet email jdecker@keelynet.com (Jerry Decker)
Files also available at Bill Beaty's http://www.eskimo.com/~billb

---

One of the most incredible discoveries in our time has been unheralded and is only known to a small group of people who study such matters. It pertains to the closest duplication of UFO flight characteristics and power sources that I have seen in many years of studies in attempts to ascertain how to duplicate them for practical use.

Go To Next Page--->

The principal players in this story;

David Hamel - inventor and contactee who built several magnetically self-driven devices which produced phenomena ranging from energy production to flight

Pierre Sinclaire - private researcher, businessman, co-author and investigator who first met and worked with Hamel to verify and try to duplicate the phenomena through Project Magnet.

Jeanne Manning - writer and author specializing in the coverage of activities relating to new energy researchers and their discoveries

For full details of the David Hamel story, I recommend you purchase a copy of the book,

'The Granite Man and the Butterfly'

written by Jeanne Manning. All income from the book is used to sponsor Project Magnet, Pierre Sinclaire's effort to carry out a controlled duplication of the phenomena first investigated by inventor David Hamel.

Each book is $11.99 plus $5.00 for shipping outside Canada ($16.99 total). NOTE: this price may not be current, BEWARE!

Order from :

Project Magnet / BOX 839 / 9037 Royal Street / Fort Langley, British Columbia / Canada V1M 2S2

Pierre's InterNet email address is magnet@smartt.com

Please understand he is quite busy working on Project Magnet, so MAKE YOUR QUESTIONS COUNT, THANKS!

Inspiration - where the idea came from

While watching television with his wife and housekeeper, David Hamel experienced a sort of 'waking trance' in which he was mentally transported to an alien ship. (The book gives a more detailed description of other observations, but for our purposes, we will stick to the magnetic drive technology.) Noticing a vibration within the ship, Hamel asked what caused it. The ship was constructed around **two large cones,** with the wide ends on the bottom. **One cone was supported on the top of the other and suspended by magnets and pinions.**

Basic Explanation of the principle

A tornado-like rushing of air moved up through the ship to produce a tremendous friction. The **cone within cone** wobbled at high speed and were kept continuously off-balance. As the cones wobbled and the air rushed between them, **lightning-like flashes appeared between them.** Hamel was shown the outside rim of the ship, where numerous openings served to allow the in and out motion of the air as it rushed between the wobbling cones. These air openings controlled not only the amount of air but the direction of flow. As the air was moved at high velocities through the gap between the wobbling cones, it became ionized to produce a stream of charged particles. The cones not only produced energy but also provided propulsion. This was accomplished by **a small weighted ball, rolling in circular path** in a restricted space. **The circular movement of this ball** appeared to have a falling motion, always seeking equilibrium. The upper area of the cones were suspended on magnetic parts which were kept unbalanced to sustain the disruption of equilibrium to produce the wobbling effect.

Imagine a horizontal disk, suspended on point, forever falling or tilting sideways as a metal ball rolled forward on its rim. This produced the graceful fluttering effect which Hamel likened to 'a butterfly above a magnetic field.' The magnets would not wear out because they were suspended on a magnetic field. **Movement of the cones produced an electro-gravitational field to cause the ship to lose its connection with gravity, thereby neutralizing it's 'weight'.** Movement of the ship could be controlled by pulling the ball out of rotation. Hamel was given the term **'weight into speed'** to help remember what he was being shown. **The aliens informed Hamel they had given this technology to our ancestors many times over history and we would find evidences of it as historical artifacts and in legends.** Hamel was also told that we humans use energy technology which produces heat as it dissipates the energy. The natural way, the aliens said, is to produce cooling by use of implosion forces, rather than explosion. [R. Crandall>> This alleged cooling effect could explain the frozen (or simply not hot) wires experienced by such "free energy" researchers as Floyd Sweet, Moray, and others.]

Initial experiment

Eventually, after much thought, consideration and research, Hamel decided to try to duplicate the cone within cone system. Using bicycle rims as the base support for his aluminum sided cones, magnets were held onto the sides with electrical tape. When the magnets were taped just right, they produced a rejection force. **A 45 gallon steel barrel was lined with magnets on the inside to create a magnetic suspension zone. The cone within cone arrangement was placed inside this barrel.** Once everything was aligned, Hamel screwed down the cover of the barrel. The cones were floating on a repelling magnetic field produced by the magnets on the lower rim of the cones. When a larger magnet was pressed down onto the top cone, a tumbling motion was created which caused the floating cones to wobble in a circular motion, in a constrained path, at an ever increasing speed. At a certain speed, the vibration stabilized, much like the smooth rotation of a properly balanced, rotating tire on your car.

Shortly after closing the barrel and due to the lateness of the hour, Hamel and his wife went to bed. Within a brief period, they were awakened by a loud bang, followed the dull red glow of what appeared to be a fire from the room with the barrel. On investigation, Hamel found the barrel had exploded into pieces strewn all over the room. (Pierre told me the barrel had IMPLODED because the barrel was caved in. This fits with the implosion theory and the idea that magnetic energy ATTRACTS TO itself while electricity REPELS FROM itself.) Further experiments with the suspended cones produced **unusual energy effects** such as **scrambling television reception, fogging photographic film or causing double exposures** when a photo was attempted.

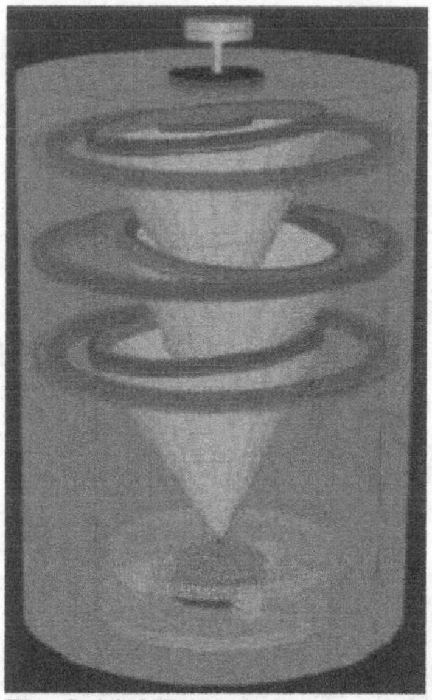

The Initial Experiment.
(3 cones in a barrel)

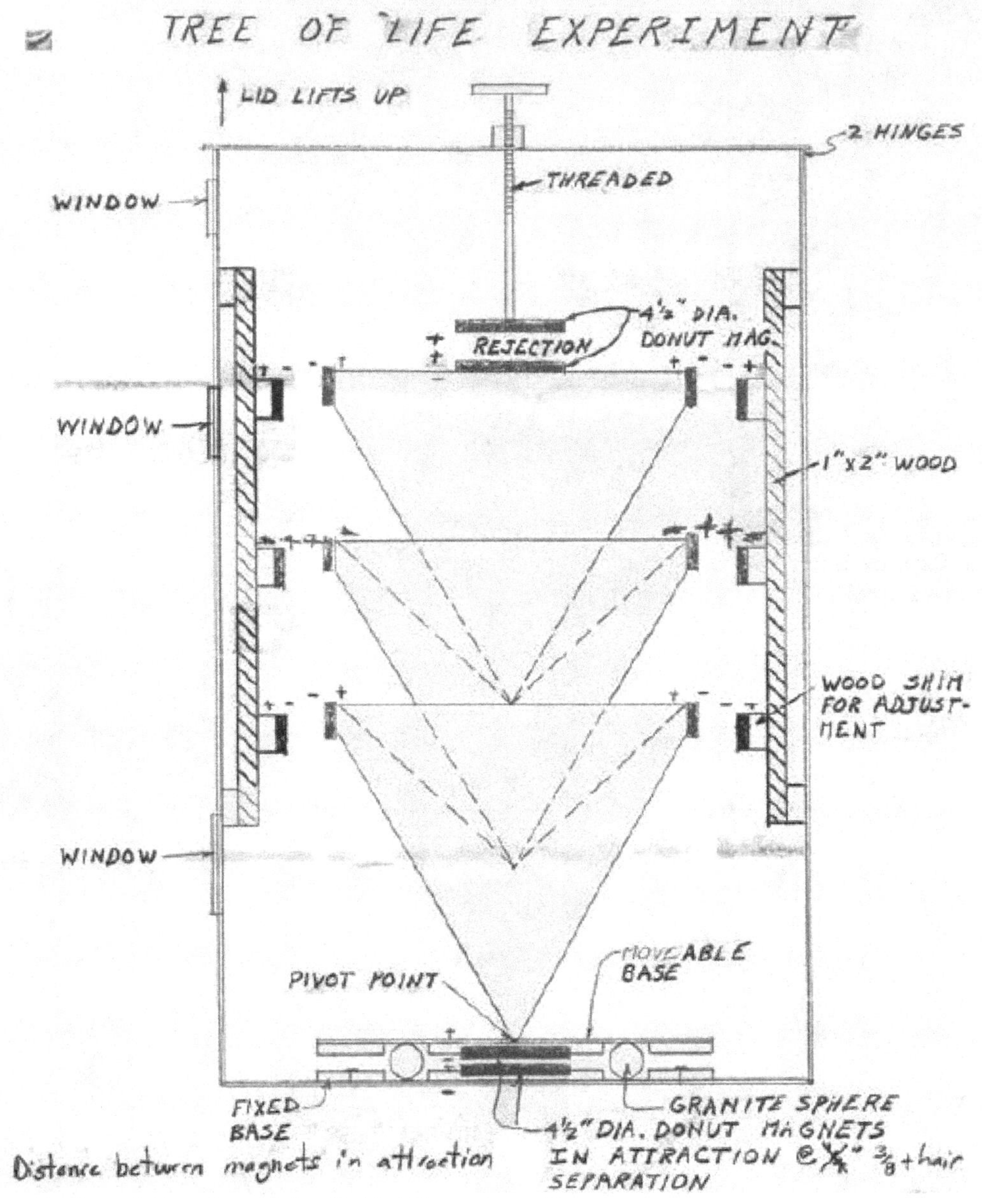

Hamel's Actual Drawing of the "Initial Experiment".

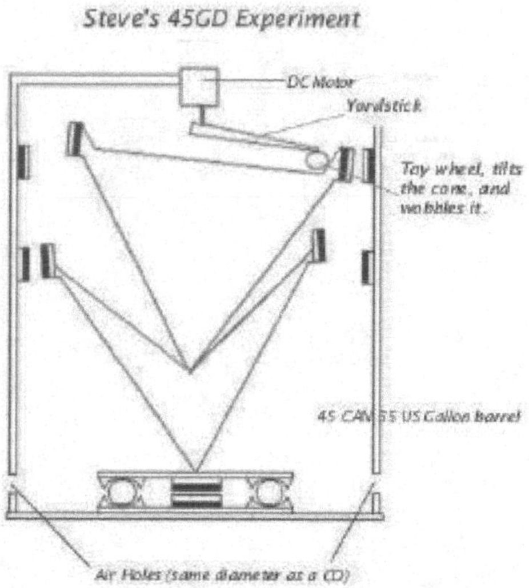

A Two-Cone Rig Built By Another Experimenter.

Yet Another Experimenter's Device.

Other Experimenter's 3-Cone Device.

Ditto. (It will be explained in greater detail, and with credit to the person, later in this book.)

Now, continuing with the "Keelynet" file, HAMEL.ASC:

Most Advanced Experiment

The next major step, after many smaller experiments, involved the construction of a **saucer shaped** cone within cone mechanism which was 7 feet, 3 inches in diameter, with a height of 3.5 feet. It was situated on a **platform** reached by a 16 foot ladder. At 11PM one evening, Hamel screwed down the garbage can lid that compressed the top magnet to make the cones wobble. He **noticed a glow and a sudden wind being sucked into the craft.** Fearing for what would happen next, Hamel climbed down the ladder and removed it for safety. **His wife yelled that the TV set had gone out again, followed by a power failure that had plunged the neighborhood into darkness.** Hamel ran into the now dark house to get his Brownie camera and as he reached the door on his way back outside, **the craft was glowing red and changing to green as it rose off the suspended platform.** As it continued to rise, **the color went to blue,** then **bright white** as it shot off up into the air and out towards space. Hamel managed to get 12 photographs as the craft rose, though the most spectacular are in a series of five.

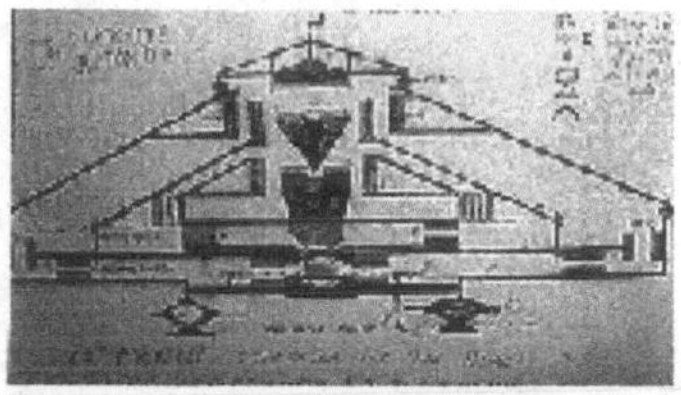
Design Drawing of Hamel's Craft (the one that got away, as explained above).

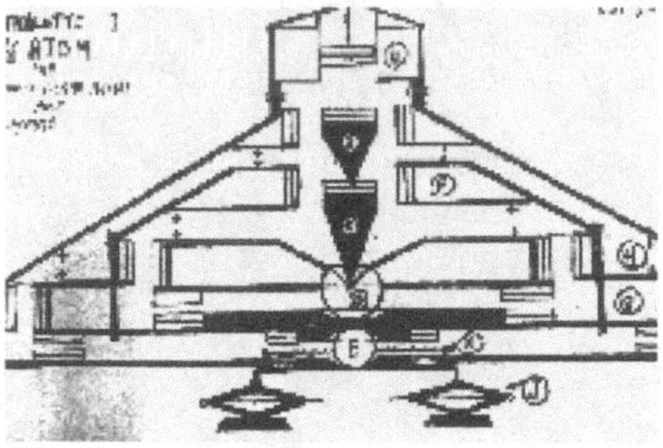
Ditto.

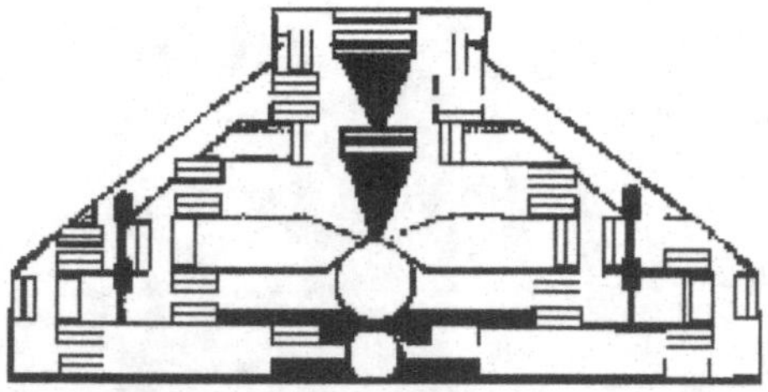
Ditto, Again.

The Final Prototype Model of the Craft, Constructed Per the Drawing.
Dave says that his working models are often confiscated by officials!

Here's the Finished Craft,
Perched High Atop a Tall Platform.
The neighbors thought he
was nuts!...

...But When the Device Started
to Vibrate, it Then Took Off Straight
Up and Was Never To Be Seen
Again. Government Officials soon
afterwards arrived.

David Showing the Base of This
Prototype Model That
Levitated and Was Never Seen
Again.

Balancing the Ring Magnet on David's Almost Confiscated Prototype Model. Government officials walked right past this model not knowing that it was what they were assigned to confiscate!

David Showing a $3000 Granite Pinion, Designed For One of His Other Prototypes.

Now, continuing with the "Keelynet" file, HAMEL.ASC:

Current Research Effort

Pierre Sinclaire has vowed to build a working model and secure a patent in David Hamel's name. Both feel this technology is critical to the survival of humankind on our world as indicated by Hamel's contact with the aliens.

Comments

It is obvious that the phenomena discovered (or shown to him, take your pick) by Hamel is identical to that of John Searl of England. Where Searl requires $10,000 to build just one magnet, Hamel used Radio Shack magnets in his drum prototype and more expensive, higher flux density magnets in the levity experiment. The entire thing seems to revolve around some mysterious phenomena which occurs when energy is condensed into a given space and not allowed to freely radiate/expand. The 'aliens' said the technology was foreign to our 'modern' science yet we would see evidence of it in archeological artifacts if we could only comprehend the technology. Many ancient records speak of flying machines, incredible weapons and other remarkable claims that we cannot see as being remotely possible based on our idea that their technological abilities were quite limited. This of course, when compared with our achievements.

For example, we have seen references to a 'liquid electricity' which was produced from water. The closest understanding we have to this using modern discoveries is the storage of high volumes of electrical energy in a Dewar flask of liquid helium. The helium sloshes around and becomes denser as more power is stored in the bottle. A KeelyNet file called POWERING.ASC speaks of using giant rings of supercooled superconducting material to hold vast amounts of electrical energy in a circulating current.

Hamel found (as did Searl) that a rapidly rotating magnetic disc would continually speed up on its own. Just as Searl claims to have done, **Hamel wanted to put a governor on the device, to prevent it from taking off into the stratosphere.** Except in Hamel's case, he wanted to embed the thing in concrete. One of the correlations Searl found that Hamel and other gravity researchers might find of use in controlling this ever accelerating rotation is the use of the horizontal scanning frequency from a video camera. Searl says his rotating disc slowed down when in the presence of a video camera, though worked fine around movie film. He realized it must be one or more frequencies from the camera which he later used to control the rotational velocity. Perhaps Hamel and others can use this observation to further accelerate (no pun intended) their research efforts. When this technology has been duplicated, in a controlled environment as Pierre Sinclaire is now attempting to do, it will lead to an entirely new world for all of us.

<u>End of the "Keelynet" file excerpt, HAMEL.ASC.</u>

Now, here are more Hamel pictures:

Time & Space, Described By Mr. Hamel.

David Hamel's "formula":

Formula: "Y" x E^2 or EEY.
Y = (1/2) the "Inductance" (L).
E = the "Mass Current" (I), Kg/sec.
Thus EEY = (1/2) x L x I^2 = (inductive) energy of a particle = effective (i.e. apparent) particle mass x C^2.
Where, C is speed of light value.
(Hamel shows "particle mass" as a rectangle above a triangle).

Thus, David's formula, as it appears, means "Inductive Energy" = "mass" (x C^2).

Reversed, it gives:
"mass of a particle" = "Stored Inductive Pure (gravitomagnetic) Energy" / C^2.
This agrees EXACTLY with my new physics theory!!

David calls this **pure energy** (which equals EEY or Y*E^2), **"Kryptonique".**

So Hamel is showing that:
Kryptonique (Y*E^2) = (a particle's routinely measured mass) * C^2.

On David Hamel's web site, he says that **Kryptonique is a Physical/Non-Physical substance,** that is a component of all atoms, that can be used and harnessed to produce Free energy.

This "physical/non-physical" character exactly agrees with my new physics theory, which identifies the routinely-measured mass of a particle with simply the "effective" mass of a small clump of pure (gravitomagnetic) ENERGY.
[This pure energy "clump" has to be an <u>extremely</u> strong and localized gravitomagnetic field].

Patent Application:

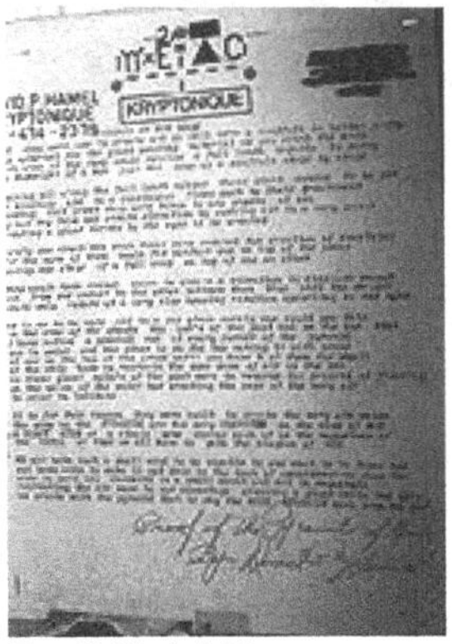

KRYPTONIQUE

An Official Document
Requesting a Patent Was
Submitted to The Ottawa
Patent Office.
Some of David's hundreds of drawings were also sent to the Patent office in Ottawa. Over the past 20 years of filing such documents, all seemed to have vanished. During his visit to the patent office, David was escorted out by several security guards when he started to complain that his file was empty.

And here's Hamel's OLD farm house:

People Driving By On the highway get upset when their cars stall in front of David's house in Vancouver, when his machines are on!

AGAIN, **I STRONGLY ENCOURAGE** __**everyone**__ **(this means YOU!) to purchase the Hamel Video, as well as the Hamel Book,** both obtainable

from http://www.world-famous.com/DavidHamel.html. There is information in these which can affect everyone in the near future. You owe it to your future life and happiness to get these and study them!!
ELSE, you could someday very soon just be left "holding the bag", so-to-speak....
Incidentally, the Video has far more technical diagrams than does the book. Most of Hamel's drawings do NOT appear in the Book.

FORENSIC ANALYSIS [David Hamel]:

A Very Complex Device, Indeed!
(look at all the shells & floating magnets).

David Hamel's story is unquestionably true, as can be proven by my new understanding of matter, the physics of matter and gravity, and of energy (magnetic and gravitomagnetic, electric and gravitoelectric). I have developed this new theory of physics only recently, and it is explained in **"The Antigravity Papers"** section **(part B of this book).** The more detailed parts of the theory are in the Appendix.

Physics doesn't lie. It is solidly mathematical and ponderable and measurable. It is also testable by experiment and repeatable. If physics says that a device which is desired to do "so-and-so" MUST be built with exactly "such-and-such" a thing in order to work, then we would expect all devices claimed to do "so-and-so" must contain at least one or more "such-and-such".

So if David Hamel is building real devices, based upon gravitomagnetism, for creating antigravity lift, **then his devices MUST contain** either triangle shapes or **spinning cone shapes,** which **MUST be made of one of a set of specific materials** as per the **requirements of this new physics theory.** The existing physics theories before this new one are not incorrect, nor is this new theory. The new theory supports and explains the old existing theories, but adds new levels of detail and additional structure. It also allows for some new effects related to gravity. This new theory is more correct in details than are the existing ones.

As is discussed later on in "The Antigravity Papers" section of this book in Chapter 7 "Next step of the Journey--the simplest model for a solid", we see that in order to produce significantly powerful antigravity effects, based upon our new understanding of gravitomagnetism, we **MUST build such devices based upon CONES and / or**

TRIANGLES. Cylindrical shapes will not work, since they can't cause the "positive" and effective "negative" mass to be pulled apart like a cone shape can do.

Antigravity devices are all about "cones" & "triangles"!
Please refer to Figures 9A&B, 11A&B, 12, and 13 in order to understand this fact.

All of Hamel's claimed devices are based upon having two or more **cones** somewhere in each device, as the new physics theory requires.

Also, see **(in Part B)** Figures 27 and 32. Figure 27 shows a device I invented which looks **amazingly similar to Hamel's 3-cone device** ("Tree of Life", as he calls it, as shown in FIG 43). Compare Figure 27 to Figure 4C and Figure A21, which show my more detailed model of the Proton (a subatomic particle) which was developed out of my new mathematical theory of antigravity physics. Notice that each ring of the Proton corresponds to each cone in the device, and note that the direction of spin for each cone must match the spin of its corresponding ring (in the Proton). **This agrees with the spin directions required of a 3-cone device, per Hamel.** Now, look at Figure 32, one of my inventions. This is a type of 1-cone device, as the "cone" is created by the spinning of the wedge (an inverted **triangle**) shape. It is effectively a **cone,** when it rotates rapidly.

Also, a similar wedge-shaped device I invented is shown in Figure 26, and it is similar to my device in Figure 32.

Now, regarding materials (ref. Chapter 5, "Next step of the Journey--models, models, and more models..."), the new theory states that only elemental materials which have an unpaired nucleon (that is, a Proton or a Neutron) can be used as parts in the device which create the gravitomagnetic field. See also Figure 6B. The list of materials includes copper, aluminum, bismuth, element 115, and even lead and titanium. Elements which cannot be used for these same field generating parts include iron, steel, gold, & silver. **This agrees with the materials chosen by Hamel for the cones, which can either be aluminum or red granite (which contains a large percent of aluminum).**

Well, I could stop right here, and what I've presented so far proves, through physics & mathematics, that David Hamel is telling the truth, and I have not had to mention anything about aliens in my proof. However, there are many more items of proof which do relate to aliens, and it would be a shame not to mention these as well. So what follows is in regard to advanced physics information that David Hamel apparently learned from his alien visitors.

So far then, we can see that Hamel **knew** to build his antigravity devices out of spinning (or vibrating) aluminum cones. And he knew what the spin directions of each cone in his "Tree of Life" device would have to be in order for it to work. So Mr. Hamel must have somehow gotten specific information about advanced antigravity physics. **Before the publishing of this book, I have never seen this antigravity physics information and theory published anywhere, in any form.** This includes books, scientific journals, the news media, and the Internet. I have certainly kept my information secret, to myself, and furthermore, I have never met Hamel or Lazar at all, and I have no ties with them whatsoever at the time of this writing (although, that may change very soon!). So, how then did Hamel get the knowledge to build antigravity devices? It seems that there are only TWO possible answers: 1) The secret government, and 2) Aliens from space. The only reasonable conclusion would be #2. The #1 reason is not probable, since Hamel has had his devices confiscated by government officials and by people who worked for Boeing "Phantomworks" (an advanced aerospace company in Seattle, Washington USA, which is very close to where Hamel resided then -- Vancouver, BC, Canada). All of these "Men in Black" (they all wore black and drove a black car!) were very nasty to Mr. Hamel, and just let themselves into Hamel's house and workshop, and usually took some of his devices and told him to "stop building and testing these things". That's all they would say. These men did wear ID badges which showed them to be employees of Boeing, in Seattle. [Ref: "The Granite Man & the Butterfly" book, by Sinclaire & Manning, and also the video "Contact From Planet Kladen"].

So then, we conclude that Hamel got his information from Aliens. What other evidence is there of this? Plenty.

In what follows, the reference material was the book and the video just mentioned. David Hamel has drawn numerous (at least twenty, or so) diagrams, in color, of devices and principles of physics which he was exposed to when he was visiting the alien spaceship. I have been able to correlate these drawings with my new physics theory.

Upon comparing Hamel's drawings with my figures A14 & A21 in my "A New Theory of Quark Substructure for Mesons and Baryons", dated 2-01-2002, I could see that some of his drawings were amazingly similar to mine. In fact, one of his drawings was identical to my new diagram of the internal structure of the electron particle. It had 4 dots arranged in a circle. Now refer to Figure 4D, in "The Antigravity Papers". **According to my new theory,** half of the circulating sub-particles must be positive mass, and half must be negative mass. **So, based-upon my new theory for the electron, which has 4 sub-particles total, we know the following: the electron has 2 sub-particles which are normal mass, and it has 2 sub-particles which are negative mass.** In the video referenced above, **Mr. Hamel says that the particle (electron) is made of 2 of "matter" and of 2 "invisible".** Clearly, Hamel was told that two were of "invisible" instead of being told that two were of "negative mass", which is a term he would probably not understand.

Also, Hamel spoke of what "they" called **the "isotope line" of the energy.** Hamel draws this as a 3-legged bent line. **It is obvious to me that this refers to the spin vectors of the three quarks in a proton.** But Mr. Hamel doesn't know what a quark or a proton is.

In addition, **Hamel said that most antigravity flying craft must be elevated a significant distance from the ground during launch,** in order for them to "take off" and fly. He even said that Stonehenge in England was used as such a landing or launching pad. If you look at pictures of Hamel's saucer which flew and got away, you can see that he launched it from a platform, with tripod, high in the air. **My new physics theory shows that it's true that devices like these indeed must be launched from a certain minimum height, or they will just be sucked back down to the ground.** For this, see Figure 17 and the end of Chapter 9 in "The Antigravity Papers".

Also, **Hamel (and others as well!) says his devices** produce bright light and **will cause total havoc, if not total destruction, of all nearby electrical / electronic equipment.** Recall his ship that "got away". The pictures he took of it when flying look simply like extremely bright blobs of light (see his book)! Now, look at the last several paragraphs of Chapter 10 and also Chapter 11 in "The Antigravity Papers" section of this book. It says "In fact, as our system continues to spin, **HUGE amounts of gravitational** energy are slowly and continually released", and also "...the total amount of energy (which emerges in the form of an ejected zero-frequency gravity shockwave..) which was (slowly!) released, is equal to the mass of the cone multiplied by the speed of light squared! This is an ENORMOUS, HUGE, STAGGERING amount of energy that was released!!!" Also, "...the total amount of energy released by running your device for an extended period of time is, in fact, the equivalent of having exploded an atomic bomb of mass m." Also, see Figure 18. Now, since this device should produce Terawatts of gravitomagnetic (gravity-like) energy, you can stand right next to it and remain unharmed, if no parts of your body have metal in or on them. My theory says that the extremely strong gravity waves travel far, and when they encounter a metal object containing copper or aluminum or lead, they will vibrate the you-know-whats out of these atoms which will cause the object, even if positioned at a long distance from the source generating device, to emit very high energy in the form of electromagnetic waves-- both bright light AND heat and radio RF. **This explains what Hamel claims,** and also incidentally explains a lot of the pictures taken of balls of very bright light flying over Area 51 in Nevada! This also explains the "Foo Fighters" (balls of light) that were seen in World War II.

In addition to all the above proof of Hamel's claims, **Hamel says that the saucer device shown in Figure 41** (also see the picture above of this same device sitting on its platform -- it has a large hole in the bottom center) **creates a "Black Hole", right below the device,** where space is bent and deformed. **Based upon my theory, I can see that such a "Black Hole" of distortion could be formed there,** particularly. **Also, recall the experiment Bob Lazar did where he could focus a device to create a small, spherical "Black Hole", where time and space are distorted.**

Also, **Hamel has a drawing of the "6 powers" or "6 poles" of a magnet.** Normally, we think of a magnet of having only 2 poles. But, if all the protons and neutrons in a magnet could be synchronized, then we indeed could detect and measure 6 poles, instead of 2, due to the fact that **there are 3 quarks in each proton, and they each create a pair of magnetic poles!**

Also, note that Hamel has design drawings of considerably more complex devices than just the 3-cone device. Things such as craft with 4 or 5 internal wings (cones), which also take advantage of air flow and ionization, and even nuclear transmutation. I can see how the combination of these wings, plus the outer shell on top, could act as mirrors which focus the gravity waves into a small "Black Hole" area below the craft.

Also, very importantly, don't forget the "KRYPTONIQUE" math formula the Aliens gave Hamel for the stored inductance energy (gravitomagnetic), a picture of which I have shown earlier in this chapter. **This formula exactly corresponds to a formula I derived in my new theory,** which is used to calculate the mass of a particle, based upon the amount of gravitomagnetic stored energy that it contains!! See Figure 5C and 5D in "The Antigravity Papers".
Hamel says the Aliens had told him that "Kryptonique" (an Alien physics term) is a Physical/Non-Physical substance, that is a component of all atoms, that can be used and harnessed to produce
Free energy.
This "physical/non-physical" character exactly agrees with my new physics theory, which identifies the routinely-measured mass of a particle with simply the "effective" mass of a small clump of pure (gravitomagnetic) ENERGY.

Also, very importantly, the Aliens told Hamel that there is actually an "aether", which they referred to as "the gases of space". This was mentioned in the Hamel video. It fits my new theory exactly, as I've found the "empty" vacuum of space to actually be a "sea" of energetic (high temperature) **vacuon particles.** These high temperature particles correlate nicely with **"the gases of space",** mentioned by the Aliens.

And last but not least, the appearance of very large amounts of mysterious tiny plant leaves. This isn't exactly scientific analysis, but here's what happened to David Hamel and his wife, Nora. **On exactly 12 different occasions,** (which were spread out over a period of time) the Hamels received a large amount of these little leaf buds. **Each one of them had 12 little points on it. The aliens told Mr. Hamel that these leaves are a sign of hope for the future** for you. They said that someday, he would finally understand what the little leaves meant. Until then, they said, you may keep these little leaves for "good luck".

Well, I can happily say that I have figured out what is the meaning of the little leaves! And I hope David Hamel someday soon may read this, as follows:

David, I have figured out what is the meaning of the little leaves! Take a look at Figure A20 of my new ANTIGRAVITY PHYSICS THEORY of matter near the back of this book. Notice that all protons and neutrons have in them exactly 12 little "clusters" of 14 "vacuons", each. Also note that the "generic structure" in the center of the page has 3 rings, with 4 little double clusters in each. Therefore, the total number of these little double clusters is 3 x 4, which equals 12. The number 12 is a VERY important number in the new physics of matter and energy and antigravity. This theory will, I firmly believe, stand the test of time, and someday, everyone will see the absolute significance and importance of the number 12, the basic number of sub-parts in all particles of matter (i.e. protons). Good luck to you David, in the near future. You will soon be having many visitors with investment money in hand, now that the theory of how your devices operate has been discovered and made public. Your long wait has now paid off and has been worth it! I hope to meet you myself later, when this book is finally published.

From all this evidence, and even some more which I saw on the video and read about which I will not mention here, **I absolutely know Hamel to be telling the truth,** and I believe **his entire story to be the truth.** Why would Mr. Hamel lie about any one small part of the story, since all parts of the whole story are just too fantastic for most people to believe? I believe that such a story would have to be either all true, or all fabricated.

This story was all true.

Notice that all of this was proven by PHYSICS ALONE. Remember, physics doesn't lie.

Incidentally, the **Boeing company** located in Seattle Washington (the source of the men who came and confiscated some of David's devices) has recently released information that it has been working on antigravity, and that the "science (of it) is valid". **An article in "Jane's Defence Weekly" told about the company's announcement about antigravity, and I have included a copy of it later on in this book!**

[ ]

[THIS PAGE IS INTENTIONALLY BLANK]

PART B

THE ANTIGRAVITY PAPERS

[ ]

This section of the book presents the reader with a reasonably easy-to-understand explanation of how antigravity fields behave, and how they can be produced by various devices. Although there are **MANY mathematical formulas** to be seen in the Figures, you can **just IGNORE them ALL, and you will still be able to completely understand ALL of the operational principles!** You can even mark out all of the math, if it upsets you to look at it! The math parts are only there for readers who happen to have the schooling background necessary to be able to understand them.

In Figures 27 through 51, and in chapters 11, 14, and 18, you will learn the principles & what's needed **to build and experiment with your very own home-made antigravity "engine".** Others have already built these, and **have actually MADE THEM WORK.**

Then, AFTER the section entitled "The Antigravity Papers: A Journey from Zero to One (And, A WARNING about asteroid....)", the reader will find the full scientific paper of THE COMPLETE DETAILS & MATHEMATICS of the new theory of antigravity, in its entirety. THIS SECTION IS ENTITLED "A New Theory of Quark Substructure for Mesons and Baryons", and IS NOT recommended for most readers of this book. It's only there for the "Ph.D." types who can understand it. You, the reader, are encouraged to show that part to people at your local universities (especially "grad" students).

Most readers will want to skip "A New Theory of Quark Substructure for Mesons and Baryons", and its FIGURES, and proceed onward to PART C, Early Pioneers, or PART D, Government & Suppression, or PART E, Lost Technology, Mercury Vortex Drive, (& The Caduceus), or PART F, Ancient Civilizations Had Antigravity.

The Appendices (Part H) contain such things as actual **Patent Office Invention Disclosure Documents (H3)** of many of my much older (cylindrical or disk shaped, and NOT cone shaped) **inventions,** and Appendix (H4) contains the old papers (very mathematical) which were on my web site.

D&R Scientific Pub. Co.; Denver, CO Jan. 2003. (originally publ. on 04-17-2002) (Figs: 2-1-02)

The Antigravity Papers: A Journey from Zero to One.

(And, A WARNING about asteroid 4179 Toutatis)

Richard P. Crandall
Dept. of Physics & Research, SPC
4-17-2002

PROLOGUE:

This book is a ***<u>must read</u>*** ***for*** ***<u>EVERYONE</u>*****!!** **<u>Every man, woman, and young adult</u> needs to at least be familiar with this information. The importance will be quite apparent a little further into this, so please keep reading.**

First, A Note of Caution for the squeamish:

This version of "The Antigravity Papers" has been written in about as simple a way as is possible. However, it does contain some very nasty-looking mathematical equations. Don't worry about this....just IGNORE THEM, if you must. It will not detract from the content of the material in any way. The equations are there so that if, per chance, you ever do see them somewhere else, you will immediately recognize them. You DO NOT have to understand how to manipulate or use the equations. All you get to do is to see what they look like!

And, you DO NOT have to know <u>anything</u> about physics, except, perhaps, that Einstein invented the equation E = M x "C squared" (i.e. energy content equals the mass of an object times <u>the speed of light squared</u> (a VERY large number)).

If you see the term **"mass"**, this is essentially the same thing as the "weight" of something.

Also, **atoms** make up the smallest units of matter which still determines all the properties of an <u>elemental</u> material (example: Copper, Oxygen, Carbon, Hydrogen, Aluminum, Iodine, Sulfur, Uranium, etc...).

And, do you remember hearing that atoms are made from three kinds of smaller particles, namely the **proton,** the **neutron,** and the **electron?**

Finally, the term "electric charge", or just **"charge"**, means how much of that "hair-raising" static electricity something contains. You won't get shocked by a material which has zero total charge.

So that's it. You're good to go.

There is also a "Ph.D. physicist" version of my new theory of antigravity, which you are encouraged to show to your local university professors! It is called "A New Theory of Quark Substructure for Mesons and Baryons."

D&R Scientific Pub. Co.; Denver, CO Jan. 2003. (originally publ. on 04-17-2002) (Figs: 2-1-02)

SO, WHAT CAN BE LEARNED BY READING THIS SHORT PAPER???

The answer is... :

Here are the most important facts you the reader will (& should!) learn:

(NOTE--For physicist and scientist readers, just skip anything that you already know, and *please* continue to read to the very end, AND, please DO NOT read anything beyond this point unless you have first read the "Ph.D." version, entitled "A New Theory of Quark Substructure for Mesons and Baryons". Else, you will probably be very skeptical about that which follows!)

- Yes, devices for antigravity propulsion really do exist. The thrust they produce can easily be made powerful enough to lift just about anything: ships, cars, even buildings!, from the ground straight upward, and as high as desired (even to Earth orbit). They can also hover "seemingly-effortlessly" and indefinitely at a fixed altitude.

- I have lately actually seen prototypes, whose operational principles are based upon my new theory, that have defeated the Earth's force of gravity, and have flown.

- This book contains easy, incremental steps (i.e. figures) in order for the ordinary, average reader to start with what is currently known about physical processes and progress to a good understanding of advanced gravitational wave generation and propulsion. DON'T WORRY about seeing all those equations of math. Just ignore them, and you'll do fine. **Don't give up. Please at least skim the entire book so you don't miss anything.**

- Antigravity engines do not consume any fuel, and they produce no heat. Quite different from today's engines.

- These machines can be used to generate electrical power. No fuel is consumed in the process of them running. In the professional scientific literature, this principle is called "tapping into the ZPE" (Zero Point Energy of space).

- Now, comes **THE WARNING:** Normally, the following things would NOT be contained in a book of this type. They include a warning about the Earth, and a call-to-action. I could have put these in a separate publication, but due to the incredible urgency, I feel I must include them in this book. [Be it known, though, that the primary reason for this booklet is for everyone to learn the basics of antigravity. It is a fun and fascinating subject, and it is fairly inexpensive to build and test these devices yourself. However, DO NOT attempt to operate such a device without a good electrical engineer present. The spinning cones can produce tens of million volts, and can spark over to YOU. Be incredibly cautious.]

- We NEED these new technologies NOW, as the Earth's supply of coal and oil and natural gas is nearly gone. Nations may someday begin warring over who gets what's left of the oil supplies! **Our lifestyles depend on getting this controversial new technology out into the mainstream, so it's NO LONGER CONTROVERSIAL!!!!**

- And, **there are far worse problems** (than the fuel shortage) lurking on the horizon. **On September 27, 2004, a large, heavy, asteroid is scheduled to come nearer to the Earth than just about any other asteroid or comet ever has.** ***In fact, it's the one object, with a predictable orbit, that will come the closest to Earth in THE NEXT 60 YEARS.*** The asteroid's name: **it's called 4179 Toutatis.** How do I know this? I got this information from one of NASA's (or JPL's) website for plotting orbits of possible Near Earth Objects (NEO's). Now, NASA says the asteroid does NOT pose any danger at all to Earth, as it's closest approach to the Earth will be about 4 times the Earth-Moon distance. **But..., that's IF the asteroid is NOT HIT by any large particles, or other "space trash", during its entire journey from where it's currently at, to here.** And it is a very **long** journey. That's what they don't tell you!! I believe that the asteroid threat is real, and such asteroid could be diverted by some impacts, and thus be caused to actually hit the Earth, therefore ending all life as we know it. Also, Toutatis' orbit is so elongated that it stays mostly outside of the asteroid belt, but will pass through the belt right before it arrives at Earth, and so it may take some "stragglers" along with it! And, it certainly will have a lot of chances to hit something as it flies through the asteroid belt. **Of course, no one can perfectly predict future events, BUT I AM ESPECIALLY DISTURBED BY THE GROWING NUMBER OF PEOPLE REPORTING DREAMS OR VISIONS OF AN OBJECT HITTING THE EARTH, USUALLY IN THE OCEAN OR SPECIFICALLY THE ATLANTIC OCEAN, IN LATE 2004. It has been my experience that mass dreams, especially if they are of the same thing, are NOT to be just ignored. I fully believe that this is God's way of warning us, especially those who believe in God. Please read about "Wormwood" in the Book of Revelation. It could be Toutatis... Also, a credible witness (authenticated forensically, by physics) to a UFO visitation was told that the surface of the Earth would be destroyed by a celestial event between the years 2000 and 2005. He was told this in 1975. PEOPLE, please do not ignore this warning!!**

- **So, what can we do about the above problem of the asteroid?** My father always said "It's better to be safe than sorry, because you just don't get a second chance after it's all over". Wise words to be heeded. **We currently have no technology to end the asteroid threat.** All we have right now is nuclear-tipped missiles which can be sent to the asteroid, and our delivery system for the missile is **very** slow, being based upon chemical rockets. So for one problem, we just can't get the missile to the asteroid in time to do something significant to it. And for the second problem, nuclear explosions are NOT THE ANSWER because they may accidentally cause the asteroid to FRAGMENT into a hundred pieces, all of which would remain on the exact same course to Earth!! Then, we'd have a far worse problem, as we would be pelted all over by big fragments, and our opportunity to divert the asteroid's orbit would be forever ENDED!

- **We can only solve our problems by making all of the new antigravity information contained herein these papers AVAILABLE to the public at large (and universities!), and QUICKLY!** Antigravity-based ships or craft can travel thousands of times faster than can chemical-rocket-based propulsion engines. They can get there (to a threatening asteroid) on time and can use artificial gravity fields to gently push the object into a different, safer orbit around the sun. I have proof (a color brochure plus testimony) that NASA will be placing many large habitation modules in space and will be connecting them to the International Space Station starting in January of 2004. Apparently, they don't plan to help save us, and the elite and the scientists are planning to live up there when & if the big one hits the Earth. I KNOW that they know possible disaster is coming. **Sadly, I think we may already be "a day late and a dollar short" in accomplishing the task of stopping the asteroid threat before it's TOO LATE.** After all, it is now early year 2003. That gives only 1 and a fraction years to **1)** spread this new knowledge among mainstream physicists at universities so they will no longer laugh at and denounce any possibility of antigravity or free energy (normally, a professor could be fired from his job for talking about such "nonsense"), and **2)** have these same people figure out how to design a device for gravity beam-forming (this is NOT easy, but is possible) and for propulsion, and **3)** give the designs for all of this to the lowest bidder, and pray that they can BUILD and RELIABLY TEST and LAUNCH before the deadline is up!!!! Hmmm, less than 2 years. Are you afraid? You sure should be. So good luck. Remember, better safe than sorry (i.e. dead). As a tidbit of sobering grim humor, if we fail, we could be on **"A Journey from One to ZERO"**, as it were.

- **It's totally up to you (the reader of this) to spread this crucial information as quickly as possible to as many other people as you possibly can!** Don't depend on me or others to do it!! I can only do so much, and yes, our very lives may depend on it!!! PLEASE, it's EXTREMELY important that you get at least one copy of this short paper entitled, "The Antigravity Papers: A Journey From Zero To One" (incl. its figures 1A - 54), and also "A New Theory of Quark Substructure for Mesons and Baryons" (including its FIGURE pages: A1 - A27) to a university or college Prof., or BETTER, to a **grad student** right away. HOWEVER, YOU ARE NOT AUTHORIZED TO COPY ANYTHING ELSE IN ANY WAY OR FORM FROM ANY OTHER PARTS OF THIS ENTIRE PAPERBACK BOOK ENTITLED "They All Told The Truth: The Antigravity Papers". **The paperback book must be purchased. NOTE: ALSO, the PAPERBACK BOOK contains MANY, MANY more PHOTOS and fascinating information than did the "OLD CD-ROM version!" Order your BOOK Today! For ordering info, see one or all of the following web sites:**

www.theyalltoldthetruth.com

http://members.tripod.com/richard_crandall

www.theantigravitypapers.com

www.antigravitypapers.com

www.trafford.com (the publisher)

- A final thought concerning the Warning, before going back to antigravity. There is always that chance that my new theory of gravity may be partially wrong. HOWEVER, if there is EVEN A SMALL CHANCE that any part of the theory works and can be implemented, WE OWE IT TO OURSELVES and LOVED ONES to go ahead and understand and develop and build & test the devices ANYWAY. Why? **Because again, our worst threat for the next 60 years occurs in Sept 2004, and we will NOT get a second chance to stop it at the "last minute". This threat is FAR, FAR WORSE than such "little" things as Bin Laden and Saddam Hussein! And it's not (YET) even discussed in the news media! As father said, "Better Safe than Sorry".**

- Now, continuing with antigravity…

D&R Scientific Pub. Co.; Denver, CO Jan. 2003. (originally publ. on 04-17-2002) (Figs: 2-1-02)

Table Of Contents ["The Antigravity Papers"]:

CHAPTER 1 "What is PHYSICS, and, The Holy Grail?":

To talk about **antigravity devices**, or any devices that defeat the Earth's gravitational force (a force which we are all familiar with), one has to include talk about a subject called "physics". **Physics** is the experimental and theoretical study of how things work (here, being careful to exclude chemistry, biology, and other studies which necessarily actually operate at a higher level than physics). Examples of areas of interest that fall under the classification of physics, are electronics (transistors, capacitors, inductor coils, etc.), electric motors & generators, fossil fuel-powered engines, nuclear reactions & radiation, the study of radio waves and light, planetary movements, rocket science, refrigeration and air conditioning, mass, momentum, and energy, and so on.

For many decades, right up until today in fact, our collective understanding of modern physics forbids certain kinds of things from occurring. This is from having developed **mathematical equations** which **model** (as best we can) and describe all the different physical processes which occur in nature, or which can be made to occur by man, under the right conditions. The problem with mathematical models, though, is that we can never completely and perfectly understand any physical process, but, we can become really, really, really, darn familiar with it. So it should not be surprising to the reader that our various models for physics have been changed and improved upon a number of times since the dawn of science. Through experimentation, we are always seeking better and more accurate mathematical models for things. Once in a while, we discover something which does not fit our current model, and we have to figure out how to revise the model.

Well, up until now, our best mathematical models for physics have clearly shown, by the power of the equal (=) sign (a VERY powerful "tool"!), that certain processes cannot occur and certain devices cannot be built. These are 1) perpetual motion machines, 2) devices that produce either totally free energy, or at least put out more energy than they take in (to operate), and 3) devices which generate **powerful** gravity waves, enough to equal or exceed the Earth's gravity. [Note that we do, however, know how to produce exceedingly weak gravity waves. That's easy to do!] These kinds of things are forbidden to operate because they violate certain mathematical models of physics, namely **"conservation of momentum"** and **"conservation of energy"**. Note that the U.S. Patent Office will not allow anyone to patent a device which violates these principles of physics. They "know" in advance that such a device will not work!

Now recently, I have created a more accurate model of matter and energy. This new model does not alter or somehow "ruin" any previous models or theories. In fact, it upholds, supports, and even explains the true root-cause for those existing theories and models. And, in addition, it adds some exciting NEW effects that are related to gravity, and these can be demonstrated. This new model of subatomic particles and photons (particles of light!) is superior, more accurate, and it even explains the underlying physical cause for certain rules in physics which have long been a mystery. In addition, this new model, which I've termed "vacuon theory" (VAC- YOU- ON), forces us to revise our previous definitions of what is "conservation of momentum" and "conservation of energy", and where does energy come from (subspace!, of course). However, it does this such that these two things, momentum and energy, are still, in fact, conserved in all physical processes, under the new definitions!! The end result of the new theory is that certain types of devices, that were once placed under categories 1), 2), and 3) above, can now be made to work!

So now, we are at least in the right ballpark. Now we must "start the game", so to speak. This new vacuon theory is explained in a document entitled "A New Theory of Quark Substructure for Mesons and Baryons", and should have been included in this package documents (if you can't find it, you should try to get a copy of it). Granted, this is a PhD level document, but the *abstract* at the top of it (which is probably the only part you will want to read) explains how essentially all of the existing theories of physics have been derived and explained by the new theory. It also shows how the long sought-after "Holy Grail" of Physics has been found.

So what is the "Holy Grail" of physics?? It is two things, which are 1) the grand unification of all forces of nature into <u>one</u> underlying force (which, we find, is gravity), and 2) the explanation of the (seemingly random!) pattern of masses (i.e. weights) of all the different subatomic particles (and, there are many of them). The new theory accomplishes both of these, but only #2 will be explained here, for brevity.

Here is an excerpt from the paper's abstract:

- Predicts the masses of 18 particles (a mix of common mesons and baryons). The error is only a few percent for most of them. For one particle, it's 0.9%.

- Finds a <u>never-before-discovered</u>, very simple pattern regarding all these masses. Amazingly, the square root of the mass of any particle, divided by the square root of the mass of the electron, is always an integer (approximately). More amazingly, the result (rounded to an exact integer) is <u>always</u> evenly divisible by either 7 or 8. See figure A_MASS, or investigate it for yourself. The value for the proton is 42, which is 7 x 6.

Now, compare that with the following.

PHYSICS BOOKS QUOTES:

From Section 26-5 of <u>Physics</u> (part two) [authors: Resnick, Robert, and Halliday, David; publisher: Wiley, 1978, ISBN 0-471-34529-6] we have:

"There exists today no theory that predicts the quantization of charge (or the quantization of and values of <u>mass</u>, that is, the existence of fundamental particles such as protons, electrons, muons, and so on)."
I have done both of these, with my theory.

And from the same book, Section 26-1, we have:

"Present interest in electromagnetism takes two forms. At the level of engineering applications Maxwell's equations are used constantly and universally in the solution of a wide variety of practical problems. At the level of the foundations of the theory there is a continuing effort to extend its scope in such a way that electromagnetism is revealed as a special case of a more general theory. Such a theory would also include (say) the theories of gravitation and of quantum physics. This grand synthesis has not yet been achieved." **I have done this, where electromagnetism is a special case of gravitation. I've also been able to couple quantum physics harmoniously with this.**

From page 19 of <u>Quarks</u> [author: Fritzsch, Harald; publisher: Basic Books, 1983, ISBN 0-465-06781-6] we have:

"It is indeed remarkable that the electric charge of the proton and that of the electron are exactly equal in magnitude because in nearly every other respect electrons and protons are entirely dissimilar. .... The proton and electron must have something in common that makes their

charges equal in magnitude but opposite in sign despite the fact that the physical properties of the two (mass, behavior, and so forth) are so very different.
Physicists have, of course, been searching for the common denominator between electrons and protons, so far without results. It appears that electrons and protons have nothing in common that would allow us to deduce the equality of their electric charge. Even today we are puzzled by this." **I have solved this connection (the "common denominator"). No longer am I puzzled.**

And from the same book, page 246, in the chapter "Does Physics Come to an End?", we have:

"The electroweak theory, however, contains a serious and as yet unresolved problem. It fails to explain why the electric charges of leptons and quarks are quantized in definite units. (The electric charge of electrons and muons is –1, and the electric charge of quarks is either +2/3 or –1/3). It appears as though physics contains a secret law that compels the various particles to contain only well-defined charges. But what is that law?" **I have discovered this.**

And from the same book, page 259, we have:

"Even Einstein spent many years of his life trying to combine the theories of electrodynamics and gravity, but without success." **I've had great successes in the solution to this problem.**

So, you see, this theory is the big one we've all been waiting for. And it opens the doors (wide open!) for building devices for defeating gravity, and for production of free electricity.

But first, the reader needs to understand, on at least a vague or non-technical level, the names of the current theories of Physics, and what they do for us, in a broad sense. What you'll see in the next chapter is a summary of the 3 major theories of Physics we have today. The theories of EVERYTHING ELSE in Physics have been based upon these three foundational ones.

Remember, what you'll see in the figures contains some <u>ominous-looking</u> **math equations**. However, you can ignore them or mark them out if you want, because they are not essential for understanding. They are just there, in case you want to know what they look like, and in case you ever see them elsewhere and want to be able to recognize them for what they are.

The idea of this book is for anyone to understand the content, and so even though you SEE lots of equations, you don't have to understand math AT ALL to understand this book!

<u>CHAPTER 2 "Beginning the Journey, starting at ZERO. What theories do we currently have?":</u>

Now, please refer to Figure 1A. It shows the 3 foundational theories of modern Physics, which have been the "state-of-the-art" of our understanding for many years. They are 1) <u>Relativity</u> (both Special and General), 2) <u>Quantum Mechanics</u>, and 3) <u>Electromagnetic Theory</u>. [The theory of Quantum Mechanics has even evolved into first, Quantum Electrodynamics (QED), and then into Quantum Chromodymanics (QCD, or quark theory).] My new Vacuon Theory does not change these theories at all. They are correct as they stand, and they are a great credit to their inventors. In #1), Einstein shows that energy and mass are somehow equivalent, and one can be converted into the other. Apparently, a very small amount of mass can be released as an incredibly HUGE amount of energy. This, unfortunately, is the principle behind the atomic bomb. Einstein's second equation shown, tells us how the very presence of matter will actually deform or warp space itself. The effect is so slight that we never see any such bending in everyday life, so don't go looking for it!! I show this second equation of Einstein's, only because it is the starting point from which I developed my entire new theory, as will soon be seen in Figure 3A.

Figure 1B tells us a little about Quantum Mechanics (or QM). The subject of QM will not show up anywhere later in this book, but it is discussed extensively in my detailed theory document. **My theory does account for and explain QM effects,** but these are beyond the scope of this book. Einstein has been known to say that he can't believe that God would play with dice with the universe, meaning that he is very disturbed that we can't ever know the precise location of a particle, but only the approximate location!

Figure 1C displays for us the third and last pillar of physics. It is Electromagnetic Theory, and the equations are called "Maxwell's" equations, after James Clerk Maxwell. The letter E represents the electric field, and the letter H represents the magnetic field. The symbols E and H will be seen throughout this book. For now, they are just the electric and magnetic field. Later, though, they will be changed to Eg and Hg, where the subscript "g" means gravity. Also seen in the equations are electric current and electric charge density.

Just for your information, the Theory of Relativity is the principle behind the operation of electric motors and generators, and of all kinds of magnets. This is because it can be shown that a magnetic field is really just an electric field which has been distorted by Relativity effects. And Relativity has applications in understanding energy, time, and space. Quantum Mechanics is the theory that explains how computer chips and transistors work. And, Electromagnetic Theory (Maxwell's Equations) is the theory that explains all our experience with radio wave transmission, light transmission, and all the usual passive electrical devices such as coils, resistors, capacitors, and so on.

Figures 2A through 2C explain, using diagrams, what is the meaning of two of Maxwell's equations and also the (supplemental) Lorentz force law equations. We see that equation #1 is the principle by which electric generators operate. A circular coil of wire is usually used to "catch" the electric field and thus produce a voltage. The power companies really like this one! Also, it is true that an increasing (opposite of collapsing) magnetic field will also produce a circular electric field. We see that equation #2 demonstrates the dual effect, namely that a collapsing or increasing electric field will produce a circular magnetic field. It is interesting to note the the combination of #1 and #2 is what allows radio waves (i.e. electromagnetic waves) to propagate through empty space. [When the electric field collapses, a magnetic field is formed. But then, when this new magnetic field collapses, a new electric field is formed. And the cycle repeats on, and on, and on...] Now, equation #3 is the simple rule regarding how like and un-like electrical charges behave (attractive or repulsive force). Ben Franklin had played with these concepts. Finally, equation #4 shows how forces are created between two current bearing wires.

CHAPTER 3 "Next step of the Journey—Discovering 'Magneto'-Gravity":

In a totally separate paper (called the "G" series), authored on 9-02-2000, I showed how we can start from Einstein's mass-warp equation (fig 1A), and ultimately "wind up with" (that's Texas talk!, meaning "get" or "create") a set of equations which look amazingly similar to Maxwell's equations! However, instead of having electric (E) fields and magnetic (H) fields and electrical charge density and electrical current, we have **ordinary gravity (Eg) fields,** and **"gravitomagnetic" (Hg) fields** (this is new!), and **mass density** and **mass flow current.** This can been seen in

pages G38 and G45 of my "G" series of papers. The gravitomagnetic, or "GM", field is new, and it is a gravity-related force which behaves similarly to how magnetism behaves, but, it's still a type of gravity!
I was able to show that these new equations will accurately describe the motions of masses in gravity fields, under special conditions called "weak field" or "linearized" conditions; later, I found that these are exactly the conditions that all atoms and particles "live" in, all the time. So, I am an independent co-discoverer of the gravitomagnetic field, and I developed equations which could be called "The Maxwellized Equations" of Gravity. I found that these equations do very well at describing the motion and behavior of the "vacuon" protoparticles which I mentioned earlier. And all atoms are composed of these vacuons. Just to clarify, from here on in this book if I talk about a gravitomagnetic field, I'm NOT talking about a magnetic field. Only a magnet could produce that. **So please just forget about electric and magnetic fields and forget about electric current throughout the rest of this book. For the rest of the book, we will only be dealing with gravitoelectric fields, gravitomagnetic fields, mass density, and mass flow (or mass current).** And, everywhere you see the terms "gravitoelectric" or "gravito-electric", these terms are exactly interchangeable with the terms "gravity" or "ordinary gravity that we are all familiar with" (it's what makes us fall down when we get tripped, and it also can cause apples in trees to fall to the ground!).

Figure 3A shows these "Maxwellized Equations of Gravity". I've added a subscript letter "g" to denote that we are talking about gravity and NOT electricity or magnetism. You may compare these with the equations on Fig. 2A; the changes are shown double underlined. Since there are 4 equations in Fig. 3A, I have given 4 example diagrams to show what they mean. The diagrams for equations #1 and #2 are not drawn, since they are identical to earlier diagrams already shown. These #1 and #2 will not be discussed further in this book.

Figures 3B and 3C describe the attraction and repulsion force laws between various masses and introduces a new kind of mass not usually seen or detected anywhere. The last case shown there is VERY important and should be committed to memory.

Figure 3D shows that mass flow currents actually can attract or repel each other. You may say, well if I place one garden hose parallel to another garden hose, and run water through them, then why don't they move or exert any forces? This is because this gravitomagnetic effect is only noticeable when both mass flows are very, very, very fast (approaching in fact, the speed of light itself!). So we don't normally observe this in our day-to-day routines.

CHAPTER 4 "Next step of the Journey—The Atom, and Smaller, and Smaller, and Smaller...":

Figure 4A shows a simplified diagram of what we call the "atom". As the reader has probably already learned in high school, everything on Earth, including ourselves, is composed of molecules or atoms, and lots of them. And molecules are composed of two or more atoms. So, let's take a closer look at the atom.

We see that any atom is composed of just three things. They are protons, neutrons, and electrons. That's all. All the protons and neutrons are clumped together in a little ball that's called the nucleus of the atom. This clump is right in the center, and it's actually quite smaller than is shown in the diagram. And, orbiting around the nucleus, are the electrons. For the rest of this book, we will not be concerned at all with electrons. They are only good for studying chemical properties of substances, and we do not need that here. So, we will zoom in on the nucleus only, for the rest of the book, as the nuclei (plural) of atoms is all that we will be concerned with, in order to understand gravity and "antigravity". So, just think of everything in nature as being made up of many billions of tiny nuclei. In fact, the proton and neutron are collectively referred to as "nucleons", as they are the only two kinds of particles found in a nucleus!

Figure 4B zooms in on nucleons, and shows that any nucleon is actually composed of 3 "quark" particles. This term was coined by Murray Gell-Mann years ago. Richard Feynman played an important role with quarks as well.

Figure 4C tells us that this is the "end of the line" for current modern physics. These current theories cannot go any "levels of detail" deeper than the quarks. We just treat them as little point-sized particles and know nothing about any possible internal structure to them. The best possible model we have that is accepted and verified is called the "Standard Model of QCD". Yes, there are theories out there that try to go beyond this, such as "string theory", but these are on very shaky ground, at best, and are not yet accepted. So, we really know little or nothing about the internal structure, if any, of quarks. Until now. Figure 4C shows my newly developed model of the proton. **It actually takes us 2 levels of detail below that of the quarks, and it shows the internal dynamic substructure of the proton. You can see that the proton is actually made up of a total of 168 vacuons (i.e. "protoparticles"). I have discovered that the vacuon, NOT the quark, is the smallest possible constituent of matter (i.e. of particles).** In fact, I've found that everything in the universe, including photons as well as material particles, are composed of different numbers of vacuons, where there is only one known type of vacuon. Vacuons are all identical, and thus all have the same size and mass. **So everything truly is made up of just one kind of little particle, and all of these little particles are identical.** How much simpler could it get!! [Actually, to be completely correct, there are two types of vacuons, but they are both the same particle. The difference is that one of these types comes in clumps of 7 vacuons, and the other is just a single free vacuon; they are the same particle, but the masses are very different, because one is bound in a clump with others, and the other is free. The difference in mass is called "binding energy".] It has often been said that the simpler and more elegant that a theory of physics is, the more chance it has of being correct, because nature, in principle, should be very simple at the lowest level. For years, physics got harder and more complicated. Now, it just got simpler!

<u>CHAPTER 5 "Next step of the Journey—models, models, and more models...":</u>

Figure 4D shows a much simpler model (remember how important models are) for either kind of nucleon (i.e. either a proton or a neutron). Yes, we lose some detail, but we don't need that level of detail to understand and design antigravity devices. The model shows just one ring of particles. Amazingly, the <u>mass</u> of half of the particles appears to us to be **negative!** You don't need to know the exact details why this is so, but just to give you a hint, it can be shown that half of the particles are moving forward in time, and the other ones actually move BACKWARD in time. And, it can be shown that particles moving backward in time will appear to have a negative mass. If you can imagine seeing two heavy masses, both moving backward in time, they would appear to be getting farther apart, instead of closer (as they would if moving forward in time). This creates the startling effect of them each having a mass value that is negative!

Figure 5A shows an even simpler model for a proton or neutron. Here we use two complete rings instead of two half-rings. This is much easier to deal with for us. The figure states that the sum of the two masses is zero. This may not seem correct, since we know that protons and neutrons have some mass, even if it is very tiny amount. Just wait, as this issue will be resolved in figure 5C.

Figure 5B shows the final and simplest model which we can possibly use. Also, just as a reminder that it is there, the gravitomagnetic field that's produced by the mass in the rings is also shown. This counter-rotating ring model is very important, and should be committed to memory.

Figures 5C and 5D explain that we are all "pure energy" beings, as we are all made up these nucleons which are nothing more than bound (or trapped) pure energy. There's no "real" mass there at all (i.e. all "real" mass cancels out to exactly zero)!! Remember, the two rings are made of "real" mass; however, these completely cancel. So atomic nuclei are just clumps of trapped pure gravitomagnetic energy. And this tremendous energy behaves as if it were real mass. We just can't tell the difference. So now we know that the long sought after nuclear binding force is actually......gravitomagnetism!! For years we believed that matter was the "parent" of energy. Now we see that actually energy is the "parent" of matter!

Figures 6A and 6B show that we can model the entire nucleus (of an atom) as a single pair of rings, instead of having a pair of rings for each nucleon, and we shall do so. There, we use a trick from electrical engineering. So now we have "THE EQUIVALENT MODEL for a NUCLEUS, (or for an ATOM)". We also see that any antigravity device must be made of only certain materials, namely those containing either copper, aluminum, bismuth, and etc., due to the fact that these elements have either an unpaired proton or an unpaired neutron.

Figure 7 shows the "EQUIVALENT MODEL for a SLICE OF MATERIAL". That model turns out to be just a single pair of counter-rotating mass rings, as we have seen so often. There is a "catch" to be aware of. This model assumes that the spins of the majority of the atomic nuclei are oriented in the same vertical direction. There are ways to achieve this unusual state of matter, which will be described later. There is also another "catch" or "gotcha", regarding the diameter of the individual "mesh elements", which are just the individual nuclei (which have a small radius). This "catch" will be discussed also, but much later on. It will then bring up the subject of "how do we dramatically widen the diameters of all the nuclei?". [yes, there is a way to do this].

CHAPTER 6 "Next step of the Journey—we're halfway there!":

Figure 8 shows why no one has detected these really strong gravitomagnetic fields before. Interestingly, we have unknowingly seen the effect of this previously unknown field in the form of what's called the Pauli Exclusion Principle for electron shell filling. But that will not be discussed here.

Figures 9A and 9B show a very important physical force effect that a gravitomagnetic field (i.e. the "Hg" field) has upon an individual mass ring (as compared to observing a parallel pair of rings, as we did before). Mass rings are, of course, a part of just about every model we have discussed. We see that a uniform field may either compress or expand a ring, but there is no overall force on the ring, either upward or downward. However, we see that a non-uniform, or converging/diverging, field does produce an overall force upon a mass ring. It is shown that any positive mass ring will always be pushed away from the more dense (or converging) flux line part of the GM field, and the other type of mass ring (i.e., the negative type) will always be pulled toward the more dense (or converging) flux line part of the GM field. So the DIRECTION of the field's flux lines DOES NOT MATTER! Admittedly, this is one of the more difficult parts of the theory to understand, but you have to go back and look at fig. 3B, and then realize that an upward directed GM field (in fig. 9B) actually behaves as if it were being produced by a HUGE external mass (i.e. having much more mass content than one of the "mass rings") which is NEGATIVE.

Thus, in an upward GM field, a (+) mass behaves as if it were a tiny (-) mass, and a (-) mass behaves normally. But in figure 9A, the downward GM field behaves as if it were created by a HUGE external (+) mass. Then, a (+) mass behaves normally, but a (-) mass behaves as if it were a tiny (+) mass, just as a tiny mass will always be attracted to the (+) Earth, regardless of whether it is (+) or (-)!!

Figure 10 shows just 4 slices ("layers of atoms") of the block of material from fig. 7. Case "A" shows how the "equivalent model" large sized mass rings are affected by a uniform GM field. They are not affected at all. Case "B" shows how they are affected by a NON-uniform GM field. All (+) rings are always pulled upwards, and all (-) rings are always pulled downward toward the converged part of the GM field. The DIRECTION ("polarity") of the GM field does not matter. The final result of applying a non-uniform field is that we have, as our model of this, only two mass rings left, overall. One positive, and one negative. They are widely separated. Between them, though, we still have effective mass currents which still do produce the same gravitomagnetic fields as they did before. The effective result of this is that we have the usual normal mass of the block of material still there, being sandwiched between two end masses. These two end masses, which are not detectable in a "normal" block of matter, are now there, and they are vastly more massive than the usual normal mass which is sandwiched in-between the two outside end-mass rings.

CHAPTER 7 "Next step of the Journey—the simplest model for a solid":

Figures 11A and 11B show that, if we know we are going to be working with particularly strong GM fields, we can model an entire solid piece of a material as being solely composed of just one pair of mass rings, which is usually located near the middle part of the solid shape. In 11A, no field is assumed to be there. But in 11B, we assume a strong internal GM field is present, which would no doubt be possible by somehow (the somehow will be discussed in Fig. 12) having most of the nuclei set up with their spins aligned in the same vertical direction, either all upward, or all downward.

Figure 11B is a very important one. **It shows that a GM field will always try to match its shape to the overall shape of the material that it is going through. This is similar to how a magnetic field will always try to follow the path of any iron or steel pieces through which it travels.** And, also important to the figure, we see that **only for a cone-shaped** (or triangular-shaped or even frustrum-shaped) solid will we get the effect that will pull the rings apart. The rings will NOT be pulled apart if we are using a cylinder or disk of material. In fact, we will later realize **that in order to have a working gravity-based device, we MUST have either some cones or triangle shapes somewhere in the device. Antigravity devices are all about "cones" and "triangles"!** These are the shapes which allow us to widely pull apart the two mass rings (of the model). And as usual, the positive mass always accumulates at the wide end of the cone, and the other mass always accumulates toward the apex or point of the cone.

CHAPTER 8 "Next step of the Journey—Cones, Triangles, and how create your 'very own' GM field":

Figure 12 shows how to induce a GM field into a cone shaped piece of material. This is unfortunately a gradual and slow process. It may take even up to 40 minutes to do so, but much less time than that is more likely the case. There are two methods shown, but actually there is a third method of placing a piece which already has a strong GM field in it close to a piece which we want to form a new field in. Eventually, in that case, the second piece will become "gravitomagnetized" by the first piece! However, this case requires us to already have created a GM field, which may or may not be practical. Anyway, the two main methods of inducing a new GM field in a cone are 1) spinning the cone, and 2) subjecting the cone to an external, axial magnetic field, which is strong. And, cold temperatures help speed up and intensify the GM field even further. The interested reader may wish to look up the (publicly available) patent by H. W. Wallace, # 3,626,606 entitled "Method and Apparatus for Generating a Dynamic Force Field". He spins disks (not cones) rapidly, and is able to detect some sort of generated field, which is no doubt actually just a gravitomagnetic field. He does measure a definite physical effect, in the form of a noticeable temperature drop in a material. Wallace did this experimenting back in the early 1970's.

Figure 13 shows that we will eventually produce, in a cone, two separate large masses, one positive and one negative, and they form at opposite ends of the cone. They are **very large** amounts of mass, as compared with the untouched cone itself. However, they are equal in magnitude but opposite in sign, so they exactly cancel and thus don't make the cone have any more or less mass than it had to start with. However, a very important and interesting effect is produced by having the two masses separated as such, and this is explained in the following figures. This effect is crucial to gravity wave powered flight.

CHAPTER 9 "Next step of the Journey—The Basic Antigravity Effect":

Figure 14 shows what happens when we have a small positive mass sandwiched within two very large equal, but opposite signed masses. It's the basic effect that takes us back to Figures 3B and 3C. The darn thing will just keep speeding up forever and ever!

The momentum balance equation required by physics for this device is:
(-M)(0) + (m)(0) + (M)(0) = (-M)(vFinal) + (m)(vFinal) + (M)(vFinal) – pEjected, where vFinal is the final velocity (to the right) and pEjected is the total momentum that is "ejected" to the left (in the form of gravity waves). YES, I'm sneaking some math into here, but it's not really all that bad.
Thus, using another standard equation,
the GravityWaveEnergyEjected = (C)(pEjected) = (C)(m)(vFinal). Also, the loss of mass, mLoss, which is the amount of mass that EACH ATOM in the device must "use" as fuel (or "burn up") in order to create the final velocity, is:
mLoss = (GravityWaveEnergyEjected) / [(C)(C)(numberOfAtoms)], by Einstein's equation.
Therefore, we have:
mLoss = (m)(vFinal) / [(C)(numberOfAtoms) = (MassPerAtom)(vFinal) / C.
So, NewMassPerAtom = MassPerAtom – [(MassPerAtom)(vFinal) / C],
Or, stated another way,

PercentMassDecreasePerAtom = 100 * (vFinal) / C %, (C is the speed of light)

AND,

GravityWaveEnergyEjected = (C)(m)(vFinal).

Therefore, we see why item #2) in Figure 14 says that, for the device to fly, the device must eject **powerful** gravity waves towards the leftward direction. It is because we can see that, from the above formula, the Gravity Wave Energy Ejected is a HUGE number, due to the "C" in the formula!! HUGE!!

So, from where can we get that huge amount of energy that's needed to make the device (the cone) fly? Easy. In order to produce that huge amount of energy, all we need to have happen is for the **mass** of each atom in the device to decrease by only **PercentMassDecreasePerAtom** %. BUT, we see, from the formula above, that this is a VERY VERY small percent. In fact, it is equal to the Final Velocity divided by the speed of light. That gives a really, really small number for the percent mass decrease required for each atom.

So, to finally answer the question, **from where can we get the huge energy required to run this thing?** Answer: it comes from just a tiny decrease in the mass of each one of the atoms in the device. It only takes a little bit!

We will find out later, in Figure 18, that **there actually is** a way to cause the mass of each one of the atoms in the device to become slightly less than the original amount. THAT'S where the needed energy comes from!!!

FORTUNATELY FOR US, we see stated in Figure 14 that since the gravity equations work just like Maxwell's equations for electricity, then an **accelerated** mass MUST emit gravity waves. It has NO CHOICE but to do this! And the amount of the energy required for those gravity waves must be the large amount calculated above. We'll later show just how easy it is to do this!

Now, you may have noticed one strange detail.
When the device is at first just sitting still on the launching pad, how can we produce a force (or a change in momentum) upon the device itself when the device is clearly **NOT accelerating**, yet?

Consider this: Let's say that initially, the device is held in place by locking clamps which are attached to the ground. Remember that the device, once spinning for a long enough time, surrounds itself with a localized gravity field which it generates (briefly, look ahead to Fig. 16 & Fig. 17). Now, let's say the locking clamps are released. Since the device is emitting (ejecting) powerful waves in ONE direction, then there must be a reaction force produced upon the device in the OTHER direction. This reaction force causes an acceleration of the object, also in that OTHER direction. Since an accelerating mass MUST emit gravity waves, as was stated above, then the device MUST emit gravity waves back in the ONE direction. Again, the device must therefore experience a reaction force in the OTHER direction, and thus the device must accelerate in that OTHER direction. And this cycle continues, on and on and on.....
So the device behaves as if it were a rocket, ejecting hot gases in one direction, and accelerating in the opposite direction. So, if the locking clamps are held in place, so that the device is clearly not accelerating, the clamps must exert a large force downward upon the device to keep it from leaving the launchpad.

So, to answer the above question, there is a force produced upon the device at all times, even if it is being held motionless on the launchpad by clamps!

Now, most readers should skip Figs 15A, 15B, 15C, and go DIRECTLY to Figure 15 D. (Unless you are a grad student or equiv.).

In figure 15D, we see the formulas for the electric and magnetic field for an accelerating electric charge, and YES they are NASTY and you DON'T have to understand them. Don't even try. Just know that if these two equations are transformed into the equivalent equations for an accelerated **mass**, they then would show how much gravitoelectric and gravitomagnetic fields are produced by that accelerating mass (as well as what direction these fields are pointing). These gravity fields which are produced & ejected, I'll call a **"Zero-Frequency Shockwave" (of gravity)**. Now, see Figure 15E. Zero-Frequency waves are not at all familiar to most electrical engineers. Now, see Figure 16. There, I have drawn a diagram of what the shockwave looks like (if we could actually see it, but we can't), which would be produced by our particular device, with its plus and minus mass separation. The shape of the diagram follows directly from the nasty equations in Figure 15D, but I have done the hard work for you! As you can probably see from Fig. 16, the gravity shockwave is ejected, generally, in the leftward direction. Thus a force (and an acceleration, if the device is not clamped down) is produced in the rightward direction! **This is the basic "antigravity" effect, which is the title of this chapter.** However, it's not actually ANTI-gravity we are producing, but rather just a powerful gravity wave which is ejected in one direction, and a consequential reaction force in the opposite direction.

The gravitomagnetic field need not be considered here, or in what follows, for reasons stated in the figure. We only need to be observing the gravitoelectric field (which is just an ordinary gravity field).

Recalling that the rod in the diagram on Fig. 16 actually represents a CONE, now please turn to Figure 17.

Figure 17 thus shows the final **Basic Antigravity Effect.** The gravity field produced from the spinning cone goes outward and also downward, at a slight angle from the horizontal. Due to the overall downward direction of the ejected shockwave, the cone experiences an upwardly directed reaction force which is consequentially being produced upon it. This force, if strong enough, can overcome the Earth's gravity, and the cone would rise upwards! Would this not be a strange looking UFO for the shocked onlookers!

However, we can now see an immediate problem which plagues all such gravity-based lifting devices. Notice that the direction of the field lines of the generated gravity field is upward. With such a situation, if a small object were placed directly within that field, the object would be pulled upward, which would cause a small equal and opposite force which tends to pull the cone itself DOWNWARD. So, if the cone were not placed atop of a table or platform, but were instead placed on the ground, it would tend to pull the ground itself upward, and since the ground cannot move, the cone would find itself sharply and strongly pulled DOWNWARD. **So the crazy problem is this: if the device is not placed high atop a platform (where the shockwave doesn't intersect with the ground anywhere near the cone's position) it will find itself being sucked down to the ground instead of being lifted upwards!!**

So an important rule is to launch the device from a tall platform or tripod, and how tall depends on experimentation.

I have noticed that a certain person, who claims to have been abducted by a UFO, said he was told that the Stonehenge structure in England, as well as numerous other such structures to be found in the area, were actually used as landing pads for UFO's thousands of years ago! I can now understand why. A photo of that person can be seen in Figure 51. Some of his fascinating story will be discussed later on in this document.

Now, we must face one final problem in order to get the lifting device to actually work. This will be discussed in the next chapter...

CHAPTER 10 "Next step of the Journey—Fixing the "Gotcha" problem, & Reducing Mass":

The device in Figure 17 will not work, as is, due to the "gotcha" problem. Look back now at Fig. 7 and Fig. 11A to understand the problem. Then re-read the end of chapter 5. Back then in Figure 7, there was a critical, but incorrect, assumption made, upon which everything that followed depends heavily on. If we look now at Figure 6A, we again can see a well-known "trick" from Electrical Engineering. The "trick", which converts a mesh of individual small currents to a single outer loop current, can be used rightly & properly only if the size of the individual small current flows are large enough so that they are either touching each other or are almost touching each other.

Now, back to Figure 7. The upper right part of that figure shows how a slice of material can be diagrammed as many small individual mass current loops, where the spin axis of each is directed upwards, out of the page. Each little current loop is just an accurate model (for our purposes) of a nucleus of an individual atom. Now here is where we made the incorrect assumption. We used the "trick" to replace all the individual small current loops with just ONE big outer current loop. However, that "trick" only can be applied if the diameters of all the small mass current loops are large enough to be touching, or nearly touching each other. But this is not the case, because in a solid material, the diameter of each nucleus (i.e. mass current loop) is incredibly small, and there is a large void of space between nuclei. So our assumption is blown, along with the model in Fig. 7, UNLESS we can somehow dramatically expand the diameters of all of the atomic nuclei in the material. If we could somehow do that (and it turns out that we can, as will be shown!), then the model in Figure 7 would be correct, and thus all the models which follow afterwards would be correct also. Before moving on, do notice one other thing. The model also assumes that the spin of most of the nuclei are aligned to approximately the same direction, namely along the axis of the block (or cylinder) of material. This assumption is already correct, due to the fact that the material will be spinning (or will have an external magnetic field applied), and thus, like shown in Figure 12, the spins of the nuclei will tend to be aligned after a long enough period of time.

So, if we can somehow dramatically expand the diameters of all the nuclei in the material, then the models in Fig. 7, Fig. 11A, and onward will work correctly, or at least much better, depending on the amount of expansion!

NOW, look at Figure 18.
If we place a rotating "exciter magnet" assembly, as shown, under our cone device, it will have an interesting effect on the nuclei throughout the cone. Since the external field coming from the assembly is at an angle (instead of straight up-and-down), there will be an interaction between the external assembly and each nucleus in the cone device.

The spinning external field torques and stretches each nucleus, causing each nucleus to have an undulating motion. And, the external field interferes with the nucleus' natural tendency to want to precess, as does a gyroscope. Anyway, the net effect is that **angular momentum** will slowly be transferred from the nuclei to the external exciter assembly.

Angular momentum is a quantity in Physics which has not yet been discussed in this document, but now it will be, because it is critical to the operation of the cone devices. Angular momentum can be defined as the rotational inertia (that's yet another definition) of an object multiplied by the angular velocity of the object. It is proven and well-known in Physics that the TOTAL angular momentum of any isolated complex system is CONSERVED (in other words, it is a constant). And, the total angular momentum of a system is defined to be equal to the sum of the angular momenta of its parts.

So, here's what happens within the complex system we are examining at the moment. The system is isolated, and consists of 1) the spinning cone, and 2) the exciter assembly. And, the cone itself is actually composed of many atomic nuclei. As angular momentum is being transferred from the nuclei in the cone to the exciter assembly, two things happen: 1) The exciter assembly begins to rotate faster, and 2) the nuclei begin to rotate slower. But, the total angular momentum is constant, as required by Physics.

As a nucleus rotates slower, its diameter must increase, due to radial force-balance equations (which are too complex to be shown here). Now NORMALLY, in any complex but **isolated** system, the total ENERGY is conserved also, meaning that the TOTAL energy of the system is a constant, as is required by Physics. However, our system is NOT a "NORMAL" system! This is because each nucleus is nothing more than a disk which contains a **HUGE** amount of trapped (confined) potential energy, in this case specifically, gravitomagnetic type energy. We can imagine, and correctly so, that the outer circumference of the nucleus is like a "rubber band" of varying tension, and thus of varying ability to confine its contents. As the diameter of a nucleus increases slightly, it releases an <u>enormous</u> amount of the trapped energy. The formula I derived from the new physics theory states that the <u>***mass of a nucleus is simply equal to a constant divided by the diameter of the nucleus***</u>. A small increase in diameter means a small percent decrease in mass, and this, in turn, means that a HUGE amount of trapped (gravitational) energy is released, according to the Einstein formula:
Energy = MassLost * SpeedOfLightSquared.

So we see that, as the device spins, total energy is NOT conserved, as it "normally" would be, because each nucleus acts as a "source" of (stored) energy, and Physicists know that a system is not "<u>isolated</u>", by definition, if any "sources" or "sinks" of energy are present in a system. Thus, well-known conventional physics says that energy does NOT have to be conserved, in this case.

So in our system (cone plus exciter assembly), we see that total angular momentum is conserved (is a constant), but energy is NOT, nor does it have to be. In fact, as our system continues to spin, HUGE amounts of gravitational energy are slowly and continually released. In theory, if the diameters of the nuclei could be made large enough, presumably after a long enough running time, then **the total mass of the cone itself would be reduced to near zero!** Talk about your weightless objects! And even more amazing, the total amount of energy **(which emerges in the form of an ejected zero-frequency gravity shockwave, of course!)** which was (slowly!) released, is equal to the mass of the cone (m) multiplied by the speed of light squared! This is an ENORMOUS, HUGE, STAGGERING amount of energy that was released!!!

Anyway, as the title of this chapter indicates, you've finally solved the "gotcha" problem, AND, as a bonus, you've figured out a real way to greatly reduce the mass of an object, an <u>enormous</u> achievement! AND, the laws of conventional physics were NOT violated!

AND, ... YOU'VE JUST BUILT A...

CHAPTER 11 "Next step of the Journey—You've Just Built an ATOMIC BOMB!!! (really!)":

Yes! It's true! You have built a SLOW-release atomic "bomb". But don't panic or worry. Yes, the total amount of energy released by running your device for an extended period of time is, in fact, the equivalent of having exploded an atomic bomb of mass m. However, the BIG difference is that you did it VERY, VERY, VERY, VERY, VERY SLOWLY. AND, you'll be happy to know that if such a device ever gets out of control, it will simply destroy itself and fall apart (it's NOT a NUCLEAR explosion, or anything nearly that bad). There is NO deadly radiation produced at all. If a device ever gets too "energetic" is can be easily turned off, or it will just fall apart or fly apart.

Actually, this is all very good news, as we've just solved the world's energy shortage problem, and in a comparatively safe-to-operate way. This type of energy release is better than what is done at nuclear power plants (our device will never "MELT DOWN"!), and is even better that fusion power or "cold-fusion" power! But I won't go off into that line of discussion just right yet...

Now, the next big question is (or should be): Ok, Now that all of this huge energy has been released and is gone off into space (in the form of gravity shockwaves, of course), and the mass of my cone has been reduced to just about zero, what happens when I physically stop all the the parts in the device from spinning (perhaps by stopping them and then even separating them)? Here's what happens...

Over a long period of time (perhaps more than 30 minutes), each nucleus in the device will **slowly** shrink its diameter back down, until it gets to its original diameter. This gradual effect is due to occasional collisions with the randomly-moving free vacuon particles which are present everywhere (but they are sparse, like a gas is). Eventually, then, everything will have returned to normal, and we may do the experiment again, if we wish. <u>This completes the second half of a big overall energy cycle</u>. The two parts of the cycle can be understood, as follows.

During the first half of the energy cycle, a huge amount of gravitational energy is released slowly, as the device operates. We will call the total amount of this lost energy "E". Now, suppose the device is shut down and stopped. It must regain this same amount of energy, "E", to return all of its nuclei back to normal. This constitutes the second half of the energy cycle. During that time, the energy "E" which comes back into the nuclei is supplied from the vacuon gas, which does have a "temperature", and thus has energy available in it (it's purely kinetic energy). Collisions with this gas restores, slowly, the potential energy normally trapped in the nuclei.

So, during the first half of the energy cycle, trapped potential energy is released from each nucleus. Then, during the second half of the energy cycle, each nucleus gradually absorbs energy from the seething vacuum of space, until its original level of trapped potential energy is restored. This cycle can be repeated over and over, to **transduce** and **free up** large amounts of energy contained in the vacuum of "empty" space. In fact, there is virtually an unlimited supply of energy in the vacuum, because the universe is very, very large.

So, if we look at the <u>entire</u> energy cycle, we see that energy actually IS conserved, and thus it doesn't just magically appear from "nowhere". The energy released does come from "somewhere", and that place is the vast kinetic energy in the infinite vacuon-filled "gas" of space, known as the "vacuum". As we have seen, it takes an ENORMOUS amount of ejected energy to run an "antigravity" engine and make it fly. However, fortunately, that enormous energy IS EASILY AVAILABLE. It comes from the first half of the above-described "energy cycle".

So, all the skeptics can be totally satisfied, as we see that Energy, Momentum, and Angular momentum are ALL conserved, just as conventional physics has always required! And, the enormous energy needed to power a device actually does come from an identifiable "somewhere", and there is an infinite amount of free energy available from that "somewhere", and it can be effortlessly "pumped out" via the use of the energy cycle.

And, it's kind of comforting knowing how to build your own "safe" atom bomb!

So getting back to figures 17 and 18, we see that a **SINGLE CONE** device can be made to operate IF it has present, underneath it, an exciter magnet assembly.

And this brings us now to the subject of my first **single cone** based (actually, TRIANGLE-WEDGE BASED) antigravity device which I invented. That is one of the subjects of the next chapter...

CHAPTER 12 "Next step of the Journey—Inventions; One, Two, and Three Cone Antigravity":

Now, take a look at Figure 32, which is my first and simplest to build invention. Note that instead of using an aluminum cone, I am using just a thin slice of a cone, which is basically a triangular slab of aluminum. However, this is still equivalent to a full cone if the cup assembly is rotating fairly fast. The reason for using a wedge and not a cone is so that the produced gravity shockwave will be emitted in only one direction, and not evenly all around in all directions like a cone would do. (This is useful, we will later see, for allowing optional placement of additional devices around the periphery which function as pulsed gravity wave **amplifiers**.) One other thing to note: I am showing a large disk with 2 embedded disk magnets. I thought at one time that such would be needed, but now I think that this part can be omitted. So don't worry about it.

This device should work, but how well? This device would have to be experimented with a lot, because we don't yet know the optimal speed to rotate the exciter magnet, nor do we know the optimal angle which it should be adjusted to. And a third variable is we don't know how fast we should rotate the cup assembly, or do we really need to try to rotate it at all? Maybe it will just start rotating naturally. So you see, there are many questions. Also, worse, we may need to use a computer or microcontroller mechanism to dynamically adjust the angle and speed in real time. So, it's easy to build, but hard to test. By the way, the reason the triangle is in a cup is so that either "dry ice" or liquid nitrogen can be added to cool down the triangle piece. The gravity effects will work from 100 to 10,000 times better if the triangle is very, very cold.

At the time of this writing, the only device I personally have tested, so far, is this particular device. I had a lot of trouble with the device falling apart during spinning, due to imbalances. But, I did get positive results on just one of the several predicted physical effects. Besides antigravity, another expected effect is that the triangle will become electrically charged. And, this effect DID ACTUALLY HAPPEN when I ran it! This charging effect CANNOT be explained by conventional physics. What happens, due to the NEW physics, is that the charge values of the protons in the atoms actually change, but the charge values of all the electrons in the atoms does not. So, now we have an imbalance, and thus the object has a NET charge. Why is beyond the scope of this book, and will not be further explained. However, note that we have identified a very significant safety hazard, and safety issues will be discussed later herein. Here, I'll just say that all these devices produce thousands or millions of volts of charge, which is unwanted, unavoidable, and is very dangerous. These high voltages can spark over a distance of one foot and ZAP you!! I did not have equipment to detect any weight changes in the device, but of course, this is another predicted effect. If designed well, the device should weigh less when running, and may even levitate!

Before moving on, I'll repeat that I classify this device as a **"SINGLE CONE"** device. (even though it's just a triangle wedge).

Now, on to the next invention:

Please see Figure 26. Here is a more technical version of the previous invention, which replaces the exciter magnet with a rotating RF (radio frequency) source, all being inside of a "waveguide" (there are books on waveguide theory). And the waveguide has been modified by flaring out the lower part of it. A normal waveguide would have been simply cylindrical and of constant diameter. Other than that, the device operates similarly to the previous one. However, this version is extremely hard to build and test due to the complications of waveguide "tuning". Now, note in the lower left corner that this device can be used in conjunction with several "gravity wave (GW) amplifiers". This was the case also for the previous invention. GW amplifiers will be discussed shortly, but they are rather advanced, and are not needed for preliminary testing.

By the way, later on, I was amazed to find out that this design is nearly identical to a design shown on Bob Lazar's website, at **www.boblazar.com**. Bob claimed to be working years ago on a government project to back-engineer UFO's supplied to us by aliens from Zeta Reticuli. There's absolutely no way Bob could have faked his information regarding the UFO engine design, because I could clearly see that his design WOULD WORK, according to the principles of my new physics theory. I may well be the first person to fully understand how the craft worked, because Bob and others apparently never did figure out all of the details of the mechanism. So, I am FORCED to believe everything on his web site, as I have verified that his reporting of the UFO hardware must be genuine and totally true. So, what does this mean? Well, for starters, we are not alone in the universe, and there really are bug-eyed looking aliens which are 3 feet tall...

At any rate, I classify this device as a **"SINGLE CONE"** device. (even though it's also just a triangle wedge).

Now, on to the next invention:

Please now see Figure 20 and 21. Here we see a **"TWO CONE"** device. And this is no doubt the second worst drawing I've ever made (you'll shortly see the first worst also!)! Now note that there is NO EXCITER MAGNET assembly. This is because each cone actually functions as the EXCITER ASSEMBLY for the other cone! How could this be? Simple. Over time, as the cones spin, they develop not only a gravito-magnetic field, but they develop an ordinary, parallel magnetic field as well. Both of these fields result when the nuclei of the atoms begin to align their spin directions. Thus, since the two cones are normally at an angle to each other, we see that each cone acts as the exciter assembly for the other cone!

There's one other thing to note about single cone and two cone devices. For both type devices, we need to keep at least one cone (or exciter) turning by externally applying a torque. This can be done by an electric motor, water jets, or even compressed air jets. Without this external stimulus, the cones will eventually slow down and stop. However, the next invention solves this problem...

Now, on to the next invention:

Please now see **Figure 27**. Here we see a "**THREE CONE**" device. Also, please refer back to **Figure 4C**. We recall that the *actual* configuration of each and every Proton in a nucleus is THREE parallel-stacked spinning rings. Up until now, we have been using a *much simplified model* of the Proton, which shows it as having only ONE spinning ring. Thus, a **much more natural** antigravity device would consist of THREE parallel-stacked spinning cones, where each cone corresponds to each of the three rings in the Proton. Also note, like the proton's rings, adjacent cones must rotate in opposite directions as well. Rings of repelling magnets are used to keep the cones from falling too far to the sides.

Due to the perfect "hand-and-glove" fit of the cones with the actual proton model, it is believed that this device will only need a "kick start", and from that point on, it should sustain itself, and the cones should keep spinning faster and faster as the atomic energy is released. This device could be classed as an infinite feedback loop device, as it is in an energy cycle that "feeds" upon itself, and continually grows. Be careful, as it may just speed up out of control.

The THREE cone device is the most reliable design, and is guaranteed to work if built properly and balanced properly. These are the type of devices we should experiment with first, to prove the new theory. You will see later in this document numerous photos and information from various sources regarding their actual experiences with attempting to build and operate 3-cone devices.

What? You mean people have already been building these things, without having first developed the new physics theory, like you did, to explain how to build them? Yes, it's true. And I suppose I shouldn't be so surprised, because after all, there are billions of people on the Earth, and that's a lot of people. Someone was bound to stumble across it, even if it was simply through years of just tinkering with different mechanisms. Has anyone actually built one that really worked and levitated? Yes, again, according to what I have read.

Well, here's what I have found out, briefly. There are apparently several people (perhaps half-a-dozen) who are working with such devices, but all of their fundamental design information & sources point right back to one man, and one man only. And, that man's name is David Hamel. He is the same person I mentioned in an earlier chapter who claims to have actually boarded a UFO and seen how its propulsion engine works. This happened back in 1975. In fact, the aliens demonstrated many advanced things to him. David said the aliens look just like regular people you'd see every day, but he said that they are "angels".

Interestingly, I am forced to believe that everything he is claiming is absolutely true. This is because David has been a carpenter by trade, and has had no "formal" education. There's absolutely no way that David could describe the particular devices and advanced physics diagrams he's seen without actually having gotten the information either from the government, or from outworlders with advanced knowledge. Although David started with testing similar 3-cone devices, his later designs are vastly more complex that anything I've come up with so far. As crazy as it is, if this part of David's story is true, which it must be, then why would he lie about any of the rest of his story? So I do completely believe his total story. And, he is not trying to make any money from his devices, which DO work. Instead, he is trying to give his technology away, for the betterment of others. Much more will be said about David, who is now almost eighty years old, at the very end of this book.

CHAPTER 13 "Next step of the Journey—Briefly, of Interest, a Few Other Related Inventions":

[This chapter may be skipped, if desired.]

And now, I go on to describe my **gravity-wave amplifier** inventions. To start with, please look now at Figure 22. Here, we see a complete, single stage gravity shockwave amplifier which I invented. Notice that it has an "electrode" which is made of either aluminum or copper. These are the two most common materials which interact strongly with gravity waves. By the way, a material which does not react or interact with gravity waves is iron or steel. A UFO whose outer skin is made of aluminum would light up at night like the very sun itself had risen in the middle of night! A UFO whose skin is made of steel or plastic would not exhibit this behavior at all. And this explains why some UFO's are reported as blindingly bright balls of light, and others are reported as just disks. Anyhow, the electrode shown is basically the shape of an inverted triangle (or frustrum). A square-shaped electrode would not work at all, just like cone shapes work but cylinders or disks do not. The triangle shape causes a separation of positive and negative mass rings, just as was the case for cones. The top part of the device is just a helix (a coil) of copper or aluminum tubing, or it can be made of solid, large cross-section wire as well. The upper coil is called the "primary", and the lower electrode is called the "secondary". The generated force of thrust on the device is in the upward direction. One uncorrected mistake in this diagram is that the "optional bar magnets" shown are NOT optional, and each of them (there are two) must actually be an "exciter magnet assembly", as was shown in Figure 32. Also, see Figure 23, which shows the simplest "building block" part which must be present in all gravity wave amplifiers. A complex, powerful amplifier would use many of these tied together.

Now, why would we ever want to amplify (increase) a gravity wave. Simple. If the wave produced by the central spinning cone assembly is insufficient to lift the device, perhaps because it is just too heavy, then we use external amplifier coils like this one to produce a stronger gravity wave, which comes out from the "secondary" of each amplifier.

How do we set up these amplifiers for use with a **one-cone or wedge device?** An example of this can be seen in Figure 26, lower left corner. As the central wedge spins around quickly, its produced gravity wave also goes around, just like a light beam coming from a lighthouse. Each time this gravitoelectric wave hits an amplifier, it causes a short pulse of gravitomagnetic energy in the primary coil, and thus produces a pulse of gravitomagnetic energy in the electrode as well. This field in the electrode is similar to the field produced inside of a spinning cone in a cone-based device. So the positive and negative mass separates, where the negative mass concentrates toward the narrow ends of the electrode. Thus, these electrodes eject an even more powerful gravity shockwave, which produces more overall lift in the device. Now, see

Figure 19. This diagram better shows the fields involved when using a system like the one in the lower left corner of Figure 26. However, in this diagram, only one of the three amplifiers is shown. Also, the central spinning wedge is shown. The primary of the coil MUST BE placed in the path of the shockwave coming from the central wedge. Note that the triangular electrodes shown separate out the + and – mass, just like the central spinning wedge does, and thus produce their own gravity shockwaves as well. teotwawki-ps100405h ell.

Here's an interesting note regarding the diagram in this Figure 19. Look at the bottom center part of the page. You will see what happens when we build an amplifier which is extremely efficient and powerful. Then, the field lines coming off either side of the triangular electrodes actually bend down far enough so that they combine into just one narrow beam of field lines, pointed straight up and down. Thus, **a gravity beam device has been created.** Such a beam, when aimed at the ground under a flying ship, would tend to levitate or suck up objects which are on the ground. This would be a useful device for collecting things. Now the flipside of this is also interesting. If the gravity beam device is set up with the field lines reversed (i.e. pointing downward), then the beam, when pointed at the ground, would act like an extremely heavy column of weight (just effective as a heavy concrete column!), which could be swept around on the ground. And this is precisely the device the alien ships use in order to make "crop circles" on the ground, in a field which is growing crops. Mystery solved.

Now, how do we set up gravity wave amplifiers for use with a **TWO-cone or a THREE-cone device?** An example can be seen in Figure 24. Now here's the absolutely WORST drawing I've ever been not-proud to make!
Also, see Figure 25. What we see is a two or three cone device encased inside of an aluminum shell which has a window opening cut out in it. So, why do we need the aluminum shell?? A spinning 2 or 3 cone device emits its gravity waves in a completely direction-changing, non-coherent, and time-varying way. The aluminum shell is there to fix that problem. The produced gravity waves can only exit through the window, thus producing a more-or-less (but not perfect) uniform and constant source of the wave. This wave is then intercepted by a copper coil which is in front of the window. Due to the motion of the copper coil, a pulsed gravitomagnetic field is produced and fed down to a "bank" of electrodes. Note that a profile view of the electrodes shows the usual required inverted triangular shape. Finally, the "bar magnets" are NOT optional. In fact, each one has to be a rotating EXCITER MAGNET ASSEMBLY like in Figure 32. The forces which lift a spaceship which uses this engine are produced inside of the electrodes in the "bank" of electrodes.

Suppose we want to produce a really powerful gravity wave? How do we do it? One way is to cascade a number of amplifier coils. See Figure 28. A pulsed gravity wave source is shown in the upper left corner. It could be like the windowed capsule which was in Figure 24. Several stationary coils are used. The output electrode (called the "secondary") of each amplifier is placed right next to the "primary" coil of the next amplifier. In theory, we can thus produce as powerful of a wave as we want, just by adding additional amplifiers. Remember that all this produced energy is free and is unlimited in supply. **In this figure, I show how to convert the gravity wave into useful free electricity. All one has to do is to add a multi-turn coil of iron wire around a part of the final amplifier or its coil, and voila, you get an infinite supply of free electricity.**

Now, on to the next invention:

Please now see **Figure 29.** Here we see a device which has nothing at all to do with gravity. It doesn't produce gravity nor does it use gravity. It is a purely electromagnetic device. So what does it do? It, also, **produces an unlimited supply of free electricity**. Instead of the device producing a long range Zero-Frequency Shockwave of GRAVITY FIELD, it actually produces a long range Zero-Frequency Shockwave of ELECTRIC FIELD. The device utilizes **ionized** mercury vapor, which is spun around inside of a drum, and at a high speed of rotation. The device's theory of operation comes from a small "offshoot" theory which was derived when I was developing the vacuon and gravity theory. This "offshoot" theory gives us a new way to understand what really creates electric fields. I found out all charged particles actually emit a steady stream of photon particles out uniformly in all directions. That is, under "normal" conditions. **However, if we take, for example, a dielectric (or even metal) disk or cylinder, which has been charged with static electricity, and rapidly rotate it about its axis, we see that the outgoing flow of photons gets bent into a flow-shape which mainly emits in the outward direction, but not as much in the upward/downward or inward direction. Since the acceleration vector (and thus the force vector) of the charges is perpendicular to the velocity vector of the charges, no work is being done upon the charges, and thus they lose NO ENERGY at all due to the rotation and the emission of zero-frequency electric shockwaves. Even if we use a coil of wire to intercept and absorb the shockwaves, the total energy of the charged disk remains constant, and the disk does not slow down (except as is due to friction).**

Now if we look at Figure 31, we can see how to tap into the emitted electric shockwave coming from the spinning drum device from Fig. 29. **The device shown generates an unlimited supply of absolutely free electric power.** All we had to do was to add a single loop of wire near the drum. For more output, we would use a multi-turn loop of wire instead. And wouldn't you know it, I found out that there is an isolated religious community in Switzerland which has been using similar devices (not mercury-based drum, but a charged rotating disk instead) for many decades to produce free electricity for their homes!! This will be discussed later, when I get to Figures 37 through 39.

The last invention is shown in Figure 30. I'm not really sure if this arrangement would work or not, but I believe that the produced electric shockwave might be used to initially start up a multi-cone device. After reading a lot of material regarding the sky ship "vimanas" that ancient India supposedly had, I believe that they were using an arrangement such as this in their designs.

This concludes the discussion of all the inventions, and I show a notarized signature page in Figure 33 as proof of date of these things.

CHAPTER 14 "Next step of the Journey—QUICK SUMMARY of Principles Of Operation: 3 CONES":

Now for all that follows, I will talk only of **3-CONE** devices, instead of 1- or 2-CONE devices, simply because the 3 cone ones should be easier to test and get working, and should produce unmistakable spectacular effects (discussed later on). In what follows, I will not again talk about where the gravity wave energy comes from, nor how to restore the nuclei to their original state, as these subjects have been discussed earlier, and were referred to as the complete energy cycle. As you will see, the mechanism provides us with an **infinite feedback** device, where the produced energy loops back on itself, and then produces even more energy, and so on. It works similarly to some chemical reactions I've seen, where you mix two reactant liquid chemicals together, and then.... Nothing happens. So you wait, and you wait, and then a minor reaction starts to occur. And so you wait some more, and then reaction rate increases. So you leave to get a cup of coffee, and come back in 15 minutes, only to find that the reaction is way out of control and is producing considerable boiling and heat, and is splattered everywhere, and is producing great quantities of poison fumes! **The lesson to be learned here is to watch your device at all times, and have a long wooden stick attached to a safety release mechanism**

on the device, so you can disengage the device if and when it gets out of control. Also, see the safety points brought out later in this book from people who have done experiments with these devices!

OPERATION PRINCIPLES OF 3-CONE DEVICES:

1) Start one cone spinning by manual intervention.

2) Some angular momentum is gradually transferred from the nuclei spins in the one cone to angular momentum in the form of overall cone spin in an adjacent cone. In other words, a neighboring cone will then start to spin because of this effect.

3) The rate of spin of cones increases as angular momentum is transferred (via collision interactions between atoms) FROM the individual nuclei in each cone TO the overall spins of the adjacent cones. The **rate of transfer depends on** the angle between the cones and also the **strength of the GM** (gravitomagnetic) **field** in the cones. The TOTAL ANGULAR MOMENTUM of the system is conserved, per conventional laws of physics. The TOTAL ENERGY is NOT conserved, as allowed by conventional physics, since there are "energy sources" present in the system, namely the trapped gravitomagnetic fields within each proton and neutron. The rest of the energy is conserved in the system, but the nuclei gradually "leak out" their enormous amount of potential mass/energy, and this appears in the form of powerful ejected gravity shockwaves.

4) As the angular momentum of individual nuclei steadily decreases, the diameter of these nuclei begins to increase, for reasons of having the inward/outward radial forces balance at all times. **As a result, the mass of individual nuclei begins to gradually decrease (since the mass is an inverse function of the diameter).** We will call "mLoss" the amount of mass that is lost by each nucleus.

5) Thus, an amount of ordinary gravity (shockwave) energy is released. The amount of this ejected energy will be called "GravityWaveEnergyEjected". Conventional physics then says that GravityWave EnergyEjected = (number_of_nuclei)(mLoss)(C)(C). A slow "atomic bomb"!

6) Conventional physics then also says that ChangeInMomentum = EjectedMomentum = GravityWaveEnergyEjected / C. Thus, we have ChangeInMomentum = (number_of_nuclei)(mLoss)(C).

7) So, by conventional physics, upward thrust force produced, thrustProduced = ChangeInMomentum / timeElapsed.

8) The **strength of the GM** (gravitomagnetic) **field** in the cones increases steadily and dramatically as more of the nuclei align their spins. (more nuclei align their spins because of the increase in the overall spin rate of the cones, an effect which occurs each time we pass through step 3, above).

9) Also, +M and –M get larger and farther apart. (due to increased GM field).

10) **Go back to step 3) and repeat, until device is shut down.**

CHAPTER 15 "Next step of the Journey—The New Theory Links Diverse Researchers Together":

So, at last, the math and physics talk is (97%!) over, and we can now concentrate on what various researchers are doing now, and what they have done. Now, take a look at Figure 34.

We see one new type of device here. It is based upon applying an external high-voltage power supply to a capacitor with plates that are different sizes. The diagram shows that this high voltage applied tends to pull back the electrons of the atoms, which then causes many of the nuclei to align their spins with the (conical-shaped) electrical field. Then, everything else in this device happens essentially the same as it did before for one-cone antigravity devices. A conical GM field forms, and the +M and –M separate, and so on, and so on... Now compare that diagram with the diagrams in Figure 35 & 36, which is a copy of part of a patent application authored by Thomas Townsend Brown. One apparatus that Brown shows is like my diagram, except he uses a "bank" of several of these, all side-by-side. So we see that Brown's devices can be explained by my new physics theory as well. History shows that Brown actually did get devices like this to levitate, back in the 1940's – 1960's. Apparently, shortly after Brown got these working, the government saw what he was doing, and they stepped in, and made all of Brown's projects classified (secret).

Now back to Figure 34. So now we see, for the first time, that three major players in the antigravity field are all linked together, as my new physics theory explains the principles of operation governing all of their devices! **Bob Lazar** speaks of a triangular wedge (apex pointing downward) surrounded by three gravitational amplifiers; **David Hamel** speaks of mechanical spinning of cones; and **Thomas Townsend Brown** (now deceased) spoke of and demonstrated using high voltage applied to devices like shown, and other shapes as well. So, these are the "big 3", and they all told the truth about what they were doing.

So, now we have a list of all the ways known to create antigravity. They are mechanical spinning, high voltage, magnetic field, and GM field. Item number 5 shown just enhances the effects of the above 4 methods. You may wish to study the history of T.T. Brown more, but I will not discuss him further, as I want to mainly concentrate on **3-cone devices** from here on.

CHAPTER 16 "Next step of the Journey—Brief Look at How Others Make Free Electricity!":

Refer now to the photos in Figures 37, 38, and 39. **This is a brief but important talk before we get back on track with learning how to build working 3-cone devices.**

What you see is a free-electricity producing device used by people currently living in Methernitha (Linden), Switzerland. The devices are a modification of well-understood existing machines called "Wimshurst" generators. A Wimshurst generator usually has two dielectric material disks, which rotate in opposite directions, and they build up a very large static electric charge as they spin. Now referring back to the earlier discussion regarding Figure 29, it was stated that either a rotating drum of ionized mercury, OR **a rotating disk** (or cylinder) containing a large **static electric charge** on it could be used to produce free electrical energy. Now the Methernitha Testatika machine does have 2 disks, but as I've said, you can make free electricity with only ONE rotating charged disk. The Testatika machine looks very nice and very, very complex.

However, to prove the concept of these machines, all you have to do is to have ONE rotating disk with static charge on it, and you also need the two really large coils of wire, which you see sitting out to the sides of the machine and slightly towards the front. The machine will not work if the coils are sitting directly to the sides, so they **must be moved forwards** just a little. Also, these coils must be of very large diameter, as you can see in the photos. IN FACT, all you really need is just ONE of the big coils. The windings on the big coil(s) may be of copper wire, if desired, as this machine does not produce any gravity waves which could seriously interact with the copper.

Try building this much simplified version of the device, and I guarantee you will measure a nice DC voltage across the two end wires from the coil. You can power anything with it, and the rate of rotation of the big disk will not decrease, except as is due to friction. So, in a practical device, you will need to have a motor or something similar to run the big disk, so it won't eventually stop due to friction. **Happy Experimenting!**

Incidentally, Professor Marinov, as pictured, is considered the inventor of the device. Unfortunately, he is now deceased, after apparently having jumped out of a window in a very tall building. Many people say, however, that he was "pushed". If you think about it, such machines could be a really big threat to the electric power companies and even oil & gas companies.

CHAPTER 17 "Next step of the Journey—Final Brief Look at a Top Secret Government Project":

Please refer to Figure 40. Here are two drawings from Bob Lazar, who was mentioned earlier. To help prove that all he claims is true, start by looking at what's inside the hemisphere in the second drawing. What you see is a waveguide, which is of the design I invented in Figure 26. We can see that the lower part of each waveguide is flared out somewhat. Bob said that inside of the top of the waveguide in the ship, there is a triangular slab of "ununpentium" material. This corresponds to my design in Figure 26, except I know that other materials such as aluminum or copper will also work. And Bob never did mention anywhere whether he knew or not that the triangle has to be spinning. I know that this is required in order to make the amplifiers work in a "pulsed" manner. Bob did at least mention that the amplifiers are "pulsed", and not on constantly. Also, Bob probably wasn't allowed to view the R.F. source, and so he did not know that it has to spin or move around, as well.

And now regarding amplifiers. Referring back to Figure 23, we see what I have designed as the basic "building" block, out of which all GW amplifiers must be made. From the side, it looks like an inverted copper triangle. Now back to Figure 40. I have seen a large poster version of the saucer shown at the top of the page. I looked closely at the cutaway view of the 3 amplifiers, and lo and behold, I could see several inverted copper triangle shapes, and basically the whole thing is made of copper plating. So there you have it. Bob Lazar is completely telling the truth. Check out his website at www.boblazar.com. This website is really worth looking at, and it is the most beautifully done website I've ever seen. You could literally spend hours looking at all its different pages.

Bob, from all of us here on Earth, we deeply thank you for coming forth with this information, for the betterment of us all. You must have put yourself at great risk by coming forth, and this is understood and very appreciated. God Bless you for doing the right thing. We all are in your debt.

CHAPTER 18 "FINAL step of the Journey—HOW TO BUILD YOUR OWN ANTIGRAVITY SHIP":

For the rest of this book, there are two people which we must give great HONOR and RESPECT to. They are **DAVID HAMEL,** and **Justin Szymanek.**

First, **David Hamel.** There is absolutely no question that all of David's story is true. As I've said before, there is NO WAY he could have drawn physics diagrams where at least some of them were just like my diagram for the electron (4 vacuons in a circle), and etc. I saw these in his video "Contact From Planet Kladen". There is no way that David would have known to try devices with 3 spinning nested aluminum cones, unless he was told to by beings who are not from this Earth. David's terminology is strange, but I could see in his video that he had clearly been told about negative mass, but he called it "invisible" instead of "negative mass". He said the electron is made of two of "matter" and two of "invisible", and indeed, my theory shows the electron is made of 2 positive mass vacuons, and 2 negative mass vacuons. David spoke of what he called an "isotope line" of the energy, which looks like a 3-legged bent line. It is obvious to me that this refers to the spin vectors of the three quarks in a proton. But David doesn't know what a quark or a proton is. David was given some designs which include red-granite spheres (pinions) as well as hollow aluminum cones. In fact, the aliens (angels) specifically TOLD David to use red-granite. David, when asked "why granite, and not some other rock such as marble?", replied that he thought it was because it weighs more. I did some checking, and found out that **red-granite** is LOADED with **aluminum!!** No other naturally occurring rock has this much aluminum in it. And, as my new theory show, materials like aluminum and copper react well with gravity waves, where all other materials, even iron, don't react with gravity waves. And, regarding granite spheres, I checked with a travel & tourist web site for Costa Rica, and it said that one region of that area contains, to this day, many, many perfectly round granite spheres, some as big as 8 feet diameter, and some only 2 inches in diameter. The aliens had told David that he could find this and other evidence all over the world to prove that some ancient civilizations had antigravity. I myself have seen a little device that the Egyptians had, called a "DJED", which consisted of three hollow, nested cones! I could go on and on, but there is not enough room here. For those who are interested, you can check out David's web site at: **www.world-famous.com**. On that site, there is an order form for David's book or his video. I have no involvement myself with David or his web site.

David Hamel is a hero. He has taken upon himself the Herculean task of spending the last fourth of his life investigating how to understand and build antigravity ships to save us, the people of Earth, from the upcoming Armageddon and destruction, which the aliens said is sure to come, between the years 2000 and 2005, according to them. And throughout the many years, he had to suffer many criticisms and taunting by people with small minds and no hearts, as well as many other hardships. David is getting very old, now, but he is still hard at work building a ship, or ships. One of them, I understand, is nearly finished. Interestingly, the Aliens told David that Noah's Ark was not actually a boat, but was an antigravity ship! And in a real sense, David is a modern Noah, building a modern "ark". He just wants to help people, and I believe that he deserves the utmost honor & respect, and now, a lot of help too. Thank you David.

I'll give some important final information regarding David at the end of this book. But now, on to the second person, **Justin Szymanek.**

Justin has been building several of David Hamel's designs, over the last several years, and has documented this on his web site **www.geocities.com/undergsci/** . This is an awesome website, filled with color photos of all his different devices he constructed, information about test runs, and a good deal of safety-of-operation information. Justin deserves honor and major recognition as well, for the great deal of work he put into building & testing the devices and for the awesome website he created to display the devices. Justin's experiences with building and testing will help us all speed up the process of making a device that works, and works well. From his reports, we can avoid problems and headaches and "blind alleys" which he encountered and wrote about. And, we can take his suggestions and solutions. **We all thank you, Justin.**

For the latest updates, do get online, and look at Justin's current site!

Now, after having given out the thank you's and other necessary credits & things, I get to the heart of this chapter, which is **How to build your own antigravity ship!**

Let's see how others have done it. All the examples and diagrams and photos I include from here on can be explained as being based upon my new theory of physics. So I know that they can all be made to actually work! **You'll see spinning (or vibrating) aluminum cones in ALL these devices!** (no surprise!).

Now, see Figure 41. At the bottom of the page, is shown the original drawing of a device that David made from memory, after having seen a drawing of the same thing aboard the alien ship. The shell is made of steel, and the two spheres are made of granite, and the two cones are aluminum. At the top of the page is shown a computer drawn & improved version of the same device supplied by Jean-Louis Naudin of France. Figure 42 shows a 3-cone device, and this drawing was also supplied by JL Naudin. Clearly shown in both drawings, is the method by which the energy output is regulated. At the top of both devices is a repulsion magnet which can be pushed down or moved back up. The output of the devices increases as the top magnet is lowered. If it is raised all the way, then the devices will slow down to a stop.

JL Naudin has one of the best web sites I've ever seen on the subject of free energy generation and antigravity. It is chock-full of many, many different devices created by many experimenters and scientists. And it has a good section on David Hamel. And, it has an animated version of the two drawings! So, I recommend looking at this site. I never can remember it's exact name, but you can find it quickly by going to a search engine, and having it look for "JLN Labs" or just JLN.

Finally, regarding Figure 41, this is a MUCH more advanced and complex device of David's than is the 3-cones in a barrel device. David did test 3-cones in a steel barrel at first, years ago, but he's since graduated to more complex designs such as what's in Figure 41. For us simple folk, however, we should initially try to just build 3-cone in a barrel devices, as there is much less chance for error in the building of these type devices.

Figures 43 through 45 show the drawings that David Hamel made, at Justin's request, several years ago. This is a **3-cone device**, and the cones are kept from hitting the sides of the barrel by using rings of magnets. The cones must be made of aluminum. The barrel is made of steel.

Figure 46 shows a very nice computer rendering that Justin made for David's 3-cone antigravity device.

Figures 47 and 48 show two 3-cone devices which Justin built. The construction is very nice. The outer steel barrel is not shown. Note that he built these back in year 2000.

Figure 49 shows a 2-cone device. This is a drawing of a device which Steve Thompson actually built, tested, and had levitate right in his living room. For safety, the air holes are critical, as the device will implode if there are none. It seems that 2-cone devices will work, but they must be driven by an external motor. A 3-cone device should spin by itself, and not require a driving motor.

Figure 50 shows another device built by Justin Szymanek.

SAFETY ISSUES:

What follows is REQUIRED READING for anyone who wishes to build and test cone devices. Also, it should be required that you read all of Justin's website, as he has very critical comments in there which might affect your safety. It wouldn't hurt to look at other websites, as they come online, as well.

I have included the following excerpts from Justin's (he's the expert in this field) and other's websites, because they may be (and probably are!) safety related, and reading them might actually save your life. Neither Justin, nor I, nor any other people are responsible in ANY WAY for any accidents or deaths, or damages to electronic or other nearby equipment. WE ARE NOT RESPONSIBLE! This is a totally new technology, and we certainly don't yet understand all of the issues. THE EXPERIMENTER TAKES FULL RESPONSIBILITY FOR HIS/HER SAFETY!!!! And, I advise you to have a good electrical engineer (not just a technician) be there to assist you. The cones develop millions of volts and can spark over to you and can be LETHAL.

I do know that these cone devices can easily (and probably WILL!) DESTROY computers and other electronic equipment within a city block or so of the device. Even high voltage road transformers down the highway have been known to explode! Also, PEOPLE WHO MAY HAVE METAL PIECES MADE OF ALUMINUM OR COPPER ANYWHERE IN THEIR BODIES (or ON their bodies, such as rings, jewelry...) COULD SUFFER SEVERE INTERNAL BURNS AND SO ON FROM THE FACT THAT GRAVITY WAVES CAN HEAT UP THESE METAL PIECES AND

MAY EVEN MELT THEM!! However, if you have embedded devices made of stainless steel or titanium or gold, these should not cause a problem. Fillings in teeth may or may not react to the devices, so use caution. Nearby objects containing any copper or aluminum may find themselves either arcing, sparking, shorted out, or melted.

Now, with all that having been said, here are the **excerpts**:

EXCERPT:

The vent hole in the bottom of the shell should be 1/2 the diameter of the cones. I made mine bigger to allow more air flow, but it reduced it a little. I would say to stay with the right geometry and problems will be reduced. You need a SINGLE hole in the center of the shell bottom, directly under the center of the base oscillator. That is how it has to be. the holes around the shell can be very small; it needs to build pressure and if you release the pressure before it can get built up, the device won't work. Think of the holes as regulators. The base hole is larger than the total area of the smaller shell holes, so the pressure has to build up to a certain level. If the holes are right, the shell will get warm at the rings as the device is starting build energy. If the holes are too large, you will never get to this point. When testing your device, you will have:

1st. air movement into the bottom vent hole in the shell; it will be cool.
2nd. warmth at the rings on the shell. This is from the friction of the magnetic fields.
3rd. a strange ozone smell. Actually noble gases being created by the friction of air through the magnetic fields.
4th. small sparks of electricity from the cone rims toward the shell.
5th. very strong ozone smell and increased air movement.
6th. slight glowing starting at the side shell holes. will be an orangish color.
7th. color intensity increases and the color changes to a bluish white.
8th. the lid will now start glowing orange and becoming bluish white.
9th. If you made it this far and no implosion, you will have a self-running electric generator; or it will go BOOM!

You can listen on the radio for static, but I did not find it a very important thing to do. You know it is doing something when it starts creating a cool breeze and warming at the rings on the shell. If it isn't doing these after a few minutes of testing, something isn't right. I feel your cones are lifting too much. The upward force is greater than the downward and sideways force. So, you really don't have a totally balanced system; XYZ. Physical and magnetic balance are KEY. If all the magnetic forces are equal on the XYZ axis, then only a small increase in downward force will be needed to get it started. If the upward force is too great, then the downward force needed to hold everything down will be too great and force the cones against the rings. ALL FORCES MUST BE EQUAL!!!! You then just increase the downward force slightly to activate it. Simple enough? Also, the repelling magnets should be equal to 1/2 the diameter of the cones; actually, slightly smaller than half the diameter of the cones. I think David used 6" magnets for his device.

I still can't stress enough how important total balance is. You are so close, but seem to be missing the big picture of how it works. Make all your forces equal. That doesn't mean to have all the spaces equal, but just the forces needed to achieve:
BALANCE...BALANCE>>>BALANCE>>>...

EXCERPT:

1. The entire device Must be lifted 1/2 it's height or more from the ground or any surface.
2. A wire should be attached to the shell and connected to a ground rod; Use your own ground rod(1/4" rod driven into the ground about 4+'.
3. DO NOT TOUCH THE SHELL AFTER ACTIVATION!!! It will be charged!
4. The device may take some time to reach resonance, so be patient.
5. After the device reaches resonance, the device will start to produce the plasma.
6. Have an adjustable vent on the side of the device to prevent implosion. If the device is glowing too intensely, open this vent with a wooden rod.
7. To tap the energy created by the standing wave, place a copper wire or coil between the 2 base magnets in attraction. It must be isolated totally from the device. Only one wire will come from the device. The other wire comes from a ground rod or difference of potential in the medium you are in at the time. Just use a ground rod.
8. The rejection magnet on the lid, must be touching the metal lid or be contained in a metal cup or can to orient the magnet field.

EXCERPT:

After you complete the parts of the M3CD, you can add the wire to the stationary plate of the oscillator. Put 2 pegs(stand-offs) on either side of the magnet 180 degrees apart. drill 2 small holes in each peg. Feed some copper wire through the top holes and stop at the bottom holes, so it makes a loop. It will cross the magnet twice. Make sure you place this wire exactly in the center of the oscillator magnets! One end will be long; this is your power line. It will be were you will tap the power from. Connect this wire to whatever device on the + and the other wire for whatever device to a good earth ground. Make sure that the oscillator wire doesn't touch any part of the shell !!!!!!!
The power that these devices produce is in relation to the care in building them. If done right, they will produce more power than you will ever be able to use. You can try to use a multimeter to test for voltage, but take it off it goes too high. Just hook the + lead to the wire going to the oscillator and the - lead to a good earth ground. Even if you don't really start it, but just wiggle the cones, you will see voltage. I had a spark jump off the cone rim and hit my finger! It hurt but was exciting, I knew at this point I was close. The voltage can go into the thousands, so be careful! **This is why David said not to touch the shell when it is running.!!!!**
Have you weighed your cones? Mine weigh almost 3lbs each, and move like silk. The oscillation in the cones and oscillator really are not that noticeable at resonance. It will look like the cones are shimmering, but the vibration produced is great. I would say they move about 1/32". It is like the bees I told you about; their bodies move vibrate in such a way that they look kinda fuzzy. This is how the cones will look before the plasma builds up.

Added 07-13-00
Shock:
Well, I had just assembled the first device (one that imploded), and had the lid on, but not in the shell. I started pushing down on the top cone rim, making the cones below rock. After I got them rocking very fast I let go of the cone rim and a blue spark jumped off the top cone and hit my finger. The spark was about 1/2" from the cone rim after the top cone inherited the motion from the cones below it. It did hurt and made me jump, but was very exciting!
My cones were/are/always will be electrically connected as were David's.

EXCERPT: (From Steve Thompson)

First off I cut 4 ventilation holes in the top and the bottom of the barrel. The ventilation holes were the size of a CD disc. I used that as a template. The holes were aligned top to bottom. The holes on the bottom were kept 100% open while the holes on the top were cover. They were riveted so I could open them to different degrees. The small motor was insulated and stabilized by 4 brackets to the inside of the barrel. A dry cell battery was wired to the motor. A dry cell would burn out this particular motor so a small rheostat (kind of like a light dimmer) was attached to regulate the flow from the battery to the motor. The motor was inside, with two wires extending outside by about 12 feet away. Stereo wiring was used and seemed to work well. The rheostat would allow to slowly increase the power to the motor and to increase the speed of the motor. Around 9:00 p.m. the set-up was complete, with the wiring, battery and rheostat about 12 feet away. A small TV was within 6 feet, which was a mistake because it blew up. Anyway, the barrel was set on 6 bricks to allow for ventilation. At 9:05 the motor was set into movement. At 9:10 nothing was happening so I increased the speed of the motor using the rheostat. At 9:15 the vibration began and at 9:17 the high pitch noise began. Since this had happened in the past I was not too surprised. At this time I decided to increase the speed of the motor using the rheostat. I also decided to open (using attached wires) the holes on the top. Upon opening the holes on the top the vibration and noise decreased by, i would guess 50%. I continued to view the barrel for the next 4 minutes. Not much was happening except for the low grade vibration and the low level high pitched noise which was getting rather annoying. I decided to increase the motor speed just slightly more. I didn't want to blow the motor by juicing it too much but I did want to increase the speed.

Which I did. After 3 minutes of increase speed the vibration stopped and the noise took on a deeper quality of humming like. Then a orangish red glow came from the vent holes. The whole drum DID NOT GLOW, just from the vent holes. Around the time the glow occurred the lights in the house began to flicker and go dull. The lights DID NOT GO OUT, but they did dim. I also noticed the neighbor behind me, back porch light dimmed. The lights remained dim, which caused my wife to come out and find out what the hell was going on. She was there to witness the event. The barrel , much to my delight and amazement rose about 6-8 ft off the bricks and hovered for about 30 seconds at which time the strain on the wire caused the battery to disconnect. When the battery disconnected the barrel elevated about another 1 foot before coming crashing to the ground. Right before the barrel started to levitate was when the TV exploded scaring the (blank out of both of us). The dim house lights returned to normal brightness. I was truly astonished by the whole event, not to mention that my wife couldn't believe it and has now promised she won't make fun of me anymore. I have gathered up everything a tucked it away in a double locked garage. Not to long after going back into the house the phone rang. It was one of our neighbors asking my wife if we had , had problems with our electricity. My wife said we had ...but left it at that. If anyone would care to share if they have had similar experiences , I would appreciate it. Right now, I am not interested in why or how it worked. I just know what I saw and what my wife saw with me, so I know I wasn't hallucinating. She had no idea as to what the barrel was suppose to do, yet she saw it rise and hover just like I did. So, I don't think it was group hallucination either. Plus, I have a destroyed TV Any help or further suggestions would be appreciated, as I don't know where to go from here. I did try to video tape the event, but the video tape did not turn out for some reason. One additional thing that I also noticed, right before the barrel rose and while it was hovering. There was kind of a weird distortion that appeared to extended about 4 inches from the barrel. I can't really describe it, except it was like looking through air when it is real hot, how things look all wavy like. BUT NOTHING WAS HOT !!! The weird distortion is what Wilhelm would call Orgone energy and is photon light. Did Hamel notice this? Has anyone else visualized this? Any one who has ever seen a UFO has seen this light. Someone let me know, please.

Thanks,

D&R Scientific Pub. Co.; Denver, CO Jan. 2003. (originally publ. on 04-17-2002) (Figs: 2-1-02)

Steve

EXCERPT:

Build BIG! Hamel says that a device small than a 45 gallon drum one is too small and won't work. It has to do with the volume of magnetic repulsion. I see it as a critical mass thing. For example, a small scale sun will not work, because for the sun to undergo nuclear fusion it must have a critical mass to do so. The Hamel devices are similar in this way, they need to have a critical amount of magnetic repulsion. The bigger, the better.

EXCERPT:

Created by Steven Dufresne (stevend@entrenet.com)
Created on 22 August 1999
Last Modified on 29 June 2000

Results from my visits with David Hamel

Combined results of my visits

The same information below appears broken up by date of visit further down. I've reordered the information so that information about related parts are close together. The items in bold are from my latest visit.

Cone Construction and Placement

- There are three cones. According to David, two will not do as many people are saying. I got a very clear statement from David on this as I wanted to be sure. I specifically asked if two would do and he said no, and that you need three cones for the isotope line.
- I asked if it was okay if the top cone had a closed top (as in Justin's design). He said yes, that was okay.
- I asked if the cones were all the same diameter. He said yes, they were all the same diameter.
- Each cone should be sitting on the cone below it, "halfway" down from that cone's height. This is in keeping with what most people are saying these days. Anything else, David was very clear on this, was wrong.
- Each cone floats, due to the rim/ring magnets' repulsion, and is just barely touching the cone below (or the oscillator as in the case of the bottom cone).
- **On David's 45GD, the bicycle rim was aluminum so that it did not affect the magnetic field.**
- **On David's 45GD, he didn't have any supporting frame in the cones.**
- **On David's 45GD, each of the three cones was actually made up of two cone skins, an outer full height cone skin and an inner cone skin that was made such that the cone above it would sit halfway down the cone below it.**
- **On David's 45GD, the cone skins were riveted to the bicycle rim. Just where this was done doesn't matter.**
- **On David's 45GD, there was no nail for the point at the bottom of the cone. Instead the cone skin was formed to a point.**
- **The cones skins are very important for moving/torqueing the air.**
- **The cones are what make the invisible (air) become visible (the air becomes plasma). (So that's what he means by invisible and visible!)**

Magnets

- He said that the top magnet and the top cone magnet which it opposes were about 10" diameter. I would like to verify this with him again, though, as he keeps referring to the magnets for the saucer that he is building when he tells me about this.
- The drum ring magnets repel the cone rim magnets with like poles facing each other and the opposite poles facing the other direction horizontally, just as in the Bob Thomas drawings. This is something David stressed to me as he had another drawing he was showing me where it was drawn wrongly and wanted me to understand that it was wrong.
- I did ask David about the pole reversal that you see in the Bob Thomas drawing, wherein the top cone's and bottom cone's rim magnets repel the ring magnets with one polarity

and the middle cone repels with the other polarity. He said that this is not necessary, but his was an offhand response so I will have to get clarification.

- I asked him about the magnet poles of all the magnets. He said that they were as follows:
  - Top magnets - North facing North
  - Top cone rim magnets/opposing magnets - South facing South
  - Middle cone rim magnets/opposing magnets - North facing North
  - Bottom cone rim magnets/opposing magnets - South facing South
  - Oscillator magnets - Attractive (I didn't get details)

  I'll try and remember to ask him how he determines North and South.
- On the drum that I showed David, my drum ring magnets were taped to an aluminum ring which was in direct contact, fitting snugly, with the inner wall of the drum. David said that this should not be so, otherwise I'd get **electrocuted**. I wasn't sure though if that was a cautionary note based on what happened with his own drum or if that was how if drum was built. He said that I should put three pieces of wood down vertically within the drum, on the drum wall, spaced out 120 degrees apart and that the rings should be attached to this wood instead. This seems to be in keeping with what is shown in the Bob Thomas drawing.
- I asked for more specs on the spacing between the cone rim magnets and their opposing ring magnets. He said that The spacing when not wobbling was very small, about 1/4 inch. The wobble was also very small, about 1/16 inch. He said that the isotope line barely moved.
- A few people had mentioned that David had embedded the top magnet in cement which was in turn in a container of some sort. I wanted to verify this with David. He said that he did not do this. He did do the following. The top magnet was a ring magnet. He needed to attach the rod to it so that it could be lowered. So he filled the hole in the ring with cement and put the rod in the cement. That's how he attached the rod to the magnet.
- **On David's 45GD, the magnets were inserted into the rim and fixed in place by wrapping tape around the magnets and rim.**
- **David gave me a few of the actual magnets that he'd used in his drum. He didn't have single magnets that fit snugly into the bicycle rims. Instead he used stacks of magnets. Starting from closest to the inside of the bicycle ring, he stacked around three rectangular 1"x3/4"x5/32" magnets and then one ring magnet od=1 1/8", id=3/8", thickness=7/32". The rectangular ones fit within the rim with the long edge (1") tangential to the curve of the rim. I say there was 'around three' because he showed me using a rim I'd brought, not his. The ring magnet would not fit inside the rim and hence the reason for stacking the smaller rectangular ones inside the rim to fill it in. My camera can't take good pictures of something that small but here are some anyway.**
- **When I visited him, David had one of those cards you can use to view a magnetic field. It's basically a translucent plastic that is made up of magnetic particles in oil (I think). He was using it to analyze the fields in my magnets. He was checking for continuity. He also mentioned that when the device was working, a third magnetic field becomes visible between the two that are made up from the cone rim magnets and the repelling ring magnets. This third field should appear between their two fields. So if you hold the card flat over the gap between the magnets, you should see three fields when the magnets are pressed close together as they would be during wobbling: one that follows the line of the cone magnets, one that follows the line of the repelling magnets, and one between those two.**

<u>Repelling Ring</u>

- **On David's 45GD, the repelling ring (the ring with magnets that repel the cone rim magnets) was made out of an aluminum bar formed into a circle.**

Drum Construction

- I wanted a clarification on whether on his original drum there was any air gap between the cover and the top of the drum. He said that there was none.
- I asked if there were any holes in the drum. He said no. However, he did suggest putting measuring devices for temperature, pressure and so on in the drum and using some sort of view ports to see them as the cones do their stuff.

Oscillator

- David described the oscillator as in the Bob Thomas drawing, with three balls in triangular form, but David said you didn't need the magnet in the middle. When I showed him mine with the magnet in the middle in attraction, he did not say it was wrong. So the presence of the magnet may be an enhancement and looks to be optional in David's mind. He did specifically ask if mine was in attraction, and when I said yes, he was happy with this.
- As I was not sure from my previous visit, I asked if his oscillator was the type that had the magnet in the middle. He said that it was and that it was attractive.
- **The oscillator barely moves, it just vibrates a bit.**

Running the Experiment

- I asked David, what liquid did he put in his cones to stabilize them (I'd heard both water and glycerine from others). David said he didn't put any. However, he did say that he put water in the drum so that he could observe the wave motions made by the wobbling cones.
- **I told David that my cones were too light to just sit in the magnetic fields and he suggested adding water into the cones to make them the correct weight. (Didn't Pierre Sinclaire say that David did this? Funny, when I asked David about this before he said he didn't. I'll have to clarify that someday. Maybe he didn't use this trick himself but was just suggesting it.)**
- I asked him how he started it up. Various people asked me to ask him this. He said that all he did was bolt on the lid.
- I asked David if the cones rotate or spin? David said no, they just wobble.
- **I asked David what he sat his drum on (bricks, the floor, ...?) He didn't really answer. Instead he said it might be a good idea to put it on rubber or even a tire so that it will be able to vibrate better. (Remember, David didn't have any holes in his drum.)**

The following is the same information as above except broken up by date of visit.

August 7, 1999 - visit to David's place

A mutual friend of David Hamel's and mine took me out to visit David. We only stayed for about 3 hours and as it was my first visit, we spent a lot of time just getting to know each other. As a result I was not able to get detailed information on the 45 gallon drum experiment, nor specifically on the construction details of the one he built. That will have to wait for a second visit in a couple of weeks.

However, We did get a little time to go over a drawing he had that someone else had made that he said was incorrect, as well as a drawing that I brought along that was much like the Bob Thomas drawing but with my construction details. I also brought along and showed him what I had built so far. He was delighted to see this and we went over it together briefly (he had to go cook dinner after).

David was working on his large device when I arrived and he gave me a quick tour of it. Amazing to actually see one in person! It was not yet completed but was a far way along. From what I saw, it matches very closely what Dan La Rochelle and Jean-Louis Naudin describe.

The following are the results of my conversations that day with David. Note I asked David if he minded if I post this information and he said, sure, go ahead but that people should wait for Bob Thomas's book as it will have detailed drawings of at least the larger flying craft.

- There are three cones. According to David, two will not do as many people are saying. I got a very clear statement from David on this as I wanted to be sure. I specifically asked if two would do and he said no, and that you need three cones for the isotope line.
- Each cone should be sitting on the cone below it, "halfway" down from that cone's height. This is in keeping with what most people are saying these days. Anything else, David was very clear on this, was wrong.
- Each cone floats, due to the rim/ring magnets' repulsion, and is just barely touching the cone below (or the oscillator as in the case of the bottom cone).
- The drum ring magnets repel the cone rim magnets with like poles facing each other and the opposite poles facing the other direction horizontally, just as in the Bob Thomas drawings. This is something David stressed to me as he had another drawing he was showing me where it was drawn wrongly and wanted me to understand that it was wrong.
- I did ask David about the pole reversal that you see in the Bob Thomas drawing, wherein the top cone's and bottom cone's rim magnets repel the ring magnets with one polarity and the middle cone repels with the other polarity. He said that this is not necessary, but his was an offhand response so I will have to get clarification.
- David described the oscillator as in the Bob Thomas drawing, with three balls in triangular form, but David said you didn't need the magnet in the middle. When I showed him mine with the magnet in the middle in attraction, he did not say it was wrong. So the presence of the magnet may be an enhancement and looks to be optional in David's mind. He did specifically ask if mine was in attraction, and when I said yes, he was happy with this.
- I wanted a clarification on whether on his original drum there was any air gap between the cover and the top of the drum. He said that there was none. This is the only statement I have as of this date where he said how something was on his original drum.
- On the drum that I showed David, my drum ring magnets were taped to an aluminum ring which was in direct contact, fitting snugly, with the inner wall of the drum. David said that this should not be so, otherwise I'd get **electrocuted**. I wasn't sure though if that was a cautionary note based on what happened with his own drum or if that was how if drum was built. He said that I should put three pieces of wood down vertically within the drum, on the drum wall, spaced out 120 degrees apart and that the rings should be attached to this wood instead. This seems to be in keeping with what is shown in the Bob Thomas drawing.

October 23, 1999 - visit to David's place

Visited David again with our mutual friend. David now has the second cone of his saucer put on top of the first cone and is starting on the third cone.

I again brought my 45 gallon drum along with a short list of questions. The following are the results of our conversation regarding the drum.

- I asked for more specs on the spacing between the cone rim magnets and their opposing ring magnets. He said that The spacing when not wobbling was very small, about 1/4 inch. The wobble was also very small, about 1/16 inch. He said that the isotope line barely moved.
- A few people had mentioned that David had embedded the top magnet in cement which was in turn in a container of some sort. I wanted to verify this with David. He said that he did not do this. He did do the following. The top magnet was a ring magnet. He needed to attach the rod to it so that it could be lowered. So he filled the hole in the ring with cement and put the rod in the cement. That's how he attached the rod to the magnet.
- I asked David, what liquid did he put in his cones to stabilize them (I'd heard both water and glycerine from others). David said he didn't put any. However, he did say that he put water in the drum so that he could observe the wave motions made by the wobbling cones.

- I asked David if the cones rotate or spin? David said no, they just wobble.
- I asked him how he started it up. Various people asked me to ask him this. He said that all he did was bolt on the lid.
- I asked if there were any holes in the drum. He said no. However, he did suggest putting measuring devices for temperature, pressure and so on in the drum and using some sort of view ports to see them as the cones do their stuff.
- I asked if the cones were all the same diameter. He said yes, they were all the same diameter.
- I asked if it was okay if the top cone had a closed top (as in Justin's design). He said yes, that was okay.
- As I was not sure from my previous visit, I asked if his oscillator was the type that had the magnet in the middle. He said that it was and that it was attractive.
- I asked him about the magnet poles of all the magnets. He said that they were as follows:
    - Top magnets - North facing North
    - Top cone rim magnets/opposing magnets - South facing South
    - Middle cone rim magnets/opposing magnets - North facing North
    - Bottom cone rim magnets/opposing magnets - South facing South
    - Oscillator magnets - Attractive (I didn't get details)

  I'll try and remember to ask him how he determines North and South.
- He said that the top magnet and the top cone magnet which it opposes were about 10" diameter. I would like to verify this with him again, though, as he keeps referring to the magnets for the saucer that he is building when he tells me about this.

February 13, 2000 - phone call

- He used ordinary bicycle rims, not aluminum (NOTE THAT ON MY JUNE 29, 2000 VISIT, THIS TURNED OUT TO BE WRONG). He took the spokes out. The magnets fit in perfectly. He did not put any lining in between the magnets and the rim.

May 8, 2000 - phone call

- His top magnets (top rejection magnet and top cone magnet) were od=6", id=4". When they are 3 or 4 inches apart they will begin to repel.

June 29, 2000 - visit to David's place

Visited David again with our mutual friend. David now has the base cone on a stand outdoors and the middle two cones (the ones that will be floating) are more or less complete. Only parts of the top cone have been made so far. He is now working on the rings that have the rejection magnets.

I again brought my 45 gallon drum along with a short list of questions. The following are the results of our conversation regarding the drum both during the visit and from a follow-up conversation. You won't like some of the information as it is different than what is commonly understood to be the case. One example is that he used aluminum bicycle wheels instead of steel - so that they don't affect the magnetic fields. Much of the other material confirms what we already know.

I called to double check that his 45GD used aluminum bicycle rims and he said yes. I can see where there was some confusion though. I started out by asking if "he used steel bicycle rims" and he said yes. Then, knowing that he has a lot of bicycle rims now and uses them as construction aids, I pressed him further - "on your 45GD that exploded, did you use aluminum or steel bicycle rims" and he said aluminum.

- On David's 45GD, the repelling ring (the ring with magnets that repel the cone rim magnets) was made out of an aluminum bar formed into a circle.
- On David's 45GD, the bicycle rim was aluminum so that it did not affect the magnetic field.

- On David's 45GD, the magnets were inserted into the rim and fixed in place by wrapping tape around the magnets and rim.
- David gave me a few of the actual magnets that he'd used in his drum. He didn't have single magnets that fit snugly into the bicycle rims. Instead he used stacks of magnets. Starting from closest to the inside of the bicycle ring, he stacked around three rectangular 1"x3/4"x5/32" magnets and then one ring magnet od=1 1/8", id=3/8", thickness=7/32". The rectangular ones fit within the rim with the long edge (1") tangential to the curve of the rim. I say there was 'around three' because he showed me using a rim I'd brought, not his. The ring magnet would not fit inside the rim and hence the reason for stacking the smaller rectangular ones inside the rim to fill it in. My camera can't take good pictures of something that small but here are some anyway.
- On David's 45GD, he didn't have any supporting frame in the cones.
-
- On David's 45GD, each of the three cones was actually made up of two cone skins, an outer full height cone skin and an inner cone skin that was made such that the cone above it would sit halfway down the cone below it.
- On David's 45GD, the cone skins were riveted to the bicycle rim. Just where this was done doesn't matter.
- On David's 45GD, there was no nail for the point at the bottom of the cone. Instead the cone skin was formed to a point.
- I told David that my cones were too light to just sit in the magnetic fields and he suggested adding water into the cones to make them the correct weight. (Didn't Pierre Sinclaire say that David did this? Funny, when I asked David about this before he said he didn't. I'll have to clarify that someday.)
- The oscillator barely moves, it just vibrates a bit.
- I asked David what he sat his drum on (bricks, the floor, ...?) He didn't really answer. Instead he said it might be a good idea to put it on rubber or even a tire so that it will be able to vibrate better. (Remember, David didn't have any holes in his drum.)
- The cones skins are very important for moving/torqueing the air.
- The cones are what make the invisible (air) become visible (the air becomes plasma). (So that's what he means by invisible and visible!)
- **When I visited him, David had one of those cards you can use to view a magnetic field. It's basically a translucent plastic that is made up of magnetic particles in oil (I think). He was using it to analyze the fields in my magnets. He was checking for continuity. He also mentioned that when the device was working, a third magnetic field becomes visible between the two that are made up from the cone rim magnets and the repelling ring magnets. This third field should appear between their two fields. So if you hold the card flat over the gap between the magnets, you should see three fields when the magnets are pressed close together as they would be during wobbling: one that follows the line of the cone magnets, one that follows the line of the repelling magnets, and one between those two.**

<u>END OF EXCERPTS.</u>

HAPPY & SAFE EXPERIMENTING & GOOD LUCK.

CHAPTER 19 "From Zero to One, or from ONE to ZERO? The Choice We Must Make NOW.":

Welcome to the final chapter of **The Antigravity Papers: A Journey from Zero to One (And... THE WARNING.).**

See Figure 51. Here, we see David showing us one of his more advanced designs, which uses multiple internal "wings" and ionized airflow. If you look close, you can see a lot of magnets, just like in some of his other designs. For all that follows, which is the short remainder of this book, we will contemplate **the world's current situation,** as relayed to David Hamel by the "angel" astronauts. Incidentally, in David's video, he said that these astronauts look just like ordinary people. He said if you saw them on the street, you couldn't tell the difference!

What we'll look at comes from several pages in the book about David, entitled "The Granite Man and the Butterfly", which was mainly written by Jeanne Manning and is published and co-authored by Pierre Sinclaire of Project Magnet Inc. Pierre can be reached through his website, www.projectmagnet.com, or at the following mailing address:

Project Magnet
Box 839, 9037 Royal Street,
Fort Langley, British Columbia, Canada V1M 2S2

What we'll look at next are some excerpts from the above book.
Since the book has a 1995 Copyright Notice, I can't show the exact wording of the excerpts, nor can I show what's on the rest of each of these pages from which the excerpts are pulled. Perhaps I will someday have permission to show the entire pages...

Now, see Figures 52, 53, and 54. These are the last three figures in "The Antigravity Papers". As you read these excerpts from "The Granite Man...", remember that I have proof, good enough for a courtroom, as you've seen in earlier chapters, that everything that David has relayed to us is absolutely true.
Now, please read all three remaining figures.

So we now have proof, through David Hamel, that the surface of the Earth will be destroyed sometime between 2000 and 2005. That much is absolute. Could it be due to 4179 TOUTATIS?? Anyway, predicting the exact time within that time-frame is another matter and is beyond the scope of "The Antigravity Papers".

It does appear that this destruction of the Earth will happen through natural means, as it will be a natural disaster which will happen at a certain time and date, just like clockwork, based upon the current position and velocity of an object or objects which is/are headed our way and will soon pass close to (or maybe impact with) Earth. **So, it would appear that the outer-world powers who want to help save us are requiring that we CARE ENOUGH TO DO OUR PART to get our act together, stop polluting the planet, develop alternate energies and antigravity ships, and help ourselves possibly by leaving the planet in ships we've built, or by other means involving antigravity physics & devices.**

So, do what YOU think is right, and soon.

Good Luck to Us ALL, and God bless.

Rick C.

D&R Scientific Pub. Co.; Denver, CO Jan. 2003. (originally publ. on 04-17-2002) (Figs: 2-1-02)

The 3 great pillars of physics, upon which all existing physical theories rest:

#1) Relativity (Einstein)

#2) Quantum Mechanics, or Q.E.D. (Dirac, and others)

#3) Electromagnetic Theory (Maxwell and Lorentz)

#1) Relativity. Einstein's two most important equations are:

A) energy/mass: $E = mc^2$

total energy → E; mass → m; $c = 3 \times 10^8$ meters/sec = the speed of light.

Thus, $c^2 = 9 \times 10^{16}$ (a very large number!).

B) How mass (or energy) warps space (and time!):

$$R_{\mu\nu} - \frac{1}{2} g_{\mu\nu} R = k T_{\mu\nu}$$

(the "effect"): resultant bending (or warping) of space (and time).

(the "cause"): mass plus energy present.

Figure 1A

D&R Scientific Pub. Co.; Denver, CO Jan. 2003. (originally publ. on 04-17-2002) (Figs: 2-1-02)

#2) <u>Quantum Mechanics</u>. Schrodinger and Dirac were two of the pioneers of this area of physics. The basic equation of Q.M, is: (in simplified form)

$$\boxed{H\Psi = E\Psi}.$$

The "Ψ" is the commonly used Greek letter psi (pronounced "sigh"). It represents a "probability wave" which describes the motion of any particle in terms of the probabilities of where the particle could be located at any specified time. We can never know for certain the <u>exact</u> location of a particle, but only where the most likely locations are!

The "H" is a differential calculus "operator" which is called the system Hamiltonian function.

The "E" is the "energy operator" function.

Figure 1B

#3) Electromagnetic Theory. The currently well-known equations, called "Maxwell's" equations, are:

a) $\vec{\nabla} \times \vec{E} = \dfrac{-\partial(\mu H)}{\partial t}$

b) $\vec{\nabla} \times \vec{H} = \dfrac{\partial(\epsilon \vec{E})}{\partial t} + \vec{J}$ (electric current)

c) $\vec{\nabla} \cdot (\epsilon \vec{E}) = \rho$ (charge density)

d) $\vec{\nabla} \cdot (\mu \vec{H}) = 0$

All of our experimental experience with light, radio waves, electric currents and electrostatic charges, and power generation/motors can be explained by these four equations.

$\vec{E}$ represents the electric field.

$\vec{H}$ represents the magnetic field.

Note that permanent magnets have a strong $\vec{H}$ field, but no $\vec{E}$ field.

Figure 1C

Maxwell's Equations for Electricity and Magnetism: (well known)

(1) $$\vec{\nabla} \times \vec{E} = \frac{-\partial(\mu \vec{H})}{\partial t}$$

(2) $$\vec{\nabla} \times \vec{H} = \frac{\partial(\epsilon \vec{E})}{\partial t} + \vec{J}$$

Also:

$$\vec{\nabla} \cdot (\epsilon \vec{E}) = \rho$$

$$\vec{\nabla} \cdot (\mu \vec{H}) = 0$$

Lorentz's Electric and Magnetic Force Equations: (well known, also)

(3) $$\vec{F} = Q\vec{E}$$ Electric force law.

(4) $$\vec{F} = Q\vec{V} \times (\mu \vec{H})$$ Magnetic force law.

Examples/Details of these:

(1) A collapsing Magnetic field ($\vec{H}$) produces a circular-like Electric field.

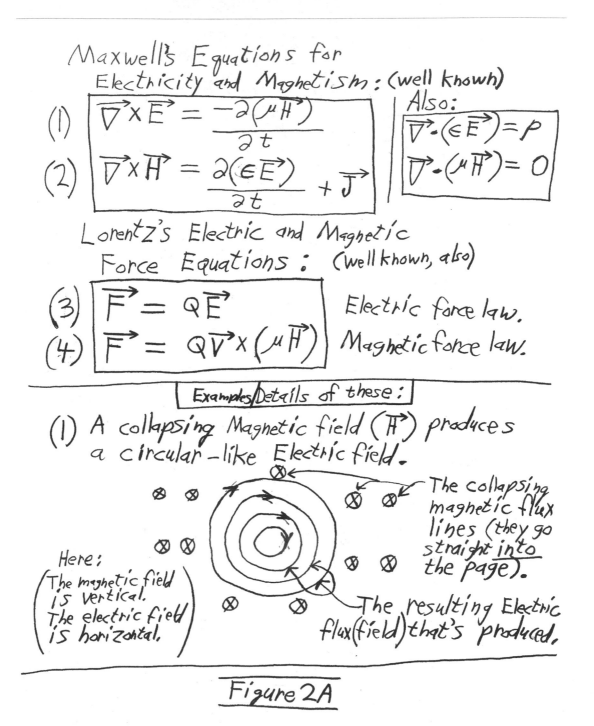

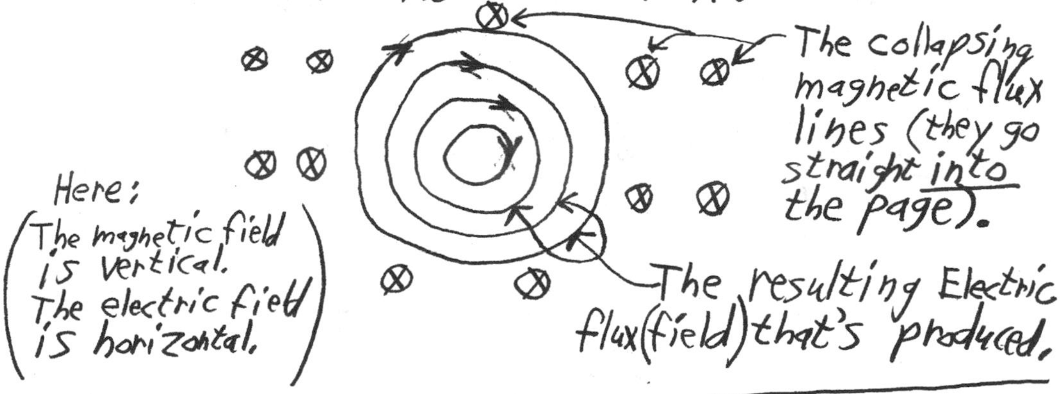
The collapsing magnetic flux lines (they go straight into the page).
Here: (The magnetic field is vertical. The electric field is horizontal.)
The resulting Electric flux (field) that's produced.

Figure 2A

(2) A collapsing Electric field ($\vec{E}$) produces a circular-like Magnetic field. (i.e. "duality")

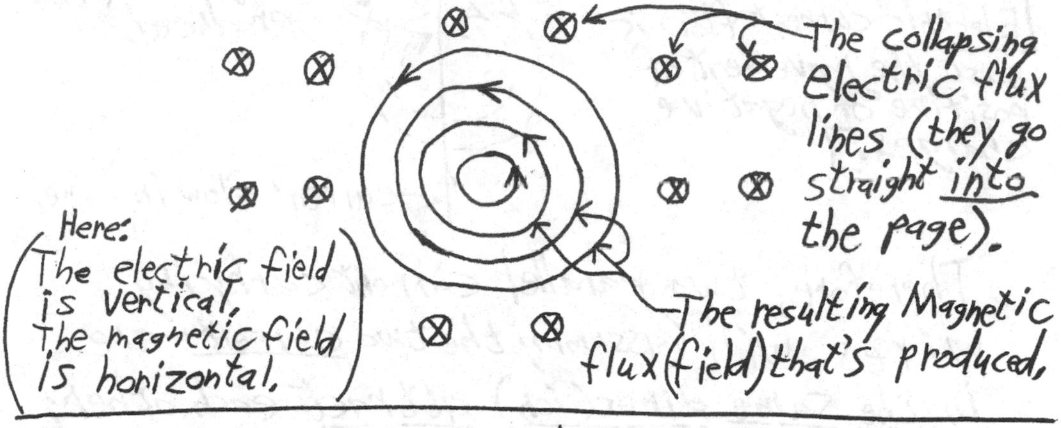
The collapsing electric flux lines (they go straight into the page).
Here: The electric field is vertical, The magnetic field is horizontal.
The resulting Magnetic flux (field) that's produced.

(3) Like charges repel each other: (always)

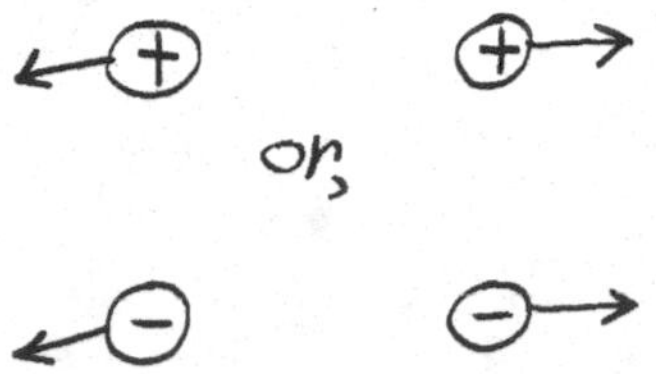
or,

Un-like charges attract each other; (always)

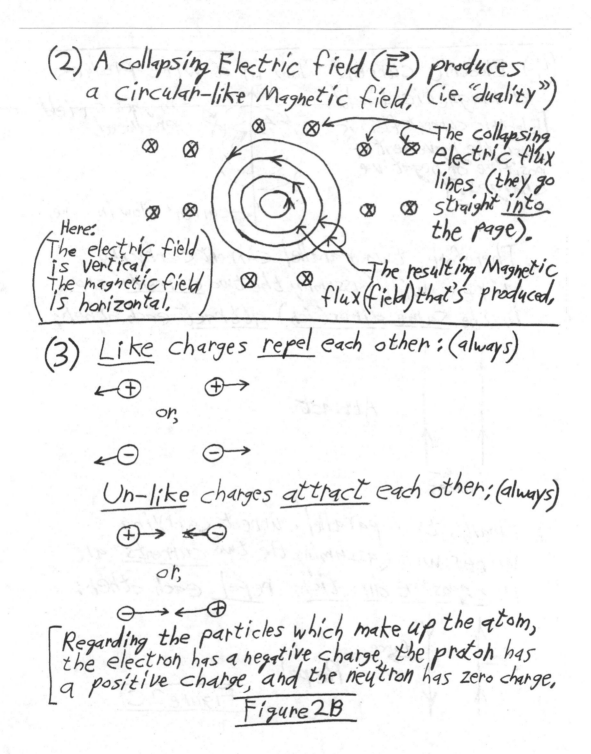

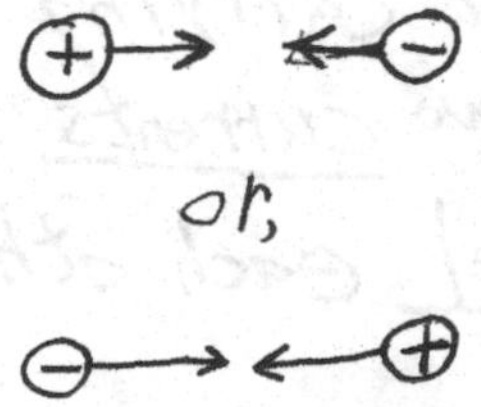
or,

[Regarding the particles which make up the atom, the electron has a negative charge, the proton has a positive charge, and the neutron has zero charge.]

Figure 2B

D&R Scientific Pub. Co.; Denver, CO Jan. 2003. (originally publ. on 04-17-2002) (Figs: 2-1-02)

(4) Electric current flow in a wire produces a magnetic field.

[Electric current flow is just the movement of positive and negative charges.]

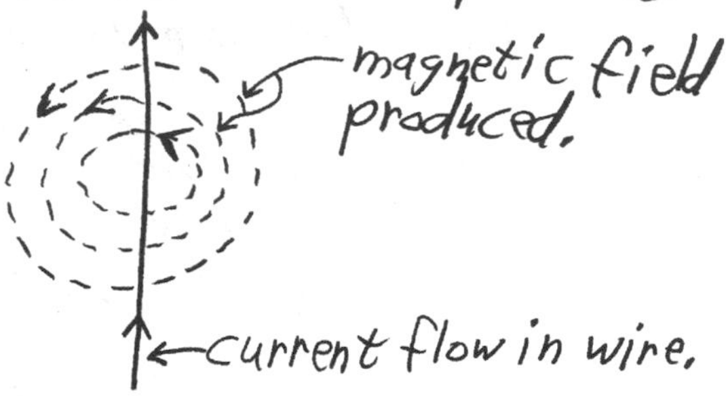

Therefore, two parallel current carrying wires will (assuming the two currents are in the same direction) attract each other:

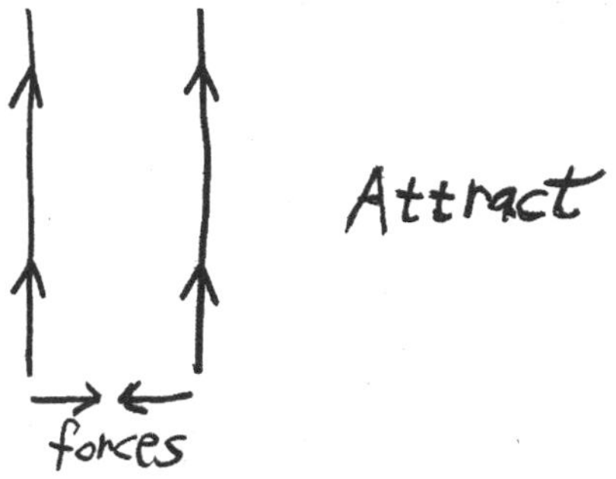

Finally, two parallel current carrying wires will (assuming the two currents are in opposite directions) repel each other:

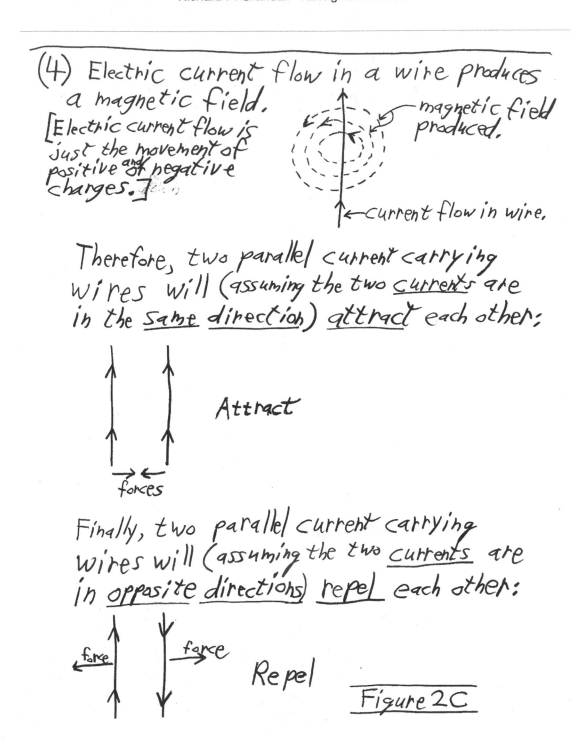

Figure 2C

My equations for gravity and magneto-gravity look very similar to Maxwell's equations. They are:

(1) $$\vec{\nabla} \times \vec{E_g} = -\frac{\partial(\mu_g \vec{H_g})}{\partial t}$$

(2) $$\vec{\nabla} \times \vec{H_g} = \frac{\partial(\epsilon_g \vec{E_g})}{\partial t} - 4\pi \rho_g \vec{u}$$

Also:

$$\vec{\nabla} \cdot (\epsilon_g \vec{E_g}) = -\rho_g$$

$$\vec{\nabla} \cdot (\mu_g \vec{H_g}) = 0$$

AND, the force upon a moving mass, m, equals:

(3) and (4) combined:

$$\vec{F}_{total} = m\vec{E_g} + m\vec{V} \times (\mu_g \vec{H_g})$$. (NO CHANGE)

[the changes are double underlined].

The subscript "g" denotes "gravity".

Examples/Details of these: (for gravity)

(1) A collapsing GravitoMagnetic (or "GM") field ($\vec{H_g}$) produces a circular-like ordinary gravity field ($\vec{E_g}$).

(2) A collapsing ordinary gravity field ($\vec{E_g}$) produces a circular-like Gravitomagnetic field ($\vec{H_g}$).

[These two will not be re-drawn, since they look exactly the same as the corresponding electric/magnetic drawings of (1) and (2), earlier.

Figure 3A

D&R Scientific Pub. Co.; Denver, CO Jan. 2003. (originally publ. on 04-17-2002) (Figs: 2-1-02)

(3) (for gravity). This is quite different from electricity. Here, we assume there is such a thing as "negative mass", as well as ordinary (positive) mass.

Positive masses attract each other: (always)

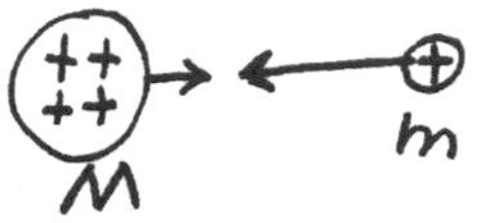

Negative masses repel each other: (always)

A larger, Positive mass attracts a negative mass:

A larger, Negative mass repels a positive mass:

Two equal, but opposite sign, masses maintain their distance of separation, and they chase each other. Their speeds continue to increase!

Figure 3B

(3 continued...)

Note that the length of the arrows (of force) is important.

The last case looks impossible, as if it continuously creates more and more energy from nowhere. However, this case is perfectly valid and correct, as a physicist will tell you, due to the fact that all the "Conservation" laws are obeyed:

> Conservation of Momentum:
>
> $-(m\gamma)v + (m\gamma)v = 0$, always.
>
> AND,
>
> Conservation of Energy:
>
> $-(m\gamma)c^2 + (m\gamma)c^2 = 0$, always.

[This last case is the one upon which practical "anti-gravity" devices are based.]

go to next page →

Figure 3C

(4) (for gravity). This is also different from its electrical/magnetic counterpart. <u>Mass flow</u> (in a tube, for instance) produces a gravitomagnetic ($\vec{H}_g$) field, which is circular.

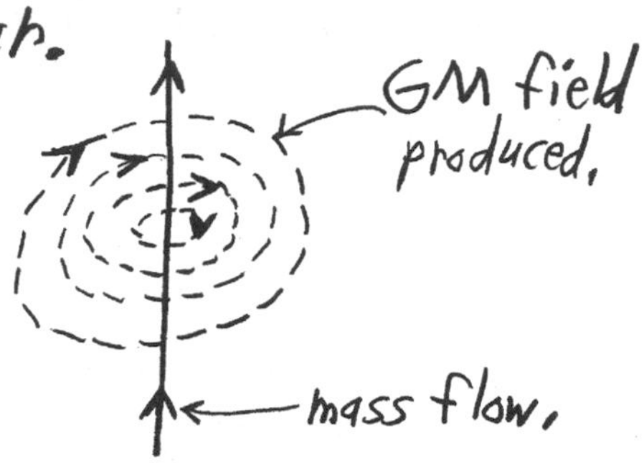

Notice the GM flux lines point clockwise, which is just the opposite of the electrical counterpart.

Thus, we see that the opposite of the electrical counterpart occurs for parallel <u>mass</u> flows:

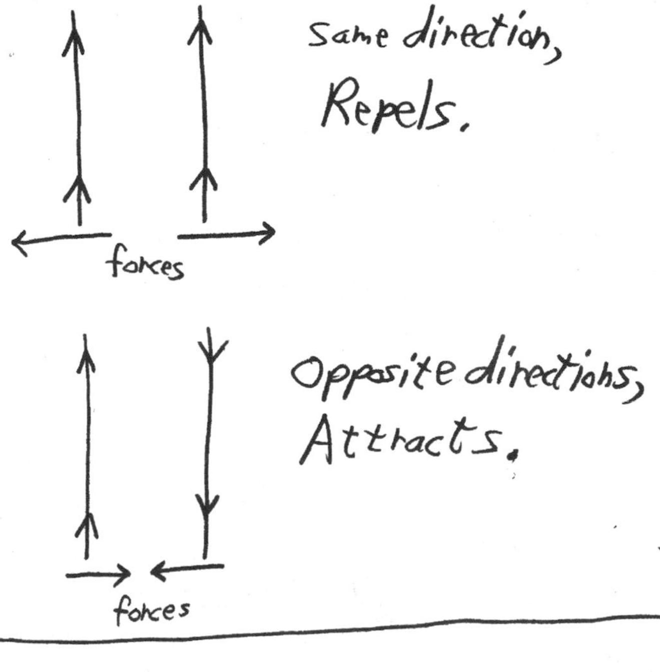

Figure 3D

The atom:

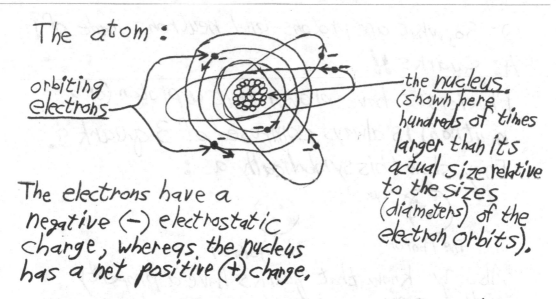

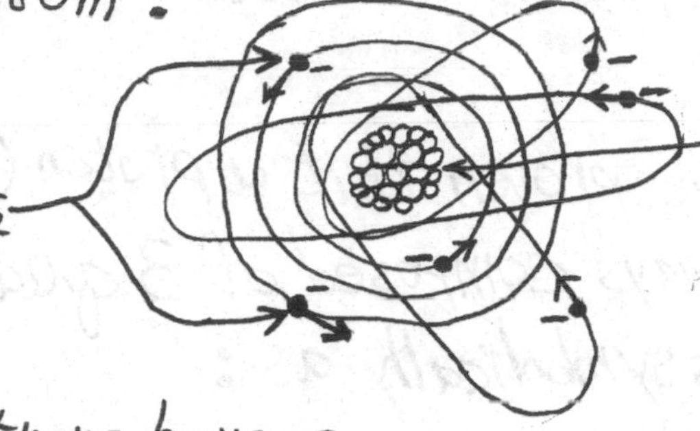
orbiting electrons
the nucleus. (shown here hundreds of times larger than its actual size relative to the sizes (diameters) of the electron orbits).

The electrons have a negative (−) electrostatic charge, whereas the nucleus has a net positive (+) charge.

The nucleus consists of two types of particles (or nucleons). They have approximately the same physical size. They are:

The Proton ⊕ (positive electric charge)
and,
The Neutron ○ (no electric charge).

The mass (i.e, weight) of a proton is about the same as the mass of a neutron.
The number of neutrons is about the same as the number of protons (in the nucleus).
[The number of electrons must equal the number of protons, to have an atom with no electric charge.

Figure 4A

D&R Scientific Pub. Co.; Denver, CO Jan. 2003. (originally publ. on 04-17-2002) (Figs: 2-1-02)

Q: So, what are protons and neutrons made of?

A: Quarks!!

Physicists have proven that a proton (or a neutron) is always composed of 3 quarks. They show this symbolically as:

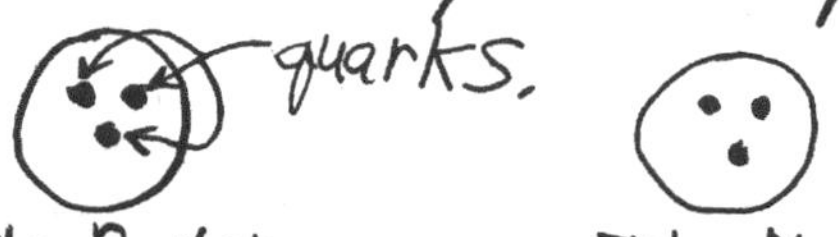

The Proton The Neutron

Also, we know that quarks have a property called spin, and it can take on only two possible values: $\frac{+1}{2}$ (spin up) or $-\frac{1}{2}$ (spin down).
[Often, the spin "direction" is denoted by an arrow: ↑ is spin up, ↓ is spin down.]
Quarks also possess fractional electric charge. Examples: $\frac{-1}{3}$, $\frac{+2}{3}$, etc...
However, the 3 charges *must* add up to +1. (for proton)
That is the net charge of a proton.
Thus, we can diagram the two nucleons as:

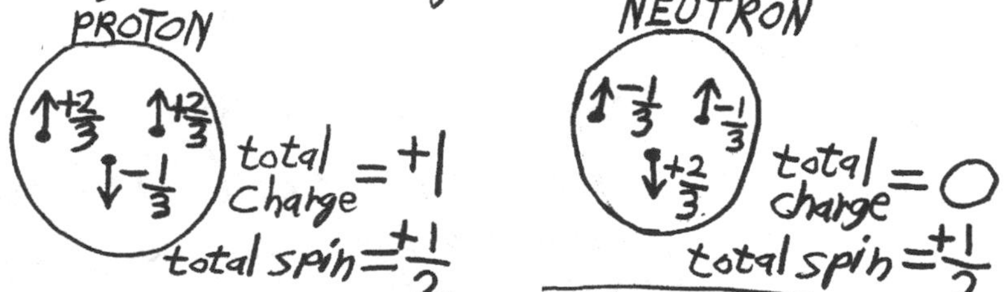

Figure 4B

Q: Do modern physicists know what the quarks are made of?

A: NO!! They have no clue, but some unaccepted theories, such as "string theory", have made so far some not-so-successful attempts at going a "level" deeper than quarks.

My Grand Unified Field Theory has managed to take us 2 levels of detail below that of the quarks:

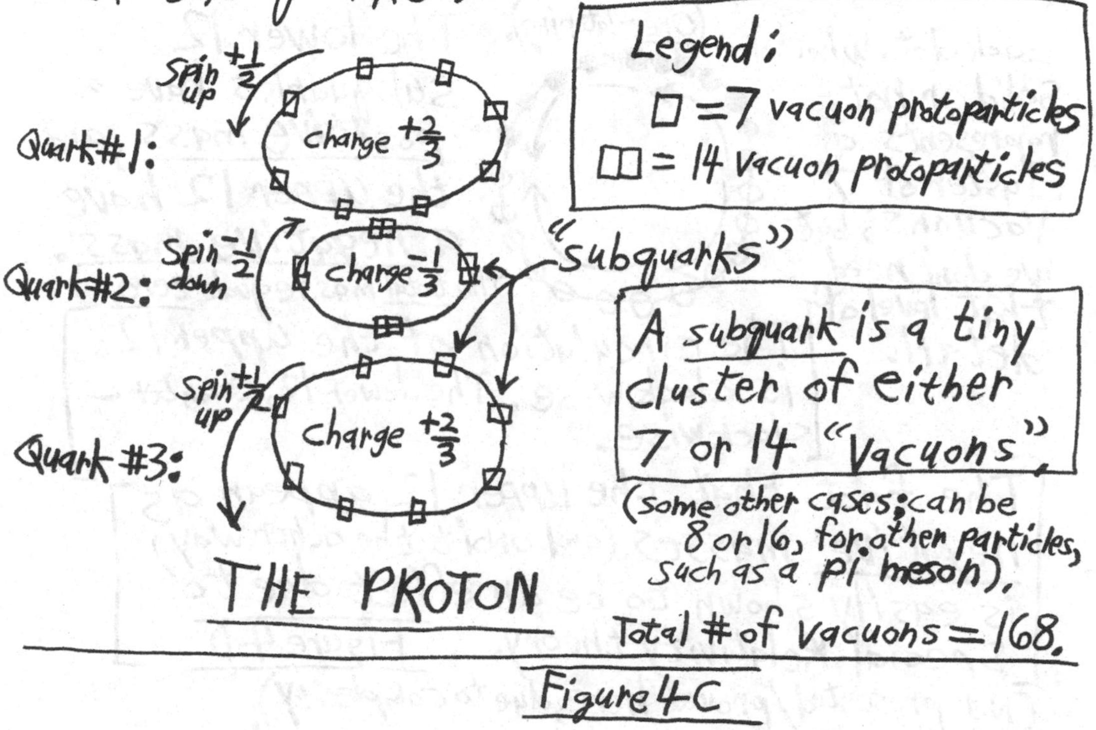

Figure 4C

D&R Scientific Pub. Co.; Denver, CO Jan. 2003. (originally publ. on 04-17-2002) (Figs: 2-1-02)

The term "vacuon" (VAC-YOU-ON) was created by me to describe what are now known to be the smallest constituent particles, of which, everything else is made of.

[For the rest of the purposes of all that follows, simpler models of the proton will be used. These are easier to understand, but are still correct enough to be used to demonstrate <u>antigravity</u> principles.

One simpler model of a proton is: (or, neutron)

(<u>One</u> flat ring):

Each dot, whether solid or not, represents a cluster of 7 vacuons, but we dont need that level of detail.

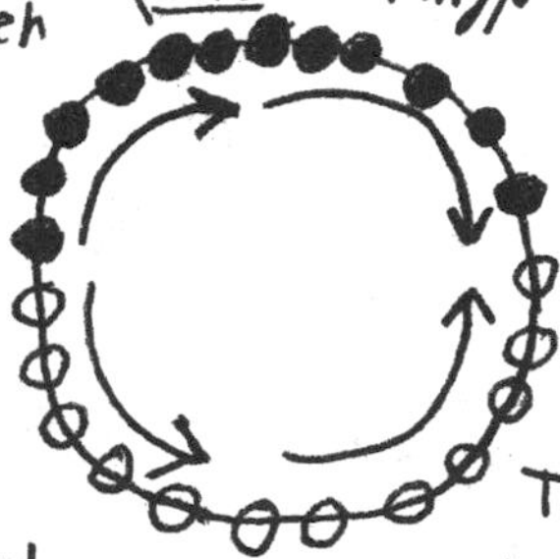

The lower 12 subquarks have a <u>positive mass</u>, and the <u>upper</u> 12 have a <u>negative mass</u>.

The <u>total</u> mass equals <u>zero</u>.

[The circulation of the upper 12 is clockwise. The lower 12, counter-clockwise.]

[The fact that the upper 12 appear as <u>negative</u> masses (and orbit the other way) is easily shown to be an effect due to Special Relativity theory.] <u>Figure 4D</u>

(Not presented/proved here, due to complexity).

An even more simple model, which is the model of a proton (or neutron) that will be used <u>from this point forward</u>, is:

A parallel stack of 2 counter-rotating rings, having thin walls.

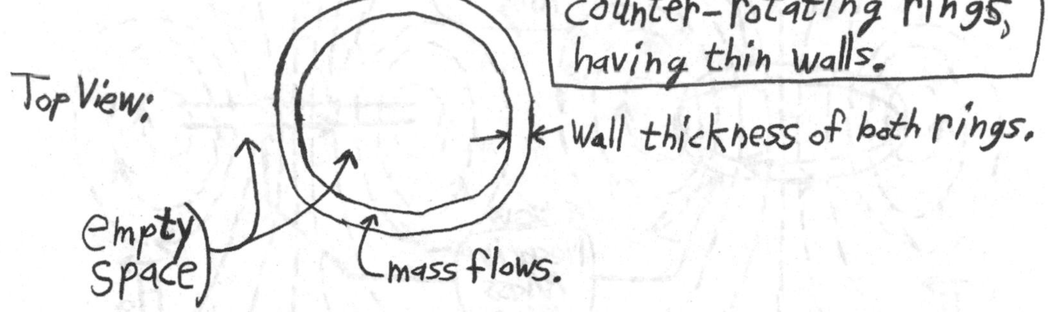

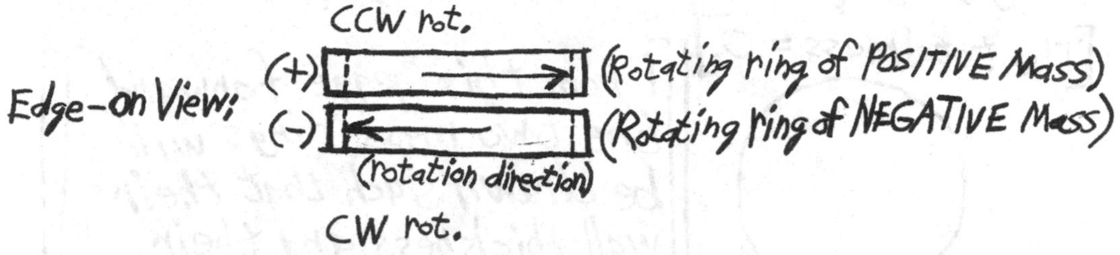

Note: the <u>total</u> mass equals zero.

[I had actually derived a model like this one many months before I derived the final model which was shown several pages earlier.

THE (PRELIMINARY) COUNTER-ROTATING RINGS (OF MASS) MODEL For a proton or a neutron.

Figure 5A

THE FINAL COUNTER-ROTATING RINGS (of Mass) MODEL: →(for a Nucleon)←

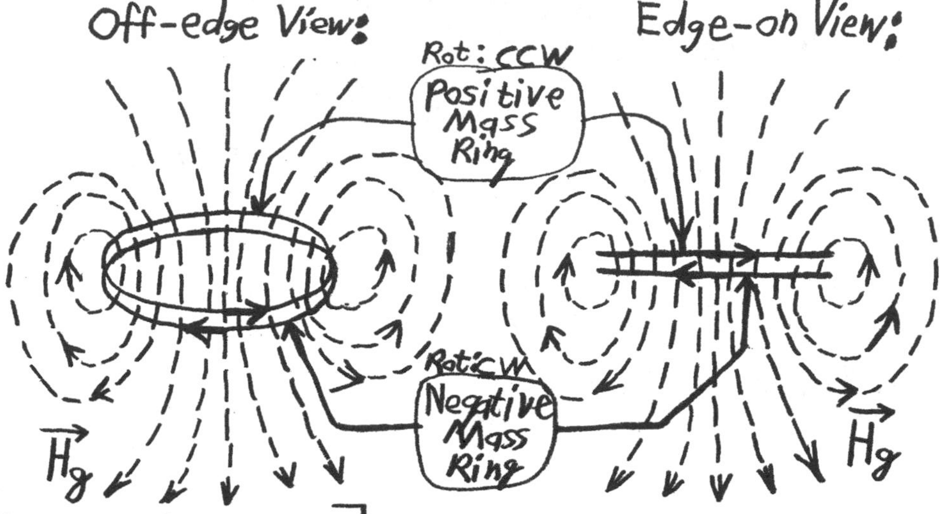

[The total mass = 0.]

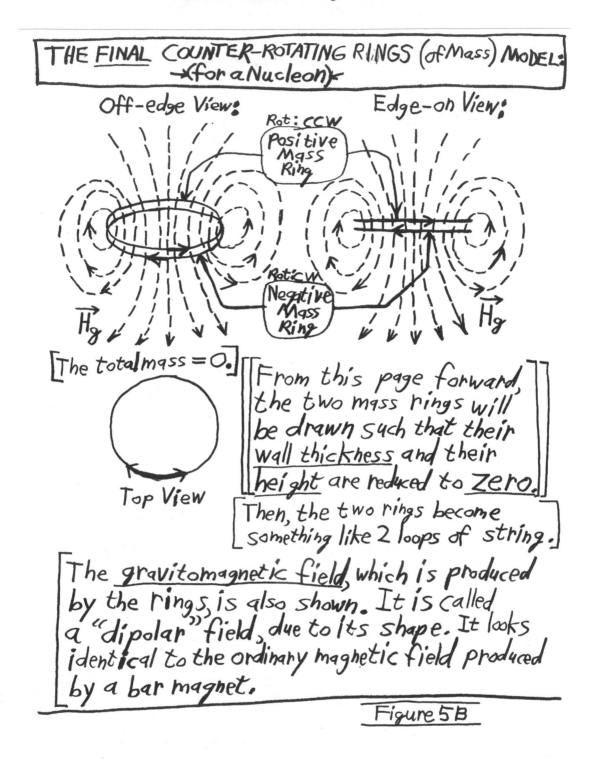

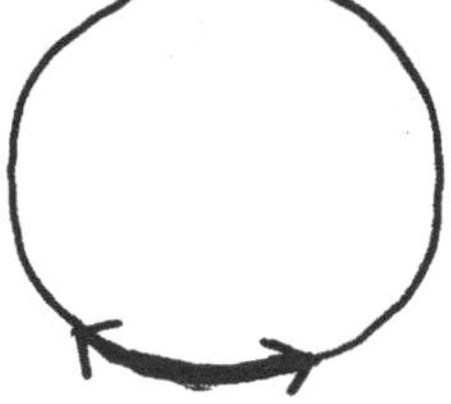

[[From this page forward, the two mass rings will be drawn such that their wall thickness and their height are reduced to zero.]]

[Then, the two rings become something like 2 loops of string.]

[The gravitomagnetic field, which is produced by the rings, is also shown. It is called a "dipolar" field, due to its shape. It looks identical to the ordinary magnetic field produced by a bar magnet.

Figure 5B

Figure 5C

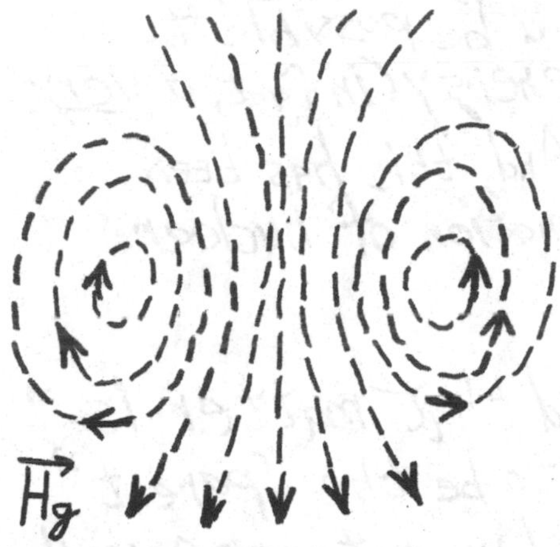

Here, we can legitimately draw the model of the nucleon without showing the 2 rings of mass, since the magnitudes of the ring masses are equal, but their signs are opposite. Thus, the ring masses exactly CANCEL each other.

So, all that remains is only the tiny dipolar gravitomagnetic field! That's ALL THERE IS. This field is extremely strong and is localized within a nucleon-sized volume.

Einstein showed that a field of any form of energy behaves as though it actually has an effective mass, even though no real mass is present to account for such.

So, the experimentally measured mass of a field is given by: $m = \frac{E}{c^2}$ where E is the total energy, contained in the field.

☆ So, what we normally measure for the mass of a nucleon (or for the mass of an entire nucleus) is just the total energy of the pure gravitomagnetic field, divided by c^2. There's no "real" mass there at all!

Until now, the Einstein formula, $E=mc^2$, has indicated to us that it should be possible to convert mass into pure energy (in fact, a very large amount of energy). And, this has been demonstrated by the detonation of nuclear explosives.

So, for years, we believed that matter is (or can be) considered to be the "parent" of energy. And we found ways to apparently convert mass into energy.

But now, we see an entirely different picture. We see that any subatomic particle consists not of matter, but of pure (gravitomagnetic) energy. So the tables turn. We now see that energy is the "parent" of matter.

Perhaps we should start re-writing Einstein's equation as: $m=\frac{E}{c^2}$.

So, the (measurable) mass of a particle is equal to its total stored gravitomagnetic energy, E, divided by the speed of light squared!

Figure 5D

Figure 6A

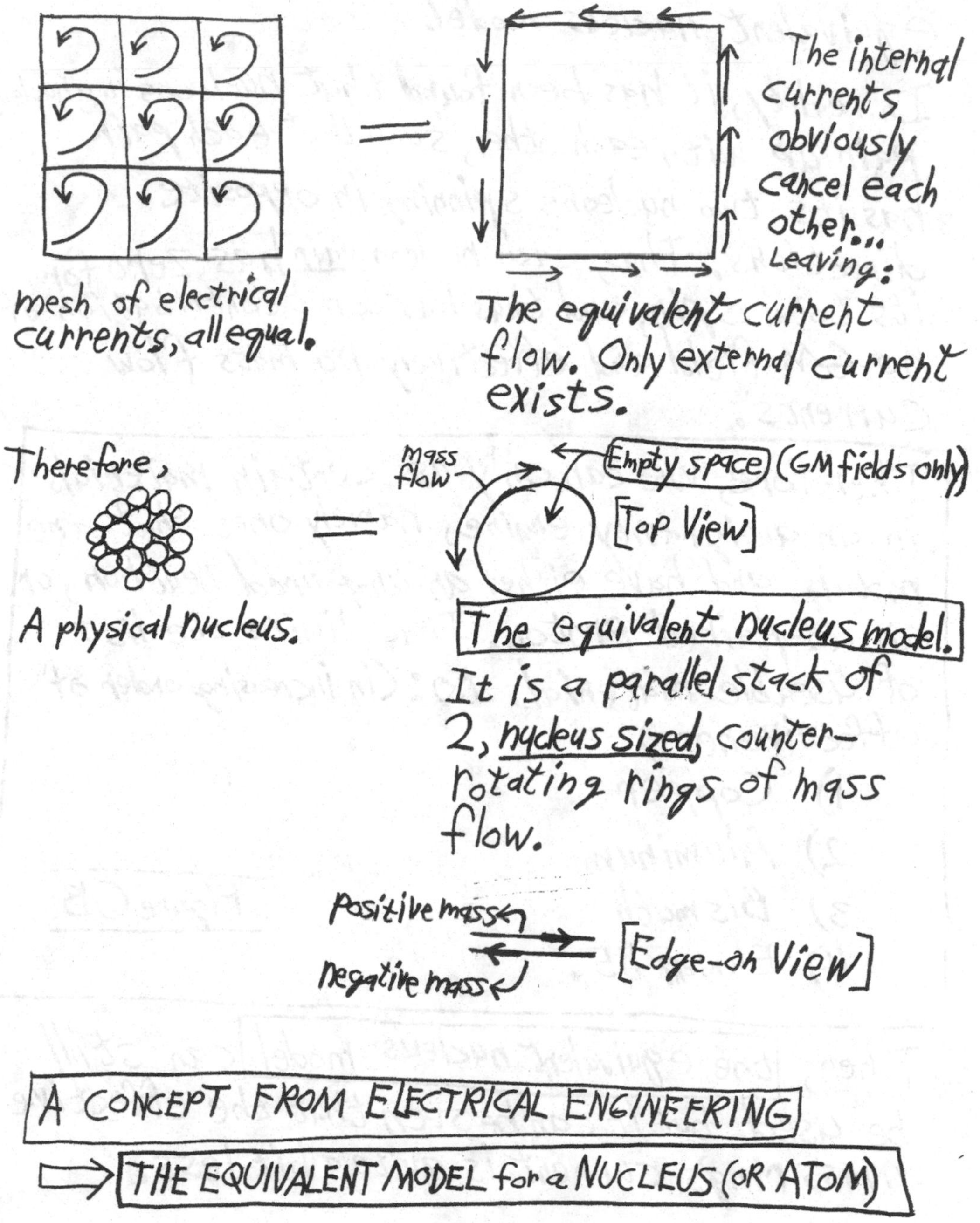
The internal currents obviously cancel each other... Leaving:
mesh of electrical currents, all equal.
The equivalent current flow. Only external current exists.
Therefore,
mass flow
Empty space (GM fields only)
[Top View]
A physical nucleus.
The equivalent nucleus model.
It is a parallel stack of 2, nucleus sized, counter-rotating rings of mass flow.
positive mass
negative mass
[Edge-on View]
A CONCEPT FROM ELECTRICAL ENGINEERING
THE EQUIVALENT MODEL for a NUCLEUS (OR ATOM)

There is a "catch", though, regarding this new equivalent nucleus model.

In reality, it has been found that nucleons in a nucleus pair up with each other, such that each pair has its two nucleons spinning in opposite directions. Thus, each nucleon <u>pair</u> has zero for its total spin, and thus has a net contribution of no GM field and effectively no mass flow currents.

Therefore, we can only use certain materials in an antigravity engine, namely ones that are metals, and have either an un-paired neutron, or an un-paired proton. That limits the list of useable materials to: (in increasing order of effectiveness)

1) Copper
2) Aluminum
3) Bismuth
4) Element 115.

Figure 6B

Then, the equivalent nucleus model can still be used, but it can be seen that the effective mass ring's current is quite a bit less.

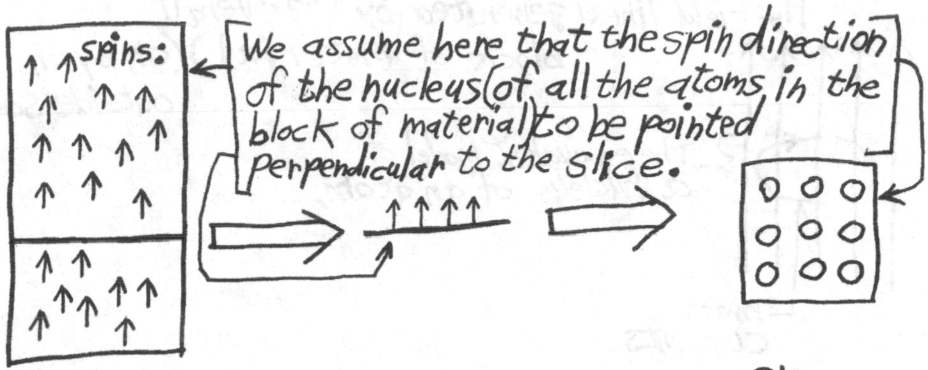
spins:
We assume here that the spin direction of the nucleus (of all the atoms in the block of material) to be pointed perpendicular to the slice.

A tall, rectangular block of material.

slice of the material.

Slice, Flipped 90°. (Only) Nine equivalent-model atoms (nuclei) are shown.

Using that same technique from Electrical Engineering, we may replace the entire slice of material with its equivalent model:

There's no mass current inside the square, Only the perimeter sides contain a mass flow (current).

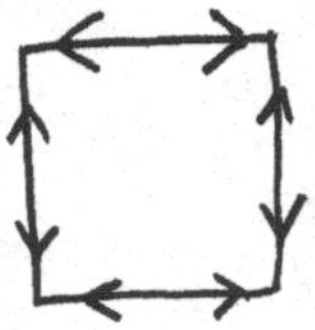

[Top View]

pos. mass

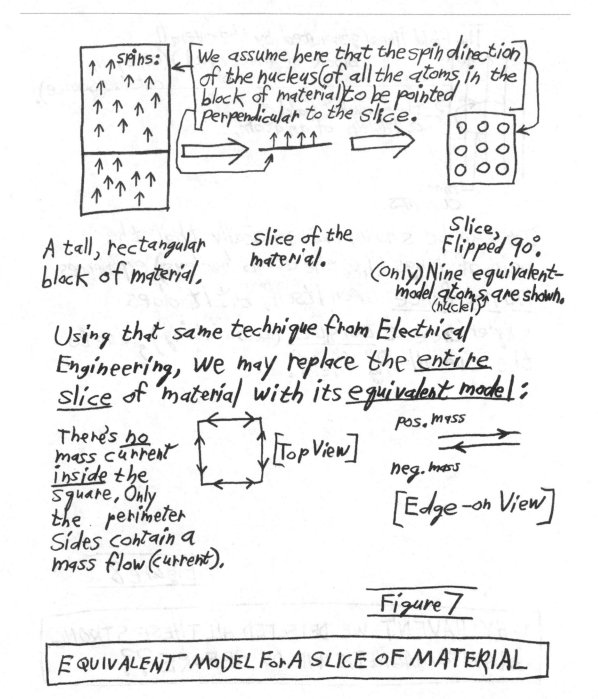

neg. mass

[Edge-on View]

Figure 7

EQUIVALENT MODEL For A SLICE OF MATERIAL

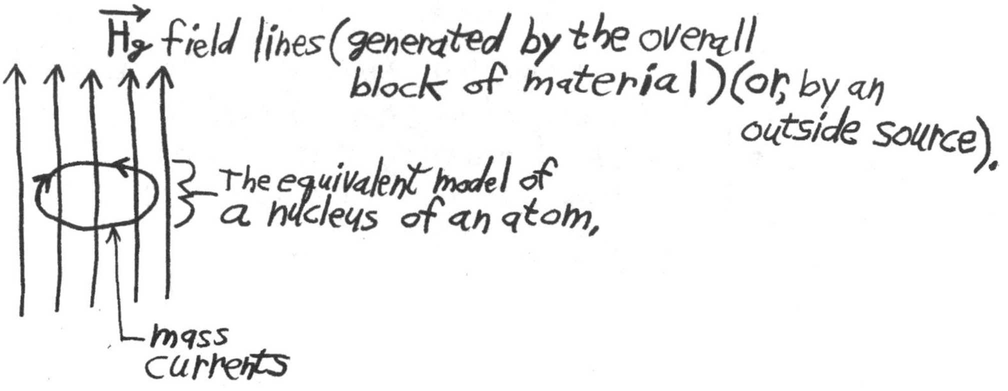

It can be shown mathematically that the pair of rings (i.e. the atom's nucleus) experiences <u>no</u> <u>net</u> <u>force</u> upon itself, but it does experience a <u>torque</u> (a twisting), due to the overall $\vec{H}_g$ field.

Figure 8

WHY HAVEN'T WE DETECTED ALL THESE STRONG GRAVITOMAGNETIC FIELDS BEFORE??

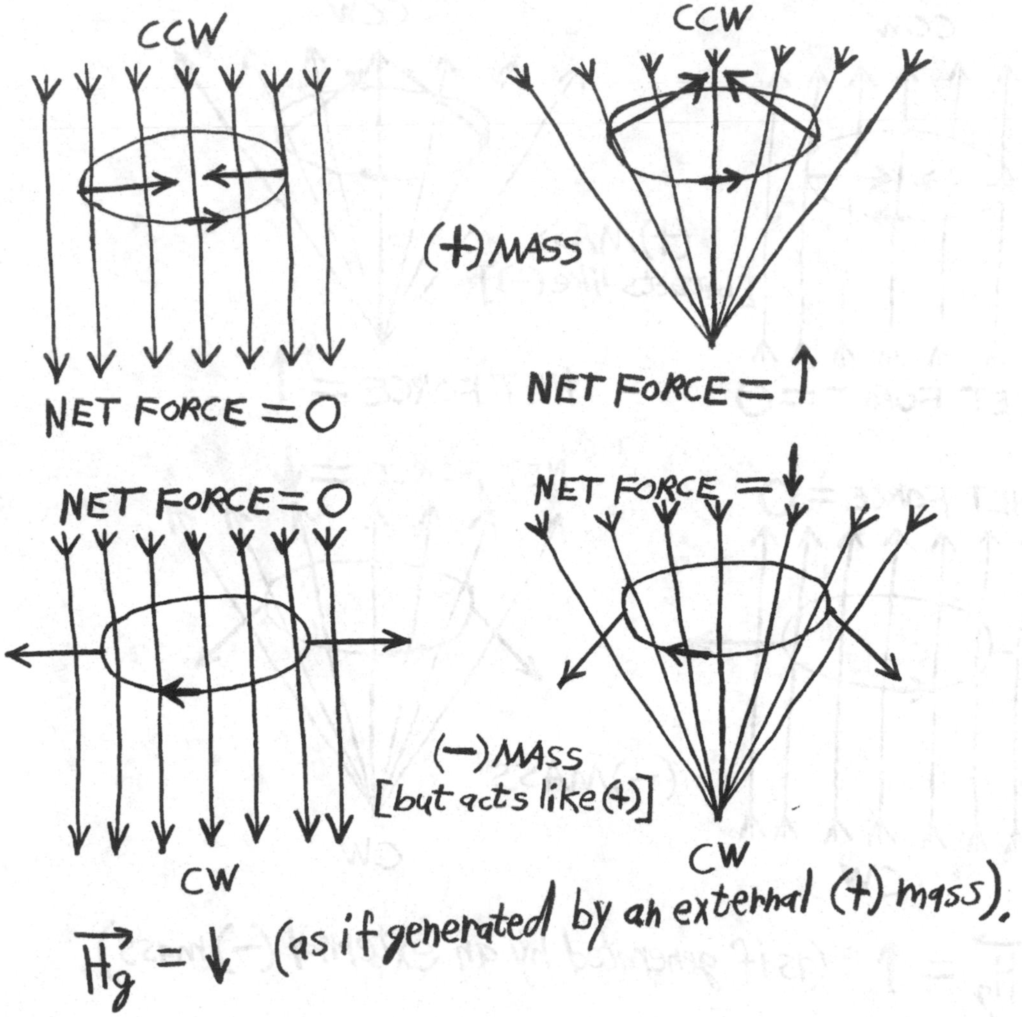

UNIFORM FIELD VS. NON-UNIFORM FIELD (CONICAL)

Figure 9A Downward GM field.

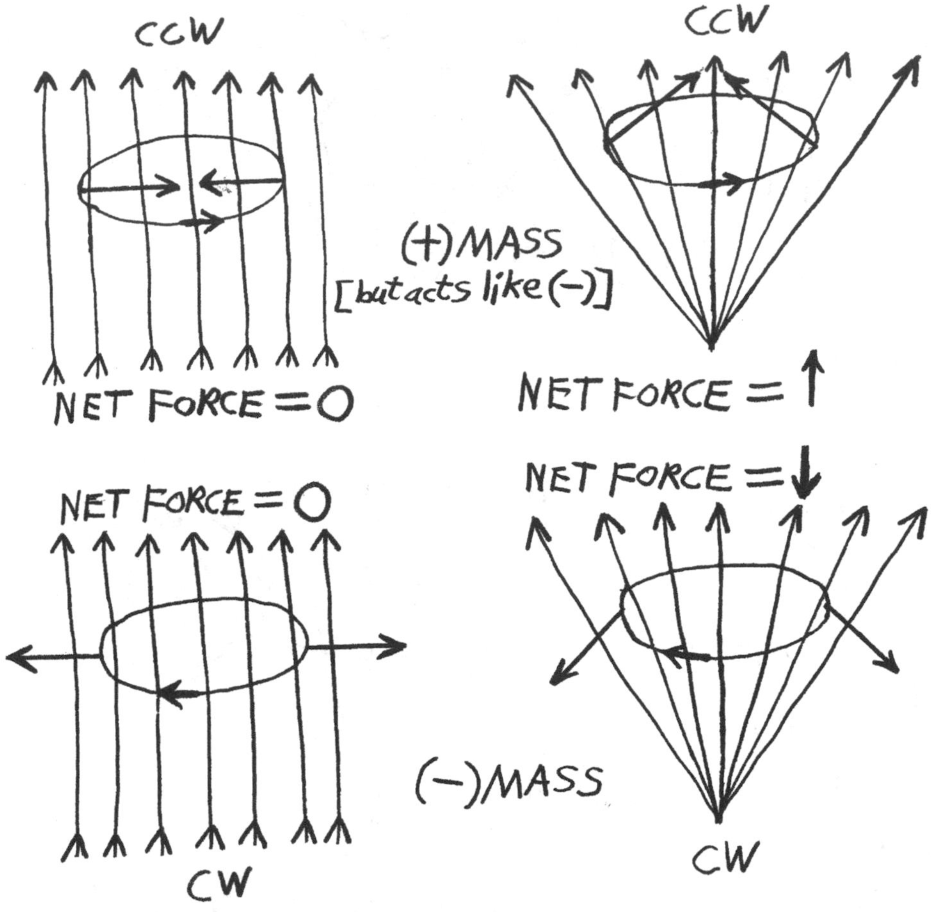

$\vec{H_g} = \uparrow$ (as if generated by an external (−) mass).

UNIFORM FIELD VS. NON-UNIFORM FIELD (CONICAL)

Figure 9B UPWARD GM field.

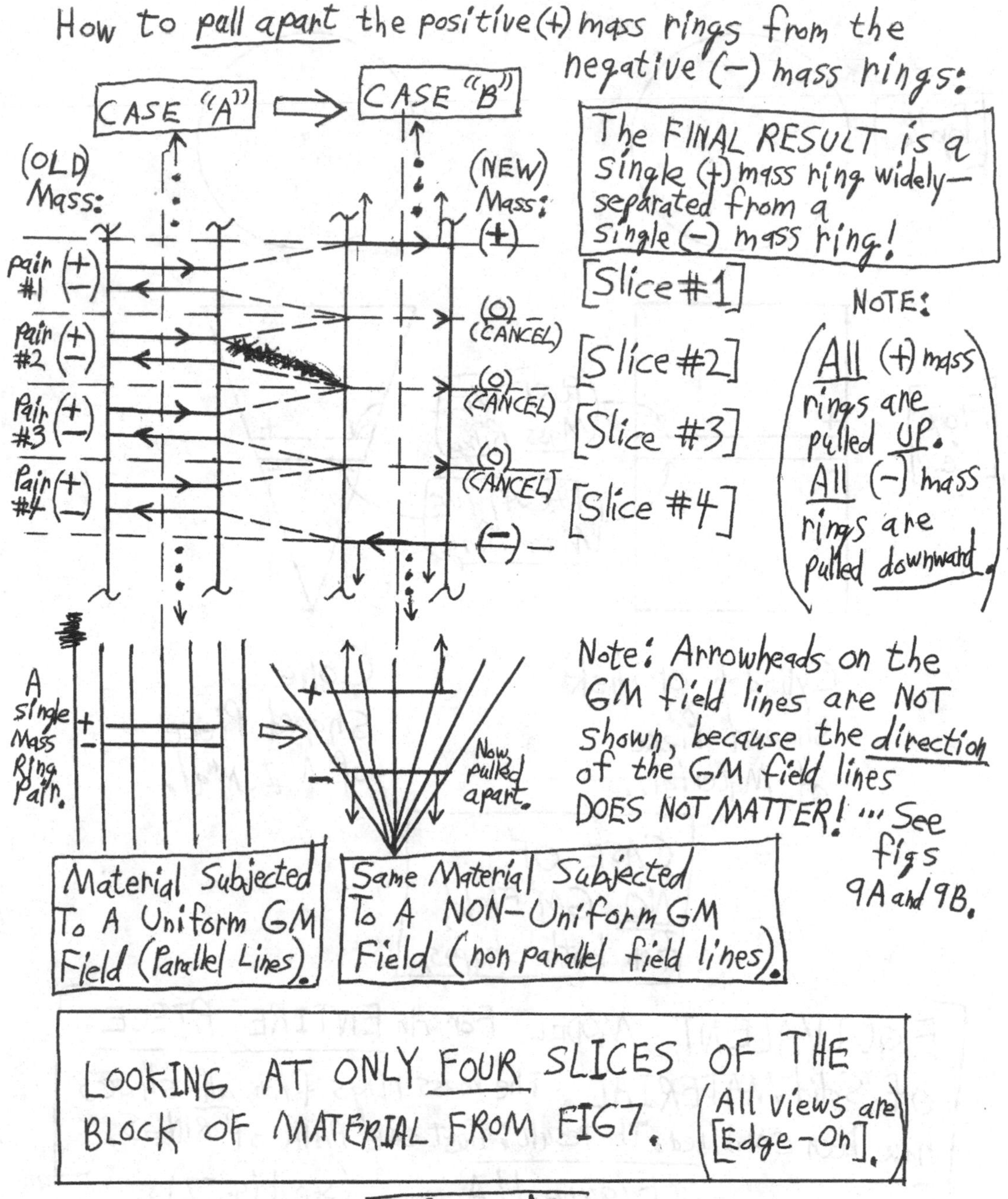
How to pull apart the positive (+) mass rings from the negative (−) mass rings:
CASE "A"
CASE "B"
(OLD) Mass:
(NEW) Mass:
The FINAL RESULT is a single (+) mass ring widely-separated from a single (−) mass ring!
pair #1
pair #2
Pair #3
Pair #4
(+)
(0) (CANCEL)
(0) (CANCEL)
(0) (CANCEL)
(−)
[Slice #1]
[Slice #2]
[Slice #3]
[Slice #4]
NOTE:
All (+) mass rings are pulled UP. All (−) mass rings are pulled downward.
A single Mass Ring Pair.
Now, pulled apart.
Note: Arrowheads on the GM field lines are NOT shown, because the direction of the GM field lines DOES NOT MATTER! ... See figs 9A and 9B.
Material Subjected To A Uniform GM Field (Parallel Lines).
Same Material Subjected To A NON-Uniform GM Field (non parallel field lines).
LOOKING AT ONLY FOUR SLICES OF THE BLOCK OF MATERIAL FROM FIG 7.
(All views are [Edge-On].)

Figure 10

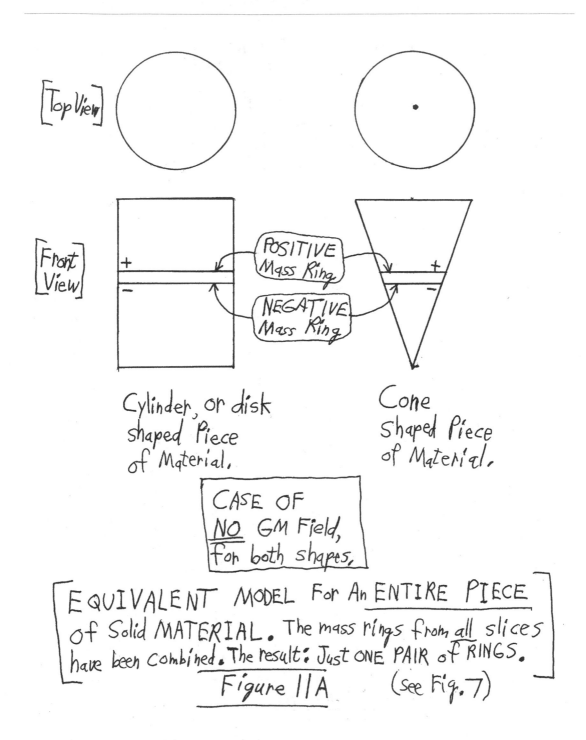

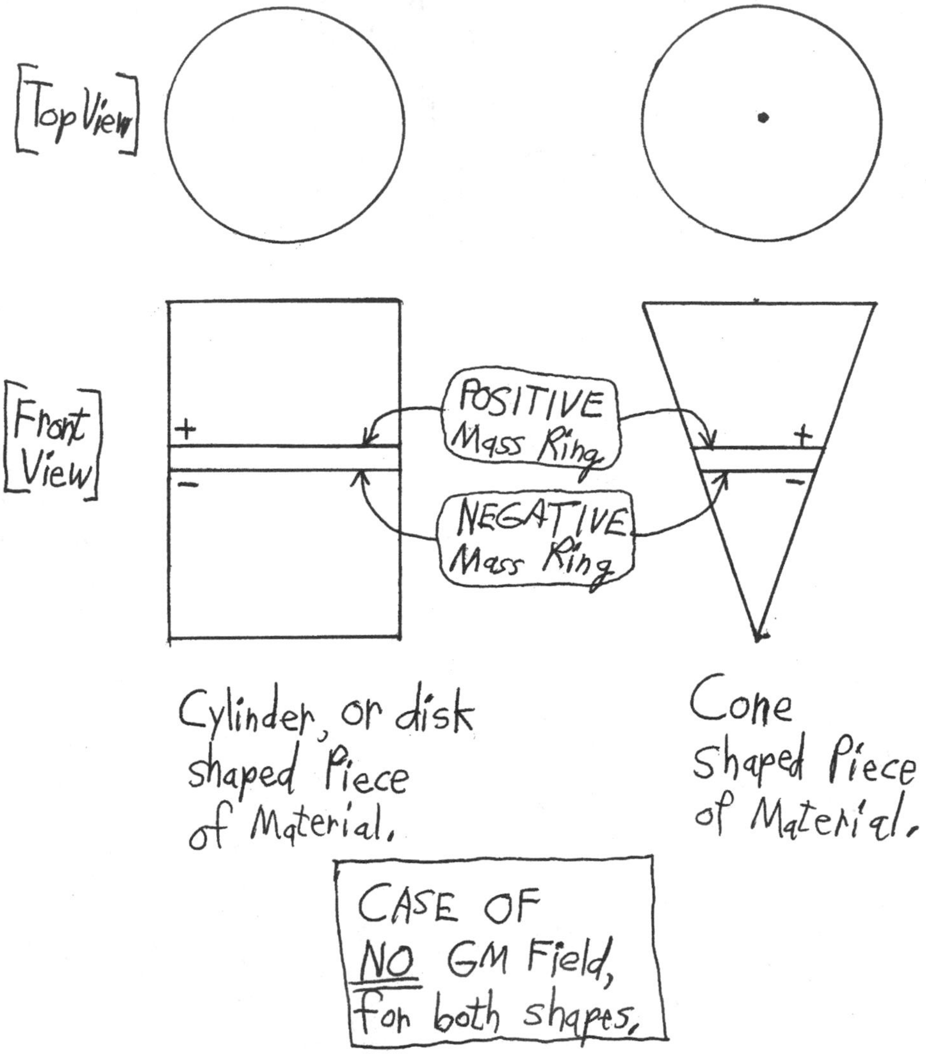

[EQUIVALENT MODEL For An ENTIRE PIECE of Solid MATERIAL. The mass rings from all slices have been combined. The result: Just ONE PAIR of RINGS.]

Figure 11A (see Fig. 7)

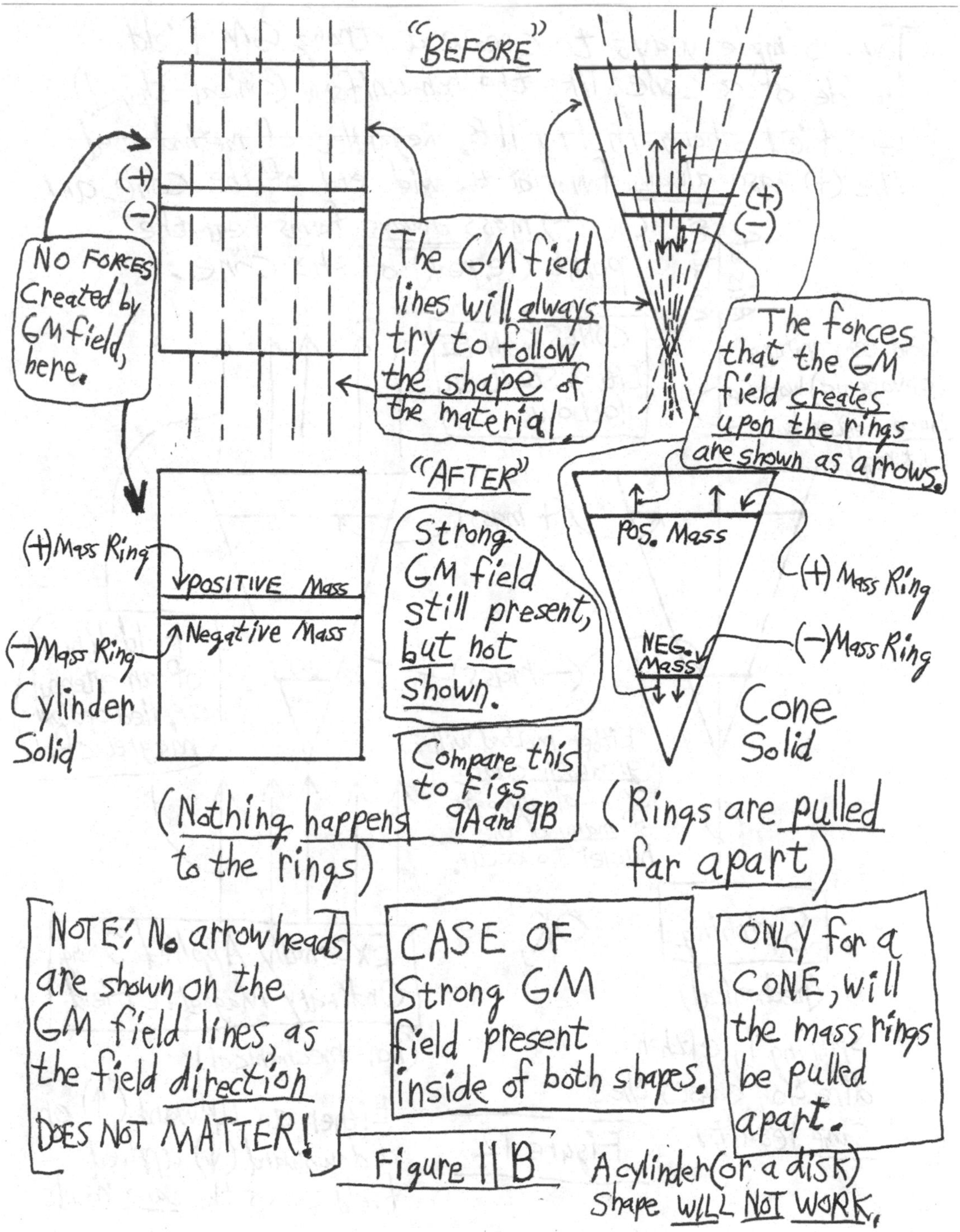
"BEFORE"
(+)
(−)
NO FORCES Created by GM field, here.
The GM field lines will always try to follow the shape of the material.
(+)
(−)
The forces that the GM field creates upon the rings are shown as arrows.
"AFTER"
(+) Mass Ring
POSITIVE Mass
Negative Mass
(−) Mass Ring
Cylinder Solid
Strong GM field still present, but not shown.
POS. Mass
(+) Mass Ring
NEG. Mass
(−) Mass Ring
Cone Solid
Compare this to Figs 9A and 9B
(Nothing happens to the rings)
(Rings are pulled far apart)
NOTE: No arrowheads are shown on the GM field lines, as the field direction DOES NOT MATTER!
CASE OF Strong GM field present inside of both shapes.
ONLY for a CONE, will the mass rings be pulled apart.
A cylinder (or a disk) shape WILL NOT WORK.

Figure 11B

Two simple ways to create a strong GM field inside of a cone, like the non-uniform (conical-shaped) GM field shown in Fig 11B. Regardless of method used, the (+) mass always forms at the wide end of the cone and the (−) mass always forms near the point (i.e. peak) of the cone.

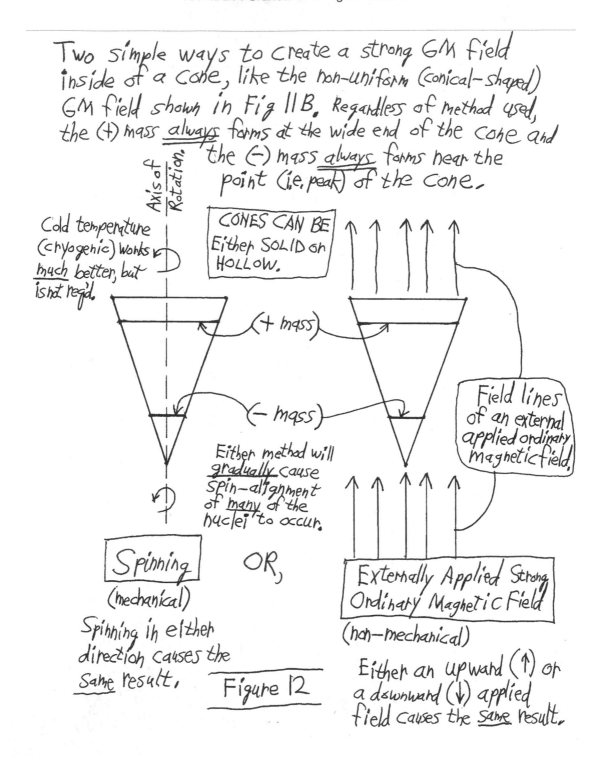

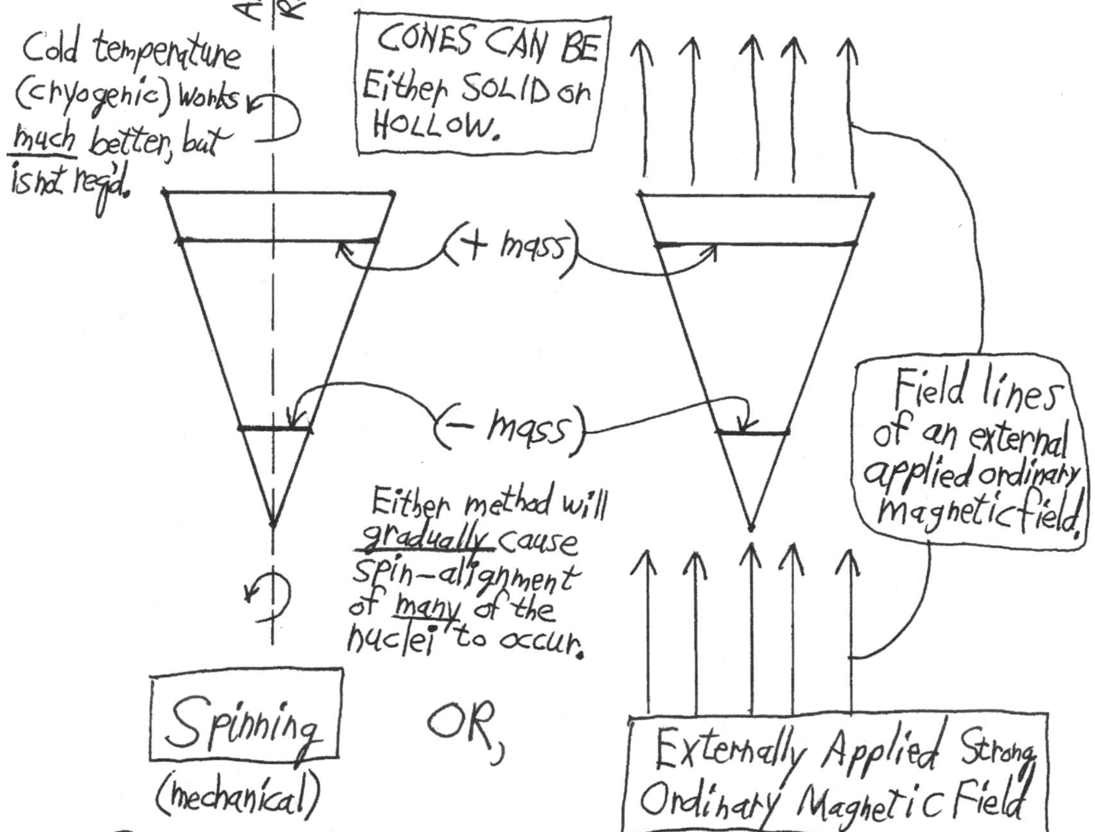

Spinning in either direction causes the same result.

Either an upward (↑) or a downward (↓) applied field causes the same result.

Figure 12

The magnitude (i.e, amount) of the widely-separated masses that form is very, very, very HUGE as compared with the measured mass of the cone, before the cone is "started up." This assumes that the spinning is fast enough or the external magnetic field is strong enough, (and enough time has elapsed).

[Why? Because the measured mass of the cone is equal to the GM field energy divided by C^2. And, The (+) and (−) mass rings have to be HUGE in amount to produce even a tiny GM field!]

So, we have this situation:

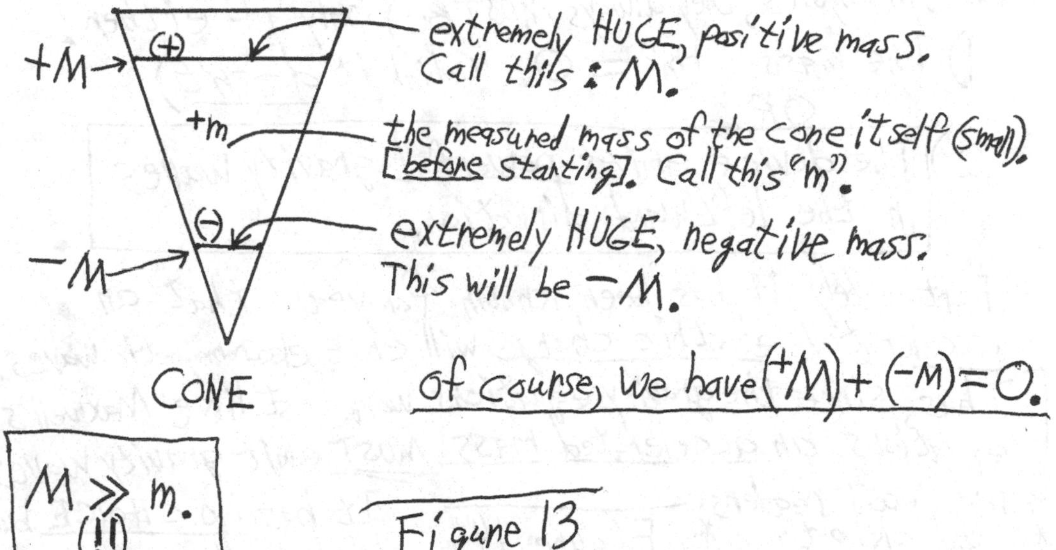
+M
(+)
+m
(−)
−M
CONE
extremely HUGE, positive mass. Call this: M.
the measured mass of the cone itself (small). [before starting]. Call this "m".
extremely HUGE, negative mass. This will be −M.
of course, we have (+M) + (−M) = 0.
M ≫ m. (!!)

Figure 13

We can symbolically diagram this situation of having a large (+) mass, separated from a large (−) mass as follows: [the cone would be laying on its side, here, the apex being on the left side]

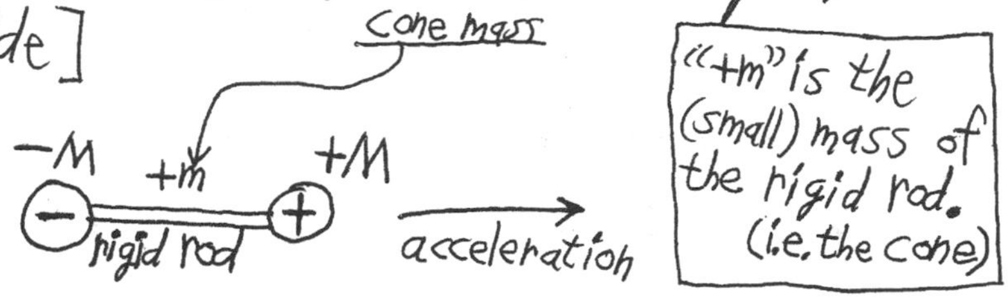

This device will speed up, forever, heading towards the right. Why?, see fig. 3B and 3C.
However, momentum and energy will be conserved (and, in Physics, they always MUST Be) only if either:

1) The mass "+m" = 0. (which it doesn't)

OR

2) The device emits powerful gravity waves in the leftward direction.

Fortunately, it has been known for years that an accelerated electric charge will emit electromagnetic waves.
[Thus, since the gravity equations work just like Maxwell's equations, an accelerated mass MUST emit gravity waves.
It has NO CHOICE but to do so!!

* Now, MOST readers*
*should skip to *
* Fig. 15D. *

Figure 14

Now, Maxwell's equations for electromagnetics can be written in the following (tensor) form:

(1): $F_{rs,s} = J_r$ (1,2,3,4 (indexes))

and,

(2): $F_{mn,r} + F_{nr,m} + F_{rm,n} = 0$

(comma means partial derivative)

(where we define:

$E_1 = iF_{14}$, $E_2 = iF_{24}$, $E_3 = iF_{34}$

and,

$H_1 = F_{23}$, $H_2 = F_{31}$, $H_3 = F_{12}$.)

Rewriting these, we find that we get:

(3): $\frac{1}{c}\frac{\partial E_\rho}{\partial t} + J_\rho = \epsilon_{\rho\mu\nu}\frac{\partial H_\nu}{\partial X_\mu}$ (1,2,3 only. (Indexes)) ⇒ Fig 1C: b)

and,

(4): $-\frac{1}{c}\frac{\partial H_\rho}{\partial t} = \epsilon_{\rho\mu\nu}\frac{\partial E_\nu}{\partial X_\mu}$ ⇒ a)

and,

(5): $\frac{\partial E_\rho}{\partial X_\rho} = J_4/i = \rho$ ⇒ c)

and,

(6): $\frac{\partial H_\rho}{\partial X_\rho} = 0$ ⇒ d)

$\epsilon_{\rho\mu\nu}$ is the "permutation" or "cross product" operator.

By changing "units" and converting (3), (4), (5), and (6) to "vector" form, we simply get Maxwell's equations (b), (a), (c), and (d) as shown in fig. 1C.

Figure 15A

Thus, equations (1) and (2) are correct.

If we define a new variable ϕ_r (the "4-potential") so that (7) and (8): $\boxed{\Box\phi_r = -J_r \text{ and } \phi_{r,r} = 0}$ is *always* true,

then Maxwell's equations ((1) and (2)) on fig. 15A ARE SATISFIED *if* (and *ONLY if*)

(9): $\boxed{F_{rs} = \phi_{s,r} - \phi_{r,s}}$.

(comma means partial derivative)

["Grad" students are familiar with (1) through (9), and all Electrical Engineers are very familiar with (a), (b), (c), and (d) in fig. 1C.]

For what follows, not too many people are familiar with it, or perhaps it's a "lost art":

Equations below come from Synge, J.L.; "Relativity: The Special Theory"; (1972) North Holland Publishing Co.; page 392 etc. See also (for radiation from an accelerated charge), Larmor; J. Phil. Mag (5); 44; (1897); 503.

[go to Fig 15C] → [Note: I throw away the "advanced" potential, as it's not needed.]

Figure 15B

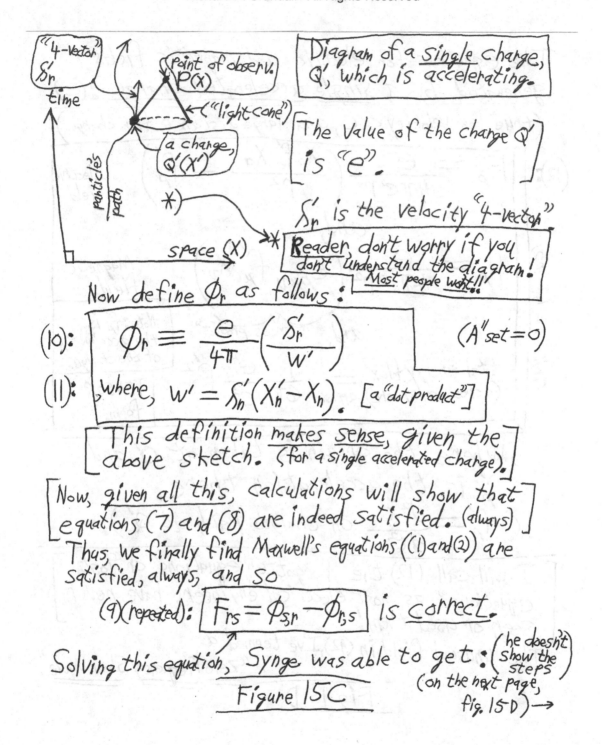

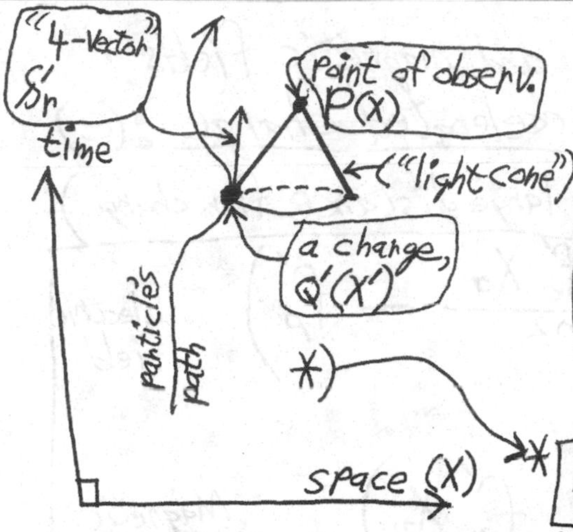

Diagram of a single charge, Q', which is accelerating.

[The value of the charge Q' is "e".]

S'_r is the velocity "4-vector".

*Reader, don't worry if you don't understand the diagram! Most people won't!!

Now define ϕ_r as follows:

$$(10):\quad \phi_r \equiv \frac{e}{4\pi}\left(\frac{S'_r}{w'}\right) \qquad (A''set=0)$$

$$(11):\quad \text{where, } w' = S'_n(X'_n - X_n). \quad [\text{a "dot product"}]$$

[This definition makes sense, given the above sketch. (for a single accelerated charge).]

[Now, given all this, calculations will show that equations (7) and (8) are indeed satisfied. (always)]

Thus, we finally find Maxwell's equations ((1) and (2)) are satisfied, always, and so

$$(9)(\text{repeated}):\quad F_{rs} = \phi_{s,r} - \phi_{r,s} \text{ is correct.}$$

Solving this equation, Synge was able to get: (he doesn't show the steps) (on the next page, fig. 15D) →

Figure 15C

The equations for electric and magnetic fields generated by a single accelerated charge: (e)
(true for velocity $\ll$ C, and large distance from charge)

(12): $$E_\rho = \frac{e}{4\pi c^2 r}\left(\frac{X_\rho f'_\sigma X_\sigma}{r^2} - f'_\rho\right)$$ Electric field

and,

$$H_\rho = \frac{e}{4\pi c^2 r^2}\left(\epsilon_{\rho\mu\nu} f'_\mu X_\nu\right)$$ Magnetic field

("cross product")

and,

total energy flux (across a sphere) $\underset{\text{approx.}}{=} \frac{1}{6\pi}\frac{e^2}{c^3} f'^2$

(indices = 1, 2, or 3)

Note: f'_ρ is the acceleration of the charge.

TENSOR form.

, where the origin is taken at Q′ and f'_ρ is the acceleration there.
$\left(f'_\rho = \frac{d^2 X'_\rho}{dt^2}\right.$ on charge's path at Q′).

[I will call (12) the "forgotten equations" or "lost equations", as most electrical engineers have never seen or used them!]

The unusual field in (12), I've termed a "Zero Frequency Shock Wave".

Fig 15D

Electrical/Electronic engineers usually only work with the concept of electromagnetic waves that have a NON-zero frequency. That is, these waves reverse polarity every "so-many-fractions-of-a-second", and are like a "sine" wave travelling (i.e. a wave with ripples in it).

[In fact, they will tell you that a zero-frequency wave won't even propagate [it won't even leave the antenna!]. Now that's true for a "sine wave", but not for a Zero-Frequency-Shock Wave.

[A Zero-frequency Shockwave actually does propagate, and can travel a long distance!] (Engineers, take note of this!!)

[Now, if we replace the "e" in equ. (12) with a (minus) mass (either (+M) or (-M)), and let f' be the acceleration of that mass, then we can calculate what Gravitoelectric and Gravitomagnetic fields will be produced and propagated. It will constitute a "Zero Freq. Gravity Shockwave".

Fig 15E

Next, is Fig 16.

Now, we return to our "train-of-thought" from figure 14 (finally!). [See figs 14 and 15 D and E]:

[Using (12), for accelerated masses, we can calculate what gravity waves will be (MUST be) emitted (or "ejected") by the two masses in fig 14. Here's what the result looks like:

(Remember, the device we are drawing actually represents a cone, with separated + and − mass.)

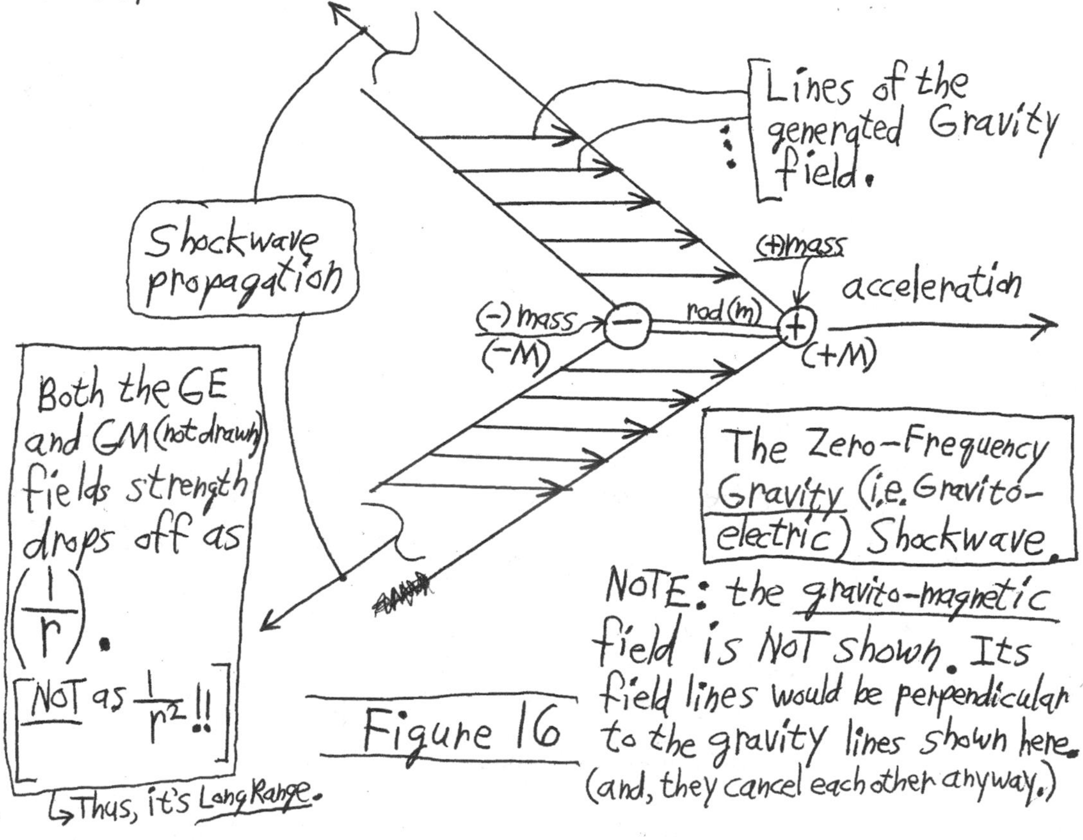
Lines of the generated Gravity field.
Shockwave propagation
(+) mass
acceleration
(−) mass
(−M)
rod (m)
(+M)
Both the GE and GM (not drawn) fields strength drops off as $\left(\frac{1}{r}\right)$.
[NOT as $\frac{1}{r^2}$!!]
↳ Thus, it's Long Range.
The Zero-Frequency Gravity (i.e. Gravito-electric) Shockwave.

NOTE: the gravito-magnetic field is NOT shown. Its field lines would be perpendicular to the gravity lines shown here. (and, they cancel each other anyway.)

Figure 16

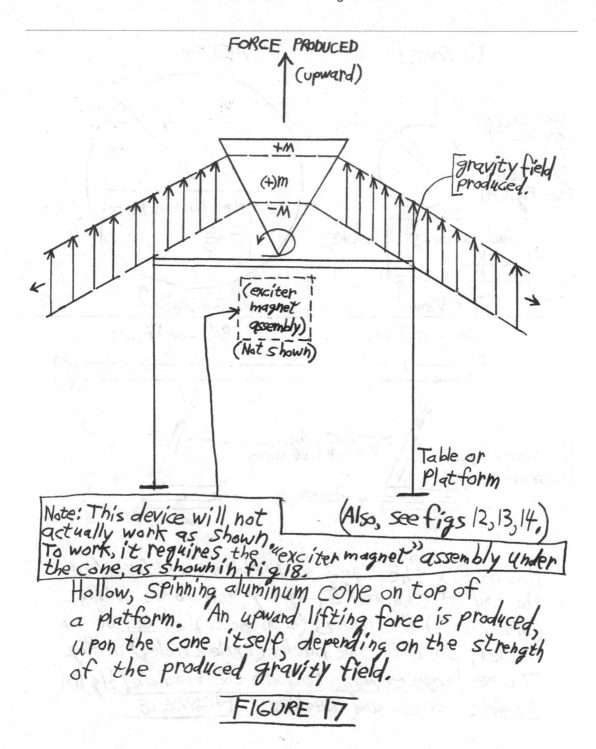

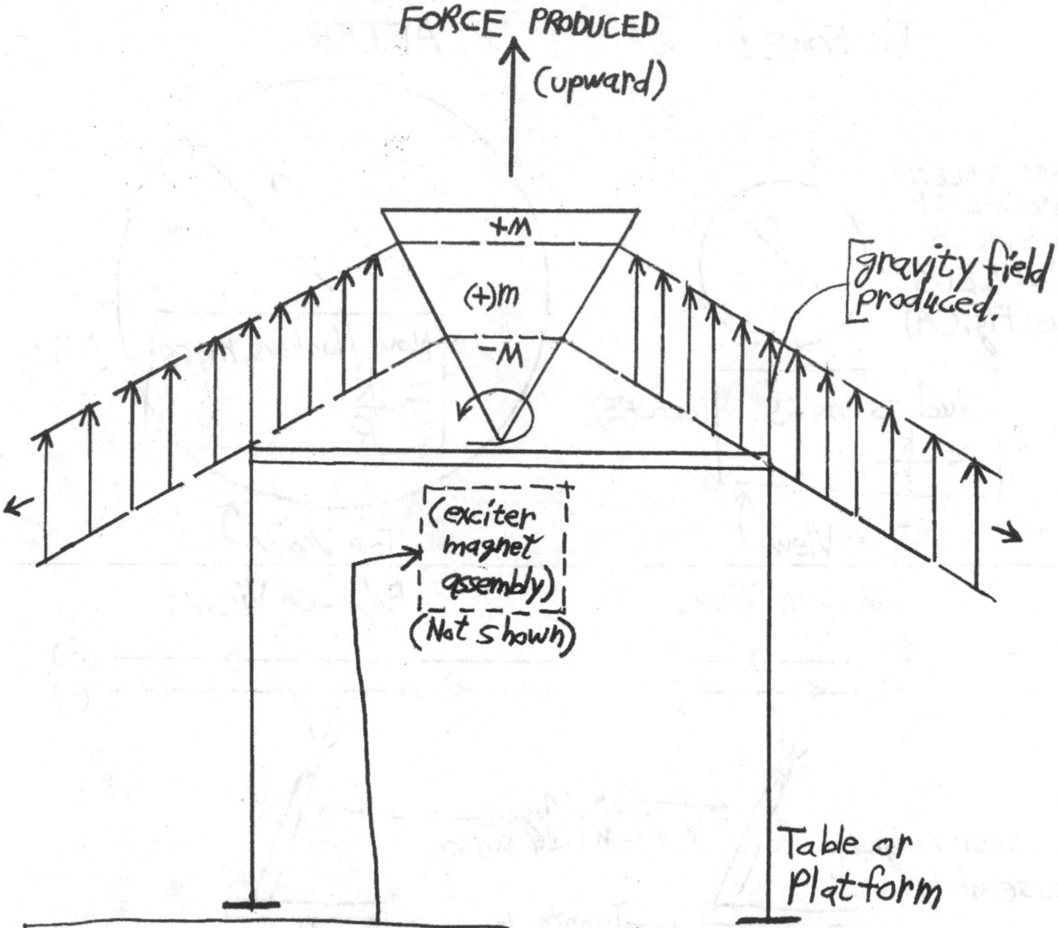

Note: This device will not actually work as shown. To work, it requires, the "exciter magnet" assembly under the cone, as shown in fig 18.

(Also, see figs 12, 13, 14,)

Hollow, spinning aluminum cone on top of a platform. An upward lifting force is produced, upon the cone itself, depending on the strength of the produced gravity field.

FIGURE 17

D&R Scientific Pub. Co.; Denver, CO Jan. 2003. (originally publ. on 04-17-2002) (Figs: 2-1-02)

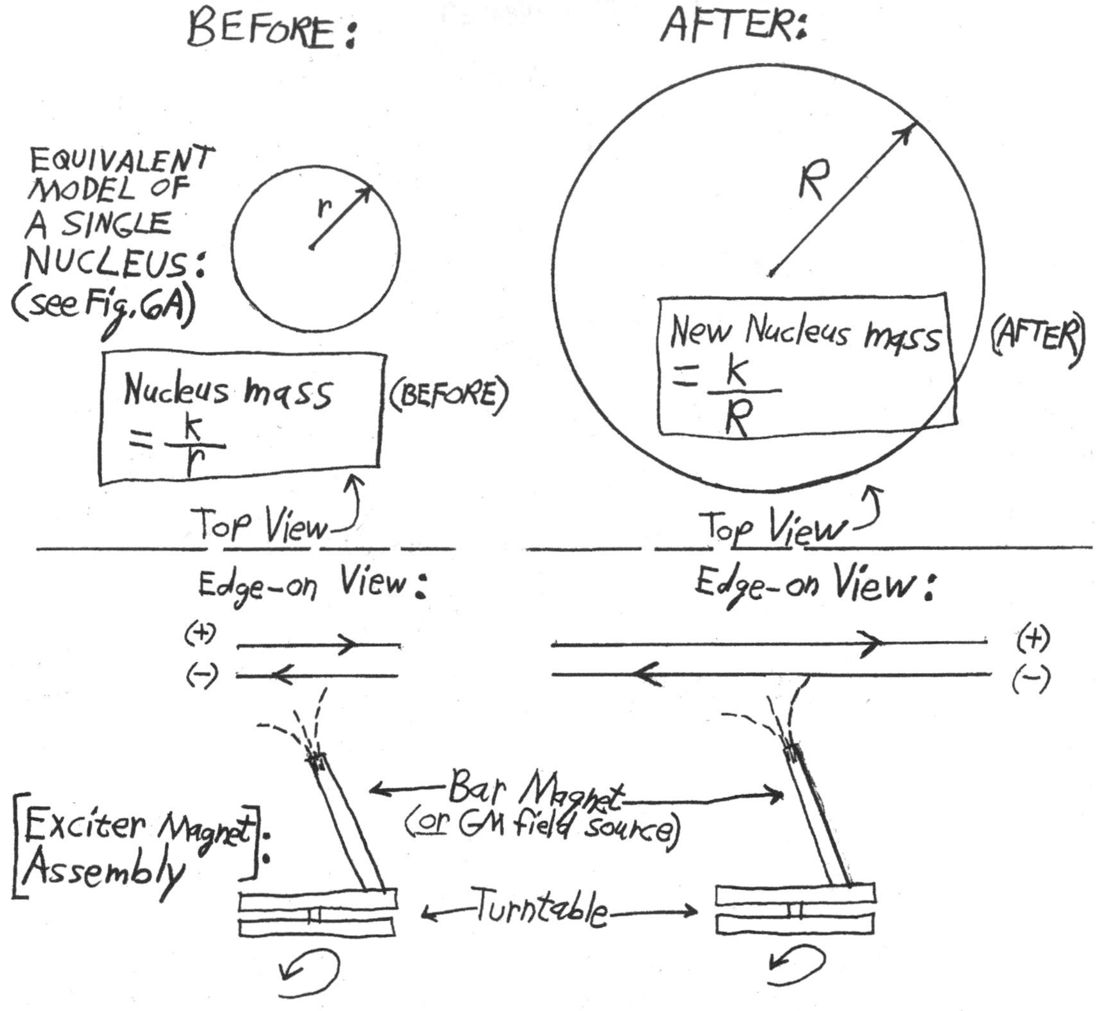

How to fix the "gotcha" problem mentioned at the end of Chapter 5, namely how do we widen (dramatically) the radius (or diameter) of all of the nuclei, so that the nuclei (little circles) in fig 7 are large enough to make the fig 7 and fig 11A models correct and useable. FIGURE 18

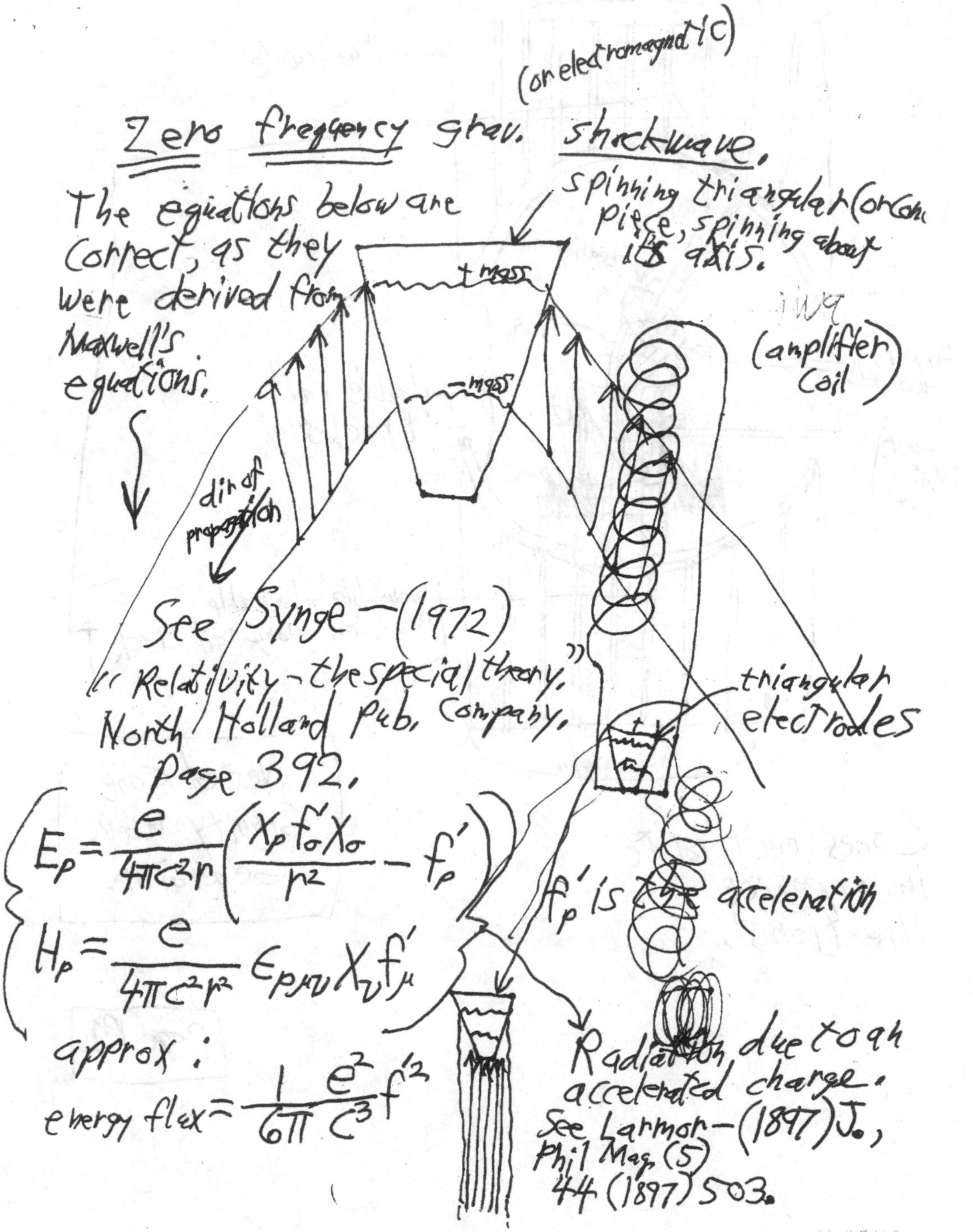
(or electromagnetic)
Zero frequency grav. shockwave.
The equations below are correct, as they were derived from Maxwell's equations.
spinning triangular (or cone) piece, spinning about its axis.
+ mass
- mass
(amplifier) coil
dir of propagation
See Synge – (1972) "Relativity - the special theory," North Holland Pub. Company, Page 392.
triangular electrodes
$E_\rho = \frac{e}{4\pi c^2 r}\left(\frac{X_\rho f'_\sigma X_\sigma}{r^2} - f'_\rho\right)$
$H_\rho = \frac{e}{4\pi c^2 r^2}\epsilon_{\rho\mu\nu} X_\nu f'_\mu$
f'_ρ is the acceleration
approx:
energy flux $= \frac{1}{6\pi}\frac{e^2}{c^3} f'^2$
Radiation due to an accelerated charge.
See Larmor – (1897) J., Phil Mag (5) 44 (1897) 503.

FIGURE 19

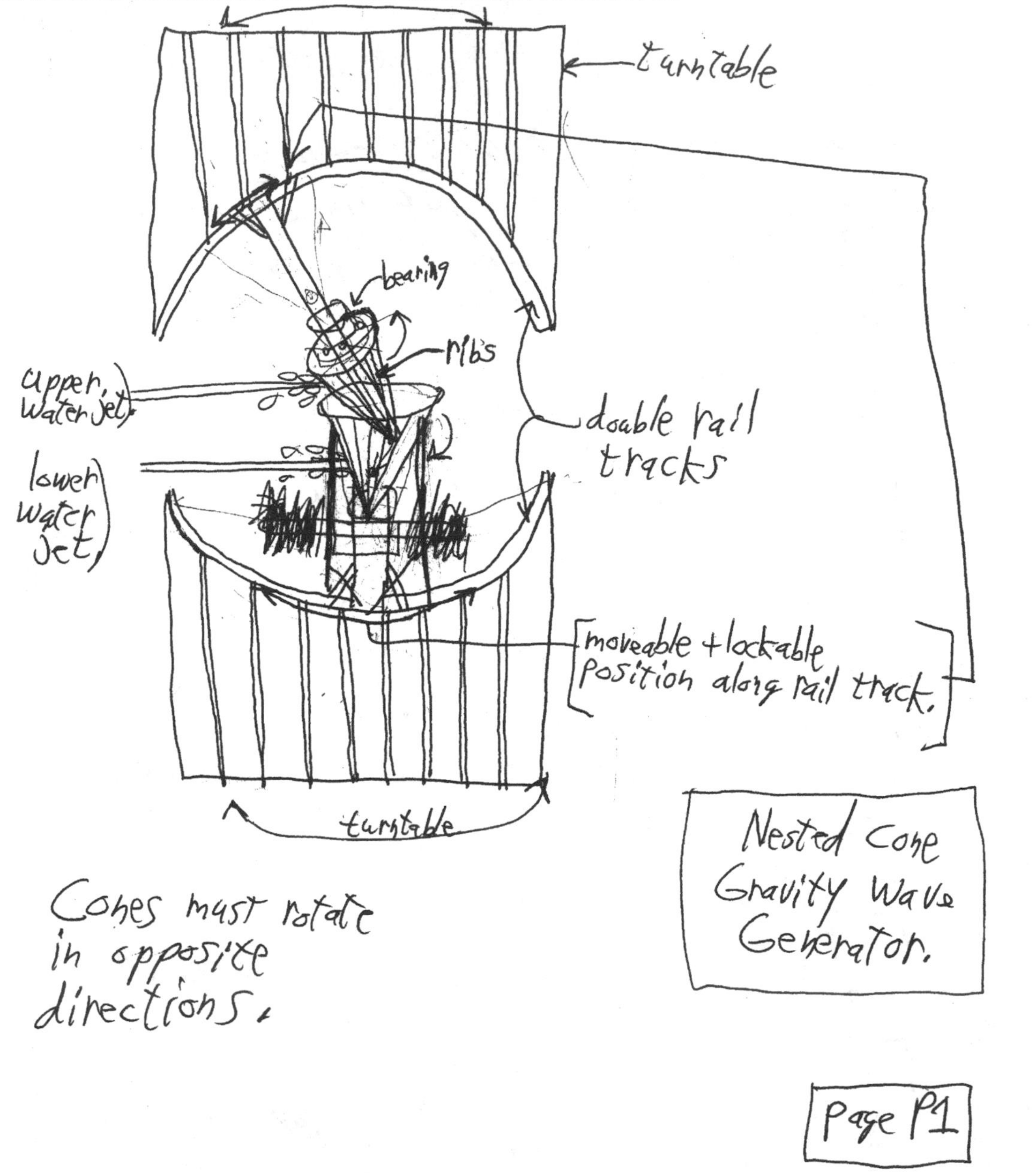
turntable
bearing
ribs
upper water jet
lower water jet
double rail tracks
moveable + lockable position along rail track.
turntable
Cones must rotate in opposite directions.
Nested Cone Gravity Wave Generator.
Page P1

FIGURE 20

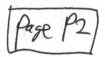

Page P2

FIGURE 21

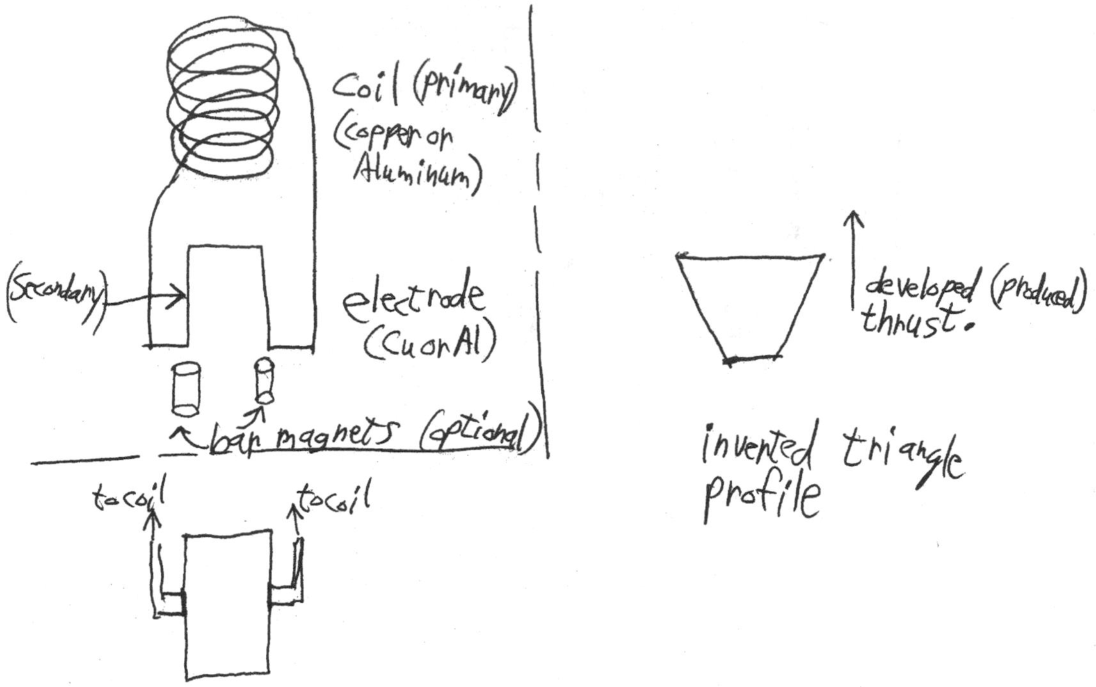

Complete, single electrode pair, single stage, Gravity wave Amplifier

Page P3

FIGURE 22

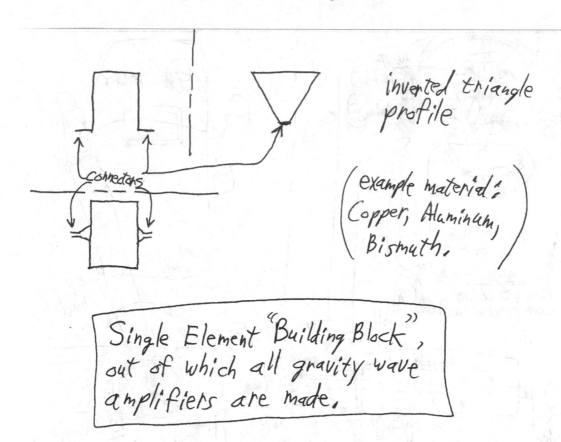

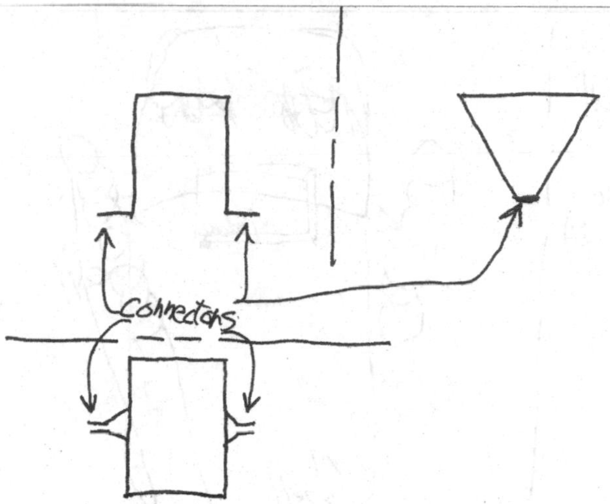

inverted triangle profile

(example material: Copper, Aluminum, Bismuth.)

Single Element "Building Block", out of which all gravity wave amplifiers are made.

Typic[redacted]al G.W. amp. would have many of these, connected together.

Page P4

FIGURE 23

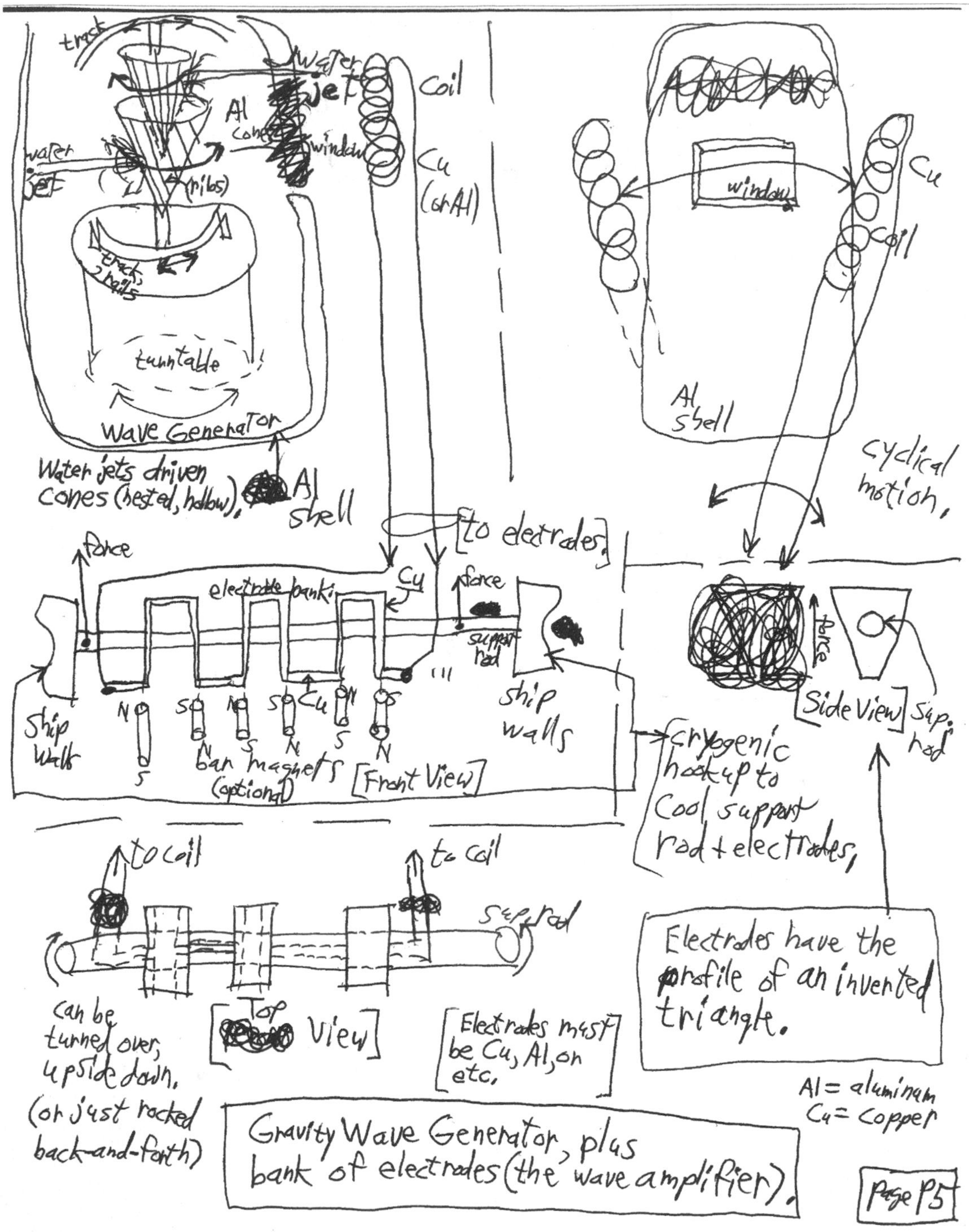
thrust
water jet
Al cones
window
Coil
Cu (or Al)
water jet
(ribs)
track, 2 rails
turntable
Wave Generator
Water jets driven cones (nested, hollow), Al shell
window
Cu
coil
Al shell
cyclical motion,
[to electrodes]
force
electrode bank
Cu
force
support rod
ship walls
Ship walls
N S N S N S
S N S N S N
bar magnets (optional)
[Front View]
force
[Side View]
Sup. rod
cryogenic hookup to cool support rod + electrodes,
to coil
to coil
sup. rod
can be turned over, upside down, (or just rocked back-and-forth)
[Top View]
[Electrodes must be Cu, Al, or etc.]
Electrodes have the profile of an inverted triangle.
Al = aluminum
Cu = copper
Gravity Wave Generator, plus bank of electrodes (the wave amplifier).
Page P5

FIGURE 24

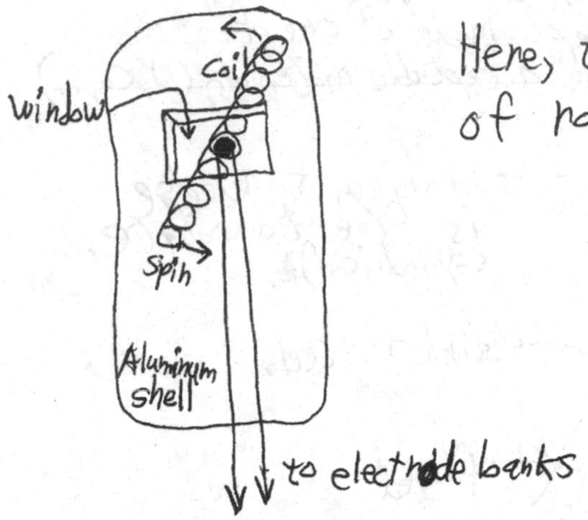

Here, the coil spins instead of rocking back and forth.

Page P6

Alternative coil mechanism for gravity wave generator.

FIGURE 25

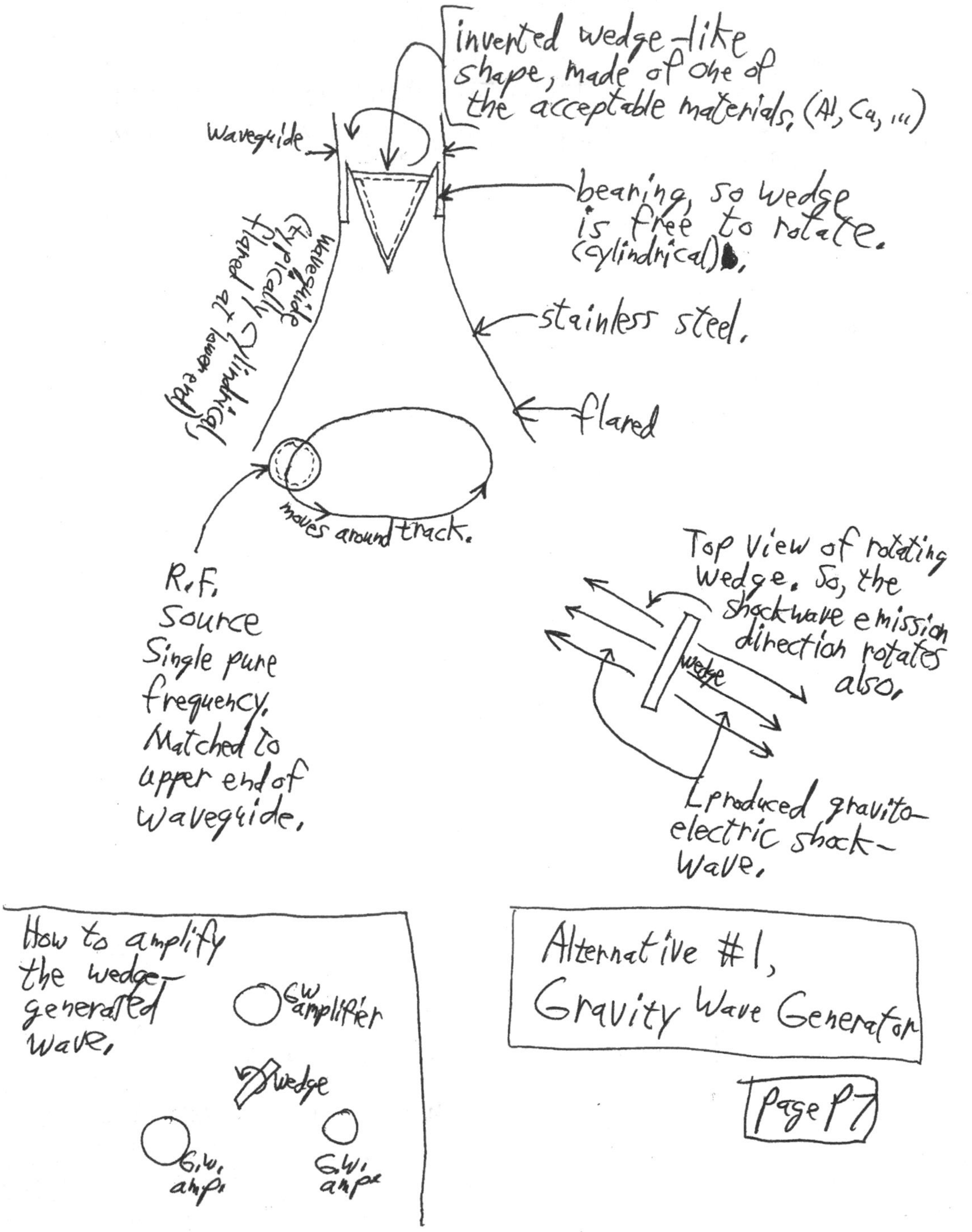
inverted wedge-like shape, made of one of the acceptable materials, (Al, Ca, ...)
waveguide
bearing, so wedge is free to rotate. (cylindrical)
waveguide typically cylindrical (flared at lower end)
stainless steel.
flared
moves around track.
R.F. Source Single pure frequency, Matched to upper end of waveguide.
Top view of rotating wedge. So, the shockwave emission direction rotates also.
wedge
produced gravito-electric shock-wave.
How to amplify the wedge-generated wave.
GW amplifier
wedge
G.W. amp.
G.W. amp.
Alternative #1, Gravity Wave Generator
Page P7

FIGURE 26

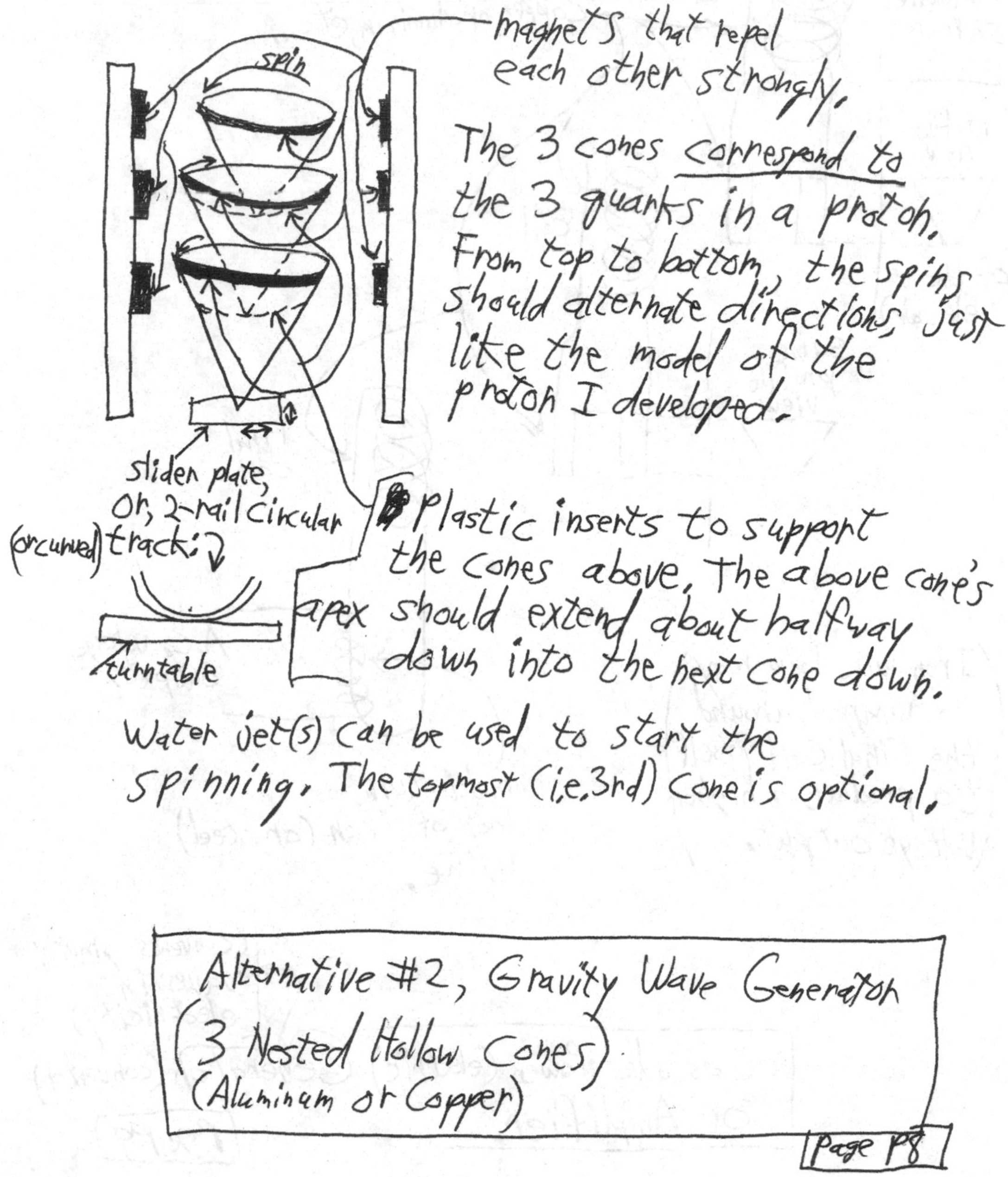
spin
magnets that repel each other strongly.
The 3 cones correspond to the 3 quarks in a proton. From top to bottom, the spins should alternate directions, just like the model of the proton I developed.
slider plate, or, 2-rail circular (or curved) track:
turntable
Plastic inserts to support the cones above. The above cone's apex should extend about halfway down into the next cone down.
Water jet(s) can be used to start the spinning. The topmost (i.e. 3rd) cone is optional.
Alternative #2, Gravity Wave Generator (3 Nested Hollow Cones) (Aluminum or Copper)
Page P8

FIGURE 27

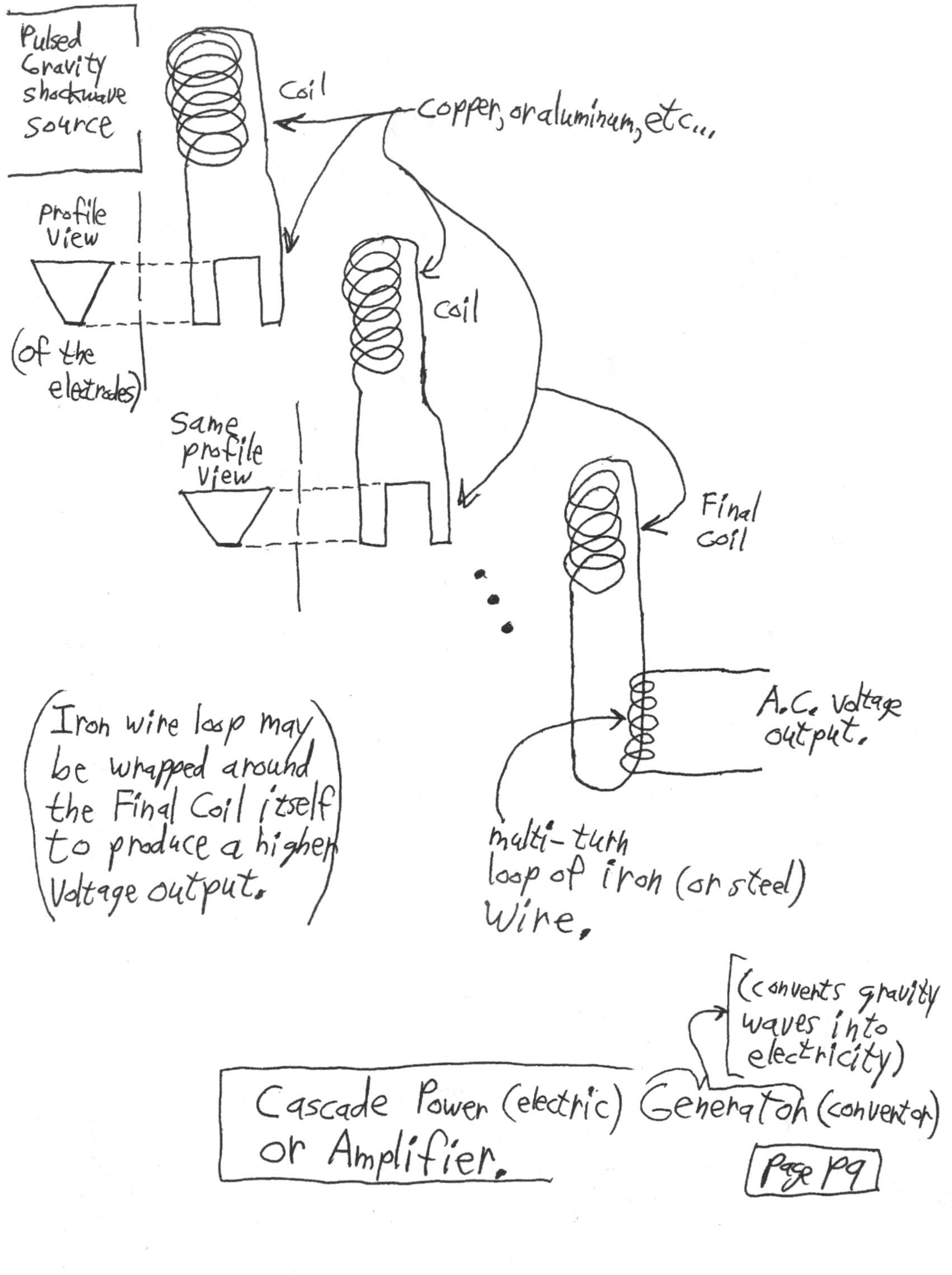
Pulsed Gravity shockwave source
Coil
copper, or aluminum, etc...
Profile view
(of the electrodes)
coil
Same profile view
Final coil
A.C. Voltage output.
Iron wire loop may be wrapped around the Final Coil itself to produce a higher Voltage output.
multi-turn loop of iron (or steel) wire.
(converts gravity waves into electricity)
Cascade Power (electric) Generator (convertor) or Amplifier.
Page P9

FIGURE 28

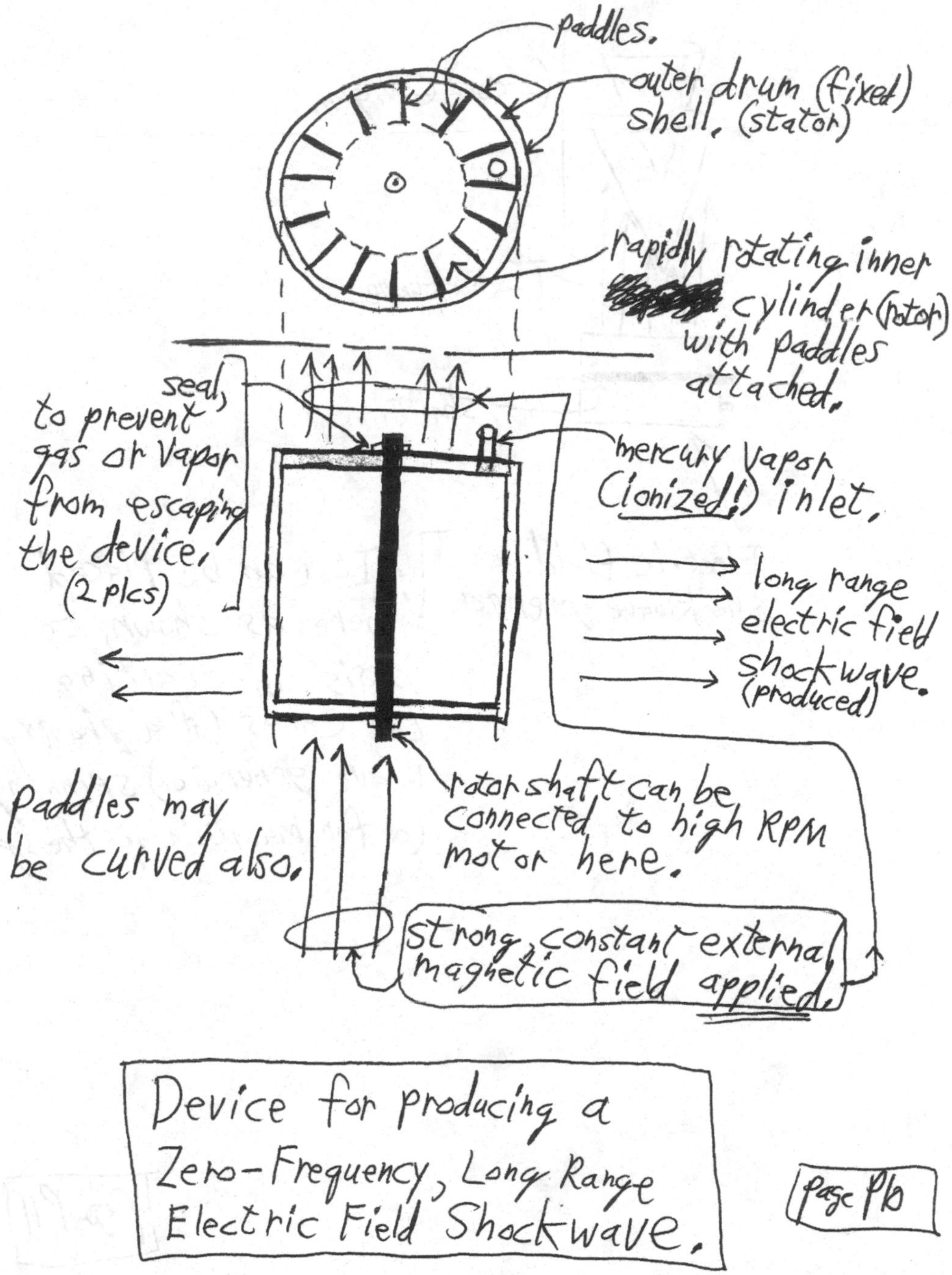
paddles.
outer drum (fixed) shell. (stator)
rapidly rotating inner cylinder (rotor) with paddles attached.
seal, to prevent gas or vapor from escaping the device. (2 plcs)
mercury vapor (Ionized!) inlet.
long range electric field shockwave. (produced)
paddles may be curved also.
rotor shaft can be connected to high RPM motor here.
strong, constant external magnetic field applied.
Device for producing a Zero-Frequency, Long Range Electric Field Shockwave.
page P6

FIGURE 29

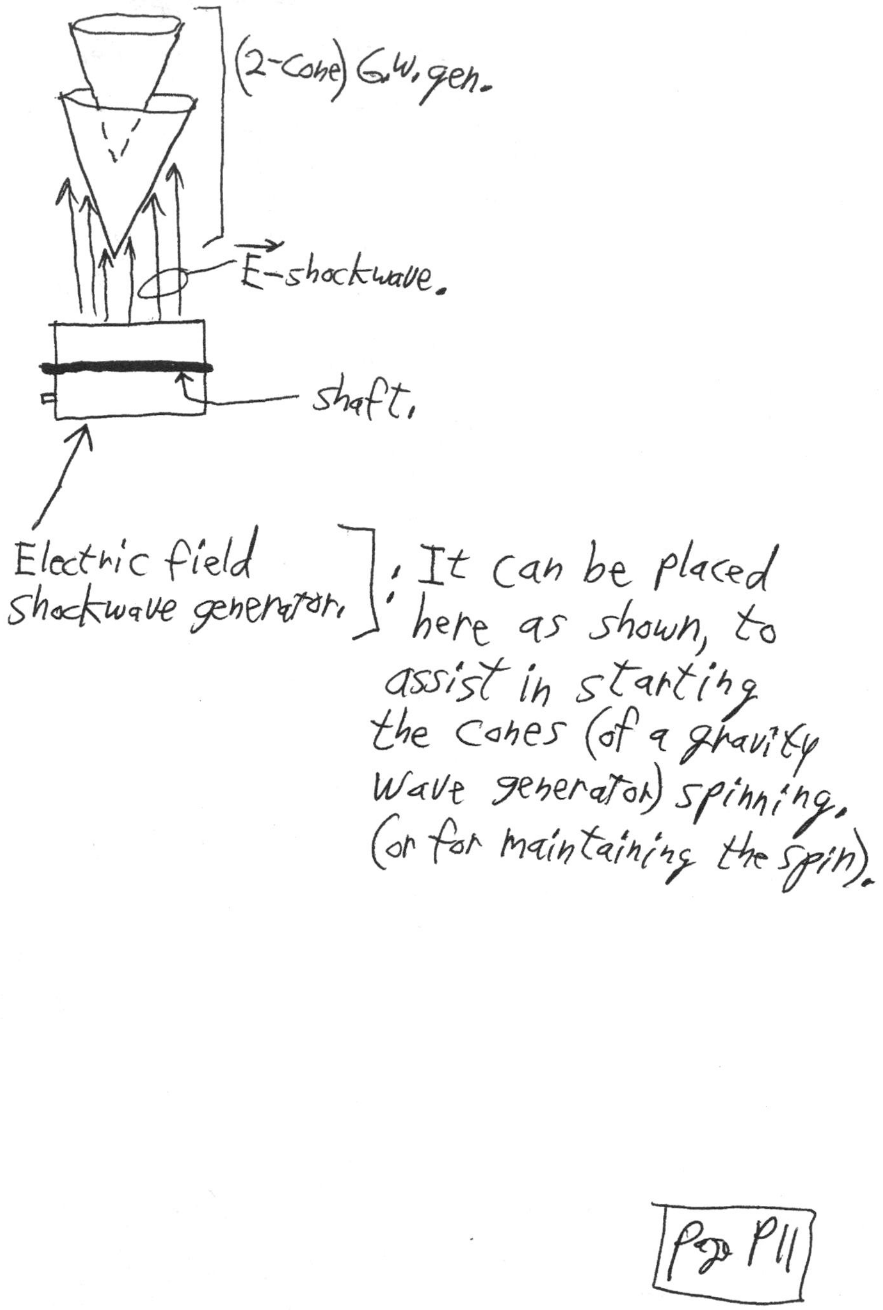
(2-cone) G.W. gen.
E-shockwave.
shaft.
Electric field shockwave generator.
It can be placed here as shown, to assist in starting the cones (of a gravity wave generator) spinning. (or for maintaining the spin).
Page P11

FIGURE 30

D&R Scientific Pub. Co.; Denver, CO Jan. 2003. (originally publ. on 04-17-2002) (Figs: 2-1-02)

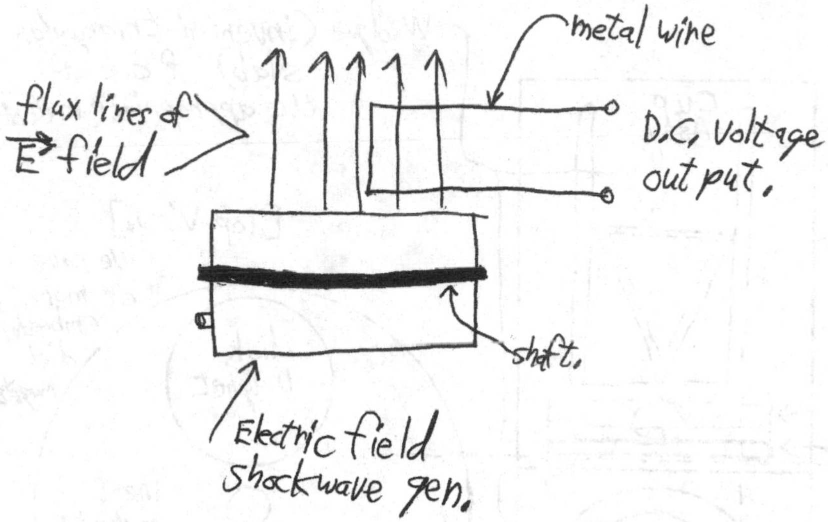

Electric Power Convertor

Page P12

FIGURE 31

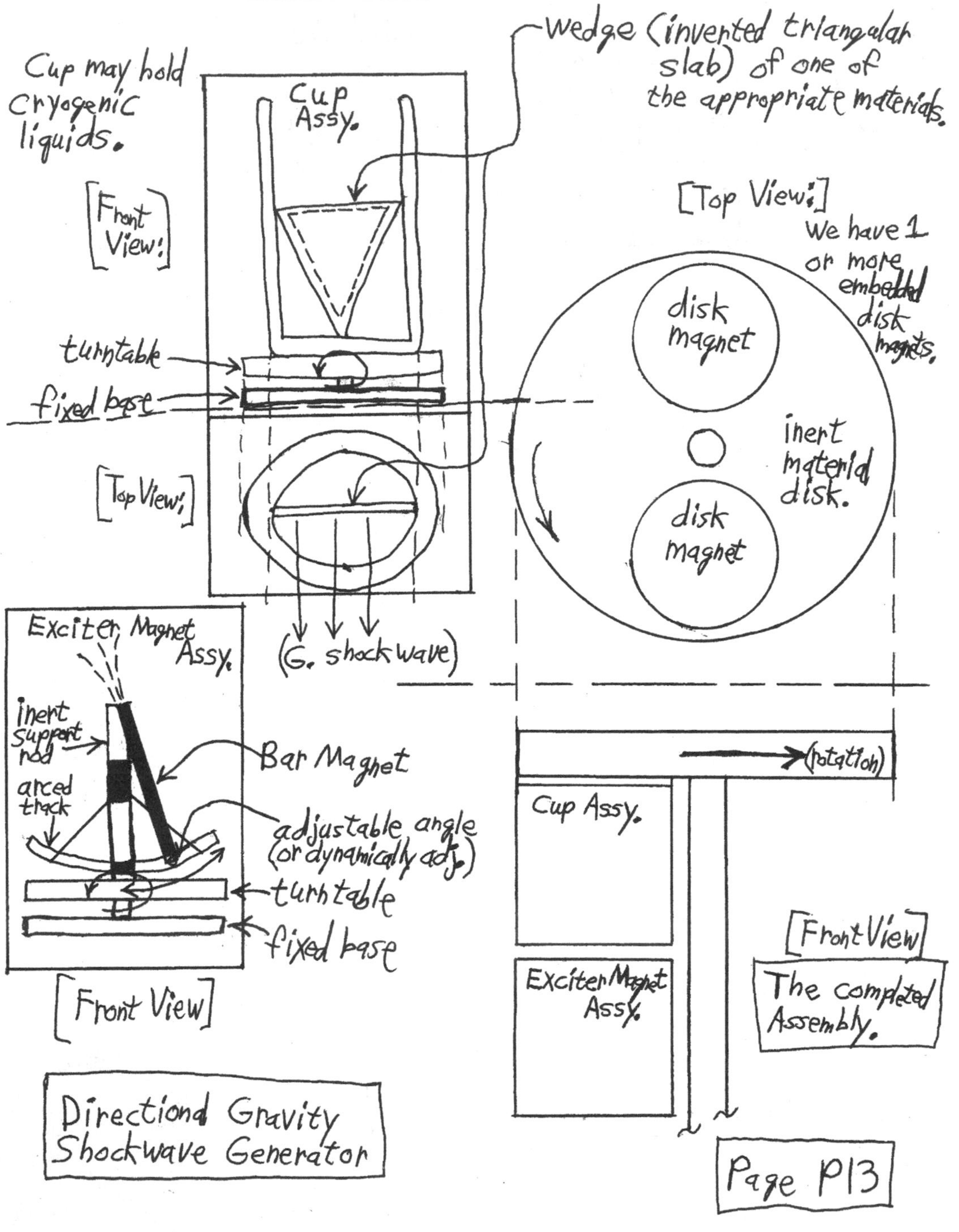
Cup may hold cryogenic liquids.
Cup Assy.
wedge (inverted triangular slab) of one of the appropriate materials.
[Front View:]
[Top View:]
We have 1 or more embedded disk magnets.
disk magnet
inert material disk.
disk magnet
turntable
fixed base
[Top View:]
(G. shockwave)
Exciter Magnet Assy.
inert support rod
Bar Magnet
arced track
adjustable angle (or dynamically adj.)
turntable
fixed base
[Front View]
(rotation)
Cup Assy.
Exciter Magnet Assy.
[Front View]
The completed Assembly.
Directional Gravity Shockwave Generator
Page P13

FIGURE 32

SIGNATURE PAGE SIGNATURE PAGE SIGNATURE PAGE

By signing below, I the INVENTOR swear and affirm that all of the information herein this complete document is complete, true, and accurate to the best of my knowledge:

Richard P. Crandall 12-31-2001

Richard P. Crandall, INVENTOR SS# 453 80 7642 Date Signed:

Notary Witness Information:
By signing and dating below, you swear and affirm and acknowledge that you were a witness to the act of the above named INVENTOR signing this document as you watched, and you have verified the identity of the above signer to be Richard P. Crandall by looking at the signer's Colorado driver's license and any additional required identification as you see fit.

SEAL OF TEXAS
COUNTY OF Randall
SUBSCRIBED & SWORN BEFORE ME
THIS 31 DAY OF Dec, 2001
Roberta Lynn Olguin
NOTARY PUBLIC

ROBERTA LYNN OLGUIN
MY COMMISSION EXPIRES
June 12, 2004

Notary Page (page P14) for Pages P1 – P13

FIGURE 33

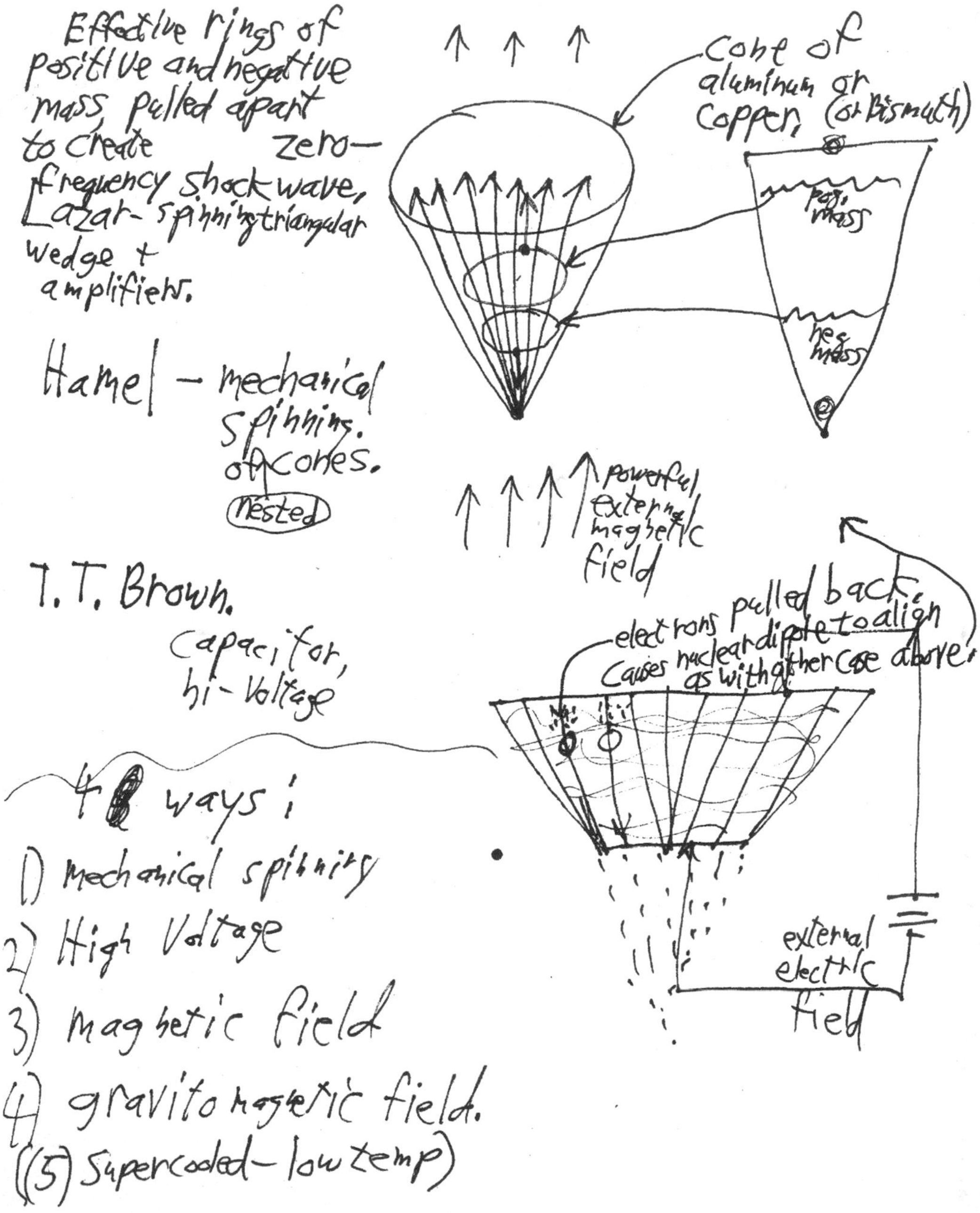
Effective rings of positive and negative mass, pulled apart to create zero-frequency shock wave, Lazar- spinning triangular wedge + amplifiers.
cone of aluminum or copper, (or Bismuth)
pos. mass
neg. mass
Hamel - mechanical spinning of cones.
nested
powerful external magnetic field
T. T. Brown. capacitor, hi-Voltage
electrons pulled back, causes nuclear dipole to align as with other case above.
external electric field
4 ways;
1) mechanical spinning
2) High Voltage
3) magnetic field
4) gravitomagnetic field.
((5) Supercooled - low temp)

FIGURE 34

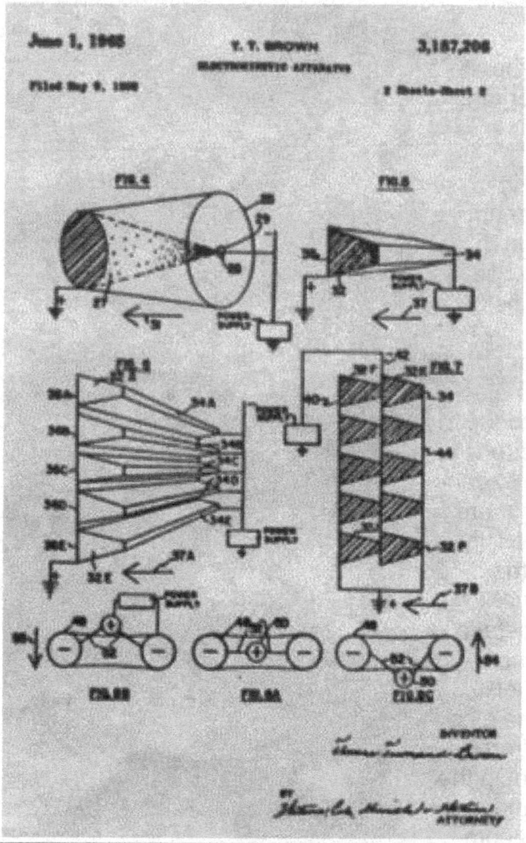

3,187,206
ELECTROKINETIC APPARATUS
Thomas Townsend Brown, Walkertown, N.C., assigner, by mesne assignments, to Electrokinetics, Inc., a corporation of Pennsylvania
Filed May 9, 1958, Ser. No. 734,342
23 Claims (Cl. 310-5)

This invention relates to an electrical device for producing a thrust by the direct operation of electrical fields.

I have discovered that a shaped electrical field may be employed to propel a device relative to its surroundings in a manner which is both novel and useful. Mechanical forces are created which move the device continuously in one direction while the masses making up the environment move in the opposite direction.

FIGURE 35

When the device is operated in a dielectric fluid medium, such as air, the forces of reaction appear to be present in that medium as well as on all solid material bodies making up the physical environment.

In a vacuum, the reaction forces appear on the solid environmental bodies, such as the walls of the vacuum chamber. The propelling force however is not reduced to zero when all environmental bodies are removed beyond the apparent effective range of the electrical field.

By attaching a pair of electrodes to opposite end of a dielectric member and connecting a source of high electrostatic potential to these electrodes, a force is produced in the direction of one electrode provided that electrode is of such configuration to cause the lines-of-force to converge steeply upon the other electrode. The force, therefore, is in a direction from the region of high flux density toward the region of low flux density, generally in the direction through the axis of the electrodes. The thrust produced by such a device is present if the electrostatic field gradient between the two electrodes is non-linear. This non-linearity of gradient may result from a difference in the configuration of the electrodes, from the electrical potential and/or polarity of adjacent bodies, from the shape of the dielectric member, from a gradient in the density, electric conductivity, electric permittivity and manetic permeability of the dielectric member or a combination of these factors.

A basic device for producing force by means of electrodes attached to a dielectric member is disclosed in my Patent 1,974,483. In one embodiment disclosed in my patent, an electrostatic motor comprises devices having a number of radially directed fins extended from one end of the dielectric body and a point electrode on the opposite end of the dielectric body. When this device is supported in a fluid medium, such as air, and a high electrostatic potential is applied between the two electrodes, a thrust is produced in the direction of the end to which the fins are attached.

Other electrostatic devices for producing thrust are disclosed and described in detail in my British Patent 300,-311, issued August 15, 1927.

FIGURE 36

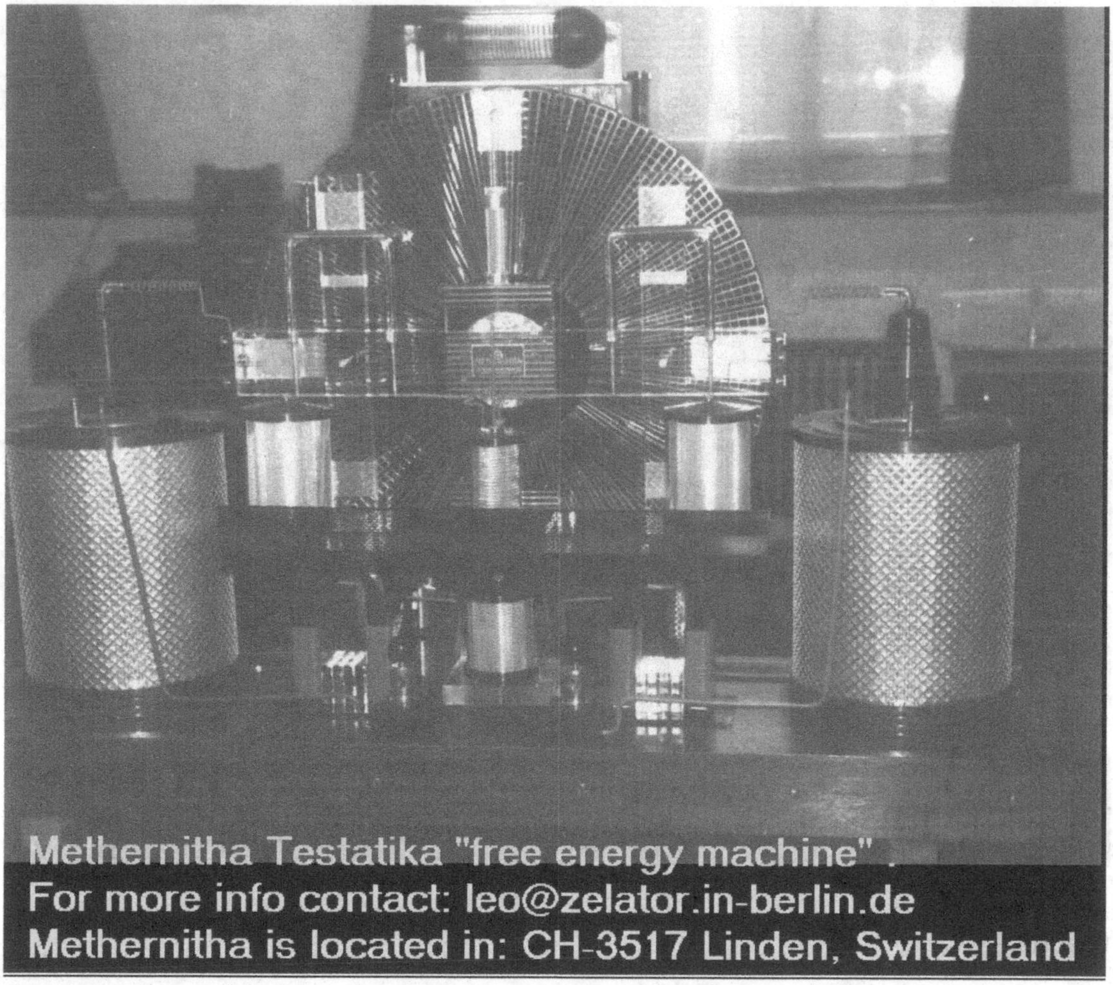
Methernitha Testatika "free energy machine" .
For more info contact: leo@zelator.in-berlin.de
Methernitha is located in: CH-3517 Linden, Switzerland

FIGURE 37

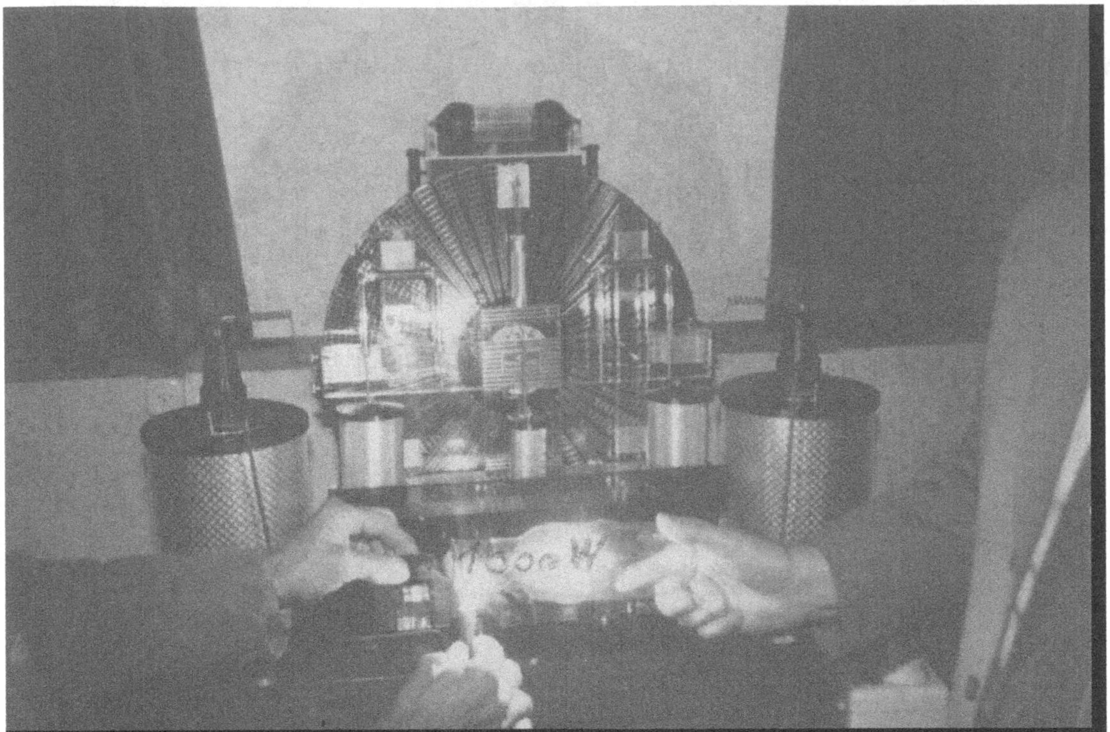

Methernitha Testatika "free energy machine" lightning a
1000 Watts bulb, for more info contact: leo@zelator.in-berlin.de

FIGURE 38

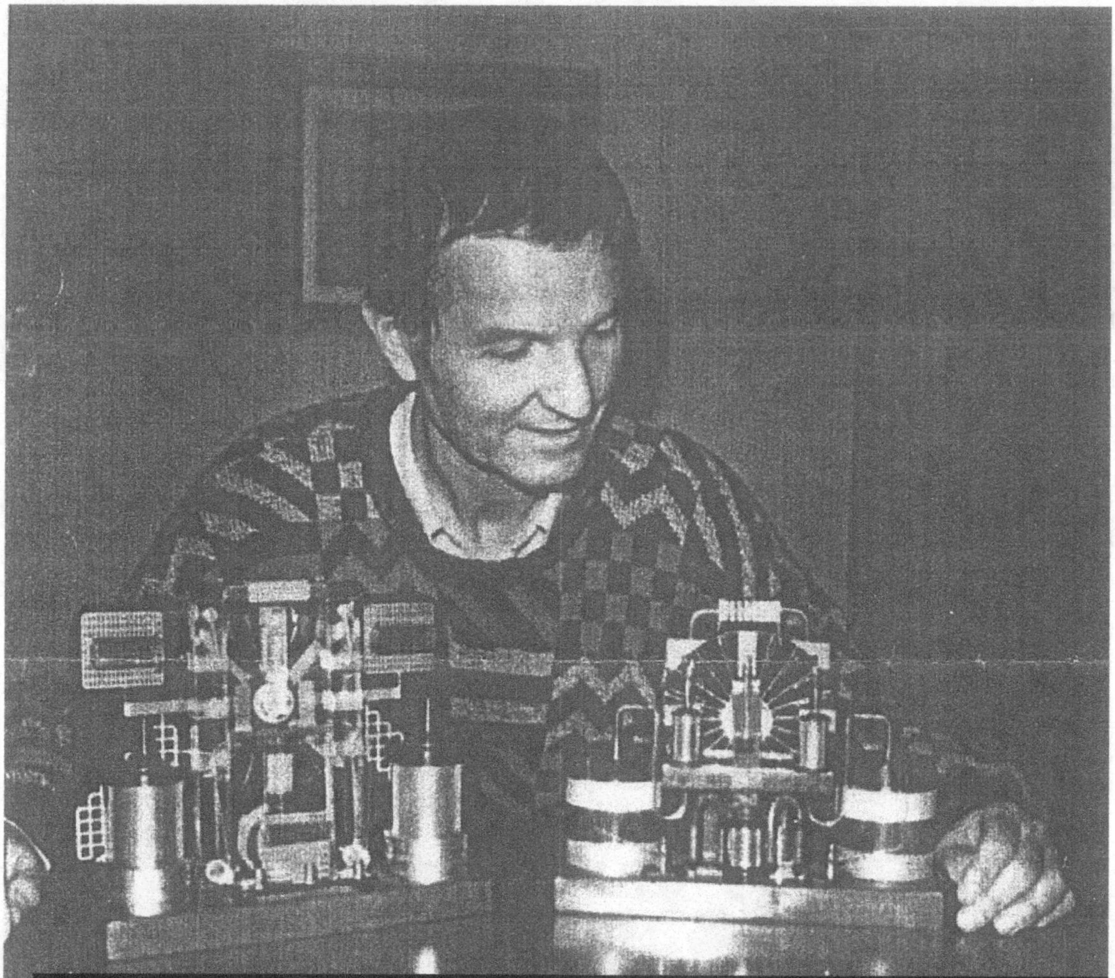

Prof. Stefan Marinov behind the 2 small 300 Watts
Methernitha Testatika "free energy machine" models.
He has tested the machines and stated that they work.
For more info contact: leo@zelator.in-berlin.de

FIGURE 39

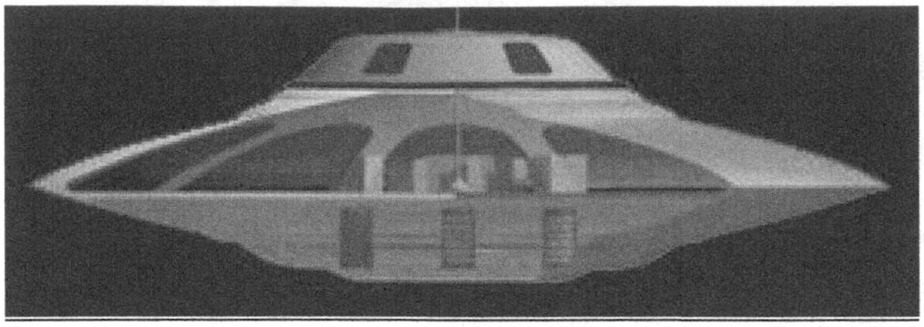

FIGURE 40

D&R Scientific Pub. Co.; Denver, CO Jan. 2003. (originally publ. on 04-17-2002) (Figs: 2-1-02)

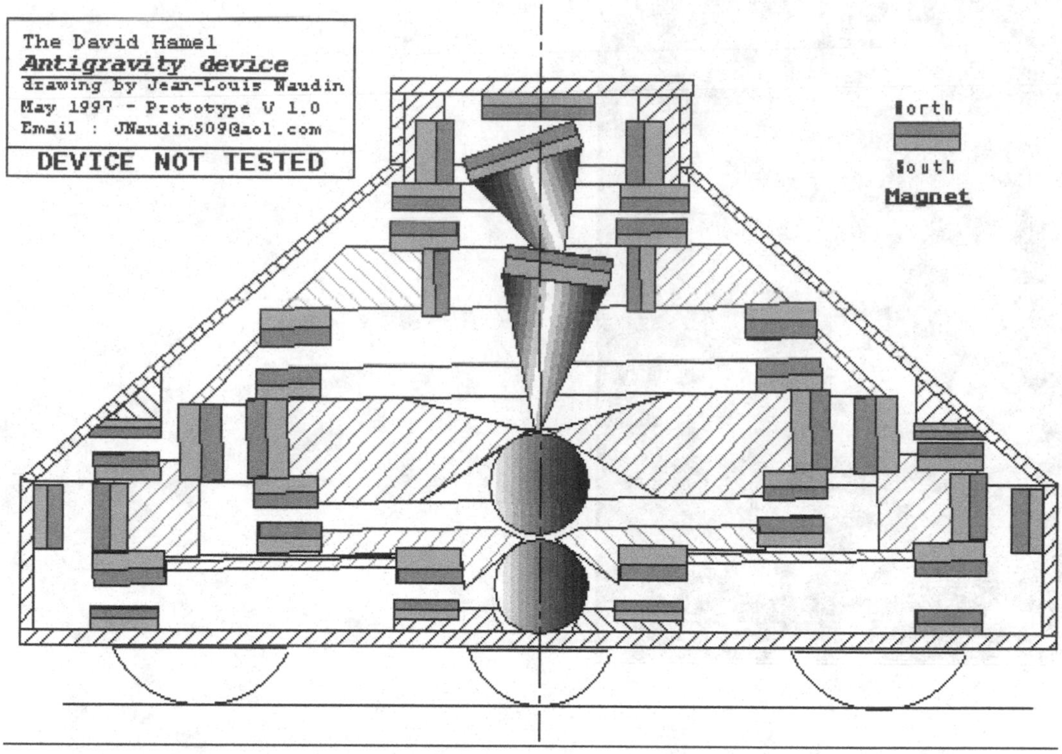

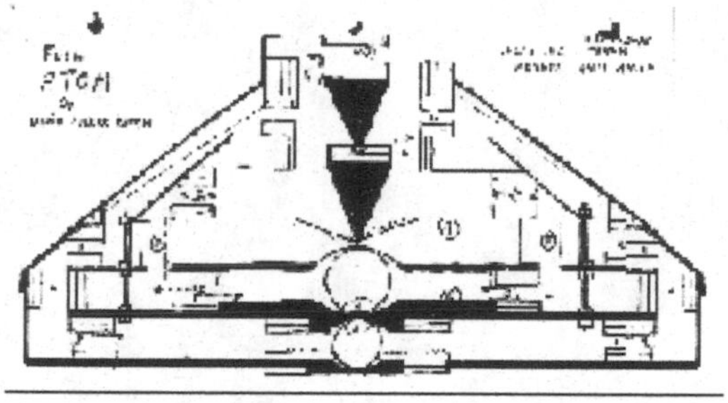

FIGURE 41

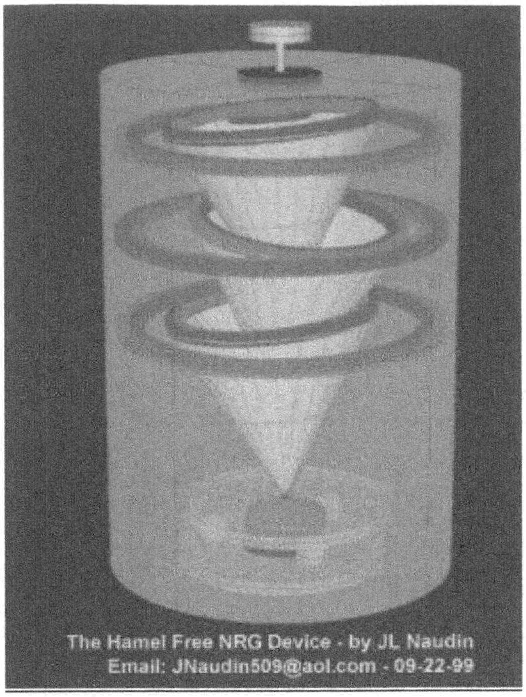
The Hamel Free NRG Device - by JL Naudin
Email: JNaudin509@aol.com - 09-22-99

FIGURE 42

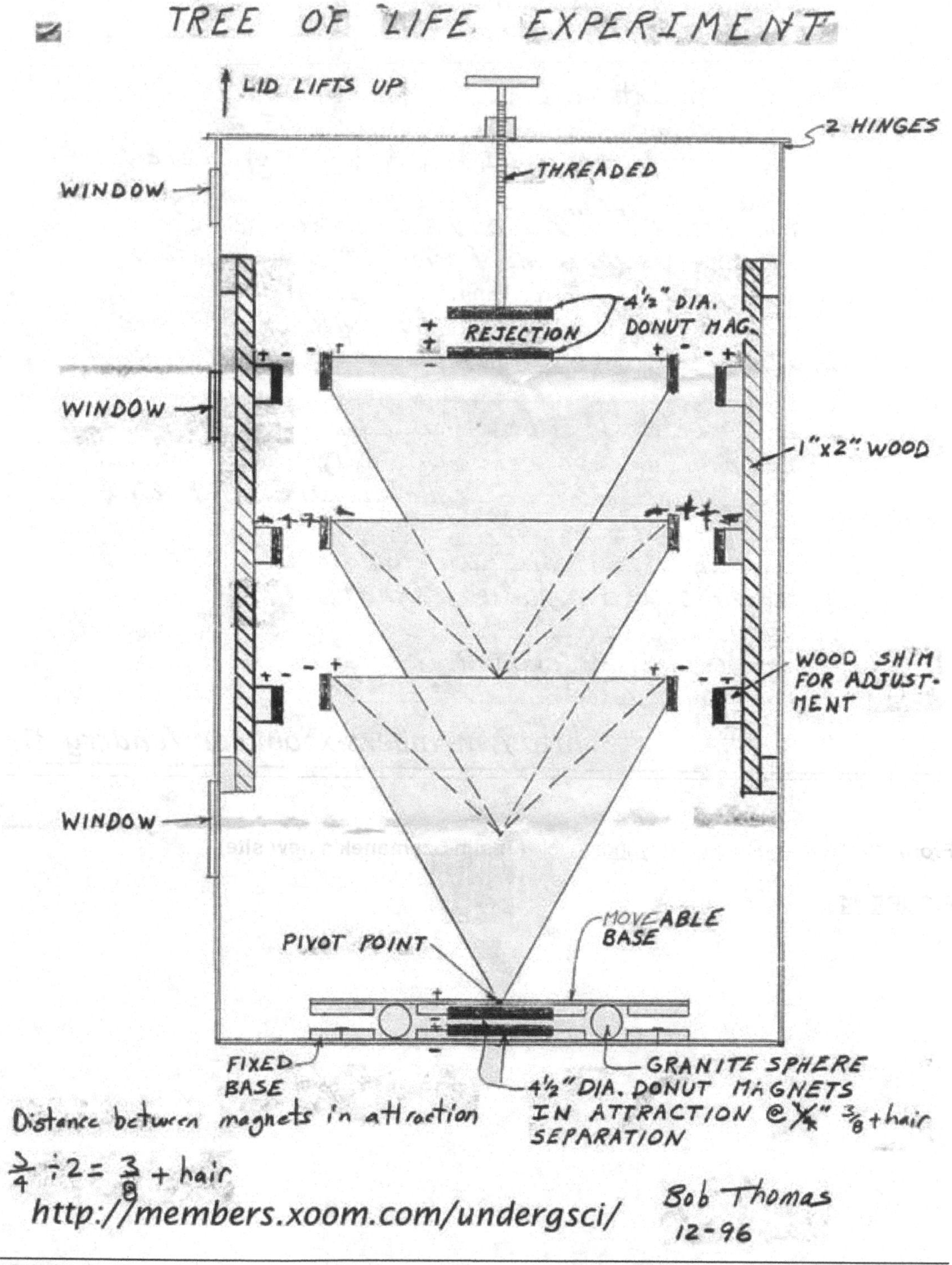

FIGURE 43 From http://www.geocities.com/undergsci/ (Justin Szymanek's new site).

Cones : Bicycle rim with metal band
14 5/8" Dia.

height of cones 13 5/8"
cone dia. with magnets 15 9/16"
center cone 1/2 height of full cone

magnets : 7/8" x 1 7/8" x 3/8" Ceramic 5
Berium Ferite @ 3/16" separation on
metal bands.
The magnets on the cones and cylinder are
attached to a 1/32" thick steel band. Glued + taped

Maximum oscillation time : 1 5/8" - separation
between mag. on cone + cylinder, donut magnet
separation - 2 1/4". Oscillation time 14 min.

Match low point of oscillating cone
magnets and cylinder magnets.

24" O.D. steel cylinder
1/4" thick steel

http://members.xoom.com/undergsci/

From http://www.geocities.com/undergsci/ **(Justin Szymanek's new site).**

FIGURE 44

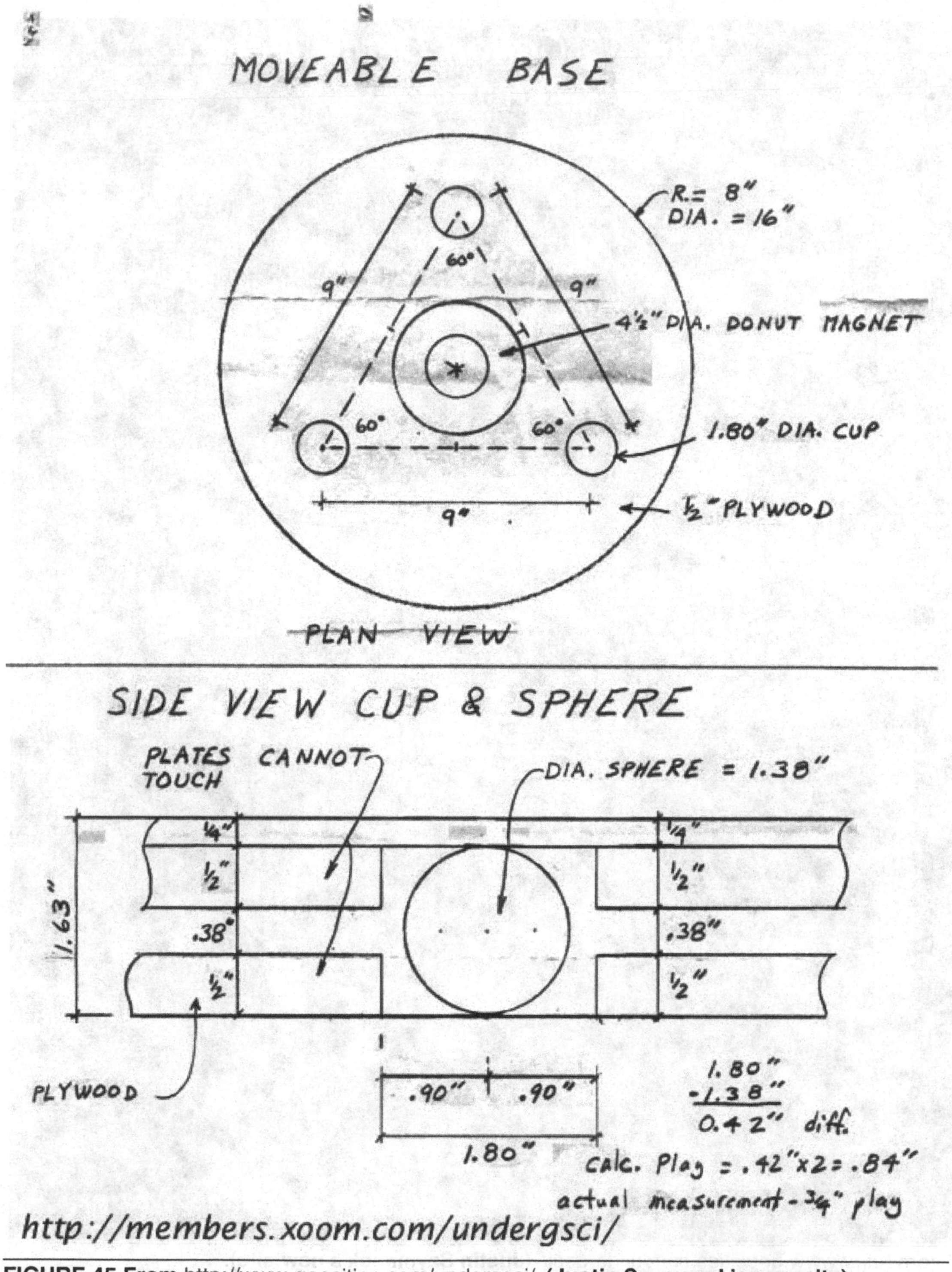

FIGURE 45 From http://www.geocities.com/undergsci/ **(Justin Szymanek's new site).**

From http://www.geocities.com/undergsci/ **(Justin Szymanek's new site).**
FIGURE 46

From http://www.geocities.com/undergsci/ **(Justin Szymanek's new site).**

FIGURE 47

From http://www.geocities.com/undergsci/ **(Justin Szymanek's new site).**

FIGURE 48

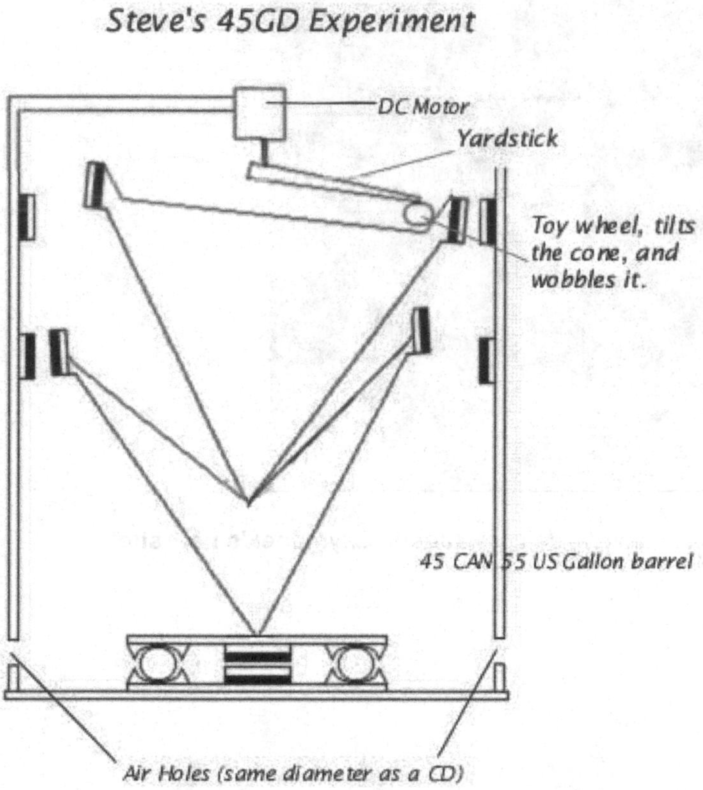

From http://www.geocities.com/undergsci/ **(Justin Szymanek's new site).**

FIGURE 49

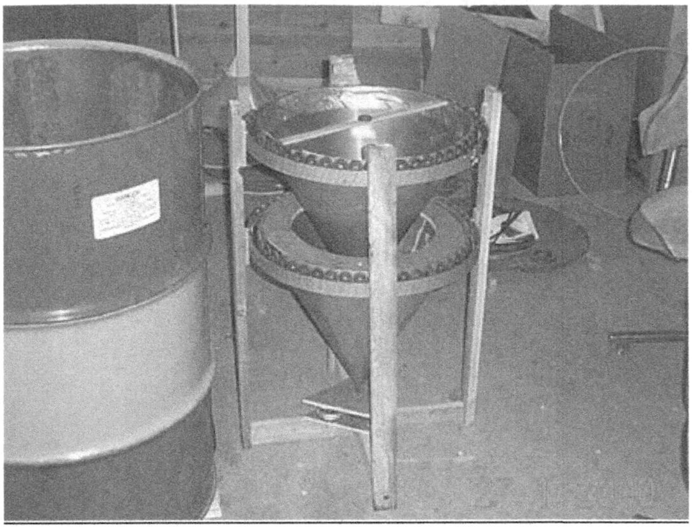

From http://www.geocities.com/undergsci/ **(Justin Szymanek's new site).**

FIGURE 50

D&R Scientific Pub. Co.; Denver, CO Jan. 2003. (originally publ. on 04-17-2002) (Figs: 2-1-02)

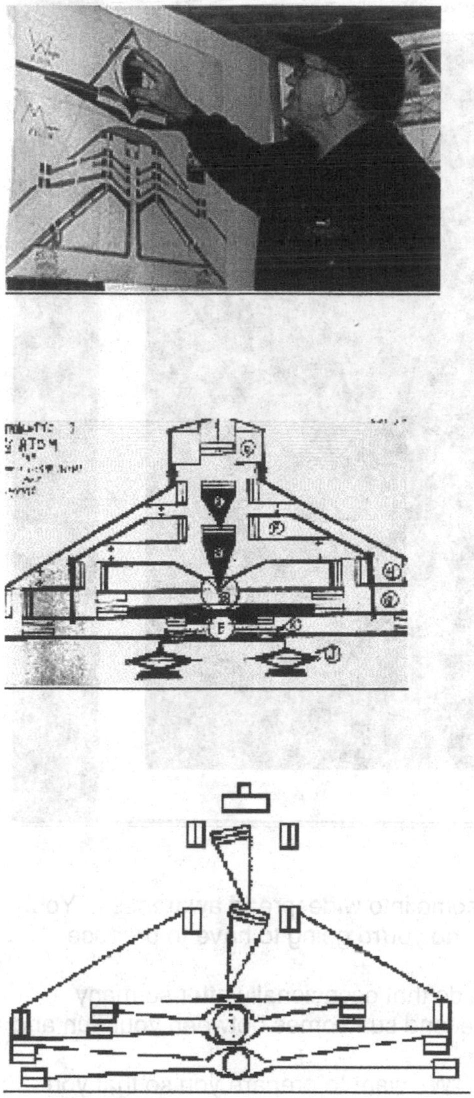

FIGURE 51

60 THE GRANITE MAN AND THE BUTTERFLY

THE GRANITE MAN AND THE BUTTERFLY 61

[redacted]

"There is a whole field of knowledge that has to come into widespread awareness. Your population has exceeded the number that you can feed, and you're going to have to produce more food."

"And there is the recycling of the planet. Planets do that occasionally after so many thousands of years. This will happen to Earth after the second sun comes between your sun and the planet."

The speaker paused to get David's full attention. "We want to prepare you so that you can build a ship to elevate people away from the planet during that recycling time."

Regardless of whether planet Earth would "recycle", an equally important reason for learning how to build energy machines, they told him, was to preserve Earth's resources and to learn how to respect nature. As well, humanity could then speed great distances in almost no time – travel to other planets and galaxies and thus preserve its species.

<u>FIGURE 52</u>

D&R Scientific Pub. Co.; Denver, CO Jan. 2003. (originally publ. on 04-17-2002) (Figs: 2-1-02)

[redacted]

"Me? Build a spaceship? Then what?"

"The ship would ferry thousands of people into space to another planet," they replied. "Don't think that Earth will stay the way it is, because everything will be changed."

Had he learned enough about the unusual propulsion system to build a ship? He understood that the energy came from magnets, and that the energy is universally available.

"It's a power that Earth people don't know about, but the knowledge has to become common, for the good of all," they said.

FIGURE 53

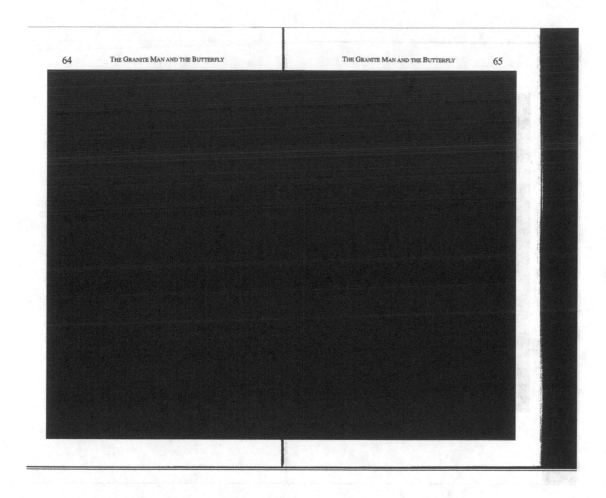

154 THE GRANITE MAN AND THE BUTTERFLY

Much as he needed financial help after so many years of taking care of an invalid wife without that help, David had his mind on bigger concerns. Regarding possible earth changes, he said, "In twenty more years everything on earth will be devastated. It's the magnetic that will change. I was told that when the lineup of planets and the second sun passes for three days and three nights in front of our sun and blacks it out, the magnetic will change. Between the years 2000 and 2005."

A number of people have asked David Hamel to speculate on the identity of the beings who gave him these prophecies and took him on the fantastic flight on October 12, 1975. In 1977, searching for the words to reply to reporter Ron Thody, David said that God had helped him through "...angels. The astronauts on board that ship."

FIGURE 54

D&R Scientific Publishing Co.; Denver, CO 02-01-2002

A New Theory of Quark Substructure for Mesons and Baryons

Richard P. Crandall
Dept. of Physics & Research, SPC
Received 01-06-02

ABSTRACT:

This new theory predicts & explains many of the major features of quantum mechanics, quantum electrodynamics, quark theory, and electromagnetic theory, by starting with simpler, lower-level principles. The entire theory is based upon a linearized (i.e. weak-field) solution of the general Einstein tensor equation for matter-induced curvature of spacetime. This set of equations looks very similar to Maxwell's equations of Electrodynamics. The equations show the existence of what are termed "gravitomagnetic" fields and "gravitoelectric" (just plain, ordinary gravity) fields. The theory accomplishes the following:

- Predicts the masses of 18 particles (a mix of common mesons and baryons). The error is only a few percent for most of them. For one particle, it's 0.9%.
- Finds a never-before-discovered, very simple pattern regarding all these masses. Amazingly, the square root of the mass of any particle, divided by the square root of the mass of the electron, is always an integer (approximately). More amazingly, the result (rounded to an exact integer) is always evenly divisible by either 7 or 8. See figure A_MASS, or investigate it for yourself. The value for the proton is 42, which is 7 x 6.
- Therefore, the mass of any particle is proportional to N squared, where N is an integer. I've discovered that N is equal to the number of "protoparticles" (a new concept; I've named them "vacuons") of which any given particle is comprised of. The reason for the mass equaling the square of N instead of N is because the measured mass of any particle is found to actually equal the amount of energy stored in a localized gravito-magnetic field (divided by C squared, of course). If a particle is visualized as a circular mass current of N protoparticles, then the amount of gravito-magnetic field energy present is proportional to N squared, just as the energy stored in an electrical current-carrying solenoid is proportional to the electrical current squared.
- **Derives (predicts) most of the major features of Quantum Mechanics (QM), Quantum Electrodynamics (QED), quark theory (QCD), Classical Electrodynamics (EM) theory, and also Special Relativity (SR), starting from simpler principles.** In this new proposed theory, though, General Relativity (GR) is a given and is assumed to be valid for these models right from the beginning. These predicted features are as follows:
- (QM and some QED) The Dirac relativistic theory of fermions. Relativistic DeBroglie waves (probabilistic). Quantum energy states diagram.
- (QM) A particle's wave (state) function squared gives the probability of finding the particle at any location. The wave function's amplitude (if a pure sinusoid) is simply equal to the radius of the "protoparticle" dynamic structure. The wave function frequency (if a pure sinusoid) is equal to the usual frequency of the particle's relativistic DeBroglie wave.

- (QM) The physical reason for the "collapse" of the wave function upon taking measurement. (This has previously been unknown).
- (QM) The true physical origin of spin, being typically one-half of "h-bar". (This has previously been unknown).
- (EM & QCD) The true physical origin of "electric charge". (Previously unknown). The true physical origin for the values of the "multiple of one third" signed charges of each of the 3 quarks in the proton and also the neutron is shown (Previously unknown).
- (EM) Maxwell's equations, derived from simpler principles. Shows what the electromagnetic field is physically composed of: a stream of "protoparticle" dynamic structures, which we see to be extremely low frequency photons.
- (QCD) **The exact radius of the Proton is predicted / calculated.** No other theory has come so close to the actual measured value.
- (EM) The physical reason for attractive photon exchanges between particles of mass. It reveals the concept of negative linear momentum. (This has previously been unknown).
- (QM and QED) The physical reason that explains the Pauli Exclusion Principle. (Previously unknown). It's due to gravitomagnetism.
- (QCD) The strong nuclear force. It is shown to be none other than a (very, very strong) gravitomagnetic field. (Previously unknown).
- (EM) The electric and magnetic force. Their physical origin, previously unknown, is explained from simpler principles.
- (SR) The Lorentz transformation, and the origin of it.

The following lists some of the discoveries and predictions of this new theory of matter:

- Gravity-based propulsion systems can be built, since we now know that nuclei of atoms contain extremely strong gravitomagnetic fields, which can be converted to intense gravitoelectric (i.e. just ordinary gravity) fields.
- Quarks can be pulled apart to large macroscopic distances.
- The mass of any particle can be reduced (even to near-zero) by forcibly enlarging the diameter of the particle's "protoparticle dynamic substructure".
- Protoparticle dynamic substructures are held together at equilibrium radius by a balance of forces, namely centripetal and centrifugal forces.
- Relativistic total energy and momentum are always conserved, as expected.

INTRODUCTION:

Before proceeding on, it's very helpful (and important) to first quickly look at the 5 excerpted scanned Figures, below.

MAIN BODY OF THE THEORY :

FORMAT:

The body of this theory is currently a **mixed document,** as follows. A few pages beyond this point (i.e., after the 5 copies of excerpted figures), the reader will find **"A New Theory of Quark Substructure for Mesons and Baryons, continued",** which is a **typed** version of the first 42 pages of my handwritten theory notebook, which I called The "UFT" (Unified Field Theory) pages. It contains mostly Postulates and then the main text of the theory. And following that, you will see the scanned images of my entire handwritten theory notebook (except for its **appendix** DL and DNSS pages) containing the theory and then all the Figures (A_MASS, & Figures A1 through A27). The reasons for this format of presentation are multiple: a document so written is not easy to alter, my signature can appear on every page, and the reader gets to see my scratchouts and deletions, so the reader can see that creating and then writing down the theory wasn't just a one-pass operation. It took months of thinking, many wrong detours, and a lot of extra "scratch paper".

However, the primary reason for the handwritten format is because of the time factor. It is intended for this to be a RUSH EDITION, version 1, of the theory. Hopefully later it will be redone as more of a publication quality work.

The first few pages of the "UFT" pages are admittedly a bit artificial, but please KEEP READING, as the much more concretely-based, mathematical equation-oriented, parts follow this.

The **appendix** material (the DL, and DNSS ,and L, and NSS pages) from the theory notebook, which is also a very important part of this theory, is presented in a later section.

The theory then ends with "FINAL CONCLUSIONS" [on the next page].

FINAL CONCLUSIONS:

Assuming the reader has read at least the entire set of typed "UFT" pages and has seen the "Figure" pages, he/she should see that this theory is certainly well-founded, and it is very much worth further investigation and extension. Also, the reader may have seen that there are the possibility of experiments which can be done to verify the theory. These experiments are related to the generation of gravitoelectric waves for propulsion, and also the possible generation of "free", or at least low-cost, electrical power. Just don't forget: relativistic momentum and energy are always conserved at the lowest level.

The following pages contain the 5 selected Figures, which have been excerpted from later parts of the theory.

A FIGURE FROM THE THEORY: (Einstein's GR equation can be solved in the context of a weak-field, linearized situation, as follows. It looks very similar to Maxwell's Equations.)

Potentials:

$$\vec{H}_g = \vec{\nabla} \times \vec{A}_g \qquad \gamma = \frac{1}{\sqrt{1-\frac{u^2}{c^2}}}$$

$$\vec{E}_g = -\vec{\nabla}\phi - \mu_g \frac{\partial \vec{A}_g}{\partial t} \qquad \rho_m = \text{mass density in } \frac{kg}{m^3}$$

"Lorentz" force law:

Force on "$m_o\gamma$" $\Rightarrow$ $\vec{P} = m_o\gamma\vec{E}_g + m_o\gamma\vec{u}\times(\mu_g\vec{H}_g)$

The linearized field equations for the electro-gravitational and magneto-gravitational fields:

$$\vec{\nabla}\times\vec{H}_g = \frac{\partial(\epsilon_g\vec{E}_g)}{\partial t} - 4(\rho_m\gamma\vec{u})$$

$$\vec{\nabla}\cdot(\epsilon_g\vec{E}_g) = -(\rho_m\gamma)$$

$$\vec{\nabla}\times\vec{E}_g = -\frac{\partial(\mu_g\vec{H}_g)}{\partial t}$$

$$\vec{\nabla}\cdot(\mu_g\vec{H}_g) = 0$$

$$G = 6.67\times10^{-11}\ \frac{N\cdot m^2}{kg^2}$$

$$G = \frac{1}{4\pi\epsilon_g}$$

$$c = \frac{1}{\sqrt{\mu_g\epsilon_g}} \cong 3\times10^8$$

$$\therefore \mu_g = 9.37\times10^{-27}\ \frac{N\cdot s^2}{kg^2}$$

Richard P. Crandall 12-24-2001

Figure A1

The governing field equations. The ordinary, or gravitoelectric, field is $\vec{E}_g$ and the gravitomagnetic field is $\vec{H}_g$.

A FIGURE FROM THE THEORY: (the explanation of QM effects, and negative mass)

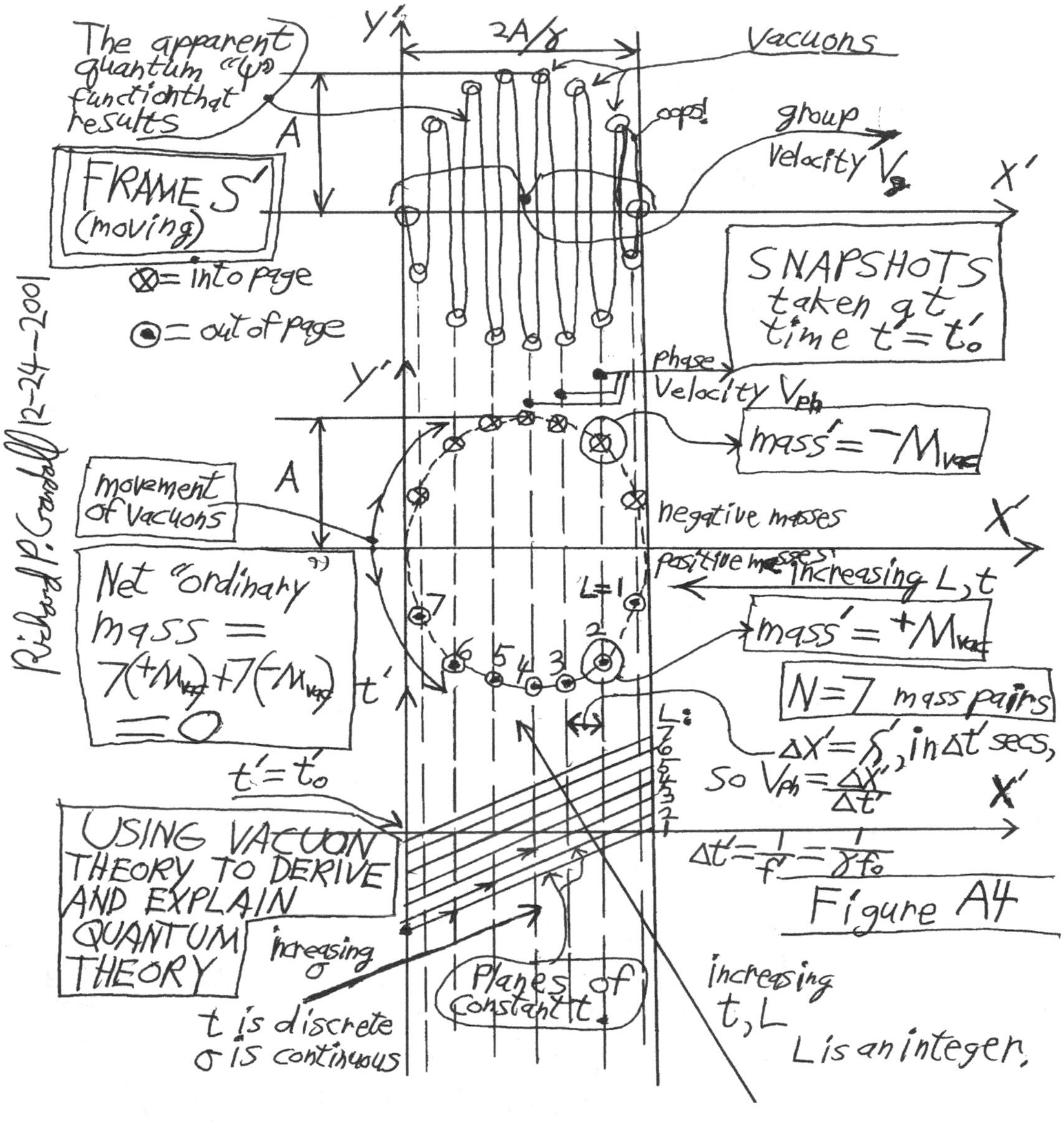

A FIGURE FROM THE THEORY: (the 3-quark substructure of the proton and neutron)

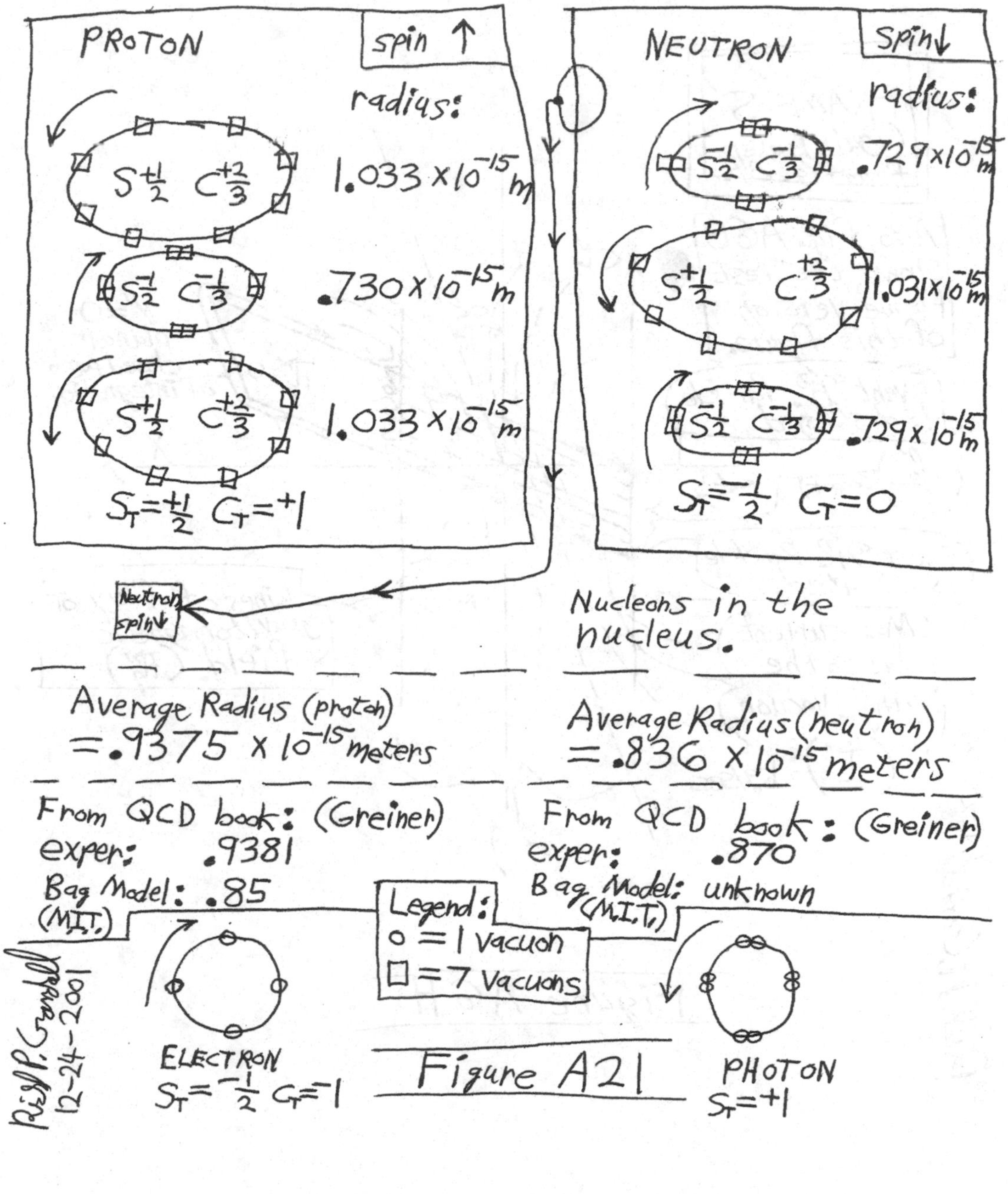

Figure A21

A FIGURE FROM THE THEORY: (gravitomagnetic field calculations)

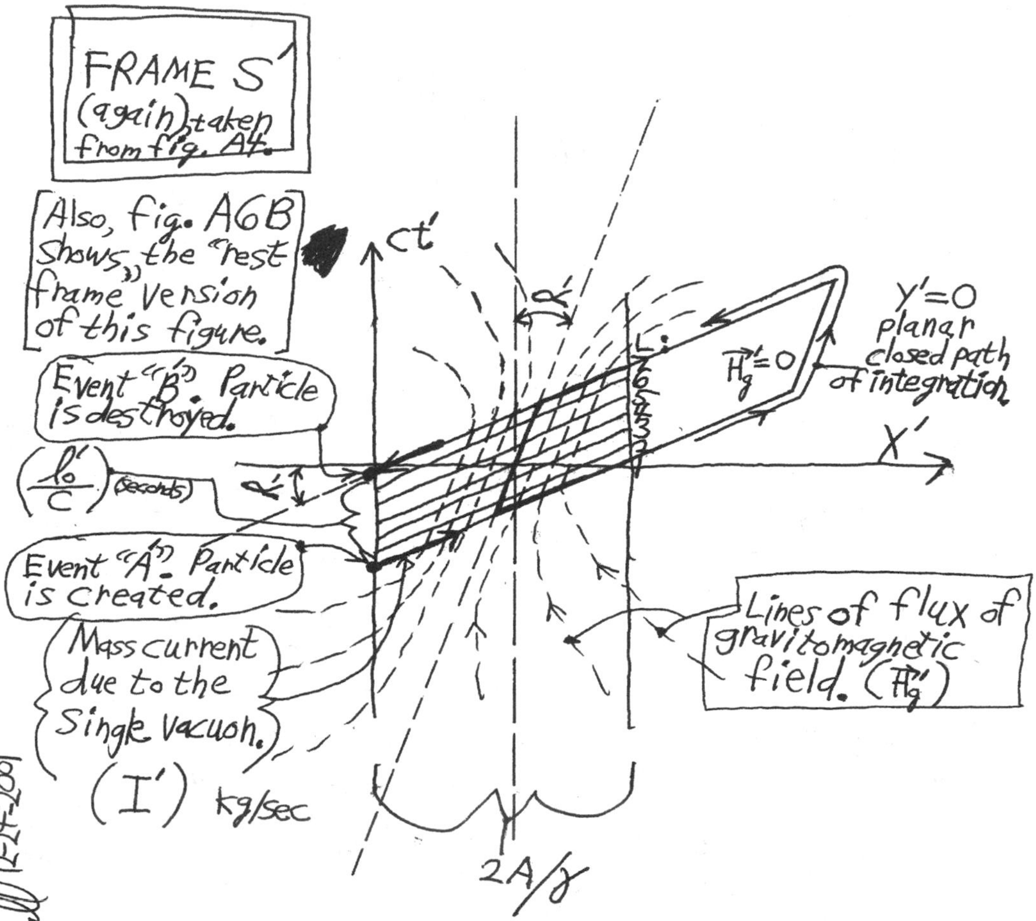

Figure A6H

Richard P. Gandall 12-24-2001

A FIGURE FROM THE THEORY: (particle mass worksheet from lab notebook)

e^- electron	MeV mass (0000.51109)	$\sqrt{m_0/m_e}$ (1;1x1x1)	formula; N_L/N_E, xQ, xN_L	total;	% err on $\sqrt{\ }$;	mass, per formula	% err on mass
μ^+	0105.7	14.38	7x2x1	14	+2.7	0100.2	+5.5
π^+	0139.6	16.53	8x2x1	16	+3.3	0130.8	+6.7
π^0	0135.0	16.25	8x2x1	16	+1.6	0130.8	+3.2
K^+	0493.7	31.08	8x2x2	32	-2.87	0523.4	-5.7
K^0	0497.7	31.21	8x2x2	32	-2.47	0523.4	-4.9
η^0	0548.8	32.77	8x2x2	32	+2.41	0523.4	+4.9
Proton	0938.3	42.85	7x3x2	42	+2.02	0901.6	+4.1
Neutron	0939.6	42.88	7x3x2	42	+2.10	0901.6	+4.2
Λ^0	1116	46.73	8x3x2	48	-2.65	1178	-5.3
Σ^+	1189	48.23	8x3x2	48	+0.48	1178	+0.9
Σ^-	1197	48.39	8x3x2	48	+0.81	1178	+1.6
Xi(avg)	1318	50.78	8x3x2	48	+5.79	1178	+11.9
Ω^-	1672	57.20	7x3x3	63	-9.21	2029	-17.6
Λ_c^+	2273	66.69	7x3x3	63	+5.86	2029	+12.0
D(avg)	1866	60.42	7x2x4	56	+7.89	1603	+16.4
J/ψ	3097	77.84	8x2x5	80	-2.70	3271	-5.3
B	5260	101.45	8x2x6	96	+5.68	4710	+11.7
Υ	9460	136.05	8x2x8	128	+6.29	8374	+13.0

Richard P. Crandall 12-24-2001

FIGURE A_MASS

↑ Halliday

Fritzsch ↓

Used these to calc. .51109 above

(6.29)

N_L turns out to be 4xQ

$c = 2.998 \times 10^8$

OVER→ (1 Joule = 6.2422×10^{12} MeV $m_e = 9.109534 \times 10^{-31}$

[This Page is Intentionally Blank]

A New Theory of Quark Substructure for Mesons and Baryons, continued

A Unified Field Theory ("UFT" Pages)

by Richard P. Crandall

POSTULATES :

I. At this level of description, the following do not exist: ordinary subatomic particles of any kind (examples: electrons, protons, mesons, baryons, quarks, antiparticles such as the positron, etc...), photons, any aspect of electromagnetism (including charge, fields, etc...), mass, force, energy, momentum, any theory of quantum mechanics (QM, QED, QCD), special relativity theory, and general relativity theory.

II. However, the following do exist: time (in an absolute sense, but this will later be disposed of, in favor of Relativity) and 4 orthogonal dimensions of space (in each of which direction, length may be measured). Our local region of the universe mainly consists of a "substrate" or "background" which will be referred to as "the vacuum." All particles, entities, and concepts exist relative to the vacuum, and will be derived and explained based upon the simple properties of the vacuum.

III. The vacuum is not empty. Even regions of it devoid of all photons, energies, or particles or any other <u>observable</u> phenomena are not truly empty. The vacuum is actually seething with proto-matter (or proto-particles) which are unobservable due to their large number density and their even distribution and their high velocities. These particles (proto-particles) will be referred to as **"vacuons"** (VAC-You-Ons).

IV. More specifically, the substrate of the vacuum will be considered to be <u>analogous</u> (for purposes of understanding) to a 4-dimensional grid of empty cubes (or other shape) (or, more specifically, 4-cubes), where the edges of such cubes are all that exists and these are made of a highly nonlinear elastic-like substance. The insides of the cubes are considered truly empty. The grid is "teaming" (i.e. extremely active) with only <u>two</u> things: violent waves (random), and the many violently moving proto-particles called vacuons. The vacuons are seen as being simply knots, kinks, or what-have-you, in this "rubbery" grid. These kinks, or twists, are assumed to be indestructible and non-changing. They are also all assumed to be identical, for the present purposes of this theory.

V. The current theory can be used to derive all properties of matter related to Relativity and Q.M. and Q.E.D. The base theory may later be amended by the inclusion of different types of vacuons as well as more dimensions of space, as may be needed to fully explain Q.C.D.

VI. The reason for the artifice of introducing such a regular grid is so that each vacuon could be visualized as the result of pressing two cubes, or sides of cubes, together. Such cubes would stick together and would then exist permanently. Such pairs of stuck-together cubes would account for all vacuons being <u>identical</u> in all respects. Such stuck pairs could also account for vacuons being un-changing and indestructible. It's quite possible that the "proto-mass" of a total of 2 cubes would, when confined to a single cube, create effectively a tiny "black-hole", which would of course last forever, but this speculative avenue will not be pursued in the present theory. The grid concept artifice will shortly be disposed with; just remember that all vacuons are assumed identical.

VII. Since the vacuum (a.k.a. the rubbery grid) contains many, many vacuons (identical knots or kinks) traveling over a distribution of very high speeds, the vacuum can be said to have inherent in it a high degree of what will be called "proto-energy" (all "kinetic" energy, there being no such thing as potential energy at this level). In fact, proto-energy will be seen to be an actual form of real energy. It will be seen that this proto-energy of the ("sea of") vacuons can act as an infinite source or sink of energy to the physical (observable) quantum particle domain. We will later see clear mechanisms by which real energy can be transferred back and forth between the vacuum (unobservable) domain and the observable physical domain of particles and photons.

VIII. Since the substance comprising the "grid" is highly non-linear (as General Relativity is non-linear), a range of velocities is available to the vacuons, instead of having all of them moving at a constant wave speed as in linear substances. It will be assumed that there exists a maximum speed for vacuons, which is >> C^2, (i.e. much, much larger than the speed of light value squared, based upon absolute time).

IX. The vacuum has three identical-like dimensions of space (except for being usually essentially orthogonal and being named differently), and one different (or "preferred") dimension of space. Therefore there is a total of 4 spacelike orthogonal dimensions. The nature of and reason for the fourth dimension of space being different from the other three is as follows. It is believed that the vacuon "sea" is actually more like a "river", whose "flow" is of very, very high velocity and is oriented in one particular direction -- that being the direction of the "preferred" or "special" fourth dimension of space. Also, it's possible that the "fabric" of the grid itself is highly and uniformly stretched in the fourth dimension's direction, but is not stretched nearly as much in the other 3 directions. Both the high-velocity directed flow of the vacuons and/or the possible stretching of the grid itself can possibly be explained as being the result of some type of large explosion or "Big Bang" which would certainly create a large directional outflux of vacuons and maybe additionally the stretching of the grid in that same direction.

An apparent illusion is created as a result of the 4th space dimension being "different." The illusion causes intelligent beings in the observable physical particle domain to attribute this fourth dimension to time itself. We beings can't see or detect or observe the fourth dimension at all, in the usual daily experiences of our lives. We only perceive 3 dimensions of space. However our modern theory of Special Relativity compels us to label the tick of our clocks as being some mysterious but physically real fourth dimension. Time itself is the fourth dimension, we say.

The mechanism, as best as it can be understood, which causes and creates the illusion will now be explained. This mechanism is such that the entire theory of Special Relativity can be derived from it.

The quantitative details are not worth detouring into, so a qualitative and simplified explanation will suffice. Due to the extreme average velocity of the vacuons in the direction of the 4th dimension, we can see that the vacuons do not interact or collide with each other very much at all in the downstream direction of the 4th dimension. This is because of the coherent alignment of the average velocity vectors of the vacuons. Particles (i.e. protoparticles (vacuons)) with the same velocity vectors (or velocity vector component in a particular direction) just aren't as likely to collide as often as particles with random velocity vectors. This can be seen upon drawing a simple diagram (now shown). So the most often of the collisions (or close interactions) are those that occur in the cross-stream direction of any of the 3 isotropic dimensions, as opposed to the downstream direction of the fourth dimension. The cross-stream direction consists of random velocity vector components, whereas the downstream direction consists of essentially equal velocity vector components.

So we see, for the most part, that collisions or interactions between the protoparticles, and thus the concepts of proto-force, proto-mass, & proto-acceleration with regard to the vacuons, occur only in the cross-stream directions, and not in the 4th dimension direction. Since very few or no interactions (proto-forces or proto-accelerations) occur in the 4th dimension direction, we cannot normally observe, view, or even detect this 4th dimension. We just see and interact with the usual three dimensions. We are constrained and confined to a narrow 3-D slice which rushes along through the 4th dimension.

So we can't see or detect this physically real and existing special 4th dimension, as we are confined to a (moving) 3-D slice of it.

The next question is why then do we physicists of Special Relativity treat this extra, but unobservable, dimension as being time itself, or as being a possible way for us to somehow travel back in time or fast forward in time? (In a physically real way?)

Here's why. Due to the gas-like nature of the vacuons (moving as they do at speeds approaching or equal to the maximum wave speed possible for the elastic grid, between "collisions" or close approaches), we see that a high speed individual vacuon will interact at a far greater rate with the other vacuons than will a slow speed individual vacuon. Due to this effect and also the non-linearity of the medium (i.e. the grid), we see that there should be a maximum average speed at which a "clump" of bound vacuons can travel, in the cross-stream direction. Also, due to a strange side-effect of the non-linearity, it may be possible for individual vacuons to travel at speeds greater than this "maximum" speed, on up to essentially infinite speeds (the large max. wavespeed of the grid, which is assumed to be >> C^2). This could be explained by the very short interaction times, which allow little effect to occur from the other nearby vacuons.

So for most particles (clumps of vacuons), there appears to be an upper maximum speed barrier, which we now recognize as being "C", the speed of light constant. Since we perceive a maximum limiting speed for most objects traveling within our little 3-D slice, then the entire theory of Special Relativity results from this one observation. We see that relative speeds of objects don't combine in an additive fashion (else they could in some cases appear to travel faster than "C"). The only explanation we have for this observation is that our rulers and clocks become changed or distorted depending on our relative velocities.

From this, the well-known Lorentz Transform formulas of Special Relativity are easily derived, in the usual way.

The net result, as is well-known, is that we are strongly compelled to view traveling back or forward (i.e. upstream or downstream) in this fourth dimension as being completely equivalent to physically traveling back or forward in time itself. The fourth dimension IS time. Thus born is the unified S.R. concept of "spacetime."

X. Again, regarding the formation of identical vacuons, we saw they could possibly be created by squeezing two cubes of the grid into one cube. Due to the postulated stretching of the grid lines in the timelike direction (i.e. in the fourth dimension), we see that a vacuon could only be created by squeezing the substrate in the 3-D spacelike direction, and NOT by squeezing the substrate along the timelike direction, as the grid lines in that direction are just too far apart.

At this point, the artifice of the grid may be deleted. We simply view our surroundings as being a substrate of elastic, nonlinear material, with many energetic vacuon knots embedded in it (analogous to a "gas" of such particles). The vacuon proto-particles are all identical, for the present theory.

XI. Thus, all the equations of Special Relativity apply to the vacuon "sea." The vacuons, being "proto-particles", can be seen as having properties such as "proto-mass", "proto-momentum" and "proto-energy." The only "forces" of interaction at all between vacuon "particles" are "proto-gravitational" in nature, following the ideas from the General Theory of Relativity, we postulate.

XII. It is possible that a simpler theory of General Relativity could be formulated such that it is applicable to having either 4 dimensions of space and absolute time, or 3 dimensions of space and absolute time where the 4th dimension is "inaccessible." However, this possibility is not worth detouring into for the purposes of this present theory.

Due to the fact we have the (illusory) Special Theory of Relativity in place which applies to vacuons, combined with this more primitive General Theory of Relativity for the vacuons, we conclude that the FULL theory of General Relativity and Special Relativity apply to the vacuons comprising the vacuon "sea." These two theories apply, exactly as written, to the vacuons, which can be viewed as very tiny particles, each of which is a tiny kink in or highly localized warping of spacetime (or "knot"). Then, according to G.R., each vacuon has an equivalent or "effective" mass which behaves in all respects just like real mass as we know it. It behaves like real mass, but it will be referred to as "proto-mass."

XIII. So the vacuons each have an "effective" mass called its "proto-mass." Therefore, Newton's law of force and acceleration can be applied, and we can define a quantity called "proto-force," which is equal to a "proto-mass" multiplied by acceleration (of a vacuon). Admittedly, we could do away with the concept of proto-force, since we postulated that all vacuons are the same and have the same proto-mass. After all, the only interaction between vacuons is gravitational, which amounts to having accelerations due to the warping of spacetime, and thus the concept of force is meaningless and unneeded. However, we will retain the concept of "proto-force," since we may later have to deal with clumps of several vacuons acting together as a single "proto-mass," and later changes to the theory may allow vacuons of different "proto-masses." We can thus define "proto-momentum" and "proto-energy" as their usual definitions as seen in the well-known theories of Newton's laws, the Special Theory of Relativity, and the General Theory of Relativity.

XIV. Since all "collisions" or interactions between vacuons are assumed to be perfectly elastic and lossless, we can state the following: the relativity formulas for conservation of momentum and conservation of energy apply to vacuons, exactly as they are written except substitute the words "proto-mass" for "mass" and "proto-momentum" for momentum and "proto-energy" for "energy." So at this unobservable level of the vacuons, which is below the level of quantum mechanics and its corresponding observables such as mass, energy, and electric charge, we see that proto-momentum and proto-energy are CONSERVED for any closed system of vacuon proto-particles, in the full sense of Special Relativity and General Relativity.

XV. Special and General Relativity must also apply at the higher levels of Quantum Mechanics and upward.

XVI. **THE PRIMARY POSTULATE.** (Yes, this is redundant with postulate XIV, but states it with less verbiage.) **At the subquantum level, which is the true vacuum, the quantity relativistic momentum IS CONSERVED involving all interactions of individual vacuons (the proto-particles of which all space and time is comprised of). It is called the "proto-momentum." Also, relativistic energy ("proto-energy") IS CONSERVED.**

XVII. In order to simplify calculations, and also understanding "collisions" between any two vacuons are defined to occur only at extremely, extremely close distances between the said two proto-particles. When vacuons are farther apart than that, we will assume no interaction occurs. Although the induced "proto-forces" and accelerations are actually continuous functions of space and time, again we can simplify understanding of further derived concepts this way.

XVIII. **THE COMPOSITION, OR SOLITON, POSTULATE.** Also at the subquantum level (which I will also refer to as the "subspace" level, or "vacuum" level, or "non-observable" level): It is postulated that all of the subatomic particles of physics (quantum electrodynamics and quantum chromodynamics) are formed from groups of two or more **co-orbiting** clusters of vacuons. This rule applies to photons as well. The simplest case would be two individual vacuons orbiting around a common point (the center of mass). Soliton theory states that stable dynamic structures can persist in a nonlinear medium. In this case, these structure would be the known observable particles of physics.

XIX. **THE ENERGY EXCHANGE, OR EQUILIBRIUM, POSTULATE.** For a given dynamic structure (i.e. particle) of vacuon protoparticles, there is **a preferred radius** and **a preferred orbital (actually co-orbital) velocity,** which represents a stable condition of **equilibrium.** These values of equilibrium tangential velocity and overall radius are measured from the rest-frame of the structure. When the structure is taken out of equilibrium by either collisions or by a gravitational field, it will **tend to return to its equilibrium condition,** once the external influence subsides. **This amounts to none-other than an energy exchange mechanism between any observable (i.e. quantum) particle of physics (in "normal" space) and "subspace" (which is the unobservable vacuon "sea" of protoparticles).**

This could also be viewed as an exchange or transfer of "proto-energy" (UNOBSERVABLE) to "real energy" (OBSERVABLE) and vice-versa.

XX. **THE GRAVITOMAGNETIC FIELD POSTULATE.** In a separate paper, I developed a weak-field, or linear, solution to the G.R. (Einstein) equation. These equations (except for a factor of -4 and -1) look identical to Maxwell's Equations of electromagnetism, where charge density and flow are replaced by mass density and flow, and where the electric and magnetic fields are replaced by the "gravitoelectric" and "gravitomagnetic" fields, respectively.

The gravitoelectric field is none-other than the usual gravity we are all familiar with. The gravitomagnetic field is the gravitational analog to the magnetic field. The familiar values of mu and epsilon are also replaced by their gravitational counterparts, denoted by the subscript small-"g". This postulate states that these "Maxwellized" linear equations of gravitation are the governing equations of motion which describe all of the interactions between the vacuons. The "proto-mass" of all the vacuons is the same, and is positive.

XXI. **THE REST MASS, OR FUNDAMENTAL, POSTULATE.** The measured rest mass (or "effective" mass) of any observable subatomic particle is equal to E divided by C squared, where E is the value of the total energy of the localized gravitomagnetic field generated by the rotation (i.e. co-orbital motion) of the particle's dynamic substructure (or soliton or "subspace" structure).

Thus, what we've always thought of as the "real" mass of a particle is not "real" at all! It's just the effective mass of the (immensely strong and localized) dipolar gravitomagnetic field which emanates from the subspace (i.e. unobservable) dynamic structure of vacuons which the particle is composed from.

Remembering how a rotating ring of static electric charge generates a dipolar magnetic field, it's easy to see how a rotating ring of massive vacuons would create a dipolar gravitomagnetic field. **The gravitomagnetic field is THE main connection we have between ordinary space and SUBSPACE.**

END OF POSTULATES.

Note, we don't normally detect the gravitoelectric fields of the vacuons, as they average out to zero.

THE MODERN THEORY OF QUANTUM MECHANICS CAN BE DERIVED FROM / PREDICTED BY THE MORE FUNDAMENTAL THEORIES OF SPECIAL RELATIVITY AND VACUONS :

Figure A1 depicts the linearized form, as was developed and derived in a separate paper, of the usual Einstein equation of General Relativity. Mass density is "rho-sub-m" and mass flow is "rho-sub-m" times the velocity vector "u". The variable "gamma" is the usual mass increase factor from special relativity. The "Lorentz" force law is also shown, as a way to detect and measure the gravitoelectric field vector "E-sub-g" and the gravitomagnetic field vector "H-sub-g". The constant "mu-sub-g" is derived from the ordinary gravitational constant "G". These equations will later be used to calculate the energy density and total energy of the gravitomagnetic field, which is postulated to account for the mass of any subatomic particle.

Figure A2 depicts an "idealized" version of the simplest possible stable dynamic structure of vacuons. The simplest case requires two vacuons, obviously, for dynamic balance with respect to the origin, but for the sake of simplicity of calculations, we use the ideal case of a single vacuon orbiting around the origin with a constant radius "A" and constant velocity. The vacuon possesses a mass (i.e. "proto-mass") which is "M-sub-VZERO", and therefore, when orbiting the origin, will produce a time-axis directed gravitomagnetic field. The rotation period is "B".

Figure A3 shows the well-known Lorentz Transform formulae of special relativity. The frame of reference "S" in that figure corresponds to the views we see in figure A2. It is a fixed frame of reference. The frame " S' " is a moving frame of reference traveling at velocity "V" in the X axis direction. The view of the orbiting vacuon that we see from that frame's viewpoint is as shown in figure A4.

In the moving frame viewpoint " S' ", instead of seeing a single proto-particle in a circular orbit, special relativity theory says we will see a flattened ring of many proto-particles. Refer to figure A4 in all of what follows, which is a detailed description of the apparent ring dynamic structure.

In figure A5, we begin the calculations which derive everything we need to know about the dynamic structure.
Refer to figure A5 through A5E now.
Note that this flattened ring, along with its wavelength, frequency, phase velocity, and group velocity is beginning to look like a relativistic DeBroglie wave (or wave packet) as is used in Dirac's relativistic theory of the electron (and other Fermion particles), which can be seen in most senior-level undergraduate texts on Elementary Quantum Mechanics. This is no accident, as we will later see also that

E' = (h) f' and P' = (h) / (lambda').

This will confirm Dirac's theory, and so vacuon theory explains Q. M. effects!

Before discussing tachyons, it is also important for the validation of Quantum Mechanics to show how the major radius "A" of the flattened ring structure can be seen as identical to the magnitude of a particle's quantum wave function "psi()". Since all we observers see is our present time slice at t-sub-zero' (since we can't normally observe the past or future at large) we can view the ring of vacuons as being like a (see fig. A4, top) wave function, which can be written mathematically as the product of an envelope function and a pure (constant amplitude and constant wavelength) sinusoidal wave. The envelope function takes care of the overall ring shape. What's left is the pure sinusoidal function. Its amplitude equals A. Imagine that we write this pure function as a complex-exponential pure wave (with amplitude A). Recall that the magnitude of a quantum wave function squared gives the probability of finding a particle there. This probability is essentially the same as the probability of an incoming (along the X' axis) particle hitting our test particle.

Well, that probability can easily be computed from seeing what cross-sectional area that our test particle presents to incoming projectiles with pure X' axis velocities. Such projectiles are assumed to arrive with random displacements from the X' axis (both above and below). Also, we must recognize that in reality, there is a third, or "Z" dimension of space. Rotating our test particle's wave function about the X' axis then gives us the correct and complete view of what cross-sectional area that our test particle presents to the incoming projectiles. That area is proportional to "A" squared. So "A" squared is the (unnormalized) probability of our test particle being hit. Thus "A" squared is the probability of finding our test particle there. Thus "A", which is the radius of our particle's ring structure, is equal to the squareroot of the probability of finding our particle at this particular position along the X' axis. Thus we find that the waveform we casually drew in Figure A4 (at the top of that page) can be considered to be identical in form and in interpretation (probabilistic) to a standard relativistic DeBroglie wave (or wave packet) of Quantum Mechanics.

Now we return to the tachyon issue. We've seen three places, denoted by asterisks "*", where it looks like vacuons must be tachyons. It turns out that this is not a problem, even though physics has not to this date ever proved the existence of tachyons. These "useless and unobservable" particles are only mentioned in a few textbooks. But, there seems to be agreement between these texts on the kinematics and kinetics of these heretofore "hypothetical" particles. These are summarized here, as follows:

1) The "rest-mass", or better yet the "base-mass" (since a tachyon can never be slowed down to a resting state) of a tachyon is an imaginary number. However, this is not a problem due to the next property:

2) Since the "momentum-energy four-vector" of tachyons is (just like it is for all known particles) equal to the product of the rest mass ("base" mass) and the "velocity four-vector", we get the following: since the components of the tachyon's velocity four-vector are imaginary (these components are real numbers for ordinary particles), and since the product of two imaginary numbers results in a real number, we see that the components of the energy-momentum four-vector are real numbers just like as for ordinary particles. In other words, both relativistic momentum and relativistic energy for tachyons are real numbers. These properties are real and should "come into play" as expected during collisions or interactions between tachyons and ordinary particles, as well as between tachyons and other tachyons.

3) For tachyons, relativistic momentum and relativistic energy are conserved in all interactions, be they tachyon-ordinary or tachyon-tachyon.

4) When tachyons gain energy, they slow down. They can never go as "slow" as the speed of light because their mass (effective, or apparent mass) starts to become infinite as they approach the speed of light from above. It would require infinite energy from an outside source to slow down a tachyon to exactly the speed of light.

5) The magnitude of the factor "gamma" can be less than unity. This occurs at speeds greater than the square root of 2 multiplied by "C", the speed of light. Due to this, the tachyon's effective mass can be less than its "base-mass", and in fact can approach zero as the speed of the tachyon approaches infinity. As a tachyon approaches infinity speed, its energy approaches zero.

6) A tachyon's effective mass is a real number between zero and infinity.

--[End of Tachyon properties list]--

We postulated earlier that the maximum speed of a vacuon is equal to the intrinsic maximum wave speed of the "rubbery grid" (or vacuum substrate). This speed is assumed to be far in excess of "C", the speed of light. **Since the maximum speed is not infinity, there will therefore be a minimum value for the effective mass and for the total energy and even the effective momentum of a vacuon (or of a dynamic structure of vacuons).**

So, due to the 3 places we've seen marked by a "*" in figs A5-A5E, and due to several such places in what is to follow, we are compelled to make the following **new postulates:**

XXII. **THE TACHYON POSTULATE.** Vacuons are probably tachyons, meaning that their speeds can only range from slightly above "C" (the speed of light) up to the maximum wave velocity of the elastic grid, or vacuum substrate, which will be called V-sub-grid or V-sub-wave. **[later, XXII (this postulate) will be refuted, except for photons].**

XXIII. **THE VACUON INTERACTION CONSERVATION LAWS POSTULATE.** The quantities total relativistic momentum and total relativistic energy are conserved for any closed system of vacuons and vacuon dynamic structures (i.e. ordinary quantum particles) and gravitoelectric and gravitomagnetic fields produced by such vacuons or structures of vacuons. This always holds true whether interactions are vacuon-vacuon, or vacuon-ordinary particle, or ordinary-ordinary.

XXIV. **THE LOWEST MINIMUM ENERGY POSTULATE.** There is a minimum possible energy that an individual vacuon can have, and that is its total relativistic energy when it is traveling in a straight path with its velocity equal to V-sub-grid or V-sub-wave. This minimum energy will be called the LOWEST MINIMUM ENERGY, or "LME".

Before moving on, we note that Figure A4 has some other interesting features. If an observer were to look at a typical ordinary subatomic particle, the particle moving at speed V-sub-g, they would see (if they could see things this tiny) a flattened ring of many points of mass, where the mass pairs move to the right at a phase velocity greater than "C". However, we know that each of these mass points is actually just a different "view" of the **same** vacuon protoparticle! The upper mass points circulate clockwise, and the lower mass points circulate counterclockwise. The mass point's effective masses are all real and have the same magnitude, or absolute value.

From figure A4, we see that the **lower** masses are coming up out of the page (i.e. they're moving **forward in time,** in fact faster than we, the observers, are) and the **upper** masses are traveling into the page (i.e. they are moving **backward in time!**). This is due to the counterclockwise motion of our vacuon in the rest frame of Figure A2. In fact, if the single vacuon in the rest frame were orbiting clockwise, then in figure A4 we would have to reverse the Y' locations of each mass point (pair) for each X' location. It will later be shown that **electric charge** of a particle is determined simply by the direction of orbit of the single rest-frame vacuon! (it can only be + or -).

And now the most interesting feature of all. We must conclude that the effective mass (from our point of view) of all the *upper* masses is negative and the effective mass of all the *lower* masses is positive. This is due to two reasons: 1) Masses traveling backward in time are thought to work in reverse and thus produce an apparently repulsive field, and 2) since we postulated that the effective mass of an observable particle is due solely to the gravitomagnetic field produced by its vacuons, we would get a net g.m. field of zero if the upper effective masses are positive. This is because the upper vacuons are rotating in the opposite sense compared to the lower vacuons.

Note that the sum of the 2N masses is therefore zero. So the effective (measurable) mass of our particle is E divided by C squared, where E is the total energy stored in the gravitomagnetic field produced by all the point masses in the ring structure. The mass of any given mass point is probably equal to " gamma' " multiplied by "M_sub_vac0". We will from here on use the positive, real number form of " gamma' " and also the positive, real number form of "M_sub_vac0". Thus, as we require, the product of these two will be real, positive. So, we say that "M_sub_vac" is equal to (hopefully) the product of " gamma' " and "M_sub_vac0". Each upper point has an effective mass of -"M_sub_vac" and each lower one of +"M_sub_vac".

In later discussions, we may for convenience refer to a particle as being composed of counter-rotating rings (or half-rings) of equal amounts of positive and negative protomass. A very interesting application of this model is that of pulling the rings apart (it's easy to do!) in order to create physically separate large amounts of positive and negative mass, for the purpose of creating gravity-based propulsion.

CALCULATION OF FIELDS AND MASS FLOW CURRENTS [Based Upon Inductance] :

This section will get us back on the path of deriving other quantum mechanical features from vacuon theory, such as intrinsic spin of fermions, and more importantly, that E' = (h)f' and P'=(h)/(lambda'), as are stated in Dirac's modern theory of relativistic Quantum Mechanics. See the "A6_" series of figures for this (i.e. A6 through A6R).

This series of figures first derives formulas for the dynamic structure, namely, the mass-flow current and the gravitomagnetic energy contained therein. This energy is what accounts for the measured rest mass of the particle. The reason for the so-called "collapse of the quantum wave function", upon taking a measurement of an observable is explained. Finally, it is shown that E'=hf' and that P'=h/lambda', completing the proof that vacuon theory predicts and explains the features of Dirac's relativistic electron theory!

[Therefore, Vacuon Theory explains/derives Quantum Mechanics.]

THE MULTI-VACUON PARTICLE MODEL. INDUCTANCE-BASED PARTICLE MASS CALCULATION AND SUBQUARKS, SPIN, PREDICTED MASSES, STEADY-STATE RADIUS, & ROTATIONAL INERTIA :

For the following, refer to **Figures A7 through A7U**.

We identify a "subquark" as being a tiny cluster (or "clump") of vacuons. (Much later, we will identify the ring structure in Fig. A7 as being a single quark.) **Thus, vacuon theory shows 2 levels of detail below what we now call quarks.** Next, we get a formula for the rest mass, m_sub_0, of a particle (specifically, a 1-quark, or simple particle) in terms of inductance correction factor, L_sub_c, permeability, mu_sub_g, structure radius, A, rest frequency, f_sub_0, single vacuon apparent protomass, M_sub_V0, number of vacuons per subquark, tilde N_sub_V, and the temporal length of the structure, l_sub_0. We see that the variables L_sub_c and l_sub_0 are troublesome, as there is currently no way to know their values, yet. Also, M_sub_V0 is troublesome for the same reason.

Then, using the angular momentum and rotational inertia of any solid disk, we are able to calculate the equilibrium (steady-state) radius of the particle. This is done by enforcing that the angular momentum must equal a half-integral multiple of h-bar.
The radius is seen to equal a function of the number of subquarks multiplied by the particle's (well-known) "Compton Wavelength". Computed radius values are seen to be "in the ballpark" of the measured value for the proton.

Now, in Fig. A7K, we show that the square-root of the ratio (rest mass of any particle divided by rest mass of the electron) is always (in theory, at least) an integer. This can be seen from Figure A_MASS, which tabulates this square-root ratio for 18 known particles of physics. Specifically, we see for the proton (and neutron) that this value = 42.

***** This has never been solved before in physics, and it is the "Holy Grail" which has been sought for so long. *****

The interesting thing to note, here, is that all of the square root mass ratios are always (approximately) equal to an integer multiple of either 7 or 8. Strange indeed. **You can see that all of the "8" type particles are the same as the usual well-known short-lived mesons, and the "7" type particles are the stable particles, such as the proton or neutron.** We will re-visit this later and will explain the meaning of the values in the "formula:" column in the table.

Therefore, we see that the troublesome product of M_sub_V0 squared, and L_sub_c squared, and [1 / l_sub_0], as was found in the mass formula in Figure A7C, is simply equal to a constant divided by V_sub_V0 squared! (Fig. A7K).

Thus, we can write a much simpler formula for h (compared to Figure A7C), which is shown in Figure A7M.

Then, we see that the assumptions in Figure A7E4-E5 were correct, and we may write a formula for steady-state (i.e. ss) velocity, V_sub_V0, and a formula for steady-state radius, A_sub_0.

Thus, the particle can be viewed as a rotating solid disk whose angular momentum = (1/2)h-bar. See Fig. A7O.

The ss velocity is seen to be a function of the number of subquarks, and the ss radius is seen to be a function of the particle's Compton wavelength and the number of subquarks.

>>> Finally, we discover that our initial guess, that vacuons are probably tachyons, is FALSE. Thus, vacuons which make up a material (mass not= 0) particle always travel slower than the speed of light ("c"). <<<

> **However, I have good arguments to show that for photons (mass = 0), the constituent vacuons ARE TACHYONS, but these will not be discussed here.**

Finally, **Figs A7P-A7U show all the formulas in corrected form, due to the above discovery.**

Figure A7P shows that we no longer have "phantom" (or apparent) vacuons, so all we ever see is tilde_N_sub_Q vacuons, total.
> **Also, we see good evidence that tilde_N_sub_Q should be an EVEN INTEGER. <**

The same effects, as were shown before, related to negative mass points and their direction of motion, can be proven to still exist, but this will be left as an exercise for the reader. This regards the non-tachyon vacuon case. This was shown to be the correct situation.

The formula for the s.s. particle rest mass, m_sub_0, is then correctly written in Figure A7S. The new correct formula for the vacuon rest mass and apparent mass is in Figure A7T.

> **Finally, we see great evidence that tilde_N_sub_Qmin equals 4. (i.e. the minimum # of subquarks). And, to recap, it should be an even integer. <**

[**For the single quark model, such as the electron.]**

All of the A7 series figures are for a single quark (i.e. one-ring) model. Protons can't be handled by this model.

E / M FIELDS AND CHARGES DERIVED, AND MULTIQUARK MODELS, & NEGATIVE MOMENTUM. :

Yes, the electron is actually a single, free quark (i.e. it is a one-ring particle, thus obeying all the formulas we've written so far).

The proton, however, has 3 quarks (as is well-known) and thus it apparently must consist of 3 parallel-stacked rings. Each "ring" of vacuons is a "quark". Each ring contains subquarks (defined earlier). Each subquark is a cluster (or clump) of one or more vacuons. Now, a precise value for the average radius of the proton can be calculated from a QCD (quantum chromo-dynamics) book (by Greiner, in this case) on it's page 128. Also, see Figure A7G for this value. It is, in units of 10^-15 meters, 0.9381. This value was experimentally determined. Deriving from the particle radius formula in Figure A7N, we get for the proton radius, the value 0.2108 x 10^-15 multiplied by the squareroot of tilde_N_sub_Q. Units are meters.
SEE FIGURE A8.

It looks like we get the right radius when the total number of subquarks = 20, which can be split up as 2 quarks of eight subquarks each plus 1 quark of four subquarks.
Figures A9 through A11 explain and show the resulting model for the proton's structure. The resulting required model for the electron and the photon are also shown in Figure A11.

>>> **Thus, the electron is a single quark consisting of 4 individual vacuons.** <<<

The photon has 2 quarks, for a total of 8 vacuons. But, it's organized as 4 subquarks having 2 vacuons each.

Figures A12-A13 discuss how a 4 vacuon quark (more correctly, a 4 subquark quark) can continually stir up (or "dig" up) the 4 subquark photons from the vacuon "sea" (just like the Dirac "sea").

The steady stream of "launched" or "ejected" photons is the phenomenon which completely accounts for electrostatic charge of a particle and the resulting (inverse-square) law of forces between such charged particles.

ALL OF ELECTRODYNAMICS THEORY CAN BE DEVELOPED FROM THIS BEING THE STARTING POINT. (Not done here, since the steps involved should be obvious to most physicists).

Negative momentum is shown to explain how photon exchange can cause attractive forces!

Also, the reason for the proton's quark charges is discussed. Also, see Fig. A14. Figures A15-A18 treat the neutron's structure to the same level of detail. For the neutron, we get the total number of subquarks = 16, which is split up as 2 quarks of 4 subquarks each plus 1 quark of 8 subquarks (i.e. just the flipside of the proton's structure).
Figure A19 calculates the total number of vacuons in the proton or in the neutron. It is 168. Note this equals 24 x 7.

Figure A20 shows how to derive either a proton or a neutron from a single generic structure, which is a ring of 24 clusters, where each cluster is a clump of 7 vacuons.

Figure A21 shows the final detailed models for the proton, neutron, electron, and photon.

Figure A22 shows that the 3-quark structures (as well as the nucleus of an atom) are cohesive and somewhat stable.

>> The "strong nuclear force" is seen to be simply the gravitomagnetic force. Also, the root-cause for the Pauli Exclusion Principle is shown. <<

Figure A23 shows the GM forces responsible for the formation of two distinct ring diameters. The compressive and expansive forces being correct depends on having each "felt-by" GM field pointing towards the temporal future, and that, in turn, depends upon the later finding that the smaller rings are significantly more massive than are the larger rings.

>> Later, in the "L" pages and the "NSS" pages, it will be shown that any **isolated** (free) ring always must contain **exactly 4 subquarks**, due to force-balance reasons. <<
(Also, see Fig. A24.)

Figure A25 explains why the powerful GM fields of this new theory have not been detected in experiments.

Figure A26 explains all of the remaining information in Figure A_MASS. If we take the ratio of any particle's rest mass to the electron's rest mass, and then square root it, and then multiply by 4, we get (APPROXIMATELY!) the total number of vacuons in the particle.

Figure A27 shows the new states/energies diagram, which replaces the Dirac electron theory diagram. The E_sub_min energy level shown is the lowest possible value that a photon can take. It is photons of this energy that are continually launched by each and every quark (and evenly in all spatial directions). [A27 is the last Figure.]

Photons, and collisions with them, will not be discussed in this paper any further. It should be known that some mathematical studies of them, done by this author, show strong evidence that photons must consist of **tachyon** vacuons!

*****> THIS CONCLUDES THE MAIN PART OF THIS PAPER. <*****

What remains is a set of pages named L0 through L25, and NSS0 through NSS6AM. Also, there is a set of pages named DL1 through DL4, and DNSS1 through DNSS5. These two latter sets describe the equations in the two former sets. Collectively, all of these sets can be considered to be an "appendix" to this overall paper (of Mesons and Baryons & Quark Substructure) just presented.

DL1-DL4, Brief Summary :

- Develop a much more accurate formula for the inductance of the helix, even for non-equilibrium.
- Calculate the formula for the inductance correction factor.
- Create formula for large radius (LR) dynamic rest mass. For large enough radius, rest mass goes to near zero.
- Based on balance of centrifugal and centripetal forces, calculate the rest mass of **both** vacuon types. **They are: M_sub_r01 = 5.56 x 10^-9 kg and M_sub_r07 = 5.106 x 10^-10 kg. The ratio = 10.89.** The formula for the total binding energy in a 7-vacuon cluster, is 7 x (M_sub_r01 - M_sub_r07) x C^2.
- Create formula for small radius (SR) dynamic rest mass.

DNSS1-DNSS5, Brief Summary :

- Create accurate formulae for inward and outward dynamic force.
- For dynamic balance, these must be equal.
- Create formula for dynamic mass of a particle (the 2 forces are =).
- Show that, for a **single isolated** quark, the # of subquarks **must be 4,** and thus vacuon velocity range is **.4 x C through .5 x C.**

- Derive a final formula for **dynamic mass of a particle**, in cases where inward force does NOT equal outward force. This is the **final, most general** answer to the question **"What is MASS?"**. We find that it is equal to a constant, multiplied by the total apparent vacuon mass squared, and divided by the structure's radius. **Rest frame, all cases of radius and rotational speed.**

[LAST PAGE, of the typed Pages]

[what follows is all hand-written]

[ ]

[THIS PAGE IS INTENTIONALLY BLANK]

D&R Scientific Publishing Co.; Denver, CO
"A New Theory of Quark Substructure for Mesons and Baryons"

Completed on 12-24-2001
(The "UFT" pages)

A Unified Field Theory
by Richard P. Crandall

UFT1
8-6-01

Postulates:

I. At this level of description, the following do not exist: ordinary subatomic particles of any kind (examples: electrons, protons, mesons, baryons, quarks, antiparticles such as the positron, etc...), photons, any aspect of electromagnetism (including charge, fields, etc...), mass, force, energy, momentum, any theory of quantum mechanics (QM, QED, QCD), special relativity theory, and general relativity theory.

II. However, the following do exist: time (in an absolute sense, but this will later be disposed of, in favor of Relativity) and 4 orthogonal dimensions of space (in each of which direction, length may be measured). Our local region of the universe mainly consists of a "substrate" or "background" which will be referred to as "the vacuum." All particles, entities, and concepts exist relative to the vacuum, and will be derived and explained based upon the simple properties of the vacuum.

Richard P. Crandall 12-24-2001
Richard P. Crandall 8-6-01

D&R Scientific Publishing Co.; Denver, CO
"A New Theory of Quark Substructure for Mesons and Baryons"

Completed on 12-24-2001
(The "UFT" pages)

UFT2
8-6-01

III. The vacuum is not empty. Even regions of it devoid of all photons, energies, or particles or any other observable phenomena are not truly empty. The vacuum is actually seething with proto-matter (or proto-particles) which are unobservable due to their large number density and their even distribution and their high velocities. These particles (proto-particles) will be referred to as "vacuons" (VAC-YOU-ONS).

IV. More specifically, the substrate of the vacuum will be considered to be analogous (for purposes of understanding) to a 4-dimensional grid of empty cubes (or other shape) (or, more specifically, 4-cubes), where the edges of such cubes are all that exists and these are made of a highly nonlinear elastic-like substance. The insides of the cubes are truly empty. The grid is "teeming" (considered extremely active) with only two things: violent waves (random), and the many violently moving proto-particles called Vacuons. The Vacuons are seen as being simply knots, kinks, or what-have-you, in this "rubbery" grid. These kinks, or twists, are assumed to be indestructable and non-changing. They are also all assumed to be identical, for the present purposes of this theory.

Richard P. Crandall 12-24-2001

D&R Scientific Publishing Co.; Denver, CO
"A New Theory of Quark Substructure for Mesons and Baryons"

Completed on 12-24-2001
(The "UFT" pages)

UFT3
8-6-01

V.

The current theory can be used to derive all properties of matter related to relativity and Q.M. and Q.E.D. The base theory may later be amended by the inclusion of different types of Vacuons as well as more dimensions of space, as may be needed to fully explain Q.C.D.

VI.

The reason for the artifice of introducing such a regular grid is so that each vacuon could be visualized as the result of pressing two cubes, or sides of cubes, together. Such cubes would stick together and would then exist permanently. Such pairs of stuck-together cubes would account for all vacuons being <u>identical</u> in all respects. Such stuck pairs could also account for vacuons being un-changing and indestructible. It's quite possible that the "proto-mass" of a total of 2 cubes would, when confined to a single cube, create effectively a tiny "black-hole," which would of course last forever, but this speculative avenue will not be pursued in the present theory. The grid concept artifice will shortly be dispensed with; just remember that all vacuons are assumed identical.

Rich'd P. Crandall 12-24-2001

D&R Scientific Publishing Co.; Denver, CO
"A New Theory of Quark Substructure for Mesons and Baryons"

Completed on 12-24-2001
(The "UFT" pages)

UFT4

8-6-01

Richard P. Crandall 12-24-2001

VII. Since the vacuum (aka the rubbery grid) contains many, many vacuons (identical knots or kinks) travelling over a distribution of very high speeds, the vacuum can be said to have inherent in it a high degree of what will be called "proto-energy" (all "kinetic" energy, there being no such thing as potential energy at this level). In fact, proto-energy will be seen to be an actual form of real energy. It will be seen that this proto-energy of the ("sea of") vacuons can act as an infinite source or sink of energy to the physical (observable) quantum particle domain. We will later see clear mechanisms by which real energy can be transferred back and forth between the vacuum (unobservable) domain and the observable physical domain of particles and photons.

VIII. Since the substance comprising the "grid" is highly non-linear (as general relativity is non-linear), a range of velocities is available to the vacuons, instead of having all of them moving at a constant wave speed as in linear substances. It will be assumed that there exists a maximum speed for vacuons, which is $\gg c^2$. (greater than speed of light squared). Based upon the absolute time.

D&R Scientific Publishing Co.; Denver, CO
"A New Theory of Quark Substructure for Mesons and Baryons"

Completed on 12-24-2001
(The "UFT" pages)

UFT5
8-6-01

IX.

The vacuum has three identical-like dimensions of space (except for being usually essentially orthogonal and being named differently) and one different (or "preferred") dimension of space. Therefore there is a total of 4 spacelike orthogonal dimensions. The nature of and reason for the forth dimension of space being different from the other three is as follows.

It is believed that the vacuon "sea" is actually more like a "river", whose "flow" is of very, very high velocity and is oriented in one particular direction -- that being the direction of the "preferred" or "special" fourth dimension of space. Also, it's possible that the "fabric" of the grid itself is highly and uniformly stretched in the fourth dimension's direction, but is not stretched nearly as much in the other 3 directions. Both the high-velocity directed flow of the vacuons and/or the possible stretching of the grid itself can possibly be explained as being the result of some type of large explosion or "big bang" which would certainly create a large directional outflux of vacuons and maybe additionally the stretching of the grid in that same direction.

Richard P. Crandall 12-24-2001

D&R Scientific Publishing Co.; Denver, CO
"A New Theory of Quark Substructure for Mesons and Baryons"

Completed on 12-24-2001
(The "UFT" pages)

UFT6
8-6-01

Richard P. Crandall 12-24-2001

...Cont'd

An apparent illusion is created as a result of the 4th space dimension being "different." The illusion causes intelligent beings in the observable physical particle domain to attribute this fourth dimension to time itself. We beings can't see or detect or observe the fourth dimension at all, in the usual daily experiences of our lives. We only perceive 3 dimensions of space. However our modern theory of Special relativity compels us to label the tick of our clocks as being some mysterious but physically real fourth dimension. Time itself is the fourth dimension, we say. The mechanism, as best as it can be understood, which causes and creates the illusion will now be explained. This mechanism is such that the entire theory of Special relativity can be derived from it.

The quantitative details are not worth detouring into, so a qualitative and simplified explanation will suffice. Due to the extreme average velocity of the vacuons in the direction of the 4th dimension, we can see that the vacuons do not interact or collide with each other very much at all in the downstream direction of the 4th dimension.

D&R Scientific Publishing Co.; Denver, CO
"A New Theory of Quark Substructure for Mesons and Baryons"

Completed on 12-24-2001
(The "UFT" pages)

E/M FIELDS AND CHARGES Derived, AND MULTIQUARK MODELS, NEGATIVE MOMENTUM.

UFT37

Yes, the electron is actually a single, free quark (i.e. It is a one-ring particle, thus obeying all the formulas we've written so far). The proton, however, has 3 quarks (as is well-known) and thus it apparently must consist of 3 parallel-stacked rings. Each "ring" of Vacuons is a "quark". Each ring contains subquarks (defined earlier). Each subquark is a cluster (or clump) of one or more Vacuons. Now, a precise value for the average radius of the proton can be calculated from a QCD book (by Greiner) on page 128. Also, see figure A7G for this value. It is, in units of 10^{-15} meters, .9381. This value was experimentally determined. Deriving from the particle radius formula on figure A7N, we get for the proton radius, the value $.2108 \times 10^{-15}$ multiplied by the square root of tilde N (sub Q). Units are meters. SEE FIGURE A8.

Richard P. Crandall 12-24-2001

D&R Scientific Publishing Co.; Denver, CO
"A New Theory of Quark Substructure for Mesons and Baryons"

Completed on 12-24-2001
(The "UFT" pages)

UFT38

It looks like we get the right radius when the total number of subquarks = 20, which can be split up as 2 quarks of eight subquarks each plus 1 quark of four subquarks. Figures A9 through A11 explain and show the resulting model for the proton's structure. The resulting required model for the electron and the photon are also shown on figure A11.

Thus, the electron is a single quark consisting of 4 individual vacuons.

The photon has 2 quarks, for a total of 8 vacuons, (But it's organized as 4 subquarks having 2 vacuons each.)

Figures A12 – A13 discuss how a 4 vacuon quark (more correctly, a 4 subquark quark) can continually stir up (or "dig" up) the 4 subquark photons from the vacuon "sea" (just like the Dirac "sea"). The steady stream of "launched" or "ejected" photons is the phenomena which completely accounts for electrostatic charge of a particle and the resulting (inverse-square) law of forces between such charged particles.

All of electrodynamics theory can be developed from this being the starting point.

Negative momentum is shown to explain how photon exchange can cause attractive forces.

Richard P. Crandall 12-24-2001

UFT39

Also, the reason for the proton's quark charges is discussed. Also, see fig. A14. Figures A15–A18 treat the <u>neutron's</u> structure to the same level of detail. For the neutron, we get the total number of subquarks = 16, which is split up as 2 quarks of <u>4 subquarks</u> each plus 1 quark of <u>8 subquarks</u> (i.e., just the flipside of the proton's structure).

Figure A19 calculates the total number of vacuons in the proton or in the neutron. It is 168. Note this equals 24×7.

Figure A20 shows how to derive either a proton or a neutron from a single generic structure, which is a ring of 24 clusters, where each cluster is a clump of 7 vacuons.

☆ Figure A21 shows the final detailed models for the proton, neutron, electron, and photon. ☆

Figure A22 shows that the 3-quark structures (as well as the nucleus of an atom) are cohesive and somewhat stable. The "strong nuclear force" is seen to be simply the gravitomagnetic force. Also, the root-cause for the Pauli Exclusion Principle is shown. ☆☆

Richd P. Crandall 12-24-2001

D&R Scientific Publishing Co.; Denver, CO
"A New Theory of Quark Substructure for Mesons and Baryons"

Completed on 12-24-2001
(The "UFT" pages)

UFT 40

Figure A23 shows the GM forces responsible for the formation of two distinct ring diameters. The compressive and expansive forces being correct depends on having each [redacted] "felt by" GM field pointing towards the temporal future, and that, in turn, depends upon the later finding that the smaller rings are significantly more massive than are the larger rings. Later, in the "L" pages and the "NSS" pages, it will be shown that any isolated (free) ring always must contain exactly 4 subquarks, due to force-balance reasons. (Also, see fig. A24.)

Figure A25 explains why the powerful GM fields of this new theory have not been detected in experiments.

Figure A26 explains all of the remaining information in Figure A_MASS. If we take the ratio of any particle's rest mass to the electron's rest mass, and then square root it, and then multiply [redacted] by 4, we get (APPROXIMATELY!!) the total number of vacuons in the particle.

Richard P. Crandall 12-24-2001

UFT41

Figure A27 shows the new states/energies diagram, which replaces the Dirac electron theory diagram. The E_{min} energy level shown is the lowest possible value that a photon can take. It is photons of this energy that are continually launched by each and every quark (and evenly in all spatial directions). [A27 is the last figure.]

[Photons, and collisions with them, will not be discussed in this paper any further. It should be known that some mathematical studies of them, done by this author, show strong evidence that photons must consist of <u>tachyon</u> vacuons!]

⇛ This concludes the main part of this paper. ⇚

[What remains is a set of pages named LØ through L25, and NSSØ through NSS6AM. Also, there is a set of pages named DL1 through DL4, and DNSS1 through DNSS5. These (2) latter sets describe the equations in the 2 former sets. Collectively, all of these sets form an <u>appendix</u> to the overall paper just presented.]

next page →

Rich P. Crandall 12-24-2001

D&R Scientific Publishing Co.; Denver, CO
"A New Theory of Quark Substructure for Mesons and Baryons"

[Last Page] UFT 42

DL1 — DL4 Briefly:

* Develop a much more accurate formula for the inductance of the helix, even for non-equilibrium.
* Calculate the formula for the inductance correction factor.
* Create formula for large radius (LR) dynamic rest mass. For large enough radius, rest mass goes to near zero.
* Based on balance of centrifugal and centrepital forces, calculate the rest mass of both vacuon types. They are: $M_{r01} = 5.56 \times 10^{-9}$ kg and $M_{r07} = 5.106 \times 10^{-10}$ kg. The ratio = 10.89. The formula for the total binding energy in a 7-vacuon cluster, is $7(M_{r01} - M_{r07})C^2$.
* Create formula for small radius (SR) dynamic rest mass.

Richard P. Crandall 12-24-2001

DNSS1 — DNSS5 Briefly: inward and outward dynamic force.

* Create accurate formulae for inward and outward dynamic force.
* For dynamic balance, these must be equal,
* Create formula for dynamic mass of a particle (the 2 forces are =).
* Show that, for a single isolated quark, the # of subquarks must be 4, and thus vacuon velocity range is .4c through .5c.
* Derive a final formula for dynamic mass of a particle, in cases where inward force does not equal outward force. This is the final, most general answer to the question "What is MASS?". We find that it is equal to a constant, multiplied by the total apparent vacuon mass squared, and divided by the structure's radius. Rest frame, all cases of radius and rotational speed. (Last Page. End of paper)

— — —

D&R Scientific Publishing Co.; Denver, CO
"A New Theory of Quark Substructure for Mesons and Baryons"

Completed on 12-24-2001
(The "UFT" pages)

FIGURE A_MASS

Richard P. Crandall 12-24-2001

e^- electron	MeV mass (0000.51109)	$\sqrt{m_o/m_e}$ (1;1x1x1)	formula; N_v x Q x N_c (N_E)	total;	% err on $\sqrt{}$	mass, per formula	% err on mass
μ^+	0105.7	14.38	7x2x1	14	+2.7	0100.2	+5.5
π^+	0139.6	16.53	8x2x1	16	+3.3	0130.8	+6.7
π^0	0135.0	16.25	8x2x1	16	+1.6	0134.8	+3.2
K^+	0493.7	31.08	8x2x2	32	-2.87	0523.4	-5.7
K^0	0497.7	31.21	8x2x2	32	-2.47	0523.4	-4.9
η^0	0548.8	32.77	8x2x2	32	+2.41	0523.4	+4.9
Proton	0938.3	42.85	7x3x2	42	+2.02	0941.6	+4.1
neutron	0939.6	42.88	7x3x2	42	+2.10	0941.6	+4.2
Λ^0	1116	46.73	8x3x2	48	-2.65	1178	-5.3
Σ^+	1189	48.23	8x3x2	48	+0.48	1178	+0.9
Σ^-	1197	48.39	8x3x2	48	+0.81	1178	+1.6
Xi (avg)	1318	50.78	8x3x2	48	+5.79	1178	+11.9
Ω^-	1672	57.20	7x3x3	63	-9.21	2029	-17.6
Λ_c^+	2273	66.69	7x3x3	63	+5.86	2029	+12.0
D (avg)	1866	60.42	7x2x4	56	+7.89	1603	+16.4
J/ψ	3097	77.84	8x2x5	80	-2.70	3271	-5.3
B	5260	101.45	8x2x6	96	+5.68	4710	+11.7
Υ	9460	136.05	8x2x8	128	+6.29	8374	+13.0

↑ Halliday
Fritzsch ↓

Used these to calc. above: .51109, (6.29)

N_Q turns out to be 4xQ.

$c = 2.998 \times 10^8$

OVER→ 1 Joule = 6.2422×10^{12} MeV $m_e = 9.109534 \times 10^{-31}$

D&R Scientific Publishing Co.; Denver, CO
"A New Theory of Quark Substructure for Mesons and Baryons"

Completed on 12-24-2001
(The "UFT" pages)

Potentials:

$$\vec{H}_g = \vec{\nabla} \times \vec{A}_g \qquad \gamma = \frac{1}{\sqrt{1 - \frac{u^2}{c^2}}}$$

$$\vec{E}_g = -\vec{\nabla}\phi - \mu_g \frac{\partial \vec{A}_g}{\partial t} \qquad \rho_m = \text{mass density in } \frac{kg}{m^3}$$

"Lorentz" force law:

Force on "$m_0\gamma$" $\Rightarrow$ $\vec{P} = m_0\gamma \vec{E}_g + m_0\gamma \vec{u} \times (\mu_g \vec{H}_g)$

The linearized field equations for the electro-gravitational and magneto-gravitational fields:

$$\vec{\nabla} \times \vec{H}_g = \frac{\partial(\epsilon_g \vec{E}_g)}{\partial t} - 4(\rho_m \gamma \vec{u})$$

$$\vec{\nabla} \cdot (\epsilon_g \vec{E}_g) = -(\rho_m \gamma) \qquad G = 6.67 \times 10^{-11} \frac{N \cdot m^2}{kg^2}$$

$$G = \frac{1}{4\pi\epsilon_g}$$

$$\vec{\nabla} \times \vec{E}_g = -\frac{\partial(\mu_g \vec{H}_g)}{\partial t} \qquad c = \frac{1}{\sqrt{\mu_g \epsilon_g}} \cong 3 \times 10^8$$

$$\vec{\nabla} \cdot (\mu_g \vec{H}_g) = 0 \qquad \therefore \mu_g = 9.37 \times 10^{-27} \frac{N \cdot s^2}{kg^2}$$

Richard P. Crandall 12-24-2001

Figure A1

The governing field equations. The ordinary, or gravitoelectric, field is $\vec{E}_g$ and the gravitomagnetic field is $\vec{H}_g$.

D&R Scientific Publishing Co.; Denver, CO
"A New Theory of Quark Substructure for Mesons and Baryons"

Completed on 12-24-2001
(The "UFT" pages)

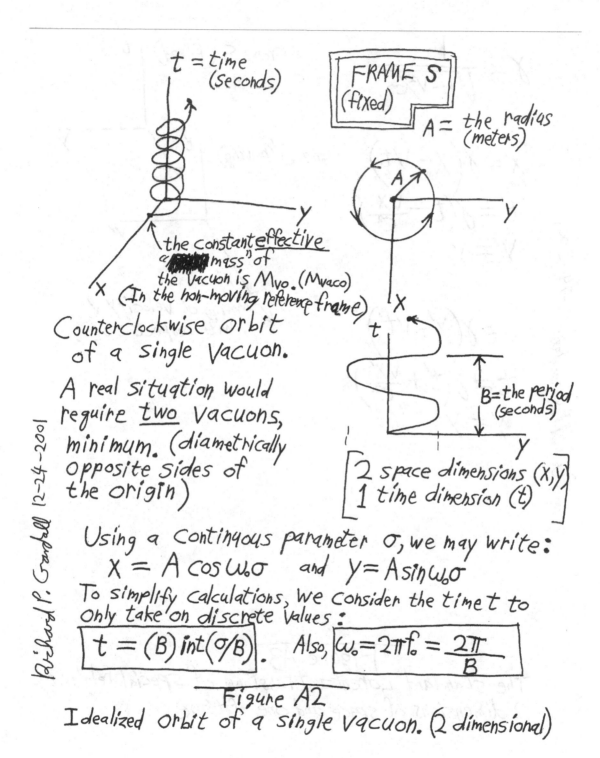

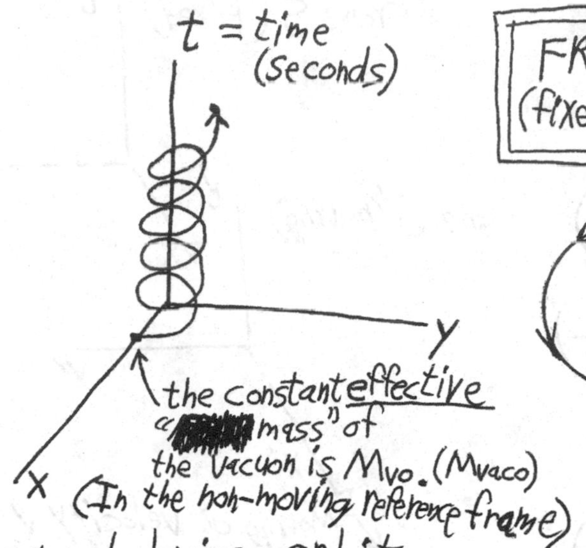

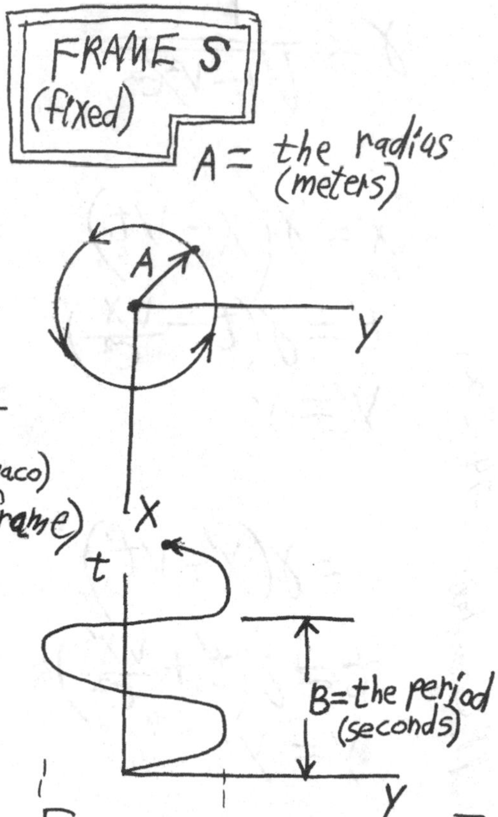

Counterclockwise orbit of a single vacuon.

A real situation would require two vacuons, minimum. (diametrically opposite sides of the origin)

[2 space dimensions (x, y)
1 time dimension (t)]

Richard P. Crandall 12-24-2001

Using a continuous parameter σ, we may write:

$$x = A\cos\omega_0\sigma \quad \text{and} \quad y = A\sin\omega_0\sigma$$

To simplify calculations, we consider the time t to only take on discrete values:

$$t = (B)\,\text{int}(\sigma/B).$$ Also, $$\omega_0 = 2\pi f_0 = \frac{2\pi}{B}$$

Figure A2
Idealized orbit of a single vacuon. (2 dimensional)

D&R Scientific Publishing Co.; Denver, CO
"A New Theory of Quark Substructure for Mesons and Baryons"

Completed on 12-24-2001
(The "UFT" pages)

$$\gamma = \frac{1}{\sqrt{1 - V^2/C^2}}$$

$$X' = \gamma(X - Vt)$$

$$t' = \gamma\left(t - \frac{VX}{C^2}\right)$$

$$Y' = Y$$

$$X = \gamma(X' + Vt')$$

$$t = \gamma\left(t' + \frac{VX'}{C^2}\right)$$

$$Y = Y'$$

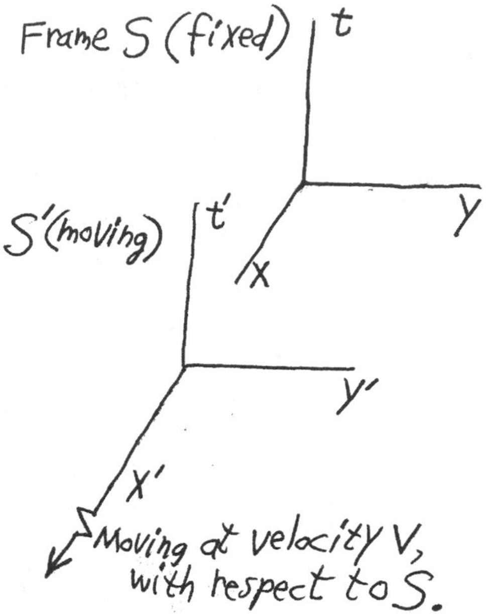

Richard P. Crandall 12-24-2001

Figure A3

The standard Lorentz transform of special relativity. 2 dimensions of space and one of time.

D&R Scientific Publishing Co.; Denver, CO
"A New Theory of Quark Substructure for Mesons and Baryons"

Completed on 12-24-2001
(The "UFT" pages)

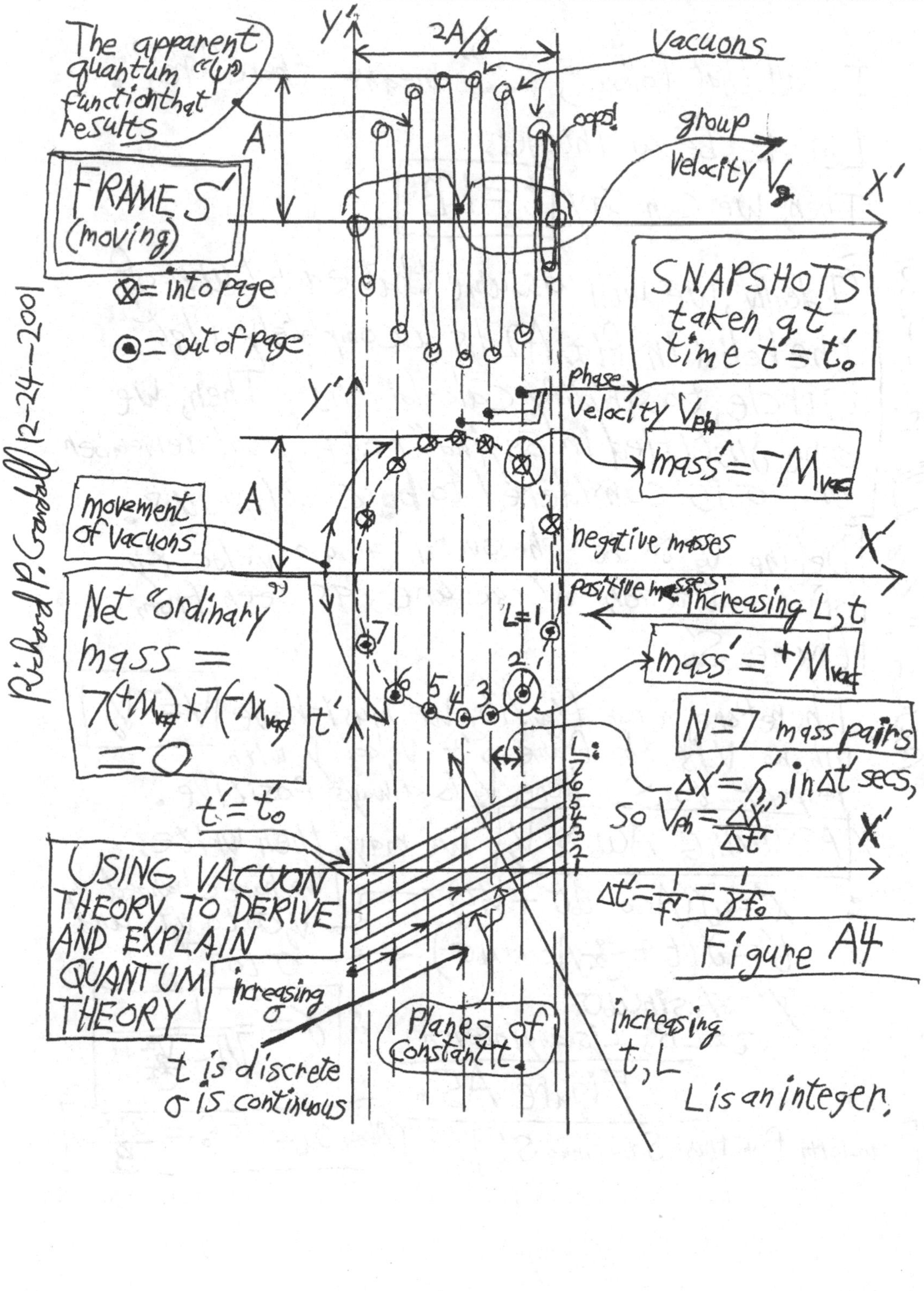

D&R Scientific Publishing Co.; Denver, CO
"A New Theory of Quark Substructure for Mesons and Baryons"

Completed on 12-24-2001
(The "UFT" pages)

In all that follows, "∴" means "therefore,".

Let L be an integer.

Then, we can say $\boxed{t = LB}$.

["Ideally", we will assume that each turn of the helix in fig. A2 is a completely flat circle, to simplify calculations. Then, we are justified in saying "$t=LB$", but remember that σ is considered to **be** continuous.

[Define V_g to be the overall group velocity of the vacuon structure, as seen from frame S'.

Therefore, from fig. A3, we must have $\boxed{V = {}^{-}V_g}$, where V is the frame S''s velocity with respect to S. Here, V_g is always positive.

* $\boxed{\text{ASSUMING } A\omega_0 \gg V_g}$, we may then write:

$$\therefore \quad x' = \gamma(A\cos\omega_0\sigma + V_g t)$$

$$t' = \gamma\left(t + \frac{V_g}{C^2}A\cos\omega_0\sigma\right)$$

$$y' = A\sin\omega_0\sigma$$

$t = LB$, L is any integer.

[We have changed $V_g\sigma$ to $V_g t$, and σ to t].

$$\therefore \boxed{\gamma = \frac{1}{\sqrt{1 - \frac{V_g^2}{C^2}}}}$$

Figure A5

[Transform from frame S to frame S'.]

$\boxed{\text{Also, } \omega_0 = 2\pi f_0 = \frac{2\pi}{B}}$

Richard P. Crandall 12-24-2001

[Everywhere in the figure 5 series, a "*" means that here's a very good reason to believe that all vacuons are "tachyons," which are particles that always travel faster than "c," the speed of light. More will be said later about this.]

Now, if we do a "snapshot" in frame S', so that $\boxed{t' = t'_o = \text{a constant}}$, we see that:

$$\therefore\ t + \frac{V_g A}{c^2}\cos\left(\frac{2\pi\sigma}{B}\right) = \frac{t'_o}{\gamma}$$

$$\therefore\ \cos\left(\frac{2\pi\sigma}{B}\right) = \left(\frac{t'_o}{\gamma} - LB\right)\left(\frac{c^2}{V_g A}\right)$$

$$\text{and}\quad LB = \frac{-V_g A}{c^2}\cos\left(\frac{2\pi\sigma}{B}\right) + \frac{t'_o}{\gamma}$$

[Defining L_{first} and L_{last} to be smallest and largest values of L within the "visible" range in S',]

$$\therefore\ L_{first} = \left(\frac{t'_o}{\gamma} - \frac{V_g A}{c^2}\right)\left(\frac{1}{B}\right),\ \text{and}\ L_{last} = \left(\frac{t'_o}{\gamma} + \frac{V_g A}{c^2}\right)\left(\frac{1}{B}\right)$$

This was arrived at by changing (varying) σ so the "cos()" ranges from +1 to −1.

FIGURE A5B

Richard P. Crandall 12-24-2001

D&R Scientific Publishing Co.; Denver, CO
"A New Theory of Quark Substructure for Mesons and Baryons"

Completed on 12-24-2001
(The "UFT" pages)

Defining "N" (an integer, positive) to be the number of mass pairs that are observed in S′ (also equals the number of wave crests, or cycles), then we have:

$$\Delta L = L_{last} - L_{First} = \frac{2V_gA}{C^2B} =$$ (large N only) (when V_g ≠ 0)

$$* \boxed{N = \frac{2V_gAf_o}{C^2}}$$

(for N>>1 only)

The minimum value for N is actually ½.

For N to be > 1, we need f_o, and thus V_{ph}, to be a very large number.

V_{ph} in many situations is > "C".

We see that for any particular $L = L_o$, in the range of L_{first} to L_{last}, there are two values for σ that satisfies the equations.

These are, (if we write $\sigma = L_oB + \alpha_oB$, where $0 \leqq \alpha_o < 1$): (derived as follows)

$$\cos(2\pi L_o + 2\pi\alpha_o) = \left(\frac{t_o'}{\gamma} - L_oB\right)\left(\frac{C^2}{V_gA}\right) = \cos(2\pi\alpha_o).$$

The two solutions for α_o, called α_1 and α_2, are such that $0 \leqq \alpha_1 \leqq \pi$, and $\alpha_2 = -\alpha_1$.

$$\therefore \boxed{\sigma_1 = (L_o + \alpha_1)B, \quad \sigma_2 = (L_o + \alpha_2)B.}$$

Of course, for both of these, $t = L_oB$.

Figure A5C

Richard P. Crandall 12-24-2001

D&R Scientific Publishing Co.; Denver, CO
"A New Theory of Quark Substructure for Mesons and Baryons"

Completed on 12-24-2001
(The "UFT" pages)

Richard P. Crandall 12-24-2001

For what follows, let $\alpha_0 = \alpha_1$ or α_2 (so that $\sigma = \sigma_1$ or σ_2, respectively):

$$\therefore X' = \gamma\left(A\cos\left(\frac{2\pi}{B}(L_0+\alpha_0)B\right) + V_g L_0 B\right)$$

Substituting, we get:

$$X' = \gamma\left(A\left(\frac{t_0'}{\gamma} - L_0 B\right)\left(\frac{c^2}{V_g A}\right) + V_g L_0 B\right)$$

$$= \frac{c^2 t_0'}{V_g} - \gamma B\left(\frac{c^2}{V_g} - V_g\right)L_0$$

$$= \frac{c^2}{V_g}\left(t_0' - B\gamma\left(1 - \frac{V_g^2}{c^2}\right)L_0\right)$$

$$\therefore \boxed{X' = \frac{c^2}{V_g}\left(t_0' - \frac{B}{\gamma}L_0\right)}$$

Note that, by varying L_0, we get equally-spaced X' locations for the mass-pairs.

L_0 is integer. Note $+\Delta L_0$ gives $-\Delta X'$.

Define δ as the wavelength.

$$\therefore |X_1' - X_2'| = |-\Delta X'| = \delta' =$$

$$\frac{c^2}{V_g}\left(t_0' - \frac{B}{\gamma}L_0\right) - \frac{c^2}{V_g}\left(t_0' - \frac{B}{\gamma}(L_0+1)\right) = \frac{BC^2}{\gamma V_g}.$$

$$\therefore \boxed{\delta' = \left(\frac{c^2}{V_g}\right)\left(\frac{1}{f_0\gamma}\right)}$$

Figure A5D

D&R Scientific Publishing Co.; Denver, CO
"A New Theory of Quark Substructure for Mesons and Baryons"

Completed on 12-24-2001
(The "UFT" pages)

Richard P. Crandall 12-24-2001

Now, we let the time t' in S' start advancing beyond the "snapshot" time t_0. We keep X' equal to some constant value X'_0. In this case it's easy to show that:

$$\boxed{t' = \frac{V_g X'_0}{C^2} + \frac{L_0 B}{\gamma}}$$

$$\therefore |t'_2 - t'_1| = |\Delta t'| = \frac{1}{f'}$$

$$= \left[\frac{(L_0+1)B}{\gamma} + \frac{V_g X'_0}{C^2}\right] - \left[\frac{(L_0)B}{\gamma} + \frac{V_g X'_0}{C^2}\right]$$

$$= \frac{B}{\gamma} = \frac{1}{\gamma f_0}.$$

$$\therefore \boxed{\boxed{f' = \gamma f_0}}$$

The 3 double-boxed formulas match the well-known Dirac/De Broglie relativistic quantum theory.

Now, we can derive a relation between phase vel "V_{ph}" and group vel "V_g":

$$\therefore f'\lambda' = V_{ph} = (\gamma f_0)\left(\frac{1}{f_0 \gamma}\right)\left(\frac{C^2}{V_g}\right)$$

$$\therefore * \boxed{\boxed{V_{ph} = \frac{C^2}{V_g}}}$$

So V_{ph} is superluminal, again strongly suggesting that vacuons must be tachyons!

Figure A5E

$$\boxed{C \leq V_{ph} \leqq \infty}$$

D&R Scientific Publishing Co.; Denver, CO
"A New Theory of Quark Substructure for Mesons and Baryons"

Completed on 12-24-2001
(The "UFT" pages)

Richard P. Crandall 12-24-2001

Per Figure A4, and standard knowledge of relativistic length contractions, we see that the distance across a particle's mass rings (actually protomass rings to be more precise) in the direction of travel is $\text{dist.} = \frac{2A}{\gamma}$.

Define τ as the total transit time for a wave crest (i.e. mass pair) to cross the flattened ring.

$$\therefore \tau = \frac{\text{dist}}{(V_{ph} - V_g)}$$

Note: this formula really always doesn't work unless [redacted] $V_{ph} > V_{gmax} = C$.

$$= \frac{2A}{\gamma\left(\frac{c^2}{V_g} - V_g\right)} = \frac{2V_g A \gamma}{C^2} = \frac{N}{f_0}\gamma$$

2N — Since Vacuum cannot be in All [redacted] places at same time [redacted]

Tot. mass in either half-ring: tot. mass = [redacted] $N M_v / 2N$.

Define "i" to be the effective mass flow rate in the entire ring (as if all the mass were positive and all flowing counterclockwise).

$$i = \frac{\text{tot. mass}}{\tau} = \frac{N M_v \gamma \left(\frac{c^2}{V_g} - V_g\right)}{2[redacted]N(2A)}$$

$$= \left(\frac{2V_g A f_0}{C^2}\right)\left(\frac{M_v \gamma \left(\frac{c}{V_g} - V_g\right)}{2A}\right)\left(\frac{1}{[redacted]N}\right)\left(\frac{1}{2}\right)$$

$$= \frac{(f_0 V_g)}{2[redacted]N}\left(\frac{1}{c^2}\right)(M_v \gamma)\left(\frac{c^2}{V_g} - V_g\right) = f_0 M_v \gamma \left(1 - \frac{V_g^2}{C^2}\right)\left(\frac{1}{[redacted]N \cdot 2}\right)$$

Figure A6

Here, M_v is the total apparent mass of the entire flattened ring. (all masses considered positive).

D&R Scientific Publishing Co.; Denver, CO
"A New Theory of Quark Substructure for Mesons and Baryons"

Completed on 12-24-2001
(The "UFT" pages)

$$\therefore \boxed{i = \frac{f_0 M_v}{2\,\text{[redacted]}\,\gamma N}}$$ = the effective mass flow current in the flattened ring. (kg/sec)

For what follows, we treat the flattened ring structure analogously to a solenoid, or tightly wrapped single layer wire helix, which carries an electric current. For this purpose, we go back to Figure A2. The structure should create an upward (or t axis) directed gravitomagnetic field, just as an inductor or solenoid creates an axial magnetic field. It's useful to remember the concept of absolute time here for measuring time (as with a clock), and then we are free to think about the "t" axis in Figure A2 as not being time but having it be just another ordinary dimension of space (such as a "z" axis). We will use "c" as the conversion factor from units of seconds to units of meters, and vice-versa. Several concepts and formalas from electrical engineering (such as inductance) will be used, but will be modified whenever necessary to make them agree with the "Maxwellized" equations of Figure A1. These are the same as the equations of electromagnetism, except for a factor of −4 and of −1. Here's the helix of figure A2, re-drawn as an idealized solenoid:

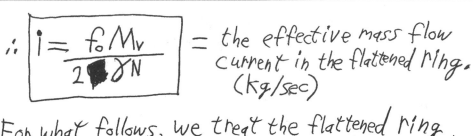

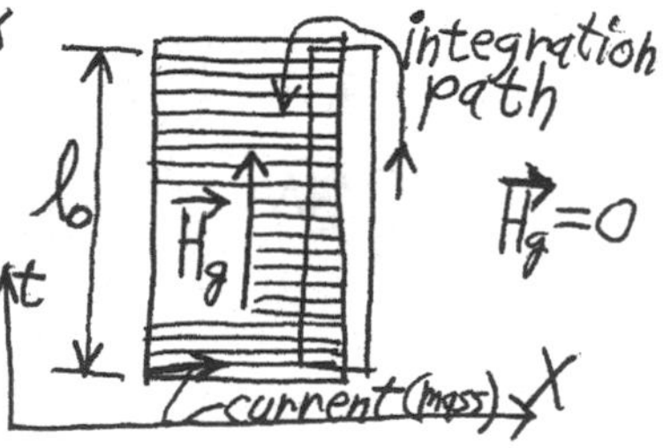

Here, l_0 is in meters (not seconds). Figure A6B

Richard P. Crandall 12-24-2001

We define l_0 to be the "width" or "thickness" of the present. It defines the thickness of that little 3-dimensional slice that we are all confined to, as was described in about the middle of postulate IX. It is in the units of meters.

$\vec{H_g}$ is the gravitomagnetic field (it's zero outside of the solenoid.) We use the closed integration path, on the right.

$$\oint \vec{H_g} \cdot d\vec{l} = \iint_S (\vec{\nabla} \times \vec{H_g}) \cdot \vec{ds} \quad \text{(by Stoke's Theorem)}$$

$$= -4 \iint_S \gamma \rho_m \vec{u} \cdot \vec{ds} \quad \text{(by Fig. A1, steady state)}$$

$$= -4 \iint_S \vec{J_g} \cdot \vec{ds} = -4 I_{g\,tot,encl.} \quad (kg/sec)$$

, where we define $\vec{J_g} = \gamma \rho_m \vec{u}$.

Define I_s to be the mass current flow (kg/sec) in any one turn of the helix (solenoid), as seen from the rest frame S, and using absolute time to measure the time.

The magnitude of the field inside will be called H_{go}, which is a constant.

Figure A6C

Richard P. Crandall 12-24-2001

D&R Scientific Publishing Co.; Denver, CO
"A New Theory of Quark Substructure for Mesons and Baryons"

Completed on 12-24-2001
(The "UFT" pages)

[Define N_w to be the total number of "windings" within the "length" l_o.]

$$\therefore \; -H_{go} l_o = +4 N_w I_{so} / N_w$$

(since the vacuon can only be in one winding at a time!)

$$H_{go} = \frac{(-4 I_{so})}{l_o}$$

(this vector actually points downward!)

$$\therefore \text{tot.energy} = \iiint_V \frac{1}{2} \mu_g H_{go}^2 \, dV$$

(per Poyntings theorem, re-written for gravity.)

$$= \left(\frac{1}{2}\mu_g\right)\left(\frac{16 I_{so}^2}{l_o^2}\right)\left(S_o l_o\right)$$

note, $S_o = \pi A^2$

$$= \frac{\frac{1}{2}\mu_g (-4 I_{so})^2 S_o}{l_o}$$

$$= \frac{1}{2} L_o (-4 I_{so})^2$$

electrical engineering formula, re-written with regard to figure A1.

["L_o" is the "inductance" of the vacuon stable dynamic structure (for a "length" $= l_o$). It will change when we go to different frames of reference (or velocity).]

$$\therefore \; \boxed{L_o = \frac{\mu_g S_o}{l_o}}$$

Figure A6D

Richard P. Crandall 12-24-2001

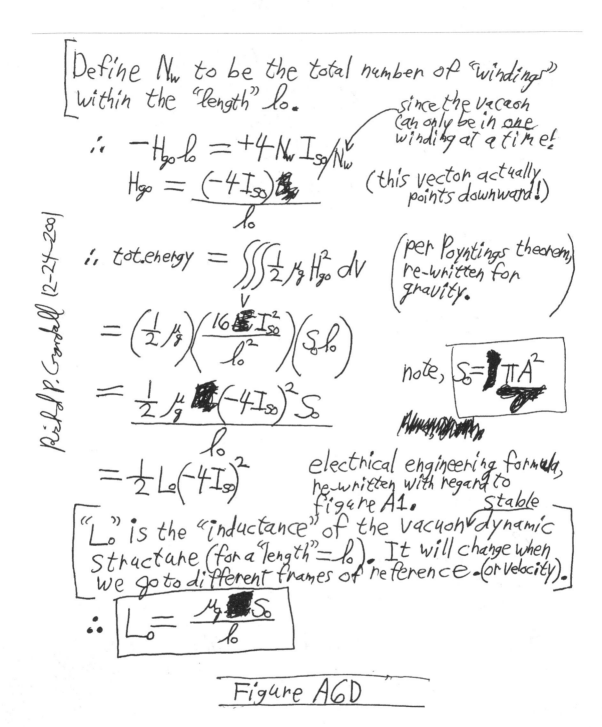

D&R Scientific Publishing Co.; Denver, CO
"A New Theory of Quark Substructure for Mesons and Baryons"

Completed on 12-24-2001
(The "UFT" pages)

$$L_o = \frac{\Psi_{toto}}{(-4I_{so})}$$

(another appropriately modified electrical formula. Ψ_{tot} is (N_w)(the simple flux Ψ).)

$$\therefore \Psi_{toto} = N_{wo}\Psi_o = (-4I_{so})L_o = \frac{-4\mu_g I_{so}\,[redacted]\,S_o}{l_o}$$

$$\therefore \boxed{\Psi_o = \frac{\mu_g\,[redacted]\,(-4I_{so})S_o}{l_o N_{wo}}}$$ = the gravitomagnetic flux.

Define "n_o" to be the number of windings (or turns) per meter of length (along the t axis).

[illegible]

Thus, we will find that N_w will vary depending on the speed of the observer. Also, l' will vary.

$$\frac{N_{wo}}{l_o} = \frac{1}{cB} = \frac{f_o}{c} = n_o$$ (in turns/meter (of time))

[illegible]

[illegible]

Figure A6E

Richard P. Crandall 12-24-2001

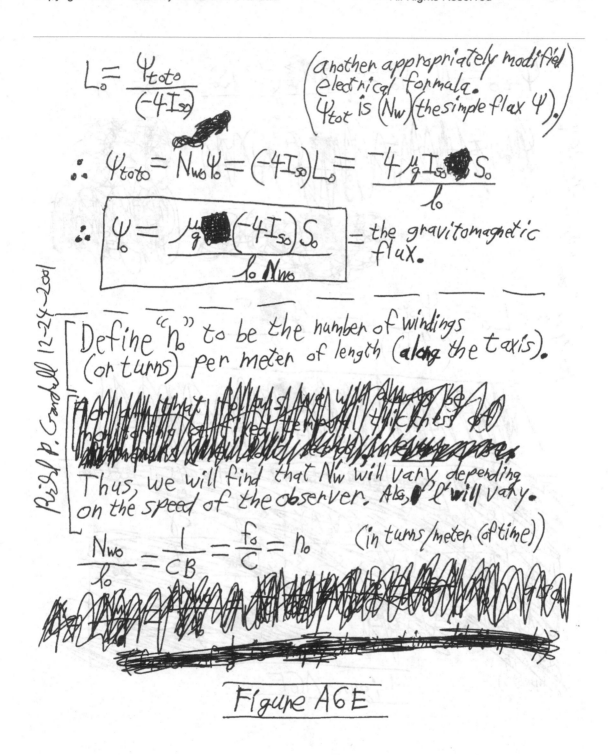

D&R Scientific Publishing Co.; Denver, CO
"A New Theory of Quark Substructure for Mesons and Baryons"

Completed on 12-24-2001
(The "UFT" pages)

$$\Psi_{toto} = N_{wo}\Psi_o = \frac{\mu_g \text{[redacted]} (-4I_{so}) S_o}{l_o} \qquad S_o = \pi A^2$$

$$\therefore \quad L_o = \frac{\Psi_{toto}}{(-4I_{so})} = \frac{\mu_g \text{[redacted]} S_o}{l_o}$$

R. P. Crandall 12-24-2001

Figure A6F

D&R Scientific Publishing Co.; Denver, CO
"A New Theory of Quark Substructure for Mesons and Baryons"

Completed on 12-24-2001
(The "UFT" pages)

other useful formulas are:

$$\boxed{\frac{N}{\tau}=\frac{f_o}{\gamma}}$$ (see fig. A6)

From fig. A2 it was stated that the effective mass of the vacuon in fixed frame S is M_{vo}. In fig. A6B it was found that the effective mass flow "i", in kg/sec, in the flattened ring, is $i = \frac{f_o M_v}{2\,\text{[redacted]}\,\gamma N}$ (moving frame S′) (2N). So, $i_o = M_{vo} f_o$ [redacted]. (fixed frame S) (because $N = 1/2$ when $\gamma = 1$)

Starting with the formula on the right, we could say that $i =$ (or should equal) [redacted] $\gamma M_{vo} f_o$ [redacted] but this is wrong because the vacuon (seen from S′) cannot be in all [redacted] pairs position [redacted] at the same time t_o, but only in one of these. (2N) Thus, the correct formula is $i = \text{[redacted]}\,\gamma\left(\frac{M_{vo}}{2\,\text{[redacted]}\,N}\right) f_o \text{[redacted]} = \gamma M_{vo} f_o$ [redacted]. (2N) ("dilated mass")

$$\therefore \boxed{i = \frac{f_o M_v}{2\,\text{[redacted]}\,N\gamma} = \gamma M_{vo} f_o}$$

$$\therefore \boxed{M_v = \gamma^2 M_{vo} N 2}$$ "un-diluted." (not apparent)

$$\therefore i = \left(\frac{M_v}{2N}\right)\left(\frac{N}{\tau}\right)$$

$$\therefore \boxed{i = \frac{M_v}{2\tau}}$$ ✓

∴ apparent mass of each apparent vacuon $= \frac{M_v}{2N} \equiv M_{VAC} = \gamma^2 M_{vo}$

Figure A6G

Richard P. Crandall 12-24-2001

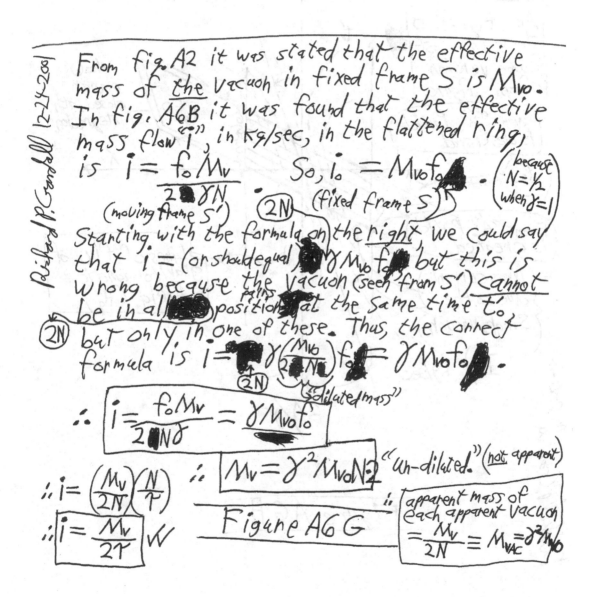

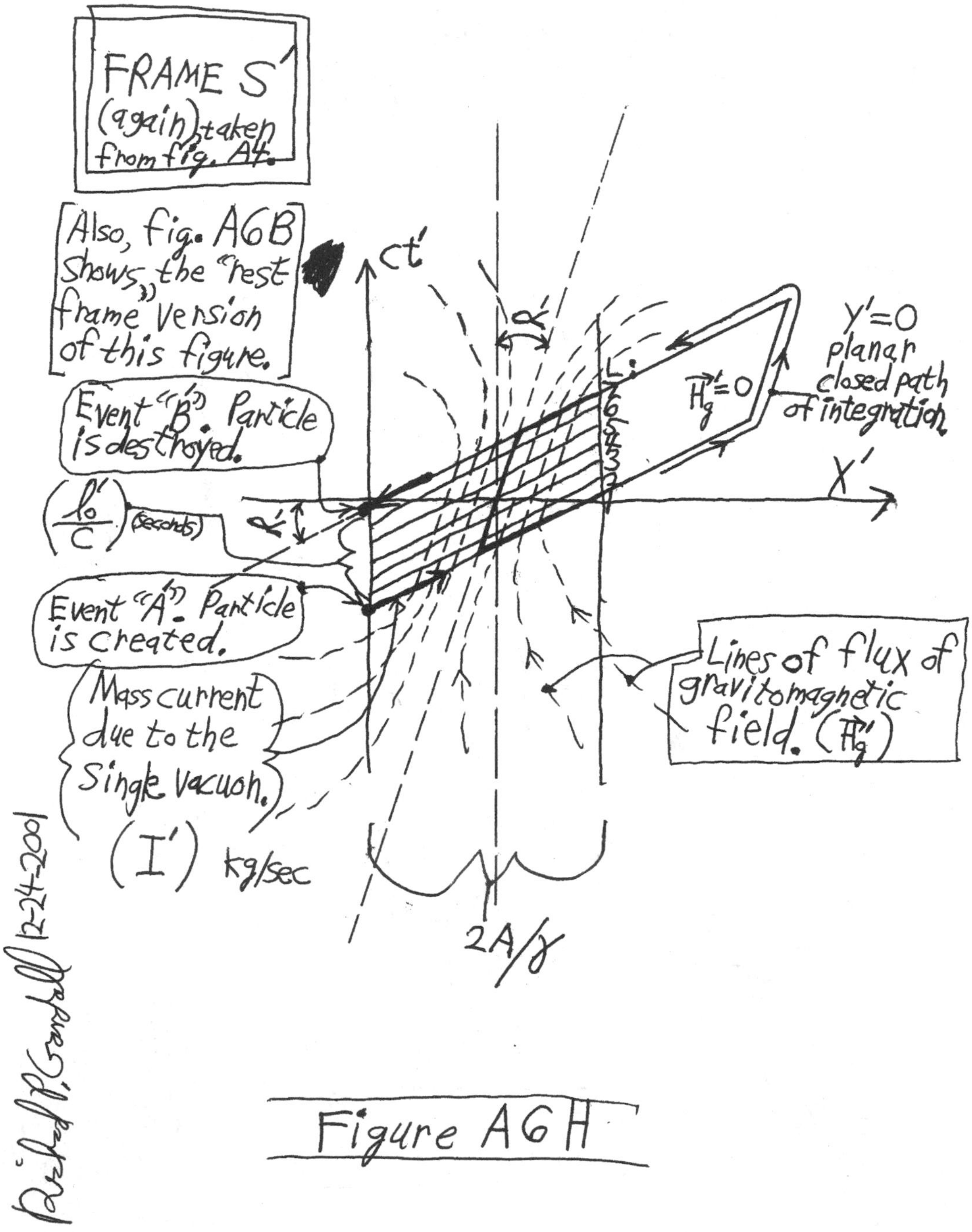
FRAME S' (again) taken from fig. A4.
[Also, fig. A6B shows the "rest frame" version of this figure.]
ct'
α'
Y'=0 planar closed path of integration
$\vec{H}_g'=0$
X'
L
Event "B". Particle is destroyed.
$(\frac{l_0'}{c})$ (seconds)
α'
Event "A". Particle is created.
Mass current due to the Single vacuon.
(I') kg/sec
Lines of flux of gravitomagnetic field. ($\vec{H}_g'$)
2A/γ
Richard P. Crandall 12-24-2001

Figure A6H

D&R Scientific Publishing Co.; Denver, CO
"A New Theory of Quark Substructure for Mesons and Baryons"

Completed on 12-24-2001
(The "UFT" pages)

If fig. A6H were re-drawn using Minkowskian coordinates ($\tilde{X}'_1 = X'$, $\tilde{X}'_2 = y'$, $\tilde{X}'_3 = z' = 0$, $\tilde{X}'_4 = ict'$), then that figure would look more like how figure A6B looks. It would be an undistorted cylinder, but would be rotated some (about the y' axis) (not shown or drawn). I will denote the use of Minkowskian coordinates with a tilde symbol (~).

[The value "l_0" is the temporal "length" (in meters) of the rest-frame solenoid, as shown in fig A6B. The corresponding apparent value in the moving frame S', in fig A6H, will be denoted by "l'_0".]

[Define N'_w to be the number of solenoid "windings" seen in S'. Define N_{wo} to be the corresponding value as seen in frame S.]

To our surprise, we will find that $N'_w \neq N_{wo}$. Only [redacted] speeds could account for this, but it should not be too surprising since we have already shown in fig A4 that we apparently see multiple copies or "rings" due to a single vacuon.

(high)

Figure A6I

Richard P. Crandall 12-24-2001

D&R Scientific Publishing Co.; Denver, CO
"A New Theory of Quark Substructure for Mesons and Baryons"

Completed on 12-24-2001
(The "UFT" pages)

Richard P. Crandall 12-24-2001

The value I′ shown in fig A6H is different from the "i" (or "i′") talked about in fig A6G. Now, I′ is defined to be the mass current as seen from a 4-dimensional frame of view. We can use the concept of having a clock to measure absolute time.

Also, we can say that $\boxed{\tilde{N}'_w = N'_w}$. ($\neq N_{wo}$)

It is important to consider the particle to begin its existence at spacetime event "A′", and to end its existence at event "B′". Alternately, we can say that a collision with another vacuon photoparticle (not shown) occurs at both events A′ and B′. The collision at event B′ can be regarded, as in quantum mechanics, as being the act of taking a measurement of the position of our observable particle. This would have to be done by using a collision with some other particle (electron, photon, etc..). THUS, WE HAVE A PHYSICAL REASON OR INTERPRETATION OF THE SO-CALLED "COLLAPSE" OF THE QUANTUM WAVE FUNCTION (Ψ), upon (or due to) the very act of taking a measurement!

***

Figure A6J

D&R Scientific Publishing Co.; Denver, CO
"A New Theory of Quark Substructure for Mesons and Baryons"

Completed on 12-24-2001
(The "UFT" pages)

The vacuon in the figure is in an equilibrium stable condition (see postulate IXX, equilibrium postulate), until it is hit at event B′. At that point, it will go careening off into an unsymmetrical spiral orbit, until it has had time to settle back into an equilibrium situation. Only then will its orbit resume being a symmetrical spiral helix. The same goes for the particle's vacuon that hit our target particle's vacuon. We will see later that this exchange of angular proto-momentum is the root cause explanation for our famous and well-known law of conservation of linear momentum and of energy!

Returning to fig. A6H,
from a well-known formula derived from Maxwell's equations, and modified per fig A1, we have: (in Minkowskian coords′)

$$\oint_{\text{closed path}'} \vec{\tilde{H}}_g' \cdot d\vec{l}' = (-4)\iint_{\text{surface}'} \vec{\tilde{J}}_g' \cdot d\vec{s}'$$

$\vec{\tilde{H}}_g'$ is the gravitomagnetic field.

$\vec{\tilde{J}}_g'$ is the mass current flux through the surface.

Figure A6K

Richard P. Crandall 12-24-2001

D&R Scientific Publishing Co.; Denver, CO
"A New Theory of Quark Substructure for Mesons and Baryons"

Completed on 12-24-2001
(The "UFT" pages)

(We are, of course, using 4-dimensions and absolute time.)

∴ We would then think we could write:

$$-\tilde{H}_g' \tilde{l}_o' = +4 \tilde{N}_w' \tilde{I}'$$

, but this is incorrect, because the vacuon cannot be in all $\tilde{N}_w'$ windings at the same time, [redacted] so it's mass is distributed and "diluted" by a factor of $(1/\tilde{N}_w')$. Thus, the correct formula is;

$$\boxed{-\tilde{H}_g' \tilde{l}_o' = +4\tilde{I}'}$$

$$\therefore \boxed{\tilde{H}_g' = -4\frac{\tilde{I}'}{\tilde{l}_o'}}$$

Note: due to the nature of time versus space, we consider the flux lines in fig A6H to be perpendicular to the current rings, in accord with special relativity. Thus, the shape of the integration path is as it is.

Also,

$$\boxed{\frac{\tilde{N}_w'}{\tilde{l}_o'} = \text{a constant, due to the non-distortion of Minkowskian space.} \therefore \frac{\tilde{N}_{wo}}{\tilde{l}_o} = \frac{N_{wo}}{l_o} = \frac{1}{CB} = \frac{f_o}{C}}$$

Since, $\tilde{N}_w' = N_w'$, [redacted] ∴ we must have $\boxed{\tilde{l}_o' = l_o'}$

Figure A6L

Richard P. Crandall 12-24-2001

D&R Scientific Publishing Co.; Denver, CO
"A New Theory of Quark Substructure for Mesons and Baryons"

Completed on 12-24-2001
(The "UFT" pages)

Now, looking only at the two events A′ and B′, and the corresponding same events A and B, as seen from frame S, we calculate the deltas between these two events as follows:

We see that $\boxed{\Delta X = 0}$ (in fixed frame S).

∴, referring to fig A3, we find that:

$$\boxed{\Delta t' = \gamma \Delta t}.$$

$$\therefore \Delta t' = \frac{l_o'}{c} = \gamma \frac{l_o}{c}, \quad \therefore \boxed{l_o' = \gamma l_o}$$

so, the length (temporal) of this "inductor coil" changes with V_g!

$$\therefore \frac{\tilde{N}_w'}{\tilde{l}_o'} = \frac{N_w'}{l_o'} = \frac{N_w'}{\gamma l_o} = \frac{f_o}{c} = \text{a constant.}$$

∴, as warned about earlier, (a tachyon? effect):

$$\boxed{N_w' = \frac{l_o f_o}{c} \gamma = \gamma N_{wo}}$$

Number of windings changes (!!) also, but the ratio of windings/meter is CONSTANT.

$$\therefore \boxed{N_{wo} = \frac{l_o f_o}{c}} \quad \text{Also, } \frac{\tilde{N}_w'}{\tilde{l}_o'} = \boxed{\frac{N_w'}{l_o'} = \frac{N_{wo}}{l_o}} \boxed{= \frac{f_o}{c} = \text{a Const.}}$$

(this makes sense!)

Figure A6M

Richard P. Crandall 12-24-2001

D&R Scientific Publishing Co.; Denver, CO
"A New Theory of Quark Substructure for Mesons and Baryons"

Completed on 12-24-2001
(The "UFT" pages)

Finally, we are ready to calculate, via the REST MASS (or FUNDAMENTAL) POSTULATE, the effective [redacted] mass of our observable quantum subatomic particle (which is made of only one vacuon). This [redacted] mass should be equal to the total energy of the gravitomagnetic field in the "solenoid" divided by the speed of light squared. Also, we calculate the inductance "L" L_s of the "solenoid."

We define E' to be the total energy of the observable particle (such as electron, photon, etc.) as seen in frame S' (not the rest frame). Then,

$$E' = \iiint_{\tilde{V}'} \frac{1}{2}\mu_g \tilde{H}_g'^2 \, d\tilde{V}' = \frac{1}{2}\mu_g \tilde{H}_g'^2 \left(\pi A^2 l_0'\right)$$

$$= \left(\frac{1}{2}\mu_g\right)\left(\frac{16\tilde{I}'^2}{l_0'^2}\right)\left(\pi A^2 l_0'\right) = \frac{1}{2}\tilde{L}_s'\left(4\tilde{I}'\right)^2$$

$$\therefore \tilde{L}_s' = \boxed{L_s' = \mu_g \frac{\pi A^2}{l_0'} = \left(\mu_g\right)\left(\frac{\pi A^2}{\gamma l_0}\right)}$$

Figure A6N

NOTE: This (when $\gamma = 1$) agrees with "L_0" in figure A6D.

Richard P. Crandall 12-24-2001

Due to the mass increase effect, we should have:
(since this "current" is due solely to the apparent proto-mass of the vacuon) (and since the "windings per meter" is a constant, regardless of the value of V_g)

$$\tilde{I}' = I' = \gamma I_0$$

(Kg/sec) (Kg/sec) (in S') (rest frame S)

Also, note that I_0 here equals I_{so} from fig. A6D. (due to a formula on A6L).

$I_0 = I_{so}$

and

$$L_s' = \left(\frac{\mu_g \pi A^2}{l_0}\right)\left(\frac{1}{\gamma}\right) = \frac{L_{so}}{\gamma}$$

$$L_{so} = L_0 = \frac{\mu_g \pi A^2}{l_0}$$

(see fig. A6D for L_0 and I_{so})

And, from A6G,

$$M_{vac} = \gamma^2 M_{vo} \quad \therefore M_{vaco} = M_{vo}$$

See definition of M_{vo} in fig. A2.

also, $i = \frac{f_0 M_{vac}}{\gamma} = \gamma M_{vo} f_0 = \gamma i_0$

where $i_0 = M_{vo} f_0$ (in non-moving frame S).

See fig. A6G.

Richard P. Crandall 12-24-2001

Figure A6O

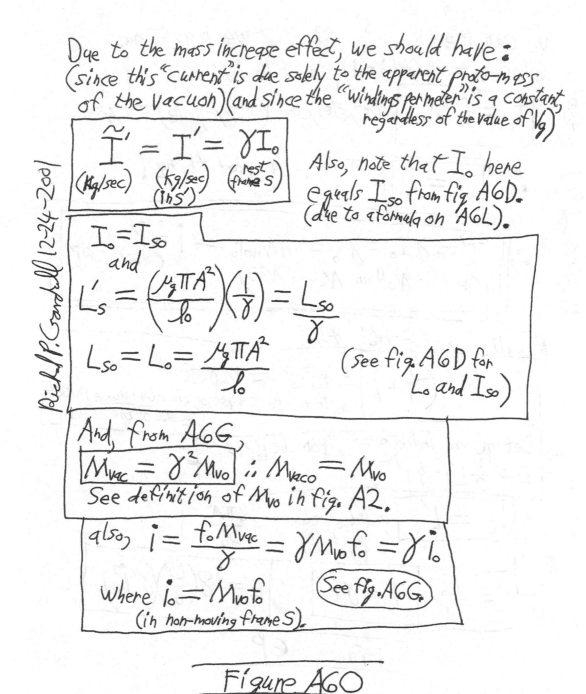

D&R Scientific Publishing Co.; Denver, CO
"A New Theory of Quark Substructure for Mesons and Baryons"

Completed on 12-24-2001
(The "UFT" pages)

We can see from fig A2 that we must have, when $V_g = 0$ (and $\gamma = 1$), the following:

$I_o = f_o M_{vo}$ (cycles/sec)(kg/cycle) = kg/sec

$\therefore$ $I_o = i_o$ This is a good thing!

$\therefore$ $I' = \gamma I_o = \gamma i_o = \gamma M_{vo} f_o = i \equiv i'$ (definition of i')
(See figs A6H and A6 and A6B)

Recall from fig A6N that

$E' = \frac{1}{2} L'_s (-4I')^2 = m^* c^2 = m_o \gamma c^2$
m_o is the rest mass of the observable particle (not of the vacuum)

Define an inductance "L'" for the i', as:
$E' = \frac{1}{2} L' (-4i')^2$.

$\therefore$ $L' = L'_s$ and $L_o = L_{so} = \frac{\mu_g \pi A^2}{l_o}$ (as was found in fig A6O)

$\therefore$ $L' = \frac{L_o}{\gamma}$ $i' = \gamma i_o$ $\therefore$ $E' = 8\left(\frac{L_o}{\gamma}\right)(\gamma^2 i_o^2)$

Figure A6P

Richard P. Crandall 12-24-2001

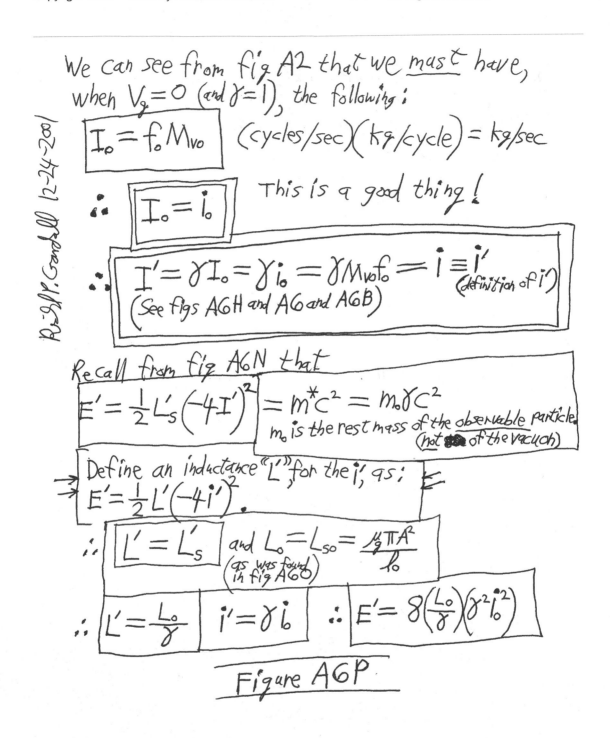

D&R Scientific Publishing Co.; Denver, CO
"A New Theory of Quark Substructure for Mesons and Baryons"

Completed on 12-24-2001
(The "UFT" pages)

$$\therefore\ m_0 \gamma c^2 = E' = \underbrace{8(L_0)(i_0^2)}_{\text{constant}}(\gamma) = \underbrace{(m_0 c^2)}_{\text{constant}}(\gamma)$$

$$\boxed{\frac{E'}{\gamma} = 8 L_0 i_0^2 = \frac{8\mu_g \pi A^2 M_{vo}^2 f_0^2}{\ell_0} = \text{a const.} = E_0}$$

[Now, let P' be the relativistic momentum of our observable particle as seen in frame S'.

$$\boxed{P' = m_0 \gamma V_g}\ ;\ \therefore\ (\gamma V_g) = \frac{P'}{m_0}$$

by definition

$\therefore$ (see fig A5D)

$$\delta' = \frac{c^2}{(V_g \gamma)(f_0)} = \frac{c^2}{\left(\frac{P'}{m_0}\right)(f_0)} = \frac{m_0 c^2}{f_0 P'}$$

$$\therefore\ \boxed{P' = \frac{m_0 c^2}{f_0 \delta'} = \frac{(E_0/f_0)}{\delta'}}$$

Define: a constant $\boxed{d_0 = \frac{8\mu_g \pi A^2 M_{vo}^2 f_0}{\ell_0}}$ = a constant

(provided we have chosen a fixed value for ℓ_0)

Figure A6Q

Richard P. Crandall 12-24-2001

D&R Scientific Publishing Co.; Denver, CO
"A New Theory of Quark Substructure for Mesons and Baryons"

Completed on 12-24-2001
(The "UFT" pages)

Richard P. Crandall 12-24-2001

$\therefore \; E_o = d_o f_o$

$\therefore \; \boxed{P' = \frac{d_o}{\lambda'}}$

and

$$E' = d_o(f_o \gamma) = d_o\left(\frac{i'}{M_{vo}}\right) = d_o f'$$

(fig A6G) (fig A5E)

and $E' = P'\lambda' f' = P' V_{ph} = \frac{P' C^2}{V_g}$

$\therefore$ we can identify d_o as Planck's constant "h"!
(as was claimed in UFT31)

$\therefore \; \boxed{E' = hf', \; P' = \frac{h}{\lambda'}}$

and $\therefore$ $\boxed{h = (8\mu_g \pi A^2 M_{vo}^2 f_o)\left(\frac{1}{l_o}\right)}$

And, to recap, $|M_{vac}| = \gamma^2 M_{vo}$ = any (apparent) Vacuon's apparent mass.

[redacted]

see fig. A4 again, and fig. A2.

Figure A6R

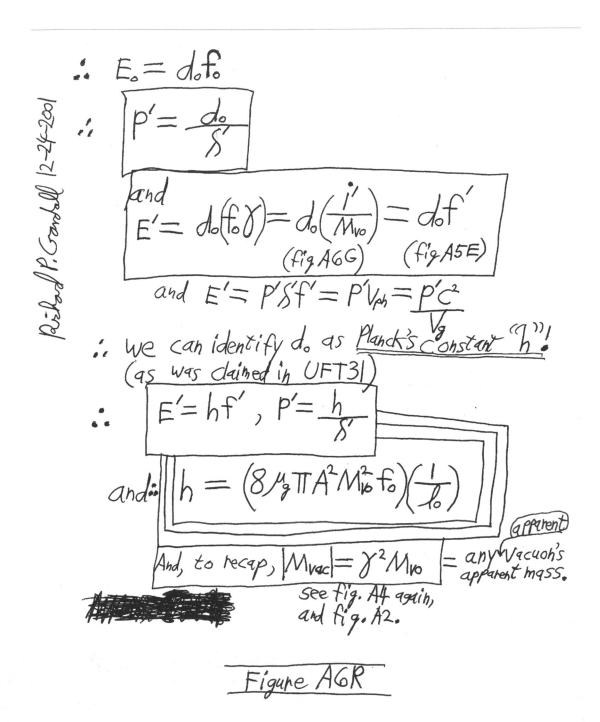

D&R Scientific Publishing Co.; Denver, CO
"A New Theory of Quark Substructure for Mesons and Baryons"
Completed on 12-24-2001
(The "UFT" pages)

Now, for multi-vacuon particles:

View from rest frame S:

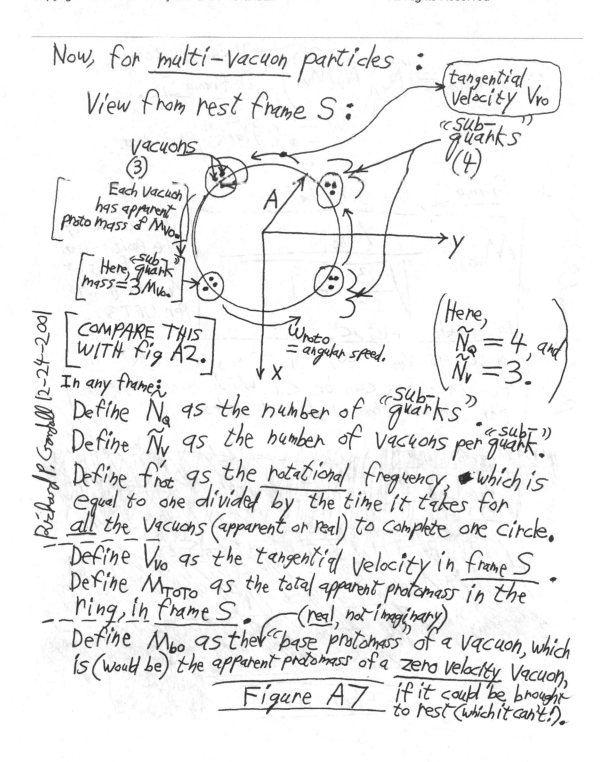

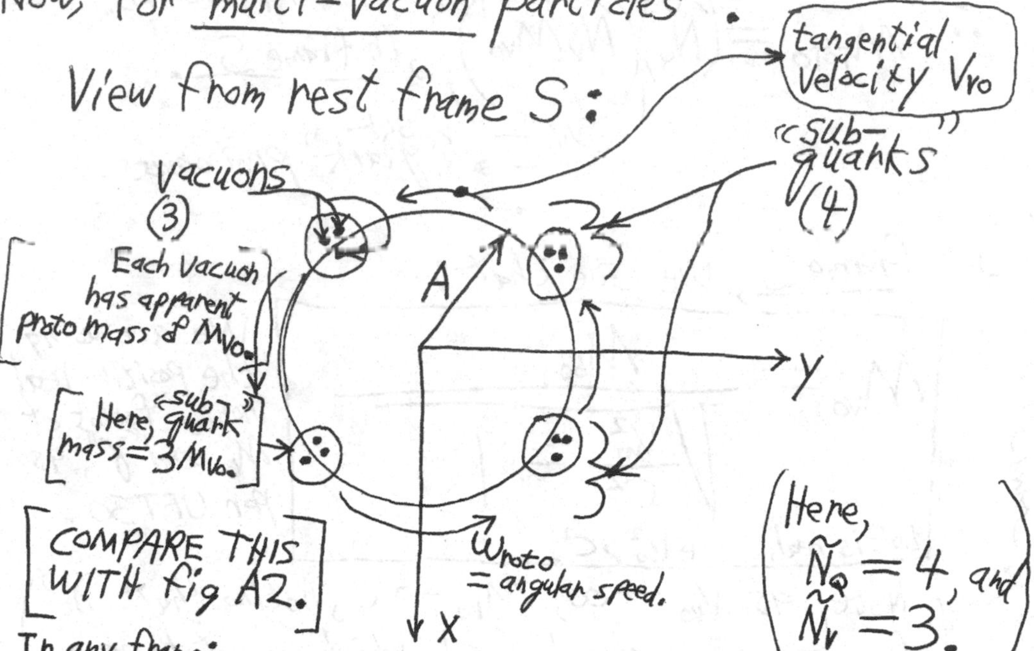

(Here, $\tilde{N}_Q = 4$, and $\tilde{N}_V = 3$.)

In any frame:

Define $\tilde{N}_Q$ as the number of "sub-quarks".

Define $\tilde{N}_V$ as the number of vacuons per "sub-quark".

Define f_{rot} as the rotational frequency, which is equal to one divided by the time it takes for all the vacuons (apparent or real) to complete one circle.

Define V_{vo} as the tangential velocity in frame S.

Define M_{TOTO} as the total apparent protomass in the ring, in frame S.

Define M_{bo} as the (real, not imaginary) "base protomass" of a vacuon, which is (would be) the apparent protomass of a zero velocity vacuon, if it could be brought to rest (which it can't!).

Figure A7

Richard P. Crandall 12-24-2001

D&R Scientific Publishing Co.; Denver, CO
"A New Theory of Quark Substructure for Mesons and Baryons"

Completed on 12-24-2001
(The "UFT" pages)

$$\therefore \; M_{TOTO} = (\tilde{N}_Q)(\tilde{N}_V M_{VO}), \text{ in frame S.}$$

($\tilde{N}_V M_{VO}$ → Vacuon protomass.)

($\tilde{N}_Q$ and $\tilde{N}_V M_{VO}$ together → "sub quark" protomass.)

In frame S, we see that:

$$M_{VO} = \frac{M_{bo}}{\sqrt{\frac{V_{VO}^2}{C^2} - 1}}$$

this is real, since $V_{VO}^2 > C^2$.

We are using the positive real number forms of M_{bo} and "γ", as per UFT30.

(Note, as $V_{VO} \to \infty$, $M_{VO} \to 0$, noting that "γ" for tachyons can be < 1, which cannot happen for slower-than-light particles.)

[illegible]

Figure A7B

Richard P. Crandall 12-24-2001

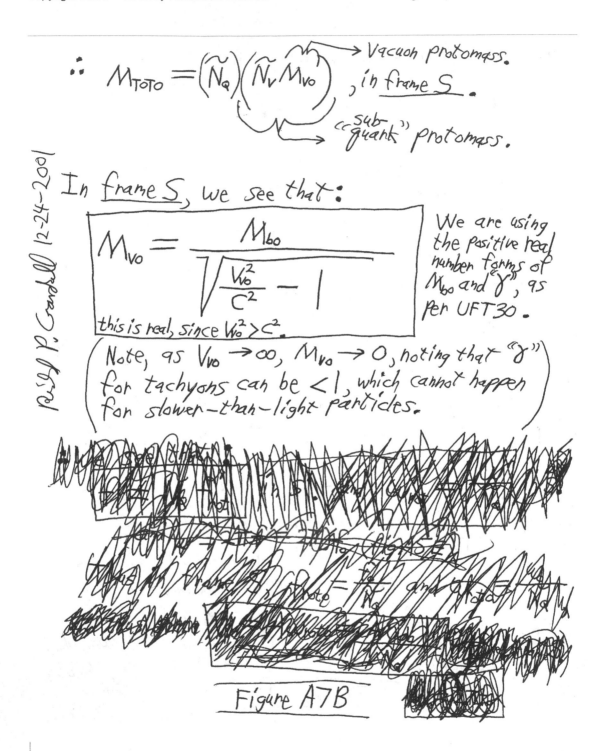

D&R Scientific Publishing Co.; Denver, CO
"A New Theory of Quark Substructure for Mesons and Baryons"

Completed on 12-24-2001
(The "UFT" pages)

Richard P. Crandall 12-24-2001

$$L_o = \frac{\mu_g \pi A^2 (L_c^2)}{l_o} \qquad L' = \frac{L_o}{\gamma}$$

Note: L_c^2 is a correction factor, as the actual solenoid is more like a helix, and does deviate from the ideal solenoid.

[we expect L_c^2 to be <1, due to flux leakage.]

We see that $\dot{i}_o = (\tilde{N}_Q \tilde{N}_v) f_{roto} M_{vo}$,

so $\boxed{\dot{i}_o = f_o \tilde{N}_v M_{vo}}$ $\boxed{i' = \gamma i_o}$.

$$m_o \gamma c^2 = E' = \underbrace{(8 L_o i_o^2)}_{const.}(\gamma) = \underbrace{(m_o c^2)}_{const.}(\gamma)$$

$$\boxed{E_o = m_o c^2 = \frac{E'}{\gamma} = 8 L_o \dot{i}_o^2 = (L_c^2) 8 \mu_g \pi A^2 f_o^2 M_{vo}^2 \tilde{N}_v^2 / l_o} = const.$$

$$p' = \frac{(E_o/f_o)}{\lambda'}$$

Then, for multi-vacuon particles,

$$\boxed{h = (L_c^2) 8 \mu_g \pi A^2 \tilde{N}_v^2 M_{vo}^2 f_o / l_o}$$

Note: $\tilde{N}_Q$ doesn't show up here!!

Then, $\boxed{E_o = h f_o}$, $\boxed{p' = h/\lambda'}$, and ∴

$$\boxed{E' = h f' = h f_o \gamma = h\left(\frac{i'}{\tilde{N}_v M_{vo}}\right)}$$

Figure A7C

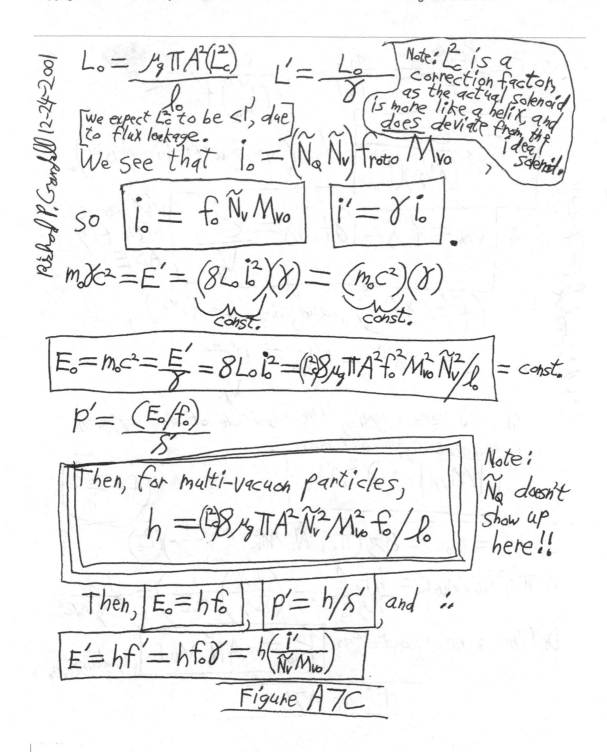

D&R Scientific Publishing Co.; Denver, CO
"A New Theory of Quark Substructure for Mesons and Baryons"

Completed on 12-24-2001
(The "UFT" pages)

$$\therefore \lambda' = \frac{(E_0/f_0)}{p'} = \frac{m_0 c^2}{f_0 p'} = \frac{c^2}{\left(\frac{p'}{m_0}\right) f_0} =$$ [redacted]

$$\boxed{\lambda' = \frac{c^2}{(\gamma V_g)(f_0)}}$$ ✓✓ (agrees with fig A6Q).

$$\therefore \boxed{V_{ph} = f'\lambda' = (\gamma f_0)\lambda' = \frac{c^2}{V_g}}$$ ✓✓ agrees with fig A5E.

$(f' = \gamma f_0)$, and, since $h = (p'\lambda')$,

$$\therefore E' = p'\lambda' f' = p' V_{ph} = \frac{p' c^2}{V_g}.$$

Also, to recap again, the magnitude of any apparent vacuon's apparent mass =

$$\boxed{|M_{vac}| = \gamma^2 M_{v0}}$$

$$f_0 = \frac{m_0 c^2}{h}$$

$$E_0 = hf_0 = m_0 c^2 = 8\mu_g (\pi A^2 \tilde{N}_v^2 M_{v0}^2) f_0^2 \left(\frac{1}{l_0}\right)(L_c^2)$$

$$\therefore \pi A^2 \tilde{N}_v^2 M_{v0}^2 L_c^2 = \frac{h f_0 l_0}{8\mu_g f_0^2} = \left(\frac{h l_0}{8\mu_g}\right)\left(\frac{h}{m_0 c^2}\right) = \frac{h^2 l_0}{8\mu_g m_0 c^2}$$

Define a new constant, as, $\boxed{K_0 = A\tilde{N}_v M_{v0} L_c}$, by definition.

Figure A7D

Richard P. Crandall 12-24-2001

D&R Scientific Publishing Co.; Denver, CO
"A New Theory of Quark Substructure for Mesons and Baryons"

Completed on 12-24-2001
(The "UFT" pages)

$$\therefore K_o = \sqrt{\frac{h^2 l_o}{8\pi \mu_g m_o c^2}\left(\frac{m_o}{m_o}\right)} = \sqrt{\left(\frac{h}{m_o c}\right)^2 (l_o m_o)\left(\frac{1}{8\pi \mu_g}\right)}$$

$$\therefore \boxed{K_o = \left(\frac{h}{m_o c}\right)\sqrt{\frac{l_o m_o}{8\pi \mu_g}}}$$

Note, $\tilde{N}_o$ doesn't show up here! (nor does $\tilde{N}_v$).

Richard P. Crandall 12-24-2001

Next page is A7E2.

Figure A7E

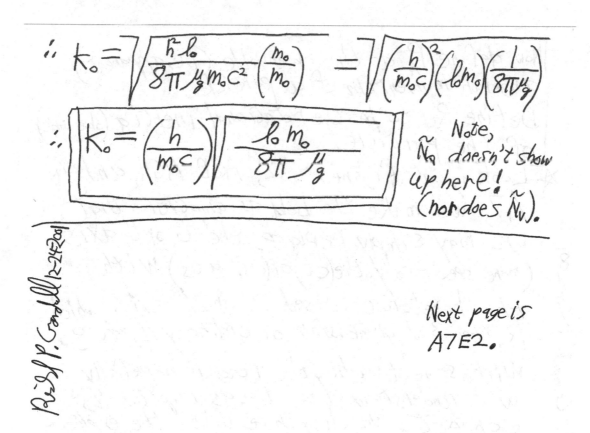

Now, define P_A as the observable (i.e. apparent) angular momentum of a particle.

Define R_I to be the rotational inertia (observable) of the particle.

☆ Looking at figures A2, A6B, A6H, and ~~A~~ A7, we make the bold assumption that we may simply replace the t or t′ axis (more specifically, the (ct) or (ct′) axis) with the heretofore unused Z or Z′ axis, which is the 3rd dimension of ordinary space. With some thought, the reader hopefully will understand that the assumption is correct. We dispense with the explicit time axis from here on, as it has served its purpose and is no longer needed. Thus, we see the particle structure as being a thin (but not zero thickness) disk containing homogeneous, Z-directed, lines of gravitomagnetic flux. Outside the disk, these flux lines are much more sparsely packed, and they exit the disk and curve back around to enter the other side of the disk, analogous to an ordinary, almost flat, solenoid of current carrying wire.

Figure A7[redacted]E2

Richard P. Crandall 12-24-2001

D&R Scientific Publishing Co.; Denver, CO
"A New Theory of Quark Substructure for Mesons and Baryons"

Completed on 12-24-2001
(The "UFT" pages)

The outer perimeter of the disk can be seen as a ring of an extremely large amount of mass flow "current", but the large mass vacuon points in the half-rings are not observable, as their actual protomass is completely shielded from our observation by the protomasses of all the other vacuons that make up the local vacuon "sea". So all we see is the powerful dipolar-like field lines which result from the large, but undetectable, mass flow (actually protomass) around the perimeter of the flat disk shaped region of space (not completely flat, but just very small thickness). So to repeat from one of the postulates, all observable particles appear to us as consisting of a (solely) tiny but powerful dipole-like gravitomagnetic field, which emanates from a tiny, thin disk-shaped region of space. That's all there is. Also, "ℓ_0" is thus the thickness, in meters, of the disk, as can be measured with a ruler which is aligned parallel to the Z axis. Thus, we start the following calculations by considering R_I to be rotational inertia of a solid, flat disk. (or cylinder) Thus, ...

Richard P. Crandall 12-24-2001

Figure A7[redacted]E3

D&R Scientific Publishing Co.; Denver, CO
"A New Theory of Quark Substructure for Mesons and Baryons"

Completed on 12-24-2001
(The "UFT" pages)

$$\boxed{P_A = R_I\,\omega'_{rot}}$$

For what follows, $\hat{A}$ may [redacted] be found to vary slightly from A, the real radius;

and, for a cylinder or disk, $R_I = \dfrac{M_{cyl}\hat{A}^2}{2}$

$$\therefore P_A = \left(\frac{M_{cyl}}{2}\right)\left(\hat{A}^2\right)\left(\frac{\omega'}{\tilde{N}_Q}\right)$$

(Neglect ring flattening along X axis due to "γ").

$$\omega' = 2\pi f' = \frac{2\pi E'}{h} = \frac{2\pi c^2 m_o \gamma}{h}$$

$M_{cyl} = m_o\gamma$ also, $m_o^2 = \dfrac{h^2 f_o^2}{c^4}$

$$P_A = \frac{2\pi \hat{A}^2 m_o^2 c^2 \gamma^2}{2\tilde{N}_Q h} = \left(\frac{\pi\hat{A}^2 c^2\gamma^2}{\tilde{N}_Q h}\right)\left(\frac{h^2 f_o^2}{c^4}\right)$$

$$P_A = \frac{\pi\hat{A}^2 f_o^2 h\gamma^2}{\tilde{N}_Q c^2}, \quad \therefore \boxed{P_{Ao} = \frac{\pi\hat{A}^2 f_o^2 h}{\tilde{N}_Q c^2}}$$

Thus, for particles with a quantum-Z directed spin # S, we must enforce this equality: (example, S = ½).

$$\frac{\pi\hat{A}^2 f_o^2 h}{\tilde{N}_Q c^2} = \frac{S h}{2\pi}; \text{ so } \frac{2\pi^2\hat{A}^2 f_o^2}{\tilde{N}_Q c^2} = S$$

$$\hat{A} = \sqrt{\frac{2S\tilde{N}_Q c^2}{4\pi^2 f_o^2}} = \left(\frac{c}{2\pi f_o}\right)\sqrt{(2S)\tilde{N}_Q}$$

Figure A7E4

Richard P. Crandall 12-24-2001

D&R Scientific Publishing Co.; Denver, CO
"A New Theory of Quark Substructure for Mesons and Baryons"

Completed on 12-24-2001
(The "UFT" pages)

$$\frac{c}{2\pi f_0} = \frac{c}{\omega_0} = \frac{c}{2\pi\left(\frac{m_0 c^2}{h}\right)} = \frac{\hbar}{m_0 c}$$

$$\therefore \hat{A} = \left(\sqrt{(2S)\tilde{N}_Q}\right)\left(\frac{\hbar}{m_0 c}\right)$$

Now, A (the actual radius), might deviate from $\hat{A}$, due to the L_c factor on fig A7C, as will be shown later.

So define a new constant K_A, such that:

$$\boxed{A = K_A \hat{A} = \left(K_A\right)\left(\sqrt{(2S)\tilde{N}_Q}\right)\left(\frac{\hbar}{m_0 c}\right)}$$

(it will finally be shown later that $K_A = 1$).

next page is A7F.

Richard P. Crandall 12-24-2001

Figure A7E5

D&R Scientific Publishing Co.; Denver, CO
"A New Theory of Quark Substructure for Mesons and Baryons"

Completed on 12-24-2001
(The "UFT" pages)

[redacted]

Note that $V_{Vo} = A\,\omega_{roto}$

, and, $\omega_{roto} = \dfrac{\omega_o}{\tilde{N}_Q} = \dfrac{2\pi f_o}{\tilde{N}_Q} = \dfrac{2\pi E_o}{\tilde{N}_Q h} = \dfrac{2\pi m_o c^2}{\tilde{N}_Q h}$

$\therefore\ A = \dfrac{\tilde{N}_Q h V_{Vo}}{2\pi m_o c^2}$] This will be used [redacted] once later, at the bottom of fig A7H.

Define A_{min} to be the minimum radius of the observable particle.

We see that $A = A_{min}$ when $V_{Vo} = c$;

$\therefore\ A_{min} = \left(\dfrac{\tilde{N}_Q}{2\pi}\right)\lambda_c$, where λ_c is the well-known compton wavelength for a particle.

$\lambda_c = \dfrac{h}{m_o c}$, by definition. (well-known)

also, $A_{min} = \left(\tilde{N}_Q\right)\left(\dfrac{1}{(m_o)\left(\frac{c}{\hbar}\right)}\right)$; where $\left(\dfrac{c}{\hbar}\right) = 2.84 \times 10^{42}$

Figure A7F

Richard P. Crandall 12-24-2001

Also, letting $\tilde{N}_q$ be it's lowest possible value, 2, we get $A_{min,min}$:

$$\therefore \boxed{A_{min,min} = \frac{2}{(m_0)\left(\frac{c}{\hbar}\right)}}$$

Richard P. Crandall 12-24-2001

Some numbers: mass of:

$h = 6.63 \times 10^{-34}$ proton 1.67×10^{-27} kg

$c = 2.99 \times 10^{8}$ m/s electron 9.11×10^{-31} kg

$\mu_g = 9.37 \times 10^{-27}$

Actual measured radius of the proton, from a QCD book (Greiner) is (page 128): ($\times 10^{-15}$ meters)

$= \sqrt{.88} = .9381$ (exactly).

For the neutron, $= \sqrt{.88 - .12?} = .87?$ (?= hard to read)

$\therefore A_{mmproton} = .42 \times 10^{-15}$ meters (looks reasonable!)

$A_{mmelectron} = 7.73 \times 10^{-13}$ meters (expected, as the electron size is still unknown!!)

Figure A7G

D&R Scientific Publishing Co.; Denver, CO
"A New Theory of Quark Substructure for Mesons and Baryons"

Completed on 12-24-2001
(The "UFT" pages)

Richard P. Crandall 12-24-2001

Define a new constant g_c, such that:

$$K_o = (g_c)\left(\frac{\sqrt{m_o}}{m_o}\right).$$

$$\therefore \boxed{g_c = \left(\frac{h}{c}\right)\left(\sqrt{l_o}\right)\left(\frac{1}{\sqrt{8\pi\mu_g}}\right)}$$

$$\therefore A\tilde{N}_v M_{vo} L_c = \frac{g_c}{\sqrt{m_o}} = (g_c)\left(\frac{\sqrt{m_o}}{m_o}\right) = K_o$$

(see fig. A7D).

$$\therefore \frac{A\tilde{N}_v M_{bo} L_c}{\sqrt{\frac{V_{bo}^2}{c^2} - 1}} = \frac{g_c\sqrt{m_o}}{m_o}$$

$$\therefore M_{bo} = \frac{(g_c)\left(\sqrt{\frac{V_{bo}^2}{c^2} - 1}\right)\sqrt{m_o}}{(L_c)\tilde{N}_v\, m_o}\left(\frac{1}{A}\right)$$

See fig. A7F for the $\left(\frac{1}{A}\right)$.

Figure A7H

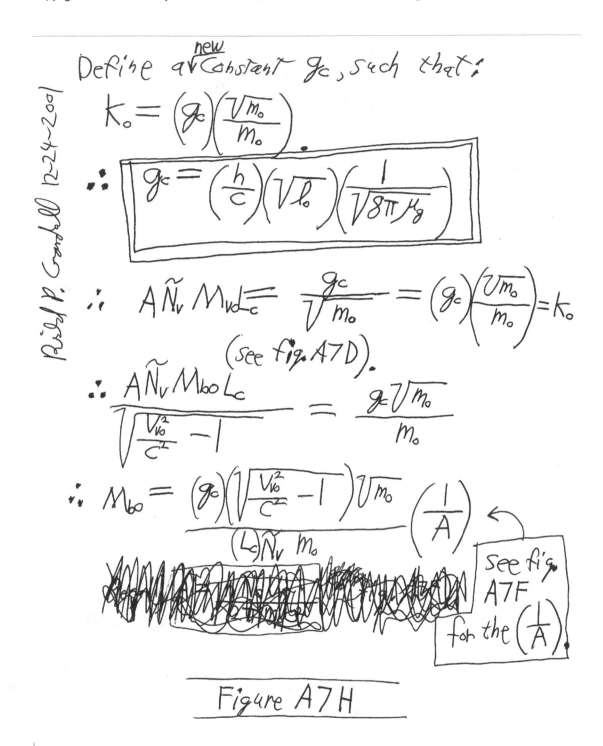

$$\therefore\ M_{bo} = \frac{(g_c)(2\pi c^2)(\sqrt{m_o})\sqrt{\frac{V_{vo}^2}{c^2}-1}}{h\tilde{N}_q\tilde{N}_v V_{vo} L_c}$$

$$M_{bo} = \frac{2\pi g_c\sqrt{m_o}\,(c)\sqrt{1-\frac{c^2}{V_{vo}^2}}}{h\tilde{N}_q\tilde{N}_v L_c}$$

Also,

$$M_{vo} = \frac{2\pi g_c\sqrt{m_o}\,c^2}{h\tilde{N}_q\tilde{N}_v V_{vo} L_c}$$

(see fig A7F for the $\frac{1}{A}$),

$$M_{bo} = M_{vo}\sqrt{\frac{V_{vo}^2}{c^2}-1}$$

$$\therefore\ M_{vo} = \left(\frac{2\pi\sqrt{m_o}\,\not{c}}{\not{h}\tilde{N}_q\tilde{N}_v L_c}\right)\left(\frac{c}{V_{vo}}\right)\left(\frac{\not{h}\sqrt{l_o}}{\not{c}\sqrt{8\pi\mu_g}}\right)$$

$$= \left(\frac{1}{L_c}\right)\left(\frac{\sqrt{m_o}}{\tilde{N}_q\tilde{N}_v}\right)\left(\sqrt{l_o}\right)\left(\sqrt{\frac{\pi}{2\mu_g}}\right)\left(\frac{c}{V_{vo}}\right)$$

$$\therefore\ \frac{M_{vo}L_c}{\sqrt{l_o}} = \left(\frac{\sqrt{m_o}}{\tilde{N}_q\tilde{N}_v}\right)\left(\sqrt{\frac{\pi}{2\mu_g}}\right)\left(\frac{c}{V_{vo}}\right)$$

Figure A7I

Richard P. Crandall 12-24-2001

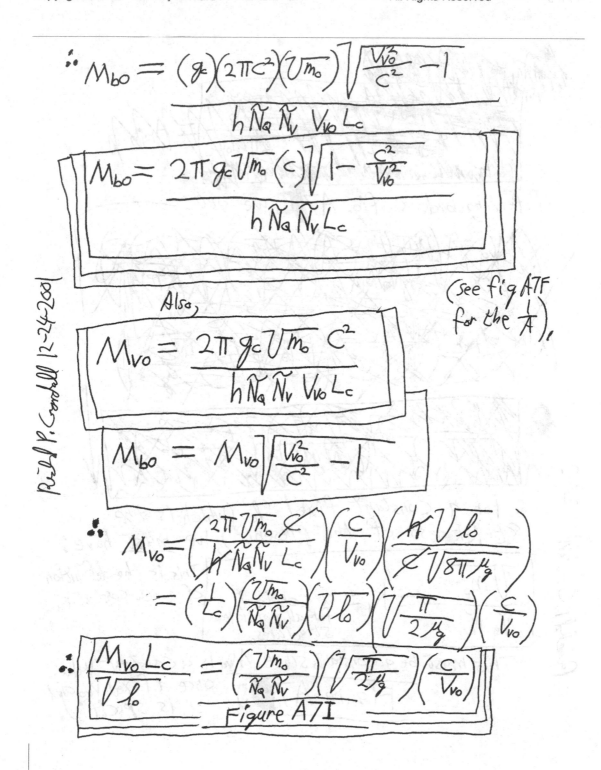

D&R Scientific Publishing Co.; Denver, CO
"A New Theory of Quark Substructure for Mesons and Baryons"

Completed on 12-24-2001
(The "UFT" pages)

Referring back to fig. A7[redacted], we get:

∴ for a constant fixed "l_o" (which is the "thickness of the present"), we must have:

$$\frac{\overline{V}m_o}{\tilde{N}_Q\,\tilde{N}_V} = K_1 = \text{a constant, for all particle structures.}$$

This is the definition of a new constant K_1.

This must be a constant, since N_{vo} is a constant value, once a fixed V_{vo} and l_o is specified.

Figure A7[redacted]J

Richard P. Crandall 12-24-2001

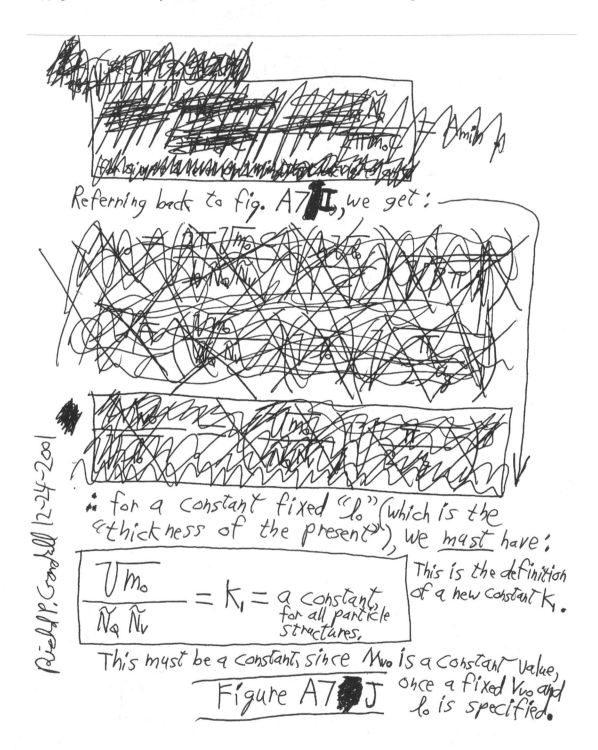

Richard P. Crandall 12-24-2001

[Define a const K_2, which is equal to the (integer) value of $\tilde{N}_Q \tilde{N}_V$ for the electron (who's rest mass is called m_{oe}).]

$\therefore$

$$K_1 = \frac{\sqrt{m_{oe}}}{K_2}$$

Later, we'll see that $K_2 = 4$.

$$\therefore \frac{\sqrt{m_o}}{\tilde{N}_Q \tilde{N}_V} = \frac{\sqrt{m_{oe}}}{K_2} \; ; \quad \sqrt{\frac{m_o}{m_{oe}}} = \frac{\tilde{N}_Q \tilde{N}_V}{K_2}$$

SEE FIG. A_MASS!!! (Proof we're on the "right track")

So, at least ideally, we should find that for any known particle, the square root of (its mass ÷ m_{oe}) should be an integer, provided that the electron is the root mass component of all particles. (which turns out to be true!).

$$\therefore \frac{L_c M_{vo} V_{vo}}{\sqrt{l_o}} = \frac{\sqrt{m_{oe}}}{K_2} \left(\sqrt{\frac{\pi}{2 \mu_g}} \right) (c) = \text{a constant.}$$

$$\text{So, } \frac{M_{vo}^2 L_c^2 V_{vo}^2}{l_o} = \frac{\pi \, m_{oe} c^2}{[\text{redacted}] \, 2 \mu_g (K_2^2)} = \text{a constant.}$$

Figure A7[redacted]K

D&R Scientific Publishing Co.; Denver, CO
"A New Theory of Quark Substructure for Mesons and Baryons"

Completed on 12-24-2001
(The "UFT" pages)

Recall for multi-vacuum particles (with $\tilde{N}_Q$ and $\tilde{N}_V$) we have, from fig A7C,

$$h = \left(8 \mu_g \pi A^2 \tilde{N}_V^2 f_0\right)\left(\frac{M_{V0}^2 C_c^2}{l_0}\right).$$

NOTE: $\tilde{N}_Q$ DOES NOT SHOW UP!!

$$\therefore \quad h = \left(8 \mu_g \pi A^2 \tilde{N}_V^2 f_0\right)\left(\frac{\pi m_{0e}}{(2 \mu_g)(K_2^2)}\right)\left(\frac{C^2}{V_{V0}^2}\right).$$

$$h = \frac{4 \pi^2 A^2 \tilde{N}_V^2 f_0 m_{0e} C^2}{(K_2^2)(V_{V0}^2)}$$

$$\therefore \quad h = \frac{\left((2\pi)\tilde{N}_V^2 A\right)(A \omega_0) m_{0e} C^2}{K_2^2 V_{V0}^2}$$

Richard P. Crandall 12-24-2001

Figure A7L

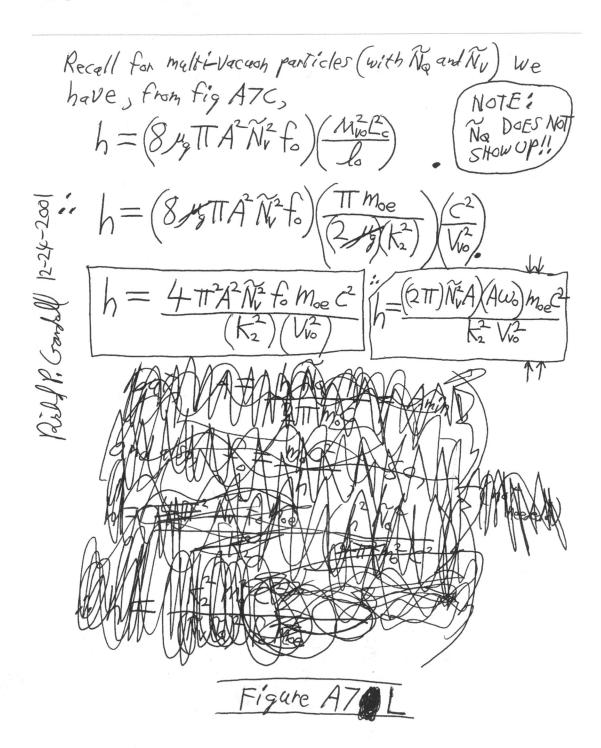

D&R Scientific Publishing Co.; Denver, CO
"A New Theory of Quark Substructure for Mesons and Baryons"

Completed on 12-24-2001
(The "UFT" pages)

Recall that (fig A7F) $\omega_o = \tilde{N}_Q \omega_{roto}$. [redacted]

$$\therefore \hbar = \left(\frac{\tilde{N}_V^2 m_{oe}}{K_2^2}\right)(A)\left(A \tilde{N}_Q \omega_{roto}\right)\left(\frac{C^2}{V_{Vo}^2}\right)$$

Now $\tilde{N}_Q$ finally shows up!

$$= \left(\tilde{N}_V \tilde{N}_Q\right)\left(\frac{\tilde{N}_V m_{oe}}{K_2^2}\right)(A)\left(A \omega_{roto}\right)\left(\frac{C^2}{V_{Vo}^2}\right)$$

$$= \left(\tilde{N}_V \tilde{N}_Q\right)\left(\tilde{N}_V\right)\left(\frac{m_o}{\tilde{N}_Q^2 \tilde{N}_V^2}\right)(A)\left(A \omega_{roto}\right)\left(\frac{C^2}{V_{Vo}^2}\right)$$

$$\therefore \boxed{\hbar = (m_o)(A)\left(\frac{A \omega_{roto}}{\tilde{N}_Q}\right)\left(\frac{C^2}{V_{Vo}^2}\right)}$$

Now, $\tilde{N}_V$ is gone.

also, $$\boxed{\frac{1}{2}\hbar = \left(\frac{m_o A^2}{2}\right)\left(\omega_{roto}\right)\left(\frac{C^2}{\tilde{N}_Q V_{Vo}^2}\right)}$$

(Compare this to top of fig A7E4).

$$\hbar = \left(\frac{m_o}{\tilde{N}_Q^2}\right)(A)\underbrace{(2\pi f_o)}_{\rightarrow\ \omega_o}(A)\left(\frac{C^2}{V_{Vo}^2}\right)$$

$$= \left(\frac{m_o}{\tilde{N}_Q^2}\right)(A)\left(2\pi A\left(\frac{1}{h}\right)m_o C^2\right)\left(\frac{C^2}{V_{Vo}^2}\right)$$

Figure A7M

Richard P. Crandall 12-24-2001

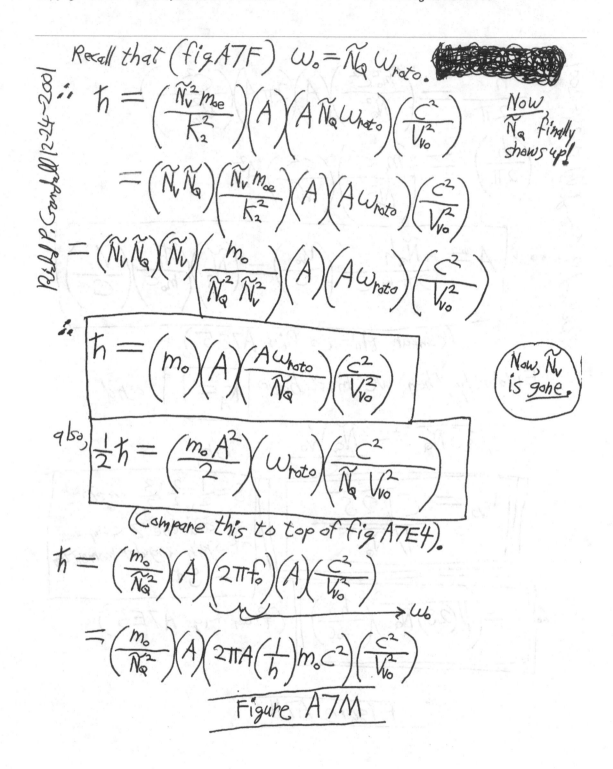

D&R Scientific Publishing Co.; Denver, CO
"A New Theory of Quark Substructure for Mesons and Baryons"

Completed on 12-24-2001
(The "UFT" pages)

Richard P. Crandall 12-24-2001

$$\therefore \frac{h^2}{2\pi} = \left(\frac{m_o^2 C^2}{\tilde{N}_Q^2}\right)(A)(2\pi A)\left(\frac{C^2}{V_{Vo}^2}\right)$$

$$\left(\frac{h}{2\pi}\right)^2 = \left(\frac{m_o^2 C^2}{\tilde{N}_Q^2}\right)\left(\frac{C^2}{V_{Vo}^2}\right)A^2$$

$$\therefore \boxed{A = \frac{\tilde{N}_Q h}{2\pi m_o C}\left(\frac{V_{Vo}}{C}\right) = \left(\tilde{N}_Q\right)\left(\frac{\hbar}{m_o C}\right)\left(\frac{V_{Vo}}{C}\right)}$$

(compare this to fig A7E5).

Evidently, then, we must have $\boxed{k_A = 1}$, and

$$\sqrt{2S\tilde{N}_Q} = \frac{\tilde{N}_Q V_{Vo}}{C}, \text{ or,}$$

$$\boxed{V_{Vo} = \frac{C\sqrt{2S}}{\sqrt{\tilde{N}_Q}}}$$

$S = \frac{1}{2}, \frac{2}{2}, \frac{3}{2}, \ldots$ etc. and the Z component of spin angular momentum $= S\hbar$

$$\therefore \boxed{A_o = \left(\sqrt{(2S)\tilde{N}_Q}\right)\left(\frac{\hbar}{m_o C}\right)} \text{ (from fig A7E5)}$$

Figure A7N

D&R Scientific Publishing Co.; Denver, CO
"A New Theory of Quark Substructure for Mesons and Baryons"

Completed on 12-24-2001
(The "UFT" pages)

Richard P. Crandall 12-24-2001

and, ∴ $\frac{1}{2}\hbar = \left(\frac{m_o A^2}{2}\right)\left(\omega_{roto}\right)\left(\frac{1}{2S}\right)$ (see fig. A7M).

$$\therefore \left(\frac{m_o A^2}{2}\right)\left(\omega_{roto}\right) = S\hbar$$

Compare to fig. A7E4 !

Thus, to our amazement, we see that our initial guess, that all vacuons are probably tachyons, is false. Apparently, these particular vacuons always have a speed less than "C", per fig A7N (V_{vo}).

✡ [However, it is a very good thing that we are now comfortable with the physics of tachyons, because it will be shown, later in this paper, that photons actually are composed of tachyon vacuons!]

Now, we summarize how the above discovery affects the previously derived formulas.
[The details change, but the main end-result is the same. But, we will see the need to change all M_{vo} to M_{vaco}.]

Figure A7O

D&R Scientific Publishing Co.; Denver, CO
"A New Theory of Quark Substructure for Mesons and Baryons"

Completed on 12-24-2001
(The "UFT" pages)

In fig A4, the apparent vacuons shown become real vacuons. Specifically, each mass point shown becomes a tiny cluster of $\tilde{N}_v$ vacuons.

From fig A7N, $f_0 A = \sqrt{(2S)\tilde{N}_q}\left(\frac{hf_0}{2\pi m_0 c}\right) = \sqrt{(2S)\tilde{N}_q}\left(\frac{c}{2\pi}\right)$

$$\therefore \quad \omega_0 A = \left(\sqrt{2S\tilde{N}_q}\right)(c)$$

Thus, the assumption of $A\omega_0 \gg V_q$, from fig A5, is still valid.

The old formula for N on fig A5C now calculates out to be: $(2V_q)\left(\sqrt{(2S)\tilde{N}_q}\right)\left(\frac{c}{2\pi}\right)/c^2 = (V_q)\left(\frac{\sqrt{2S\tilde{N}_q}}{\pi c}\right)$, which is $\ll 1$. (i.e. the phantom vacuons will never appear).

Thus, we must replace that formula with:

$$N = \frac{\tilde{N}_q}{2}$$

(independent of V_q!) (see fig A5C) ***

So, it's suggestive that $\tilde{N}_q$ should be an even integer.

The formulas for λ', f', and V_{ph} in figs A5D, A5E, should not change at all.

Figure A7P

Richard P. Crandall 12-24-2001

D&R Scientific Publishing Co.; Denver, CO
"A New Theory of Quark Substructure for Mesons and Baryons"

Completed on 12-24-2001
(The "UFT" pages)

Richard P. Crandall 12-24-2001

In the fig A6 series, the following are changed:

M_V = mass of whole ring $= 2N\tilde{N}_V M_{vac}$

and $2N = \tilde{N}_Q$.

$$i = \frac{N M_{vac}\tilde{N}_V}{\tau} = \frac{N M_V \gamma\left(\frac{c^2}{V_g} - V_g\right)}{2N2A} = \frac{(NM_V)(\gamma)\left(\frac{c^2}{V_g}\right)\left(1-\frac{V_g^2}{c^2}\right)}{2N2A}$$

$$= \boxed{\frac{M_V c^2}{4AV_g\gamma}}$$ (diff) And, we must have $\boxed{i = \gamma i_o}$ (same), also.

$L_o = \frac{\mu_g S_o}{l_o}$ (same)

In fig A6G: (note $\frac{N}{\tau} \neq \frac{f_o}{\gamma}$, anymore).

[In what follows, we must seek the limit as $V_{go} \to 0$.

So, $M_{vo} = \frac{i_o(4)AV_{go}}{c^2}$, $\therefore \frac{M_V c^2}{4AV_g\gamma} = \frac{\gamma c^2 M_{vo}}{4AV_{go}}$

$$\therefore \boxed{M_V = \gamma^2\left(\frac{M_{vo}}{V_{go}}\right)V_g} \text{ (diff)}$$

$$\therefore i_o = \frac{M_V c^2}{4AV_g\gamma^2} = \left(\frac{c^2}{4AV_g\gamma^2}\right)(\gamma^2)\left(\frac{M_{vo}}{V_{go}}\right)V_g$$

$$\therefore \boxed{i_o = \left(\frac{c^2}{4A}\right)\left(\frac{M_{vo}}{V_{go}}\right)} \text{ (diff)} \qquad \boxed{M_{vac} = \frac{M_V}{2N\tilde{N}_V}}, \text{ also.}$$

Figure A7Q

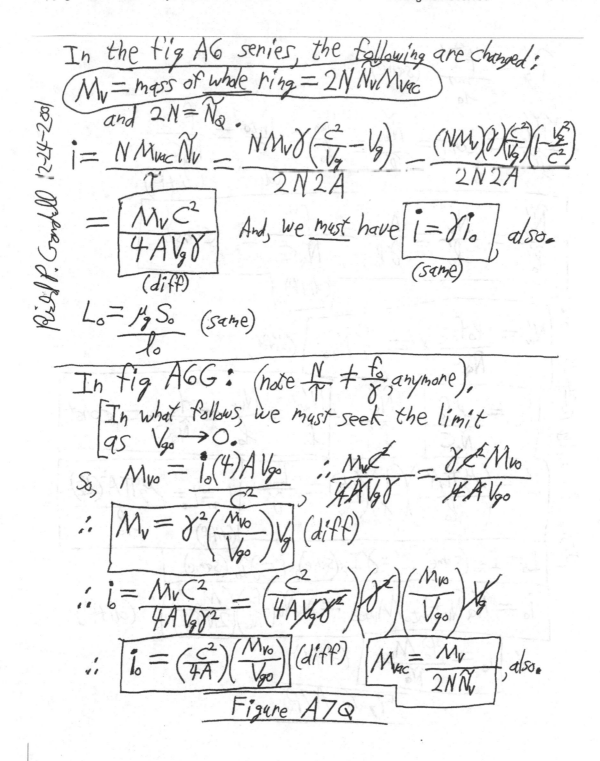

D&R Scientific Publishing Co.; Denver, CO
"A New Theory of Quark Substructure for Mesons and Baryons"

Completed on 12-24-2001
(The "UFT" pages)

$$\tilde{H}'_g = \frac{-4\tilde{I}'}{\tilde{l}'_o} \quad \text{(same)}$$

$$\frac{\tilde{N}'_W}{\tilde{l}'_o} = \text{const} = \frac{\tilde{N}_{Wo}}{\tilde{l}_o} = \frac{N_{Wo}}{l_o} = \frac{f_{roto}}{c} = \frac{f_o}{\tilde{N}_Q c}$$

(diff) (diff)

$$\boxed{\frac{\tilde{N}'_W}{\tilde{l}'_o} = \frac{N'_W}{l'_o} = \frac{N'_W}{\gamma l_o} = \frac{f_o}{\tilde{N}_Q c} = \text{const.}}$$

(diff)

$$\therefore \boxed{N'_W = \frac{l_o f_o}{\tilde{N}_Q c}\gamma = \gamma N_{Wo}} \quad \text{(diff)}$$

$$\boxed{N_{Wo} = \frac{l_o f_o}{\tilde{N}_Q c}} \quad \boxed{\frac{\tilde{N}'_W}{\tilde{l}'_o} = \frac{N'_W}{l'_o} = \frac{N_{Wo}}{l_o} = \frac{f_o}{\tilde{N}_Q c} = \text{const.}}$$

$$\boxed{L' = \left(\frac{\mu_g \pi A^2}{l_o}\right)\left(\frac{1}{\gamma}\right)\left(L_c^2\right) = \frac{L_o}{\gamma}} \quad \boxed{L_{so} = L_o = \frac{\mu_g \pi A^2 (L_c^2)}{l_o}}$$

(diff)

$$\boxed{I_o = I_{so} \text{ (same)},\ I' = \gamma I_o \text{ (same)},\ i' = \gamma i_o \text{ (same)}}$$

$$\boxed{i_o = \tilde{N}_Q \tilde{N}_V f_{roto} M_{vaco} = \tilde{N}_Q \tilde{N}_V \left(\frac{f_o}{\tilde{N}_Q}\right)\left(\frac{M_{vo}}{2N\tilde{N}_V}\right)} \quad \text{(diff)}$$

$$\therefore \boxed{i_o = \frac{f_o M_{vo}}{\tilde{N}_Q}} \quad \text{(diff)}$$

Figure A7R

Richard P. Crandall 12-24-2001

D&R Scientific Publishing Co.; Denver, CO
"A New Theory of Quark Substructure for Mesons and Baryons"

Completed on 12-24-2001
(The "UFT" pages)

Richard P. Crandall 12-24-2001

$I_o = \tilde{N}_Q \tilde{N}_V f_{roto} M_{vaco}$, ∴ $\boxed{I_o = i_o}$ (same)

∴ $I' = \gamma I_o = \gamma i_o = \boxed{\dfrac{\gamma f_o M_{vo}}{\tilde{N}_Q}} = i = i'$
(diff)

note, now M_{vaco} (one) $\neq M_{vo}$ (all)
(diff)

so, $\boxed{M_{vaco} = \dfrac{M_{vo}}{\tilde{N}_Q \tilde{N}_V}}$
(diff)

$$|M_{vac}| = (\gamma^2 M_{vo})\left(\frac{V_g}{V_{go}}\right)\left(\frac{1}{\tilde{N}_Q \tilde{N}_V}\right) = \gamma^2\left(\frac{V_g}{V_{go}}\right)|M_{vaco}|$$
(diff)

$$E_o = m_o c^2 = \frac{E'}{\gamma} = 8 L_o i_o^2 = (L_c^2)\, 8\mu_g \pi A^2 \left(\frac{1}{\tilde{N}_Q^2}\right) M_{vo}^2 f_o^2 / \ell_o$$
(same) (same) (diff) = a constant.

Note $\tilde{N}_V$ doesn't show up.

new

$$\text{So } h = (L_c^2)\, 8\mu_g \pi A^2 \left(\frac{1}{\tilde{N}_Q^2}\right) M_{vo}^2 f_o / \ell_o$$
(diff)

∴ $\left(\dfrac{1}{\tilde{N}_Q^2}\right) M_{vo}^2 = \tilde{N}_V^2 M_{vaco}^2$

∴ $h = (L_c^2)\, 8\mu_g \pi A^2 \tilde{N}_V^2 M_{vaco}^2 f_o / \ell_o$; $E_o = h f_o$

(same as fig A7C, IF we replace M_{vo} with M_{vaco}).

Figure A7S

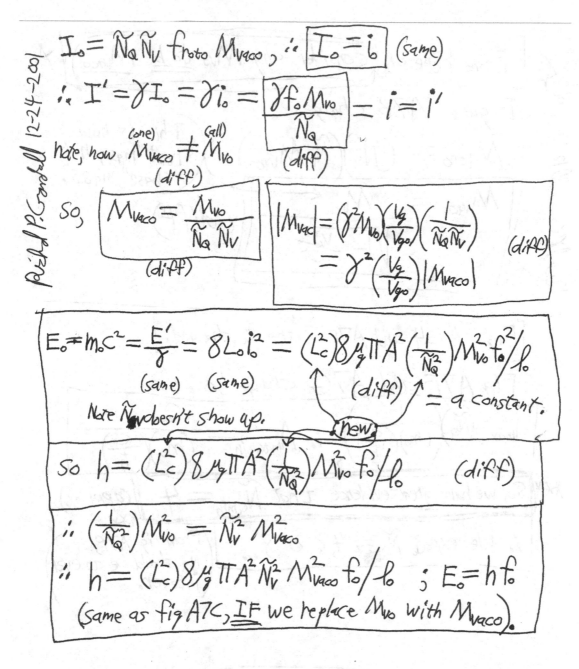

D&R Scientific Publishing Co.; Denver, CO
"A New Theory of Quark Substructure for Mesons and Baryons"

☆ [From here on, consider any M_{vo} to be M_{vaco}] ☆

Richard P. Crandall 12-24-2001

Figure A7B changes:

$$M_{TOTO} = (\tilde{N}_Q)(\tilde{N}_V M_{vaco})$$

$$M_{vaco} = \frac{M_{ro}}{\sqrt{1 - \frac{V_{vo}^2}{C^2}}}$$

(diff)

(This is how "rest" mass, not "base" mass.)

[Figs A7E4 and A7E5 are unchanged.]

Fig A7F and A7G change to:

$$A_{max} = (\tilde{N}_Q)\left(\frac{1}{(m_o)\left(\frac{c}{\hbar}\right)}\right), \quad \therefore A_{max,min} = \frac{2}{(m_o)\left(\frac{c}{\hbar}\right)}$$

*** So, we have great evidence that $\tilde{N}_{Qmin} = 4$ (approx.) ***

∴ We expect $\tilde{N}_Q$ [redacted] $= 4, 6, \text{or } 8 \ldots$

(From fig A7P, $\tilde{N}_Q$ should be an even integer.)

Figure A7T

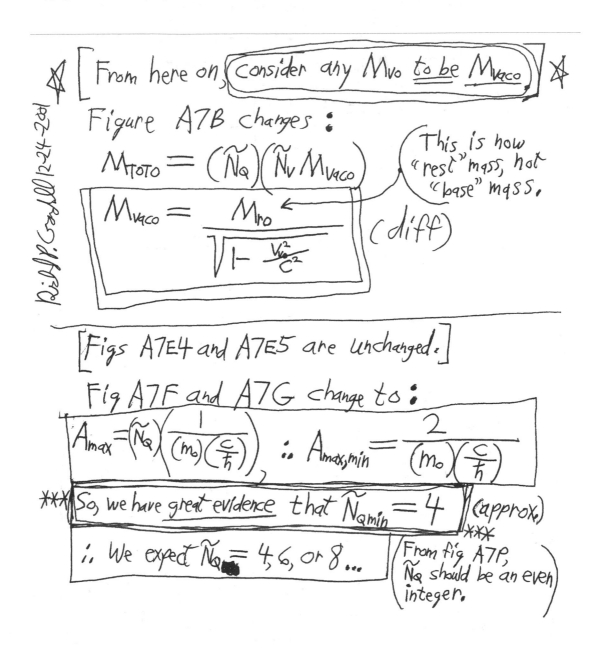

D&R Scientific Publishing Co.; Denver, CO
"A New Theory of Quark Substructure for Mesons and Baryons"

Completed on 12-24-2001
(The "UFT" pages)

Changes to fig A7I :

$$M_{ro} = \frac{(q_c)(2\pi c^2)(\mho m_o)\sqrt{1-\frac{V_{vo}^2}{c^2}}}{h \tilde{N}_Q \tilde{N}_V V_{vo} L_c}$$ (diff)

$$M_{vaco} = \frac{2\pi q_c \mho m_o c^2}{h \tilde{N}_Q \tilde{N}_V V_{vo} L_c}$$ (same)

apparent mass of any one vacuon.

$$M_{ro} = M_{vaco}\sqrt{1-\frac{V_{vo}^2}{c^2}}$$ (diff.)

"rest" mass.

$$\frac{M_{vaco} L_c}{\mho l_o} = \left(\frac{\mho m_o}{\tilde{N}_Q \tilde{N}_V}\right)\left(\sqrt{\frac{\pi}{2\mu_q}}\right)\left(\frac{c}{V_{vo}}\right)$$ (same)

Figs A7J, A7K, A7L, A7M : no changes.

Figs A7N, A7O : NO changes

(end of changes.)

(last AZ fig.)

Remember, always replace any M_v with M_{vac}, and any M_{vo} with M_{vaco}.

Figure A7U

Richard P. Crandall 12-24-2001

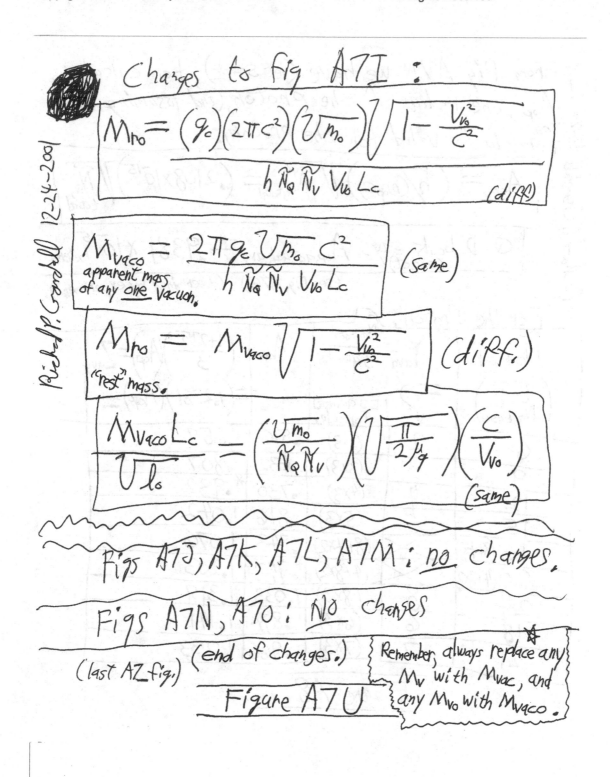

Richard P. Crandall 12-24-2001

From fig A7N we have (let $S = \frac{1}{2}$), for calculating A_{op}, the radius of the proton, (and assuming this formula is valid for the proton):

$$A_{op} = (\hbar/(m_{op}c))\sqrt{\widetilde{N}_{Q,total}} = (.2108 \times 10^{-15})\sqrt{\widetilde{N}_{Q,total}}$$

QCD book says (exact) $A_{op,actual,avg} = .9381 \times 10^{-15}$ meters.

(see Fig. A7G)   (determined by experiment).

For the Proton:

$\widetilde{N}_{Q,total}$ (or $\widetilde{N}_{Q,base}$)	$Y_{avg} = \dfrac{\widetilde{N}_{Q,total}}{3\,quarks} = ?$	[3 quarks] Formula:	A_{op}	$\left(\dfrac{1+2\sqrt{2}}{3}\right)(A_{op}) = Z_{avg} = (1.2761)(A_{op}) =$
6	2	(2×3)	.516	.659
9	3	(3×3)	.632	.807
12	4	(4×3)	.730	*.932
15	5	(5×3)	.816	1.042
16 (8+4×2)	✗	(8+4×2)	.843	1.076
20 (4+8×2)	✗	(4+8×2)	*.943	1.203
24	8	(8×3)	1.033	1.318
18	6	(6×3)	.894	1.141
21	7	(7×3)	.966	1.233

Figure A8

D&R Scientific Publishing Co.; Denver, CO
"A New Theory of Quark Substructure for Mesons and Baryons"

Completed on 12-24-2001
(The "UFT" pages)

Let Y_n = the number of subquarks in quark #n, where $n = 1, 2,$ or 3.

$\therefore \tilde{N}_{Q,total} = \sum_{n=1}^{3} Y_n$; also, $(3\,Y_{avg}) = \tilde{N}_{Q,total}$.

Assume $Y_1 = Y_2 = Y_3$ $\therefore \tilde{N}_{Q,total} = 20$,

but 20 is not a multiple of 3. Now 21 is the closest, but $Y_{avg} = Y_n = 7$, and 7 is not even. This case is no good. (Also, .966 isn't close enough.)

Now, Assume $Y_1 = 2Y_2 = Y_3$ [For the Proton, only.]

$\therefore \tilde{N}_{Q,total} = 2Y_2 + Y_2 + 2Y_2 = 5Y_2 = 20$, still.

$\therefore Y_2 = 4$, and, $Y_1 = Y_3 = 8$. All 3 are even.

This gives $A_{op} = .943$. ←use some particular

Now, suppose we "base" $\tilde{N}_Q$. The average radius of 3 separate, 1-quark, particles (Y_1', Y_2', Y_3') is thus given by $(A_{op,base})\left(\frac{1+2\sqrt{2}}{3}\right) = Z_{avg} = .932$, when "base" $\tilde{N}_Q = 12$. Then, the doubled $\tilde{N}_Q$ = base*2 = 24.

$(24 = (2)(12) = 24)$ Figure A9 $[Y_1' = 2Y_2' = Y_3']$

Richard P. Crandall 12-24-2001

D&R Scientific Publishing Co.; Denver, CO
"A New Theory of Quark Substructure for Mesons and Baryons"

Completed on 12-24-2001
(The "UFT" pages)

Note that $\frac{12+24+24}{3} = 20$, and

Avg radius $= \frac{.730 + 1.033 + 1.033}{3} = .932 = Z_{avg}$, when $\tilde{N}_{Q,base} = 12$.

Richard P. Crandall 12-24-2001

> If, instead, we assume, say, $Y_1 = (\frac{1}{2}) Y_2 = Y_3$, then we'd have $\tilde{N}_{Q,total} = \frac{1}{2}Y_2 + Y_2 + \frac{1}{2}Y_2 = 2Y_2 = 20$, so that $Y_1 = Y_3 = 5$ and $Y_2 = 10$. This case is no good, since Y_1 and Y_3 are <u>not even</u>. However, we will later see this as <u>good for neutrons</u>. So, we don't use this for the proton.

Thus, we see the 3 cases of interest as: [proton]

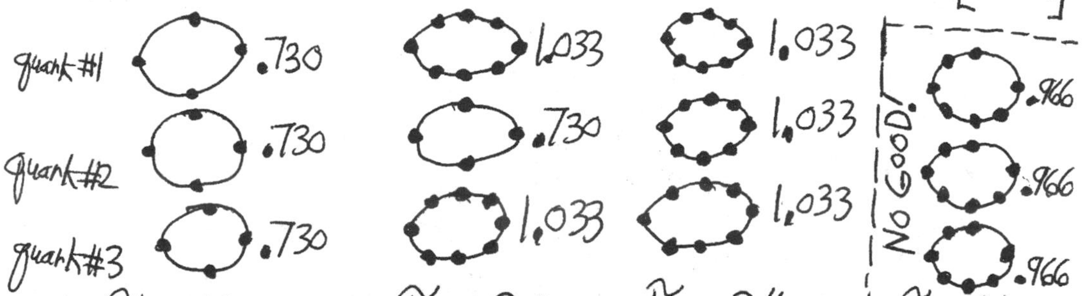

$\tilde{N}_Q = 12$	$\tilde{N}_Q = 20$	$\tilde{N}_Q = 24$	$\tilde{N}_Q = 21$
$A_{avg} = .730$	$A_{avg} = .932$ ✓	$A_{avg} = 1.033$	$A_{avg} = .966$

Note, the <u>average</u> of $(.943, .932) = .9375 =$ very close to $.9381$ (actual measured)

Figure A1Ø

D&R Scientific Publishing Co.; Denver, CO
Completed on 12-24-2001
"A New Theory of Quark Substructure for Mesons and Baryons"
(The "UFT" pages)

Incidentally, if we let $Y_1 = 3Y_2 = Y_3$, then we get $\tilde{N}_{Q,total} = 3Y_2 + Y_2 + 3Y_2 = 7Y_2 = 20$. This is NO GOOD, as Y_2 is not an integer.

Thus, we have for the Proton:

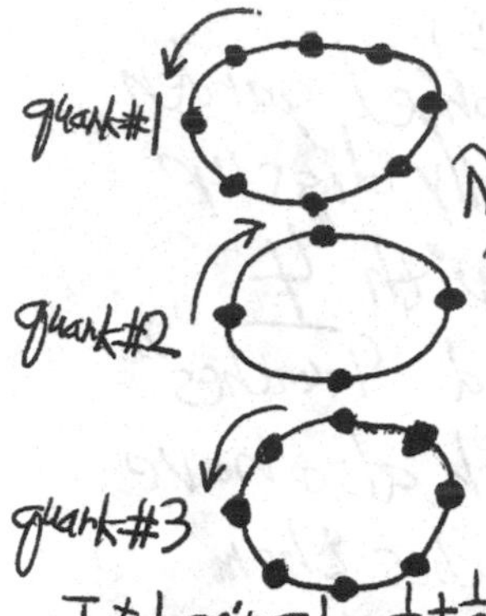

$\tilde{N}_Q = 20$
$\tilde{N}_V = ?$

Total spin $= \frac{1}{2} - \frac{1}{2} + \frac{1}{2} = {}^{+}\frac{1}{2}$.

Now, if we model the electron as

$\tilde{N}_Q = 4$ $\tilde{N}_V = 1$ spin $= \frac{-1}{2}$, then we can explain all of electromagnetic theory, including charges. Here's how:

Assumed rotation directions are shown with an arrow.

∴ It is thus believed that a photon is actually:

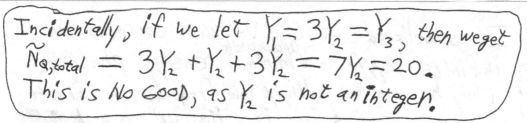

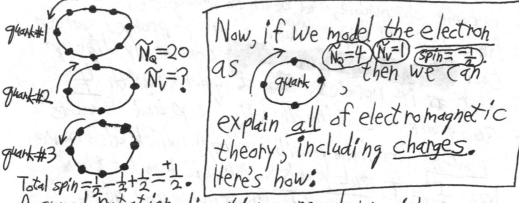

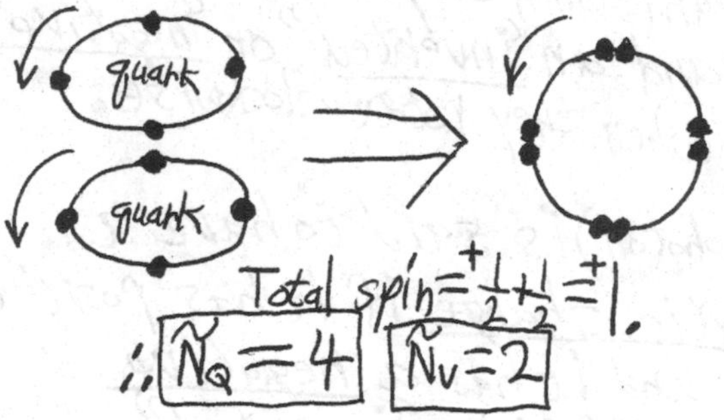

It is believed that these stick together as pairs, instead of interleave as above, since spin directions are the same.

Total spin $= {}^{+}\frac{1}{2} {}^{+}\frac{1}{2} = {}^{+}1$.

∴ $\tilde{N}_Q = 4$ $\tilde{N}_V = 2$

Richard P. Crandall 12-24-2001

Figure A11

D&R Scientific Publishing Co.; Denver, CO
"A New Theory of Quark Substructure for Mesons and Baryons"

Completed on 12-24-2001
(The "UFT" pages)

It is believed that the vacuum contains an infinite supply of unobservable minimum energy photons. They have an exceedingly low frequency, but finite. As any quark spins, it "digs up" these [redacted] photons from the unobservable vacuum (and then, they become observable, and begin to propagate). An analogy will help. Much as a 4-pronged garden tool [redacted] continually digs up dirt as it rotates, [redacted] a quark with 4 subquarks continually digs up and "launches" photons (minimum possible energy) which also have exactly 4 subquarks. And it launches them evenly in all spatial directions, and at a fixed rate. Quarks, or photons, are said to have a positive linear momentum when they rotate counterclockwise, and an "inverted" or "negative" linear momentum when they rotate clockwise.

A quark (but not a photon) is said to have a positive electrostatic charge if it has positive linear momentum, and it has a negative electrostatic charge if it has "inverted" linear momentum.

Richard P. Crandall 12-24-2001

FIGURE A12

D&R Scientific Publishing Co.; Denver, CO
"A New Theory of Quark Substructure for Mesons and Baryons"

Completed on 12-24-2001
(The "UFT" pages)

Richard P. Crandall 12-24-2001

Now, if a [redacted] <u>normal</u> momentum object collides with another <u>normal</u> momentum object, then the linear momentum of the second object will <u>increase</u>.

If an <u>inverted</u> momentum object collides with a <u>normal</u> momentum object, then the linear momentum of the second object will <u>decrease</u>. Similar for the other 2 cases.

[[There fore, like charges repel due to photon exchanges between the two particles, and unlike charges attract.

Now, looking at the upper and lower quarks in the proton, we see that they each have <u>twice</u> as many subquarks as does the middle quark. Now, we can explain the [redacted] total charge for both the electron and proton:

[Having eight subquarks is the same as having <u>two</u> of the 4-pronged "photon diggers" working at the same time, and thus, the effective rate of photon launching doubles for quarks #1 and #3.]

Figure A13

D&R Scientific Publishing Co.; Denver, CO
"A New Theory of Quark Substructure for Mesons and Baryons"

Completed on 12-24-2001
(The "UFT" pages)

Therefore, we can say that the quark charges in a proton are as follows:

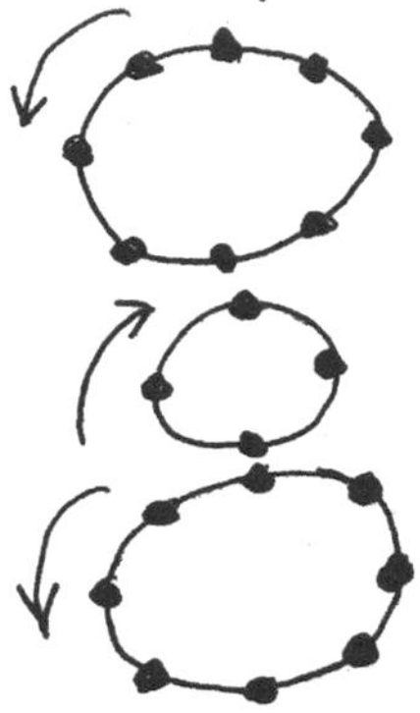

#1 charge = $+\frac{2}{3}$

#2 charge = $-\frac{1}{3}$

#3 charge = $+\frac{2}{3}$

∴ total charge = +1 ✓

For the electron:

total charge = −1

Thus, totally explained/derived, are electric charges and forces between them. What we normally observe as an Electric field (or a potential field) is just the field of force exerted due to the steady stream of launched photons.

Richard P. Crandall 12-24-2001

Figure A14

Now, for the neutron (see fig. A_MASS):

$$m_{op}/m_{on} = \frac{938.3}{939.6} \quad \text{(see Fig. A8)}$$

$$\therefore \boxed{A_{on} = (A_{op})\left(\frac{m_{op}}{m_{on}}\right) = (.2108\times10^{-15})\left(\frac{938.3}{939.6}\right)\sqrt{\tilde{N}_{q,total}}}$$

The (same) QCD book value is confusing, but I believe the intended value for A_{on}^2 is $.88 - .12$,

$$\boxed{\text{So } A_{on,actual,avg} = \sqrt{.88 - .12} = .87\times10^{-15}\ \text{meters.}}$$

(see fig. A7G) (determined by experiment).

Rich P. Crandall 12-24-2001

For the Neutron:

$\tilde{N}_{q,total}$ (or $\tilde{N}_{q,base}$)	Formula:	A_{on}	$\left(\frac{2+\sqrt{2}}{3}\right)(A_{on}) = Z_{avg} = (1.1381)(A_{on}) =$
9	(3x3)	.632	.719
12	(4x3)	.729	* .830
15	(5x3)	.815	.928
18	(6x3)	.893	1.016
16 (8+4x2)	(8+4x2)	* .842	.958
17	X	.868	X (too large)
24	(8x3)	1.031	(too large)

Figure A15

D&R Scientific Publishing Co.; Denver, CO
"A New Theory of Quark Substructure for Mesons and Baryons"

Completed on 12-24-2001
(The "UFT" pages)

Let Y_n = the number of subquarks in quark #n, where $n = 1, 2,$ or 3.

$\therefore \tilde{N}_{Q,total} = \sum_{n=1}^{3} Y_n$; also, $(3Y_{avg}) = \tilde{N}_{Q,total}$.

Assume $Y_1 = Y_2 = Y_3$ $\therefore \tilde{N}_{Q,total} = 16, 17,$ or 18.

Now 16 and 17 are ruled out, since not multiple of 3. Thus, we get $Y_{avg} = Y_n = 6$.

However, given the proton structure, we know that this solution for the neutron can't be valid. (recall fractional changes) The total charge would not be zero, but zero is the total neutron charge.

Now,

Assume $Y_1 = 2Y_2 = Y_3$ [for the neutron]

$\therefore \tilde{N}_{Q,total} = 2Y_2 + Y_2 + 2Y_2 = 5Y_2 = 16, 17, 18,$ or 19.

This case is no good, since none of these are exactly divisible by 5.

Figure A16

Richard P. Crandall 12-24-2001

D&R Scientific Publishing Co.; Denver, CO
"A New Theory of Quark Substructure for Mesons and Baryons"

Completed on 12-24-2001
(The "UFT" pages)

Richard P. Crandall 12-24-2001

Now, we try:

$$\boxed{\text{Assume } Y_1 = \left(\tfrac{1}{2}\right) Y_2 = Y_3}$$ [For the Neutron, only.]

$\therefore \tilde{N}_{Q,total} = \frac{1}{2}Y_2 + Y_2 + \frac{1}{2}Y_2 = 2Y_2 = 16, 17, 18, \text{ or } 19.$

Thus, we can only have $Y_2 = 8$ or 9. (must be integer).

However, if Y_2 is 9, then Y_1 is not an integer.

$\therefore Y_2 = 8$, and $Y_1 = Y_3 = 4$. All 3 are even.

This gives $A_{on} = .842$.

Now, suppose we use some particular "base" $\tilde{N}_Q$. The average radius of 3 separate, 1-quark, particles (Y'_1, Y'_2, Y'_3) is thus given by $(A_{on,base})\left(\frac{2+1/\sqrt{2}}{3}\right) = Z_{avg} = .830$, when "base" $\tilde{N}_Q = 12$. Then, the doubled $\tilde{N}_Q = \text{base} * 2 = 24$.

$\left(Y'_1 = \left(\tfrac{1}{2}\right) Y'_2 = Y'_3\right)$ or, $\left(12 = \left(\tfrac{1}{2}\right)(24) = 12\right)$.

Note that $\frac{24+12+12}{3} = 16$, and

Avg radius $= \frac{1.031 + .729 + .729}{3} = .830 = Z_{avg}$, when $\tilde{N}_{Q,base} = 12$.

Figure A17

D&R Scientific Publishing Co.; Denver, CO
"A New Theory of Quark Substructure for Mesons and Baryons"

Completed on 12-24-2001
(The "UFT" pages)

Thus, we see the 3 cases of interest as: [neutron]

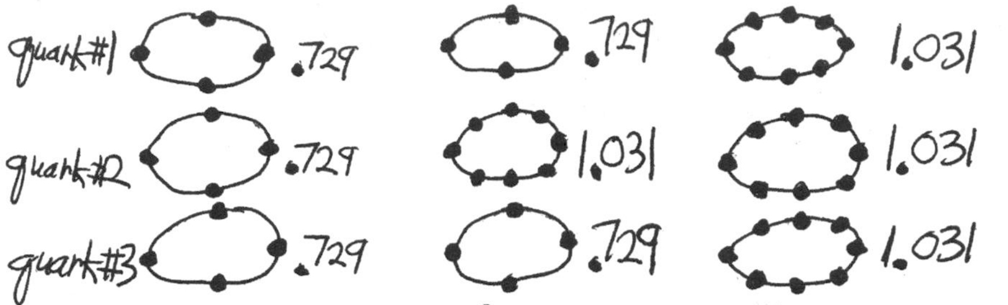

$\tilde{N}_Q = 12$ $\quad$ $\tilde{N}_Q = 16$ $\quad$ $\tilde{N}_Q = 24$

$A_{avg} = .729$ $\quad$ $A_{avg} = .830$ ✓ $\quad$ $A_{avg} = 1.031$

In Summary, for the Neutron, we have the calculated radius values as .842 and .830, and the <u>measured value</u> as .87.

Thus, we have for the <u>Neutron</u>:

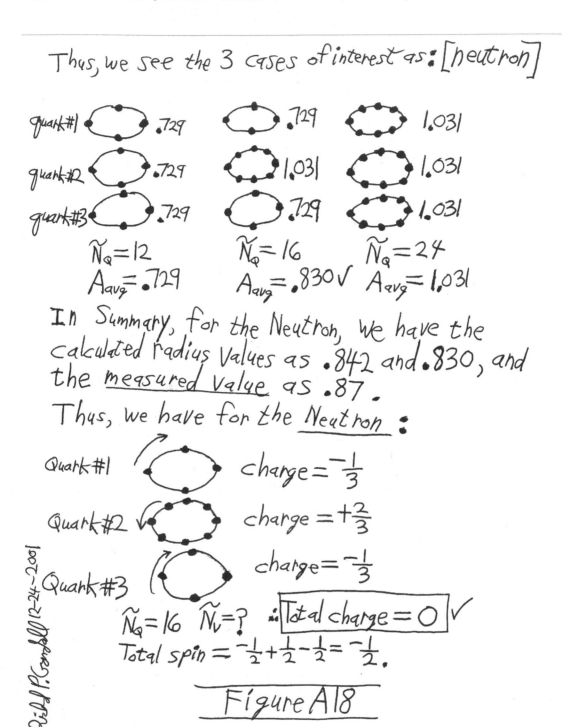

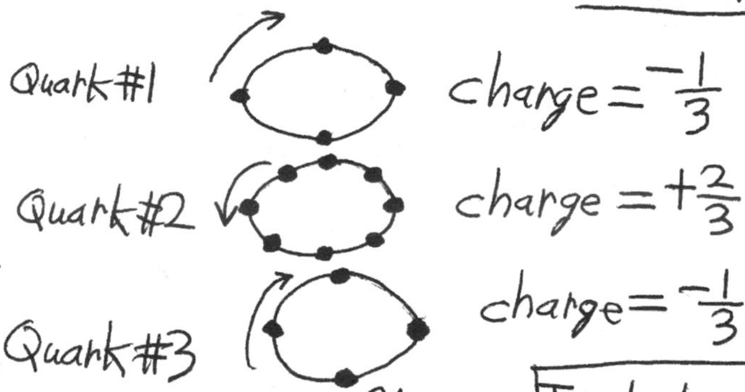

$\tilde{N}_Q = 16$ $\quad$ $\tilde{N}_V = ?$ $\quad$ ∴ **Total charge = 0** ✓

Total spin $= -\frac{1}{2} + \frac{1}{2} - \frac{1}{2} = -\frac{1}{2}$.

Figure A18

Richard P. Crandall 12-24-2001

D&R Scientific Publishing Co.; Denver, CO
"A New Theory of Quark Substructure for Mesons and Baryons"

Completed on 12-24-2001
(The "UFT" pages)

Richard P. Crandall 12-24-2001

Recall from figure A7K that:

$$\boxed{\sqrt{\frac{m_o}{m_{oe}}} = \frac{\tilde{N}_Q \tilde{N}_V}{K_2}}$$

, where $K_2 = \tilde{N}_{Qe} \tilde{N}_{Ve}$

"e" is electron.

From figure A11, we see that, for the electron, we have: $\tilde{N}_{Qe} = 4$, $\tilde{N}_{Ve} = 1$.

$$\therefore \boxed{K_2 = 4}$$

[From figure A_MASS, we see that for the proton or neutron, made into an integer, we have $\sqrt{\frac{m_o}{m_{oe}}} = 42$.

$$\therefore \tilde{N}_Q \tilde{N}_V = (4)(42) = \boxed{168}$$ [Proton, or Neutron].

Now, both Proton and Neutron can be diagrammed as:
[redacted]

Actually, think of this as a generic nucleon.

$\tilde{N}_{Q,total} = 24$.

$$\therefore \boxed{\tilde{N}_V = 168/24 = 7}$$ for a "generic" nucleon.

[Now, note from fig A_MASS that several particles, including proton and [redacted] neutron, are divisible by 7.

Figure A19

D&R Scientific Publishing Co.; Denver, CO
"A New Theory of Quark Substructure for Mesons and Baryons"

Completed on 12-24-2001
(The "UFT" pages)

A little insight then shows the true nature of both the proton and neutron, and how they can be formed from one generic structure:

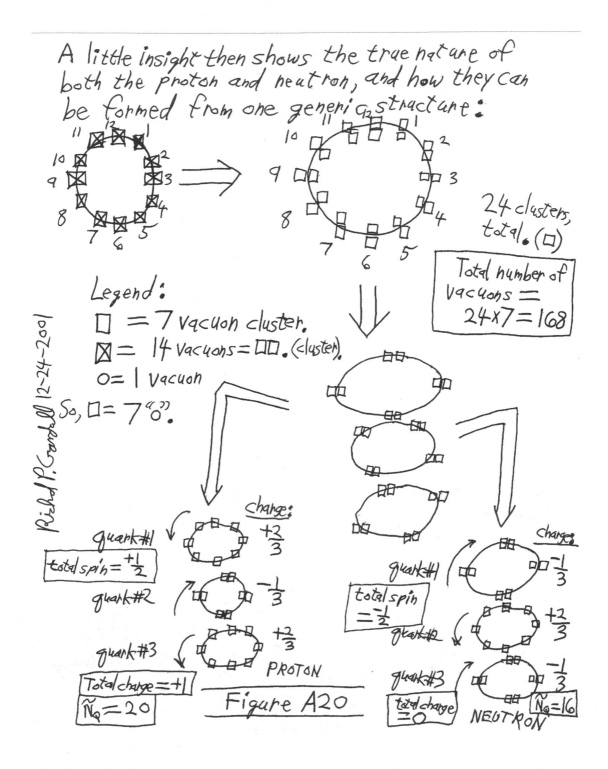

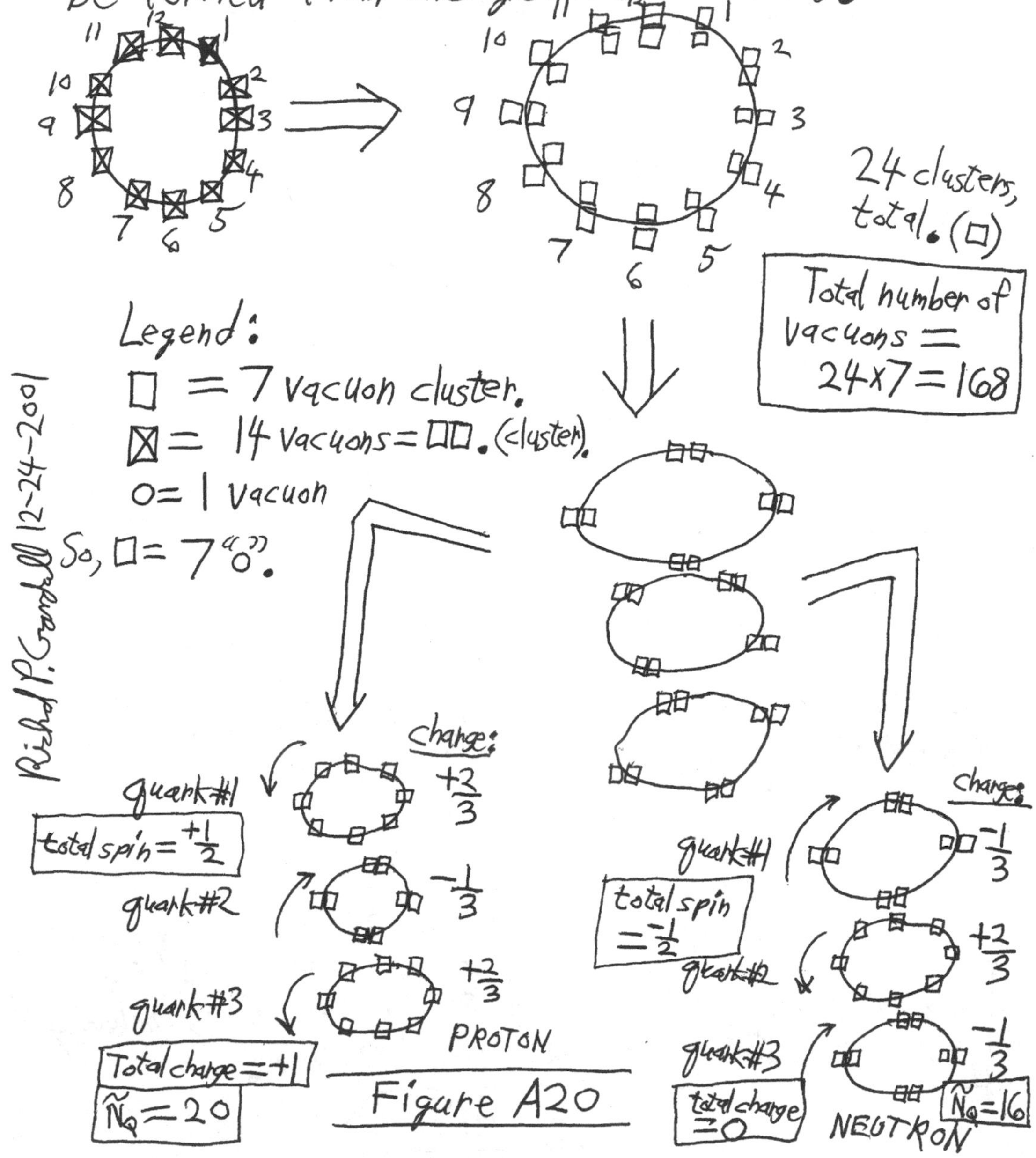

Figure A20

D&R Scientific Publishing Co.; Denver, CO
"A New Theory of Quark Substructure for Mesons and Baryons"

Completed on 12-24-2001
(The "UFT" pages)

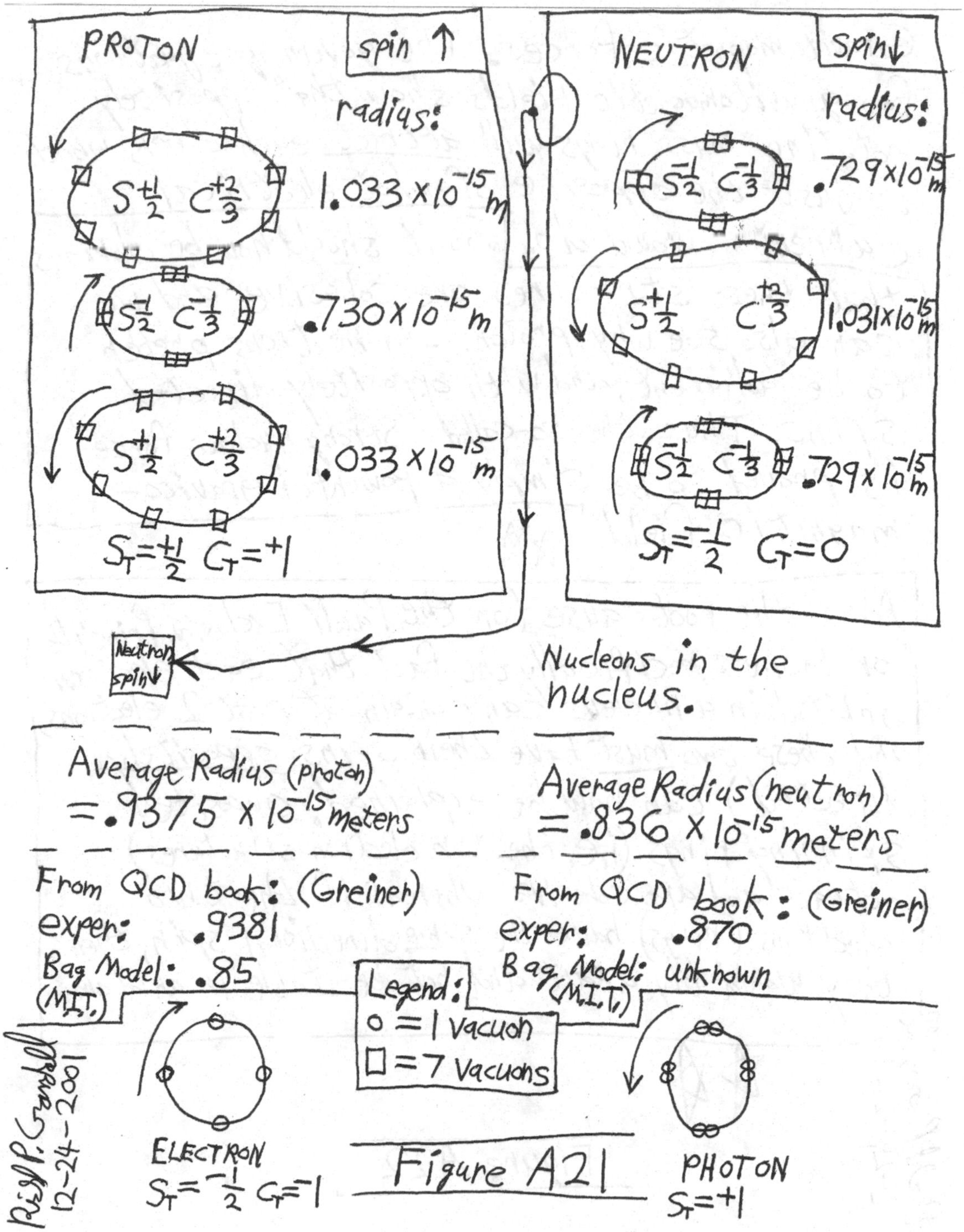

Figure A21

D&R Scientific Publishing Co.; Denver, CO
"A New Theory of Quark Substructure for Mesons and Baryons"

Completed on 12-24-2001
(The "UFT" pages)

Gravitomagnetic forces. The governing equations for gravitomagnetic fields show that oppositely rotating mass rings will attract each other, which is just the opposite of what electrical currents would do. So it should now be obvious that these structures are cohesive, and we can also see why protons and neutrons prefer to be adjacent, and with oppositely directed spins. Thus, the so-called "Strong nuclear force" is proved to be simply a powerful gravito-magnetic field. ☆☆

Also, the root cause for the Pauli Exclusion Principle (or more specifically the fact that each electron orbital in a nucleus can contain at most 2 electrons, and these two must have their spins oppositely directed) can now be explained. Oppositely spinning rings (i.e. the two electron structures) attract and are cohesive, whereas if the two electrons (rings) have the same directional spin, then they would fly apart and not be stable in an atom.

☆☆

Richard P. Crandall
12-24-2001

Figure A22

D&R Scientific Publishing Co.; Denver, CO
"A New Theory of Quark Substructure for Mesons and Baryons"

Completed on 12-24-2001
(The "UFT" pages)

Later (the "L" and "NSS" pages) it will be shown that any isolated (ie free) quark (ring structure) always must contain exactly 4 subquarks, no more, no less. This is due to force balance reasons.

Thus, we see that the nucleon rings which have 8 subquarks are very unusual. The only explanation for this is that the overall (net) gravitomagnetic field must be causing forces (outward) that tend to separate each clump of 14 vacuons into 2 separate clumps of 7 vacuons. This can be seen as follows:

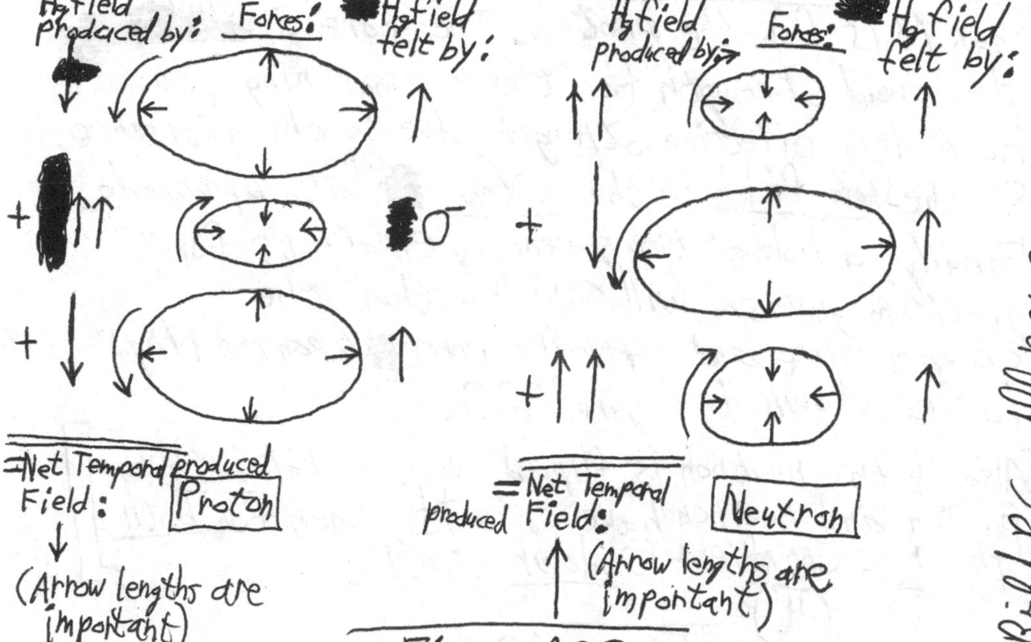

Figure A23

Richd P. Crandall 12-24-2001

D&R Scientific Publishing Co.; Denver, CO
"A New Theory of Quark Substructure for Mesons and Baryons"

Completed on 12-24-2001
(The "UFT" pages)

[Also, later (the "L" and "NSS" pages), it will be shown that the rest-mass of a ring is equal to a constant, divided the ring diameter. Thus, the smaller rings are [redacted] more massive than are the larger rings. Thus for the proton, the smaller ring is shown as producing a field which is greater than [redacted] the other two. Thus, the 2 "felt" [redacted] fields in the <u>proton</u> [redacted] are upwards, as shown. However, for the neutron, the two small rings tend to push each other farther apart due to their increased strength (as compared to two large rings). Thus, the <u>temporal length</u> of the neutron is greater than it is for the proton. Therefore, I know [redacted] the field strength for these small rings, to have less effective strength, due to the distance. So the 3 "felt" [redacted] fields in the <u>neutron</u> [redacted] are also upwards. Finally, a look at the governing equations for gravitomagnetism will confirm that the compressive and expansive force[redacted] upon the rings is as shown, in figure A23.

Richard P. Crandall 12-24-2001

[[Also, if the neutron is flipped over, so that <u>both</u> the proton and neutron have spin $+\frac{1}{2}$, then for <u>both</u> the <u>net</u> GM field is <u>down</u> (i.e. ↓).]]

(H_g)

Figure A24

D&R Scientific Publishing Co.; Denver, CO
"A New Theory of Quark Substructure for Mesons and Baryons"

Completed on 12-24-2001
(The "UFT" pages)

[So why haven't these powerful gravity related forces been detected in experiments already?

Firstly, there are no net detectable gravitoelectric forces due to the presence of the background vacuon "sea". So that leaves gravitomagnetic (GM) forces only. If we look at any generic free particle, it appears to us to only consist of a powerful, local, dipolar GM field. If this

[tiny dipolar field is immersed in an overall net [redacted] gravitomagnetic field, the force equations tell us it will experience a net torque, but it will not experience a net force. And also, in most experiments, there is no GM field present, anyway! So, we have the answer to the above question.

[It [redacted] can easily be seen that a large diameter ring (which actually can be created) can experience a net force as well, especially if immersed in a divergent or non-uniform external GM field!]

Richard P. Crandall
12-24-2001

Figure A25

D&R Scientific Publishing Co.; Denver, CO
"A New Theory of Quark Substructure for Mesons and Baryons"

Completed on 12-24-2001
(The "UFT" pages)

Now, we're in a position to explain those variables in the "formula" column in figure A_MASS. Apparently, the quantity $(\tilde{N}_v/\tilde{N}_c)$ can only take on 3 possible values: 1, 7, or 8. These represent the 3 possible sizes of a vacuon cluster. The quantity (N_c) is the number of clusters in a subquark. Therefore, the total number of vacuons in a subquark is equal to $\tilde{N}_v$. (BEFORE THE CLUSTERS SPREAD OUT OR SEPARATE)! Q is the number of quarks, or rings. It can be 1, 2, or 3. The number of subquarks in a quark is always 4. Thus, we see that the (total) $\tilde{N}_Q = 4 \times Q$, always. (BEFORE THE CLUSTERS SEPARATE)!

Therefore, for any particle, we see that the total number of vacuons $= \tilde{N}_Q \tilde{N}_v = \frac{\tilde{N}_v}{N_c} \times (4 \times Q) \times N_c$
$= 4 \times \left(\frac{\tilde{N}_v}{N_c} \times Q \times N_c \right). \cong 4 \times \sqrt{m_o/m_{oe}}$.

Richard P. Crandall 12-24-2001

Figure A26

D&R Scientific Publishing Co.; Denver, CO
"A New Theory of Quark Substructure for Mesons and Baryons"

Completed on 12-24-2001
(The "UFT" pages)

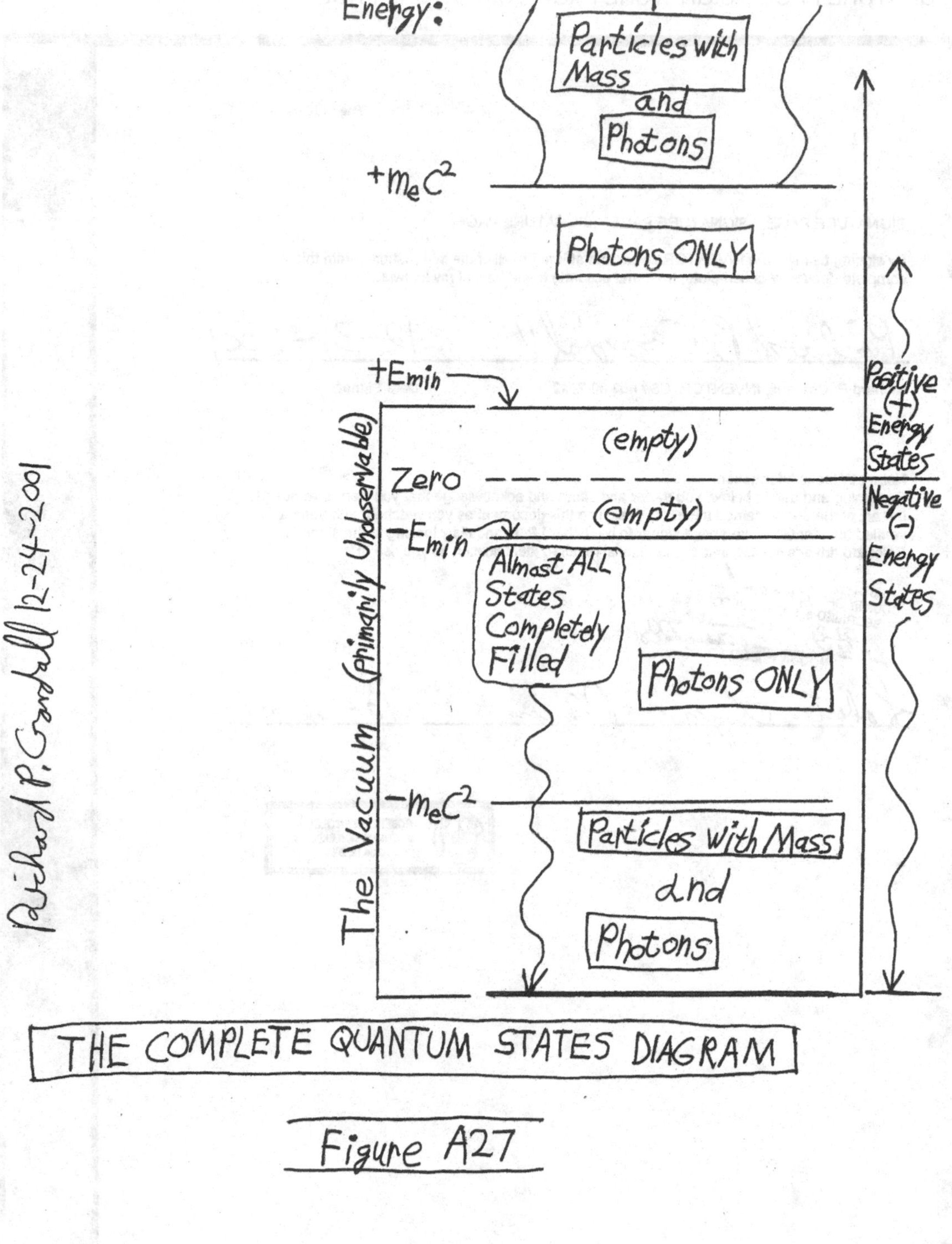

THE COMPLETE QUANTUM STATES DIAGRAM

Figure A27

SIGNATURE PAGE SIGNATURE PAGE SIGNATURE PAGE

SIGNATURE PAGE SIGNATURE PAGE SIGNATURE PAGE

By signing below, I the INVENTOR swear and affirm that all of the information herein this complete document is complete, true, and accurate to the best of my knowledge:

Richard P. Crandall 12-31-2001

Richard P. Crandall, INVENTOR SS# 453 80 7642 Date Signed:

Notary Witness Information:
By signing and dating below, you swear and affirm and acknowledge that you were a witness to the act of the above named INVENTOR signing this document as you watched, and you have verified the identity of the above signer to be Richard P. Crandall by looking at the signer's Colorado driver's license and any additional required identification as you see fit.

COUNTY OF Crandal
SUBSCRIBED & SWORN BEFORE ME
THIS 31 DAY OF Dec, 2001
Roberta Lynn Olguin
NOTARY PUBLIC

Roberta Lynn Olguin 12-31-01

ROBERTA LYNN OLGUIN
MY COMMISSION EXPIRES
June 12, 2004

PART C

EARLY PIONEERS

[]

This section of the book discusses the efforts and experiments of people who could be called the "Pioneers Of Antigravity". They did some very revealing experiments which demonstrated <u>some</u> of the aspects of antigravity fields. All of their experiments involved spinning disks of various materials at various temperatures. Had they tried **spinning cone shapes,** instead of **spinning disks,** they would have probably achieved much more spectacular results. However, they did not, in their time, know the importance of using conical shapes.

Henry W. Wallace (patent):

During the 1960s through the mid 1970s, Henry William Wallace was a scientist at **GE Aerospace** in Valley Forge PA, and **GE Re-Entry Systems** in Philadelphia. In the early 1970s, Wallace was issued patents for some unusual inventions relating to the gravitational field. Wallace developed an experimental apparatus for generating and detecting a secondary gravitational field, which he named the kinemassic field, and which is now better known as the gravitomagnetic field.

Wallace's experiments were based upon aligning the nuclear spin of material **isotopes of elements which have a NON-ZERO TOTAL NUCLEAR SPIN.** Such materials always have a total nuclear spin which is an odd integral multiple of one-half, resulting in **one** (or sometimes more) **nucleon with an unpaired spin. This is the same method that I have used to determine which materials are affected by gravitomagnetic fields and which ones are not.**

Examples of elemental materials which CAN be used for antigravity effects: copper, aluminum, lead, titanium, barium, neodymium, cobalt, mercury, bismuth, and (predictably, due to its location on the chart) element 115.

Examples of elemental materials which are NOT affected by (nor can they produce) gravitomagnetic fields: iron, steel, gold, silver, and tin.

Wallace drew an analogy between the un-paired angular momentum in these materials, and the un-paired magnetic moments of electrons in ferromagnetic materials. Wallace created nuclear spin alignment by **rapidly spinning a brass disk,** of which essentially all isotopes have an odd number of nucleons. Nuclear spin becomes aligned in the spinning disk due to precession of nuclear angular momentum in inertial space -- a process similar to the magnetization developed by rapidly spinning a ferrous material (known as the Barnett effect). The gravitomagnetic field generated by the spinning disk is tightly coupled (0.01 inch air gap) to a gravitomagnetic field circuit composed of material having half integral nuclear spin, and analogous to the magnetic core material in **transformers** (such as like a "C" shaped electromagnet core with the usual gap in it) and motors. The gravitomagnetic field is transmitted through the field circuit and focused by the field material to a small space where it can be detected. In Wallace's patent, he said that the most preferable material from which to make the spinning disk would be **Bismuth,** as it has the highest number of nucleons of any stable known element and has the property of having an odd multiple of one-half for its nuclear spin.

The first few pages of the Wallace patent are shown here.
Figure 2 is perhaps the easiest to visualize, having the horizontal spinning (toothed) brass disk on the right side, and having the dual-cone shaped (!!) detector on the left side, whose function is to focus the gravitomagnetic field onto a small chip of crystalline material. Figures 6 and 7 show a close-up of the detector arrangement, which **measures the temperature** of the crystal material upon which the field is focused by the cones. Figure 8 shows the gradual temperature change that was measured, indicating that something unusual was going on. Thus, some new and not-yet-understood field must be acting upon the crystal material.

DRAWING CAPTIONS FOR WALLACE PATENT (first page of which is shown later below):

Figure 1 is an overall perspective view of equipment constructed according to the invention, this equipment being designed especially for demonstrating the useful applications of kinemassic force fields.

Figure 2 is an isolation schematic of the apparatus comprising the kinemassic field circuit of the apparatus of the (Figure 1), showing the field series relationships of the generator and detector units.

Figures 3, 4, and 5 show the generator (of Figures 1 and 2) in greater detail.

Figure 6 is an enlarged view of the detector working air gap area of the apparatus in Figures 1 and 2.

Figure 7 is a sectional view of Figure 6 showing associated control and monitoring equipment.

Figure 8 represents measured changes in the operating characteristics of a crystalline target subject to a kinemassic force field generated in the apparatus of Figures 1 and 2.

Dec. 14, 1971 **H. W. WALLACE** **3,626,606**

METHOD AND APPARATUS FOR GENERATING A DYNAMIC FORCE FIELD

Filed Nov. 4, 1968 4 Sheets-Sheet 1

FIG-1

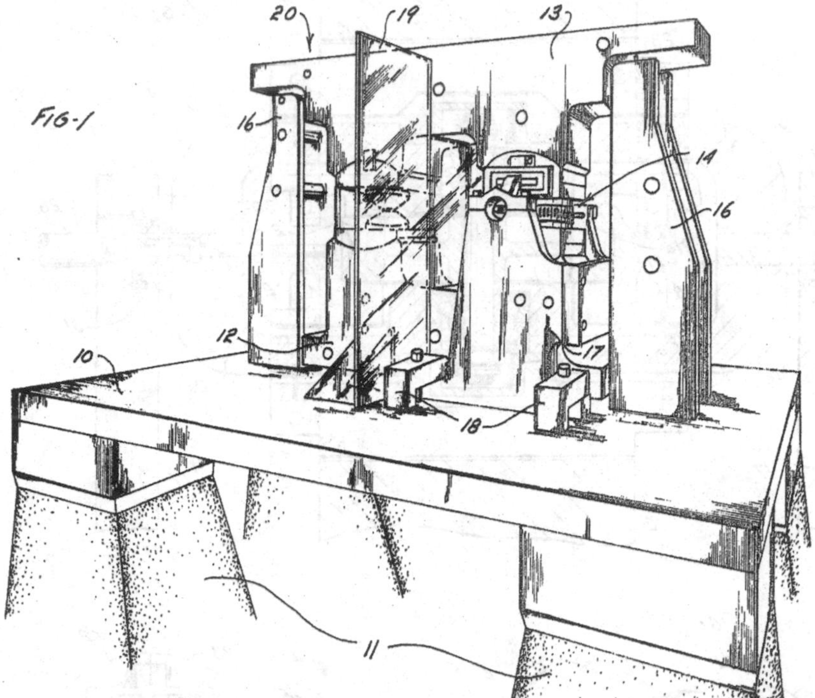

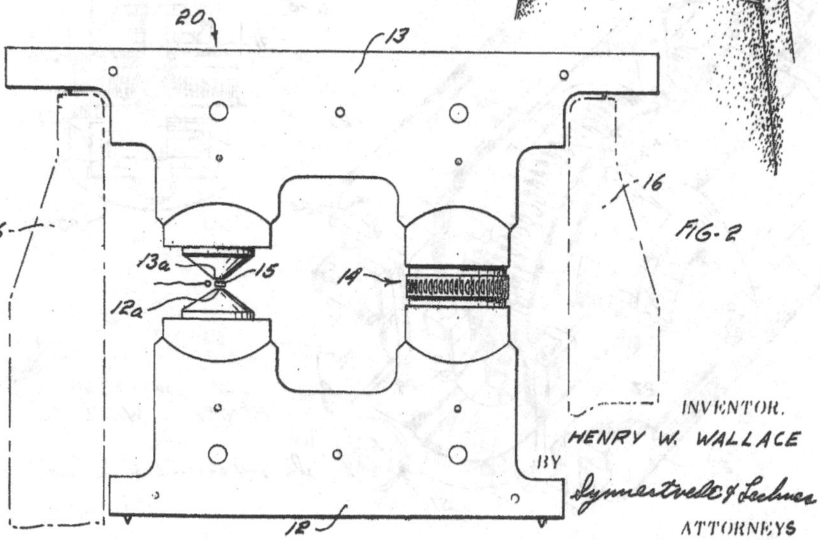

FIG-2

INVENTOR.
HENRY W. WALLACE
BY
Symmestvedt & Lechner
ATTORNEYS

METHOD AND APPARATUS FOR GENERATING A DYNAMIC FORCE FIELD

Filed Nov. 4, 1968 4 Sheets-Sheet 2

FIG-3

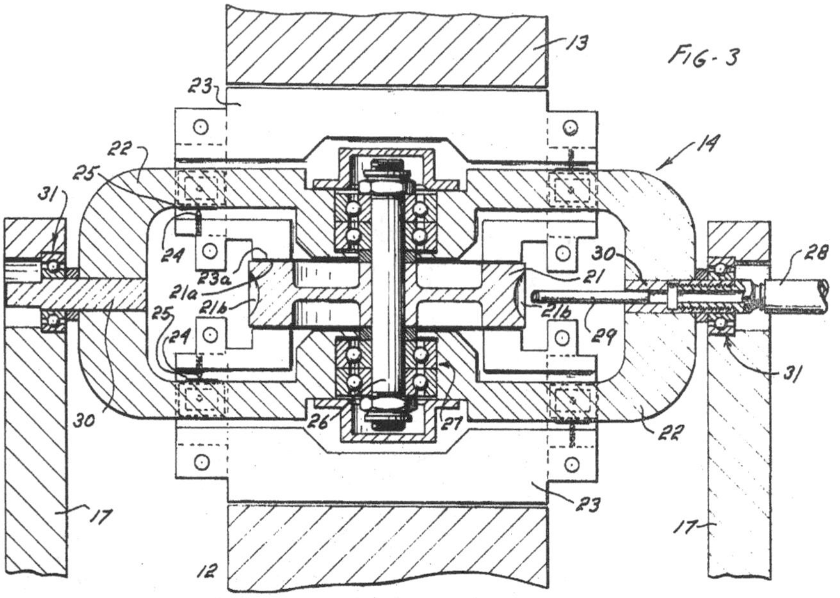

FIG-4

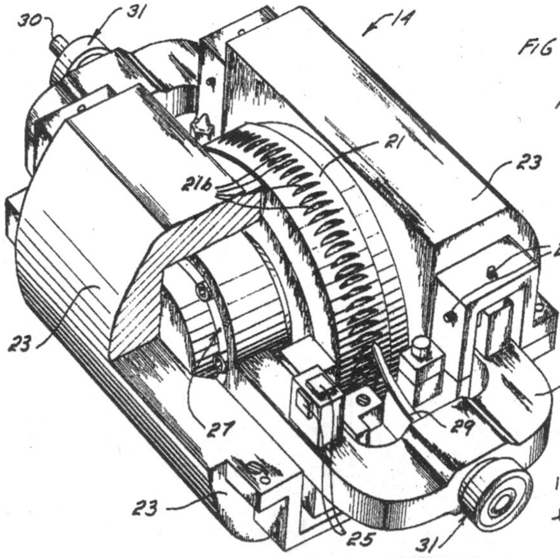

FIG-5

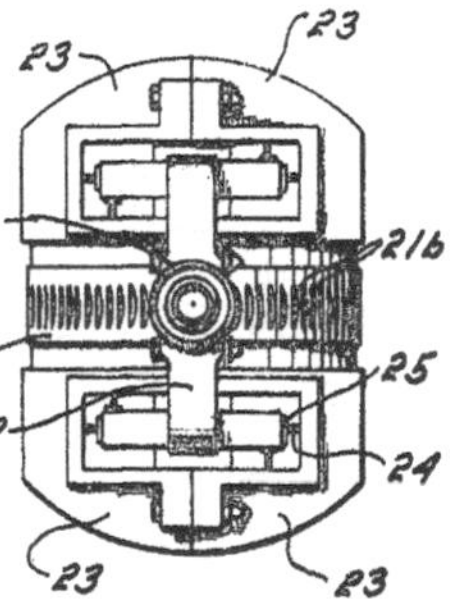

INVENTOR.
HENRY W. WALLACE
BY
Synnestvedt & Lechner
ATTORNEYS

METHOD AND APPARATUS FOR GENERATING A DYNAMIC FORCE FIELD

Filed Nov. 4, 1968 4 Sheets-Sheet 3

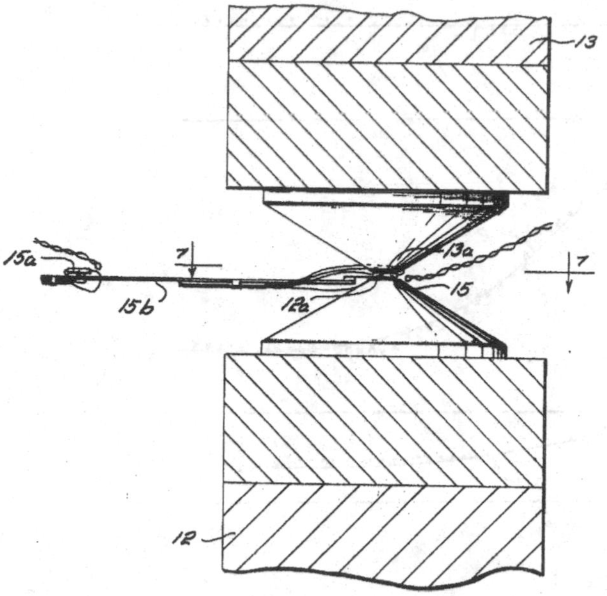

FIG-7

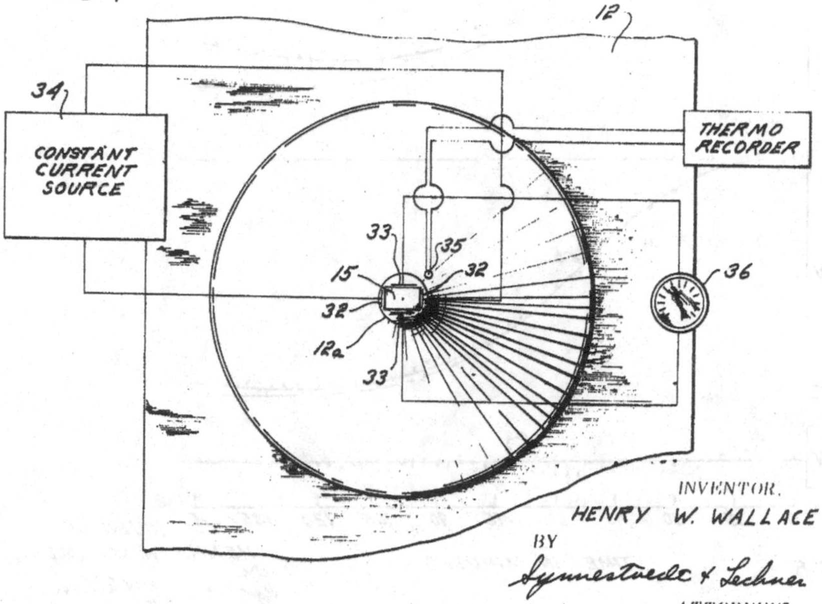

INVENTOR.
HENRY W. WALLACE
BY
Synnestvedt & Lechner
ATTORNEYS

METHOD AND APPARATUS FOR GENERATING A DYNAMIC FORCE FIELD

Filed Nov. 4, 1968 4 Sheets-Sheet 4

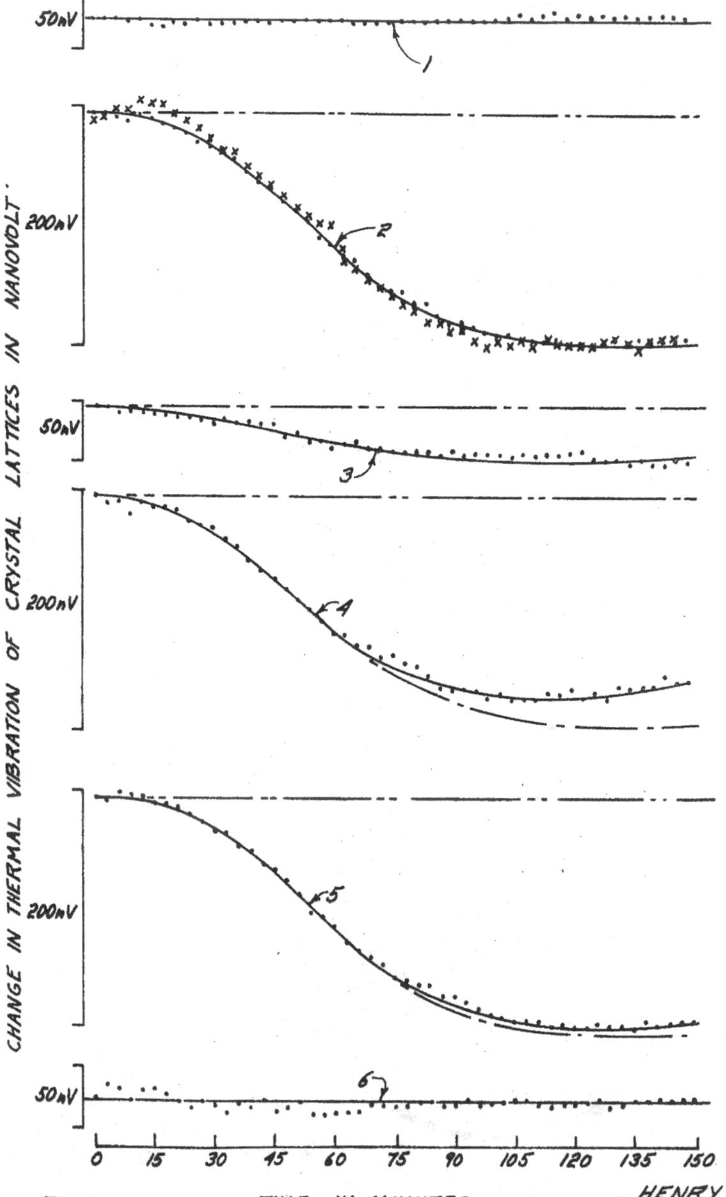

FIG-8

INVENTOR
HENRY W. WALLACE
By Lynnestvedt & Lechner
ATTORNEYS

United States Patent Office

3,626,606
Patented Dec. 14, 1971

1

3,626,606
METHOD AND APPARATUS FOR GENERATING A DYNAMIC FORCE FIELD

Henry W. Wallace, Ardmore, Pa.
(803 Cherry Lane, Laurel, Miss. 39440)

Filed Nov. 4, 1968, Ser. No. 773,116
Int. Cl. G09b 23/06
U.S. Cl. 35—19 10 Claims

ABSTRACT OF THE DISCLOSURE

Apparatus and method for generating a non-electromagnetic force field due to the dynamic interaction of relatively moving bodies through gravitational coupling, and for transforming such force fields into energy for doing useful work.

The method of generating such non-electromagnetic forces includes the steps of juxtaposing in field series relationship a stationary member, comprising spin nuclei material further characterized by a half integral spin value, and a member capable of assuming relative motion with respect to said stationary member and also characterized by spin nuclei material of one-half integral spin value; and initiating the relative motion of said one member with respect to the other whereby the interaction of the angular momentum property of spin nuclei with inertial space effects the polarization of the spin nuclei thereof, resulting in turn in a net component of angular momentum which exhibits itself in the form of a dipole moment capable of dynamically interacting with the spin nuclei material of the stationary member, thereby further polarizing the spin nuclei material in said stationary member and resulting in a usable non-electromagnetic force.

This invention relates to an apparatus and method for use in generating energy arising through the relative motion of moving bodies and for transforming such generated energy into useful work. In the practice of the present invention it has been found that when bodies composed of certain material are placed in relative motion with respect to one another there is generated an energy field therein not heretofore observed. This field is not electromagnetic in nature; being by theoretical prediction related to the gravitational coupling of moving bodies.

The initial evidence indicates that this nonelectromagnetic field is generated as a result of the relative motion of bodies constituted of elements whose nuclei are characterized by half integral "spin" values; the spin of the nuclei being synonymous with the angular momentum of the nucleons thereof. The nucleons in turn comprise the elemental particles of the nucleus; i.e., the neutrons and protons. For purposes of the present invention, the field generated by the relative motion of materials characterized by a half integral spin value is referred to as a "kinemassic" force field.

It will be appreciated that relative motion occurs on various levels, i.e., there may be relative motion of discrete bodies as well as of the constituents thereof including, on a subatomic level, the nucleons of the nucleus. The kinemassic force field under consideration is a result of such relative motion, being a function of the dynamic interaction of two relatively moving bodies including the elemental particles thereof. The value of the kinemassic force field, created by reason of the dynamic interaction of the bodies experiencing relative motion, is the algebraic sum of the fields created by reason of the dynamic interaction of both elementary particles and of the discrete bodies.

For a closed system comprising only a stationary body, the kinemassic force, due to the dynamic interaction of

2

the particles therein, is zero because of the random distribution of spin orientations of the respective particles. Polarization of the spin components so as to align a majority thereof in a preferred direction establishes a field gradient normal to the spin axis of the elementary particles. The present invention is concerned with an apparatus for establishing such a preferred orientation and as a result generating a net force component capable of being represented in various useful forms.

According, the primary object of the present invention concerns the provision of means for generating a kinemassic field due to the dynamic interaction of relatively moving bodies.

A further object of the present invention concerns a force field generating apparatus wherein means are provided for polarizing material portions of the apparatus so as to reorient the spin of the elementary nuclear components thereof in a preferred direction thereby generating a detectable force field.

The kinemassic force field finds theoretical support in the laws of physics, being substantiated by the generalized theory of relativity. According to the general theory of relativity there exists not only a static gravitational field but also a dynamic component thereof due to the gravitational coupling of relatively moving bodies. This theory purposes that two spinning bodies will exert force on each other. Heretofore the theoretical predictions have never been experimentally substantiated however, as early as 1896, experiments were conducted in an effort to detect predicted centrifugal forces on stationary bodies placed near large, rapidly rotating masses. The results of these early experiments were inconclusive, and little else in the nature of this type of work is known to have been conducted.

It is therefore another object of the present invention to set forth an operative technique for generating a measurable force field due to gravitational coupling of relatively moving bodies.

Another more specific object of the present invention concerns a method of generating a non-electromagnetic force field due to the dynamic interaction of relatively moving bodies and for utilizing such forces for temperature control purposes including the specific application of such forces to the control of lattice vibrations within a crystalline structure thereby establishing an appreciable temperature reduction, these principles being useful for example in the design of a heat pump.

The foregoing objects and features of novelty which characterize the present invention as well as other objects of the invention are pointed out with particularity in the claims annexed to and forming a part of the present spection. For a better understanding of the invention, its advantages and specific objects allied with its use, reference should be made to the accompanying drawings and descriptive matter in which there is illustrated and described a preferred embodiment of the invention.

In the drawings:

FIG. 1 is an overall perspective view of equipment constructed according to the present invention, this equipment being designed especially for demonstrating the useful applications of kinemassic force fields;

FIG. 2 is an isolation schematic of apparatus components comprising the kinemassic field circuit of the apparatus of FIG. 1, showing the field series relationship of generator and detector units;

FIGS. 3, 4 and 5 show the generator of FIGS. 1 and 2 in greater detail;

FIG. 6 is an enlarged view of the detector working air gap area of the apparatus of FIGS. 1 and 2;

FIG. 7 is a sectional view of FIG. 6 showing associated control and monitoring equipment; and

FIG. 8 represents measured changes in the operating characteristics of a crystalline target subject to a kine-

The remainder of the patent can be downloaded from the US patent office, if the reader desires to do so.

LATER EXPERIMENTAL CORROBORATION:
In 1997, a person named Harvey Morgan published, in an I.E.E.E. journal, the results of an experiment involving two parallel spinning disks made of **lead.** Here's a quote from that article:

"A mechanical experiment confirmed that momentum is indeed a field phenomena. A 2 pound **lead** flywheel was mounted on the shaft of a small, very high speed (26,500 RPM advertised) electric motor. Another flywheel was mounted on a ball-bearing shaft aligned with the motor shaft. The two flywheel's parallel faces were separated by about 1/16"... When the motor was energized, it accelerated the lead flywheel towards its top rated speed. The other flywheel, in response to the changing angular velocity and momentum of the lead flywheel, started turning briskly - in the opposite direction! The changing momentum field of the lead flywheel induced a torque in the other flywheel across an air gap. Newtonian mechanics does not predict that reaction.

When the electric motor was turned off before reaching top speed, the other flywheel stopped turning. It then started turning slowly in the same direction as the lead flywheel, urged by the collapsing momentum field and the air coupling between flywheels."

What Morgan refers to as momentum is in fact more specifically angular momentum. It is already well-known physics that angular momentum of matter generates a field known as the gravitomagnetic field. Since there were no ordinary magnetic fields present in Morgan's experiment (lead is non-magnetic), nor electric fields, **then the only fields which could possibly remain** and thus account for the phenomenon would have to be **gravity-related** (or gravitomagnetic).

In Wallace's experiments, he observed in the laboratory a gravitomagnetic type field which he named the "kinemassic" field. Here's a quote from Wallace:

"In the practice of the present invention, it has been found that when bodies composed of certain material are placed in relative motion with respect to one another there is generated an energy field therein not heretofore observed. This field is not electromagnetic in nature, being by theoretical prediction related to the gravitational coupling of relatively moving bodies.

The initial evidence indicates that this non-electromagnetic field is generated as a result of the relative motion of bodies constituted of elements whose nuclei are characterized by half-integral "spin" value, the spin of the nucleus being associated with the net angular momentum of the nucleons thereof."

Also, Wallace says:

"The [kinemassic] field strength is apparently a function of the density of spin nuclei material comprising the field circuit members. Whereas the permeability in magnetic field theory is a function of the density of unpaired electrons, the kinemassic permeability is a function of the density of spin nuclei and the measure of magnitude of their half-integral spin values."

And:

"...when the generator wheel is made to spin at rates upward of 10 or 20 thousand revolutions per minute, effective polarization of spin nuclei within the wheel structure gradually occurs."

Wallace's invention is of extreme importance when compared with the next two devices, one of which is a spinning disk (actually ring) tested by a Mr. Podkletnov, and the other is a spinning disk tested by NASA. Both of these disks were composed of what's called YBCO superconductive material. The "YBCO" means "Yttrium, Barium, Copper Oxide".

Wallace had discovered that **you <u>do not</u> need to use superconducting materials** in the spinning disks. We will shortly see that Podkletnov's disks and NASA's disks were superconducting materials, kept very cold. The theories put forth by those two parties were derived based-upon a novel theoretical link between gravitomagnetism and superconductive materials.

The <u>real</u> <u>reason</u> that those spinning superconductor disks worked so well to produce gravity-related effects is, as Wallace and I have discovered, that they contained barium and copper (two of the proper materials, as stated above). I think Yttrium may be one of the proper materials, as well, but I can't remember.

The spinning disks DO NOT HAVE TO BE superconductors or even be very cold! The cold environment of the disks only ENHANCES an already-existing effect!

It will be very interesting indeed when someone, hopefully soon, tries using a 3-cone-in-a-barrel device which is kept very cold by cryogenic means (such as using liquid air or solid carbon dioxide). I predict at least a thousand-fold multiplication in energy output, due to the fact the nuclei will have statistically better numbers of alignment.

[ ]

[THIS PAGE IS INTENTIONALLY BLANK]

Podkletnov & Tampere University:

Dr. Eugene Podkletnov, a researcher at the Tampere University of Technology, Finland, reported that during a superconductivity experiment, tests showed a small drop in the weight of objects placed over a device consisting of a **rotating** superconducting ceramic **disk** suspended over a magnetic field produced by three electric coils enclosed in a cryostat.
The experiments by the Finnish researcher(s) reportedly registered a maximum 2 percent drop in the weight of objects suspended over the cryostat.

The following excerpt is from The Sunday Telegraph (UK), **1st of September, 1996,** page 3:

Scientists in Finland are about to reveal details of the world's first anti-gravity device. Measuring about 12 inches across, the device is said to reduce significantly the weight of anything suspended over it. The claim -- which has been rigorously examined by scientists, and is due to appear in a physics journal next month -- could spark a technological revolution.

....

The Sunday Telegraph has learned that NASA, the American space agency, is taking the claims seriously, and is funding research into how the anti-gravity effect could be turned into a means of flight.

....

According to Dr. Eugene Podkletnov, who led the research, the discovery was accidental. It emerged during routine work on so-called "superconductivity", the ability of some materials to lose their electrical resistance at very low temperatures. The team was carrying out tests on a rapidly spinning disc of superconducting ceramic suspended in the magnetic field of three electric coils, all enclosed in a low-temperature vessel called a cryostat. "One of my friends came in and he was smoking his pipe," Dr. Podkletnov said. **"He put some smoke over the cryostat and we saw that the smoke was going to the ceiling all the time. It was amazing -- we couldn't explain it."** Tests showed a small drop in the weight of objects placed over the device, as if it were shielding the object from the effects of gravity - an effect deemed impossible by most scientists. "We thought it might be a mistake," Dr. Podkletnov said, "but we have taken every precaution." Yet the bizarre effects persisted. **The team found that even the air pressure vertically above the device dropped slightly, with the effect detectable directly above the device on every floor of the laboratory. ....** [END OF EXCERPT].

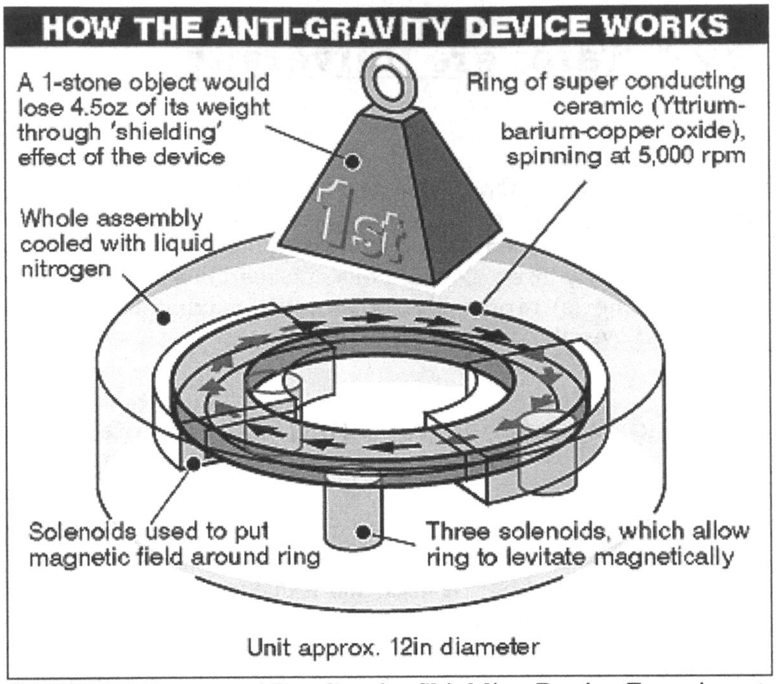

Simplified Diagram of The Gravity-Shielding Device Experiment, Done By E. Podkletnov.

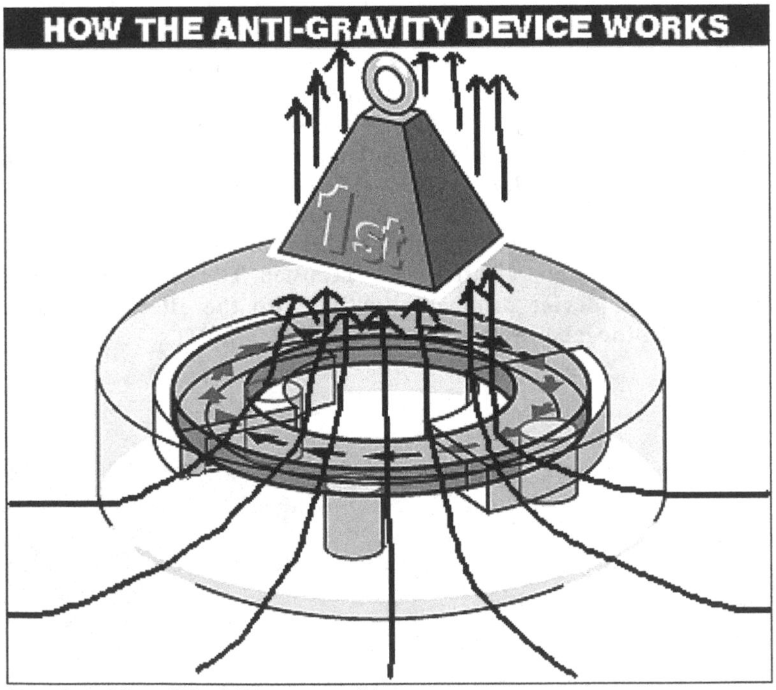

Another Simplified Diagram of it.

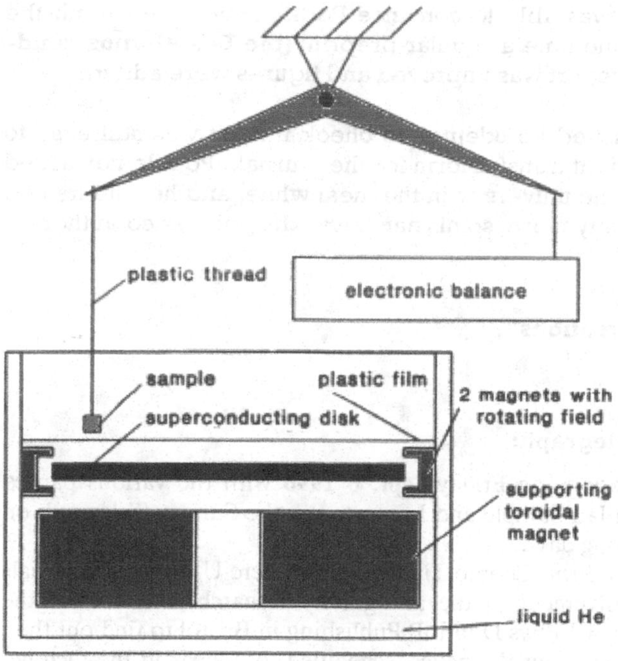

Yet Another Diagram of The Experimental Setup.

A scientific paper carrying the names of Eugene Podkletnov and Petri Vuorinen of Tampere University in Finland was accepted for publication in Journal of Physics D: Applied Physics, after proper peer review, in early or mid 1996. According to Robert Matthews of New Scientist, the paper has now been withdrawn. It appears that this work was done in the early 1990s and has only recently received a flurry of attention. <u>Tampere University denies any knowledge of the work</u> and Vuorinen stated that he had never worked on the paper with Podkletnov. Podkletnov appeared surprised and suggested that it must have been another Petri Vuorinen who worked with him -- but no such other Petri has been located.
When Podkletnov **withdrew his paper on September 9, 1996,** he stated: **"This is an important discovery and I don't want it to disappear"**, to New Scientist magazine.

Going back in time a little, a report named **"report MSU-chem 95"** was finished by Podkletnov in **January of 1995.** "MSU" stands for "Moscow State University". Levit (a co-author) was one of those who helped in the experiment, but only for technical support. The same applies to Nieminen, in an earlier work. The only one responsible for the scientific aspects was Podkletnov. Podkletnov also submitted the paper to J. Phys. (Or maybe to some other journal before J. Phys., but it was rejected). Probably the name of Vuorinen was added there, because he had made some technical contribution too, and Podkletnov wanted to acknowledge this.
As it is known, later co-author Vuorinen denied any contribution, **after the University of Tampere (apparently in reaction to the Sept-1-1996 article in the U.K. Daily Telegraph) "disowned" Podkletnov.**
Podkletnov was very sad and bitter for all this and withdrew the paper and forgot it for a while. **After the ensuing controversy with Tampere University, Podkletnov was not**

allowed to enter "his" laboratory any more (he was kind of "guest" there) and the experiment was dismounted.

At **the end of 1996,** a Mr. G. Modanese was able to convince Podkletnov to re-submit the paper for publication and make at the same time a regular preprint **(the Los Alamos cond-mat/9701074).** To this end, the previous report was improved and figures were added.

Before submitting the paper Modanese asked Podkletnov to check if Levit was still easy to reach in order to have him sign the copyright transfer form for the journal. Podkletnov asked Levit. Levit had begun a new job outside the university in the meanwhile, and he said he was not interested in being cited as co-author any more, so his name was dropped as co-author.

From: "Follow-Up Investigative Conversations",

By Robert Matthews,

Science Correspondent, The Sunday Telegraph:

What follows is a summary of my conversations on Friday **Sept. 6 1996** with the various parties involved in the anti-gravity claims, on which Ian Sample and I reported in the Sunday Telegraph on Sept. 1 1996. It's rather terse, but it's been a long day.....

1. Following the posting of the statement from Prof. Tuomo Tiainen of Tampere U. Inst. of Materials Science disclaiming any knowledge or involvement in the anti-gravity research, I contacted the Institute of Physics in London and the offices of J Phys D in IoP Publishing in Bristol to find out their reaction. Neither were aware of any problems about the paper, scheduled to appear in the October issue of the journal, and were taken aback by the reaction of Tampere. Richard Palmer, Managing Editor of J Phys D said he would investigate matters further (of which more below)

2. Following the statement by Petri Vuorinen, whose name appears on the paper as co-author, disclaiming any involvement, I faxed a set of questions to Dr. Eugene Podkletnov (lead author) asking for an explanation. While waiting for a response of the paper, I contacted PV, whose immediate reaction was to insist that I talk to Prof. Tiainen, who was apparently fielding all inquiries. However, PV did say that he had worked with EP some years ago, but had no idea why his name had appeared on the paper as co-author on this latest paper. He was anxious to distance himself from the research.

3. EP then responded to my fax by telephone. He said that the denial of any (recent) involvement by Tampere stemmed from the fact that (a) the key experiments were indeed done some years ago, in 1992; (b) that Prof. Tiainen has only been director of the Institute for four months, and was not in a position to know about the experiments before then. EP insisted that the results stated in the forthcoming paper are reliable and genuine, and that Tampere has full knowledge, with a (Finnish?) patent being applied for in its name. (I wasn't able to confirm this latter point in the time available). On the matter of PV's denial of all involvement in the paper, and mystification of his involvement, EP insisted that there must have been some mix-up over names at Tampere, and that there must be a second Petri Vuorinen working on superconductivity at the Institute, and that the one involved in anti-gravity was now working in Japan. When challenged on the sheer implausibility of this, EP said that the name was a common one in Finland. He finished by saying that he did not want to cause any trouble with Tampere, with which he appears to still have some relationship (Tampere's statement says he no longer works there; however, one researcher said he had seen EP visiting the Institute last week; the discrepancy may revolve on the question of full-time salaried staff and others like EP who appear to be on more informal arrangements - see below).

4. I went back to PV, putting to him EP's claim that there must be two researchers with the same name. PV said there are only about 60 people in the Institute, and that he was sure he would know if there was another researcher with the same name there. He added that he had indeed been in Japan - three years ago. He ended by saying that he hoped the controversy did not damage relationships between Finnish institutes and the British academic journals. Later I discovered that there is indeed another PV - Petri Vuoristo - at the Institute, but he denied all involvement in the research too.

5. Prof. Tiainen responded to a call placed AM, and began by repeating the original statement denying any involvement - except some years ago - in anti-gravity research. He re-iterated that he did not have any views on whether the claims being made in the forthcoming JPhysD were valid or not (he said he was not qualified to do so). He added, however, that "We don't want to get the credit for the result if it is good or bad". He said that EP had done good work at Tampere on thin films and S/conductivity, and that EP still came into the university, but had no official position. TT said: "I was completely surprised when I learnt these things were going on". He finished by saying that "If this turns out bad" he would consider banning EP from the Institute. He said that there had been claims that part of the work was funded by the Finnish military, but denied that this was the case.
6. I then contacted the editorial offices of JPhys D again, and was told by Richard Palmer, managing editor, that he had been contacted by TT. In the light of the conversation, RP said he and his staff were looking at the paper again, and had not ruled out the possibility of holding the paper out of the journal until EP had been contacted for clarification of various issues. Among these was the fact that documents relating to the paper's publication carry the signature of PV - who denies involvement in the research.
7. Despite repeated requests, the IoP head office did not issue a position statement during the day.
8. MONDAY **September 9:** Dr. Podkletnov has today contacted the IoP editorial offices, and requested that his paper be withdrawn from publication in JPhysD next month. His request has been accepted, and the IoP is taking no further action on this matter.

Robert Matthews,
Science Correspondent, The Sunday Telegraph
[END EXCERPT]

Later, Jane's Defence Weekly (UK) reported that Moscow was blocking attempts by anyone to collaborate with Mr. Podkletnov. Unfortunately, I don't recall the date of this report.

Here is a partial version of the paper that was finally submitted by Podkletnov, at the request of G. Modanese, at the end of 1996. Omitted parts are denoted by "....". I have **boldfaced** parts which are particularly of the greatest importance.

Weak gravitational shielding properties of composite bulk $YBa_2Cu_3O_{(7-x)}$ superconductor below 70 K under e.m. field

Report MSU-chem 95. Los Alamos database nr. cond-mat/9701074 (LaTeX).
This is an informal ASCII version.
E.E. Podkletnov
Moscow Chemical Scientific Research Centre

ABSTRACT

A high-temperature $YBa_2Cu_3O_{(7-x)}$ bulk ceramic superconductor with composite structure has revealed weak shielding properties against gravitational force while in a levitating state at temperatures below 70 K. A toroidal disk was prepared using conventional ceramic technology in combination with melt-texture growth. Two solenoids were placed around the disk in order to initiate the current inside it and to rotate the disk about its central axis. Samples placed over the rotating disk initially demonstrated a weight loss of 0.3-0.5%. When the rotation speed was slowly reduced by changing the current in the solenoids, the shielding effect became considerably higher and reached 1.9-2.1% at maximum.

P.A.C.S. 74.72.-h High-T_c cuprates.

1. INTRODUCTION

....According to public information, a NASA group in Huntsville, Alabama, is now "cloning" our experiment. ....

2. EXPERIMENTAL

....

2.1 CONSTRUCTION OF THE DISK

The shielding superconducting element was made of dense, bulk, almost single-phase YBa_2Cu_3O_{7-x} and had the shape of a toroidal disk with an outer diameter of 275 mm, an inner diameter of 80 mm, and a thickness of 10 mm. The preparation of the 123-compound and fabrication of the disk involved mixing the initial oxides, then calcining the powder at 930 C in air, grinding, pressing the disk at 120 MPa, and sintering it in oxygen at 930 C for 12 hours with slow cooling to room temperature. After that, the disk was put back in the furnace at 600 C, and the upper surface was quickly heated to 1200 C using a planar high-frequency inductor as shown in Figure 1. During this last heating, the gap between the disk and the inductor was chosen precisely so that heating would occur only in the top 2 mm-thick layer of the disk, ...

2.2 OPERATION OF THE APPARATUS

....

3. RESULTS

The levitating disk revealed a clearly measurable shielding effect against the gravitational force even without rotation. In this situation, the weight-loss values for various samples ranged from 0.05-0.07%. As soon as the main solenoids were switched on and the disk began to rotate in the vapors of liquid helium, the shielding effect increased, and **at 5000 rpm, the air over the cryostat began to rise slowly toward the ceiling. Particles of dust and smoke in the air made the effect clearly visible. The boundaries of the flow could be seen clearly and corresponded exactly to the shape of the toroid.**

The weight of various samples decreased no matter what they were made from. Samples made from the same material and of comparable size, but with different masses, lost the same fraction of their weight. However, the shape and position of the sample did affect the weight loss, causing slight variations (about 10%) in the shielding effect. The maximum loss of weight was achieved when a sample was positioned with its largest surface parallel to the surface of the disk, so that the effect being projected by the disk impinged on the maximum area. **The best measurement gave a weight loss of 0.5% while the disk was spinning at 5000 rpm,** with typical values ranging from 0.3 to 0.5%. Samples placed above the inner edge of the toroid (5-7 mm from the edge) were least affected, losing only 0.1 to 0.25% of their weight.

During the time when the rotation speed was being decreased from 5000 to 3500 rpm, using the solenoids as braking tools, the shielding effect reached maximum values: the weight loss of the samples was from 1.9 to 2.1%, depending on the position of the sample with respect to the outer edge of the disk. These peak values were measured during a 25-30 seconds interval, **when the rotational speed was decreasing to 3300 rpm.** Because of considerable disk vibration at 3000-3300 rpm, the disk had to be rapidly braked in order to avoid unbalanced rotation, and further weight measurements could not be carried out.

The maximum shielding properties were coincident with maximum current inside the superconducting disk. According to preliminary measurements, the upper layer of the disk was able to carry over 15000 A/cm^2. The samples' maximum weight loss was observed only when the magnetic field was operating at high frequencies, on the order of 3.2 to 3.8 MHz. ....

Remarkably, the effective weight loss was the same even when the samples, together with the balance, were moved upward to a distance of 3 m, but still within the vertical projection of the toroidal disk. No weight loss at all was observed below the cryostat.

The shielding effect decreases slightly the gravitational force within a **vertical projection** above the disk. **The projection creates a kind of vertical tunnel in the air, within which the air pressure is slightly reduced.** (The observed effect also works in other gases and even in liquid media.)

The difference between the atmospheric pressure over the cryostat and the pressure below it was measured with high precision using a mercury barometer. It was equal to 8 mm for the maximum shielding effect. Such

a pressure differential produces a lifting force on the cryostat (of the order of 10^2 Kg/m^2). However, in the present case, this is of no practical interest.

4. DISCUSSION

••••

5. CONCLUSIONS

A levitating superconducting ceramic disk of $YBa_2Cu_3O_{(7-x)}$ with composite structure demonstrated a stable and clearly measurable weak shielding effect against gravitational force, but only below 70 K and under high-frequency e.m. field. The combination of a high-frequency current inside the rotating toroidal disk and an external high-frequency magnetic field, together with electronic pairing state and superconducting crystal lattice structure, apparently changed the interaction of the solid body with the gravitational field. This resulted in the ability of the superconductor to attenuate the energy of the gravitational force and yielded a weight loss in various samples of as much as 2.1%.

Samples made of metals, plastic, ceramic, wood, etc. were situated over the disk, and their weight was measured with high precision. All the samples showed the same partial loss of weight, no matter what material they were made of. Obtaining the maximum weight loss required that the samples be oriented with their flat surface parallel to the surface of the disk. **The overall maximum shielding effect (2.1%) was obtained when the disk's rotational speed and the corresponding centrifugal force were slightly decreased** by the magnetic field.

It was found that the shielding effect depended on the temperature, the rotation speed, the frequency and the intensity of the magnetic field. At present it seems early to discuss the mechanisms or to offer a detailed analysis of the observed phenomenon, as further investigation is necessary. The experimentally obtained shielding values may eventually prove to have fundamental importance for technological applications as well as scientific study.

ACKNOWLEDGMENT

The author is grateful to the Institute for High Temperatures at the Russian Academy of Sciences for their help in preparing the unique superconducting ceramic disks and for being permitted to use their laboratory equipment. The effect was first observed and studied at Tampere University of Technology.

REFERENCES

••••

[END OF PODKLETNOV PAPER]

So, we see that there was very strange behavior of most of the people and the institutions involved, when Podkletnov first started to release his material and his paper. We can imagine that one or more parties, unknown, expressed a deep interest in having the information suppressed. Probably, it was government-related.

Finally, recall the following statements which are in the above paper:

The team found that even the air pressure vertically above the device dropped slightly, with the effect detectable directly above the device on every floor of the laboratory.

....and **at 5000 rpm, the air over the cryostat began to rise slowly toward the ceiling. Particles of dust and smoke in the air made the effect clearly visible. The boundaries of the flow could be seen clearly and corresponded exactly to the shape of the toroid.**

Remarkably, the effective weight loss was the <u>same</u> even when the samples, together with the balance, were <u>moved upward</u> to a distance of 3 m, but still within the vertical projection of the toroidal disk. No weight loss at all was observed below the cryostat. The shielding effect decreases slightly the gravitational force within a **vertical projection** above the disk. **The <u>projection</u> creates a kind of <u>vertical tunnel in the air</u>, within which the air pressure is slightly reduced.**

These statements indicate something astounding. Something incredible. They built a device which could clearly form a BEAM OF "GRAVITY", which does not diminish with distance from the device! Wow! In the later chapter on NASA, there is a section included there regarding USC (the University of South Carolina). They claimed to have invented a similar device as well, and perhaps their device was even better. However, their claims mysteriously and quickly disappeared after a few weeks, for unknown reasons. I myself remember seeing the announcement on the Internet, on their web site, and I also remember that it disappeared very soon thereafter.

Part D

Government & Suppression

[ ]

[THIS PAGE IS INTENTIONALLY BLANK]

NASA Marshall SFC, & Univ. Ala. Huntsville:

In 1989, Dr. Ning Li of UAH (University of Alabama, Huntsville) predicted that if a time varying magnetic field were applied to a superconductor, charged and deformed lattice ions within the superconductor could absorb enormous amounts of energy via the magnetic moment effect. This acquired energy would cause the lattice ions to spin rapidly about their equilibrium positions and create a minuscule gravitational field. Dr. Li's calculations showed that if these charged, rotating, lattice ions were aligned with each other by a strong magnetic field, the resulting change in local gravity would be measurable.
Podkletnov and Nieminen (1992) made the accidental discovery that a single-phase, dense, bulk, high T_c, superconducting, ceramic disk **spinning at 5,000 rpm** can produce a 2 percent reduction in the weight of non-conducting, non-magnetic objects placed over the spinning disk. UAH and MSFC (NASA Marshall Space Flight Center, in Huntsville Alabama) have cooperated on a joint research project to independently confirm the results of the Podkletnov experiment and to validate Dr. Li's theory of gravity modification via superconductor. On March 26th, 1997, as part of this project, the joint UAH-MSFC research team produced the largest high temperature superconducting disk ever manufactured in the USA. This disk measured 12 inches in diameter and is 0.5 inches thick.

From my website, The Antigravity Papers: (under the heading "NASA Goofed!")

The two scientists who did most of the work for NASA (both theoretical and experimental) were Ning Li and Douglas Torr. I have read and understood several of the scholarly papers they published in various physics journals, and I must say I was very impressed. I rank these two people in the category of "brilliant and accomplished thinkers."

Li and Torr had already done a certain amount of theoretical work in the area of antigravity when the world got the news that someone in Finland named Podkletnov had created a device which would shield gravity, and thus would cause objects to weigh less when they were suspended above the device. His device consisted of a rotating superconductor disk which was magnetically suspended. So apparently NASA commissioned Li and others to try to replicate the experiment of Podkletnov, and this is what the online paper found at the above link is all about.

*However, what the Physica C paper by Li, and others, discussed was an experiment using a NON-rotating superconductor disk. The overall results of the experiment were **negative** (or at least not anywhere near the magnitude of the effect Podkletnov had achieved). They stated in the paper **that they would do later research using a rotating superconductor disk** and would publish the results. As far as I am aware, the rotating disk experiment was either never done, or there were never any publications which resulted from it. The last published papers by Ning Li, as far as I can tell, were done in the 1997 time frame. If I am wrong about this, please let me know.*

After looking at the writeups about the NASA experiment and the Podkletnov experiment, I could see that both of the experiments were severely BLUNDERED! Yes, you heard that right. I'm sure both parties will be rather embarassed when you tell them the following, and see what their reaction is. Using the "Maxwellized" equations of gravity, and knowing the fact that the vortical motion (or spin vector) of the lattice ions in the disk must be aligned in a direction perpendicular to the plane of the disk, then the field lines of the "gravitomagnetic field" must be perpendicular to the plane of the disk also. This "gravitomagnetic field" would change with time, due to either the RF electric and magnetic fields applied, or to the spinning of the disk, or both. Indeed, in order to either create ordinary gravity or shield ordinary gravity, you must have a changing "gravitomagnetic field" in order to create that ordinary gravity field (per the "Maxwellized Equations"). Fact is, according to the "Maxwellized Equations", that the ordinary gravity field created by a changing (in strength), but always vertically oriented, gravitomagnetic field, would have to be HORIZONTAL, not vertical. In fact, the field lines of the generated ordinary gravity field are both horizontal and circular in nature. As an analogy, since the two types of gravity fields (i.e. ordinary gravity fields and "gravitomagnetic" fields) interact with each other exactly like electric and magnetic fields interact with each other, then consider the following. Any degreed electrical engineer will tell you that a collapsing vertical magnetic field will induce a circular horizontal electric field. This is the principle by which electrical generators work. Similarly, if a permanent magnet is thrust vertically through a horizontal circular coil of wire, then an electrical voltage is induced in the coil, due to the circular electric field which is produced by the moving magnet.

So, both NASA and Podkletnov were measuring for a change in gravity in THE WRONG DIRECTION!! They should have looked for any changes in gravity in the HORIZONTAL direction. Thus, the hanging weight in those experiments must have been pulled slightly either to the left or right (but not enough that they could visually see it).

The reason that Podkletnov got any results at all in the VERTICAL direction was simply due to the residual bending of the fields from the perfect horizontal, due to the finite size of the disk producing the fields. Then, there would have been a small but measurable vertical component to the generated gravity field.

I somehow suspect that Li and others knew about this huge mistake, but did the experiment anyway just like Podkletnov had done it (except with no rotation, oddly enough). After all, they were commissioned to reproduce the same effects as Podkletnov had observed. So why embarass Podkletnov when they could instead confirm and validate him?

But why didn't NASA go ahead and do the rotating version of the experiment? Maybe NASA and crew know alot more about antigravity than they want us to know....

Why do you think they have drastically cut spending (their new motto is "smaller, cheaper, better") on space programs which use expensive chemical rocket technology? Perhaps they know that inexpensive and efficient propulsion methods such as antigravity are at the doorstep, and so any more spending on conventional rockets would be a waste of money.

Hmmmm.......

For what follows, my comments are in brackets [->...], and I have made bold any parts that are of special importance.

Here's part of the

NASA Marshall Space Flight Center and University of Alabama Huntsville Physica-C Paper:

Static Test for A Gravitational Force Coupled to Type II YBCO Superconductors

Ning Li*, David Noever, Tony Robertson, Ron Koczor, and Whitt Brantley

Physica C 281 (1997) 260-267

NASA Marshall Space Flight Center Huntsville, AL 35812 and The University of Alabama in Huntsville, Huntsville, AL, 35804

received 8/25/97

ABSTRACT:

As a Bose condensate, **superconductors** provide novel conditions for revisiting previously proposed couplings between electromagnetism and gravity. Strong variations in Cooper pair density, large conductivity and low magnetic permeability define superconductive and degenerate condensates without the traditional density limits imposed by the Fermi energy (~10^{-6} g cm^3). **Recent experiments [-> by Podkletnov]** have reported anomalous weight loss for a test mass suspended above a **rotating** Type II, YBCO superconductor, with the percentage change (0.05-2.1%) independent of the test mass' chemical composition and diamagnetic properties. A variation of 5 parts per 10^4 was reported above a **stationary** (non-rotating) superconductor. In experiments using a sensitive gravimeter, bulk YBCO superconductors were stably levitated in a DC magnetic field. Changes in acceleration were measured to be <u>less than 2 parts in 10^8 of the normal gravitational acceleration</u>. This result puts new limits on the strength and range of the proposed coupling between static superconductors and gravity.

Extending the early experiments on gravity and electromagnetic effects by Faraday [1] and Blackett [2], Forward [3] first proposed unique gravitational tests for superconductors in an electromagnetic field: "Since the magnetic moment and the inertial moment are combined in an atom, it may be possible to use this property to convert time-varying electromagnetic fields into time-varying gravitational fields." For comparison, Forward's electromagnetic analogy shares many features at the atomic scale (e.g. ion precession) with nuclear magnetic resonance (NMR) devices and at the laboratory scale with superconducting bearings or flywheels.

Recent experiments [4-5] **[-> Podkletnov]** have reported that for a variety of different test masses, a Type-II, high temperature (YBCO) superconductor induces anomalous weight effects (0.05-2% loss). A single-phase, dense bulk superconducting ceramic of $YBa_2Cu_3O_{7-x}$ was held at temperatures below

60 K, levitated over a toroidal solenoid, and **induced into rotation** using coils with rotating magnetic fields. **Without superconductor rotation, a weight loss of 0.05% was reported,** a relatively large value which has been attributed to buoyancy corrections [6] or air currents [7] until further details of the experiment elaborated upon measurements in closed glass tubes encased in a stainless steel box. A subsequent simplified apparatus **without rotation** [-> Don't believe this] has reported transients of up to 5% weight loss [5] with lower strength magnetic fields. Three theoretical explanations have been put forward to account for a possible gravitational cause: shielding [4], absorption via coupling to a Bose condensate [5, 8] and a gravito-magnetic force [9-11]. The symmetry requirements of each explanation are different, as are the need for magnetic fields or superconductor rotation; most notably an absorption mechanism (based on an instability in the quadratic part of the Euclidean gravitational action in the presence of a Bose condensate [5,8]) may not require an external EM field (except to generate density fluctuation in the Cooper pairs), while gravitomagnetic effects in the ion lattice [9-11] depend on a time-varying gravito-magnetic potential, dA_g/dt. Careful experiments must identify and isolate the relative importance of thermal, magnetic, and any gravitational components.

••••

Discussion

Any apparent gravitational contribution of the superconductor can be derived by subtracting the contribution of the magnet and superconductor together from the magnet alone; however, since the relative gravimeter responds (weakly, $<2\text{-}5\times10^{-6}$ cm s^{-2}) to the magnetic field, the uniquely superconductive contribution must combine any gravitational effect with the diamagnetic shielding of the magnets by the YBCO superconductor itself (~20-90% shielding of the field depending on hysteresis during cooling and magnetization). In any case, the maximum contribution to a change in gravity of a **static [-> Non-rotating]** superconductor in a constant magnetic field was measured as less than 2 parts in 10^8 of the normal gravitational acceleration. **[i.e. NO results!]**
This measurement extends an approximately 4-5 order of magnitude improvement over that previously obtained with the use of an opto-electronic balance [4-5] instrumented without either thermal or magnetic compensation. Relative to a gravito-magnetic force [9-11; 18] which depends on an AC magnetic drive or source term, dA_g/dt, the static case more strongly constrains interpretations based on either simple material shielding [4-5] or absorption of gravity [8]; regardless of the relative orders of magnitude, a coupling term (quadratic) to Euclidean gravity based on the Bose condensate and radial absorption does not necessarily require either rotation or a magnetic field to induce density fluctuations in the Cooper pairs, particularly in the limit of infinite conductivity. **The rotating verion of this experiment will be reported in subsequent work. [->Never happened].** In addition to superconductors, other Bose condensates such as superfluid helium have been investigated for gravitomagnetic field exclusion [19], but the low thermal conductivity of helium limits measurable power transfer from an AC magnetic field by several orders of magnitude below a YBCO superconductor.

••••

One of the references cited at the end of Ning Li's paper:

[4] Podkletnov E. and Nieminen R., Physica C 203 (1992) 441; Podkletnov E., 1995, report MSU-chem 95, Tampere Univ. Finland; improved version (cond-mat/9701074)

[end of Physica C Paper]

Other Related Papers:
Ning Li and D. G. Torr, Phys. Rev., 43D, 457, 1991.
Ning Li and D. G. Torr, Phys. Rev., 46B, 5489, 1992.
Ning Li and D. G. Torr, Bull. Am. Phys. Sco., 37, 948, 1992.
E. Podkletnov and R. Nieminen, Physica C, 203, 441, 1992.
D. G. Torr and Ning Li, Found. Phys. Lett., 37, 948, 1993.

Mysterious USC Gravity Generator Announcement:

NOTICE: This document has since been **withdrawn** by USC.

The University of South Carolina
Available Technology
Gravity Generator

Abstract:

The gravitational field generator comprises a stationary superconductor surrounded by a special configuration of RF solenoids. Certain combinations of RF fields will result in the generation of a DC gravitational field in any desired direction in the range of about 10^-22 g m^-2 per ion, depending on ion mass, where g is the acceleration due to gravity. When directed against gravity it acts as an anti-gravity device. When directed horizontally it acts to accelerate or brake a free mass. Appropriate choice of the relative phases of the RF magnetic field causes the gravitational field to form a beam, which will not exhibit the usual inverse square dependence with distance from the source. The production of fields equal to one g is anticipated with this simple arrangement. The theoretical expertise exists to understand controlling mechanisms and to optimize the field for specific applications.

Potential Areas of Application:

The invention proposed here can be used to generate a force field in any direction whether it be for propulsion or any other similar purpose, thus it has many applications wherever a force is needed. For example, the Gravity Generator could be used both to replace the wheels of a car so that the motion would not be terrain dependent and to accelerate and brake the vehicle. This invention could also be used to lift and propel aircraft, drive generators more efficiently, and produce gravity free environments on Earth for precision manufacturing and scientific research in numerous fields (e.g., chemical engineering, medicine and pharmaceuticals).

Main Advantages:

In addition to producing the above-mentioned gravitational effects, the Gravity Generator also has the advantages of being inexpensive, due to its simplicity and low power consumption, as well as being non-polluting.

Licensing Potential:

University seeks licensee and/or joint development.

USC ID Number:

96140

Contact Info:

William F. Littlejohn or Daniel J. Antion, Ph.D.
University of South Carolina
Office of Technology Transfer
Byrnes International Building, Suite 501
Columbia, SC 29208
(803) 777-9515 or Fax (803) 777-4136

NOTES:

- Pete Skeggs heard a rumor from a reliable source that this document existed. However, that source would not send him a copy because it was issued under nondisclosure. However, the information the source gave him corresponds with the information in this document, so he believes they are one and the same.
- This document was printed inside the **back cover** of James E. Cox's "Antigravity News and Space Drive Technology" journal, July-August 1997. No comments accompanied the text. No copyright notice or other restriction accompanied this document.
- A person contacted William F. Littlejohn on 6 Oct 1997. This person confirmed that the document **is real** and is not a hoax. However, it has since been **withdrawn** by USC.

[End of USC Gravity Generator Announcement]

Boeing Phantom Works:

Recently **Jane's Defence Weekly** has reported that **Boeing,** the world's largest aircraft maker, is working on an antigravity research project at the **top-secret Phantom Works in Seattle,** entitled GRASP - gravity research for advanced space propulsion.

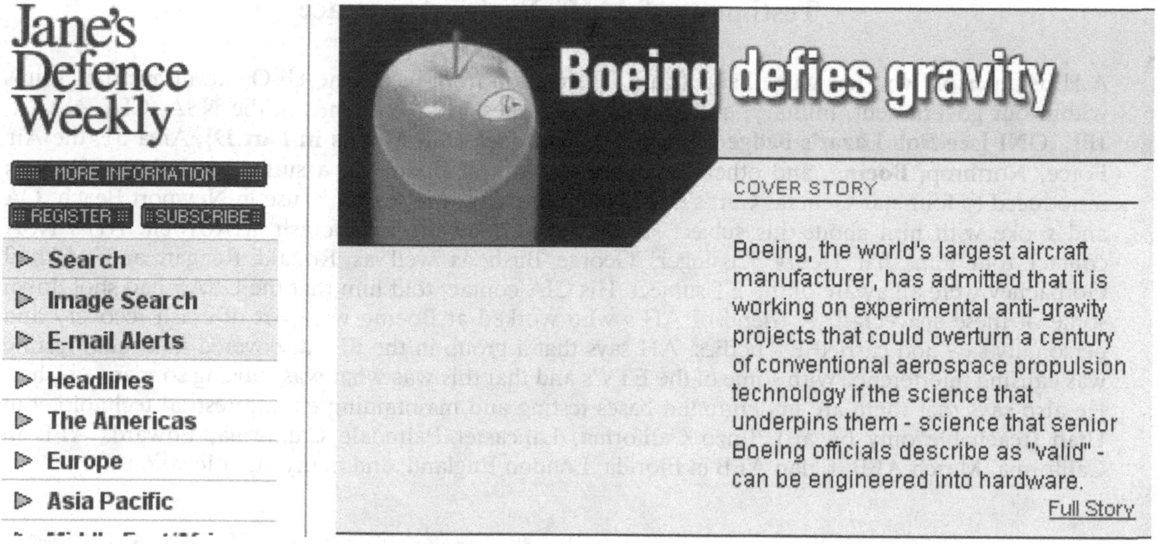
Jane's Defence Weekly
MORE INFORMATION
REGISTER
SUBSCRIBE
Search
Image Search
E-mail Alerts
Headlines
The Americas
Europe
Asia Pacific
Boeing defies gravity
COVER STORY
Boeing, the world's largest aircraft manufacturer, has admitted that it is working on experimental anti-gravity projects that could overturn a century of conventional aerospace propulsion technology if the science that underpins them - science that senior Boeing officials describe as "valid" - can be engineered into hardware.
Full Story

The Web Page From Jane's Where Article Began. "Full Story" Was Only Available To Paid Subscribers, Of Which This Author Was Not One Of.

The following contains very interesting information regarding Boeing, from a person who was obviously a "deep-insider".

Excerpt From "Disclosure":

I've **bold**-faced words which are important, and
my comments are in brackets [ ]:

Testimony of A. H., Boeing Aerospace

December 2000

A.H. is a person who has gained significant information from inside the UFO extraterrestrial groups within our government, military, and civilian companies. He has friends at the NSA, CIA, NASA, JPL, **ONI [see Bob Lazar's badge in Part A], NRO [see Dan Morris in Part D],** Area 51, the Air Force, Northrop, **Boeing**, and others. He used to work at Boeing as a surface technician. He was introduced to four-star General Curtis LeMay and one day went to his house in Newport Beach, CA and spoke with him about this subject. LeMay confirmed the ETC crash at Roswell. AH's NSA contact told him that Henry Kissinger, George Bush as well as Ronald Reagan and Mikhail Gorbachev were all aware of the ET subject. His CIA contact told him that the USAF had shot down some of these spacecraft. A friend of AH's who worked at Boeing was part of crash recovery and personally saw and carried ET bodies. AH says that a group in the FBI discovered that radar testing was causing interference with some of the ETV's and that this was what was causing so many crashes. He also says that there are underground bases testing and maintaining extraterrestrial technology in Utah (reachable only by air), Enzo California, Lancaster/Palmdale California, Edwards AFB in California, March AFB, Eglan AFB in Florida, London England, and many other locations.

AH: A. H. SG: Dr. Steven Greer

AH: ...And I asked Curtis LeMay and I said Curt, when you were in the Air Force out of the 100 percent of the sightings [of UFOs] that came into the Air Force how many of them remained not identified? And he said out of the 100 percent that came in only 35 percent of them could not be identified. And I asked well, why couldn't they have been identified. And he said because the craft were too damn fast. We can't catch them. In fact, he used some foul language in there too in his explanation.

And then I asked him about the Roswell crash, if that really occurred. And he looked over at me and he shook his head, yes, that it did occur. And I said that I was particularly interested in the strange writings that were found inside. I asked if they were ever deciphered. And he said as far as he understood they were not deciphered when he was in office...Then Curtis LeMay started to open up and talk about Roswell and the crash that occurred. And he told me that he was aware that it wasn't ours, it wasn't the military's, the Air Force or the Army or the Navy, and that it was a highly strange vehicle that crashed. ...

And by the way, **Majestic 12** is real.

[Recall again Bob Lazar's badge, in Part A. It denotes "MAJ".]

MJ12 did exist. But it does not exist today. The name was changed. The positions are still the same. Henry Kissinger is very knowledgeable about what's going on. Henry Kissinger was in the loop, I was told.

SG: Who told you this?

AH: A friend of mine who worked at the NSA [National Security Agency] told me this. He saw Henry Kissinger's name on documents. He saw George Bush's name on some of the documents. He was made aware of what's going on. About 1978 Reagan was fully briefed on the alien presence. Reagan told Mikhail Gorbachev of Russia about 75 percent of what's going on. And then Gorbachev became very, very close to us. ...

.....

.....was a national intelligence agent. He said that we hit several of them in Nevada over Ely, Nevada, and other portions of New Mexico, near Albuquerque.

SG: Did he indicate what time frame these things happened?

AH: This was in the late 60s and 70s...

Yes, I have. I met a gentleman who was in the Army who was involved with several crash recoveries when he was in the CID. In 1947 when he was inducted in the United States Army several of his friends were in the CIC, counter intelligence, and they became very close to one another. And he hand picked a friend of mine that worked at **Boeing** Aircraft Company in Long Beach to become a part of ET crash recovery/ retrieval in northern New Mexico.

He saw some of the aliens and witnessed the craft guarded this crashed disc that had crashed and they gave him an M1 carving rifle to guard against anybody who tried to get close to it. And they were given orders to shoot anybody without authorization. They recalled him on several other crashes. They had to fly him with blackened out windows to several of the crash sites. ...

.....

But he did tell me that he worked on the craft to back engineer them. He's aware of their existence. He knows that I met the Chief of Staff of the Air Force, General LeMay. He told me several times of the testing that took place at Area 51, that some of them come up behind the mountain. And they just blank out and disappear and then they appear about 15, 20 miles away within about 20 seconds and just reappear. And they do that back and forth ...

287

SG: These are made by humans?

AH: No, these are the real things. These are the craft that he has been working on. This is our military testing these ET vehicles up at Area 51...

He was working for Northrop. Directly for Northrop. He was not working for the military. He was receiving his paycheck from Northrop.

SG: So Northrop has an involvement with Area 51.

AH: Oh, yes. So does Lockheed Martin, **Boeing** Aircraft. **Boeing** does trucking for Area 51. Also Hughes is involved. Most of the major defense contractors have something to do with Area 51. Boeing trucks for Area 51. I do know that for sure. I found that out from a friend at Boeing that told me that they were hiring up there for trucking for some of the military sites. And I found out from him that he had personally gone onto Area 51, escorted to Area 51, from the main highway.

This witness is still alive. In fact, he knows where the aliens are originating from. As far as he understands there is an underground area there and there are a lot of extraterrestrial debris down below Area 51 and stored in some sort of a containment area. But he has never seen any extraterrestrials at Area 51 walking around and conversing with anybody. He just saw the craft and the technology that they're trying to extract from these vehicles to incorporate that technology into our fighters and possibly some of our space programs.

.....

Project Red Light is a program to test these vehicles and extract as much information as possible from the alien connected projects to find out how these things operate. They want to gain as much information for our fighters and bombers and for our space connected projects. I was told by people who were in the Army and the Air Force and people that worked at Area 51, all of whom corroborate what was going on regarding Project Red Light. And Project Red Light is still continuing as of this day.

So, Northrop hired him to go up there and help the cause up at Area 51 to try to find out how these things work. And I believe that he started in 1980. And he left there about 1997.

And by the way, Project Red Light and Project Grudge, they are all intermixed. The director of Project Blue Book, Robert Friend, told me this. He was fully aware of Project Red Light and Project Grudge/ Blue Book Report No. 13 was written by him. Robert Friend used to work at Fairchild in Redondo Beach on Aviation and Rosecranz.

......

.....

But **Carter** was largely left out of the loop -- the control group didn't trust the guy for some reason. They were afraid that he would come out and make blanketed statements to the news media and they just didn't trust him. They left him out of the loop ... This NSA witness entered in 1974 and left the military, the NSA, in about 1985. He said that **Henry Kissinger** was involved with the study group back in the 50s to study the ramifications of this information, and to determine what would happen if the information was leaked through a credible source. They were to do a study and pass off classified information to certain outside study groups such as the **Rand Corporation** and other think tanks of this nature. Basically, these projects were controlled by the Majestic 12 group, which is no longer called MJ12. I'm trying to find out the new name of this group. My contact that worked at Area 51 knows the name of the group, but he's refusing to tell me the name. Basically it's an oversight group intermingled with the National Security Council and the National Security Planning Group in Washington, D.C.

......

The **NSA** also collaborates back and forth with the **NRO,** the National Reconnaissance Organization, regarding tracking and intercepts, along with NORAD, the Air Force and the Army. They are all in on it and it's all connected with the top secret group called MAJI Control.

SG: What did he tell you about **MAJI Control** and how that operates?

[recall again Bob Lazar's badge, in Part A]

AH: MAJI Control is controlled by the **Office of Naval Intelligence.** It's a top secrete collection group, like the Central Intelligence Agency and the NSA. In fact, the ONI is just like the CIA. It's a top-secret organization within the Navy. It's similar to the NSA and the CID. It's all encrypted information. They have agents out in the field, just like the CIA, collecting information. It's all very, very top-secret...

.....

[END OF EXCERPT]

<u>Bill Uhouse:</u>

The following contains very interesting information, from a person who was obviously a "deep-insider".

Excerpt From "Disclosure":

I've **bold**-faced words which are important, and
my comments are in brackets [ ]:

Testimony of Captain Bill Uhouse, USMC (ret.)

October 2000

Bill Uhouse served 10 years in the Marine Corps as a fighter pilot, and four years with the Air Force at Wright-Patterson AFB as a civilian doing flight-testing of exotic experimental aircraft. Later, for the next 30 years, he worked for defense contractors as an engineer of antigravity propulsion systems: on flight simulators for exotic aircraft — and on actual flying discs. He testifies that that the first disc they tested was the **re-engineered ET craft that crashed in Kingman, Arizona in 1958.** He further testifies that **the ET's presented a craft to the US government;** this craft was taken to **Area 51,** which was just being constructed at the time, and the **four ET's that accompanied the craft** were taken to Los Alamos. Mr. Uhouse's specialty was the flight deck and the instruments on the flight deck — he understood the gravitational field and what it took to get people trained to experience antigravity. He **actually met several times with an ET that helped the physicists and engineers** with the engineering of the craft.

I spent 10 years in the Marine Corps, and four years working with the Air Force as a civilian doing experimental testing on aircraft since my Marine Corps days. I was a pilot in the service, and a fighter pilot; [I] fought in ... after the latter part of WWII and the Korean War Conflict, I was discharged as a Captain in the Marine Corps.

I didn't start working on flight simulators until about - well the year was **1954,** in September. After I got out of the Marine Corps, I took a job with the **Air Force at Wright Patterson** doing experimental flight-testing on various different modifications of aircraft.

While I was at Wright Patterson, I was approached by an individual who — and I'm not going to mention his name — [wanted] to determine if I wanted to work in an area on new creative devices. Okay? And, that was a flying disc simulator. What they had done: they had selected several of us, and they reassigned me to A-Link Aviation, which was a simulator manufacturer. At that time they were building what they called the C-11B, and F-102 simulator,
B-47 simulator, and so forth. They wanted us to get experienced before we actually started work on the flying disc simulator, which I spent 30-some years working on.

I don't think any flying disc simulators went into operation until the early 1960s — around 1962 or 1963. The reason why I am saying this is because **the simulator wasn't actually functional until around 1958.** The simulator that they used was **for the extraterrestrial craft they had, which is a 30-meter one that crashed in Kingman, Arizona, back in 1953 or 1952. That's the first one that they took out to the test flight.**

This ET craft was a controlled craft that the aliens wanted to present to our government — the USA. It landed about 15 miles from what used to be an army air base, which is now a defunct army base. But that particular craft, there were some problems with: number one - getting it on the flatbed to take it up to **Area 51.** They couldn't get it across the dam because of the road. It had to be barged across the Colorado River at the time, and then taken up Route 93 out to Area 51, which was just being constructed at the time. There were four aliens aboard that thing, and those aliens went to **Los Alamos** for testing. They set up Los Alamos with a particular area for those guys, and they put certain people in there with them — people that were astrophysicists and general scientists — to ask them questions. The way the story was told to me was: there was only one alien that would talk to any of these scientists that they put in the lab with them. The rest wouldn't talk to anybody, or even have a conversation with them. You know, first they thought it was all ESP or telepathy, but you know, most of that is kind of a joke to me, because they actually speak — maybe not like we do — but they actually speak and converse. But there was only one who would [at Los Alamos].

The difference between this disc, and other discs that they had looked at was that this one was a much simpler design. The disc simulator didn't have a reactor, [but] we had a space in it that looked like the reactor that wasn't the device we operated the simulator with. We operated it **with six large capacitors** that were charged with **a million volts each,** so there were six million volts in those capacitors. **They were the largest capacitors ever built. These particular capacitors, they'd last for 30 minutes, so you could get in there and actually work the controls and do what you had to — to get the simulator, the disc to operate. [recall the end of the T.T. Brown chapter!]**

So, it wasn't that simple, because we only had 30 minutes. Okay? But, in the simulator you'll notice that there are no seat belts. Right? It was the same thing with the actual craft — no seat belts. You don't need seat belts, because when you fly one of these things upside down, there is no upside down like in a regular aircraft — you just don't feel it. There's a simple explanation for that: you have **your own gravitational field right inside the craft,** so if you are flying upside down — to you — you are right side up. I mean, it's just really simple, if people would look at it. I was inside the actual alien craft for a start-up...

There weren't any windows. The only way we had any visibility at all was done with cameras or video-type devices. *[See the testimony of Mark McCandlish. SG]* My specialty was the flight deck and the instruments on the flight deck. I knew about the gravitational field and what it took to get people trained.

Because the disc has its own gravitational field, you would be sick or disoriented for about two minutes after getting in, after it was cranked up. It takes a lot of time to become used to it. Because of the area and the smallness of it, just to raise your hand becomes complicated. You have to be trained — trained with your mind, to accept what you are going to actually feel and experience.

Just moving about is difficult, but after a while you get used to it and you do it — it's simple. You just have to know where everything is, and you [have] to understand what's going to happen to your body. It's no different than accepting the g-forces when you are flying an aircraft or coming out of a dive. It's a whole new ball game.

Each engineer that had anything to do with the design was part of the start-up crew. We would have to verify all the equipment that we put in — be sure it [worked] like it [was] supposed to, etc.

I'm sure our crews have taken these craft out into space. I'm saying it probably took a while to train enough of the people, over a sufficient time period. The whole problem with the disc is that it is so exacting in its design and so forth. It can't be used like we use aircraft today, with dropping bombs and having machine guns in the wings.

The design is so exacting, that you can't add anything — it's got to be just right. There's a big problem in the design of where things are put. Say, where the center of the aircraft is, and that type of thing. Even the fact that we raised it three feet so the taller guys could get in — the actual ship was extended back to its original configuration, but it has to be raised.

We had meetings, and **I ended up in a meeting with an alien.** I called him **J-ROD** — of course, that's what they called him. I don't know if that was his real name or not, but that's the name the linguist gave him. I did draw a sketch, before I left, of him in a meeting. I provided it to some people and that was my impression of what I saw. **The alien used to come in with [Dr. Edward] Teller and some of the other guys, occasionally, to handle questions that maybe we'd have. You know? But you have to understand that everything was specific to the group. If it wasn't specific to the group, you couldn't talk about it. It was on a need-to-know basis. And [the ET] he'd talk. He would talk, but he'd sound just like as if you spoke — he'd sound like you. You know, he's like a parrot, but**

he'd try and answer your question. A lot of times he'd have a hard time understanding, because if you didn't put it on paper and explain yourself, half the time he couldn't give you a good answer.

The preparation we had before meeting this alien was, basically, going through all of the different nationalities in the world. Then they got into going into other forms of life, even down to animals and that type of thing. And...this J-ROD — his skin was pinkish, but a little bit rough — that kind of stuff; not horrible-looking, you know — or to me, he wasn't horrible-looking.

Some of the guys who were in the particular group that I was in — they never even made it. You know, when they gave you the psychological questions, I just answered them the way I felt and I had no problem. That's what they wanted to know — if you'd become upset — but it never bothered me. It didn't amount to much.

So basically, the alien was only giving engineering advice and science advice. For example, I performed the calculations but needed more help...I spoke of a book that — well it's not a book; it's a big assembly with various divisions dealing with gravitational technology, and the key elements are in there, but all the information wasn't there. Even our top mathematicians couldn't figure some of this stuff out, so the alien would assist. Sometimes you'd get into a spot where you [would] try and try and try, and it wouldn't work. And that's when he'd [the alien] come in. They would tell him to look at this and see what we did wrong.

Over the last 40 years or so, not counting the simulators — I'm talking about actual craft — there are probably two or three-dozen, and various sizes that we built.
I don't know much about the [ET] ones that they brought here. I know about that one [craft] out of **Kingman** but that's about it. And, I know the company that hauled it out of there — who is out here now — but... But, there's one that operates with certain chemicals.

I think these triangles that people are seeing are two or three 30-meter craft, that are in the center of it [the triangle]. And, the outside perimeter — well you could put anything you want, as long as these particular ones meet the design criteria, and they'll operate.

You know, there were certain reasons for the secrecy. I could understand that; it was no different than the first atomic bomb that they built. But they are getting so far ahead now with aircraft design. **And, like I told you gentlemen earlier — that by 2003, most of this stuff will be out for everybody to look at. Maybe not the way that everybody expects it, but in some manner they determine...appropriate to show everybody. You know, a big surprise. The reason why I said that is because the document I signed ends in 2003 and I'm not the only one who signed those.**

But, that gravitational manual — if you ever get one of these volumes of documents, you'd be on top of the world. You'd know everything.

[See the testimony of Mark McCandlish that verifies human-made antigravity craft and also the use of cameras instead of windows for imaging. SG]

[END OF EXCERPT]

[ ]

[THIS PAGE IS INTENTIONALLY BLANK]

Dan Morris, NRO Operative:

The following contains very interesting information, from a person who was obviously a "deep-insider".

Excerpt From "Disclosure":

I've **bold**-faced words which are important, and
my comments are in brackets [ ]:

Testimony of Master Sergeant Dan Morris, USAF (Retired)/ NRO Operative

September 2000

Dan Morris is a retired Air Force career Master Sergeant who was involved in the extraterrestrial projects for many years. After leaving the Air Force, he was recruited into the super-secret **National Reconnaissance Organization,** or **NRO,** during which time he worked specifically on extraterrestrial-connected operations. He had a **cosmic top-secret clearance** (38 levels above top secret) which, he states, no U.S. president, to his knowledge, has ever held. In his testimony, he talks of **assassinations** committed by the NSA; he tells how **our military deliberately caused the 1947 ET craft crashes near Roswell, and captured one of the ETs, which they kept at Los Alamos for 3 years, until he died.** He talks about the intelligence teams that were charged with **intimidating, discrediting, and even eliminating witnesses to ET/UFO events.** He talks about Germany's re-engineering of UFOs, even prior to WWII. He talks about our current energy crisis — and **the fact that we haven't needed fossil fuels since the 1940s, when free energy technologies were developed** — but have been kept from humanity. This is the real reason for the secrecy of the ET/UFO subject. "What the people in power right now don't want us to know is that this free energy is available to everybody." In conclusion, he warns against the weaponization of space and the shooting down of ET craft — this could force them to retaliate, and that would be our destruction.

I had a clearance 38 levels above top secret, which is cosmic top-secret -it is the top of all of those clearances. It is for UFOs, and aliens, etc. No president has had that level, has ever been cleared for that level. **Eisenhower** was the closest. Well, there are several intelligence agencies- the Army had it, the Air Force had it, the Navy had it. And then there were several secret intelligence agencies. One that did not exist, it was so secret, was the NRO. **You couldn't mention NRO.** It is the National Reconnaissance Organization. If you're on that level, then there's an organization worldwide called **ACIO,** that's Alien Contact Intelligence Organization. If you pay your dues and you follow the rules, your government is allowed to benefit from that organization's information. Now some people call it the high frontier. The Navy Intelligence refer to themselves that way sometimes. And they all work together. Air Force intelligence, **Naval intelligence,** and **the NRO** were at one time all in a certain part of Langley Air Force Base in Virginia. And most of the satellite interpreters were there, most of the intelligence interpreters from the Air Force, the Army, the Navy were there, that's where they worked and interpreted.

......

What happened was that we found out that **high-powered radar would interfere** with their stability, because we could watch the **AMPLIFIERS** and stabilizers come down when they were low and slow. Radar affected the UFOs when they were low and slow. We already knew that, we knew it before '47, before they came down. And where was most of our radar? **White Sands, and down at Roswell.** Who was stationed at Roswell? The only nuclear bomb squadron in the world. So they were interested and we had a lot of radar there, because we were going to protect as much as we could. Well, **we focused several, big, powerful radars on them, and it caused two of them to run together.** One of them was the one that went down and landed on the ranch, the other was the one that went down into the bank and it had two aliens on it lying outside when we got there. One of them was wounded or hurt, and the other was alive then, but before we could get him anywhere, he had already passed on. But the other we kept at **Los Alamos** over here for about three years. He got sick. We sent out on every frequency range we could and everything that he was sick, that we didn't do it, that they could come and get him if they wanted to. But he died before they got here. But they came and got his body, and that's when they went to Washington and had that formation over Washington. So they retrieved his body.

.....

[END OF EXCERPT]

Majestic-12 (MAJ, T.S. MAJIC), & Interplanetary Phenomenon Unit:

This is just a part of the many "leaked out" documents (by govt. insiders, no doubt) which have been on the Internet (and elsewhere) for the last 7 years, or so. What you see here is just the "tip of the overall iceberg". These documents speak for themselves. The reader is encouraged to seek out more information regarding documents such as these seen here!

As a reminder, recall that Bob Lazar's badge had a "MAJ" designation on it. And, it also had an "S4" designation. Bob's workplace was within the Area 51 complex, in Nevada.

Also recall, from Part A - Prolog & Introduction, that the officially-obtained (by FOIA request) document there proves that there indeed did exist an "Interplanetary Phenomenon Unit of the Scientific and Technical Branch, Counterintelligence Directorate, Department of the Army", and it existed until the late 1950's. This Unit is referenced in the following documents.

No. [redacted] - 9 July 1947

By Auth. A. C. of S., G-2
Date 22 July 1947
Initials [illegible]

WAR DEPARTMENT
Office of A. C. of S., G-2

SECRETARY'S OF

SECRET

~~TOP SECRET~~ F0662

001947122-A 1224

[redacted]

Received
July 16, 1995

THIS DOCUMENT [illegible] DE-CLASSIFIED AS [illegible] AFFECTING [illegible] NATIONAL SECURITY AND HAS BEEN UPGRADED AS ABOVE [illegible] WITH A "NEED TO KNOW" BASIS. ONLY THOSE [illegible] PERSONS WITH MAJIC ACCESS MAY HAVE ACCESS. [illegible] SCHEDULE TO COMMENCE ONLY WITH AN [illegible] PRESIDENTIAL E.O. WITH APPROVAL OF U.S. MAJESTIC [illegible] AND SECURITY

By authority of: Lt. General John A. Samford, Dir., NSA

Approved by: Allen Dulles, Director, CIA

Date: [illegible] 1960

~~TOP SECRET~~

~~TOP SECRET~~
ULTRA

J. C. S. FILE COPY

C.C.S. FILE [illegible]

27

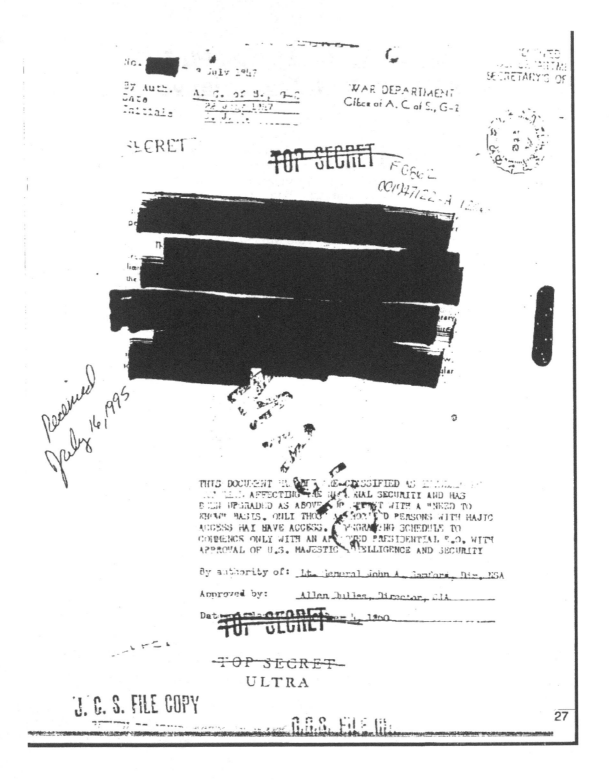

~~TOP SECRET~~

ULTRA

[redacted]

INTERPLANETARY PHENOMENON UNIT SUMMARY

INTELLIGENCE ASSESSMENT File ref. 001947122-A.1206

1. The extraordinary recovery of fallen airborne objects in the state of New Mexico, between 4 July - 6 July 1947: This Summary was prepared by Headquarters Interplanetary Phenomenon Unit, Scientific and Technical Branch, Counterintelligence Directorate, as requested by A.C. of S., G-2, at the expressed order of Chief of Staff.

2. At 2332 MST, 3 July 47, radar staions in east Texas and White Sands Proving Ground, N.M., tracked two unidentified aircraft until both dropped off radar. Two crash sites have been located close to the WSPG. Site LZ-1 was located at a ranch near Corona, approx. 75 miles northwest of the town of Roswell. Site LZ-2 was located approx. 20 miles southeast of the town of Socorro, at Lat. 33-40-31, Long. 106-28-29, with Oscura Peak being the geographic reference point.

3. The AST personnel were mainly interested in LZ-2 as this site contained the majority of structural detail of the craft's airframe, propulsion and navigation technology. The recovery of five bodies in a damaged escape cylinder, precluded an investigation at LZ-1.

4. On arrival at LZ-2, personnel assessed the finds as not belonging to any aircraft, rocket, weapons, or balloon test that are normally conducted from surrounding bases. First reports indicated that the first crash investigators from Roswell AAF that LZ-1 was the remains of a AAF top secret MOGUL balloon project. When scientists

~~TOP SECRET~~

ULTRA

[redacted]

[redacted]

from the Los Alamos Scientific Laboratory arrived to inspect IZ-2, it became apparent to all concerned that what had crashed in the desert was something out of this world.

5. Interviews with radar operators and officers from the Signal Corps Engineering Laboratories, Fort Monmouth, N.J., who were tracking these objects on-and-off since June 29 from Station "A", all indicated that these targets had periodically remained stationary for minutes at a time, then would resume thier original course, flying from the southeast to northwest. SCEL antennas had locked onto a flight of three objects on 3 July and lost them around 2330 MST on 4 July (a V-2 was scheduled for launch which is why SCEL Station "A" was able to do a track). It has been learned that at least six radar stations in east Texas (see detailed report in attchment A), and radar stations at Alamogordo AAF and Kirtland AAF, had also picked up these objects on the 4th as well. Using topographical maps and triangularisation, a last known position and bearing was calculated which helped search parties to locate the general area. Detachment 3 of the 9393rd Technical Services Unit, assigned to Alamogordo AAF, was responsible for the locating and transportation of the large sections of the craft.

6. A special radiobiological team accompanied by a SED and security detail from Sandia Base under orders from Colonel S. V. /HASBROUCK/ USA, Armed Forces Special Weapons Project, secured the immediate area surrounding the crash site. Select scientists from

[redacted]

 [redacted]

the General Advisory Committee of the Atomic Energy Commission, most notably DR. /J. ROBERT OPPENHEIMER,/was identified at LZ-2 as well as other members. Among PAPERCLIP specialist identified at LZ-2 were DR. /WERNHER VON BRAUN/(Fort Bliss); DR. /ERNST STEINHOFF/(AMC); and DR. /HUBERTUS STRUGHOLD/(AEROMEDICAL LAB, RANDOLPH FIELD).

7. Because of the stringent security measures that were in place at both crash sites, the team was not able to gain access to the several locations were wreckage and bodies are being held. CIC member of the team was able to learn that several bodies were taken to the hospital at Roswell AAF and others to either Los Alamos, Wright Field, Patterson AAF, and Randolph Field for security reasons. It is believed that this dispersion was on the orders of General Thomas Handy, Fourth Army Hdqrs. Remains of the powerplant were taken to Alamogordo AAF and Kirtland AAF. Structural debris and assorted parts were taken to AMC, Wright Field. Other remains were transported across the WSPG to the storage facilities of the NRL. All this was accomplished by 1730 MST 7 July.

8. On 7 July, Lt. General Nathan Twining arrived at Alamogordo AAF for a secret meeting with AAF Chief of Staff Spaatz and to view recovered remains of craft from LZ-2. On 8 July, Twining visited Kirtland AAF to inspect parts recovered from powerplant. On 9 July, Twining and staff flew to WSPG to inspect pieces of craft being stored there and on 10 July, made inspection of R&D facilities at Alamogordo and then returned to Wright Field. It is believed that Twining and staff is preparing a detailed report of both incidents and briefings later to follow. It is also the belief of CIC that General Eisenhower will see a showing of recoveries sometime in late August this year. The President was given a limited briefing

 [redacted]

at the Pentagon by AC/S AAF General Hoyt Vandenberg on 10 July. FBI interest was curtailed and access to stored craft at Los Alamos was denied upon request of Deputy COS Lt. General Collins. Inquiries to bases involved was restricted by General Vandenberg during the duration of recovery efforts. All teletype, telephone and radio transmissions were monitored for any disclosures of the finds. To maintain secrecy of site LZ-2, the CO of Roswell AAF was authorized to give a brief press release to local paper in which 8th AF Hdqrs. promptly denied rumors that the Army had flying saucers in their possession which effectively killed press interest. Civilians who might have seen or handled some of the wreckage, or viewed bodies were detained under the McNab law until all remaing evidence was secured in restricted bases. Witnesses were debriefed by CIC and warned of the consequences of talking to the press. So far, secrecy seems to be working.

9. All civilian and military personnel involved with the recovery operations had "need to know" access with proper security clearances. Though several MPs suffered nervous breakdowns resulting in one committing <u>suicide</u>, MP details from Alamogordo and Kirtland performed security functions very well. Ground personnel from Sandia experienced some form of contamination resulting in the deaths of 3 technicians. The status of the fourth technician is unknown. Autposies are scheduled to determine cause of death. CIC has made appropriate security file entries into dossiers with cross references for future reviews.

Liaison Committee Chairman approved on 28th.

10. With the pending approval of JAMES FORRESTAL as new Secretary of Defense, it is certain that he will be briefed on certain aspects of the discoveries. The only Cabinet member to date that may know of the details is Secretary of State Marshall. It has become known to CIC that some of the recovery operation was shared with Representative

[redacted]

/JOHN F. KENNEDY,/Massachusetts Democrat elected to Congress in 46. Son of JOSEPH P. KENNEDY, Commission on Organization of the Executive Branch of the Government. KENNEDY had limited duty as naval officer assigned to Naval Intelligence during war. It is believed that information was obtained from source in Congress who is close to Secretary for Air Force.

11. As to the bodies recovered at LZ-2, it appeared that none of the five crew members survived entry into our atmosphere due to unknown causes./DR. DETLEV BRONK/has been asked to assist in the autopsy of one well preserved cadaver to be done by MAJOR CHARLES B. REA. From what descriptions the team was able to learn and from photographs taken by intelligence photographers, the occupants appear in most respects human with/some anatomical differences/in the head, eyes, hands and feet. They have a slight build about five feet tall, with grayish-pink skin color. They have no hair on their bodies and clothed with a tight fitting flight suit that appears to be fire proof (some of the bodies looked as if they had been burned on head and and hands). Their overall stature reminds one of/young children./It is believed that there were male and female genders present, but was hard to distinguish.

12. The most disturbing aspect of this investigation was-- there were other bodies found not far from LZ-1 that looked as if they had been disected as you would a frog. It is not known if army field surgens had performed exploratory surgery on these bodies. Animal parts were reportedly discovered inside the craft at LZ-2 but this cannot be confirmed. The team has reserved judgement on this issue.

13. Our assessment of this investigation rests on two assumptions: 1) Either this discovery was an elaborate and well orchestrated hoax (maybe by the Russians), or; 2) Our country has played host to

[redacted]

[redacted]

beings from another planet.

14. Until more data can be acquired from other intelligence sources, it is the opinion of the team, that the investigation be expanded to include sources that might elucidate other possibilities not found by contemporary science. It is also recommended that appropriate budgets be allocated to facilitate future assignments that the unit may be called upon to perform. Until further orders, this investigation will continue.

Received 11-19-94
Timothy Good

TS-1-2736
TOP SECRET
MAJIC

Director
Dep. Director
Adm. Officer
Prod. & Control
Records & Pub.
[illegible] Asst.
Visual
Statistical
Correspondence
Copies [illegible]
Filed [illegible]

MAJESTIC TWELVE PROJECT

1st Annual Report 9/48 ?

A-1762.1-J1

A REVIEW OF THE PRESIDENT'S SPECIAL PANEL TO INVESTIGATE THE CAPTURE OF UNIDENTIFIED PLANFORM SPACE VEHICLES BY U.S. ARMED FORCES AND AGENCIES

PANEL

CHAIRMAN, Dr. Vannevar Bush, 4901 Hillbrook lane (Phone, REpublic 6700, branch 5484)

General J. Lawton Collins, Deputy Chief of Staff, United States Army
Major General Luther D. Miller, Chief of Chaplains, United States Army
General Hoyt S. Vandenberg, Vice Chief of Staff, United States Air Force
Lt. General Lewis H. Brereton Chairman, Military Liaison Cmt. AEC
Maj. General George C. McDonald, Director of Intelligence, United States Air Force
Brigadier General George F. Schulgen, Director, Plans and Policies, United States Air Force
Rear Admiral Paul F. Lee, Chief, Office of Naval Research, United States Navy
Admiral John Gingrich, Director, Security and Intelligence, AEC
Dr. J. Robert Oppenheimer, Chairman, General Advisory Committee, AEC
Jerome C. Hunsaker, MIT, National Academy of Sciences
Detlev W. Bronk, Chairman, National Research Council
Dr. Hugh L. Dryden, Director of Aeronautical Research, NACA
Dr. James H. Doolittle, Shell Oil

FOR
INCLUSION

TOP SECRET Central Intelligenc Agency Information Report CIA/SI 28-55 entitled: A DIGEST OF WORLD WIDE UNIDENTIFIED FLYING OBJECT INTELLIGENCE MATERIAL AS CONTAINED IN THE ARMED FORCES SECURITY AGENCY SIGNALS, RADAR, COMMUNICATIONS, AND HUMAN INTELLIGENCE OPERATIONS IN THE FIRST FIVE YEARS

Approved for Rele
Date

Note: This report has been coordinated with the Joint Intelligence Committee, the Air Technical Intelligence Center, [redacted] and the Intelligence Advisory Committee

A95[illegible]

WARNING: This document contains sensitive intelligence information affecting the national defense of the United States, within the meaning of Title 18, Sections 793 and 794, of the U. S. Code, as amended. Its transmission or revelation of its contents to or receipt by unauthorized person is prohibited.

[redacted]

OCS [illegible]-20
OCS 21347-86

INTERPLANETA[illegible] PHENOMENON UNIT

CSA

COPY

TOP SECRET

TOP SECRET

March 5, 1942.

ME[illegible] PRESIDENT :

As indicated in my [illegible] 2[illegible] memorandum to you regarding the ai[illegible] over Los Angeles [illegible] learned by Army G2 that Rear Admiral Anderson, [illegible] Intelligence, has informed the War Department of a naval [illegible] unidentified airplane off the coast of California, unlike th[illegible] no bearing on conventional explanation. Further in[illegible] revealed that the Army Air Corps also recovered a similar [illegible] the San Bernardino Mountains east of Los Angeles which cannot [illegible] as conventional aircraft. This Headquarters has come to a [illegible] that the mystery airplanes are in fact not earthly and accor[illegible] secret intelligence sources they are in all probability of interplan[illegible]. As a consequence, I have issued orders to Army G2 that [illegible] intelligence unit be created to further investigate the phe[illegible] and report any significant connection between recent inc[illegible] and those collected by the director of the Office of Coordinator [illegible].

I have further orde[illegible] investigation of all War Department files regardi[illegible] aerial phenomenon reported since 1897 and what ex[illegible] on the subject. At present, GHQ has no further in[illegible] would invalidate this conclusion. Pending any fu[illegible] investigation into this matter shall be limited to those [illegible] by you.

[illegible] G. Marshall
[illegible] Staff

TOP SECRET

Rec'd on SEP 0[illegible] 2000 [illegible]

REPORT BY THE JOINT LOGISTICS PLANS COMMITTEE

to the

JOINT CHIEFS OF STAFF

on

JOINT LOGISTIC PLAN FOR "MAJESTIC"

References:
a. J.C.S. 1844/126
b. J.C.S. 1844/127
c. "Joint Logistics Policy and Guidance" - FM 110-10, JANALP, AFM 400-4, June 1952

THE PROBLEM

1. Pursuant to the decision by the Joint Chiefs of Staff on J.C.S. 1844/126, to prepare the Joint Logistic Plan in support of MAJESTIC*.

FACTS BEARING ON THE PROBLEM AND DISCUSSION

2. MAJESTIC and its implementing directive** contain the planning and implementing procedures including assignment of responsibilities.

3. Appendix "A" to the Enclosure hereto constitutes the "Logistic Plan for MAJESTIC" and follows the standard form for a joint logistic plan to support a war plan, as set forth in reference c, Chapter 1, Section XV.

4. It is the policy of the Joint Chiefs of Staff to include, as a part of the Joint Logistic Plan, the logistic implications inherent in MAJESTIC. (See Appendix "B" to the Enclosure hereto.)

RECOMMENDATION

5. That the memorandum in the Enclosure with its Appendices be forwarded to the Chief of Staff, U.S. Army; the Chief of Naval Operations; and the Chief of Staff, U.S. Air Force.

---

* J.C.S. 1844/126
** J.C.S. 1844/127

DECLASSIFIED
Authority JCS [signature] 26 May 76
BY ____ NARS, Date ____

1	
2	
3	
4	
5	

Op-[illegible]31/at
Ser: 0001[illegible]30

TOP SECRET
SECURITY INFORMATION

TOP SECRET
SECURITY INFORMATION

2 OCT 1952

From: Chief of Naval Operations
To: Commander in Chief, Pacific
Commander in Chief, Atlantic
Commander in Chief, U.S. Naval Forces, Eastern Atlantic and Mediterranean
Commander, Second Fleet (alt. hq, CINCLANT/CINCLANTFLT)
Commander, Military Sea Transportation Service

Subj: Joint Outline Emergency War Plan "MAJESTIC"; forwarding of

Encl: (1) SM-2[illegible]-52 (Distribution of MAJESTIC), dated 25 September 1952
(2) Memorandum Receipt Form

1. Enclosure (1), with attached copies of Joint Outline Emergency War Plan "MAJESTIC", is forwarded herewith for appropriate action.

2. Please execute enclosure (2) and return to this office.

3. It is directed that each addressee maintain a chronological record of the physical sighting of all copies of "MAJESTIC" on the last day of each quarter. No report of this sighting will be required by CNO, except when requested. Attention is further invited to paragraph 5 of enclosure (1).

4. Note that paragraph 1 of enclosure (1) authorizes retention or destruction of the superseded plan at the discreation of the addressee, if retained they must be sighted as set forth in paragraph 3 above.

Copy to:
Secy JCS
[illegible]
Op-03C

AUTHENTICATED

M. W. HORTON
Lieutenant, USN

TOP SECRET
SECURITY INFORMATION

TOP SECRET
SECURITY INFORMATION

DECLASSIFIED
JCS Declass Br. 26 May 76
BY ______ NARS, Date ______

Green Copy.

TOP SECRET SECURITY INFORMATION

TOP SECRET - SECURITY INFORMATION COPY NO. ____

THE JOINT CHIEFS OF STAFF
Washington 25, D.C.

SM-2242-52
25 September 1952

SPECIAL HANDLING REQUIRED;
NOT RELEASABLE TO FOREIGN NATIONALS

MEMORANDUM FOR: Chief of Staff, U.S. Army
Chief of Naval Operations
Chief of Staff, U.S. Air Force
Commander in Chief, U.S. European Command
Commander in Chief, Far East
Commander in Chief, Pacific
Commander in Chief, U.S. Army, Europe
Commander in Chief, Atlantic
Commander in Chief, Caribbean
Commander in Chief, U.S. Naval Forces, Eastern Atlantic and Mediterranean
Commander in Chief, Alaska
Commander in Chief, U.S. Northeast Command
Commanding General, Strategic Air Command
Commanding General, U.S. Forces in Austria
Commander in Chief, U.S. Air Forces in Europe
U.S. Liaison Officer, Headquarters, Supreme Allied Commander, Atlantic
U.S. Representative, North Atlantic Military Committee
U.S. Representative, Standing Group, North Atlantic Military Committee
Secretary, U.S. Section, Canada-U.S. Military Cooperation Committee
Director, Continental U.S. Defense Planning Group
Chief, Army Field Forces (alternate headquarters, Chief of Staff, U.S. Army)
Commander, Second Fleet (alternate headquarters, CINCLANT/CINCLANTFLT)
Commanding General, Tactical Air Command (alternate headquarters, Chief of Staff, U.S. Air Force)
Commanding General, Military Air Transport Service
Commander, Military Sea Transportation Service
Director, Armed Forces Security Agency
Secretary, Joint Strategic Plans Committee
Secretary, Joint Communications-Electronics Committee

381 USSR (3-2-46) S. 65 - S.O.

Subject: Joint Outline Emergency War Plan for a War Beginning 1 July 1952.

References: a. Joint Outline Emergency War Plan MASTHEAD
b. SM-1197-51, dated 14 May 1951

1. Forwarded herewith is a copy of the Joint Outline Emergency War Plan for a War Beginning 1 July 1952 MAJESTIC.* This plan supersedes Joint Outline Emergency War Plan MASTHEAD, which was forwarded by SM-1197-51, dated 14 May 1951, copies of which ~~may be retained or destroyed by burning as desired~~ ## will be either returned or destroyed by burning.

* Enclosure filed in B.P. Part 8 by 9-25-52 date

TOP SECRET - SECURITY INFORMATION

- 1 -

See change per SM-432-53 dtd 3-2-53

9-25-52

TOP SECRET SECURITY INFORMATION

2. This plan, approved by the Joint Chiefs of Staff on 19 September 1952, is forwarded for your information and guidance in connection with your planning responsibilities.

3. The following plans in support of MAJESTIC are now under preparation:

a. A psychological warfare plan.

b. An unconventional warfare plan.

c. Cover and deception plans.

d. A civil affairs/military government plan.

e. A command plan.

f. A logistic plan.

g. Transportation guidance, to be included in the logistic plan.

h. A map and chart plan.

i. A communications plan.

4. The estimate of the Soviet Union's capability to execute campaigns and her probable courses of action contained in the Enclosure does not take into consideration the effect of opposition by any forces now in position and operational, or of unfavorable weather or climatic conditions. The purpose in emphasizing this statement, which is included in the body of the plan (page 16), is to avoid the danger of an interpretation which might accord to the Soviets capabilities unwarranted by a realistic appraisal of the facts.

5. Distribution or circulation of this plan, or portions thereof, will be restricted to those agencies and personnel whose duties specifically require knowledge of the plan and is made on a "Special Handling Required - Not Releasable to Foreign Nationals" basis.

For the Joint Chiefs of Staff:

W. G. LALOR,
Rear Admiral, U.S. Navy (Ret.),
Secretary.

Copies to:
~~Asst C/S, G-3~~
~~Secy to CNO (JCS)~~
~~Director/Plans, AF~~
Director J/S

JCS 1844/126 - Approved as amended - 19 Sep 52)

Enclosure

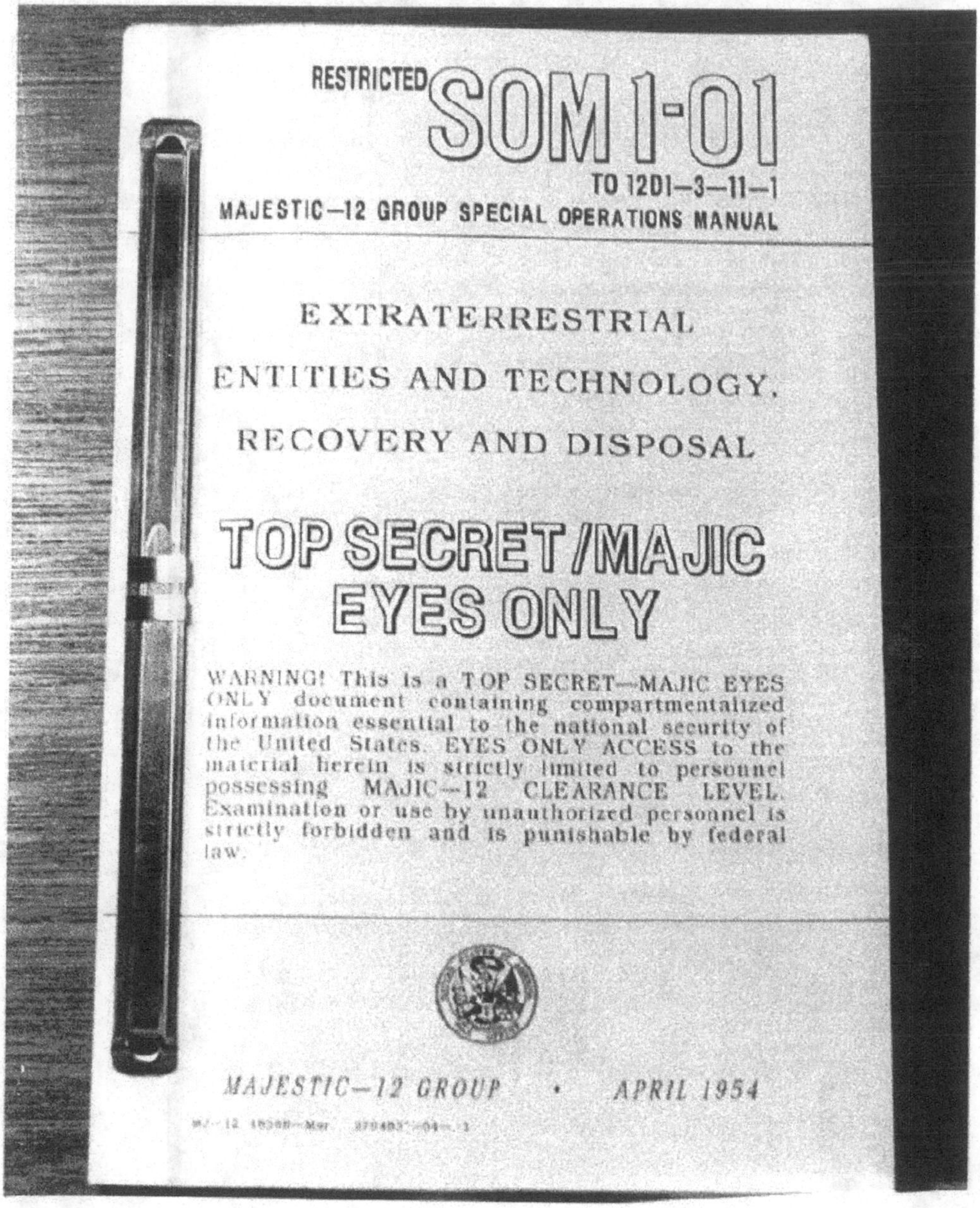
RESTRICTED SOM 1-01
TO 12D1-3-11-1
MAJESTIC-12 GROUP SPECIAL OPERATIONS MANUAL
EXTRATERRESTRIAL
ENTITIES AND TECHNOLOGY,
RECOVERY AND DISPOSAL
TOP SECRET/MAJIC
EYES ONLY
WARNING! This is a TOP SECRET—MAJIC EYES ONLY document containing compartmentalized information essential to the national security of the United States. EYES ONLY ACCESS to the material herein is strictly limited to personnel possessing MAJIC-12 CLEARANCE LEVEL. Examination or use by unauthorized personnel is strictly forbidden and is punishable by federal law.
MAJESTIC-12 GROUP • APRIL 1954

Special Operations Manual No. 1—01 } MAJESTIC — 12 GROUP
Washington 25, D. C., *7 April 1954*

EXTRATERRESTRIAL ENTITIES AND TECHNOLOGY, RECOVERY AND DISPOSAL

CHAPTER 1
OPERATION MAJESTIC—12

Section I. PROJECT PURPOSE AND GOALS

1. Scope

This manual has been prepared especially for Majestic—12 units. Its purpose is to present all aspects of Majestic—12 so authorized personnel will have a better understanding of the goals of the Group, be able to more expertly deal with Unidentified Flying Objects, Extraterrestrial Technology and Entities, and increase the efficiency of future operations.

2. General

MJ—12 takes the subject of UFOBs, Extraterrestrial Technology, and Extraterrestrial Biological Entities very seriously and considers the entire subject to be a matter of the very highest national security. For that reason everything relating to the subject has been assigned the very highest security classification. Three main points will be covered in this section.

a. The general aspects of MJ—12 to clear up any misconceptions that anyone may have.

b. The importance of the operations.

c. The need for absolute secrecy in all phases of operation.

3. Security Classification

All information relating to MJ—12 has been classified MAJIC EYES ONLY and carries a security level 2 points above that of Top Secret. The reason for this has to do with the consequences that may arise not only from the impact upon the public should the existence of such matters become general knowledge, but also the danger of having such advanced technology as has been recovered by the Air Force fall into the hands of unfriendly foreign powers. No information is released to the public press and the official government position is that no special group such as MJ—12 exists.

4. History of the Group

Operation Majestic—12 was established by special classified presidential order on 24 September 1947 at the recommendation of Secretary of Defense James V. Forrestal and Dr. Vannevar Bush, Chairman of the Joint Research and Development Board. Operations are carried out under a Top Secret Research and Development - Intelligence Group directly responsible only to the President of the United States. The goals of the MJ—12 Group

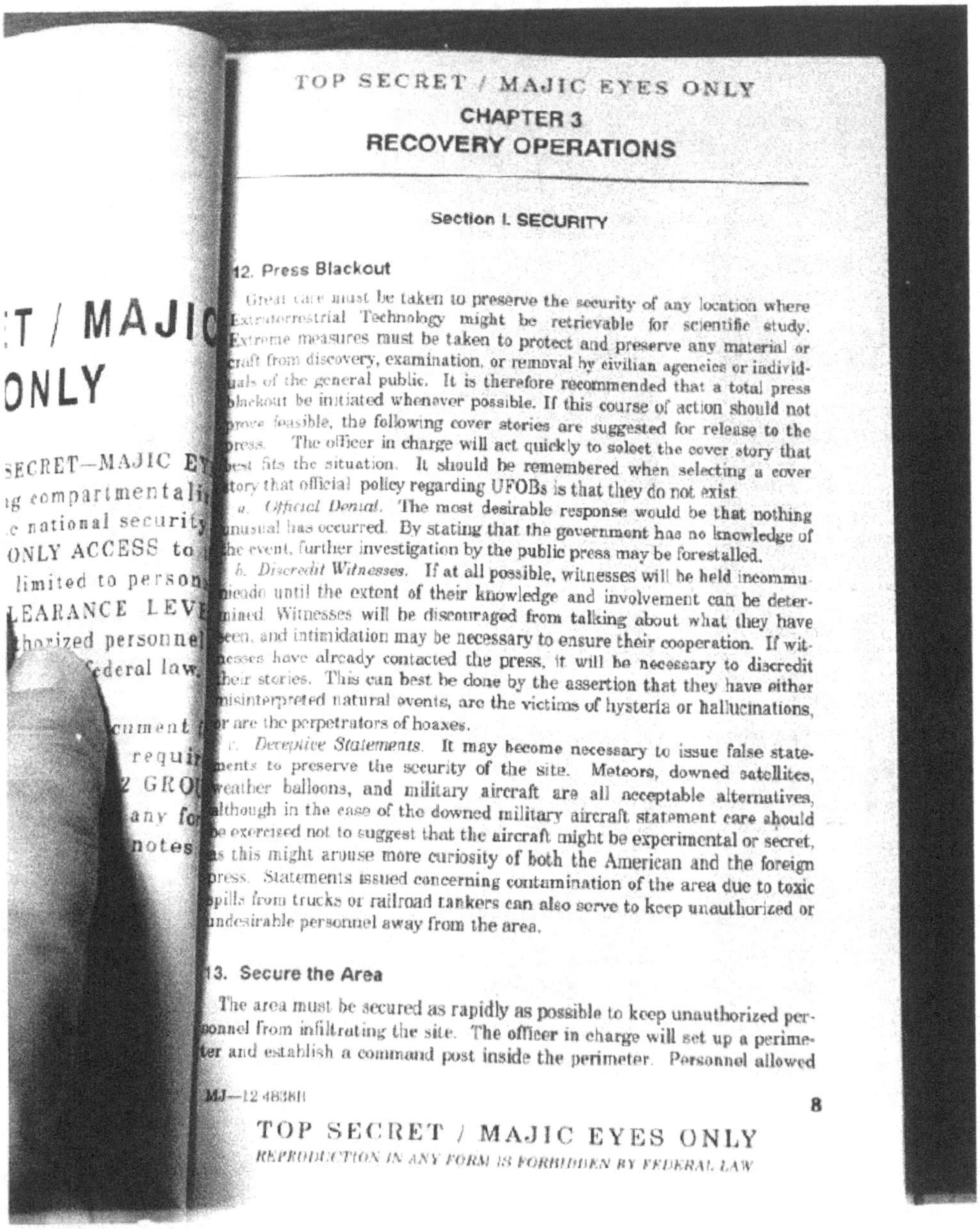

CHAPTER 3
RECOVERY OPERATIONS

Section I. SECURITY

12. Press Blackout

Great care must be taken to preserve the security of any location where Extraterrestrial Technology might be retrievable for scientific study. Extreme measures must be taken to protect and preserve any material or craft from discovery, examination, or removal by civilian agencies or individuals of the general public. It is therefore recommended that a total press blackout be initiated whenever possible. If this course of action should not prove feasible, the following cover stories are suggested for release to the press. The officer in charge will act quickly to select the cover story that best fits the situation. It should be remembered when selecting a cover story that official policy regarding UFOBs is that they do not exist.

a. Official Denial. The most desirable response would be that nothing unusual has occurred. By stating that the government has no knowledge of the event, further investigation by the public press may be forestalled.

b. Discredit Witnesses. If at all possible, witnesses will be held incommunicado until the extent of their knowledge and involvement can be determined. Witnesses will be discouraged from talking about what they have seen, and intimidation may be necessary to ensure their cooperation. If witnesses have already contacted the press, it will be necessary to discredit their stories. This can best be done by the assertion that they have either misinterpreted natural events, are the victims of hysteria or hallucinations, or are the perpetrators of hoaxes.

c. Deceptive Statements. It may become necessary to issue false statements to preserve the security of the site. Meteors, downed satellites, weather balloons, and military aircraft are all acceptable alternatives, although in the case of the downed military aircraft statement care should be exercised not to suggest that the aircraft might be experimental or secret, as this might arouse more curiosity of both the American and the foreign press. Statements issued concerning contamination of the area due to toxic spills from trucks or railroad tankers can also serve to keep unauthorized or undesirable personnel away from the area.

13. Secure the Area

The area must be secured as rapidly as possible to keep unauthorized personnel from infiltrating the site. The officer in charge will set up a perimeter and establish a command post inside the perimeter. Personnel allowed

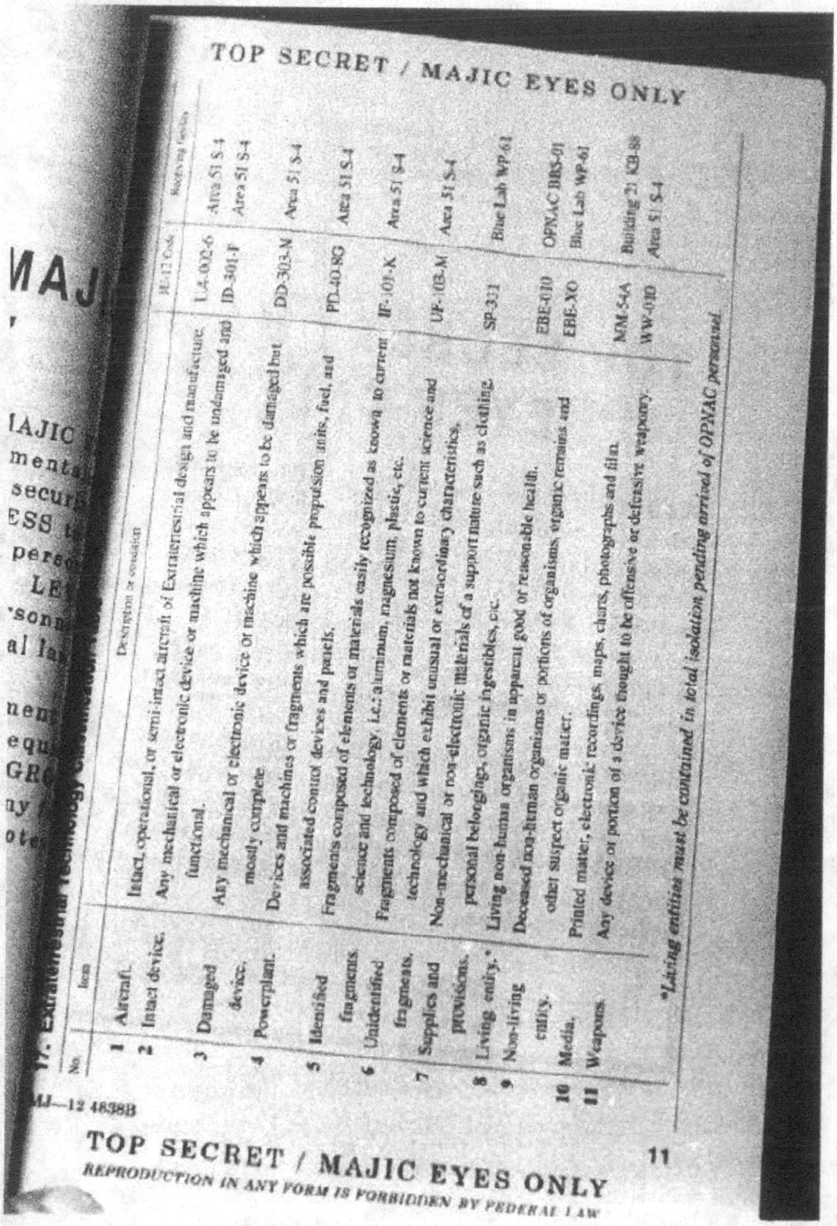

TOP SECRET / MAJIC EYES ONLY

17. Extraterrestrial Technology Classification

No.	Item	Description or condition	MJ-12 Code	Receiving facility
1	Aircraft.	Intact, operational, or semi-intact aircraft of Extraterrestrial design and manufacture.	UA-002-6	Area 51 S-4
2	Intact device.	Any mechanical or electronic device or machine which appears to be undamaged and functional.	ID-301-F	Area 51 S-4
3	Damaged device.	Any mechanical or electronic device or machine which appears to be damaged but mostly complete.	DD-303-N	Area 51 S-4
4	Powerplant.	Devices and machines or fragments which are possible propulsion units, fuel, and associated control devices and panels.	PD-40-8G	Area 51 S-4
5	Identified fragments.	Fragments composed of elements or materials easily recognized as known to current science and technology, i.e.: aluminum, magnesium, plastic, etc.	IF-101-K	Area 51 S-4
6	Unidentified fragments.	Fragments composed of elements or materials not known to current science and technology and which exhibit unusual or extraordinary characteristics.	UF-103-M	Area 51 S-4
7	Supplies and provisions.	Non-mechanical or non-electronic materials of a support nature such as clothing, personal belongings, organic ingestibles, etc.	SP-331	Blue Lab WP-61
8	Living entity.*	Living non-human organisms in apparent good or reasonable health.	EBE-010	OPNAC BBS-01
9	Non-living entity.	Deceased non-human organisms or portions of organisms, organic remains and other suspect organic matter.	EBE-XO	Blue Lab WP-61
10	Media.	Printed matter, electronic recordings, maps, charts, photographs and film.	MM-54A	Building 21 KB-88
11	Weapons.	Any device or portion of a device thought to be offensive or defensive weaponry.	WW-010	Area 51 S-4

*Living entities must be contained in total isolation pending arrival of OPNAC personnel.

MJ—12 4838B

TOP SECRET / MAJIC EYES ONLY 11

REPRODUCTION IN ANY FORM IS FORBIDDEN BY FEDERAL LAW

The Above Table Shows Where To Take The Different Pieces of any UFO Wreckage To.
As a reminder, recall that Bob Lazar's badge had a "MAJ" designation on it. And, it also had an "S4" designation. Bob's workplace was within the Area 51 complex, in Nevada.
Note the references to "Area 51 S-4" on the above document!

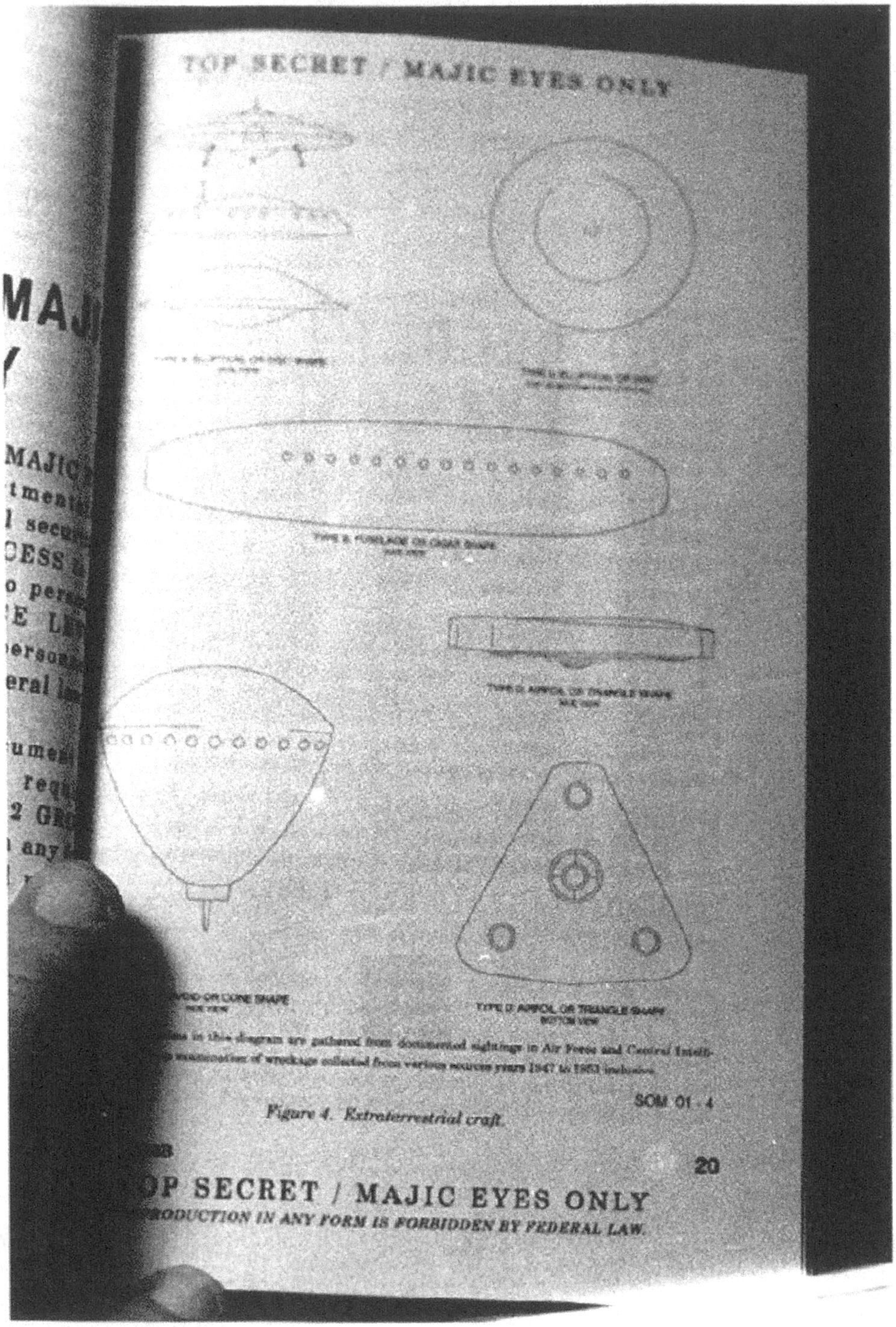
TOP SECRET / MAJIC EYES ONLY
MAJI
MAJIC
tment
l secu
CESS
o pers
E L
ersons
eral l
umen
req
2 GR
any
SOM 01 - 4
Figure 4. Extraterrestrial craft.
20
P SECRET / MAJIC EYES ONLY
RODUCTION IN ANY FORM IS FORBIDDEN BY FEDERAL LAW.

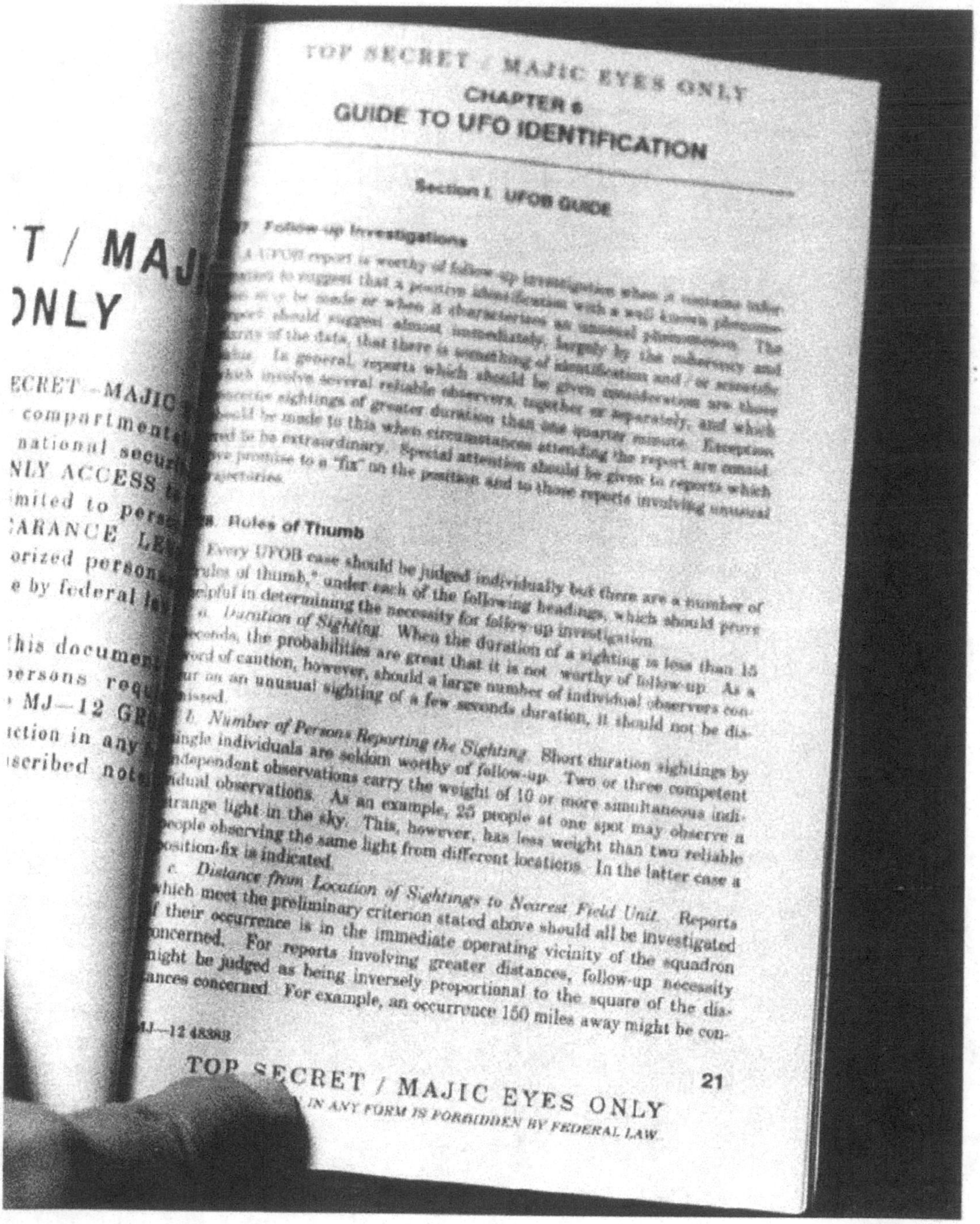

CHAPTER 6
GUIDE TO UFO IDENTIFICATION

Section I. UFOB GUIDE

7. Follow-up Investigations

A UFOB report is worthy of follow-up investigation when it contains information to suggest that a positive identification with a well known phenomenon may be made or when it characterizes an unusual phenomenon. The report should suggest almost immediately, largely by the coherency and clarity of the data, that there is something of identification and/or scientific value. In general, reports which should be given consideration are those which involve several reliable observers, together or separately, and which concern sightings of greater duration than one quarter minute. Exception should be made to this when circumstances attending the report are considered to be extraordinary. Special attention should be given to reports which give promise to a "fix" on the position and to those reports involving unusual trajectories.

8. Rules of Thumb

Every UFOB case should be judged individually but there are a number of "rules of thumb," under each of the following headings, which should prove helpful in determining the necessity for follow-up investigation.

a. Duration of Sighting. When the duration of a sighting is less than 15 seconds, the probabilities are great that it is not worthy of follow-up. As a word of caution, however, should a large number of individual observers concur on an unusual sighting of a few seconds duration, it should not be dismissed.

b. Number of Persons Reporting the Sighting. Short duration sightings by single individuals are seldom worthy of follow-up. Two or three competent independent observations carry the weight of 10 or more simultaneous individual observations. As an example, 25 people at one spot may observe a strange light in the sky. This, however, has less weight than two reliable people observing the same light from different locations. In the latter case a position-fix is indicated.

c. Distance from Location of Sightings to Nearest Field Unit. Reports which meet the preliminary criterion stated above should all be investigated if their occurrence is in the immediate operating vicinity of the squadron concerned. For reports involving greater distances, follow-up necessity might be judged as being inversely proportional to the square of the distances concerned. For example, an occurrence 150 miles away might be con-

MJ—12 4838B

[ ]

[THIS PAGE IS INTENTIONALLY BLANK]

PART E

LOST TECHNOLOGY, MERCURY VORTEX DRIVE, (& THE CADUCEUS)

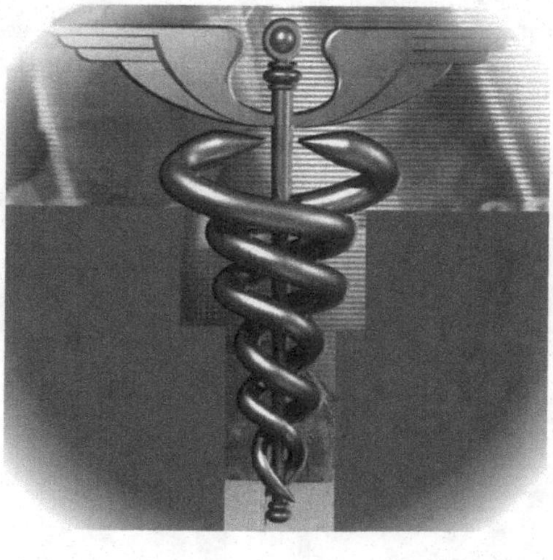

[ ]

[THIS PAGE IS INTENTIONALLY BLANK]

Lost Dwgs & Tech, Ancient India, & "Four"

THE LOST HAMEL DRAWINGS:

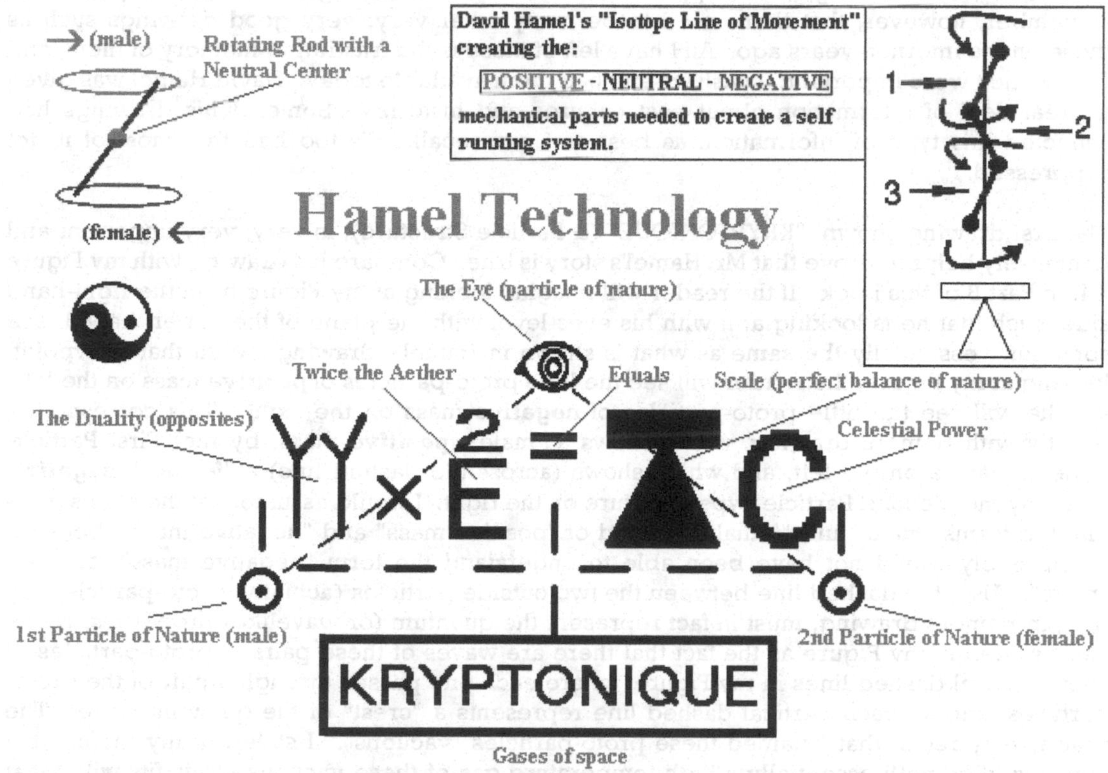

Particle Structure. [The MOST IMPORTANT Hamel Diagram of All.]

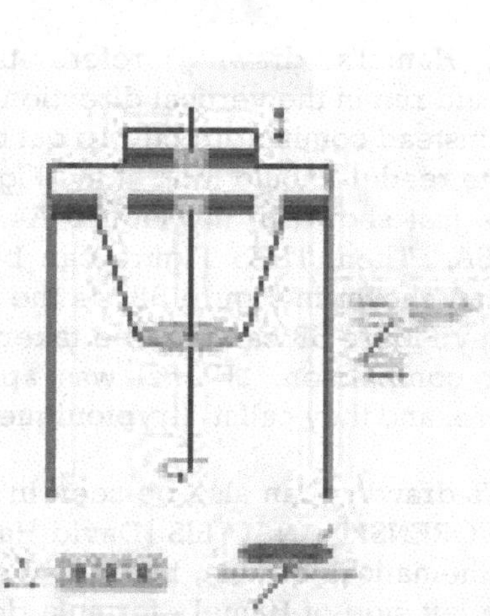

Unknown. But Notice, To The Right, The SAME Small Diagram, of Male & Female Rings With Rotating Rod And Neutral Center, as Was In The Previous Drawing, Above.

This is an unusual chapter which brings together several different (but still related) concepts, some of which are from the distant past, and some which come from the present. This first section is about the "lost" Hamel drawings. These few drawings were originally drawn by Mr. Hamel himself, and then were apparently re-drawn and improved by "someone", and that someone is unknown. I had to dig very hard to find these. All of these I was able to get from some so-called Internet archives (such as the "Way Back Machine"). The original web sites which had these are long gone (suppressed, no doubt). I found eight "lost" drawings, in all. I remember, however, that there used to be many other very, very good drawings such as these on the Internet, years ago. All I have left of these is the frustrating memory of them, and I know they were important; I wish we still had them available today. David Hamel was given a great deal of information about past cultures and histories. Some of his drawings had reflected this type of information, as best as I can recall. It's too bad that most of it got suppressed.

The first drawing shown, "KRYPTONIQUE" (& Particle Structure), is very, very important and immensely helps to prove that Mr. Hamel's story is true. Compare his drawing with my Figure A4, in Part B of this book. If the reader will imagine looking at my Figure from the right-hand side, such that he is looking at it with his eyes level with the plane of the paper, he will see something essentially the same as what is shown in Hamel's drawing. With that viewpoint, looking at my Figure, the reader will see the little proto-particles of **positive** mass on the left, and he will see the little proto-particles of **negative** mass on the right. This corresponds exactly with Hamel's drawing, which shows a "male" (**positive** mass, by me) First Particle Type of Nature on the left, and which shows (across the dashed line) a "female" (**negative** mass, by me) Second Particle Type of Nature on the right. I would assume that the aliens gave him the terms "male" and "female" instead of "positive mass" and "negative mass", because he probably would not have been able to understand the term "negative mass", or even "mass". Also, the dashed line between the two outside particles (actually, proto-particles, by me), in Hamel's drawing, must in fact represent the quantum (or wavelike) properties, which can be seen in my Figure as the fact that there are waves of these pairs of proto-particles. I show vertical dashed lines in my Figure, where each one passes through a pair of the proto-particles, and so each vertical dashed line represents a "crest" of the quantum wave. The reader may recall that I named these proto-particles "vacuons". I stated in my theory that space is filled with essentially a high-temperature gas of these vacuons. This fits with what Hamel's drawing shows, where he depicts "Gases of Space", at the bottom of his drawing.

The word "Kryptonique" in Hamel's drawing refers the powerful, but localized, gravitomagnetic field which would run in the vertical direction, on his drawing. In my Figure (A4, again), this field would be instead coming directly up out of the paper at you, the reader. If it will help clear things up, the reader should look at my Figure 4D in Part B. This is just a simplified version of what was just shown in my Figure A4. And, my Figure 4D can be redrawn, as shown in Figure 5A. Then, THAT Figure can be re-drawn after having been flipped on its side, which is then shown in Figure 5B, as the "Edge-on View". The dashed gravitomagnetic field lines in my Figure 5B can then be taken from my drawing and placed right on Hamel's drawing, for comparison. Hamel was apparently told that this "field" (gravitomagnetic, by me) is there, and they call it "Kryptonique".

The **"math formula"** in Hamel's drawing can also be seen in a figure of mine located just 2 pages before the start of the "FORENSIC ANALYSIS [David Hamel]:" section, in Part A. This formula is NOT STRICTLY a mathematical formula, but is also somewhat symbolic, especially on its right-hand side. But, the left side of Hamel's formula does fit my theory exactly. The right hand side of it was apparently to denote the "equilibrium" condition, which is the state that most particles are normally in. The "non-equilibrium" condition, or reduced-mass state (with its much larger radius), can be seen in my Figure 18 in Part B. Actually, this is where my theory predicts a totally new state of matter, in addition to the usual known ones: solid, liquid, gas, or plasma (ionized). This new 5th state of matter can be called the "reduced-mass" state, and a particle can remain in this state for a very long time (seconds, or even minutes). The fact that the large-radius ("reduced-mass") state can persist for a long time was discovered, as is explained on my Page "DNSS4" in Appendix H1. This same information is also repeated on handwritten page "UFT42", which is located just 2 pages before Figure A1 in Part B. The same is of course repeated at the end of the "typed UFT pages" in Part B. As is stated in these several locations, I found that the speed of the vacuons in a large (i.e. reduced-mass) ring is .4C, whereas the speed of the vacuons in a normal ring in equilibrium is .5C. I found that the equilibrium, average speed of the background vacuons of space is .5C also, which makes sense, and this is what keeps normal matter in the equilibrium (small radius ring) state. **Since the orbit speed in the large ring state is almost the same as it is for the normal ring state, it takes a long time for the random hits of the background free vacuons to bring a large-radius (i.e. reduced-mass) ring back down to a normal-radius (normal mass) ring!** So, where David Hamel denotes "Scale (perfect balance of nature)" in his drawing, this apparently refers to the equilibrium condition (i.e. equilibrium diameter) of the ring of proto-particles, as I just discussed. **The whole right-hand side of Hamel's equation is simply there to denote the equilibrium balance of energy (or "Celestial Power", as he calls it).**

Finally, regarding Hamel's formula in this chapter, he refers to "The Duality (opposites)". All this means is that there are, in nature, both "positive" masses and "negative" masses, for which he was given the terms "male" and "female". And, where he denotes "Twice the Aether", the formula really means **E squared,** and NOT two times E. Mr. Hamel may have not known what a "squared" number was, so they must have told him to just remember it as "twice", but to still write the "2" ABOVE the "E", instead of to the left of the "E". This "E squared" in Hamel's formula fits perfectly with my theory, and, by me, it refers to the mass current squared. The mass current (E) is in units of "Kilograms per Second", and represents the speed of flow of the circulating proto-particles in the ring. My theory shows that the value obtained by **squaring** the mass current (or "Ether Current, E", as Hamel would call it) is **proportional to** (i.e. is essentially equal to) the **actual experimentally-measured mass** of a normal particle.

Now, take a look at the upper left part of Mr. Hamel's drawing at the beginning of this chapter. He shows two parallel spinning rings, one called "male" and the other "female". The rings, as he shows them, **counter-rotate** (i.e., they rotate in opposite directions). This little diagram VERY OBVIOUSLY matches a drawing from my theory, which is in Figure 5B, in Part B, and is called "The Final Counter-Rotating Rings Model for a Nucleon". Also, see my Figures 11B and 12, in Part B also, which show how to cause the overall positive and negative mass rings (in a cone) to separate (i.e. to be pulled apart).

And lastly, the reader should look at the upper right part of Hamel's drawing (beginning of this chapter). This corresponds to my model of the Proton, which I diagram in my Figure 4C in Part B. Remember, Figures 4D, 5A, and 5B are actually just HIGHLY SIMPLIFIED VERSIONS of the MASTER MODEL which I developed for the Proton, in Fig. 4C. In Hamel's model, we can see three spinning "legs", where every other one must rotate in the opposite direction. These three "legs" correspond to the 3 quark rings shown in my model in Fig. 4C. My three rings have alternating directions of rotation also. One detail which I DID NOT SHOW in my Figure 4C is the fact, known to me, that these rings are not normally parallel. Normally, they are canted at a significant angle with respect to one another. I purposely didn't show this extra detail in my figure in order to not further confuse the reader, who is probably already confused enough! So in my own understand of the structure of the Proton, these 3 rings are slanted at an angle, and their **lines of direction** are very much the same, in reality, to the three lines (or "legs") shown in Mr. Hamel's drawing.

In summary, there can be no doubt that David Hamel's story is true. He WAS abducted by alien beings, and he was shown the CORRECT advanced physics drawings. There REALLY ARE alien beings, somewhere out there....

Oh, and notice in Hamel's drawing where it says "...mechanical parts needed to create a self running system". The US Patent Office would certainly call this a "perpetual motion machine", but we now know that it really is possible to build a 3-cone device which can output a large amount of energy, in various forms, as we have seen. And, it can do so continuously.

My ending comment for this "lost" Hamel drawing is that the yin-yang symbol shown (and possibly the eye symbol, as well) probably refers also to the fact that there is both positive and negative mass. And now, we must move on to look at Mr. Hamel's second "lost" drawing, which the reader will find on the same page as is the first drawing which we have just been discussing at length.

Hamel's **second drawing** is of some unknown device; I do not know what it is. The drawing is blurred and very hard to read. But I did notice that to the device's right, there is a little replica of the male/female spinning rings drawing, which we looked at earlier at the top-left of the above "Particle Structure" drawing. Perhaps the central rod (with the disk magnet on its top end) is supposed to move (or vibrate) in a circular fashion, just like the fictitious "neutral rod" shown between the counter-rotating rings (male & female). I only wish I knew.

Now, take a look at the third and fourth lost drawings, which appear on the next page:

[Go To Next Page]

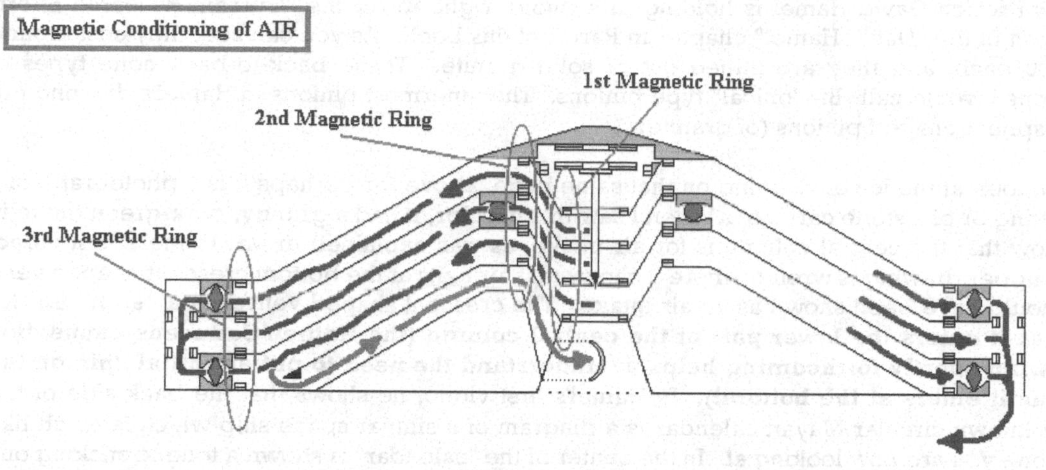

The AIR is changed 3 times as it flows through the UFO motor. Through the geometry of the two oscillating "wings", the AIR moves at different speeds as it flows past these 3 points.

One of The Better Diagrams of Hamel's MOST ADVANCED UFO Engine. Note That All Parts Float on Magnetic Fields, And The Pinions (Which Are Made of Granite) Are of a Mix of The Bi-Conical type, AND the Spherical type, as Shown.

An Amazing Drawing (or Artifact), Original Source Unknown,
But It Was Found By David Hamel.
Shows The Ancients Must Have Built These Things.
The Central Column is For Air Intake.
Compare This With The Advanced
UFO Diagram, Shown Above!

So, look at the upper drawing. I can see that the outermost-located pinions are like the one shown which David Hamel is holding in a photo, right above the "Current Research Effort" section in the "David Hamel" chapter in Part A of this book. As you can see, they cost around $3000 each, and they are milled out of solid granite. These back-to-back cone types of pinions I would call "Bi-Conical" type pinions. The innermost pinions in Hamel's drawing are just sphere shaped pinions (of granite).

Now, look at the lower drawing on that same page, above (or perhaps it is a photograph of a drawing or of a stone carving). When I saw this in color, it had a grungy, dark-green tint to it. I know that the central column is for air intake, as was explained in Mr. Hamel's first video. The upper drawing is wrong, where it shows that air <u>exits</u> at the bottom crescent-shaped vent. It should have been shown as an air <u>intake</u>. The crescent shaped vent **puts a "spin" on the air as it enters the lower part of the central column (the Hamel/Caduceus connection drawing shortly forthcoming helps us understand the need to put an initial spin on the air as it enters at the bottom!).** In Hamel's first video, he shows that the back side of the well-known circular Mayan calendar is a diagram of a similar space ship which is much like the one you are now looking at. In the center of the "calendar" is shown a tongue sticking out, denoting the need for air intake at that central point of the ship, or so he was told.

Anyway, what we see here (on the previous page) is apparently an artifact which shows an antigravity ship which was no-doubt used by ancient peoples (somewhere on this Earth). **Note that the overall device <u>almost</u> looks like a bird with wings!** COMPARE THIS TO THE ANCIENT DRAWINGS SHOWN IN THE "SUMERIAN & ASSYRIAN CARVINGS" CHAPTER IN PART F OF THIS BOOK!!

Now, look at the next two "lost" Hamel drawings, on the next page. >>>>

The two internal surfaces are actually gravitomagnetic lifting-surfaces, and Mr. Hamel rightly calls them "wings", even though they are on the INSIDE of the flying craft. A voltage is built up between these wings, per Hamel, and so they could also be called "capacitor-wings". The high voltage build-up is explained by my theory, which shows that the charge of the protons, in the nuclei, either decreases or goes to near-zero, whereas the charge on the electrons in the atoms stays the same. This huge charge unbalance manifests itself as a high-voltage net charge on the metal "wings". The charge of each proton changes (decreases), because the charge that a proton has is actually a function of the proton's "ring-radius", a concept which was discussed above. As a proton's ring-radius increases, which it does do during device operation, the effective charge of that proton decreases. Also, this is why the 3-cone device is very dangerous and usually possesses a high-voltage charge on it.

Now, the reader should proceed on to the following page, which shows two photographs of large metal cones. >>>

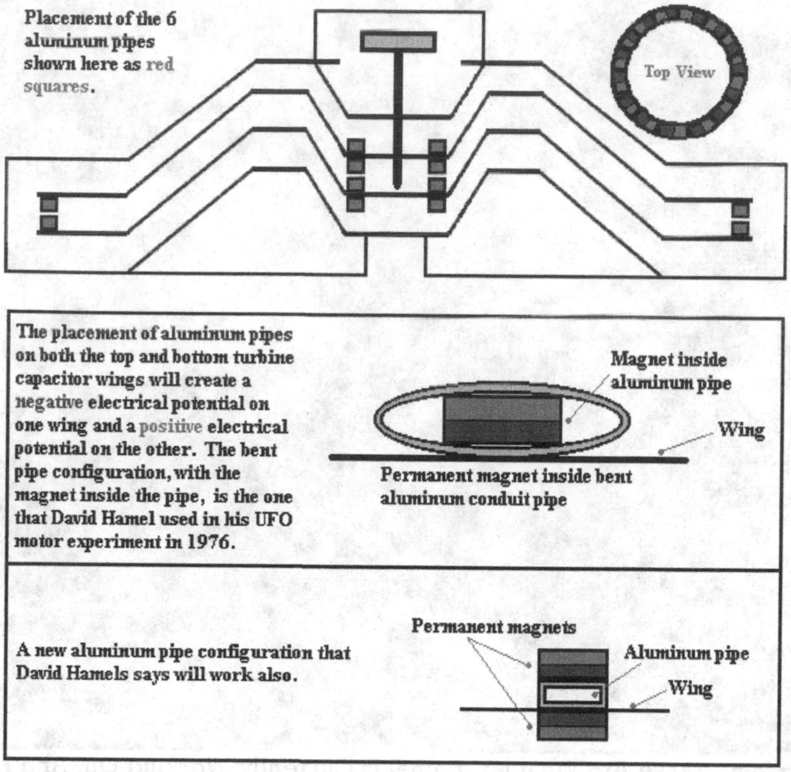

Placement of the 6 aluminum pipes shown here as red squares.
Top View
The placement of aluminum pipes on both the top and bottom turbine capacitor wings will create a negative electrical potential on one wing and a positive electrical potential on the other. The bent pipe configuration, with the magnet inside the pipe, is the one that David Hamel used in his UFO motor experiment in 1976.
Magnet inside aluminum pipe
Wing
Permanent magnet inside bent aluminum conduit pipe
A new aluminum pipe configuration that David Hamels says will work also.
Permanent magnets
Aluminum pipe
Wing

Using Aluminum Pipes, For Additional Shielding. Note: David Built This Way Back in 1976.

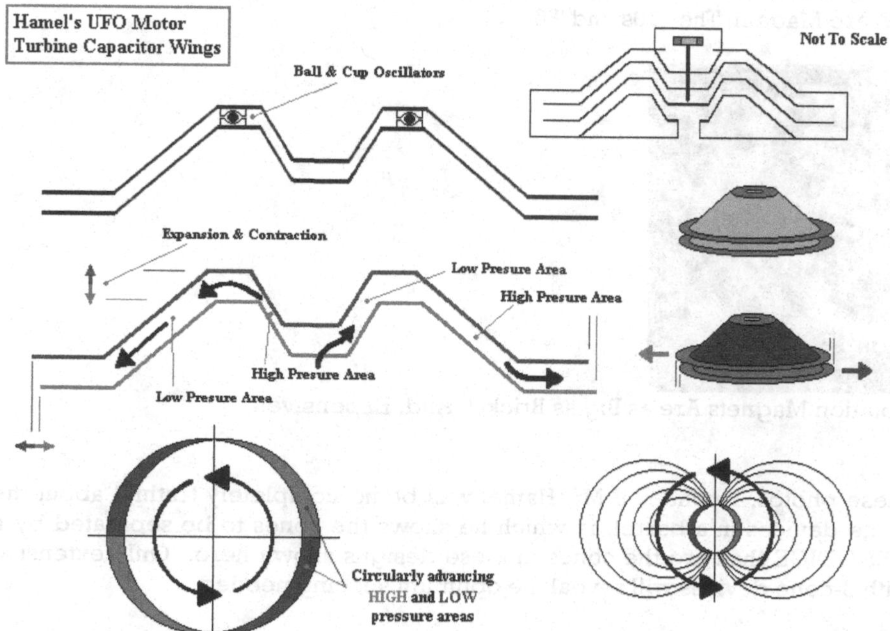

Hamel's UFO Motor Turbine Capacitor Wings
Not To Scale
Ball & Cup Oscillators
Expansion & Contraction
Low Presure Area
High Presure Area
High Presure Area
Low Presure Area
Circularly advancing HIGH and LOW pressure areas

Air Flow Action In The 2 Internal Metal Wings of The Craft.

NOTE: THESE WINGS DO NOT SPIN. RATHER, THEY OSCILLATE SLIGHTLY, BUT AT HIGH FREQUENCY, IN A CIRCULAR-MOVEMENT FASHION.

Yep, he's actually making 'em. These Are What Mr. Hamel is Currently Working On, And He Has Not Actually Flown a Device For At Least 10 Years, Due To Work on These Big-Type Devices! That's Why David Has Not Had A Working Model For Many Years Now. All His Working Models Were Made in The '70s and '80s.

Close Up. The Flotation Magnets Are as Big as Bricks! And, Expensive!!

When I look at these photos, I wonder if Mr. Hamel was being completely truthful about the design of the 3-cone devices in a barrel, in which he shows the cones to be separated by a MUCH LARGER DISTANCE than are the cones in these designs shown here. Only extensive experimenting with 3-cone devices will reveal the optimum spacing needed.

And now, I proceed on to the next subject, namely the "Caduceus" symbol.

ENTER THE CADUCEUS:

Mercury, who was the Messenger of the Gods, way back in very ancient history (or legends), carried with him his magic wand, or caduceus (pronounced Ka-DUE-see-us, or just Kuh-DUE-shus for the lazy). This winged staff allowed him to perform many wondrous feats. This ancient symbol has appeared throughout the world, but its ultimate origin is a mystery. The caduceus is usually shown as a rod (or staff) entwined by two snakes, and topped with a sphere and wings. Of course, the caduceus is used today very commonly as a medical symbol. It was, back in very ancient times, said that whenever the gods wanted to move goods (or people) from one place to another, over a long distance, they made use of **Mercury** (now is this the MAN (or God), Mercury, or could it be the liquid metal, mercury?) to accomplish such transportation. There will be more said on this possibility (of the use of the liquid metal, mercury, for antigravity flight) in an upcoming section, shortly.

THE APPARENT HAMEL/CADUCEUS CONNECTION:

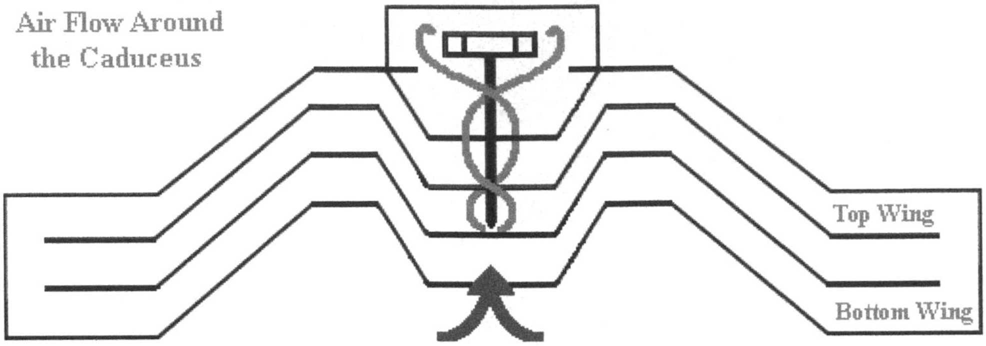

An Unknown Diagram. Perhaps It Means That The Air Flows Like This, or It May Mean That This Is How The Central Rod Vibrates (its horizontal displacement), Circularly. NOTE: The Central Rod Might Actually Be FLEXIBLE, and a Totally Stiff Rod Might Not Work?

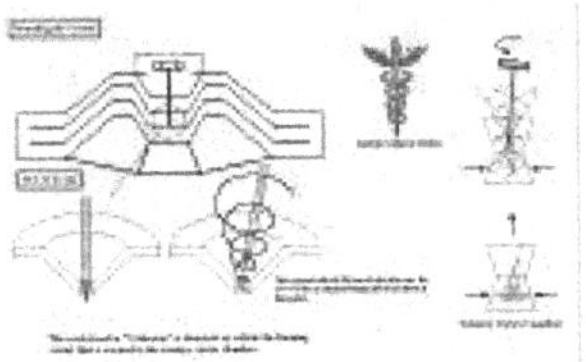

An <u>Extremely Important</u> "Lost Drawing", Originally By Mr. Hamel, Second Only In Importance To The First Drawing of This Chapter, Above, Which Shows The Male/Female Rings And The Yin/Yang Symbol.

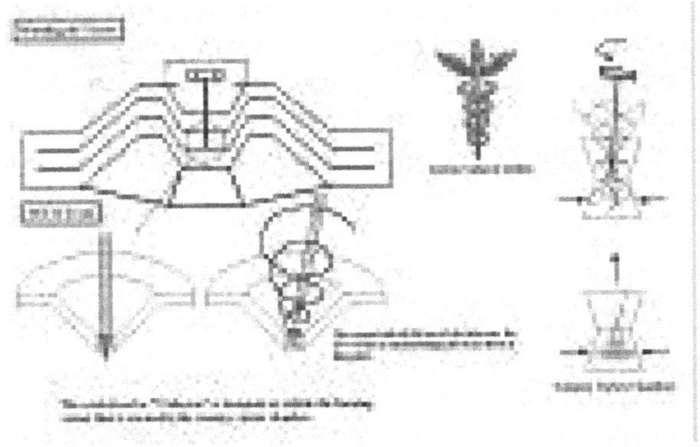

An Attempted Enlargement of Same, But Not Legible Due to The Very Poor Resolution of The Original, Which is Quite Small, To Begin With. **Here We See, at The Far Right, Both The Origin of The Caduceus, And, <u>Directly Below</u> That, The Origin of The "Rx" Symbol, Which Is Used Commonly by Our Pharmacies of Today.** What I Wouldn't Give To Get a Larger Drawing of This, With Full Explanation! Oh, Well....

As we can see from those final two "lost" Hamel drawings, there is supposed to be some sort of connection between the antigravity craft's functionality and the ancient caduceus symbol. That connection could be air flow pattern and/or (circular) vibrational pattern. The only way we'll ever know what these mean, exactly, will be if Mr. Hamel comes forward again to explain them, or if someone who has spoken to Mr. Hamel about these matters would come forward with the information. Only time will tell.

VORTEX ENGINES AND MERCURY VORTEX ENGINES:

I recall reading about antigravity engines of ancient past which were said to have worked based-upon a "vortex" principle of some sort. SOME of the engines apparently used mercury (the liquid metal, whose symbol is "Hg") and some definitely DID NOT use it. But most of these references which I read would refer to these antigravity engines as being VORTEX ENGINES. They would talk about "starting up the (powerful) vortex", or they would say that "once the vortex was started, a man could fly on the device in the sky, lit up like the sun at night", and so on. The vortex was often described as very powerful, making a fierce circulation of wind and the roar of a lion (from the vibration). Also, a glowing effect was sometimes reported, such as could be plasma or other electromagnetic phenomenon. Finally, the reader should know that in Mr. Hamel's first video ("Contact From Planet Kladen"), his cone-within-cone devices were also called the "vortox" (sic.), and there was talk of starting up the vibration (or "vortox"). I laughed a bit when I heard this, because apparently no one involved in making the video knew about the word "vortex" (spelled & pronounced correctly, here), let alone how to correctly pronounce it!

The idea of ancient civilizations having antigravity capability is fascinating, and will be covered more fully in the next chapter. But here, in this chapter, we need only recall that Mr. Hamel stated quite clearly that many ancient civilizations had antigravity, and that there is plenty of evidence for this, through artifacts found around the world. Also, there were hundreds of texts, written in Ancient India mostly, which described all aspects of antigravity flight as if it were just a normal part of daily routines. These old scrolls talked about take-off and landing, and steering & piloting, and use in warfare, and rules of traffic, and engine maintenance, and so on. Since there were hundreds of texts, fully detailed, then either the people of Ancient India were telling truthful stories, or they had an unbelievable imagination (or were taking too much peyote!). The Ancient Indian name for a vortex engine was "vimana" (more on this, shortly).

And now, a thought or two regarding the NON-mercury vimanas. The very word ***vimana*** is an old Sanskrit word, purportedly derived from ***vamana***: "he who is able at **<u>three</u>** strides to take measure of <u>the entire earth and heavens</u>." When we consider that many vimanas must have been built based-upon the **<u>3-cone</u>** design, then the original meaning of the word vimana makes sense. But, as we shall now see, they had even more powerful and advanced designs, than the 3-cone designs, for vimanas which worked based upon a sort of "mercury (Hg) vortex". However, it should be finally noted that a sufficiently large 3-cone device could very easily create the above mentioned effects, namely that of air circulation, plasma (glowing), and even a vibrational "roaring" sound.

A <u>MERCURY</u> vortex engine vimana would have used the liquid metal, mercury, in them. When I look up the information on the element mercury (symbol "Hg", coming from the old Latin word "hydrargyrum", meaning "liquid silver"), I see that mercury is indeed affected by gravitomagnetic fields, because it does have isotopes which have a net nuclear spin (specifically, 199-Hg has a 1/2 spin, and 201-Hg has a 3/2 spin). Therefore, mercury is a viable material for use in creating antigravity effects, as per earlier discussions.

A mercury-vortex engine, to function, would have to cause the mercury itself to form some sort of geometrical cone-like pattern (i.e. wide at one end, narrow at the other), ACCORDING TO MY THEORY. Well, wouldn't you know it, this brings us right back to the Caduceus concept (remember, a personage named Mercury used this "device"). The reader should now look again at the caduceus drawing (& photo) shown several pages earlier. If we imagine the coiled snakes to actually be hollow metal pipes, and if we were to force liquid mercury (or even mercury VAPOR) to flow up through these expanding-helical pipes, then we would have created a device which satisfies the physics theory requirement above, namely that of forcing the mercury into some sort of cone-like pattern! After all, the outline of an expanding HELIX is a cone! **As the reader will soon see below, this is indeed the way that the mercury-based vimanas were evidently built by the ancients, as described in some of those hundreds of very old texts.**

I now believe that the caduceus is a symbol of antigravity-flight, both mercury-based and even non-mercury based. Amazingly, there are other authors out there who also believe that the caduceus is such a symbol, or even that it is a simplified DIAGRAM for a mercury vortex engine (such as W. D. Clendenon believes). Look again at the symbol. The lower sphere could a mercury boiler. The upper sphere could be a mercury condenser. The whole thing would be a closed system, much like the system in refrigerators. The mercury is heated, in the lower sphere, until it becomes a vapor. The mercury vapor then travels up through the expanding copper coils, and loses its heat along the way, through heat exchange. Then, the mercury condenses back to a liquid, and collects in the upper sphere, for re-use. The wings are of course shown as part of the caduceus symbol, to show that the whole device actually flies. My personal belief is that these ancient mercury-vortex antigravity vehicles, as diagrammed by the caduceus, were long ago probably used to transport injured or sick people to far-away hospitals, among many other uses as well, much like the "flight-for-life" helicopters do in today's world. **I think that the exact technology used to build these very useful medical transport devices was somehow lost long ago, for whatever reasons, but the caduceus symbol STILL REMAINS WITH THE MEDICAL PROFESSION, and it is a symbol of caring and healing.**

One final note regarding the symbol. The EXPANDING coils of the snakes could also represent the fact that protons (& entire nuclei) actually do expand, over time, as an antigravity engine runs, as has been shown earlier. The expanded nuclei of the atoms have a considerably reduced mass (or weight) in the expanded state. Thus the entire antigravity engine weighs much less during its operation.

THE MERCURY-VORTEX ENGINES OF ANCIENT INDIA:

The information in this section will detail some of what's known about the antigravity craft of one ancient civilization, namely that of India. This will serve as an introduction to the next part of the book, which is Part F, "Ancient Civilizations Had Antigravity."

In a text from Ancient India called the "Samar", vimanas were "iron machines, well-knit and smooth, with a charge of mercury that shot out of the back in the form of a roaring flame." Other ancient Indian texts containing stories about vimanas are the "Samarangana Sutradhara", the "Mahabarata", the "Ramayana", and many others. My interpretation of the above is that it is NOT the mercury itself (which is VERY poisonous) that shoots out of the back as a flame, but rather the onlooker was seeing highly charged air plasma coming out, and this could very understandably look like a blue flame emerging. Also the overall vehicle structure (or shell) was, as said above, made of **iron.** This is the **one** common and useable metal which is NOT affected by the powerful gravitomagnetic waves. Other metals (such as aluminum or copper) would light up like a very bright light if used as the shell of the ship. Actually, I think some UFO's are actually PURPOSELY built with aluminum shells in order to create such an effect (which would look frightening but awesome to an unwary onlooker).

Also of great importance, there were described, elsewhere, flying crafts known as "vailx". These vailx were built, starting 20,000 years ago, supposedly by the Atlanteans (called the "Asvins" by the Ancient Indians), and these Atlanteans fought in many hideous, fierce wars against the vimana pilots of Ancient India. These vailx crafts were saucer-shaped with a generally **trapezoidal cross-section,** and with three hemispherical engine pods on the underside. And, they supposedly used a **mechanical antigravity device,** driven by engines, developing approximately 80,000 horse-power. To me, the "trapezoidal cross-section" reminds me of either the "frustrum" of a cone, or perhaps the asymmetrical capacitors used by Townsend Brown. Anyway, I'm sure that the shape of the device was critically important to its operation.

So, where are these vimanas and vailx today, and where are the descendants of the people who flew them? Well, it seems that the Ancient Indians and the Atlanteans must have extinguished themselves almost totally through horrible wars, and so all the technology to build flying craft was lost (or purposely destroyed!) in the aftermath. Their final war occurred about 12,000 years ago. The ancient "Mahabarata" describes the awesome destruction during the final war: "....(the weapon was) a single projectile charged with all the power of the Universe. An incandescent column of smoke and flame as bright as a thousand suns rose in all its splendor.... An iron thunderbolt, A gigantic messenger of death, Which reduced to ashes the entire race of the Vrishnis and the Andhakas. ....the corpses were so burned as to be unrecognizable. Their hair and nails fell out; pottery broke without apparent cause, and the birds turned white. ...After a few hours all foodstuffs were infected... ...to escape from this fire, the soldiers threw themselves in streams to wash themselves and their equipment..."

When the Rishi City of Mohenjo-Daro was excavated by archeologists, many years ago, they found skeletons just lying in the streets, some of them holding hands, as if some great doom had rapidly overcome them. These skeletons are among the most radioactive ever found, similar to the ones found at Hiroshima and Nagasaki. Ancient cities whose brick and stone walls have been apparently fused or melted, can be found in India, Turkey, and other places. I leave it up to the reader to guess what kind of weapon caused all this, so far back in the past!

Mercury vortex engines are described also in the ancient Indian text, the "VYMAANIKA-SHAASTRA", which is more commonly spelled "Vimaanika Shastra". Chapter 5 in that text describes something about how to build **a mercury vortex engine:** (my comments are in brackets "[ ]")

"Prepare a square or circular base of 9 inches width with wood and glass, mark its center, and from about an inch and a half thereof draw lines to the edge in the 8 directions, fix 2 hinges in each of the lines in order to open and shut [I read this as being a way to open and close **valves** in the copper pipes, to regulate the flow of the **mercury**]. In the center, erect a 6 inch pivot [I read this as being a 6" diameter vertical **pole**, probably at least 12 feet long, which is set upon a **pivot** so it can **tip slightly** towards any direction] and **four tubes,** made of *vishvodara* metal, [I read this as **four** (probably copper) tubes, which make up the "snakes", being the **expanding helix tubes** which must go up the central pole] equipped with **hinges** [so they can vibrate at high frequency in a small circular fashion, just like other devices we've examined earlier] and bands [for support] of iron, **copper**, **brass** or **lead**, and attach to the pegs in the lines in the several directions. The whole is to be covered."

Note that "copper, brass, or lead" are all metals which are gravitomagnetically active, but I'm curious why "iron" was mentioned. Well, I just don't know why.

The ancient text "Samarangana Sutradhara" says that (regarding mercury engines & vimanas) they were made of light material, with a strong, well-shaped body. **Iron**, **copper**, and **lead** were used in their construction. They could fly to great distances and were propelled by air. This particular text devotes 230 stanzas to the building of these machines.
Here's a quote from the text:

"Strong and durable must the body be made, like a great flying bird, of light material. Inside it one must place the **Mercury Engine** with its **iron** [Iron makes a good container for the mercury, as it is not gravitomagnetically affected (i.e. inert)] **heating apparatus** [for boiling the mercury to a vapor state] beneath. By means of the power latent in the mercury which **sets the driving <u>whirlwind</u> in motion,** a man sitting inside may travel a great distance in the sky in a most marvelous manner."

"Similarly by using the prescribed processes one can build a vimana as large as the temple of the God-in-motion. **<u>Four</u>** strong **mercury containers** must be built into the interior structure. When these have been heated by controlled fire from **iron** containers, the vimana develops **thunder-power** through the mercury. And at once it becomes a **pearl** in the sky. Moreover, if this iron engine with **properly welded joints** [so the **<u>FOUR</u>** attached **copper helix tubes** ("snakes") can vibrate properly!] be filled with mercury, and the **fire be conducted to the upper part** [just heat flow, up the **4** helices] it develops power with **the roar of a lion** [due, no doubt, to the acoustic-range frequency of the helix vibrations]."

<u>Analysis of these statements:</u>

In the first quote above, it talks about the **<u>four</u>** (expanding helix) **tubes** which go up the <u>pole</u> (called a 6 inch (diameter) "pivot"). The second quote mentions a <u>single</u> iron heating apparatus for boiling the mercury into a vapor. This may mean that the device described in the FIRST quote above would have had only **<u>one</u>** mercury boiler **container,** which would thus have **<u>four</u>** copper **tubes** spiraling up out of it.

And finally, in the last quote above, it would appear that larger vimanas need more mercury boiling capacity, and so <u>each</u> <u>one</u> of the **<u>four</u>** expansion-helix (spiraling) **tubes** has its OWN, <u>individual</u>, mercury boiler **container** attached to it.

So, what's all this about needing **<u>FOUR</u>** upwardly-expanding helix tubes around the central pole (or staff)? After all, the caduceus symbol only shows TWO of these outwardly spiraling snakes (tubes) going up the pole. Does my physics theory explain why we need **<u>FOUR</u>?** Well, the answer to that question is YES!

Look now at **Figure A21, in Part B** of this book. It is the **MASTER DIAGRAM** for the structure of ALL matter and of ALL electromagnetic energy. Every subatomic particle, even the photon (which is the quantum of electromagnetic energy), is composed of either 1, 2, or three "quarks" (a "quark" is just a ring of <u>proto-particles</u> which I call "vacuons"). We see that the electron is made up of only 1 quark, and the photon is made up of 2 quarks, and both protons and neutrons are always made up of 3 quarks. There also exist in modern physics particles which are called "mesons", and these (not shown) each have 2 quarks.

The most important thing to notice from the diagrams is that a quark (i.e. ring) usually contains exactly **<u>FOUR</u>** vacuon clusters (or just **four** individual little vacuons). I say "usually", because something strange does happen in the proton and neutron, where in <u>one or two</u> of the quark rings, the vacuon clusters will **<u>split in half</u>,** thus forming eight vacuon clusters in each of <u>those particular</u> quark rings. Specifically, in the proton, two of the quark rings do this, and in the neutron, only one of the three quark rings does this splitting-up.

But, by-and-large, we see that a so-called "generic", or temporarily isolated, quark ring will always have exactly **<u>FOUR</u>** little vacuon clusters (or just four <u>individual</u> vacuons, where each "cluster" only has <u>one</u> vacuon in it).

Said another way, <u>every physical particle</u> of <u>matter</u>, OR <u>energy</u>, is composed only of quarks, and each quark is always generically composed of exactly **FOUR** sub-quarks (a sub-quark is just a vacuon cluster, or an individual vacuon).

Now, we can see why evidently all the ancient mercury vortex engines had to have exactly <u>FOUR</u> expanding-helix tubes around the central pole. The entire device, consisting of the four expanding helix tubes containing mercury vapor, acts as a single macroscopic, coherent, quantum system. In other words, each <u>one</u> of the <u>four</u> helices corresponds with just <u>one</u> of the <u>four</u> sub-quarks which is located within each and every quark ring making up the mercury vapor. And the whole thing is synchronized such that all of the quarks in the mercury are lined-up with each other and act as a unit. So the device-as-a-whole acts as if it is ONE large (mercury) quark ring (with its four little sub-quarks), which expands its diameter as it is forced to go further up the pole. And, as we know, as any quark ring is forced to expand, its mass decreases and it gives up HUGE amounts of gravitomagnetic energy, which is used to propel the craft upwards.

Recall above that a device was built having eight supporting pegs, spaced around the "eight directions". There were metal bands connecting from each of these support pegs to the helix tubes. Now this makes sense. Since there are 4 helix tubes, and since we would need a support connected to opposite sides of EACH helix, then we would need a total of EIGHT support pegs (one for each of the "eight directions", as it was stated).

We will also see the importance of the number **<u>FOUR</u>** in the next section, as well, below.

And finally, as an interesting but related tangent to all this, Soviet scientists found what they call "age-old instruments used in navigating cosmic vehicles" in caves in Turkestan and the Gobi Desert. The devices are hemispherical objects of glass or porcelain, connected to a **cone**, of the same material, and the **cone** portion has a drop of **mercury** in it. I believe that these devices must have been used to detect gravitomagnetic fields. They were cone-shaped in order to concentrate the gravitomagnetic field and cause it to separate the positive and negative mass rings within the drop of mercury contained inside. This would probably cause the mercury drop to glow, thus emitting light. When it lights up, it means you are detecting gravitomagnetic fields in the vicinity!

THE "ALIEN ANTENNA", AND THE NUMBER FOUR (4):

Yes, this last section of this part of the book has nothing to do with either lost technology or ancient technology, but it is included because it is a future & alien technology (which we too may have someday), and it is ALSO BASED UPON THE NUMBER FOUR, and for the SAME REASONS as were the 4-helix mercury devices discussed in the previous section. And never mind that the following information is great evidence that alien beings exist, and that we are not alone in this galaxy!

Take a look at this!

Field Close To Chilbolton, England on The Morning of 08/14/2001.

Close-Up View of Same.

We used to call these unusual formations "crop circles", but now they are usually much more complex than they once were, and now should properly be called "crop designs". Crop designs historically have appeared most commonly in England. And, they are definitely on the increase, both in frequency of appearance and in complexity!

Now, look at this:

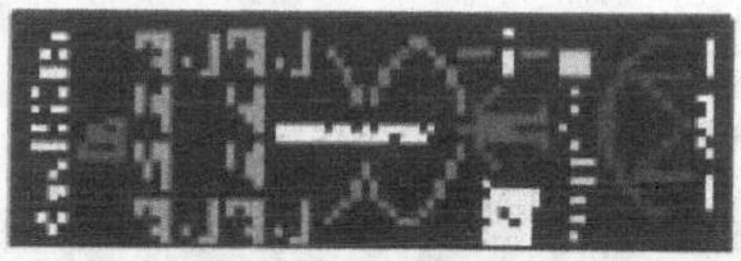

Here's a Digital Message We Sent Into Deep Space,
on 11/16/1974,
by The Arecibo Radio Telescope.
Note The Little "Man" Figure And Also
The "DNA" Molecule.

ALSO, Note The Diagram of Our **Sending**
(Radio) **Antenna** (a dish-type),
at The Right-hand End of The Picture.

Straight-Down View of the Digital "Answer",
That We Received, in a Crop Field
In Chilbolton, England.

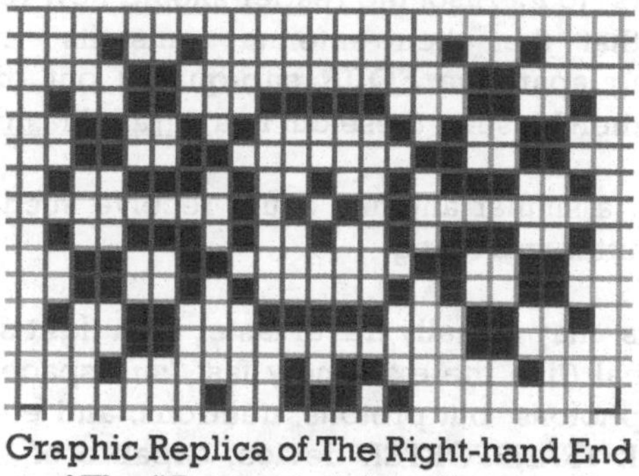

Graphic Replica of The Right-hand End
of The "Answer". It is a diagram of
an Alien Transmission Antenna,
Apparently.

[Now, Go To Next Page>>]

This Was Found in The Same Field,
on The Morning of 08/14/2000,
Exactly ONE YEAR Earlier.
This Shows a Much More Detailed Diagram of
That Same Alien Communication Antenna!

Evidently, the aliens, who responded to our outgoing message/greeting, are using a large communication antenna (or device), which is evidently BASED UPON THE number **FOUR**.

Still can't see it? Take a closer look at the graphic replica diagram on the previous page. Their communication antenna is built as a **hierarchy** of smaller and yet smaller modules, each of which has **FOUR** "corners" to it. Also, the reader should now refer again to my Figure A21 in Part B. I believe that the alien antenna transmits "signals" by a method of gravitomagnetically ripping apart the **FOUR** sub-quarks out of many protons, neutrons, electrons, or photons, and then blasting these out in a parallel fashion as a sort of "beam".

And, at the receiver's end, a similar antenna could receive such a signal by just doing the opposite of what the transmitter was doing.

These beams of sub-quarks may actually travel faster than light speed. This communication antenna is thus superluminal (like the extremely fast "sub-space" radio of Star Trek). And even better, if it uses not photons, but protons, neutrons, and electrons, then this "antenna" could constitute a MATTER TRANSPORTER device, capable of "beaming" physical objects across vast expanses of space, and at a speed exceeding that of light! Look out for Star Trek! Here it comes!

But now, let's travel back to the **PAST,** again, in the next Part, "F":

Part F

Ancient Civilizations Had Antigravity

[ ]

[THIS PAGE IS INTENTIONALLY BLANK]

Ancient Egypt:

The following contains very compelling information to show that Egypt once had some advanced technology in their midst.

Egyptian Djed

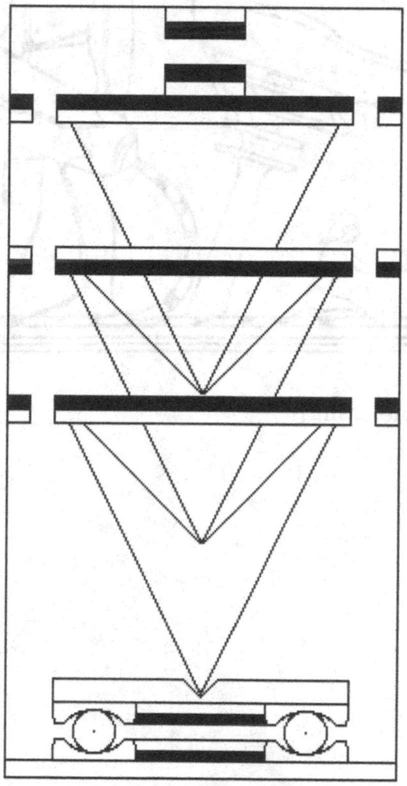

3-Cone Device

Egyptian Djed Column Side-By-Side With a 3-Cone Device.

What Are These People Doing?

[ ]

[THIS PAGE IS INTENTIONALLY BLANK]

Spheres of Costa Rica:

Diquis Spheres:
Of all the existing remnants of pre-Columbian culture, none are more mysterious than the stone spheres of the Diquis region. This region covers the southern half of Costa Rica. Dotted throughout the area are perfectly shaped spheres of **granite**, some as large as a tall person and others as small as a grapefruit. They can be seen in the Museo Nacional and various parks and gardens in San José, as well as throughout the Diquis region. Some have been found, undisturbed for centuries, on the Isla del Caño, 20 kms west of the southern Pacific coast. Who carved these enigmatic orbs? What was their purpose? How did they get to Isla del Caño?

No one has the answers to these questions.

The concept of using a small, restricted circular path which is traversed by a rolling sphere (usually granite) can also be seen in the "Inspiration - where the idea came from" section of the account of David Hamel (one of the "Big 3" witnesses).

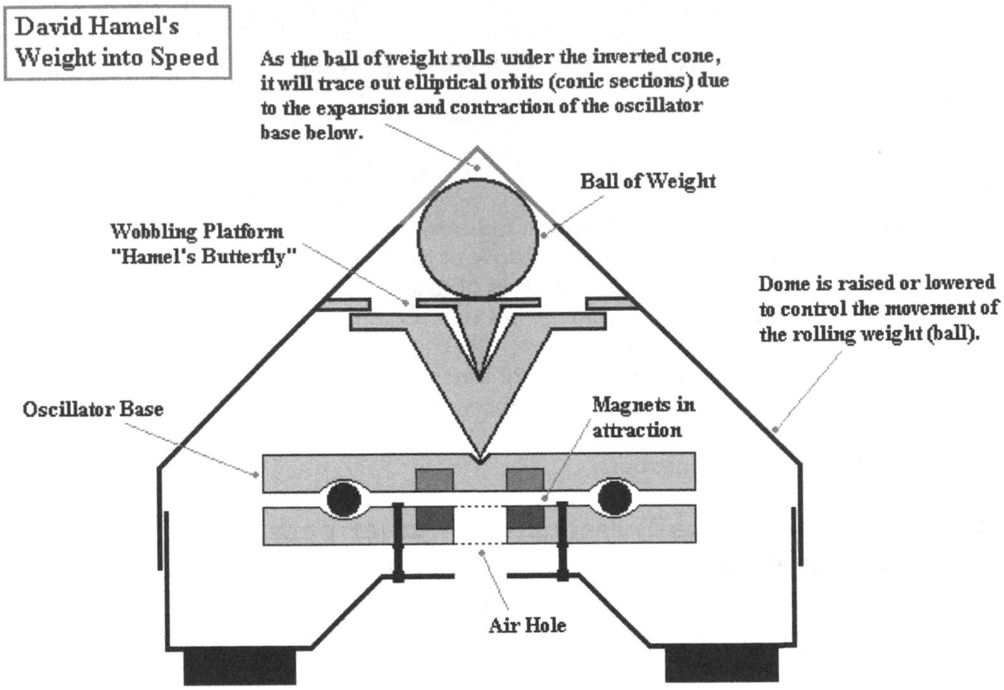
David Hamel's
Weight into Speed
As the ball of weight rolls under the inverted cone, it will trace out elliptical orbits (conic sections) due to the expansion and contraction of the oscillator base below.
Ball of Weight
Wobbling Platform
"Hamel's Butterfly"
Dome is raised or lowered to control the movement of the rolling weight (ball).
Oscillator Base
Magnets in attraction
Air Hole

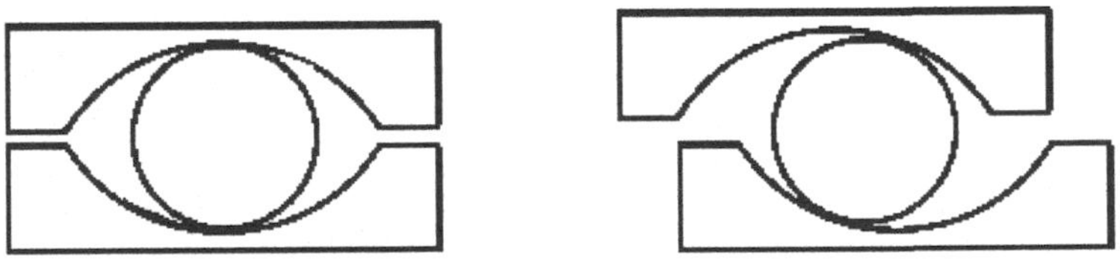

The ball and cup oscillator arrangement consists of a sphere sandwiched between two cups. The geometry used for the ball and cup is the vesica with eye. The cup's geometry (radius) can be altered to modify the horizontal and vertical movements of the sphere if needed. When each cup is oscillated 180 degrees out of phase in relation to each other, the sphere will ride up and down on the walls of the cups providing expansion in rotation and alternately contraction in rotation.

Diagram (Redrawn) From David Hamel Showing Use of Granite Ball Rolling In a Restricted Circular Path, In Order To Generate Heat & Light (& Gravity Waves!).

David Hamel States That This Device, Uncovered By Digging In Mexico City, Was Once A Rolling-Granite-Ball Device For Generating Electric Power & Heat & Light. The Large Granite Ball is Now Missing, For Whatever Reason.

[ ]

[THIS PAGE IS INTENTIONALLY BLANK]

Sumerian & Assyrian Carvings:

The Sumerians (Assyrians) are about the oldest known civilization on Earth. By "civilization", I mean a group of people who built cities of stone and who lived with rules and civilized practices. They lived in the areas which are now known as Iran and Iraq. They have left us with many awe-inspiring carvings in stone and in clay. Exhibited below are a few of these carvings, which now make it seem like they were involved somehow with antigravity flying devices, based upon my new theory of physics, of course. To understand these, refer back to "Thomas Townsend Brown" in Part A of this book. These are all "Electro-gravitic" type devices.

Note: Many of the photographs below come originally from the "Oriental Institute of Study", and they were widely available on the Internet, in multiple places.

ORINST. P 24975 PERSEPOLIS, IRAN
COUNCIL HALL. AHURAMAZDA SYMBOL
ON THE NORTH JAMB OF THE EASTERN
DOORWAY OF THE MAIN HALL.
C12

"Ahuramazda Symbol". From Persepolis, Iran. Also Known as "ZoroAster's Vehicle". Note The "Steering Wheel" For This (Imaginary or Real?) Vehicle.

ORINST. P 57464 PERSEPOLIS, IRAN.
CLAY TABLET WITH A CYLINDER SEAL
IMPRESSION. TWO SUSIAN? GUARDS
FLANKING A DATE PALM UNDER A
WINGED DISK SYMBOL. FOUND IN THE
TREASURY IN ROOM 33.
E9

A Similar Winged Disk Device, With No Rider (This Time).

ORINST. P 57472 PERSEPOLIS, IRAN.
IMPRESSION OF A CYLINDER SEAL ON
A CLAY TABLET. HERO TRIUMPHANT.
XERXES INSCRIPTION. FOUND IN THE
TREASURY.
E7

Another Such Device.

C5
ORINST. P 56563 PERSEPOLIS, IRAN. TREASURY. CLOSE-UP OF A RELIEF IN THE EASTERN PORTICO OF A COURTYARD DEPICTING KING DARIUS SEATED ON A THRONE WITH HIS SON XERXES STANDING BEHIND HIM.

The Sumerian King Darius and His Son Xerxes. From Persepolis, Iran.

ORINST. P 58722 NAQSH-I-RUSTAM, IRAN. TOMB OF XERXES. KING, GOD AND FIRE ALTAR IN THE TOP REGISTER.
E11

Tomb of Xerxes, in Naqsh-I-Kustam, Iran. Notice That This Time, The WINGS & FEATHERS Have Been Replaced by What Could Be a Stack of Many (metal?) Plates (with sections). Also, Note The "Chicken Feet" Instead of The Usual Ball-Shaped Piece.
Notice Again The "Steering Wheel".

War Scene, With Riderless Flying Device Shown. Note The Parallel Stack of Plates, In Apparently in 8 or more Sections. Notice the "Power Generator" in The Center.

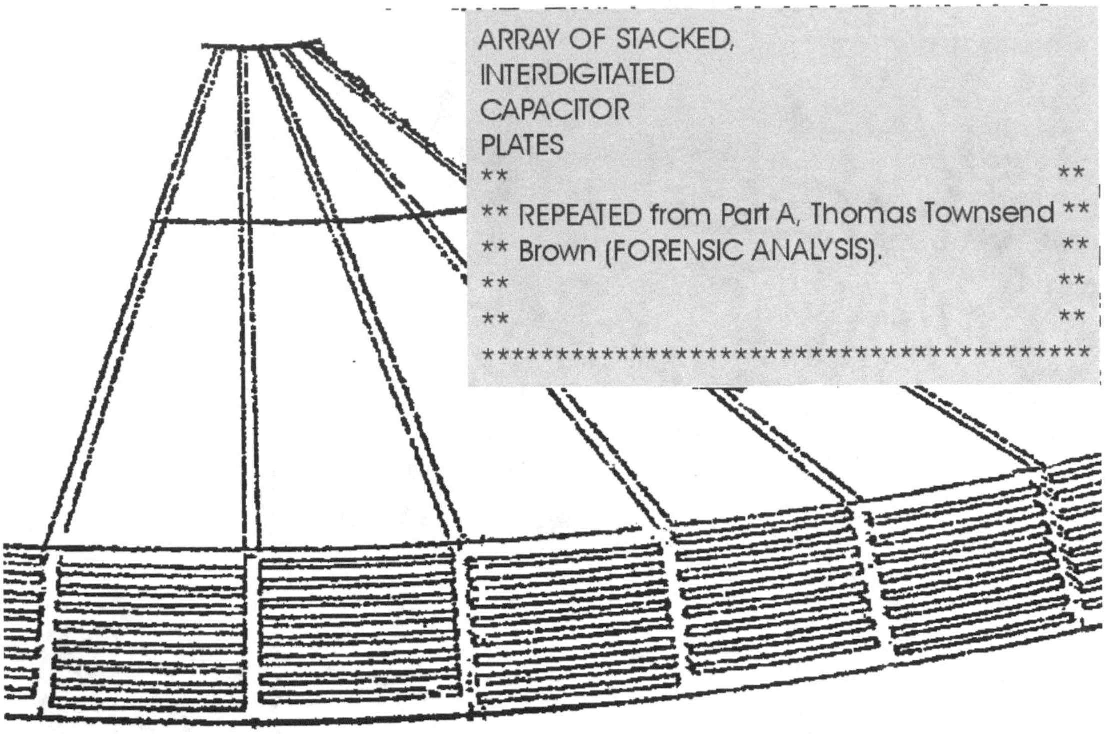

Stack Of Metal Capacitor Plates, in Many Small Radial Sections. Remember, This Device Was Built by S.A.I.C., and was Actually Flown By The U.S. Government!

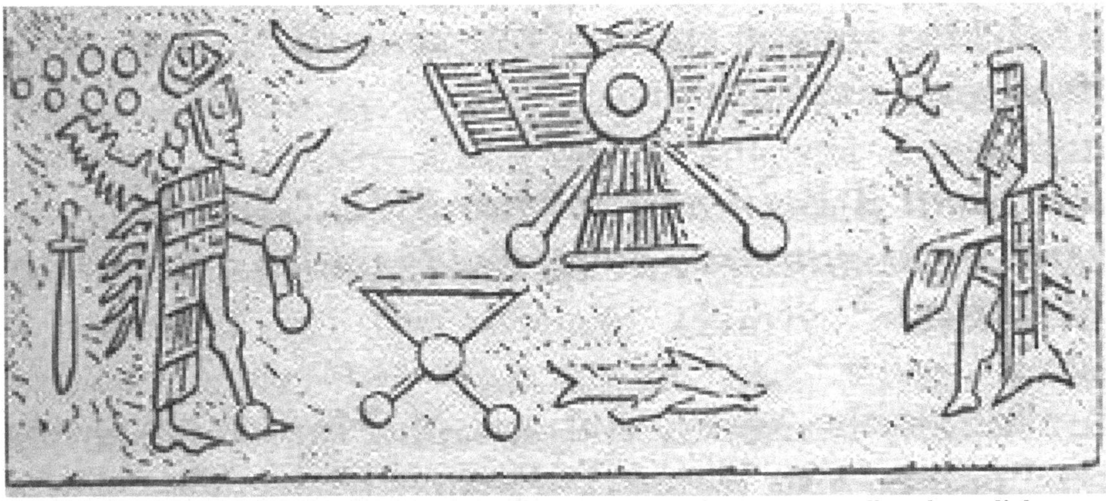

Another Sumerian Drawing. Note the Interdigitated Stack of Plates, as well as the radial sections. Interestingly, The Shape of The Device is Part of a Larger INVERTED Conical (Cone) Shape! [Remember Cones and Triangles From Antigravity Theory...]

Three Riders In This Flying Wing!

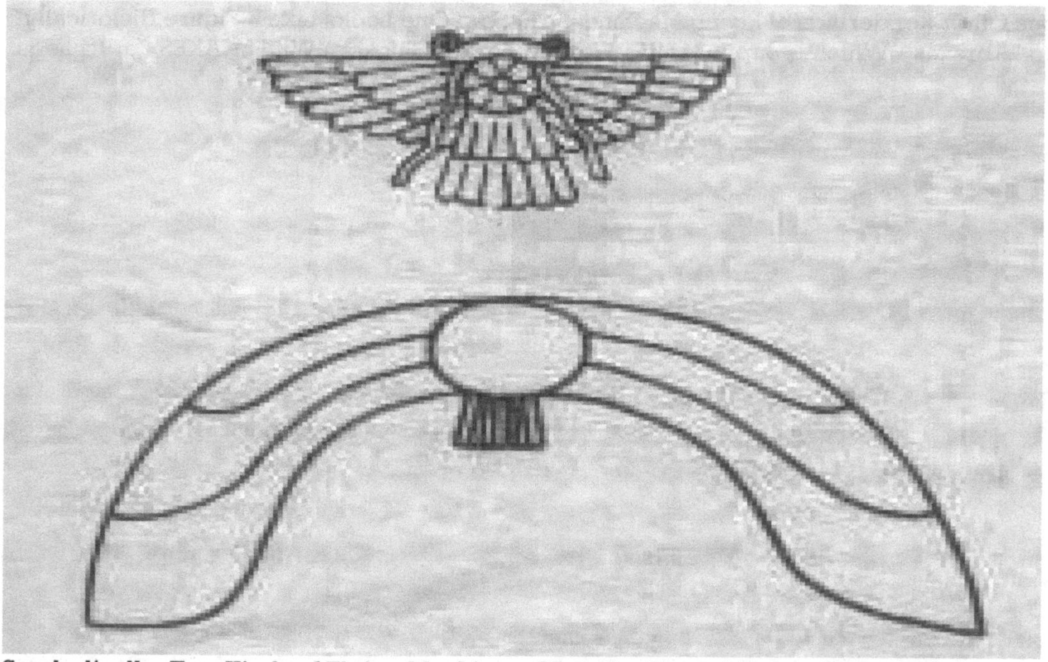

Symbolically, Two Kinds of Flying Machines. Note the Central Power Generator.

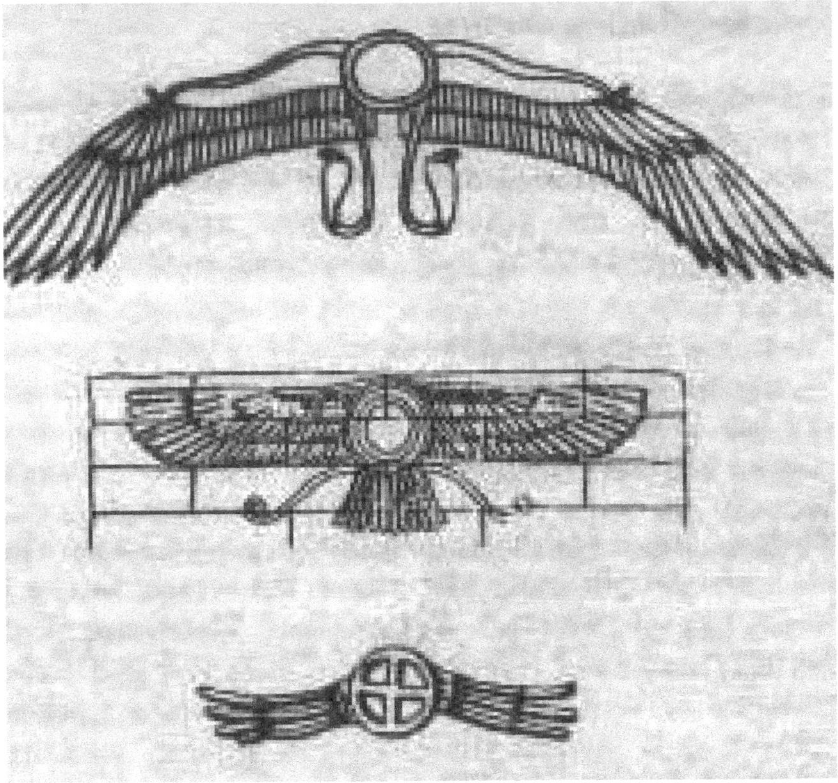

Various Other Renderings of Flying Machines. The Top One Looks Like a Figure Historically Called "Abraxas", Which is Supposedly a Winged Bird (a Chicken) With SNAKES For Its Legs.

Another Depiction of The Winged Device.

The "Mechanical Looking" Version of the Flying Disk Platform.

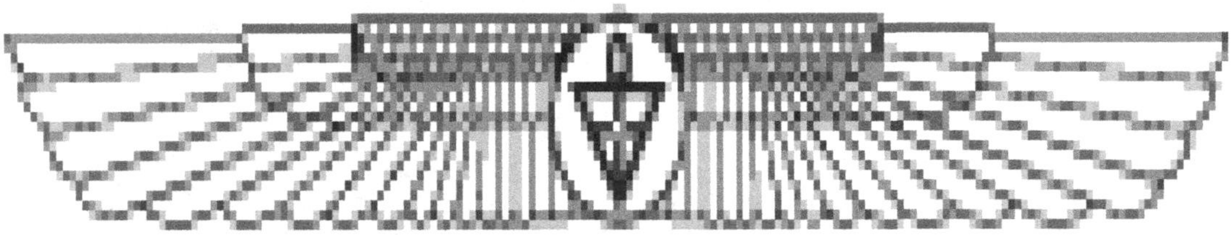

This is The Modern Symbol For the Group Calling Themselves The "Rosicrucians". They Claim They are Related To Descendants of Ancient Egypt.

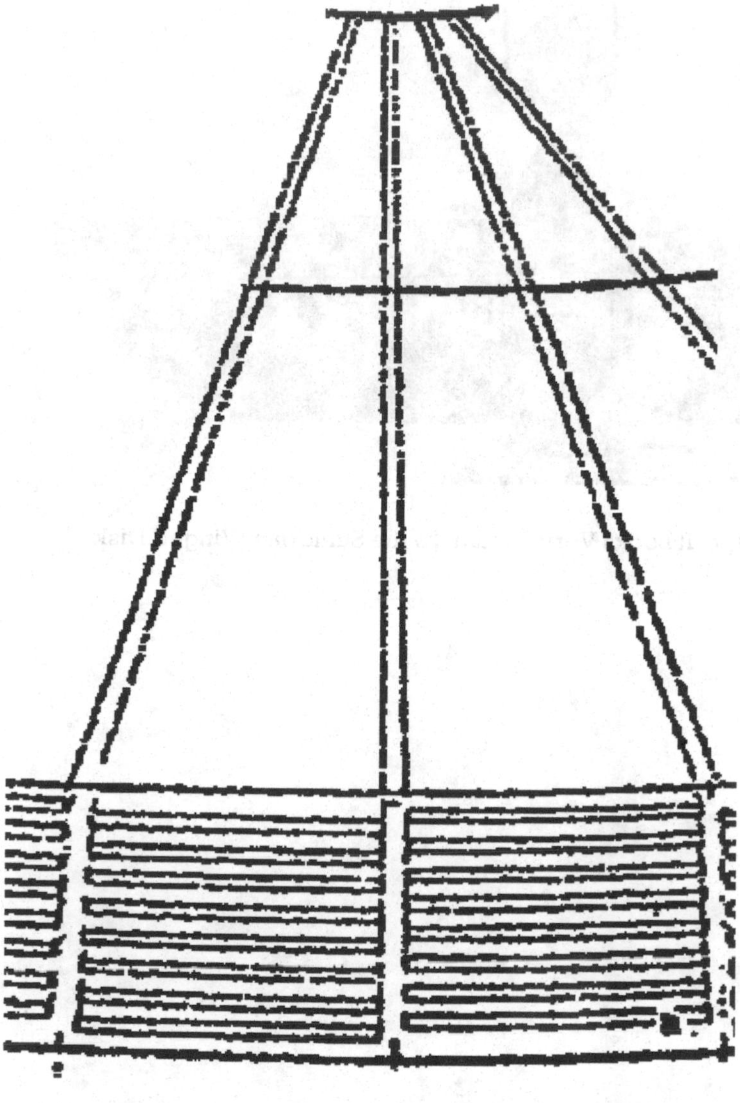

Some of The Sumerian Winged Disks Seem To Have 8 Sections, Like These.

A Symbol Found in Ancient Egypt. It Looks Very Similar To the Sumerian Winged Disk.

The Dogon Tribe:

The Dogon tribe existed many thousands of years ago. See in the next figure a design painted on one of the masks that they used.

A Ceremonial Mask of the Dogon Tribe.

From my new Unified Field Theory of physics, it is readily seen that certain numbers and ratios are extremely significant.

The numbers 7, 12, 24, 42, 1/3, 3, 4, and the ratio 10 are exactly the significant physical quantities that I've discovered during the lengthy mathematical development of my new theory of matter and antigravity.

The number 24 is the number of "vacuon" clusters in the proton. The number 7 is the number of vacuons in a cluster. Twelve is the number of clusters that behave as negative mass. The number 42 is the square root of the ratio of the proton's mass to the electron's mass (check for yourself!!). Three is the number of quarks in a proton. The charges of quarks in a proton are +/- integer multiples of 1/3. Four is the number of "subquarks" in a typical or generic quark. Also, the electron has 4 subquarks, as does the photon. Finally, 10 is the ratio of the masses of the two types of vacuons (there are only two types).

In the picture, this ceremonial mask design they drew obviously indicates either knowledge of antigravity physics, or at least the tribe was influenced by aliens. The design shows the importance of the numbers: 4, 7, 12, 3 (7 x 3 = 21 = 7+7+7), and also 1/3. If you look closely at the fourth obelisk, there are 7 vertical white stripes in every other block. The fourth obelisk also shows that there is both positive and negative mass (white vs. black).

Also, notice the positions of the round orbs, showing snapshots of a circular, rolling movement. This concept of using a small, restricted circular path which is traversed by a rolling sphere can also be seen in the "Inspiration - where the idea came from" section of the account of David Hamel (one of the "Big 3" witnesses). ALSO, SEE THE CHAPTER ON "SPHERES OF COSTA RICA" IN PART F. That chapter shows a diagram of one way that such a device was built!

The Masons:

The following is a drawing that appears in most modern Masonic Temples.

This scene is found in most Masonic Lodges. It contains information, in encrypted form, regarding advanced Antigravity Physics.

From my new Unified Field Theory of physics, it is readily seen that certain numbers and ratios are extremely significant.
The numbers 7, 12, 24, 42, 1/3, 3, 4, and the ratio 10 are exactly the significant physical quantities that I've discovered during the lengthy mathematical development of my new theory of matter and antigravity.

The number 24 is the number of "vacuon" clusters in the proton. The number 7 is the number of vacuons in a cluster. Twelve is the number of clusters that behave as negative mass. The number 42 is the square root of the ratio of the proton's mass to the electron's mass (check for yourself!!). Three is the number of quarks in a proton. The charges of quarks in a proton are +/- integer multiples of 1/3. Four is the number of "subquarks" in a typical or generic quark. Also, the electron has 4 subquarks, as does the photon. Finally, 10 is the ratio of the masses of the two types of vacuons (there are only two types).

In the picture you can see the numbers 12, 7, 4 and 3. [there are 3 support columns in front and 4 in the back, totaling 7]. Also, 4 stars are grouped in the top center. Note the abundance of triangle shapes (some with a horizontal line drawn across them) also. The triangle shapes remind one of the inverted metallic triangle used in Bob Lazar's device.

The modern Masons are actually descendants of the Knights Templar. The Knights Templar were supposedly organized in order to keep a "Great Secret" (antigravity?).

Part G

Conclusions & Epilog

[ ]

[THIS PAGE IS INTENTIONALLY BLANK]

Conclusions:

ANTIGRAVITY FLIGHT:

Antigravity Flight: what is it? How can we comprehend it? Indeed, how can we even begin to believe that it really can exist, and how could we have missed it in all of our many classic, well-known experiments of physics, which have been used to explain just about everything else? I will answer these questions with a brief summary of my own sequence of thoughts which helped me to accept that such things as "antigravity" (and gravitomagnetism) are possible. Some of them are, amazingly, simple enough to be understood from a standpoint of modern high-school physics.

Many details regarding my "chains of thought" can be discovered in the Appendices of this book. But, here, I now present these thoughts in a very concise and straight-to-the-point form. Many people have thought that if such a thing as powerful gravity, powerful gravity waves, or powerful gravitomagnetic waves could be created, it would surely require some sort of process which could yield the energy equivalent of an atomic explosion. In fact, they thought it would require either multiple sequential detonations of nuclear explosions, or a controlled but fierce output of some kind of atomic reactor (perhaps fission, perhaps fusion) which would be the equivalent of continuous nuclear explosions. If one really looks at this, however, it will readily be seen that such conclusions are fallacy, as follows.

To counteract/escape the ordinary gravity of the Earth, we would want to create an opposing gravity field (yes, just simple, ordinary gravity) to cancel or overcome the Earth's field (unless, of course, we are talking about airplanes, jets, rockets, and the like). Suppose we imagine having two atomic bombs, sitting side-by-side. Neither one of them creates a powerful gravity field of force. In fact, the gravity they produce, which is due to simply the mass of these devices, is extremely small. If the gravity produced by the two bombs was anything significant, we would witness the two exert a force upon each other, and they would pull (or try to) each other together. They would *noticeably* attract each other!

Now, let's say that we detonate one of the bombs. It makes a huge fireball, as we expect. Now, we ask the following. How much of a gravity field of force does this fireball produce (while it lasts, of course)? The answer is very, very little. In fact, it produces exactly the same force of gravity that either one of the two unexploded bombs did. This fact, that the energetic explosion occurred, did not help us in any way whatsoever to produce more gravity in the vicinity of our experiment. Why is this? This is true because of Einstein's law which states that the total of "mass-energy" in conserved (kept constant) at all times. This is because mass and energy are really just different views (or forms) of the SAME THING. In other words, when the bomb exploded, a predictable amount of its mass was converted to energy, but the overall gravity field produced by the now-exploded device, and fireball, is exactly the same because the gravity field produced by any object is always equal to the total mass/energy of the object, multiplied by a certain constant value (a number which is well-known in modern physics). And, the "mass-energy" of the fireball, and all, is equal to (again, per Einstein's law) the "mass-energy" of the unexploded bomb. Thus, the gravity output is the same, either for an unexploded atomic bomb, or one that is in the process of exploding!

So we see that NO amount of atomic energy release will EVER be able to produce powerful enough gravity (or negative gravity) fields, or even gravitomagnetic fields! **It's a hopeless cause, no matter how fast and powerful the motion of a device's constituent matter and**

<u>energy</u> components are. So how, then, do we explain how UFO's fly? We are now certain that they do exist, and they DO FLY, somehow, and that "somehow" must be related to the generation/production of gravity, gravity waves, or perhaps gravitomagnetic waves or fields. So, we must theorize possible alternatives to our current understanding of atoms and the matter which comprises them. And, recall that atoms are chiefly made up of proton and neutron particles in the central nucleus region, and there are some lightweight, and thus insignificant, electrons buzzing about, as well (far from the nucleus).

After much thought, I came to the (perhaps ridiculous, perhaps not) conclusion that there **<u>must be</u>** <u>extremely</u> powerful (and localized) gravity-like (or gravity-related) types of fields present NATURALLY (or naturally occurring) in **<u>all</u>** atoms and thus probably in all particles of matter, such as protons and neutrons. However, these powerful fields must, under "normal" conditions, either all cancel each other out, or simply be somehow "dormant" or "inaccessible", due to the obvious experimental facts that no such "fields" have ever been detected, nor have we detected any effects which could be caused by such "fields".

The reason that these tiny and extremely powerful fields must exist, and must be **<u>naturally produced</u>** somehow by nuclei of atoms, is that there is **<u>no other possible way</u>** to generate sufficiently strong <u>gravity</u> (or gravitomagnetic) waves for vehicle propulsion purposes, as we have seen so vividly and painfully in the thought experiment above with the two atomic bombs.

This was one of my <u>first clues</u> that real vehicles which produce real antigravity propulsion effects do not require any nuclear reactors, or the like, devices at all. They could just be simple, electromagnetic and/or electromechanical machines! No nasty nuclear radiation, pollution, or danger in handling hazards need be present in an antigravity vehicle.

One of my <u>other clues</u> to this enigmatic puzzle of antigravity propulsion was seen when I first completed the derivation of my gravity force formula (see page "G38" in the H4 appendix), which shows that gravity forces can manifest in four different ways. This is because there are 4 terms in the equation for the total force, Zr. Now, look at the <u>last</u> equation on that page. The "u" (velocity) to the zero power is just equal to 1. Thus, the first term, which is of order (c, or, 1/c, depending), is a "gravity"-type effect term, whereas the last two terms , which are each of order (1/c), are the "gravitomagnetic"-type effect terms. It is easily shown that the first term is always of order c as long as there is ANY ordinary static mass present AT ALL. Thus, the gravitomagnetic field terms are totally "drowned out" by a factor of c squared, in that case, and thus gravitomagnetism is just too small to ever detect or study or put to any practical use. It is only when there is NO ordinary static mass present, at all, that the first term becomes of order (1/c), and thus is on the same order as is the gravitomagnetic terms. Then, in that case, a gravitomagnetic field could be measured, studied, and also put to any sort of use.

So that was my other clue. It told me that best and most sensible situation would have to be **one in which there <u>could be</u> "negative mass", and in which the total sum of all the positive mass and the "negative" mass is ZERO.** <u>That particular situation</u> does two things for us: 1) it makes the gravitomagnetic fields present have a numeric value which is reasonable (instead of infinitesimally small or zero), and 2) it makes all the velocity-squared terms go to zero, and we didn't want them in there anyway, as they are not <u>strictly</u> a "gravitomagnetic"-type field, but something rather strange altogether.

So, to summarize, the first two clues are 1) there must exist extremely powerful, highly localized gravity fields or gravitomagnetic fields in each particle of matter (example: the proton), and 2) the total, conventional mass, which is probably the sum of some positive mass and some negative mass, should be zero. And, regarding #1), we can rule out "gravity fields", because if we don't, then #2) couldn't be satisfied, and we would have probably measured and noticed any such localized gravity fields, long ago.

<u>My only possible explanation and conclusion based upon these 2 items, was as follows. The proton (and also any other subatomic particle) must be comprised of extremely massive, counter-rotating rings of positive mass and negative mass, such that the sum of the positive and negative equals zero (i.e. they are equal in amount, but opposite in sign). This conclusion was reached on June 1, 2000.</u> Thus, there is leftover only a resulting extremely strong gravitomagnetic field to be measured there (if we could do such a thing). This is why the two rings must counter-rotate instead of co-rotate. Then, by the law of: Energy = Effective-Mass * c^2, we see that the <u>Effective Mass</u> (i.e. the usual measured mass), of a proton, is produced by simply the stored, localized (gravitomagnetic) Energy, divided by the speed of light squared. Also see item #14 on page 9 of the "Explanatory Supplement to IDD", in the H3 appendix, for a similar discussion. Also, see item #4 on page 10 of that same document. And finally, see Figures 4C through 5D (especially 5A), in Part B. We once thought that matter was the "parent" of energy, but now we see that, indeed, energy (i.e. gravitomagnetic energy) is the "parent" of matter!

Finally, this counter-rotating ring model of the proton was, months later, replaced by the model of the proton shown in Figure 4C, in Part B. This was done in order to include and explain the quantum-mechanical effects normally present in the proton and in all other subatomic particles. Also the model in Figure 4C explains all the electromagnetic effects rather nicely, such as electric charge laws and force laws and-the-like.

So there you have it. Then, as is seen earlier in this book, we can start with that model of the proton and proceed through a sequence of steps until, ultimately, we can cause a separation of the positive and negative mass, and thus emit a very powerful gravity wave (or field), which is in fact of zero frequency. It is a "shockwave", so to speak.

The Two Modes of Antigravity Flight:
We have thus achieved an antigravity engine which can cause a craft to levitate and fly. It does this by emitting an **ordinary gravity wave** (i.e. it's NOT gravitomagnetic). There are actually **2 modes** of operation. **Mode #1** is operation out in deep space. There, the large amount of momentum carried away by the ejected (emitted) gravity wave is what causes a "push" on the craft. This is analogous to a situation where a rocket expels a high-velocity gas. The ejected gas carries away momentum, and so to counter that, the craft experiences a force in the opposite direction. This is due to the well-known conservation of momentum law. **Mode #2** is operation by hovering close to the ground, on Earth, or on any other planet. Here, the emitted ordinary gravity wave field (which is d.c., not a.c., since it's a zero-frequency-shockwave) simply counteracts the already-existing local gravity field of the Earth, either exactly negating it, or, by making the craft's output field stronger, the net sum push then occurs in the upward direction, instead of in the usual downward direction.

Epilog:

THE ASTEROID THREAT SITUATION:

And here, I show the reader the **new** face of death, in fact, possibly the face of **your** soon-to-be killer. It is a large, heavy, mindless and uncaring rock, far out in space. It is slowly, but steadily making its way to the Earth. It will come very, very close to the Earth, as you shall see, because its orbit is a fairly predictable one. It's the one object, which has a predictable orbit, that will come the closest to Earth in the next 60 years, as per the NASA/JPL NEO (Near Earth Objects) web site. Its name is 4179 Toutatis. And here's what it looks like.

ASTEROID TOUTATIS:

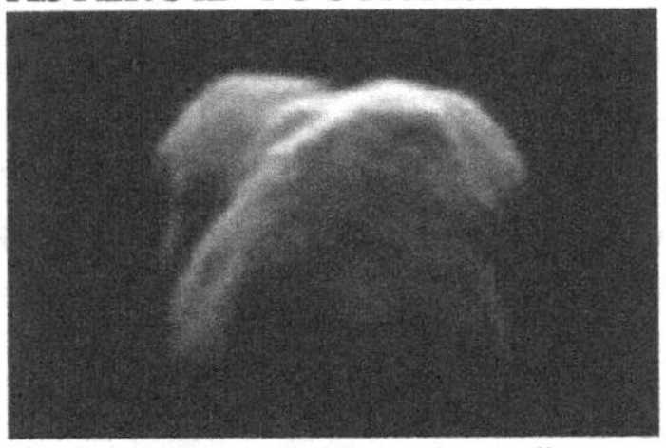

"I am become Death, ..."

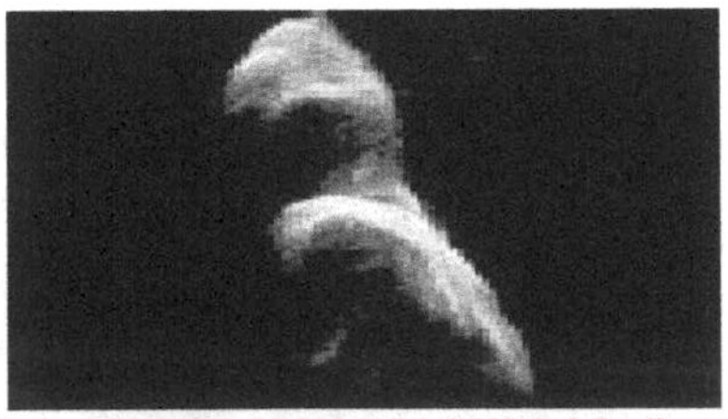

"....The Destroyer of Worlds."
-- J. R. Oppenheimer,
right after the detonation of the
first man-made atomic bomb.

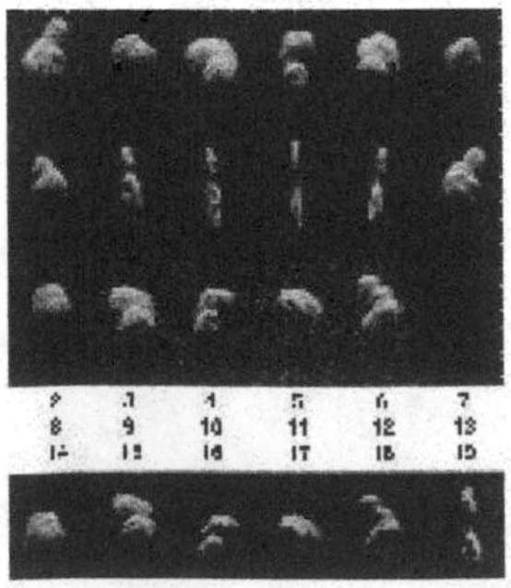

Time Lapse Photography of
Toutatis.

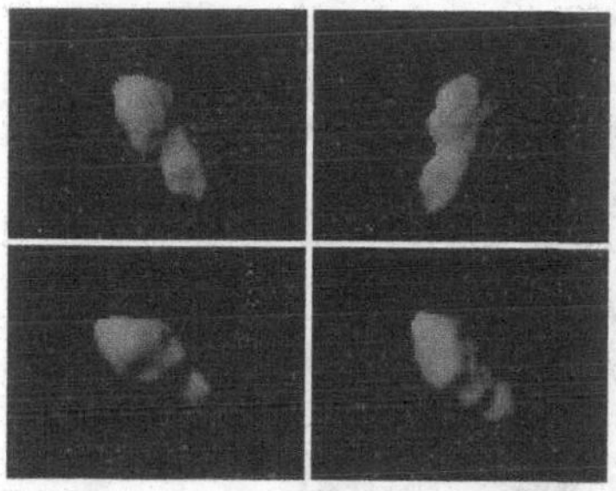

Ditto.

What follows is a set of computer plots of the orbit of 4179 Toutatis for 2004. To see them for yourself, just go to the following Internet address:

http://neo.jpl.nasa.gov/orbits/

OR, just go straight to:

http://neo.jpl.nasa.gov/cgi-bin/db_shm?rec=4179

Either way, you will be able to see what follows on your computer screen.

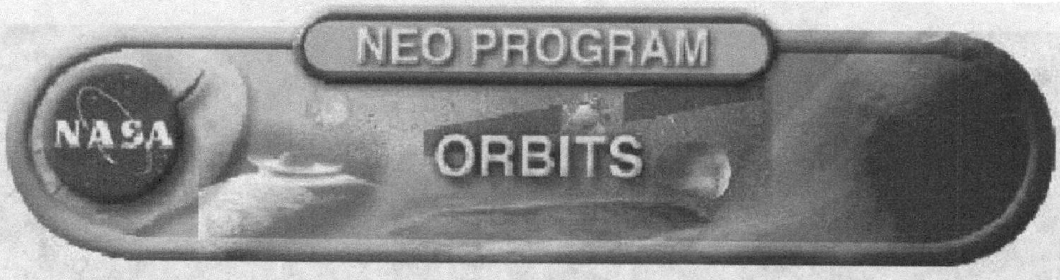

Asteroid 4179 Toutatis (1989 AC)

[go to next page to see first graphic plot]

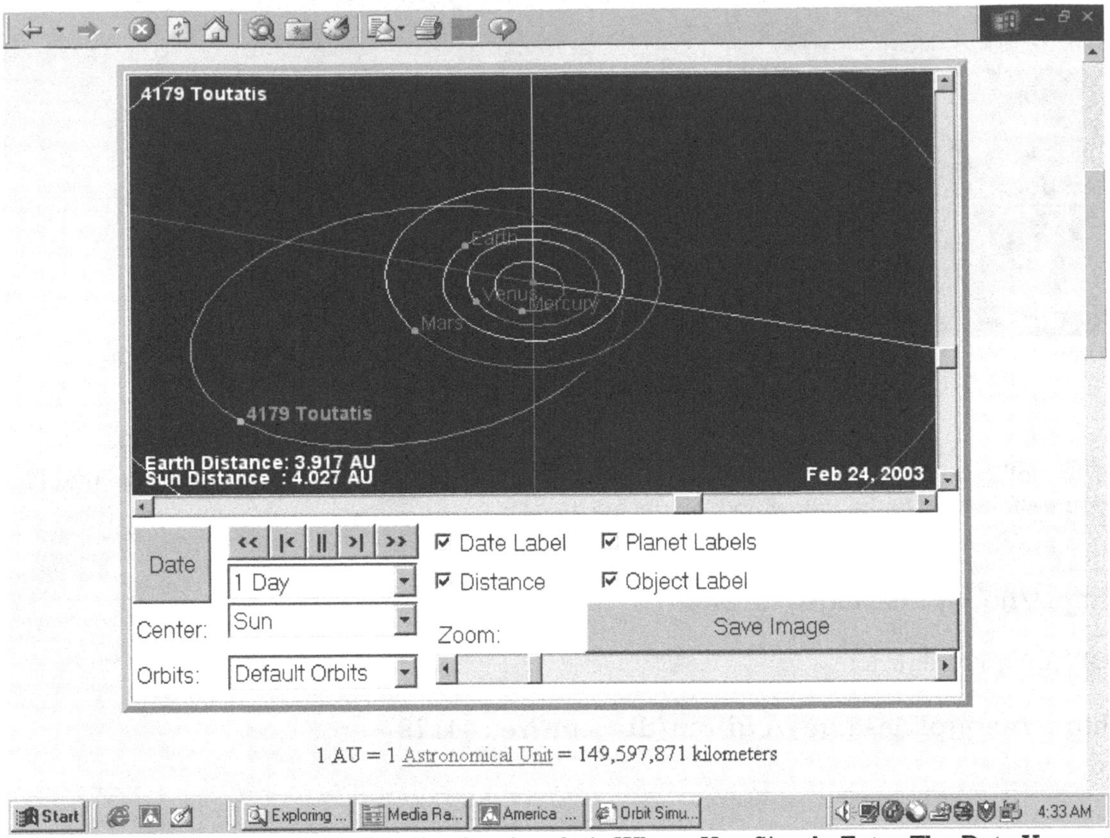

This is a NASA/JPL Interactive Application (applet), Where You Simply Enter The Date You Desire, And Then The Orbital Plot is Automatically Drawn For You.

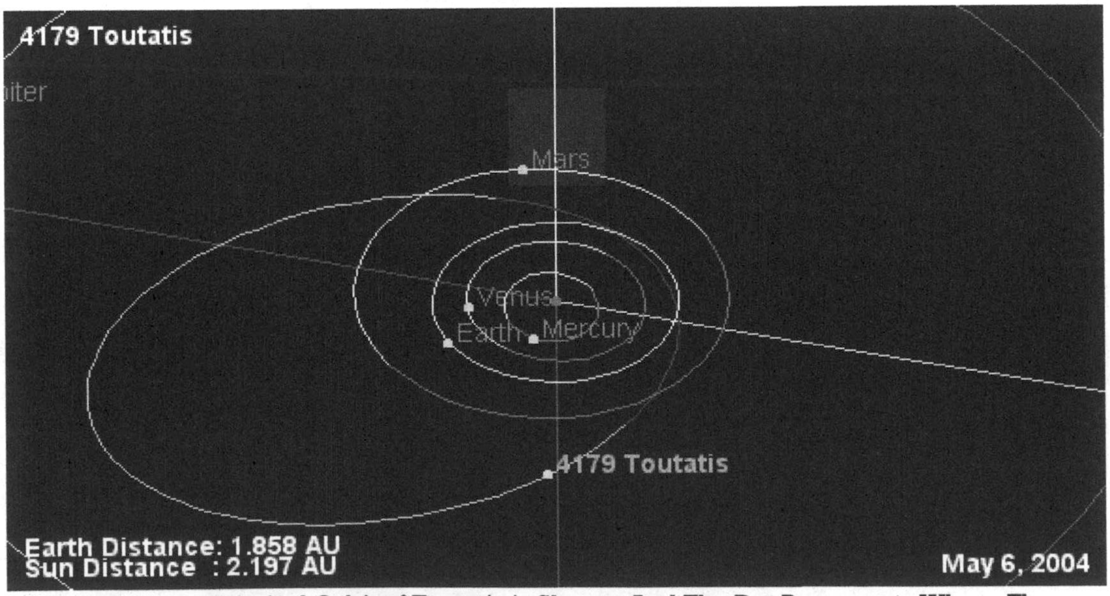

Here, The Large Elliptical Orbit of Toutatis is Shown, And The Dot Represents Where The Asteroid is on May 6, 2004.

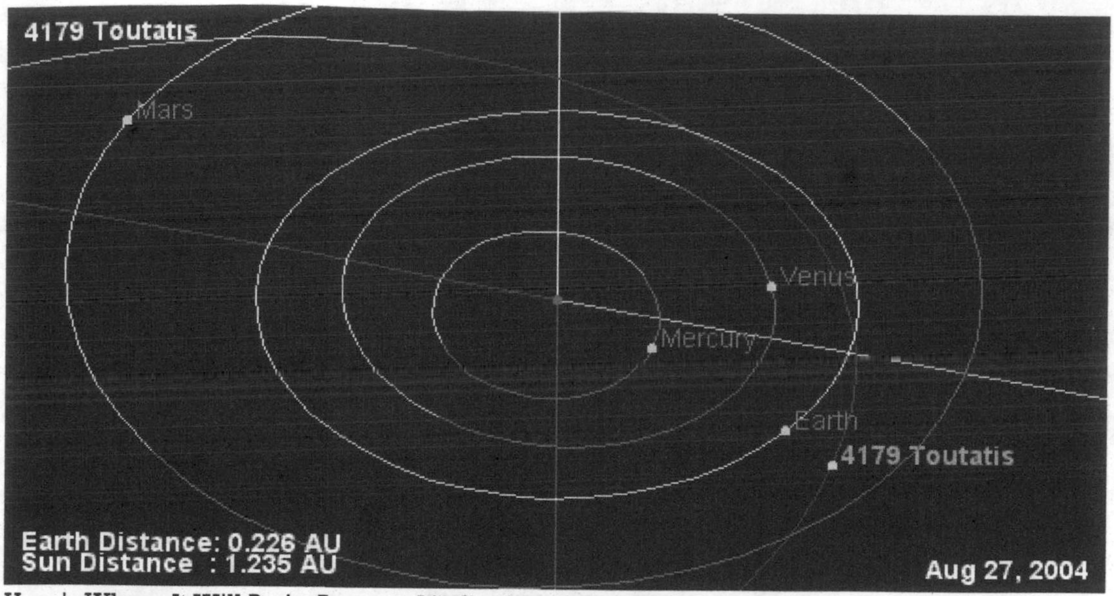

Here's Where It Will Be in August. It's Getting Much Closer.

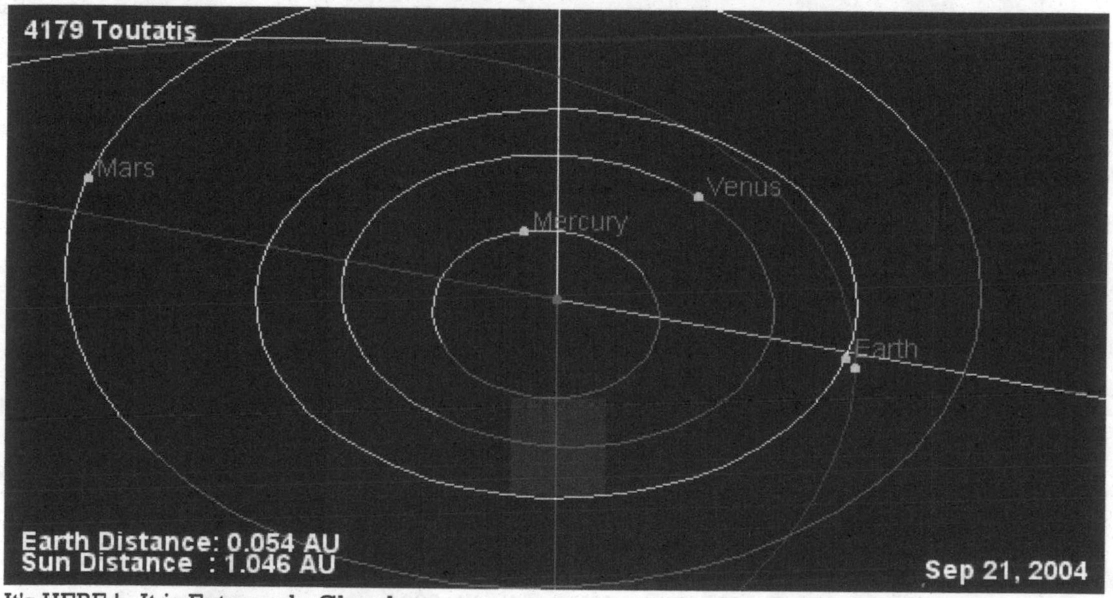

It's HERE ! It is Extremely Closeby.

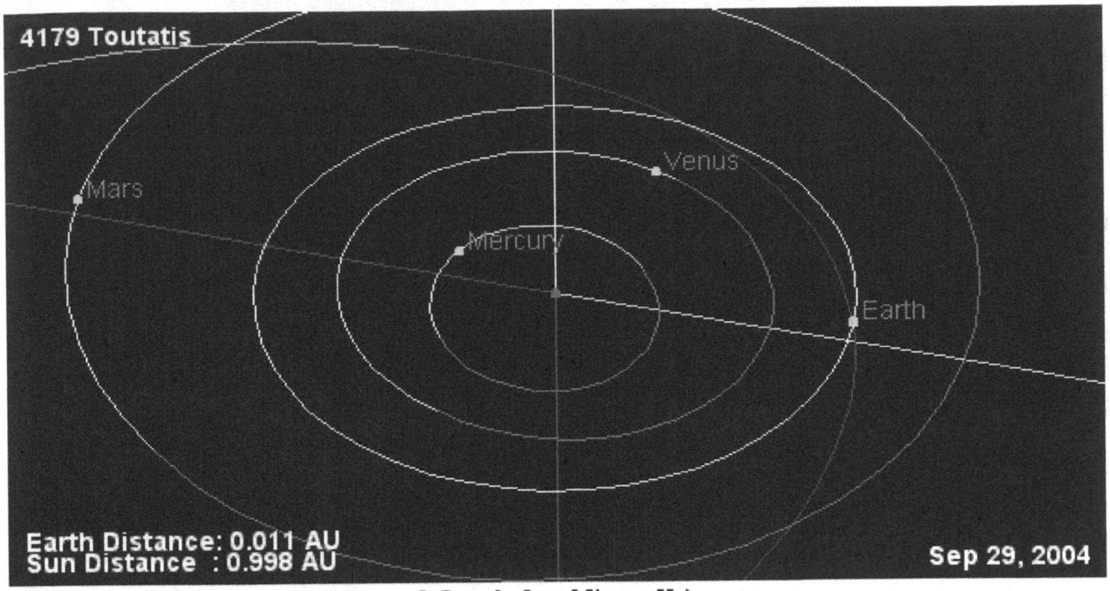

IMPACT (Unless, By Good Fortune It Barely Just Misses Us).

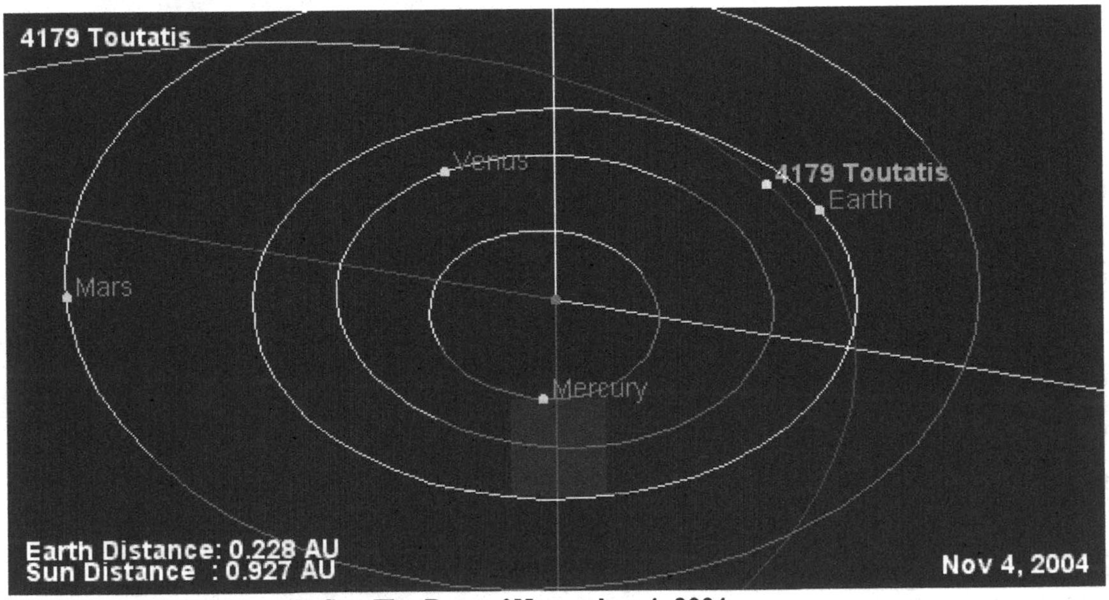

We May Not Be Around To See The Date of November 4, 2004.

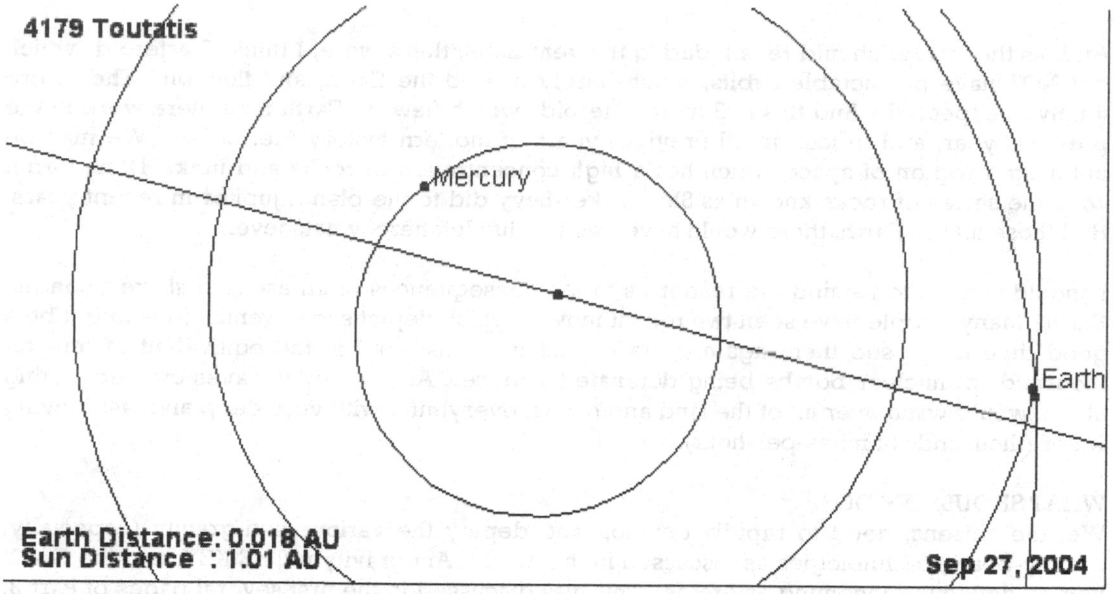

Close-Up View #1, in Inverse Video, For Clarity.

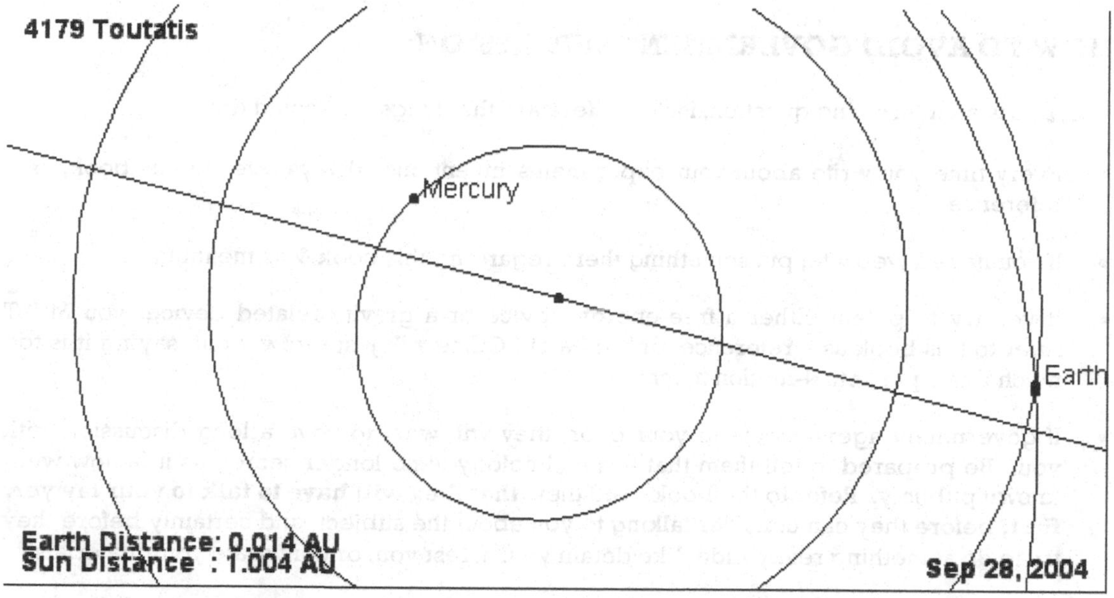

Close-Up View #2. IMPACT DATE. (or, perhaps on the 29th)

The above date, September 28, 2004, might just be your last. I think that everyone should ask NASA the question, "Can you 100% guarantee that Toutatis will not hit the Earth"? Also, we should ask NASA the question, "Are you 100% sure that Toutatis will not collide with anything, during its long journey to Earth, which could alter its course"? Or, perhaps we would do better by asking them simply "Are you 100% sure that Toutatis will not collide with anything during its long journey to Earth"?

And, as the reader should recall, during the year 2002, there were, I think, 3 asteroids which did NOT have predictable orbits, which barely missed the Earth, and flew on. They were totally unexpected. And this is 3 more asteroids which flew by Earth than there were in the previous year, and in fact, in all previous years of modern history (i.e. none). We may be entering a region of space which has a high concentration of rocks and junk. Don't forget what the series of rocks known as Shoemaker-Levy did to the planet Jupiter in recent years. Had those hit the Earth, there would have been no life left here, whatsoever.

I shouldn't have to remind the reader as to the consequences of an asteroid strike upon the Earth. Many people have seen two recent movies which depict such events. Now might be a good time to go see them again. The impact of an asteroid is the equivalent of tens-of-thousands of nuclear bombs being detonated at once. And the tidal waves created, acting alone, would wash over all of the land and drown everything with very deep and fast-moving water (thousands of miles-per-hour).

WHAT SHOULDWE DO?
We, the citizens, need to rapidly develop and deploy the various anti-gravity (especially, gravity-beam) technologies as discussed in this book. Antigravity is THE KEY, and the ONLY key, to deflecting incoming asteroids. This was discussed in the first several pages of Part B. Don't simply trust the government to do it. I will elaborate on why not in a later section herein. And, this brings us to the next problem...

HOW TO AVOID GOVERNMENT SUPRESSION:

Yes, this is an interesting question, isn't it? Here are the things we should do:

- Every time you write about your experiments/inventions, always refer to this book, as a reference.

- If you have a web site, put something there regarding this book & its meaning.

- If you try to patent either a free-energy device or a gravity-related device, you MUST refer to this book as a reference, or the Patent Office will just throw it out, saying it is too much like a perpetual-motion machine.

- If government agents come to your door, they will want to have a long discussion with you. Be prepared to tell them that this technology is no longer secret, as it is now well-known publicly. Refer to this book. Tell them **that they will have to talk to your lawyer, first,** before they can consider talking to you about the subject, and certainly before they try to do something really rude, like detain you, arrest you, or confiscate your materials!

- If things begin to get really nasty, just call the police and explain to them the same thing, when they get there. Tell the police that the agents have no authority to do anything to you, because this knowledge is now public.

- If you don't yet have a lawyer, might I suggest talking to one of the Disclosure Project's lawyers? Get in touch with Steven M. Greer, or someone at www.disclosureproject.org and ask to talk to one of their attorneys. Remember, dozens of government insiders, still with top-secret clearances, risked everything by revealing all that they knew, with the understanding that Greer's lawyers could keep them out of trouble by citing that the secrecy rules, done as they were, were illegal and thus not enforceable. So far, after about 2 years, NOTHING WHATSOEVER has happened to any of those people who disclosed what they know!

- As you develop an invention or even a theory, you should publicly disclose some, or all, of it on the Internet, perhaps a step-at-a-time. If new knowledge or a design is soon made public, then the agents won't have any reason to make you sign any non-disclosure agreements.
- Finally, tell any agents that bother you that your invention will save many lives (possibly including theirs, as well), and also tell them that all of your neighbors know what you are working on, and tell them that many, many people have copies of your work, and you never did keep a list of who those people are. And, of course, you can't remember most of them!

I can give you specific examples of people who I am sure were silenced. They will never admit to it, because they are told not only to not disclose the information any further, but they are told also to NOT TELL ANYONE that they were silenced, or even visited by agents! I would expect a violation of the secrecy is punishable by a $10,000 fine and perhaps jail time.

Here are people whom I'm sure it DID happen to:

- Jerry Bayles, for doing experiments, publicly on the Internet, involving using wedge-shaped or triangle-shaped waveguides in order to create negative mass, and then use a balance beam to measure that negative mass. His stuff was only on the Internet for 2 weeks, and then they "got him". He now has a message on his site that he found an error in his experiment, and that the results were thus null. However, the CAUSE he gave for the error was ridiculous, and would have endowed a simple high-voltage charge with intelligence enough to know the difference between "up" and "down", which obviously, it couldn't have. My cousin wrote an e-mail to Jerry with a question about "negative mass", and he got a very nasty reply back, which in essence said "mind your own damn business, and never contact me again".

- Dan LaRochelle, for putting up some very, very good improvements of David Hamel's drawings on the Internet. I remember some of them, and I sorely miss them. They had a tremendous potential to help our cause. You may ask, why then doesn't the government shut down David Hamel's website. I hate to be rude in any way, but I believe it is because that particular web site (which I will not name) actually makes David look like a complete nut/looney bird/possessed crazy man. So his web site is causing MUCH more dis-belief that it creates belief, which is exactly what the government wants! Most engineers and physicists would look at the site and see him as some kind of "witch-doctor", and nothing more (sorry David).

There were others, as well, and it was always the ones who put up such things as "negative mass" or "cones", and the like. I also noticed that sites which did not mention these things were never shut down or went off-the-air.

PROPAGANDA & DISINFORMATION:

I'm sorry to have to say the following, but it's true, and I can clearly see what is going on. The government is putting out stories (probably false ones), mainly on CNN and on AOL, in order to confuse us and to make us think that nothing is going to actually HIT the Earth for hundreds of years. Yes, CNN and AOL are the most governmentally-influenced news outlets, but I'm sure there are others.

As examples, I present the following, which are not exact and to-the-letter, but are my remembrances of the gist of what sorts of things I've heard throughout 2002.

"There is so-and-so asteroid which was recently discovered, and its calculated orbit says it will impact the Earth! But don't worry, as this will happen in the year 2025. By then, we'll have a solution to the problem..."

"It is predicted that, based upon the current and past data, no asteroids will hit the Earth, at least for decades to come."

"It could be the End Of The World! But don't worry, as this so-and-so event could only happen in the year 2,900, at the very earliest, and so it won't affect our current generation."

"A large, and very strange, asteroid, which we've actually been following since 1975, will soon come only 10 earth-moon distances from the Earth in 2007, by our calculations. But there is no danger. (I've heard the statement "but, there is NO DANGER" countless times in 2002, and always in regards to some astronomical phenomenon.) It will not hit the Earth, but we have extrapolated its orbit, and it could possibly hit the Earth in the year 2079."

"The scientists at xyz (and, I've checked, and xyz is on military government property and is not accessible to just anyone) have discovered that we may eventually expect to contact alien life, but not until the year 2017, at the earliest (for some stupid reason), and that such alien life would only be microbes, at most."

"Government scientists say that if there were intelligent life out there, that it would take them until at least the year 3000 until they could ever travel far enough to get to Earth."

"We have discovered possible microbial life on Jupiter's moons, but there is no possibility of ever discovering intelligent alien life anytime soon."

And so on, and so on....

They've even done a new expose on Roswell, and have "proven, beyond doubt" that the Roswell crash was not an alien craft, but was wreckage from a project Mogul. After I had seen all these kinds of things announced on CNN and AOL more than about 20 times, I realized that they are desperately trying to cover-up something. This is because these types of announcements have been quite rare over the last several years (indeed, decades). But now, I am seeing them all over the place. And each time I check out where the quoted scientist works, I found in every case that he/she is working on a military or government reservation, and that organization conducts either all secret projects, or at least more secret projects than normal, civilian-type projects. PEOPLE, wake up and smell the *?*$%^& !!

In addition, the director of the United States Naval Observatory recently died under very strange circumstances. He reportedly had seen "something significant" in his telescope, and he scheduled a trip for both him, and a very large telescope, to go to New Zealand for better viewing and study. The story goes that the telescope had barely gotten half-way to its destination, when the director was discovered dead, under mysterious circumstances. Several people said they thought it was murder, disguised as suicide. The telescope never made it to New Zealand, either.

And, NASA is planning to put in orbit a LARGE number of human habitation modules, which will be connected to the International Space Station. I'm sorry, but I cannot reveal my sources for this, but in addition to hearing about it, I have actually seen the brochures that were made for internal use which detail the particulars of the modules.

Couple that with the statements in "Disclosure" by more than one witness that NASA has photographed buildings on the Moon (and Mars!), and also the fact that NASA deported Ning Li back to China, before she or anyone could do a rotating superconductor experiment to demonstrate antigravity. And also bring in the news that certain people, who talked to former astronauts (in confidence) about Aliens and UFO's, have indicated that the astronauts said that they **routinely** see alien ships, and they are under orders to not reveal any of these sightings or contacts. And some of the older, braver, astronauts are now starting to talk...... Why do you think NASA would not recently allow a (multi-millionaire) US citizen to go up to visit the International Space Station? He had to go to the Russians, instead! They let him do it, but he said he had to sign secrecy and security papers of various sorts. What was NASA's excuse to not let him go up on the Shuttle? It was that "the public has NO IDEA of what extreme measures the astronauts would have to take to ensure the man's **safety** aboard the space station". I believe what they really were saying was that they would have to watch the visitor all the time to make sure he never looks out the window and sees a UFO (for his **safety**, of course!), because the UFO's are VERY commonly seen by astronauts, as witnesses have said. NASA just thought the security risk would be too high. Safety had nothing to do with it.

The bottom line is this. There is something going on, probably asteroid-related and maybe even alien-related, and they don't want us to know about it. It very well could be an impact by Toutatis. So they use propaganda releases to keep us confused. I've also read that there has been a big increase in underground activity, specifically in large underground bases built by the military. These could be used as places for people (the rich, elite, of course) to hole-up in and survive an upcoming big disaster, very possibly. But that is a story unto itself.

DANGERS TO OUR FREEDOM. EXECUTIVE ORDERS & F.E.M.A.:

In case of National Emergency (and, an asteroid impact would certainly qualify as being one), the following principal "civilian concentration-camps"-to-be, indeed, were established under the Rex '84 program:
Ft. Chaffee, Arkansas;
Ft. Drum, New York;
Ft. Indian Gap, Pennsylvania;
Camp A. P. Hill, Virginia;
Oakdale, California;
Eglan Air Force Base, Florida;
Vandenberg AFB, California;
Ft. McCoy, Wisconsin;
Ft. Benning, Georgia;
Ft. Huachuca, Arizona; &
Camp Krome, Florida.
These (and others) are referred to as "internment" camps in the Presidential Executive Orders (EOs) and related literature. Interestingly, note the word "internment" is very close to the word "interment" (a word meaning "burial"). Check out the following:

APPLICABLE EXECUTIVE ORDERS:

The following Presidential Executive Orders, now recorded in the Federal Register, and therefore accepted by Congress as the law of the land, can be put into effect at any time an emergency is declared:

10995 All communications media seized by the Federal Government.

10997 Seizure of all electrical power, fuels, including gasoline and minerals.

10998 Seizure of all food resources, farms and farm equipment.

10999 Seizure of all transportation, including your personal car, control of all highways and seaports.

11000 Seizure of all civilians for work under Federal supervision.

11001 Federal takeover of all health, education and welfare.

11002 Postmaster General empowered to register (conscript) every man, woman and child in the USA.

11003 Seizure of all aircraft and airports by the Federal Government.

11004 Housing and Finance authority may shift population from one locality to another. Complete integration.

11005 Seizure of railroads, inland waterways, and storage facilities.

11051 The Director of the Office of Emergency Planning authorized to put Executive Orders into effect in "times of increased international tension or financial crisis". He is also to perform such additional functions as the President may direct.

According to Professor Diana Reynolds, of the Fletcher School of Diplomacy at Boston's Tufts University, when Bush (the first one) declared a national emergency he "activated one part of a contingency national security emergency plan." That plan is made up of a series of laws passed since the presidency of Richard Nixon, which Reynolds says give the president "boundless" powers. According to Reynolds, such laws as the Defense Industrial Revitalization and Disaster Relief Acts of 1983 "would permit the president to do anything from seizing the means of production, **to conscripting a labor force,** to **relocating** groups of citizens." Reynolds says the net effect of invoking these laws would be the suspension of the Constitution.

She adds that national emergency powers "permit the stationing of the military in cities and towns, closing off the U.S. borders, freezing all imports and exports, allocating all resources on a national security priority, monitoring and censoring the press, and warrantless searches and seizures."

The measures would allow military authorities to proclaim martial law in the United States, asserts Reynolds. She defines martial law as the "federal authority taking over for local authority when they are unable to maintain law and order or to assure a republican form of government."

FEMA (the Federal Emergency Management Agency) is the agency which takes over and handles nearly all governmental control at all levels, in event of emergency. It can take and put into effect any rules and orders which the president wants or desires (without him having to consult with the Congress). FEMA has its roots in the World War I partnership between government and corporate leaders who helped mobilize the nation's industries to support the war effort. The idea of a central national response to large-scale emergencies was reintroduced in the early 1970s by Louis Giuffrida, a close associate of then-California Gov. Ronald Reagan and his chief aide Edwin Meese. Reagan appointed Giuffrida head of the California National Guard in 1969. With Meese, Giuffrida organized "war-games" to prepare for "statewide martial law" in the event that **Black nationalists** and **anti-war protesters** "challenged the authority of the state."

Aware of the bad publicity **FEMA** was getting because of its role in organizing for a post-nuclear world, Reagan's **FEMA** chief Giuffrida publicly argued that the 1865 Posse Comitatus Act **prohibited the military from arresting civilians**. However, Reynolds says that Congress eroded the act by giving the military reserves an exemption from Posse Comitatus and allowing them to arrest civilians. The **National Guard,** under the control of state governors in peace time, **is also exempt from the act and can arrest civilians.**

FEMA Inspector General John Brinkerhoff has written a memo contending that the government doesn't need to suspend the Constitution to use the full range of powers Congress has given the agency. **FEMA** has prepared legislation to be introduced in Congress in the event of a national emergency that would give the agency sweeping powers. **The right to "deputize" National Guard and police forces is included in the package.** But Reynolds believes that actual martial law need not be declared publicly. Giuffrida has written that "Martial Rule comes into existence upon a determination (not a declaration) by the senior military commander that the civil government must be replaced because it is no longer functioning anyway." He adds that "Martial Rule is limited only by the principle of necessary force." According to Reynolds, it is possible for the president to make declarations concerning a national emergency secretly in the form of a National Security Decision Directive. Most such directives are classified as so secret that Reynolds says "researchers don't even know how many are enacted."

Stated simply: the dictatorial power of the Executive rests primarily on three bases: **Executive Order 11490,** Executive Order 11647, and the Planning, Programming, Budgeting System which is operated through the new and all-powerful Office of Management and Budget.

Executive Order 11490 is a compilation of some 23 previous Executive Orders, signed by Nixon on Oct. 28, 1969, and outlining emergency functions which are to be performed by some 28 Executive Departments and Agencies whenever the President of the United States declares a national emergency (as in defiance of an impeachment edict, for example).

Under the terms of E. O. 11490, the President can declare that a national emergency exists and the Executive Branch can:

take over all communications media,
seize all sources of power,
....
commandeer all civilians to work under federal supervision,
....
shift any segment of the population from one locality to another,
....
regulate the amount of your own money withdrawn from your banks.

All of these and many more items are listed in 32 pages incorporating nearly 200,000 words, providing an absolute bureaucratic dictatorship whenever the President gives the word.

Executive Order 11647 provides the regional and local mechanisms and manpower for carrying out the provisions of E. O. 11490. Signed by Richard Nixon on Feb. 10, 1972, this

Order sets up Ten Federal Regional Councils to govern Ten Federal Regions made up of the fifty still existing States of the Union.

The 10 Federal Regions:
REGION I: Connecticut, Massachusetts, New Hampshire, Rhode Island, Vermont. Regional Capitol: Boston
REGION II: New York, New Jersey, Puerto Rico, Virgin Island. Regional Capitol: New York City
REGION III: Delaware, Maryland, Pennsylvania, Virginia, West Virginia, District of Columbia. Regional Capitol: Philadelphia
REGION IV: Alabama, Florida, Georgia, Kentucky, Mississippi, North Carolina, Tennessee. Regional Capitol: Atlanta
REGION V: Illinois, Indiana, Michigan, Minnesota, Ohio, Wisconsin. Regional Capitol: Chicago
REGION VI: Arkansas, Louisiana, New Mexico, Oklahoma, Texas. Regional Capitol: Dallas-Fort Worth
REGION VII: Iowa, Kansas, Missouri, Nebraska. Regional Capitol: Kansas City
REGION VIII: Colorado, Montana, North Dakota, South Dakota, Utah, Wyoming. Regional Capitol: Denver
REGION IX: Arizona, California, Hawaii, Nevada. Regional Capitol: San Francisco
REGION X: Alaska, Oregon, Washington, Idaho. Regional Capitol: Seattle

It has been said, again and again, "It can't happen here, this is AMERICA!"

Now, understand what CAN and WILL happen today, if the President of the United States decides to irrationally (or rationally as far as HE believes) put pen to paper and do something similar.

To wit, I present Executive Order #6260 - written in 1933 where this great unleashing of "governmental power" occurred for one of the first times. All boldfacing and underlining is by me.

Executive Order No. 6260 - April 5, 1933
"Executive Order: By virtue of the authority vested in me by Section 5(B) of the Act of Oct. 6, 1917, as amended by section 2 of the Act of March 9, 1933:, in which Congress declared that a serious emergency exists, I as ***President do declare that the national emergency still exists;*** *that the continued hoarding of gold and silver by subjects of the United States poses a grave threat to the peace, equal justice, and well-being of the United States; and that appropriate measures must be taken immediately to protect the interests of our people."*

"Therefore, pursuant to the above authority, I hereby proclaim that such gold and silver holdings are prohibited and that all such coin, bullion, and other gold and silver be tendered within fourteen days to agents of the Government of the United States for compensation at the official price, in the legal tender of the Government. ***<u>All of your safe deposit boxes in banks or financial institutions have been sealed,</u>*** *pending action in the due course of the law. All sales, or purchases, or movements of such gold and silver, within the borders of the United States and its territories, and all foreign exchange transactions or movements of such metals across the border are hereby prohibited."*

"Your possession of these proscribed metals and/or your maintenance of a safe deposit box to store them <u>is known to the government</u> from bank and insurance records. *Therefore, be advised that* ***your*** *vault box must remain sealed, and* ***may only be opened in the***

presence of an agent of the INTERNAL REVENUE SERVICE." *"By lawful Order given this day, the President of the United States."*

SO WHY am I including this information in this book? It is because of a serious concern I have developed regarding our past president's uses & abuses of Executive Orders, which seem to be developing an alarming trend, and other (horrifying, actually) information as well. Further discussion of this subject is beyond the scope of this book, but if the reader wants to see what I've been seeing, just do the following:

- Read all you can about the "N.W.O." (New World Order). Find out which high-ranking officials of which countries are reportedly involved. Study what is their ultimate plan.

- Check out the background of people such as Bill Clinton, George Bush Sr., Alan Greenspan, and major people that you find to be involved with UN (United Nations) plans to bring about a ONE-WORLD government, and even a ONE-WORLD religion. Notice who Clinton appointed to lead the UN (Kofi Annan). Study his background as well. Check out the billionaire friends of some of the high-ranking UN officials, one of which lives in Canada. Note that many of the recent presidents are involved with Masonry. Note that the ultimate head of all Masons, who presides over a round table of many "33rd Degree" Masons, is a well-known name in England, and is in the Royal Family.

- Check out the stories that the Royal Family has bought up large tracts of land in Colorado, and they did so under false names.

- Check out the stories of witnesses who believe that Denver, Colorado is to become the new headquarters of the NWO, after the "big disaster" hits the Earth.

- Look at the website www.flydenver.com, which is the official web site for the new Denver International Airport (D.I.A.). Look under the heading of "Public Art" displays, especially carefully at the murals and paintings (by "Leo T."). And look at other web sites regarding these murals, as well. You will be shocked beyond belief that such horrible things would ever be put in such a public place. Depicted clearly in these paintings are their plans (whoever "they" are) of what they intend to do to us, the people. Remember, Denver is supposedly going to be the new NWO WORLD headquarters! The paintings show the intended deaths (extinction, in fact--just like the "Do - Do bird" became extinct) of all Blacks, Indians & Mexicans, and Jews & Christians. People are actually shown lying dead, in coffins!!

- Upon studying the NWO deeper, you will find that 1) they are just as racist as were the Nazi Germans, and 2) they have members who have such a staggering amount of wealth (we're talking thousands of Trillions of dollars), that they are easily able to buy off many leaders of many countries to do their requested bidding.

- And, finally, note that Bill Clinton, George Bush Sr., and Alan Greenspan have all been knighted by the Queen of England. And when a person is knighted, they MUST declare a solemn oath of allegiance to the Queen and the Royal Family of England!

Need I go on? Oh, and there's one other thing. A person was interviewed who said that the government had commissioned some companies to build many hundreds of thousands of "special" railroad cars, and these are all now stored in underground tunnels. I will let the reader research and find out what makes these "special", and of this awful subject, I will say no more.

My concern is only that, in the event of an asteroid strike, we may find ourselves stripped of all our Constitutional Rights, and people could be forcibly relocated. In fact, after all that I (and many others) have learned about the secret plans of the NWO, I would not be at all surprised if they have somehow attached thruster rockets or other devices to the asteroid Toutatis in order to actually GUIDE it, to MAKE SURE that it HITS THE EARTH!! Then, they will have that "state of emergency" declaration that they need to impose martial law, and to enact all of their dictatorial powers, against the Constitution.

Moral: don't trust the government, because leaders CAN BE, and HAVE BEEN, bought off by the evil & extremely rich. Maybe the current president is a good guy, but I wouldn't necessarily count on it.

THE LAST WORD:

And so ends this book. But I do have some final advice to give, first. If you should ever cross paths with an actual alien, I know of some good advice on how to deal with them. According to a number of significant contactees, you will always find that any given alien is always ONE of TWO POSSIBLE TYPES: 1) Trinitarian (also called "service-to-others"-type), or 2) Luciferian (also called "service-to-self"-type). There is no in-between; it's either 1) or 2). Looked at another way, they either are working for God, or they are working for Satan. But the good news is this: when asked, they MUST tell you which camp they are in. My advice is to ask them to go away if they are Luciferian. Just mention the names God or Jesus, and they will run away, if they are of the Luciferian type. They are allegedly really, really afraid of the wrath of God, if they bother someone who does not desire contact with them! And, it's this Luciferian group which has been doing all the abductions and mutilations of both animals and humans. I've never personally seen an alien, but I will surely try this out if I ever meet one!

And finally, a word about GOD. I have found that God is a really great and important entity to know. He has helped me, and there's no question of this, many, many times. I have had no less than miracles occur in my life, for my benefit. He has even saved my life several times. God has deep, infinite love for every one of us (even the bad people, as they may someday learn their errors and reform). He knows us better than even WE know us. And the reader might be interested to know that the aliens worship the SAME God that we do (they usually call Him the Supreme Being). And, there's only ONE God, and he's the one who created (and still maintains) the ENTIRE UNIVERSE. I think God is very under-rated, and under-talked-about in today's society. I think God deserves MUCH better from us, his creations, than THAT. And so, this is why I bring up the subject of God, the creator, here in this book. Without Him, I would have never been able to complete this book.

I know for a fact that God wants the information contained in this book to get out to all of the people, and quickly. I believe I understand all of His reasons why all this information needs to become public, at this time, but I will not impose my opinion in place of God's true opinion and reasons, which we may never totally know or understand. All we can do is to try to understand His master plan.

I believe that we human beings, in about a decade from now, will finally learn and understand one very huge lesson, and that is that money in this world has caused UNBELIEVABLE, STAGGERING amounts of corruption, especially amongst our leaders, which we are not yet completely or even mostly aware of (but we later will be!). That's my prediction, anyway.

And lastly, what should we do, in these troubled and increasingly frightening times? My advice is to believe and TRUST in God, the Creator, and acknowledge Him every day, and thank Him for what you have, and what you are going to have. And also, help others. We are all part of God and are actually within God. Helping others is equivalent to helping yourself. Do help others, and do pray for the well-being and safety of others in the World, especially the ones who are suffering the greatest.

[ ]

[THIS PAGE IS INTENTIONALLY BLANK]

PART H

APPENDICES
(H1 - H4)

[ ]

[THIS PAGE IS INTENTIONALLY BLANK]

Appendix H1

The L & NSS Pages For The UFT

[ ]

[THIS PAGE IS INTENTIONALLY BLANK]

The inductance (ie "L" series of pages): **DL1**

<u>L0 – L2</u>, develop formula for the inductance, based upon E.E. helix formula, for any situation, including that of non-equilibrium.

<u>L3</u>, develop steady state inductance and a formula for the steadystate inductance <u>correction factor</u> $L_{c,ss}$.

<u>L4</u>, verify small radius (S.R.) dynamic rest mass.

<u>L5</u>, develop formula for large radius (LR) dynamic rest mass, and show that, for large enough radius, the rest mass of a particle can be reduced to near-zero!

<u>L6 – L14</u>, develop formula for rest mass of the vacuon which is approximate, <u>but</u> is independent of temporal length l_0^* (or Ksubl*). For now, we can ignore pages L13A and L13B, as they jump ahead and use formulas from a later part of the document to derive an expression, <u>exact</u>, for Ksubl*. Also on these 2 pages, we calculate, based upon a value of Ksubl* of 1/6.4234 (which is <u>exactly</u> correct for the electron), Mroe and Mrop. Mroe is the <u>exact</u> <u>rest mass of a single free vacuon</u>, whereas Mrop is the <u>rest mass of a vacuon while in a "clump"</u> of vacuons.

⟹ Exact single vac. rest mass $= 5.56 \times 10^{-9}$ kg. The ratio is 10.89.

Paul P. Grandall 12-24-2001

DL2

L15–L17, Rewrote the exact formula from page L13 as two slightly different methods. Method "A" uses rest mass computed as an exact multiple of the electron's mass, and method "B" uses M_o, the actual measured mass (rest) of the particle.
Using temporal length = the Compton wavelength of the electron, we get $M_{roe} = .947 \times 10^{-8}$ kg, and using temporal length = the Compton wavelength of the proton, we get $M_{rop} = .0715 \times 10^{-8}$ kg. The values are to be regarded as the absolute maximums for the values of the mass of the Vacuon (free (M_{roe}), and trapped within a multi-vacuon clump of 7 vacuons (M_{rop})). The ratio of $M_{roe}/M_{rop} = 13.24$.
And, to review, the ratio of M^*_{roe}/M^*_{rop}
[= 10.956. These "*" values are more realistic.
$M^*_{roe} = .517 \times 10^{-8}$ kg and $M^*_{rop} = .047 \times 10^{-8}$ kg.]

L18, I develop 2 very crude formulas for calculating the dynamic rest mass of the proton and electron. The dynamic mass is seen to be inversely proportional to the radius (A_{dy}). Valid only for LR (Large Radius) states.

Richard P. Gardell 12-24-2001

DL3

<u>L19–L22</u>, We compute absolute minimum (using temporal length = 0) values $M_{roe,min}$ and $M_{rop,min}$. In the minimum case, ratio = 9.74. Then, we calculate the "best" values, M_{ro1} and M_{ro7} (same as M_{roe} and M_{rop}, just a change in notation), using a temporal length of (1/40) Lambda,compton. These are $M_{ro1} = 4.7 \times 10^{-9}$ kg and $M_{ro7} = 4.7 \times 10^{-10}$ kg. The ratio of these is 10.000.

However, the "<u>EXACT</u>" values, using the balance of centrifugal and centripetal forces method, are derived as seen on page L13B.

They are:

[we use temporal length = (1/6.4234) Lambda,compton]

$M_{roe,exact} = M_{ro1,exact} = 5.56 \times 10^{-9}$ kg,
and $M_{rop,exact} = M_{ro7,exact} = 5.106 \times 10^{-10}$ kg.
The ratio is 10.89.

Lambda,cproton was used for the M_{rop} calculation, and Lambda,celectron was used for the M_{roe} calculation.

Richard P. Gardell 12-24-2001

DL4

L23–L25 Now, back to non-equilibrium (but still "SR" (Small Radius)). Formula no good for "LR".

I write a formula to compute m(sub dyox) and also m'(sub dyx) [this becomes just m(sub dyox) if we set gamma(sub gr) = 1].

Also written is the formula to calculate Ldycx (squared) as a function of the steady state radius A_{ox} and of the actual (non-S.S.) current radius A_{dy} (i.e. the dynamic radius).

[The final formula matches the one on fig A7S, if we set gamma = 1 and $A_{dy} = A_{ox}$.

[This is the last of the "L" series of pages. We have calculated dynamic mass using inductance concepts.

Next, is the "non-steady state" series of pages (ie "non-SS" or "NSS"). It uses inward/outward force balance concepts to render a more accurate dynamic mass.

Ricky P. Gardell 12-24-2001

DNSS1

The following "non-steady state" pages use inward/outward force balance to calculate a very accurate dynamic mass.

NSSØ, I calculate the total outward force in a spinning disk, which represents our physical particle. It's a function of mass (M sub dyox), outer rim velocity, and radius.

NSS1–NSS5A, I calculate the gravitomagnetic field produced by each subquark (vector B_{gn}), and the corresponding inward force produced upon a subquark by each of the other subquarks (Vector F_n). The vector i components of all these are summed to give the magnitude of the total force (inward) upon any subquark (i.e. compute the sum of F_n, for n=1 to tilde N sub Q).

On page NSS5A, I calculate the exact value for M_{ro1} (e-type (isolated) Vacuon) and M_{ro7} (p-type (in a clump) Vacuon).

[I take M_{ro1} to be the final and most accurate value for the rest-mass of the isolated Vacuon, which is 5.56×10^{-9} kg.] The value for M_{ro7} is below the minimum (ie it's totally wrong), but this expected, as the proton has multiple rings (quarks), and the force model only works for 1 ring.

Richard P. Campbell 12-24-2001

DNSS2

NSSG, I move the nasty summation to a separate function (of tilde N(sub Q)), which I have named the Z( ) function.

Then, I write the formula for F_{indyo} and F_{outdyo}, which are the inward and the outward forces acting upon each subquark, respectively. For stability (dynamic balance) to occur, we must have $F_{indyo} = F_{outdyo}$. These formulae are very accurate.

Richard P. Campbell 12-24-2001

NSSGAA, I develop the formula for dynamic mass of a particle, where $F_{indyo} \overset{\text{must}}{=} F_{outdyo}$. I call it "m(sub dyox)".

NSSGAB–NSSGA[redacted]F, By equating the inductance based m(sub dyox) formula with the balance-of-forces-based one on page NSSGAA, we are able to solve for those two pesky unknowns, namely I(sub ody) and L(sub dycx) squared. They are written as functions of V_{vody} and A_{dy}.

DNSS3

NSS6AG, We solve for K(sub ϕ), which is the fraction of the particular particle's Compton wavelength, which, when multiplied by that Compton wavelength, gives the temporal length, called l(sub ϕx). For the electron, where tilde N(sub Q)=4, we calculate k(sub ϕ) to be 1/6.4236. For any particle, we calculate the maximum [redacted] value allowed for the Z() function, which is 5.8997.

[Thus, we can conclude, for any "1-ring" particle (i.e. 1 quark), that tilde N(sub Q) can only range from 2 through 5.]

The 1/X column gives the "effective" value for 1/K(sub ϕ), due to the effect of the inductance correction factor, (L ^2) sub (CX, SS).

NSS6AH – NSS6AI, Now going back to the LR (Large Radius) case:

Richard P. Gambill 12-24-2001

DNSS4

I equate the inductance-based dynamic mass with the balance-of-forces-based dynamic mass. I derive a formula for VvodyLR, the large-radius dynamic velocity. Amazingly, it is a function solely of tilde N(sub Q)! It is not a function of the radius!

A table for tilde N (sub Q) is created. The tilde N (sub Q) value ranges from 2 through 8.

Due to a finding on page Figure A7T, and due to the finding on page NSSGAG, we

> see that the only two possible values for tilde N (sub Q), in the 1-ring (1 quark) model, [redacted] are 4 and 5, with 5 not likely.
> The corresponding velocities range, in terms of "c", are .4028 thru .5 and .4320 thru .4472 respectively.

Richard P. Gambold 12-24-2001

DNSS5

NSS6AJ It is shown that, for tilde N(sub Q) equal to 4, in the 1-ring model, that the expression $E' = hf'$ remains approximately true in the LK case (with an error of only 11% or 12%).

NSS6AK – NSS6AM, We add another subscript, "u", to indicate the most general case of all (still assuming the vacuons form a nice ring), which is such radius and velocity that the inward and outward forces are NOT equal. This is called the "unbalanced" or "unstable" case (or "u" case).

Finally, we get an answer, to the most general case, of what is mass? We see that it is equal to a constant, Alpha[redacted], multiplied by the total apparent vacuon mass squared, and divided by the structure's radius. Rest frame, all cases of radius and rotational speed.

We also derive formula for angular momentum.

Richard P. Gauthier 12-24-2001

LØ

Helix

$$L = \frac{(NR)^2}{9R + 10H} \quad (\mu H)$$

H = height, R = radius (inches), N = # turns

or,

$$\frac{.394\, R^2 N^2}{9R + 10L} \quad (\mu H)$$

also helix

(R and L: cm)

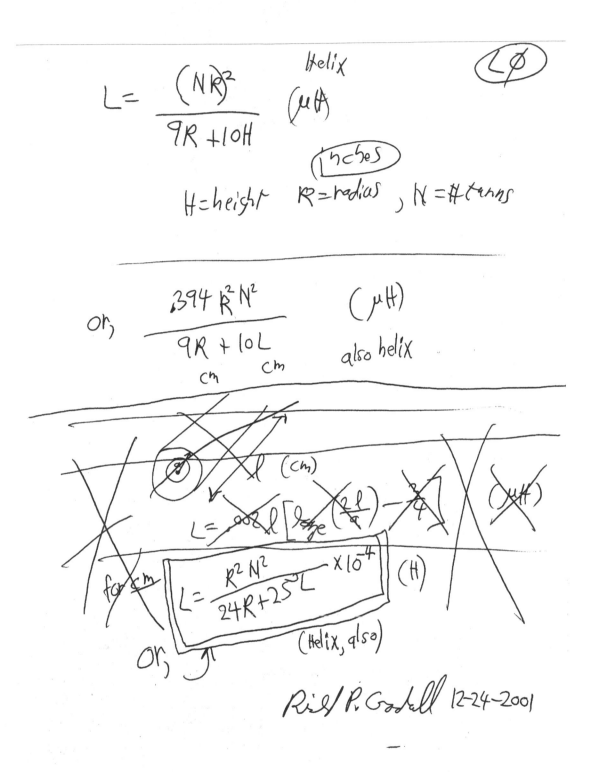

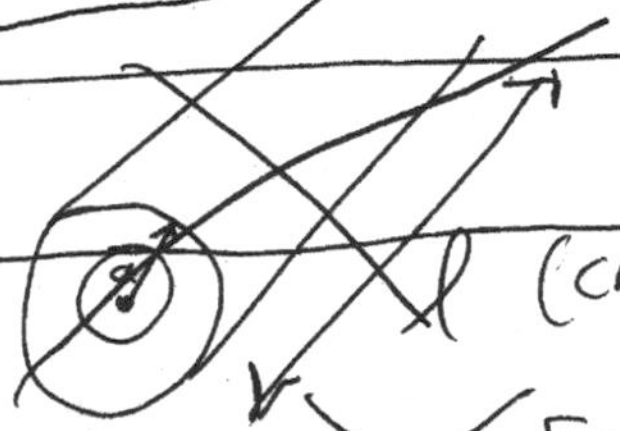

$$L = .002\, l \left[\log_e\left(\frac{2l}{a}\right) - \frac{3}{4}\right] \quad (\mu H)$$

for cm

$$L = \frac{R^2 N^2}{24R + 25L} \times 10^{-4} \quad (H)$$

(Helix, also)

or,

Rich P. Goodell 12-24-2001

L1

$$L = R^2 N^2$$

$$\text{induc} = \left(\frac{(100R)^2 N^2}{24(100R) + 25(100L)}\right) \times 1 \times 10^{-6} \quad (H)$$

R, L meters

$$\therefore \text{induc} = \frac{R^2 N^2}{24R + 25L} \times 1 \times 10^{-4} \quad (H)$$

N = # turns, R = radius, L = length or height

$$N \Rightarrow N_w', \quad L \Rightarrow l_0', \quad R \Rightarrow A_0 \swarrow !$$

$$L'_{temp} = \frac{A_0^2 \tilde{N}_w'^2 \left(\frac{1}{\tilde{N}_w^R}\right)}{22.86 A_0 + 25.3 l_0'} \times 10^{-4} = \frac{A_0^2}{22.86 A_0 + 25.3 l_0'} \times 10^{-4}$$

since current only flows in one turn at a time, (a const.)

$$\mu_e = 1.26 \times 10^{-6}$$

$$\mu_q = 9.37 \times 10^{-27} \qquad \frac{1 \times 10^{-4}}{25.3} = \mu_e \pi$$

$$\therefore L' = \frac{\pi}{\pi}\left(\frac{\mu_q}{\mu_e}\right)\left(\frac{A_0^2 \times 10^{-4}}{22.86 A_0 + 25.3 l_0'}\right) = (\pi \mu_q)\left(\frac{25.3 A_0^2}{22.86 A_0 + 25.3 l_0'}\right)$$

Rick P. Goodell 12-24-2001

(L2)

$$\therefore\ L' = \mu_g \pi \left(\frac{A_o^2}{\left(\frac{22.86}{25.3}\right) A_o + l_o'} \right)$$

$$\boxed{L' = \mu_g \pi \left(\frac{A_o^2}{(.904) A_o + \gamma l_o} \right)}$$

equilib. OR Non-equilib.

$$\frac{l_o}{N_{w_o}} = \frac{l_o'}{N_w'} = C\,T_{rot_o} = \frac{C}{f_{rot_o}} = \frac{C\,\tilde{N}_Q}{f_o} = \text{a const.}$$

SS, only $= \dfrac{C\,\tilde{N}_Q}{f_o}$

but,

$$\boxed{A_o = \frac{\left(\sqrt{(2S\tilde{N}_Q^2)}\right) C}{2\pi f_o}}$$

(see fig A7P)

SS only

Rick P. Grandfield 12-24-2001

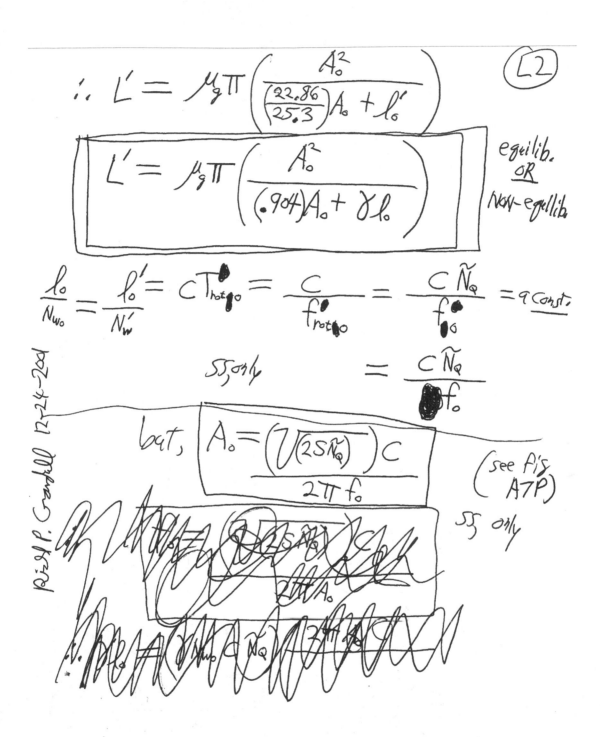

L3

Richard P. Crandall 12-24-2001

$$\therefore \quad \frac{1}{f_0} = \frac{l_0}{C \tilde{N}_Q N_{wo}}$$

$$\therefore \quad A_0 = \frac{\cancel{C}\sqrt{2S\tilde{N}_Q}}{2\pi} \; \frac{l_0}{\cancel{C}\,\tilde{N}_Q N_{wo}}$$

$$\therefore \quad (.904)A_0 = \left(\frac{(.904)(\sqrt{2S})}{2\pi N_{wo}\sqrt{\tilde{N}_Q}}\right) l_0$$

Steady state equilib. ONLY!!

$$\therefore \quad L'_{ss} = \frac{\mu_g \pi A_0^2}{\left(\frac{(.904)(\sqrt{2S})}{2\pi(\sqrt{\tilde{N}_Q})N_{wo}} + \gamma\right) l_0} = \frac{L_0}{\gamma}$$

(steady state)

$$L_0 = \frac{\mu_g \pi A_0^2 (L^2_{c,ss})}{l_0}$$

$$N_{wo} = \frac{l_0 f_0}{C\tilde{N}_Q}$$

$$N'_w = \gamma N_{wo}$$

$$\therefore \quad L^2_{c,ss} = \frac{1}{1 + \frac{(.904)(\sqrt{2S})}{2\pi(\sqrt{\tilde{N}_Q})N'_w}}$$

from fig A7S, (L4)

$$8L_oI_o^2 = m_oC^2 = \left(L^2_{c,ss}\right) 8\mu_g \pi A_o^2 \tilde{N}_V^2 M_{vaco}^2 f_o^2 / l_o$$

and

$$f_o = \tilde{N}_Q f_{roto}$$

$$f_{roto} = \frac{\omega_{roto}}{2\pi} = \frac{V_{Vo}}{2\pi A_o}$$

$$\therefore m_oC^2 = \overbrace{\left(\frac{1}{1+\frac{k_3}{\gamma N_{Wo}}}\right)}^{=L^2_{c,ss}} 8\mu_g \pi A_o^2 \tilde{N}_V^2 \left(\frac{\tilde{N}_Q^2 V_{Vo}^2}{4\pi^2 A_o^2}\right)\left(\frac{M^2_{vaco}}{l_o}\right),$$

$$\text{and } \left(\frac{M^2_{vaco} L^2_{c,ss}}{l_o}\right) = \left(\frac{m_o}{\tilde{N}_Q^2 \tilde{N}_V^2}\right)\left(\frac{\pi}{2\mu_g}\right)\left(\frac{C^2}{V^2_{Vo}}\right),$$

(See fig A7I)

$$\therefore m_oC^2 = 2\mu_g \pi \tilde{N}_V^2 \left(\frac{\tilde{N}_Q^2 V_{Vo}^2}{\pi \pi}\right)\left(\frac{m_o}{\tilde{N}_Q^2 \tilde{N}_V^2}\right)\left(\frac{\pi}{2\mu_g}\right)\left(\frac{C^2}{V^2_{Vo}}\right),$$

$$= m_oC^2 \checkmark\checkmark$$

(Small radius dynamic mass) rest

Independent of: $L^2_{c,ss}$, V_{Vo}, and A_o !

David P. Crandall 12-24-2001

L5

Large Radius Dynamic ~~Mass~~ rest Mass:
(LRdy)

(ring has been gradually forced to a <u>much</u> wider diameter)

approximately:

$$L'_{LR} = \overset{\bullet}{L}_{LRo} = \left(\frac{1}{.904}\right)\mu_g \pi A_{\underset{LRdyo}{\bullet}}$$

(power of "A" is one smaller)

And there is no $\underline{L_c}$ ($\underline{\text{or } L_{c,ss}}$), nor l_o.

$$8 L_{\underset{LRo}{\bullet}} I^2_{\bullet dyo} = m_{\underset{LRdyo}{\bullet}} c^2 = \frac{8\mu_g \pi}{.904} A_{\bullet LRdyo} \tilde{N}_V^2 M^2_{vacdyo} f^2_{\bullet LRdyo}$$

(No $\underline{l_o}$, also!)

,and, $$f_{LRdyo} = \frac{\tilde{N}_Q V_{vady}}{2\pi A_{LRdyo}}$$

Large radius state, rest mass:

drastic reduction of mass: (!)

$$\therefore\; m_{LRdyo} c^2 = \left(\frac{1}{.904}\right) 8\mu_g \pi A_{LRdyo} \tilde{N}_V^2 \left(\frac{\tilde{N}_Q V_{vady}}{2\pi A_{LRdyo}}\right)^2 M^2_{vacdyo}$$

(can be [redacted] ≈0 if A is large!)

and, $$M^2_{vacdyo} = \frac{M^2_{ro}}{\left(1 - \frac{V^2_{vady}}{c^2}\right)}$$

depends on both V_{vady} and A_{LRdyo}!

R. P. Gandall 12-24-2001

Back to steady state: (L6)

$$\left(\frac{\sqrt{m_o}}{\tilde N_Q \tilde N_V}\right) = k_1 = \frac{\sqrt{\frac{m_o}{m_{oe}}}}{\left(\frac{\tilde N_Q \tilde N_V}{\sqrt{m_{oe}}}\right)} = \frac{\left(\sqrt{\frac{m_o}{m_{oe}}}\right)\left(\sqrt{m_{oe}}\right)}{\tilde N_Q \tilde N_V}$$

$$= \frac{\sqrt{m_{oe}}}{k_2}$$

∴ (see fig A70)

$$\left.\frac{M_{vaco} L_{c,ss}}{\sqrt{l_o}}\right|_{\gamma=1} = \overbrace{\frac{\left(\sqrt{\frac{m_o}{m_{oe}}}\right)\left(\sqrt{m_{oe}}\right)}{\tilde N_Q \tilde N_V}\left(\sqrt{\frac{\pi}{2\mu_g}}\right)\left(\frac{\sqrt{\tilde N_Q}}{\sqrt{2S}}\right)}^{k_4}$$

$$L^2_{c,ss} = \frac{1}{1+\frac{k_3}{\gamma N_{wo}}}, \quad N_{wo} = \frac{l_o f_o}{C \tilde N_Q}$$

$$\left.\frac{M^2_{vaco} L^2_{c,ss}}{l_o}\right|_{\gamma=1} = k_4^2$$

$$\frac{M^2_{vaco}}{l_o}\left(\frac{1}{1+k_3\left(\frac{C\tilde N_Q}{f_o l_o}\right)}\right) = k_4^2$$

$$\frac{M^2_{vaco}}{l_o}\left(\frac{1}{1+\frac{k_3 \tilde N_Q}{l_o}\left(\frac{h}{m_o c}\right)}\right) = k_4^2$$

Richard P. Crandall 12-24-2001

∴ steady state, rest frame cont'd: (L7)

Richard P. Crandall 12-24-2001

$$\tilde{M}^2_{vaco} = l_0\left(1 + \frac{\tilde{N}_Q}{l_0}\left(\frac{h}{m_0 c}\right)\left(\frac{(.904)\sqrt{2S}}{2\pi(\sqrt{\tilde{N}_Q})}\right)\right) \cdot \left(\frac{\left(\frac{m_0}{m_{oe}}\right)(m_{oe})}{\tilde{N}_Q^2 \tilde{N}_V^2}\right)\left(\frac{\pi \tilde{N}_Q}{4 \mu_q S}\right)$$

next is L13

Let $l_0 = K_2 K_\ell \left(\frac{h}{m_e c}\right) / \tilde{N}_Q \tilde{N}_{Ve}$ — K_ℓ = [illegible] # of compton wavelengths; reflection, [illegible]

∴

$$\tilde{M}^2_{vaco} = K_\ell\left(\frac{h}{m_0 c}\right)\left(1 + \frac{\tilde{N}_Q (.904)(\sqrt{2S})}{K_\ell 2\pi(\sqrt{\tilde{N}_Q})}\right)\left(\frac{\left(\frac{m_0}{m_{oe}}\right)(m_{oe})\pi}{\tilde{N}_Q \tilde{N}_V 4 \mu_q S}\right)$$

$$M^2_{vaco} = \frac{M^2_{ro}}{\left(1 - \frac{2S}{\tilde{N}_Q}\right)}$$

NEXT PAGE = L13

$$M^2_{ro} = K_\ell\left(\frac{h}{m_0 c}\right)\left(1 - \frac{2S}{\tilde{N}_Q}\right)\left(1 + \frac{\tilde{N}_Q (.904)(\sqrt{2S})}{K_\ell 2\pi(\sqrt{\tilde{N}_Q})}\right)\left(\frac{\left(\frac{m_0}{m_{oe}}\right)(m_{oe})\pi}{(\tilde{N}_Q)\tilde{N}_V^3 4 \mu_q S}\right)$$

NEXT PAGE = L13

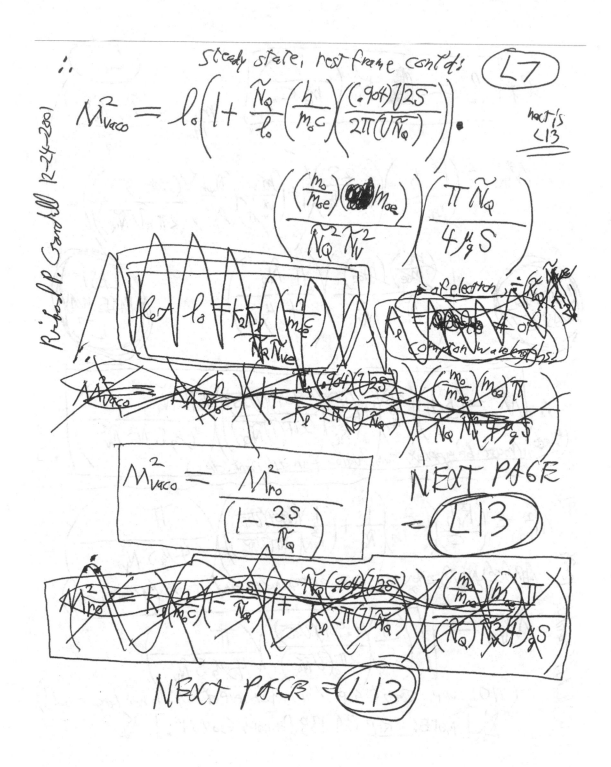

(L13)

let $\boxed{l_0^* = K_\ell^* \left(\frac{h}{m_0 c}\right)}$

$$\frac{m_0}{m_{0e}} = \frac{\tilde{N}_Q^2 \tilde{N}_V^2}{K_2^2}$$

$$\therefore M_{r0}^{*2} = \left(\frac{K_\ell^* h}{m_0 c}\right)\left(1 - \frac{2S}{\tilde{N}_Q}\right)\left(1 + \left(\frac{m_0 c}{K_\ell^* h}\right)\left(\frac{\tilde{N}_Q h}{m_0 c}\right)\left(\frac{(.904)(\sqrt{2S})}{2\pi(\sqrt{\tilde{N}_Q})}\right)\right) \cdot$$

$$\left(\frac{\left(\frac{m_0}{m_{0e}}\right)(m_{0e})}{\tilde{N}_Q^2 \tilde{N}_V^2}\right)\left(\frac{\pi \tilde{N}_Q}{4\mu_q S}\right).$$

(LAST PAGE WAS L7)

(LAST PAGE WAS L7)

$$\therefore \boxed{M_{r0}^{*2} = \left(\frac{K_\ell^* h}{c}\right)\left(1 - \frac{2S}{\tilde{N}_Q}\right)\left(1 + \left(\frac{\tilde{N}_Q}{K_\ell^*}\right)\left(\frac{(.904)(\sqrt{2S})}{2\pi(\sqrt{\tilde{N}_Q})}\right)\right)\left(\frac{\pi \tilde{N}_Q}{4\mu_q S \tilde{N}_Q^2 \tilde{N}_V^2}\right)}.$$

(~~that~~) (must be approx. same value for all particles)

$$\therefore M_{r0}^{*2} = \left(\frac{K_\ell^* h}{c}\right)\left(1 - \frac{2S}{\tilde{N}_Q}\right)\left(\frac{1}{\tilde{N}_Q} + \left(\frac{1}{K_\ell^*}\right)\left(\frac{(.904)(\sqrt{2S})}{2\pi(\sqrt{\tilde{N}_Q})}\right)\right)\left(\frac{\pi}{4\mu_q S \tilde{N}_V^2}\right)$$

~~Since $\tilde{N}_Q \gg 1$~~, we can approximate and say: (proton only)

$$M_{r0}^* = \sqrt{\left(\frac{h}{c}\right)\left(\frac{(.904)(\sqrt{2S})}{2\pi(\sqrt{\tilde{N}_Q})}\right)\left(\frac{\pi}{4\mu_q S \tilde{N}_V^2}\right)}$$

(NO!, we can't use this for electron; 4 not large enough!)

✡ [NOTE: <u>Skip</u> 13A, 13B for now. Go to 14.] ✡

Richard P. Crandall 12-24-2001

LBA

exact formula: [illegible]

SS:

$$M_{rox\,(force,\,exact)} = \sqrt{\frac{\sqrt{\tilde N_q}\,h}{3\tilde N_q C}\left(\frac{4\tilde N_q}{Z(\tilde N_q)}\right)\left(\frac{1}{\tilde N_q \tilde N_v^2 \mu_\ell}\right)\left(1-\frac{1}{\tilde N_q}\right)}$$

and from LB:

$$M^*_{rox\,(exact)} = \sqrt{\left(\frac{\pi h}{2\mu_\ell \tilde N_v^2 C}\right)\left(\frac{K_\ell^*}{\tilde N_q}+\frac{(.904)}{2\pi\sqrt{\tilde N_q}}\right)\left(1-\frac{1}{\tilde N_q}\right)}$$

And these should be equal. Also, $l_0^* = K_\ell^*\left(\frac{h}{m_0 C}\right)$

Solve for K_ℓ^* needed to make 'em equal:

$$\therefore \frac{1}{3\sqrt{\tilde N_q}}\left(\frac{4\tilde N_q}{Z(\tilde N_q)}\right)\left(\frac{1}{\mu_\ell \tilde N_q \tilde N_v^2}\right) = \left(\frac{\pi}{2\mu_\ell \tilde N_v^2}\right)\left(\frac{K_\ell^*}{\tilde N_q}+\frac{(.904)}{2\pi\sqrt{\tilde N_q}}\right)$$

$$\left(\frac{2}{\pi}\right)\left(\frac{4}{3\sqrt{\tilde N_q}}\right)\left(\frac{1}{Z(\tilde N_q)}\right) = \frac{K_\ell^*}{\tilde N_q}+\frac{(.904)}{2\pi\sqrt{\tilde N_q}}$$

$$\frac{8\sqrt{\tilde N_q}}{3\pi}\left(\frac{1}{Z(\tilde N_q)}\right) = K_\ell^* + \frac{(.904)\sqrt{\tilde N_q}}{2\pi}$$

Richard P. Crandall 12-24-2001

L13B

KEEP!

$$K_\ell^* = \left(\frac{\sqrt{\tilde{N}_Q}}{\pi}\right)\left(\left(\frac{8}{3}\right)\left(\frac{1}{Z(\tilde{N}_Q)}\right) - \frac{(.904)}{2}\right)$$

if $\tilde{N}_Q = 4$, (electron) then $Z(\tilde{N}_Q) = 3.82843$, and

$\therefore K_{\ell 4}^* = .1557 = \frac{1}{6.4234}$ Compare to page NSSGAG

$$\therefore \ell_{oe}^* = \left(\frac{1}{6.4234}\right)\left(\frac{h}{m_{oe} C}\right)$$

(Looks Good)

$\therefore M_{roe} = 5.56 \times 10^{-9}$ kg ✓

force exact and exact

$.958\bar{3}$

using the same $K_\ell^* = \frac{1}{6.4234}$ but, let $\tilde{N}_Q = 24$, $\tilde{N}_V = 7$;

$$M_{rop}^* \text{ (exact)} = \sqrt{\left(\frac{\pi h}{2 \mu_g \tilde{N}_Q \tilde{N}_V^2 C}\right)\left(K_\ell^* + \frac{(.904)\sqrt{\tilde{N}_Q}}{2\pi}\right)\left(1 - \frac{1}{\tilde{N}_Q}\right)}$$

$= 5.106 \times 10^{-10}$ ✓ $\therefore$ ratio $= 10.89$ ✓ THE END.

Richard P. Gandhi 12-24-2001

So,

$$M^*_{ro} = \sqrt{\left(\frac{h}{8\mu_g S c}\right)\left(\frac{(.904)(\sqrt{25})}{(\tilde{N}_V^2)(\sqrt{\tilde{N}_Q})}\right)}$$ (L14)

rest mass of the vacuon.

Independent of K_ℓ^* or ℓ_0^*!

where,

$$\ell_0^* = \left(\frac{h}{m_0 c}\right) K_\ell^*$$

better, 'cause we have variables $(\tilde{N}_V, \tilde{N}_Q)$, and no K_ℓ^*! (or ℓ_0^*)

letting $S = \frac{1}{2}$,

$$M^*_{ro} = \sqrt{\left(\frac{h}{4\mu_g c}\right)\left(\frac{(.904)}{(\tilde{N}_V^2)(\sqrt{\tilde{N}_Q})}\right)}$$

for electron: $\tilde{N}_Q = 4$, $\tilde{N}_V = 1$, $K_\ell^* = 1$ (or any value!)

$$M^*_{roe} = .517 \times 10^{-8} \text{ kg}$$

both for ℓ_0^* = anything.

for proton: $\tilde{N}_Q = 24$, $\tilde{N}_V = 7$, $K_\ell^* = 1$

$$M^*_{rop} = .047 \times 10^{-8} \text{ kg}$$

$$\frac{M^*_{roe}}{M^*_{rop}} = 10.956$$

I will call this a "best" value.

Richard P. Crandall 12-24-2001

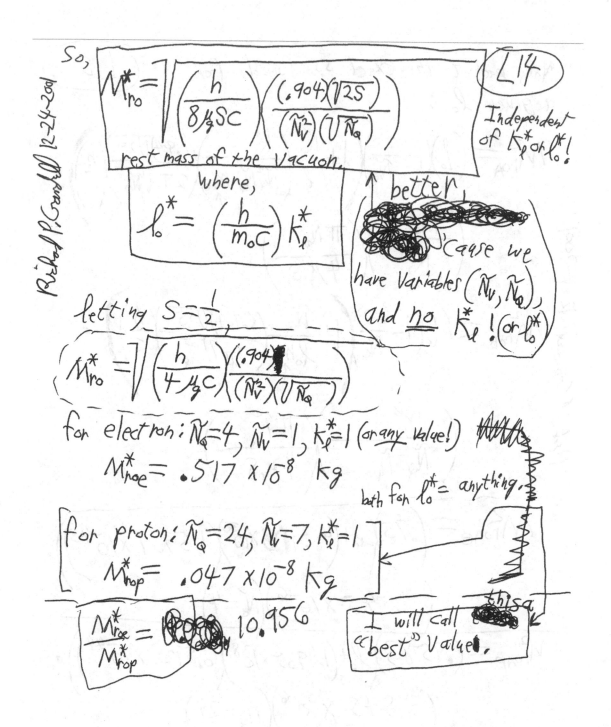

L15

Now, back to the exact S.S. formula for a given l_o:

$$M_{roA}^2 = (l_o)\left(1-\frac{2S}{\tilde{N}_Q}\right)\left(1+\frac{\tilde{N}_Q}{l_o}\left(\frac{K_2^2}{\tilde{N}_Q^2\tilde{N}_V^2 m_{oe}}\right)\left(\frac{(.904)\sqrt{2S}}{2\pi\sqrt{\tilde{N}_Q}}\right)\left(\frac{h}{c}\right)\right)\cdot$$

$$\left(\frac{m_{oe}}{K_2^2}\right)\left(\frac{\pi\tilde{N}_Q}{4\mu_q S}\right)$$

$$M_{roB}^2 = (l_o)\left(1-\frac{2S}{\tilde{N}_Q}\right)\left(1+\frac{\tilde{N}_Q}{l_o m_o}\left(\frac{(.904)\sqrt{2S}}{2\pi\sqrt{\tilde{N}_Q}}\right)\left(\frac{h}{c}\right)\right)\cdot$$

$$\left(\frac{m_o}{\tilde{N}_Q^2\tilde{N}_V^2}\right)\left(\frac{\pi\tilde{N}_Q}{4\mu_q S}\right)$$

Richard P. Gandell 12-24-2001

$$M_{roAe}^2 = (.75)\left(l_o+(4.3908\times10^{30})(1.5951\times10^{-43})\right)\cdot$$

$$(9.545\times10^{-6})(\tilde{N}_Q=4)$$

$$M_{roAp}^2 = (.9583)\left(l_o+(1.4935\times10^{28})(6.5122\times10^{-44})\right)\cdot$$

$$(9.545\times10^{-6})(\tilde{N}_Q=24)$$

L16

$$M^2_{roBe} = (.75)\left(\ell_0 + (4.3908 \times 10^{30})(1.5951 \times 10^{-43})\right) \cdot (9.545 \times 10^{-6})(\tilde{N}_Q = 4) = M^2_{roAe} \text{ exactly.}$$

$$M^2_{roBp} = (.9583)\left(\ell_0 + (1.4371 \times 10^{28})(6.5122 \times 10^{-44})\right) \cdot (9.919 \times 10^{-6})(\tilde{N}_Q = 24)$$

$$\therefore M_{roAe\,exact} = \sqrt{(2.8635 \times 10^{-5})(\ell_0 + 7.0038 \times 10^{-13})}$$

AND

$$\rightarrow M_{roAp\,exact} = \sqrt{(2.1953 \times 10^{-4})(\ell_0 + 9.7260 \times 10^{-16})}$$

$M_{roBe} = M_{roAe}$, exactly.

AND finally,

$$\rightarrow M_{roBp\,exact} = \sqrt{(2.2813 \times 10^{-4})(\ell_0 + 9.3587 \times 10^{-16})}$$

Richard P. Crandall 12-24-2001

L17

Phillip P. Crandall 12-24-2001

electron:

Using $l_o = \lambda_{ce} = \frac{h}{m_{oe} c} = 2.4340 \times 10^{-12}$ m

$\therefore$ $M_{roAe} = M_{roBe} = .947 \times 10^{-8}$ kg (exactly), given the l_o.

proton:

Using $l_o = \lambda_{cp} = \frac{h}{m_{op} c} = 1.3278 \times 10^{-15}$ m

$\therefore$ $M_{roAp} = .071 \times 10^{-8}$ kg

and $M_{roBp} = .072 \times 10^{-8}$ kg (exactly, given the l_o). Regard these as absolute MAXIMUMS.

ratio = 13.24

And, to re-cap:

electron: $M^*_{roe} = .517 \times 10^{-8}$ for l_o = any value, not exact, but close.

proton: $M^*_{rop} = .047 \times 10^{-8}$ for l^*_o = any value. not exact but close.

[illegible]

(again) ratio = 10.956

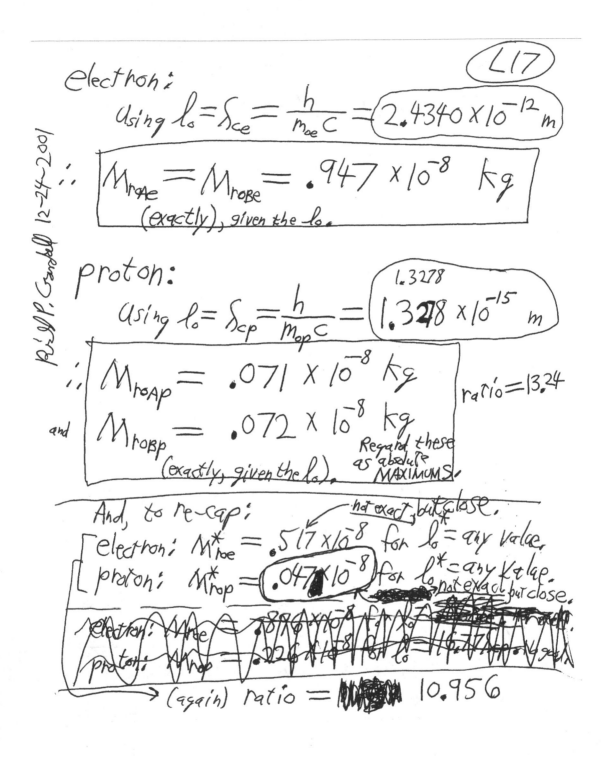

[redacted]

LRDM again: (valid for large "A" only)
(non-SS)

L18

$$M_{LRdyo} = \frac{\left(\frac{8}{.9d4}\right) \mu_q \pi \tilde{N}_v^2 \text{[redacted]} \left(\frac{\tilde{N}_q}{2\pi}\right)^2 \left(\frac{1}{A_{LRdyo}}\right) \left(\frac{V_{vody}}{c}\right)^2 M_{ro}^2}{\left(1-\left(\frac{V_{vody}}{c}\right)^2\right)}$$

For what follows, assume $V_{vody} = \frac{c}{2}$,
and $M_{ro} = .07 \times 10^{-8}$ ← (not actually correct)
(proton: $\tilde{N}_q = 24$, $\tilde{N}_v = 7$)

$\therefore$ non SS $\quad M_{LRdyop} \cong \frac{3.04 \times 10^{-4}\text{[redacted]}1}{A_{dy}}$ (kg) $\quad A_{dy}$ is meters,

Next, $V_{vody} = \frac{c}{2}$, $M_{ro} = .8 \times 10^{-8}$;
(electron: $\tilde{N}_q = 4$, $\tilde{N}_v = 1$)

non SS $\quad M_{LRdyoe} \cong \frac{2.25 \times 10^{-42}}{A_{dy}}$ (kg) $\quad A_{dy}$ is meters,

Drastic reduction of mass due to enlarged ring radius. "Large Radius dynamic mass."

[redacted]

12-24-2001

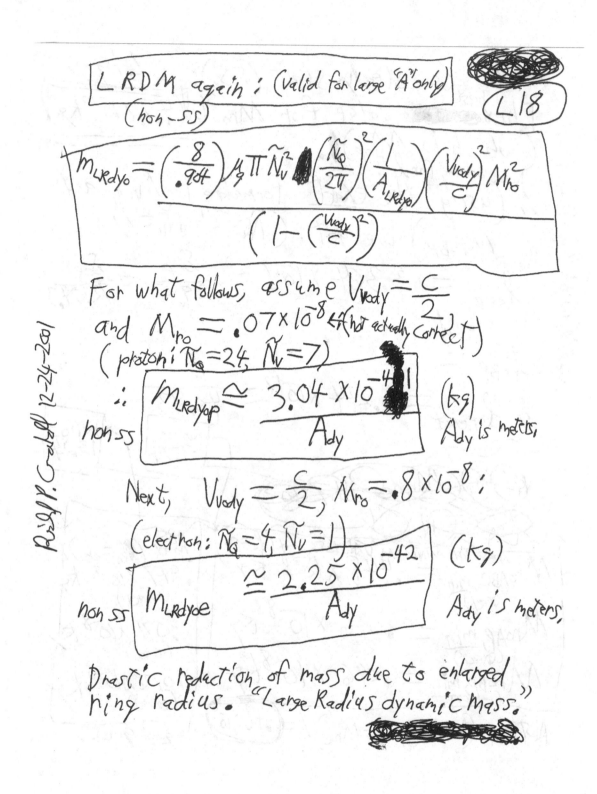

back to steady state (S.S.): (L19)

The best value for M_{ro} is $\boxed{4.7 \times 10^{-10}\ kg}$ (which is "M^*_{rop}").

Now solving for l_o:

∴ (using the exact formulae for M_{roAP} and M_{roBP}) we get: ($\tilde{N}_b = 24$, $\tilde{N}_v = 7$)

$$l_{oAP} = 3.364 \times 10^{-17} = \frac{\delta_{cp}}{39.47} = \frac{\delta_{ce}}{12{,}354.3}$$

and

$$l_{oBP} = \;\; \therefore \;\; l_{op,exact} = 3.244 \times 10^{-17} = \frac{\delta_{cp}}{40.93}$$

ratio = 9.74

ratio = 13.24

ABSOLUTE MINIMUM VALUES: ($l_o = 0$)

$M_{roABE,min}$ (exact) $= .448 \times 10^{-8}$ Kg

$M_{roAP,min}$ (exact) $= .046 \times 10^{-8}$ Kg

$M_{roBP,min}$ (exact) $= .046 \times 10^{-8}$ Kg

ABS. max ($l_o = \delta_c$):

$.947 \times 10^{-8}$ Kg (exact)

$.071 \times 10^{-8}$ Kg (exact)

$.072 \times 10^{-8}$ Kg (exact)

NOTE!: $(M_{roABE,min} + M^*_{roe})/2 = \boxed{.48 \times 10^{-8}}$ = a good guess.

Richard P. Crandall 12-24-2001

L20

Obviously, there are two distinct values for M_{ro}, due to the unbinding energy needed for separating a clump of 7 vacuoms

So, we have M_{ro}^1 and M_{ro}^7 (corresponding to M_{roe} and M_{rop})

$N_o N_v = 24 \times 7 = 12 \times 14 = 168$

$\therefore$ if $N_o = 20$, $N_v = \frac{168}{20} = 8.4$

$N_v^2 = 70.56$

N_o	N_v	$N_v^2 \cdot \Sigma N_o$	X factor (corr.)
24	7	240.85	1
12	14	679.	.595
20	8.4	315.6	.872

Rich M. Goodall 12-24-2001 Not ours!

exact values: $\ell_{op} = \frac{\ell_{cp}}{40.0}$ (exact)

letting M_{roe} = ...

then, $\frac{M_{rope}}{M_{roe}} = 9.73$ (exact) and ℓ_{oe} = ...

$\therefore \ell_{op}$ (exact) $= \frac{\ell_{cp}}{40.0}$ and ℓ_{oe} (exact) $= \frac{\ell_{ce}}{40.0}$ (exact)

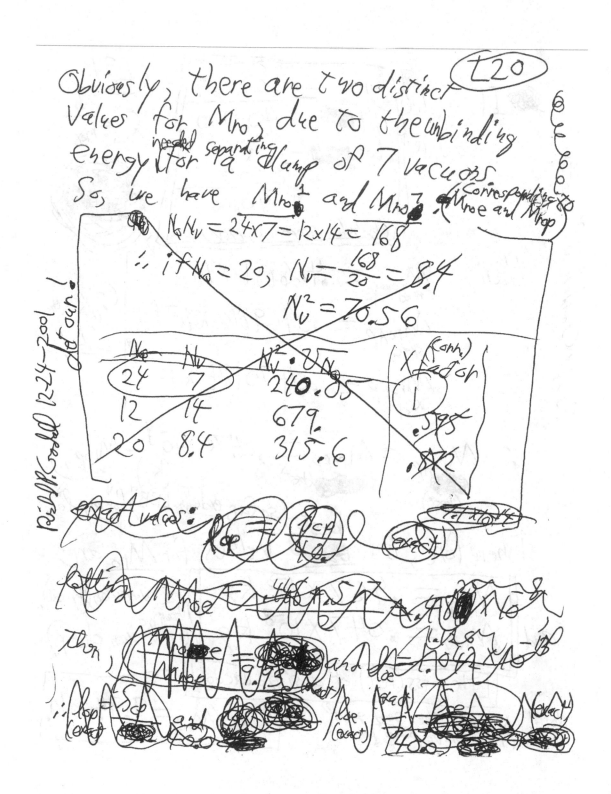

(L24)

If we let $l_{op} = \dfrac{\lambda_{cp}}{40{,}000}$

and let $l_{oe} = \dfrac{\lambda_{ce}}{40{,}000}$

(by helix proportionality)

$\lambda_{cx} = \dfrac{h}{m_{ox} C}$

then, $M_{ro\,Ae\,exact} = .466881 \times 10^{-8}$

$M_{ro\,Ap\,exact} = .0469896 \times 10^{-8}$

$M_{ro\,Bp\,exact} = .0470184 \times 10^{-8}$

(given the l_o!)

$\therefore M_{ro\,Ap\,(best)} = M_{ro\,Bp\,(best)} = .0470 \times 10^{-8}$ (kg) $= M^{*}_{rop}$, for any l_o^{*}.

Therefore, the best (so far, excluding L13B) values for M_{ro} are:

$M_{ro1} = .467 \times 10^{-8}$ kg (1 vacuon) (by itself),

and $M_{ro7} = .047 \times 10^{-8}$ kg (vacuing a 7 vacuon clump)

[The mass (rest) of a single vacuon]

NOTE! $M_{ro1} \approx$ the "good guess," from prev. page! $\therefore$ $M_{ro1}/M_{ro7} = 9.94 \approx 10.$

Richard P. Grondahl 12-24-2001

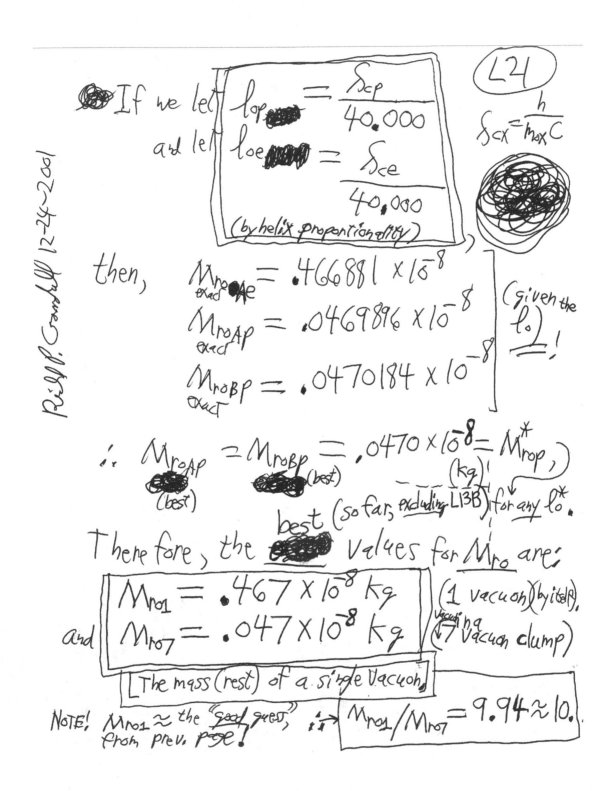

L22

re-writing,

"best" values (but not yet exact),

$$M_{ro1} = 4.7 \times 10^{-9}\ kg \quad \text{(i.e. electron-like)}$$

$$M_{ro7} = 4.7 \times 10^{-10}\ kg \quad \text{(i.e. proton-like)}$$

ratio:

$$\therefore \quad M_{ro1} / M_{ro7} = 10.$$

$$\text{and } l_{ox} = \frac{\delta_{cxo}}{40} = \frac{h}{40 m_{ox} c} = \frac{c}{40 f_{ox}}$$

The "40" here must later change to "6.4234", which is EXACT. (based upon balance of forces). (see page L13B). There, ratio = 10.89.

Now, NON-EQUILIB:

$$L_{dyox} = \frac{\mu_q \pi A_{dy}^2}{(.904) A_{dy} + \frac{\delta_{cxox}}{40}}$$

(All cases of radius) (non-equilib.)

$\frac{\delta_{cxox}}{40}$ → Constant, no matter what.

$$i_{dyox} = \tilde{N}_q \tilde{N}_v f_{rotdyo} M_{vacdyox}$$ (see fig A7R)

"X" means either: "e" = electron, "p" = proton.

$$\therefore \quad m_{dyox} c^2 = 8 L_{dyox} i_{dyox}^2$$

$$\text{and } f_{rotdyo} = \frac{\omega_{rotdyo}}{2\pi} = \frac{V_{rody}}{2\pi A_{dy}}$$

Richard P. Crandall 12-24-2001

NON-S.S. (NON-EQUILIB.) (L23)

$$m_{dyox} = \frac{\left(\frac{8}{C^2}\right) \mu_g \pi \cancel{A_{dy}^2} \tilde{N}_Q^2 \tilde{N}_V^2 V_{vody}^2 \quad M_{rox}^2}{\left((.904)A_{dy} + \frac{\lambda_{cxo}}{40}\right) 4\pi^2 \cancel{A_{dy}^2} \left(1 - \frac{V_{vody}^2}{C^2}\right)}$$

dynamic rest mass non equilib. (ANY A, V) for particle "x". All cases. (Just a function of "A" and "V"!)

$$m_{dyox} = \frac{2 \mu_g \tilde{N}_Q^2 \tilde{N}_V^2 \left(\frac{V_{vody}^2}{C^2}\right) M_{rox}^2}{\left((.904)A_{dy} + \frac{\lambda_{cxo}}{40}\right)(\pi)\left(1 - \frac{V_{vody}^2}{C^2}\right)}$$

dynamic rest mass. Valid for any radius and velocity

note, (S.S.), $A_{ox} = \frac{\left(\sqrt{2S\tilde{N}_Q}\right) C}{2\pi f_{ox}} = \frac{\left(\sqrt{2S\tilde{N}_Q}\right) \hbar}{m_{ox} C}$

and, $\lambda_{cxo} = \frac{h}{m_{ox} C}$, so

$$A_{ox} = \frac{\left(\sqrt{2S\tilde{N}_Q}\right) \lambda_{cxo}}{2\pi}$$

equilibrium radius

Richard P. Crandall 12-24-2001

(L24)

$$\delta_{cxo} = \frac{h}{m_{ox} c} \quad \text{(again)}$$

"gr" = group velocity

$$m'_{dyx} = \frac{2 \mu_q \tilde{N}_Q^2 \tilde{N}_V^2 \left(\frac{V_{vady}^2}{c^2}\right) M_{rox}^2 \gamma_{gr}^2}{\left[\left(.904\right)\left(40 A_{dy}\right)\left(\frac{\sqrt{25 \tilde{N}_Q}}{2\pi A_{ox}}\right) + 1\right]\left(\gamma_{gr} \frac{\delta_{cxo}}{40}\right)\left(\pi\right)\left(1 - \frac{V_{vady}^2}{c^2}\right)}$$

$\therefore$

$$L_{dycx}^2 = \frac{1}{1 + \frac{(.904)(40)\left(\sqrt{25 \tilde{N}_Q}\right) A_{dy}}{2\pi A_{ox}}}$$

inductance correction factor definition.

and,

$$m'_{dyx} = \frac{2 \mu_q \tilde{N}_Q^2 \tilde{N}_V^2 \left(\frac{V_{vady}^2}{c^2}\right) M_{rox}^2 \gamma_{gr}^2 \left(L_{dycx}\right)^2}{\left[\gamma_{gr}\left(\frac{\delta_{cxo}}{40}\right)\right]\left(\pi\right)\left(1 - \frac{V_{vady}^2}{c^2}\right)}$$

dynamic mass, exact, for any V_{vady} and A_{dy}. (and γ).

and $V_{vady}^2 = A_{dy}^2 \omega_{rotdy}^2$

$$\gamma_{gr}^2 = \left(\frac{1}{\sqrt{1 - \frac{V_g^2}{c^2}}}\right)^2, \quad V_g = V_{group} \neq V_{vady}!$$

Richard P. Crandall 12-24-2001

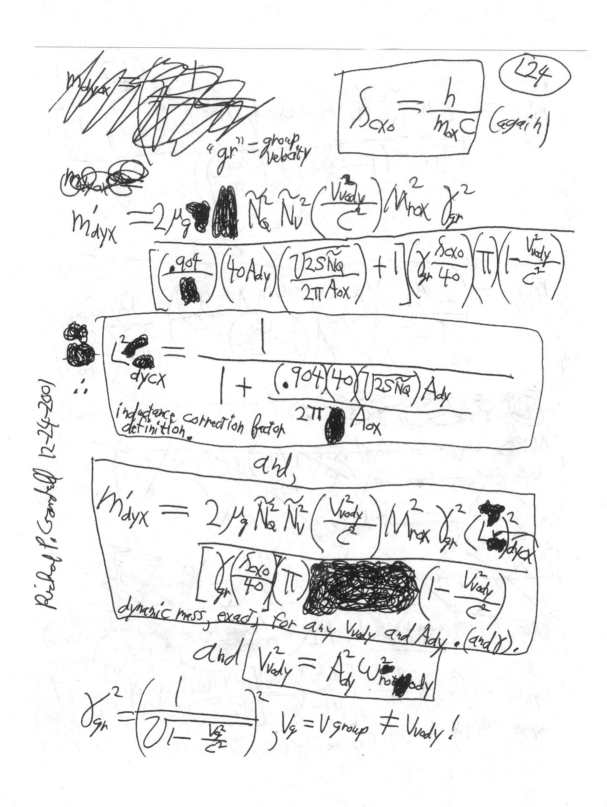

(L25)

$\therefore$

$$m'_{dyx} = \frac{2\mu_q \tilde{N}_Q^2 \tilde{N}_V^2 A_{dy}^2 \omega_{rotody}^2 M_{rox}^2 \gamma_{gr}^2 \left(L_{dycx}^2\right)}{\pi C^2 \left(1 - \frac{V_{vody}^2}{C^2}\right) \underbrace{\left(\gamma_{gr}\left(\frac{\lambda_{cxo}}{4o}\right)\right)}_{= l_o'}}$$ (Last page)

Richard P. Crandall 12-24-2001

$$m'_{dyx} = \left(\frac{2\mu_q}{\pi C^2}\right)\left(L_{dycx}^2\right)\left(\frac{\tilde{N}_Q^2 \tilde{N}_V^2 M_{rox}^2 \gamma_{gr}^2}{\left(1 - \frac{V_{vody}^2}{C^2}\right)}\right)\left(\frac{A_{dy}^2 \omega_{rotody}^2}{\gamma_{gr}\left(\frac{\lambda_{cxo}}{4o}\right)}\right)$$

~~[illegible]~~

~~Note that rotational K.E. (R_{KE}) in terms of rotational Inertia (R_I) is~~

~~$R_{KE} = \frac{1}{2} R_I \omega_{rot}^2$.~~

~~[(Also, angular momentum, P_A, is:) $P_A = R_I \omega_{rot}$]~~

~~[(Also, for ring, $R_I = M_{tot} r^2$]~~

Formula NOT GOOD for Large A_{dy}.

$$m'_{dyx} C^2 = \left(L_{dycx}^2\right)\left(\frac{8\mu_q}{\pi}\right)\left[\frac{\tilde{N}_Q \tilde{N}_V \gamma_{gr} M_{rox}}{\sqrt{1 - \frac{V_{vody}^2}{C^2}}}\right]\left[\frac{1}{2}\left(\frac{\tilde{N}_Q \tilde{N}_V \gamma_{gr} M_{rox}}{2\sqrt{1 - \frac{V_{vody}^2}{C^2}}}\right)\left(A_{dy}^2\right)\left(2\pi f_{rotody}\right)^2\right] \Big/ \left(\frac{\gamma_{gr}\lambda_{cxo}}{4o}\right)$$

(compare with fig A7S!) They are identical. (if $\gamma = 1$ and s.s.)

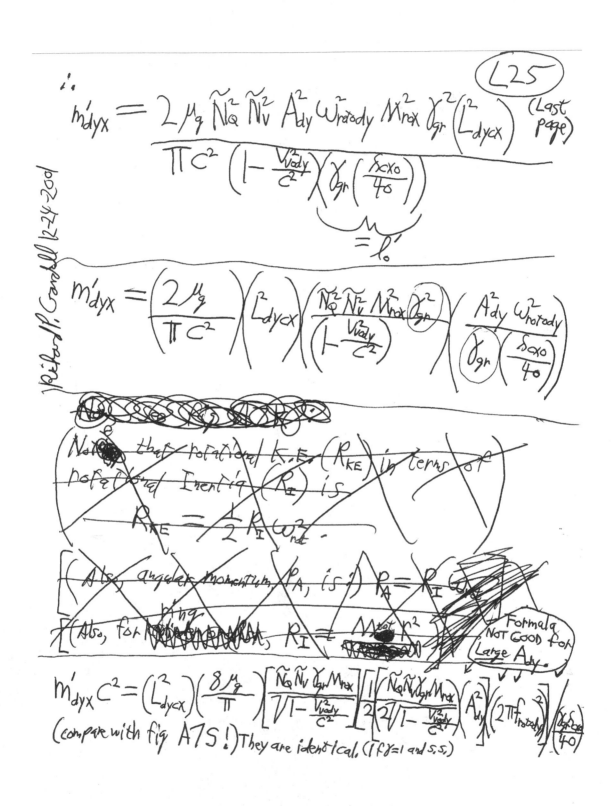

[OUTWARD Force] (on a disk)

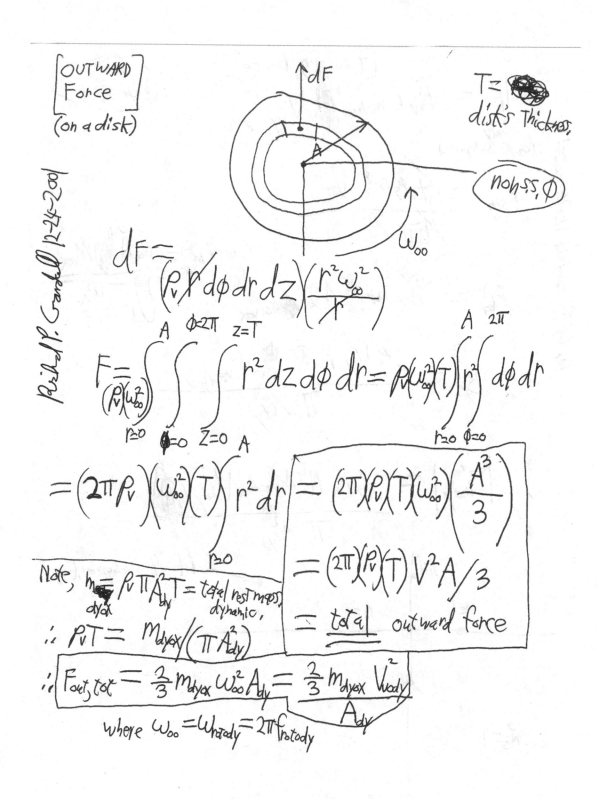

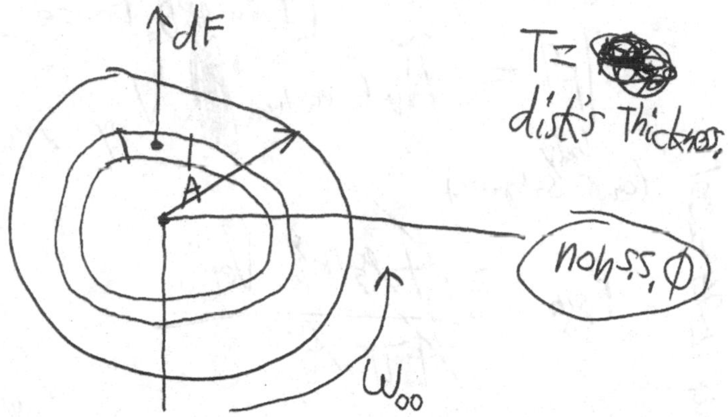

T= disk's Thickness,

Richard P. Crandall 12-24-2001

$$dF = (\rho_v \not{r}\, d\phi\, dr\, dz)\left(\frac{r^2\omega_{oo}^2}{\not{r}}\right)$$

$$\frac{F}{(\rho_v)(\omega_{oo}^2)} = \int_{r=0}^{A}\int_{\phi=0}^{\phi=2\pi}\int_{z=0}^{z=T} r^2\, dz\, d\phi\, dr = \rho_v(\omega_{oo}^2)(T)\int_{r=0}^{A} r^2 \int_{\phi=0}^{2\pi} d\phi\, dr$$

$$= (2\pi\rho_v)(\omega_{oo}^2)(T)\int_{r=0}^{A} r^2\, dr = (2\pi)(\rho_v)(T)(\omega_{oo}^2)\left(\frac{A^3}{3}\right)$$

$$= (2\pi)(\rho_v)(T)\, V^2 A/3$$

= **total** outward force

Note, $m_{dyox} = \rho_v \pi A_{dy}^2 T$ = total rest mass, dynamic,

$\therefore \rho_v T = m_{dyox}/(\pi A_{dy}^2)$

$$\therefore F_{out,tot} = \frac{2}{3} m_{dyox}\, \omega_{oo}^2 A_{dy} = \frac{\frac{2}{3} m_{dyox} V_{body}^2}{A_{dy}}$$

where $\omega_{oo} = \omega_{rotody} = 2\pi f_{rotody}$

Richard P. Crandall 12-24-2001

[INWARD Force]:

no. 55

$$|F'_{indy}| = \tilde{N}_V \, M_{vacdyox} \, V_{vody} \, B_{gin}$$

(on a subquark)

$$B_{gin} = \frac{\not{4} \mu_g m^*_{tot} V_{el}}{\not{4} \pi A_{dy}^2}$$

and $m^*_{tot} = \dfrac{\tilde{N}_Q \tilde{N}_V M_{vox}}{\sqrt{1 - \frac{V_{vody}^2}{c^2}}}$

$$m^*_{tot} = \tilde{N}_Q \, \tilde{N}_V \, M_{vacdyox}$$

$$\therefore \; B_{gin} = \frac{\mu_g V_{vody} \tilde{N}_Q \tilde{N}_V M_{vacdyox}}{\pi A_{dy}^2}$$

$$\therefore \; |F'_{indy}| = \frac{\mu_g \tilde{N}_Q \tilde{N}_V^2 M_{vacdyox} V_{vody}^2}{\pi A_{dy}^2}$$

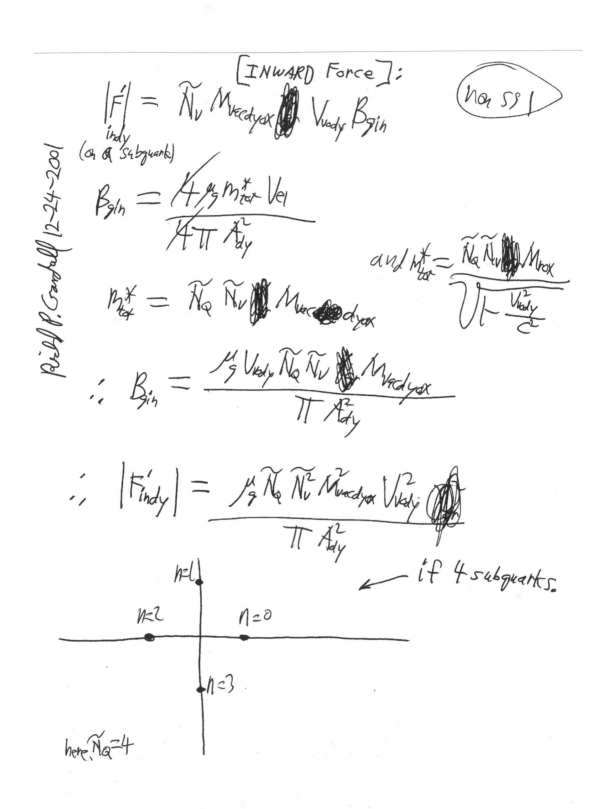

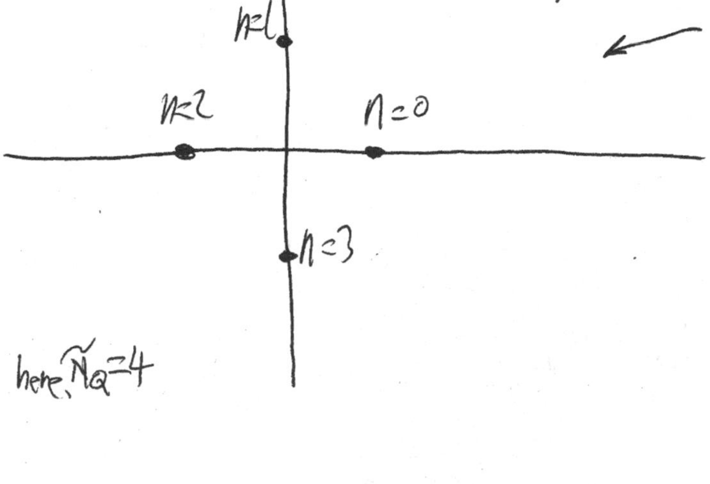

← if 4 subquarks.

here, $\tilde{N}_Q = 4$

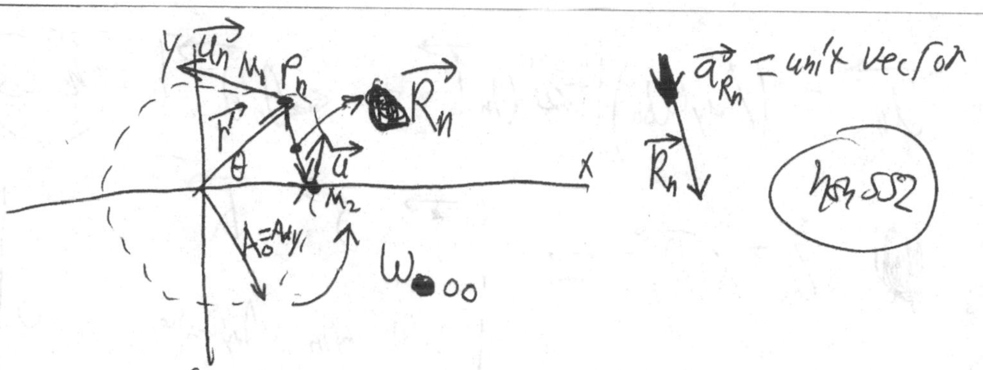

Reid D. Crandall 12-24-2001

$$\theta_n = \left(\frac{2\pi}{\tilde{N}_q}\right) n \quad , \quad n = 0,..,\tilde{N}_q - 1$$

$$\vec{r}_n = \left(A_{dy}\cos\theta_n\right)\vec{i} + \left(A_{dy}\sin\theta_n\right)\vec{j}$$

$$\vec{R}_n = \left(A_{dy} - A_{dy}\cos\theta_n\right)\vec{i} - \left(A_{dy}\sin\theta_n\right)\vec{j}$$

$$\vec{B}_{gn} = -4\mu_g \vec{U}_n \times \frac{M_1 \tilde{N}_v}{4\pi R_n^2}\vec{a}_{R_n} \quad , \quad \vec{F}_n = M_2 \vec{U} \times \vec{B}_{gn}$$

and $M_1 = M_2 = M_{ucdyox}$

and $\vec{U} = A_{dy}\,\omega_{oo}\,\vec{j} + 0\vec{i}$

$$\vec{a}_{R_n} = \frac{1}{|\vec{R}_n|}\vec{R}_n \quad , \quad \vec{F}_n = \frac{-4\mu_g M_1 M_2 \tilde{N}_v^2}{4\pi R_n^2}\left(A_{dy}\,\omega_{oo}\,\vec{j} \times \left(\vec{U}_n \times \frac{\vec{R}_n}{|\vec{R}_n|}\right)\right)$$

Richard P. Gandolf 12-24-2001

hcns553

$$\vec{U_n} = (A_{dy}\omega_{oo})\left[-\sin\theta_n \vec{i} + \cos\theta_n \vec{j}\right]$$

$$\vec{U_n} \times \vec{R_n} = \begin{vmatrix} \vec{i} & \vec{j} & \vec{k} \\ -A_{dy}\omega_{oo}\sin\theta_n & A_{dy}\omega_{oo}\cos\theta_n & 0 \\ A_{dy} - A_{dy}\cos\theta_n & -A_{dy}\sin\theta_n & 0 \end{vmatrix}$$

$$= \left(A_{dy}^2\omega_{oo}\sin^2\theta_n - A_{dy}^2(1-\cos\theta_n)(\omega_{oo}\cos\theta_n)\right)\vec{k}$$

$$= (A_{dy}^2)(\omega_{oo}\sin^2\theta_n + \omega_{oo}\cos^2\theta_n - \omega_{oo}\cos\theta_n)\vec{k}$$

$$= (A_{dy}^2\omega_{oo})(1-\cos\theta_n)\vec{k}$$

$$(A_{dy}\omega_{oo}\vec{j}) \times (\vec{U_n} \times \vec{R_n}) = \begin{vmatrix} \vec{i} & \vec{j} & \vec{k} \\ 0 & A_{dy}\omega_{oo} & 0 \\ 0 & 0 & (A_{dy}^2)\omega_{oo}(1-\cos\theta_n) \end{vmatrix}$$

$$= (A_{dy}\omega_{oo})\left((A_{dy}^2)\omega_{oo}(1-\cos\theta_n)\right)\vec{i}.$$

$$\vec{F_n} = -\frac{\mu_q M_{vacdyox}^2 \tilde{N}_v A_{dy}^3 \omega_{oo}^2 (1-\cos\theta_n)}{\pi R_n^3}\vec{i} + 0\vec{j} + 0\vec{k}$$

Item 54

$$R_n = (A_{dy})\sqrt{(1-\cos\theta_n)^2 + (\sin\theta_n)^2}$$

$$= (A_{dy})\sqrt{1 - 2\cos\theta_n + \cos^2\theta_n + \sin^2\theta_n}$$

$$= (\sqrt{2})(A_{dy})\sqrt{1-\cos\theta_n}$$

Richard P. Campbell 12-24-2001

$$\therefore F_n = [redacted] |\vec{F_n}| =$$

$$F_n = \frac{\mu_q M_{vacdyox}^2 \tilde{N}_v^2 \omega_{oo}^2 (1-\cos\theta_n)}{\pi\left((\sqrt{2})\sqrt{1-\cos\theta_n}\right)^3}$$

$n \neq 0$
$n \neq \tilde{N}_q$

$$\theta_n = \frac{2\pi n}{\tilde{N}_q}$$

[redacted]

$$F_{findyo\ [redacted]on\ (Each\ [redacted]\ Subsegment)} = \sum_{n=1}^{\tilde{N}_q - 1} F_n$$

$$= \frac{\mu_q M_{vacdyox}^2 \omega_{oo}^2 \tilde{N}_v^2}{\pi(\sqrt{2})^3}\sum_{n=1}^{\tilde{N}_q - 1}\left(\frac{1}{\sqrt{1-\cos\left(\frac{2\pi n}{\tilde{N}_q}\right)}}\right)$$

Richard P. Grandall 12-24-2001

$$2\left(\sin^2 A\right) = 1 - \cos(2A)$$

from tables, no. 555

$$\therefore \left(\frac{1}{\sqrt{1-\cos\left(\frac{2\pi n}{\tilde{N}_q}\right)}}\right) = \left(\frac{1}{\sqrt{2}}\right)\left(\frac{1}{\sin\left(\frac{\pi n}{\tilde{N}_q}\right)}\right)$$

$$\sum_{n=1}^{\tilde{N}_q - 1} \frac{1}{\sin\left(\frac{\pi n}{\tilde{N}_q}\right)} = ?$$

$$\sum_{n=1}^{\tilde{N}_q - 1} e^{-j\left(\frac{2\pi n}{2\tilde{N}_q}\right)} = \sum \cos\theta - j \sum \sin\theta$$

$$\sum \frac{\quad}{\sin(\,)\sin(\,)\sin(\,)\cdots}$$

$$\frac{1}{\sin\theta_n}$$

$$\frac{1}{e^{j\theta_n}} = \frac{1}{\cos\theta_n + j\sin\theta_n} =$$

$$= \frac{\cos\theta_n - j\sin\theta_n}{\cos^2\theta_n + \sin^2\theta_n} = \cos\theta_n - j\sin\theta_n$$

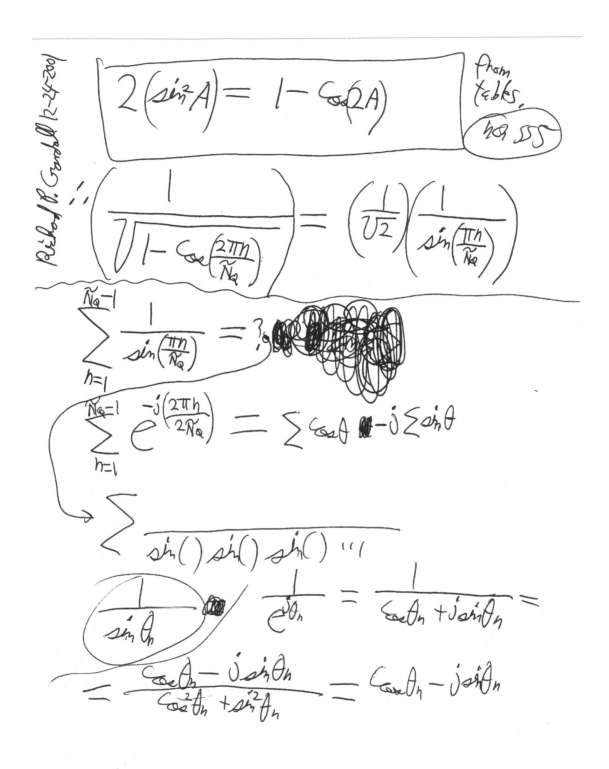

---

s.s. electron: $(\tilde{N}_q = 4, \tilde{N}_v = 1)$ $V_{vody} = c/\sqrt{\tilde{N}_q}$

$= 1.495 \times 10^{8}$

$F_{indyo} = F_{outdyo} = 4.38 \times 10^{-3}$ $A_o = \sqrt{\tilde{N}_q}\left(\frac{\hbar}{m_o c}\right)$

$= 7.748 \times 10^{-13}$

$\therefore M_{vacodye} = .642 \times 10^{-8}$

5A non s.s. [redacted]

keep!

$\therefore M_{ro1} = (M_{vacodye})\sqrt{1 - \frac{1}{\tilde{N}_{qe}}}$

$= \boxed{5.56 \times 10^{-9}\ kg}$ (exact)

(see pg. Non SS 6)

Richard P. Grandall 12-24-2001

s.s. proton: $(\tilde{N}_q = 24, \tilde{N}_v = 7)$ $V_{vody} = c/\sqrt{\tilde{N}_q}$

$= .6103 \times 10^{8}$

$F_{indyo} = F_{outdyo} = 166.94$ (no exponent)

$A_o = \sqrt{\tilde{N}_q}\left(\frac{\hbar}{m_o c}\right)$

$\therefore M_{vacodyp} = 1.6136 \times 10^{-10}$ $= 1.035 \times 10^{-15}$

$\therefore M_{ro7} = (M_{vacodyp})\sqrt{1 - \frac{1}{\tilde{N}_{qp}}}$

$\frac{M_{ro1}}{M_{ro7}} = 35.19$

don't use proton. $= \boxed{1.58 \times 10^{-10}\ kg}$ way off!

Some Values:

$\tilde{N}_Q$	$Z(\tilde{N}_Q)$	$\frac{Z(\tilde{N}_Q)}{4\tilde{N}_Q}$	$\frac{4\tilde{N}_Q}{Z(\tilde{N}_Q)}$
24	50.473	.52576	1.90201
12	19.9358	.415329	2.40773
8	11.2195	.350608	2.85219
4	3.82843	.239277	4.17926
2	1	.125	8
3	2.3094	.19245	5.19615
48	122.132	.636106	1.57206

Richard P Crandall 12-24-2001

THE "Z( )" FUNCTION (see next page)

define function

$$Z(\tilde{N}_Q) = \sum_{n=1}^{\tilde{N}_Q - 1} \frac{1}{\sin\left(\frac{\pi n}{\tilde{N}_Q}\right)}$$

non 55,6

next page is (6AA)!

$$\therefore \quad F_{indyo} = \frac{\mu_q M_{vacdyox}^2 \omega_{oo}^2 \tilde{N}_V^2}{4\pi} Z(\tilde{N}_Q)$$

(on each subquark)

$\omega_{oo} = \omega_{rotdyo}$

$$\therefore \quad F_{indyo} = \frac{\tilde{N}_Q \tilde{N}_V^2 M_{vacdyo}^2 \mu_q V_{vsdy}^2}{\pi A_{dy}^2} \left(\frac{Z(\tilde{N}_Q)}{4\tilde{N}_Q}\right)$$

(on each subquark)

$$F_{outdyo} = \left(\frac{2}{3}\right)\left(\frac{1}{\tilde{N}_Q}\right)\left(\frac{m_{dyox} V_{vsdy}^2}{A_{dy}}\right)\left(\frac{A_{dy}}{A_{dy}}\right)$$

(on each subquark)

Both valid for any radius, Large or Small. (and any V).

Richard P. Gandell 12-24-2001

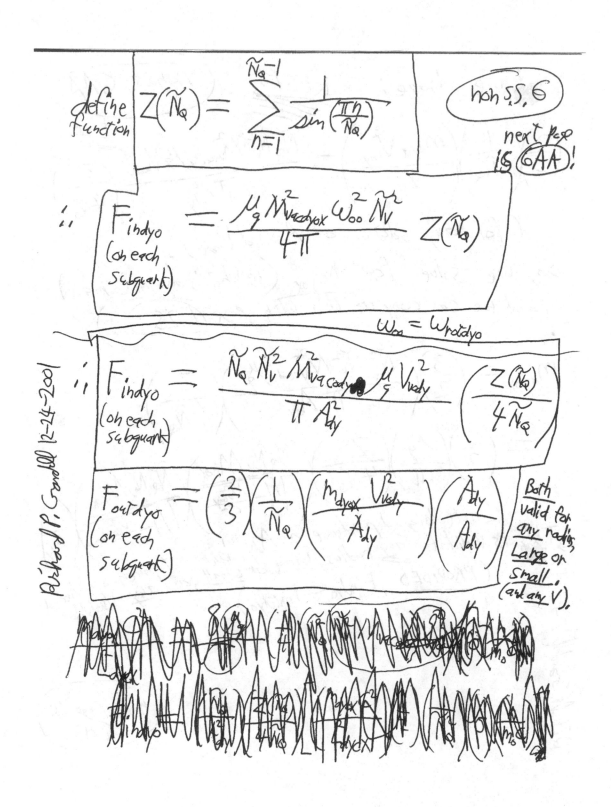

∴ we have: KEEP! (nonS.S. 6AA)

Keep!

$$\left(\frac{2}{3}\right)\left(\frac{1}{\tilde{N}_Q}\right)\left(\frac{m_{dyox} V_{vody}^2}{A_{dy}}\right) = \frac{\tilde{N}_Q \tilde{N}_V^2 M_{vacody}^2 \mu_q V_{vody}^2}{\pi A_{dy}^2}\left(\frac{Z}{4\tilde{N}_Q}\right)$$

(for any solution of A_{dy} and V_{vody})

So, we solve for m_{dyox} (ie what is dynamic mass):
(and we can compare this with page (L25).)

∴

$$m_{dyox} = \left(\frac{3}{2}\right)\left(\frac{\tilde{N}_Q^2 \tilde{N}_V^2 M_{vacody}^2 \mu_q}{\pi A_{dy}}\right)\left(\frac{Z}{4\tilde{N}_Q}\right)$$

$$= \left(\frac{3}{2}\right)\left(\frac{\mu_q}{\pi}\right)\left(\frac{Z}{4\tilde{N}_Q}\right)\left(\frac{\tilde{N}_Q \tilde{N}_V M_{rox}}{\sqrt{1-\frac{V_{vody}^2}{c^2}}}\right)\left(\frac{\tilde{N}_Q \tilde{N}_V M_{rox}}{\sqrt{1-\frac{V_{vody}^2}{c^2}}}\right)\left(\frac{1}{A_{dy}}\right)$$

[note big dependence on A_{dy}, but not much on V_{vodye}]

⇒ This is valid for any radius, Large or small, and any V_{vody}!!
(PROVIDED, $F_{indyo} = F_{outdyo}$).

Richard P. Crandall 12-24-2001

(next page is 6AB)

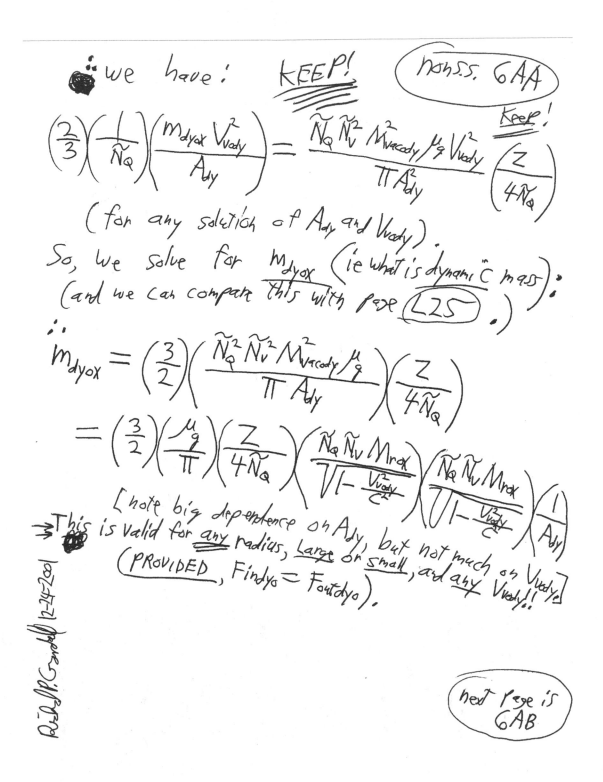

KEEP !!

(non 5.5. 6AB)

from page L25, we may write:

[this is valid for any A_{dy} and any V_{vady} such that A_{dy} is SMALL]

[and, we replaced $\left(\frac{\gamma_{gn} \delta_{cxo}}{4o}\right)$ with l_o]; (and, let $\gamma_{gn} = 1$) (then, $m'_{dyx} = m_{dyox}$)

Note: $f_{ody} \neq f_o$!

$$m_{dyox} = \frac{\left(L^2_{dycx}\right)\left(\frac{8\mu_q}{\pi}\right)}{l_o}\left[\frac{\tilde{N}_Q \tilde{N}_V M_{nox}}{\sqrt{1-\frac{V^2_{vady}}{c^2}}}\right]\left[\left(\frac{1}{2}\right)\left(\frac{1}{2}\right)\frac{\tilde{N}_Q \tilde{N}_V M_{nox}}{\sqrt{1-\frac{V^2_{vady}}{c^2}}}\right]\left(\frac{4\pi^2 A^2_{dy} f^2_{ody}}{c^2 \tilde{N}^2_Q}\right)$$

(inductance based)

$$\overset{(must)}{=} \left(\frac{3}{2}\right)\left(\frac{8\mu_q}{\pi}\right)\left(\frac{Z}{4\tilde{N}_Q}\right)\left(\frac{\tilde{N}_Q \tilde{N}_V M_{nox}}{\sqrt{1-\frac{V^2_{vady}}{c^2}}}\right)\left(\frac{\tilde{N}_Q \tilde{N}_V M_{nox}}{\sqrt{1-\frac{V^2_{vady}}{c^2}}}\right)\left(\frac{1}{8A_{dy}}\right)$$

(balance of forces based)

∴ [and, we ultimately want to solve for "l_o"];

$$\frac{\left(L^2_{dycx}\right)}{l_o}\left(\frac{\pi^2 A^2_{dy} f^2_{ody}}{c^2 \tilde{N}^2_Q}\right) = \left(\frac{3}{2}\right)\left(\frac{Z}{4\tilde{N}_Q}\right)\left(\frac{1}{8A_{dy}}\right)$$

$$\therefore \boxed{l_o = \left(\frac{64}{3}\right)\left(L^2_{dycx}\right)\left(\frac{\pi^2 A^3_{dy} f^2_{ody}}{c^2 \tilde{N}_Q}\right)\left(\frac{1}{Z}\right)}$$

Richard P. Grandall 12-24-2001

Keep!

non s.s. 6AC

from pg L24, (and L23):
(set $\gamma=1$)

$$L^2_{dycx} = \frac{1}{1 + \frac{(.904)A_{dy}}{\ell_o}} = \frac{1}{1 + \frac{(.904)A_{dy}}{k\,\delta_{cxo}}}$$

where $\ell_o = k\,\delta_{cxo}$

~~$$= \frac{1}{1 + \frac{(.904)\ell_{dy}}{k} \cdot \frac{V_{25}/V_{\tilde{N}_Q}}{3\pi A_{ox}}}$$~~ (not needed).

Since $A^2_{dy} f^2_{ody} = \left(\frac{1}{2\pi}\right)^2 A^2_{dy} (2\pi f_{ody})^2$

$$= \left[\frac{A_{dy}\tilde{N}_Q}{2\pi} \cdot \frac{2\pi f_{ody}}{\tilde{N}_Q}\right]^2, \left(\text{and since } \frac{\omega_{ody}}{\tilde{N}_Q} = \omega_{rotody}\right)$$

$$= \left[A_{dy}\,\omega_{rotody}\right]^2 \left(\frac{\tilde{N}_Q^2}{4\pi^2}\right) = \left(\frac{\tilde{N}_Q^2}{4\pi^2}\right) V^2_{vody}$$

$$\therefore \quad k\,\delta_{cxo} = \left(\frac{64}{3}\right)\left[\frac{1}{1 + \frac{(.904)A_{dy}}{k\,\delta_{cxo}}}\right]\left(\frac{\pi^2 A_{dy}}{C^2 \tilde{N}_Q}\right)\left(\frac{\tilde{N}_Q^2 V^2_{vody}}{4\pi^2}\right)\left(\frac{1}{Z}\right)$$

(we must solve this for k)

Richard P. Campbell 12-24-2001

Keep!

non S.S. 6AD

Richard P. Campbell 12-24-2001

∴

$$l_o = \left(\frac{16}{3}\right)\left[\frac{1}{\frac{1}{A_{dy}} + \frac{(.904)}{l_o}}\right]\left(\frac{\tilde{N}_Q V_{vody}^2}{C^2 Z}\right)$$

(solve for l_o)

define: $\left(\frac{1}{\beta}\right) = \frac{16\,\tilde{N}_Q V_{vody}^2}{3\,C^2 Z()}$

must be ≥ 1 !! (so V_{vody} must be close to C)

∴

$$\frac{1}{l_o} = \beta\left(\frac{1}{A_{dy}} + \frac{(.904)}{1}\frac{1}{l_o}\right)$$

∴

$$\left(1 - \left[(.904)\beta\right]\right)\left(\frac{1}{l_o}\right) = \frac{\beta}{A_{dy}}$$

∴

$$l_{ody} = A_{dy}\left(\left(\frac{1}{\beta}\right) - (.904)\right)$$

$$= \left[\left(\frac{16\,\tilde{N}_Q V_{vody}^2}{3\,C^2 Z()}\right) - (.904)\right] A_{dy}$$

good for any SMALL A_{dy}.

Keep !

Vsh S,S, 6A[redacted]E

Richard P. Crandall 12-24-2001

$$\therefore \left(\frac{A_{dy}}{l_0}\right) = \frac{\beta}{1-(.904)\beta}$$

$$\therefore L^2_{dy\alpha} = \frac{1}{1+\frac{(.904)\beta}{1-(.904)\beta}} = 1-(.904)\beta$$

$$\therefore \boxed{L^2_{dy\alpha} = 1-\left[(.904)\left(\frac{3C^2 Z()}{16\tilde{N}_Q V^2_{vody}}\right)\right]}$$

$$\boxed{\frac{L^2_{dy\alpha}}{l_{ody}} = \frac{1}{l_0+(.904)A_{dy}} \simeq \left(\frac{3C^2 Z()}{16\tilde{N}_Q V^2_{vody}}\right)\left(\frac{1}{A_{dy}}\right)}$$

(Good for any small A_{dy}). (and any V_{vody})

$$L^2_{\alpha,ss} = 1-(.904)\left(\frac{3C^2 Z}{16\tilde{N}_Q}\right)\left(\frac{\tilde{N}_Q}{C^2}\right)$$

$$\therefore \boxed{L^2_{\alpha,ss} = 1-\left(\left(\frac{3}{16}\right)(.904)Z()\right)}$$

Richard P. Gardell 12-24-2001

Keep!!

non S.S. GAF

$$\left(\frac{16\tilde{N}_Q V_{Vody}^2}{3c^2 Z()}\right) = \left(\frac{16\tilde{N}_Q\left(\frac{V_{Vody}}{V_{Vo}}\right)^2}{3c^2 Z()}\right)\left(\frac{(2S)c^2}{\tilde{N}_Q}\right)$$

$$= \left(\frac{16(2S)}{3}\right)\left(\frac{1}{Z()}\right)\left(\frac{V_{Vody}}{V_{Vo}}\right)^2$$

and,

$$A_{dy} = \left(\frac{A_{dy}}{A_{ox}}\right)\left(\frac{\sqrt{2S\tilde{N}_Q}\;\lambda_{cxo}}{2\pi}\right)$$

(See pg. ∠23)

$$\lambda_{cxo} = \frac{h}{m_{ox}C}$$ by definition (∠23)

$$\therefore\; l_{ody} = \left[\left(\frac{16(2S)}{3}\right)\left(\frac{1}{Z()}\right)\left(\frac{V_{Vody}}{V_{Vo}}\right)^2 - (.904)\right]\left(\frac{\sqrt{2S}\sqrt{\tilde{N}_Q}}{2\pi}\right)\left(\frac{A_{dy}}{A_{ox}}\right)\lambda_{cxo}$$

$$= K_{dy}\lambda_{cxo}$$

letting $S=\frac{1}{2}$: Keep!! (non s.s. 6AG)

$\therefore\ l_{ox} = K_o\,\delta_{cxo}$, and (since $V_{wdy}=V_{vo}$ and $A_{dy}=A_{ox}$),

$$K_o = \left[\left(\frac{16}{3}\right)\left(\frac{1}{Z()}\right) - (.904)\right]\left(\frac{\sqrt{\tilde{N}_Q}}{2\pi}\right)$$

for $\tilde{N}_Q = 4$, $\frac{1}{Z()} = .2612$

Now, please go back and see page L13B.

$\therefore \left(\frac{16}{3}\right)\left(\frac{1}{Z()}\right) = 1.393$

$$\therefore\ K_o = \frac{1}{6.4236} \quad \text{for } \tilde{N}_Q = 4.$$

Richard P. Gandolfi 12-24-2001

$\therefore$ The largest allowable value for Z() is 5.8997.

$\therefore$

	$\tilde{N}_Q$	Z()	$1/K_o$	$x = \frac{K_o}{L^2_{cx,ss}}$	$\frac{1}{x}$	$L^2_{cx,ss}$
.3448 (OK)	5	5.50553	43.445	(but not likely) .3448	2.900	.0668
(No Good)	~~6~~	7.3094	X	Too HIGH of Z().		X
	4	3.82843	6.4236	.4434	2.2[illegible]5	.3511
	3	2.3094	2.5812	.6366	1.571	.6086
	2	1.	1.0031	1.200	.833	.8305

from pg. L5, (non S.S., GAH)

the Large radius dynamic mass:

$$m_{dyox} = \left(\frac{1}{c^2}\right)\left(\frac{1}{.904}\right) 8 \mu_g \pi A_{LRdy} \tilde{N}_V^2 \left(\frac{\tilde{N}_Q V_{vody}}{2\pi A_{LRdy}}\right)^2 \frac{M^2_{rox}}{\left(1 - \frac{V^2_{vody}}{c^2}\right)}$$

(inductance based)

$$\overset{(must)}{=} \left(\frac{3}{2}\right)\left(\frac{8\mu_g}{\pi}\right)\left(\frac{Z()}{4\tilde{N}_Q}\right)\left(\frac{\tilde{N}_Q^2 \tilde{N}_V^2 M^2_{rox}}{\left(1 - \frac{V^2_{vody}}{c^2}\right)}\right)\left(\frac{1}{8 A_{LRdy}}\right)$$

(balance of forces based)

$$\therefore \frac{\left(\frac{1}{c^2}\right)\left(\frac{1}{.904}\right)\pi V^2_{vody}}{4\pi^2} = \left(\frac{3}{2}\right)\left(\frac{1}{\pi}\right)\left(\frac{Z()}{4\tilde{N}_Q}\right)\left(\frac{1}{8}\right)$$

$$\therefore V^2_{vodyLR} = \left(c^2(.904)(4)(\pi)\right)\left(\frac{3\,Z()}{(16)(4)(\pi)\tilde{N}_Q}\right)$$

$$V_{vodyLR} = \left(\sqrt{\frac{(3)(.904)\,Z()}{\tilde{N}_Q}}\right)\left(\frac{c}{4}\right) = \text{func}(\tilde{N}_Q) \text{ and NOT } A_{LRdy}!!$$

dynamic vel for LARGE A_{dy} ONLY.

Richard P. Gandell 12-24-2001

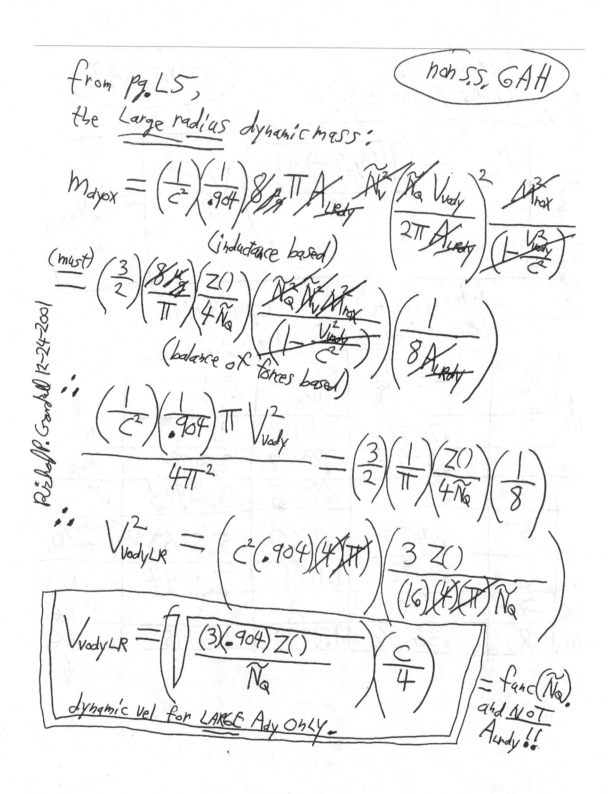

Letting $S=\frac{1}{2}$,
(LARGE Radius only):

nonss. 6AI

Richard P. Gyngell 12-24-2001

$$\therefore \quad V_{vodyLR} = \left(\frac{\sqrt{(3)(.904)Z()}}{4}\right) V_{vo,ss} = f_{VLR} V_{vo,s.s.}$$

where, $V_{vo,s.s.}$ is the equilibrium (or steady state) Value.

	$\tilde{N}_Q$	$\frac{V_{vo,ss}}{c} = \frac{1}{\sqrt{\tilde{N}_Q}}$	f_{VLR}	$Z()$	
N.G.	2	.7071	.4117	1.	N.G.
N.G.	3	.5774	.6257	2.3094	N.G.
	4	.5	.8056	3.82843	✓
	5	.4472	.9660	5.50553	(unlikely, since odd)
N.G.	6	.4082	1.1131	7.3094	N.G.
N.G.	~~7~~	X	X	X	X N.G.
N.G.	8	.3536	1.3790	11.2195	N.G.

Keep!

Richard P. Crandall 12-24-2001

Proof that we still have $E' \cong hf'$ (+12%) in the dynamic mass case.

$$m'_{dy} = \gamma' m_{ody}$$

$$m_{ody} = \frac{\alpha_o M^2_{vactoto}}{A_{dy}}$$

recall: (for any A_{dy}): $.4 \leq \frac{V_{vody}}{c} \leq .5$ (for $\tilde{N}_Q = 4$).

$$V_{vody} \cong V_{oo} = .5c \text{ (constant).}$$

$$V_{oo} = A_{dy}\,\omega_{rotdy} = 2\pi A_{dy} f_{rotdy}$$ (non S.S., GAJ)

$$= \frac{2\pi A_{dy} f_{ody}}{\tilde{N}_Q}, \quad \therefore A_{dy} = \frac{\tilde{N}_Q V_{oo}}{2\pi f_{ody}}$$

$$E_{ody} = m_{ody}c^2 = \frac{\alpha_o c^2 M^2_{vactoto}}{A_{dy}}$$

(α_o: a constant, recall. $M^2_{vactoto}$: constant, if V_{vody} = constant.)

$$\cong \frac{\alpha_o c^2 M^2_{vactoto} 2\pi f_{ody}}{\tilde{N}_Q V_{oo}} \cong \alpha_{oo} f_{ody}$$

(α_{oo}: a constant, regardless of A_{dy}.)

$$\therefore \boxed{\alpha_{oo} = h}$$

$$\text{So } \boxed{E_{ody} \cong h f_{ody}} \;!!$$

$$\therefore m_{ody} \cong \frac{h f_{ody}}{c^2}$$

$$m'_{dy} \cong \frac{h(\gamma' f_{ody})}{c^2}$$

$$\therefore \boxed{m'_{dy} \cong \frac{h f'_{dy}}{c^2}}$$

or $\boxed{E' \cong hf'}$!!

Richard P. Gandall 12-24-2001

$$M_{ody} = \frac{\alpha_o M^2_{vactotody}}{A_{dy}} = \frac{\alpha_o \left(\frac{\tilde{N}_Q \tilde{N}_V M_{rot}}{\sqrt{1-\frac{V^2_{dy}}{c^2}}}\right)^2}{A_{dy}}$$

non S.S. GAK

Keep!

$$I_{ody} = \frac{M_{vactotody} W_{rotody}}{2\pi}$$

("U" means Unstable (or un-balanced) (i.e. any A and V)) (i.e. $F_{outdy} \neq F_{indy}$)

$$\therefore M_{ody} C^2 = \alpha_o \left(\frac{2\pi I_{ody}}{W_{rotody}}\right)^2 \left(\frac{C^2}{A_{dy}}\right) = \frac{1}{2} L_{dy} I^2_{ody}$$

$$H_{odyu} = \left(\frac{1}{2}\right) M_{odyu} A_{dyu} A_{dyu} W_{rotodyu}$$

KEEP!

$$\text{So, } L_{dyu} = 2\alpha_o \left(\frac{2\pi}{W_{rotodyu}}\right)^2 \left(\frac{C^2}{A_{dyu}}\right)$$

$$\therefore M_{odyu} C^2 = \frac{1}{2} L_{dyu} I^2_{odyu}, \quad I_{odyu} = \frac{M_{vactotodyu} W_{rotodyu}}{2\pi}$$

Below, the LAST IMPORTANT Formula in this last set of papers.

$$\text{So, } M_{odyu} \cancel{C^2} = \alpha_o \left(\frac{\cancel{2\pi}}{\cancel{W_{rotodyu}}}\right)^2 \left(\frac{\cancel{C^2}}{A_{dyu}}\right) \left(\frac{M^2_{vactotodyu} \cancel{W^2_{rotodyu}}}{\cancel{4\pi^2}}\right)$$

Q: What is mass?

Answer:

$$\therefore M_{odyu} = \frac{\alpha_o M^2_{vactotodyu}}{A_{dyu}}$$

for ALL CASES.

Does NOT depend (much) on ω or V!

<THE BIG RESULT>!

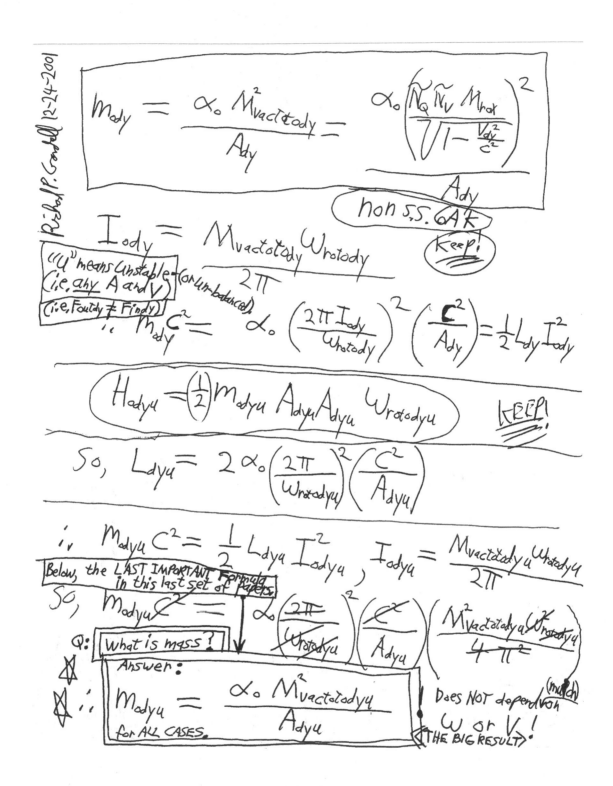

Paul P. Gambill 12-24-2001

(non S.S. GAL)

$$H_{odyu} = \frac{1}{2}\left(\frac{\alpha_o M^2_{vactotodyu}}{A_{dyu}}\right) A^2_{dyu}\, W_{rotodyu}$$

$$= \left(\frac{1}{2}\right) \alpha_o M^2_{vactotodyu}\, A_{dyu}\, W_{rotodyu}$$

$$= \boxed{\left(\frac{1}{2}\right) \alpha_o M^2_{vactotodyu}\, V_{vodyu}}$$

Keep!!

also,

$$H_{odyu} = \frac{1}{2}\left(M_{odyu}\right)\left(\frac{\alpha_o M^2_{vactotodyu}}{M_{odyu}}\right)^2 W_{rotodyu}$$

$$= \frac{\left(\frac{1}{2}\right) \alpha_o^2 M^4_{vactotodyu}\, W_{rotodyu}}{M_{odyu}}$$

$$\boxed{M_{odyu}\, C^2 = \frac{\left(\frac{1}{2}\right) \alpha_o^2 M^4_{vactotodyu}\, W_{rotodyu}\, C^2}{H_{odyu}}}$$

$\boxed{E_{ody} \cong h f_{ody}}$ but, $\boxed{E_{odyu} \neq h f_{odyu}}$

$$P'_{dyu} = \gamma'(m_{odyu}\,[redacted])u' = \gamma' u' \left(\frac{E_{odyu}}{c^2}\right) [redacted]$$

non S.S., 6AM (Last page)

$$= \gamma' u' \left(\frac{\left(\frac{1}{2}\right) \alpha_o^2 M^4_{vactotodyu} \omega_{rotodyu}}{H_{odyu}} \right)$$

but, $$(m_{odyu}) \overset{also}{=} \frac{2 H_{odyu}}{A^2_{dyu} \omega_{rotodyu}}$$

KEEP!

Richard P. Grandall 12-24-2001

$$\therefore \frac{\left(\frac{1}{2}\right) \alpha_o^2 M^4_{vactotodyu} \omega_{rotodyu}}{H_{odyu}} = \frac{2 H_{odyu}}{A^2_{dyu} \omega_{rotodyu}}$$

$$\therefore H^2_{odyu} = \left(\frac{1}{4}\right) \alpha_o^2 M^4_{vactotodyu} A^2_{dyu} \omega^2_{rotodyu}$$

so, $$H_{odyu} = \left(\frac{1}{2}\right)\left(\alpha_o M^2_{vactotodyu}\right) \underbrace{A_{dyu} \omega_{rotodyu}}_{V_{vodyu}}$$

(again)

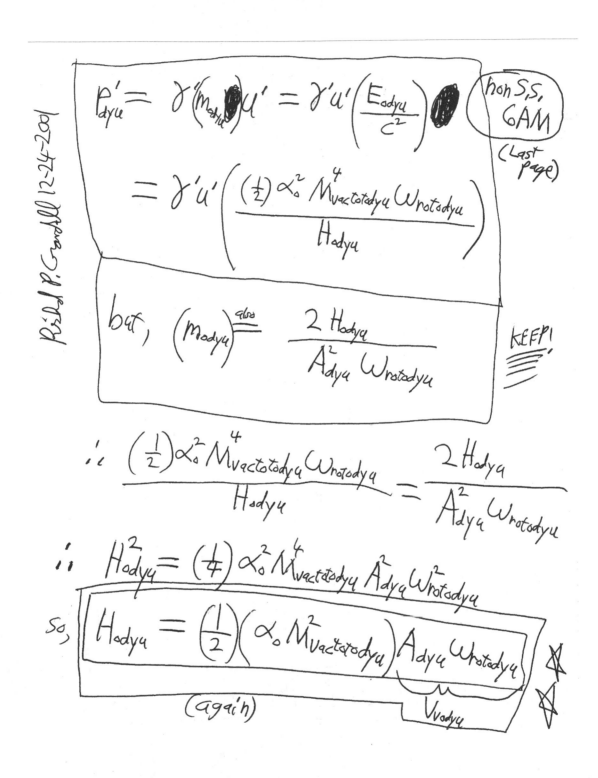

Page 55 of 55

SIGNATURE PAGE SIGNATURE PAGE SIGNATURE PAGE

By signing below, I the INVENTOR swear and affirm that all of the information herein this complete document is complete, true, and accurate to the best of my knowledge:

Richard P. Crandall 12-31-2001

Richard P. Crandall, INVENTOR SS# 453 80 7642 Date Signed:

Notary Witness Information:
By signing and dating below, you swear and affirm and acknowledge that you were a witness to the act of the above named INVENTOR signing this document as you watched, and you have verified the identity of the above signer to be Richard P. Crandall by looking at the signer's Colorado driver's license and any additional required identification as you see fit.

SEAL OF TEXAS
COUNTY OF
SUBSCRIBED & SWORN BEFORE ME
THIS 31 DAY OF DEC 2001
Roberta Lynn Olguin
NOTARY PUBLIC

ROBERTA LYNN OLGUIN
MY COMMISSION EXPIRES
June 12, 2004

The inductance (ie "L" series of pages): **DL1**

L0–L2, develop formula for the inductance, based upon E.E. helix formula, for any situation, including that of non-equilibrium.

L3, develop steady state inductance and a formula for the steady state inductance correction factor $L_{c,ss}$.

L4, verify small radius (S.R.) dynamic rest mass.

L5, develop formula for large radius (LR) dynamic rest mass, and show that, for large enough radius, the rest mass of a particle can be reduced to near-zero!

L6 – L14, develop formula for rest mass of the Vacuon which is approximate, but is independent of temporal length l_0* (or Ksubl*). For now, we can ignore pages L13A and L13B, as they jump ahead and use formulas from a later part of the document to derive an expression, exact, for Ksubl*. Also on these 2 pages, we calculate, based upon a value of Ksubl* of 1/6.4234 (which is exactly correct for the electron), Mroe and Mrop. Mroe is the exact rest mass of a single free Vacuon, whereas Mrop is the rest mass of a Vacuon while in a "clump" of Vacuons.

⟹ Exact single Vac. rest mass = 5.56×10^{-9} kg. The ratio is 10.89.

Revised [illegible] 12-24-2001

Appendix
H2

Operational Princ.
(For The I.D.D.)

[ ]

[THIS PAGE IS INTENTIONALLY BLANK]

Gravity Wave Propulsion Device Physical Principles of Operation

The entire device consists of several components, as can be seen in **FIGURE ??**. There is a large, approximately square "C" shaped electromagnet (which can be switched on or off), and there are two parallel metallic waveguides of rectangular cross-section. The waveguides are centered in the gap of the electromagnet. A disk of special material is placed inside the end of each of the two waveguides. Each disk can independently be rotated at ultra high RPM by the two electric motors present. Each waveguide is connected by coaxial cable to its own private microwave oscillator source. Finally, each waveguide has an inlet port and an outlet port for liquified gas to be pumped through, in order to cool the disks to below 100 degrees Kelvin. Now, I will describe the physics of the device operation.

It has been thought for many years now, since Einstein first published his equation for General Relativity, that matter itself may simply be a "twist" or "knot" in the fabric of spacetime (space and time are unified in Relativity theories). As such, subatomic particles of matter (such as an electron or a quark) could simply be described strictly in terms of geometry, where such geometry describes the extreme localized twisting (or knotting or bending) of spacetime. However, all attempts to date to describe particle mass, or particle matter itself, in terms of such localized warping of spacetime have been futile. This is probably due to the fact, at this small scale, the theory of Quantum Electrodynamics (QED) plays an important role in the description of any material or energy phenomena. A complete and true model of matter at this small scale would ultimately require some kind of higher order theory, which harmoniously combines quantum mechanics theory with relativity theory. Such a higher order theory, usually referred to as a "Grand Unified Field Theory", does not yet exist, and would probably be quite complex.

I propose herein (for the first time anywhere, to the best of my very informed knowledge), as <u>one</u> of the main basic physical principles of operation for my new <u>gravity wave propulsion device</u>, a completely new first order (or first attempt at a) physical model of subatomic particles as a <u>geometrically-specified localized warping of spacetime</u>. I am, in this first order model of particles, completely neglecting any quantum effects. This is a bold, but approximately justifiable, step. Perhaps someday someone will be able to fuse this first order model with quantum theory, and arrive at a better combined theory which more closely approaches a true "Grand Unified Theory". I note here, without further explanation, that my new first order model could be the first step in solving the long-standing "renormalization" problem of QED, which is basically an unjustified trick to eliminate the infinities found at the center of all particles. This is because, as will be seen below, that my first order model shows that a particle is a tiny ring of a finite size, rather than a discrete infinitesimal point. This ringlike property of particles which I propose here also adds further support to the so-called "string-theories" of matter.

Starting with the basic Einstein equation, **FIGURE 1**, I was able to derive an equation, using a weak-field approximation, which describes the gravity-related forces imposed upon (or felt by) a small moving test-mass which are caused by (or produced by) the mass (and motion) of a gravity-field generating mass. This force equation is shown in **FIGURE 2**. The left side of the equation is Z (sub r) multiplied by c squared. Here, the subscript r ranges from 1 to 4, due to the four dimensional nature of spacetime. However, we will ignore the r=4 case for the present purposes. The letter "c" represents the speed of light (a VERY large number: about 3x10^8 meters/sec). The variable Z (sub r) was defined such that Z (sub r) * c^2 equals (gamma * Force (sub r)) / m. Here, gamma is a unitless velocity-related number which is usually approximately equal to 1, and will thus be ignored everywhere it appears. Little "m" is the mass of the small, moving test-mass (the units are Kilograms). Also, "Force (sub r)" is the vector force felt by the test-mass (the units are Newtons). Therefore, Z (sub r) * c^2 has units of Newtons per Kilogram.

Note that the right side of this force equation, which I call the "Crandall Force Equation for Weak Gravity Fields" has **four** terms. This means that the total force felt by the small test mass is due to the sum of four different kinds of gravity-related forces. I thus discovered that there are four types of gravity! These I call gravity "B", gravity "E", gravity "C", and gravity "D", respectively, according to the order in which the terms appear in the equation. Gravity "E" is just the ordinary static gravity force we're all familiar with. It is the gravity force which attracts all objects towards the center of the Earth. **Gravity "B"** is now semi-widely known as the "gravitomagnetic field". It is a form of gravity force which is **generated** by the **motion** of the gravity-field generating mass, and is **felt** due to the **motion** of the small test mass. In contrast, the familiar **gravity "E"** is **generated** by the **amount of mass present** in the gravity-field generating mass, and is **felt** due to the **amount of mass present** in the small test mass. **Gravity "C"** is a form of gravity, discovered by me alone, which is **generated** by the **amount of mass (gradient) present** in the gravity-field generating mass, and is **felt** due to the **motion** of the small test-mass. **Gravity "D"** is a form of gravity, also discovered by me alone, which is also **generated** by the **amount of mass (gradient) present** in the gravity-field generating mass, and is also **felt** due to the **motion** of the small test-mass. Finally, I found that familiar **gravity "E"** can also be **generated** by the time derivative of the momentum density present in the gravity-field generating mass, but this is not discussed further. So in summary, there are four types of gravity. I suspect that other people were only able to discover the gravity "B", but not the gravity "C" and "D", because they took a different approach, and started out with the so-called "PPN" or "post-newtonian" approximation equations of relativity. I, however, started out with the general Einstein equation itself.

I invented (or discovered) my unique first order geometrical model of particles, as follows. Looking at the equation and the relative orders of magnitude of each of the terms, I could see that if there is any NET generating mass amount present at all (i.e. the "large static mass term") then the small test mass cannot detect the gravity "B", "C" and "D" fields, as they are "drowned out" by the "E" (ordinary gravity) field of the generating mass by a factor of "c" squared, where "c" is the (huge) speed of light. **So the only way to have physically detectable, non-trivial, and meaningful results for the "B" field is to have the algebraic sum of all of the generating masses to be zero.** This of course implies the existence of "negative" mass, which I will assume is reality. So, if we model a subatomic particle (which is a gravity-field generating mass) as only consisting of counter-rotating, superimposed, rings of equal amounts of "positive" and "negative" mass, then we have a situation where there is no generated gravity "E", "C", or "D" fields (since the total generating mass present is ZERO). However, there is a strong localized gravity "B" field present, due to the rotation of each of the signed mass components.

Therefore, for my first order geometrical model of electrons and quarks, there are two equivalent ways of looking at such a particle. 1) The particle is ring shaped (i.e. it is NOT a discrete infinitesimal point), and consists solely of superimposed counter-rotating rings of equal amounts of "positive" and "negative" mass. OR, 2) The particle consists solely of an extremely strong dipole-shaped gravitomagnetic (gravity "B") field, and there is no net mass present (i.e. it adds up to zero), in the usual sense of the term "mass". **The gravity "B" field is dipole shaped** since that is the shape of field created by a ring of current. This is the same shape of field as that of the magnetic field produced by a bar magnet or electromagnet, or simply a multi turn coil of wire carrying electric current. This equivalent second way of looking at a particle is seen as being a first order **PURELY GEOMETRIC** model, since there is no "actual" mass (in the usual sense) present, and all that is present is just a strong localized gravity "B" field. But ANY energy field composed purely of just any combination of gravity "E", "B", "C" and "D" fields can be viewed as being simply a localized geometrical warping (bending) of spacetime itself. There are no physical fields (such as the electromagnetic field) present; there is only a warping (i.e. a geometrically specified bending) of the fabric of spacetime itself.

So, one of the main basic physical principles of operation for my new gravity wave propulsion device is that quarks, and thus any atomic nucleus having a net spin (defined later), can be described by a first-order purely geometrical model (i.e. a geometrically-specified localized warping of spacetime). This can be better viewed, alternately, as having each particle, or nucleus, as being composed **solely** of a strong dipolar gravitomagnetic field. Therefore, there is no "actual" mass present. All that's there is the strong dipolar gravitomagnetic field. Therefore, what we normally measure as the "mass" of a particle is really just the "effective mass" of the pure gravitomagnetic energy field. This is because any energy field has an effective mass defined by the familiar equation Energy (of a field) = effective Mass * c^2. So what we ordinarily measure as the value of the mass of a particle is equal to the localized gravitomagnetic field strength divided by the speed of light squared! Therefore, even for particles with a small measured mass (i.e. the particle's effective mass), the Energy (or field strength) of the localized gravitomagnetic field must be very huge!! It has long been known that subatomic particles have a constant spin, and this is consistent with my model of counter-rotating rings. It has also been theorized, and for good reasons, that electrons and other small particles may be very similar to a tiny "black hole". This is also consistent with my model, where the amount of mass contained in each of the rings of positive and negative mass would have to be extremely huge, in order to generate the huge localized gravitomagnetic field. A "black hole" is just a very, very large amount of mass confined to a small space. The new feature present here, though, is the presence of negative mass as well.

The net overall thrust, or force of propulsion, produced by my propulsion device is the direct result of having all the atomic nuclei, in each one of two separate disks, generate an overall composite dipolar gravitomagnetic field which, in turn, produces a gravity "B" type force upon all the atomic nuclei in the other disk, and therefore produces a force upon the entire other disk. This can best be seen by viewing each nucleus as simply being composed of the two superimposed counter-rotating rings of equal simple positive and negative mass. The two opposed disks are parallel, and share a common axis. They can be seen in **FIGURE 3**. If the nuclei in each disk are spin-aligned in a certain direction, then each disk will produce a gravitomagnetic force upon the other. How the spin-alignment (of all the nuclei) can be accomplished will be described in detail later.

Given such a radical new theory of matter, with obviously strong fields produced, one may well ask "why hasn't this been detected before in all the usual experiments of physics?" "Could it have gone unnoticed?" I can, in fact, see how such fields could have easily been unnoticed, and the reason they can go unnoticed is also the root cause of one serious problem with making the propulsion device work properly. With gravity "C" and "D" absent, the Einstein equation can ultimately be re-written (still in the weak-field case, of course) in a form analogous to Maxwell's well-known equations of electromagnetism (which describes how electric fields, magnetic fields, electric charge, and flow of electric charge all interact with each other). These "Maxwellized" equations I derived for gravity "E", gravity "B", mass, and flow of mass can be seen in **FIGURE 4**. Also shown is an equation for calculating the force on a test-mass m* whose velocity vector is "u". In these equations, vector "H (sub g)" is simply equal to vector "B (sub g)" (the gravitomagnetic field vector) divided by a constant called "mu (sub g)" (the so-called free-space permeability constant for gravitomagnetic fields). The constant "epsilon (sub g)" is the so-called free-space permittivity constant for static ordinary gravity fields. The vector "E (sub g)" is the ordinary gravity field vector. The quantity "rho star (sub m)" is the density of any field-generating mass present, and the quantity "rho star (sub m) times a U vector" is the momentum density of the same field-generating mass. Now, suppose we take **for an example** two quarks or two electrons, where each one is modeled, as before, by two very, very small diameter counter-rotating (and superimposed) rings of mass (as usual, one being positive and the other negative mass, for cancellation purposes). It can be shown (after lengthy, but well-known and well-understood calculations, as in the analogous theory of static magnetic fields and currents), using the "Maxwellized equations for gravity" and the test-mass force equation, that there is NO NET VECTORED FORCE produced upon either one of the particles. However, there is A NET TORQUE produced upon each one of the particles as a result of the gravitomagnetic field emitted by the other particle. THIS IS SO, BECAUSE THE "RINGS" OF EACH PARTICLE ARE SO SMALL IN DIAMETER AND THE DISTANCE FROM THE OTHER PARTICLE IS SO GREAT

THAT ALL THE "LINES OF FLUX" OF THE "B" FIELD (emanating from the <u>other</u> particle) WHICH PENETRATE THE "RINGS" OF <u>THIS</u> PARTICLE ARE PARALLEL AND UNIFORMLY DISTRIBUTED. **So, particles can create TORQUES on other particles, but particles cannot normally create a (gravity-related) FORCE on other particles at a distance. This is why the effects predicted by this new model of matter have typically gone unnoticed before.** In beginning-level quantum mechanics books, spin of electrons is discussed, as well as a "spin-orbit" coupling. However, these books apparently never discuss anything called a "spin-spin" coupling effect between particles. The Pauli exclusion principle takes care of what would be considered to be any spin-spin coupling by rigidly enforcing the rule that in each electron orbital, there can only be two electrons, where one is spin "up" and one is spin "down". End of story. So it seems reasonable that this new gravitomagnetic spin-spin coupling (or torque coupling) could have gone unnoticed. So in atoms with an even number of electrons, the average of the gravitomagnetic fields due to the electrons would cancel. But what about nuclei? I will cover that subject next.

So now we can see the serious problem with the gravity wave propulsion device described thus far, which has the two opposing disks of some material. If we consider each atom's nucleus (in the disks) to be composed of neutrons and protons, where each neutron or proton is made up of three quarks (as is now well-known), and where each quark, and each electron orbiting around the nucleus, is modeled as being composed of the two counter-rotating rings of positive and negative mass, then the problem becomes clear. The particles (being the quarks and electrons) in one disk cannot produce any forces on the particles in the other disk, but only torques on those particles in the other disk. Thus, one disk cannot produce any overall directional force upon the other disk. **The only way to solve this apparent problem is to consider that each atomic nucleus in the disks, being composed of many closely spaced quark particles, <u>effectively</u> acts as ONE LARGE DIAMETER set of counter rotating rings of equal positive and negative mass. When I say "LARGE DIAMETER", I mean that the diameter of the "effective" or "composite" set of counter-rotating rings is the same is the diameter of the nucleus itself.** Assuming the extended gravitomagnetic field which is produced by each disk is non-uniform (which is probably the case, due to the finite size of the disks, and also whose non-uniformity could be augmented by having non-flat surfaces for the faces of the disks), then each nucleus of one disk would be penetrated by a gravity "B" field (from the other disk) whose "lines of flux" are slightly non-parallel. Obviously, for larger diameter atomic nuclei, the total divergence of the "B" field passing through a nucleus is greater. **So here it can be seen, that for maximum efficiency and workability of the propulsion system, we must use materials for the disks whose atomic nuclei are of the largest diameter possible.** This is because it turns out that a **<u>divergent</u>** gravity "B" field passing through a ring (or rings) of rotating mass WILL create a net vector force on the ring(s), and the net force produced upon the ring will be in the average direction of the lines of flux of the gravity "B" field.

One other restriction on the disk material is as follows. The majority of the atomic nuclei in each disk should be such that there is one un-paired proton, OR one un-paired neutron. This is because of the tendency of protons to pair themselves up such that the two protons comprising each pair have oppositely directed spins. The same goes for neutrons. Therefore, in all element isotopes (of the "periodic table") which have even numbers of protons AND neutrons, the gravitomagnetic fields will cancel out. Isotopes of elements which have unpaired nucleons have a net angular momentum of the nuclei. It is hoped that such elements will have a net gravitomagnetic field, such that each nucleus effectively acts as one large diameter set of counter-rotating rings of positive and negative mass (where the rings' diameter is effectively that of the nucleus).

It will be noted that in **<u>FIGURE XX</u>** that each of the two disks are coupled to a separate ultra-high RPM electric motor, which includes an ultra-high ratio gear assembly. The reasons for having the disks spin at a constant ultra-high RPM are twofold as follows. Reason #1) It is hoped that this overall rotation will increase the ability of each atomic nucleus to act effectively as one composite, large diameter (effective diameter same as the nuclear diameter) set of counter-rotating rings. The overall disk rotation would hopefully accomplish this goal by imparting a large degree of spin angular momentum to each nucleus, which should help the individual quarks' "rings" to better couple with adjacent quarks' "rings". This is the only purpose of the overall disk rotation. The disk rotation is NOT used to align the spins of the nuclei, since we have the external electromagnet to accomplish that (which will be described later, below). Admittedly, this use of disk spin to increase or improve the overall ring-like behavior of each nucleus is highly speculative, and may not work as expected. That is why we have reason #2, which represents a different mode of operation which has an equal or better chance of working properly as needed. Reason #2) In this mode of operation, the upper disk will have a net upward force produced upon it as a result of having the gravitomagnetic field produced by the lower disk passing through the upper disk. However, the force produced in the upper disk is not in this case due so much to interactions with nuclei-considered-as-rings in the upper disk, but is more due to looking at the overall disk as being a large ring of spinning positive mass, and when this large ring is pierced by a divergent or non-uniform gravity "B" field from the lower disk, then the disk will experience a net vertical force upon it. Of course, the terms "upper" and "lower" can be interchanged, as these effects work both ways.

In **FIGURE XX** it can be seen that each disk is located inside the end of a rectangular waveguide, where the usual definition of a "waveguide" is a hollow metallic tube of constant cross-section (in this case, rectangular). Each waveguide is connected by coaxial cable to it's own private microwave oscillator source (of constant frequency). The two parallel waveguides are surrounded by an electromagnet (iron core, laminated layers to reduce eddy current losses), which can be switched on and off. The disks (inside the waveguides) must be located in the gap of the electromagnet such that the lines of magnetic field produced by the electromagnet are parallel to the axis of each of the disks. Each waveguide has an inlet port and an outlet port for liquid nitrogen (or perhaps some other liquified gas) to flow through. It is important to keep the two disks at very low temperature (say, below 100 degrees Kelvin), as will be explained later.

The reason for having the microwave (electromagnetic wave) sources and for having the large electromagnet is because the device is intended to operate based upon the principles and practices of Nuclear Magnetic Resonance (commonly called "NMR"). The principles of NMR are used to first align the spins of the majority of the nuclei in each disk to a vertical direction (i.e. perpendicular to the disks, but parallel to the electromagnet's magnentic lines of flux), and then to sinusoidally and continuously flip the spin directions of the nuclei. During the first half-cycle, the spin direction of the nuclei in each disk will have been rotated by 180 degrees. During the second half-cycle, the spin directions of the nuclei are returned to their original direction. This sinusoidal flipping of the spin directions of the nuclei is accomplished by exposing the disks to continuous, sinusoidal, microwave electromagnetic waves of a particular frequency, which can be calculated. An example of how to calculate, using the principles of NMR, the required strength of the fixed magnetic field (of the electromagnet), the frequency of the microwaves, and the field intensity of the microwaves can be seen in **FIGURE ??**. These calculations are based upon the so-called measured "gamma" value for the particular type of nuclei being used as the disk material. The external fixed magnetic field is often referred to in the literature as the B0 ("B zero") field, and the electromagnetic wave source as the B1 ("B one") field. The strength of the B0 field is commonly, say, ten times that of the B1 field. Now, regarding the need for a cryogenic system for cooling the disks. There is an inlet and outlet port in the waveguides for the flow of a liquified gas. This low temperature (perhaps under 100 degrees Kelvin) operation of the disks is necessary for there to be a very large fraction of the nuclei having their spins stay aligned in the same direction. There is a formula in NMR which allows one to calculate what fraction of the spins will stay aligned, given the temperature of the apparatus.

The final physical principle of operation of the propulsion device will now be described. For the moment, assume that all the nuclei in both disks are aligned vertically (and parallel to the lines of the electromagnet's magnetic field). This is accomplished by turning on the electromagnet, keeping the microwave oscillators turned off for the moment, and having the cryogenic cooling of the disks working properly. Also, the disks should both be spinning at a constant rate. Due to the gravitomagnetic fields produced by both disks, there will be a vertical gravitomagnetic force created upon both the upper and lower disk. The forces upon the two disks will be equal in magnitude and opposite in direction, according to Newton's ??? law. Therefore, the net force produced upon the device-as-a-whole will be zero. The situation is entirely analogous to having two colinear permanent bar magnets with a small gap between the ends of the two magnets. Due to the non-uniform magnetic field produced from the ends of each bar magnet, the magnets will either repel each other, or will attract each other, depending on whether we place "like" or "unlike" ends of the magnets close together. For magnets, as well as for electric charges, "likes" repel, and "unlikes" attract. In the analogous case of the two bar magnets, the total (net) force is also zero, since the force upon a magnet is always equal in magnitude but opposite in direction to the force upon the other magnet. Therefore, using the two disks as specified, there is no net force, and so the device does not function as a propulsion system, because with a propulsion system, there must be a net overall force produced upon the device.

However, it is possible to "cheat" Newton's ??? law by use of a speed-of-light delay effect. This effect is perfectly admissable in modern physics and is well-understood. Using this method, there will be an equal force produced upon each disk, but the direction of the forces will now be in the SAME direction. Therefore, there will be an overall NET force upon the device. It can be shown that, if we can sinusoidally flip the directions of the nuclei spins in each disk (or, for the bar magnet example, if we can sinusoidally flip the orientation of both the bar magnets) at a very great frequency, then there will be a net average force produced upon the device-as-a-whole. It can also be shown that the phase difference between the spin alignments in the two disks (or the two bar magnets, in that example) must be 90 degrees, for optimum operation. Also, there is a formula which allows one to calculate the required frequency of the source microwave oscillators, once given the separation (air-gap) distance between the adjacent parallel faces of the two waveguides. As an example of what physically is happening, consider again the example of the two, initially colinear, bar magnets. If one were to suddenly flip the direction of one of the magnets, say the upper magnet, (thereby instantaneously reversing the direction of force upon it due to the field of the other magnet), then, for a very small fraction of a second, the force upon both magnets would be in the SAME direction. This is because it takes a finite, albeit very small, amount of time for the change in the field produced by the (now flipped) upper magnet to propagate down to the lower magnet, since any change in any energy field can only propagate through space at the speed of light, maximum. Once the field change from the upper magnet has propagated down to the lower magnet, then the force produced upon the lower magnet reverses, and we are finally left with essentially the same situation we had when we started. There will then be equal and opposite forces produced upon the magnets. So for a fraction of a second, there was a NET force upon the device-as-a-whole. It can be shown that if we continuosly flip the directions of the magnets, in a sinusoidal manner, such that the relative phases of the magnet positions are kept at 90 degrees, then there will be a time-averaged NET force upon the device-as-a-whole.

So we see that if we turn on the sinusoidal microwave oscillator sources in the propulsion device, and set them so that they are electrically 90 degrees out of phase, and provided the frequency of them is calculated based upon the formula that relates frequency to separation distance, then there will be a time-averaged NET force upon the device-as-a-whole. This seems to violate Newton's ?? law, but this law must be amended once the principles of special relativity are brought into the picture. It can be shown that, in special relativity theory, that conservation of momentum of mass/energy must always hold. Relativity shows that a pure energy field itself must possess an effective mass and momentum. It can be shown that, using the two disk arrangement as described, under the conditions described, there will be powerful gravity waves emitted in one overall direction (which is downward, in this case). This happens due to the already mentioned 90 degrees phase shift. These downwardly directed gravity waves, being a form of pure energy, possess a momentum, and so the device is constantly ejecting large amounts of momentum in the downward direction. Due to the conservation of momentum principle of relativity, this downward ejected momentum must be countered by having the device experience an upwardly directed average force upon it. Thus, this wave and momentum analysis agrees with the simple speed-of-light delay analysis in the previous paragraph. It can be seen that the device can be used as an antenna capable of emitting powerful, directed gravity waves, as well as be used as a propulsion (or net thrust-producing) device.

This concludes the physical principles of device operation.

[ ]

[THIS PAGE IS INTENTIONALLY BLANK]

Appendix H3

The I.D.D. & Explanatory Suppl.

[ ]

[THIS PAGE IS INTENTIONALLY BLANK]

Today's date:
April 9, 2001

DISCLOSURE DOCUMENT NO.
492109
RETAINED FOR 2 YEARS
THIS IS NOT A PATENT APPLICATION
PTO-1652 (8/99)

OIPE
APR 1 0 2001
PATENT & TRADEMARK OFFICE

To:

Disclosure Document Receiving Dept. ("Box DD")
Of The
United States Patent And Trademark Office

From Inventor:

Richard Perry Crandall, SS# 453-80-7642
6229 Zenobia Court
Arvada, CO 80003-6635
(303) – 429-3939

Subject: The three enclosed documents, which I am requesting to be filed.

Dear Invention Disclosure Document Receiving Department:

Under your "Disclosure Document Program", I the undersigned, being the inventor of the disclosed inventions, request that the all of the enclosed papers be accepted under the Disclosure Document Program, and that they be preserved together, for a period of two years.

The three documents enclosed with this cover letter are as follows:

1) "Invention Disclosure Document" dated 02-22-2001.
2) "IDD Addendum- Attach to the IDD" dated 04-09-2001.
3) "Explanatory Supplement to IDD" dated 04-04-2001.

There are NO DRAWINGS included with any of these documents.

Item 1) is the main Disclosure Document. It contains a record of most of my inventions, dating back to year 1999 and 2000. Item 2) contains inventions, and variations on my prior inventions of item 1 above, which were discovered recently, this year (2001).

Item 3) contains very important information which makes the content of my "lab notebook" and its connections to my IDD understandable. I request that this document be kept in addition to, and in the same "file" (however you define "file") with items 1) and 2) above. It is critical for understanding the connections my inventions have to my lab notebook.

Thank you, and sincerely,

Richard P. Crandall 4-9-2001

Richard P. Crandall, Inventor Date signed. [Enclosures (3), and a $10 check]

Richard P. Crandall, Inventor Date signed. [Enclosures (3), and a $10 check]

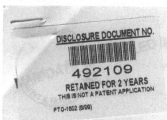

[ ]

[THIS PAGE IS INTENTIONALLY BLANK]

Today's date:
February 22, 2001

Audience: NO UNAUTHORIZED ACCESS to this document is allowed. PENALTY OF LAW. The only individuals who may read or possess this document are responsible employees of the U. S. Patent and Trademark Office, and my duly authorized agents. If you are one of my authorized agents (and not a member of the Patent Office), you must have obtained prior permission and authorization from me personally as well as having signed a non-disclosure / non-co-inventor / non-compete / confidentiality document, before you can read this document. I must retain a copy of your signed authorization. The U. S. Patent and Trademark office may read and make copies of this document without any prior permission.

Author and Inventor:
Richard Perry Crandall, SS# 453-80-7642
6229 Zenobia Ct.
Arvada, CO 80003
(303) – 429-3939

Subject:
Summary and history of my inventions, theory and claims, and all significant related activity and dates of invention, for my impending patent application.

Also this document serves as my original submission copy of a so-called "INVENTION DISCLOSURE" document, as defined by the U. S. Patent Office, which I request be retained by the U. S. Patent Office for however maximum period as it can be kept on file.
<u>[PROCEED TO NEXT PAGE, IF DULY AUTHORIZED]</u>.

Predicted title for the final patent, or for the patent search application (may change):

Devices for Gravity Wave Propulsion, Generation of Gravity Waves, and Directional Gravity Wave Antennas. Additional related keywords: Antigravity, Levitation, NMR.

Author and Inventor:
Richard Perry Crandall, SS# 453-80-7642

Dated History of, theory of, and specification of patentable inventions:

DATED RECORD OF INVENTIONS, #1: On June-01-2000, I invented a gravity wave propulsion device (which also serves as a levitation device and as a gravitational wave producing antenna) consisting of two opposing parallel coils (or loops) of wire, each one of which is wrapped around a separate cylindrical piece of material. The cylinders are placed co-axially (i.e. the central symmetry axis of each one is aligned along a common line), with their end faces separated by a small gap. In many cases, the height of a cylinder is less than its diameter. The end faces are typically not flat, in order to purposely create a non-uniform field. The cylinders must each be made of a material (not necessarily solid, but at least confined inside a cylinder shell) which has either a physical property such that the electrons of the atoms constituting the material are free enough so that the spin direction (vector) of a majority of the electrons can easily be aligned and manipulated by application of a high frequency electromagnetic field, or has a physical property such that the nuclei of the atoms constituting the material have a net angular momentum due to the contributing (i.e. non-canceling) individual spins of the protons and neutrons of each atom and such that the nuclei are free enough so that the overall spin directions of the nuclei can easily be aligned and manipulated by application of a high frequency electromagnetic field. These requirements on the material are thus because at that time (of 6-01-2000) I had formed a new theory (a theory which is still unknown to any other persons whatsoever as of the original date of my first writing this disclosure document [Feb. 20, 2001], as far as I know and to the best of my knowledge) that electrons and other lowest-order constituent (i.e. non-divisible, as we know it) particles must mainly consist of a strong dipolar Gravity "B" field (a terminology I invented). The existence of Gravity "B" fields in general had been developed (or should I say "co-developed") independently by me as of 5-07-2000, with no knowledge at that time of any co-development or similar work by others. However, other persons have co-discovered this same type of field (which I call Gravity "B") earlier than I did, and this type of field is now semi-widely known as a "gravitomagnetic" field.

It is well known that protons and neutrons are made of 3 quarks, where a quark is considered to be a lowest-order constituent particle. Therefore, protons and neutrons and atomic nuclei should also possess a strong Gravity "B" field as well as the electron. I found the local magnitude of a particle's Gravity "B" field must be incredibly strong, as per General Relativity, it takes quite a large amount of confined energy to have an equivalent effective measurable mass, according to the commonly known formula $E=M*c*c$. Therefore, by inverting, equivalent mass equals energy divided by the speed of light squared. So it takes an extremely large energy field to "weigh" as much as even a small amount of mass, such as electrons and quarks. The coils of wire are used to supply the above mentioned high-frequency electromagnetic field. Each coil is connected to a separate source of high-frequency current (or voltage) (typically in the microwave frequency region of the spectrum).

[PROCEED TO NEXT PAGE].

Each source should be preferably sinusoidal, and for optimal results, one source should be 90 degrees (ninety degrees) out of phase with the other source (as measured right at the loops). The loops must have a minimal number of turns, since microwave frequencies are used. This optimal (and approximate) 90 degree phase difference is a primary part of this patent specification and claims. Other phase shifts could be used, but would result in a drastically reduced net force produced by the overall system. There is a calculation one can perform to find the optimum spacing for the gap between the two cylinders as a function of the frequency of the radio frequency sources (not shown).

One variation on this design theme, which is itself a new design and was also invented by me on **June-01-2000,** is the alternate use of microwave electromagnetic radiation sources (typically the output of these sources is a horn-shaped end part of a waveguide). Instead of using the two coils to align the nuclei or electrons in the cylinders, we use two separate microwave sources, such that each cylinder has its own separate microwave source directed at it (at very close range, for maximum effect). As before, each source is approximately plus or minus 90 degrees out of phase (for optimum performance) with the other one. This, and all similar-in-concept configurations are also intended to be protected by this patent. As examples, the cylinders could be square or other shape, and the microwave sources may each be some other kind of waveguide or even just a small dipole antenna. When the entire device is tuned and operating properly, there will be a net force on the device as-a-whole in the direction of the central axis of the device. This force, if strong enough, will propel the device in that direction.
The device works based upon a conservation of total momentum principle. Specifically, the device creates strong gravity waves in one overall direction (in fact, it is actually a gravity wave producing "antenna", which in itself is also a patentable claimed invention), which amounts to expelling momentum in that same direction, and therefore requiring the physical device to accelerate in the opposite direction to make up for the loss of momentum to the first direction. The gravity waves produced are the result of the coherent sinusoidal (continual) turning over to one direction and then the opposite direction of the spin directions of the material's nuclei, which are in essence tiny gravitomagnetic dipoles (and as such, tiny antennas). Gravity waves are produced in one overall direction and not very much in the opposite direction because of the 90 degree phase difference of the sources, and due to the calculated proper width of the gap. Another patentable claim that I have is that this device is an efficient gravity wave producing antenna, and can be used for that purpose alone, if desired.

On 1-28-2001, I published a web site, heavily censored, regarding my development of Maxwell-like equations of gravity to help establish the significant dates and times. These type equations have been co-developed by others. I also had my father put printouts and compact disks (CD's) into envelopes and mail them to me, for me to leave unopened so I would have some proof of date (albeit about 8 months after the actual invention date) as stamped on the envelopes by the post office. The dates of the envelopes he sent me are Feb 5, 2001. All of the pages of formulas & documentation that I had created earlier (the equivalent of my "lab notebook" as defined by the Patent Office) were signed and dated by me on 9-2-00. It was on that date that I realized I should have been documenting and proving dates of when things were being done. I wanted to scan the documents and possibly publish them at that time, but at the time I did not have the means to do so. From September 2000 to January 2001, I was sick quite often with a chronic illness, and I was struggling with such major problems as finding a new job, and then keeping a job and doing well at it. Finally, in January 2001, I was able to scan in all the documents, and then publish select pages (NOT CONTAINING ANY OF THE INVENTIONS DISCLOSED IN THIS DOCUMENT WHATSOEVER) to the web site. [PROCEED TO NEXT PAGE].

DATED RECORD OF INVENTIONS, #2: **On February-06-2001,** I invented some major improvements to the devices which were mentioned above and which were invented on 6-1-2000. I do not know if this constitutes a completely new invention category, or if it would be considered part of the same invention discussed above. The descriptions of all of these improved devices is as follows. Regarding the opposed cylinders device, I decided to add a large magnet (with perhaps, but not limited to, the shape of a large letter "C") around the device. The magnet could be a permanent magnet, or an electromagnet, or better yet, a superconducting electromagnet. The magnet must be very powerful, but I will not here suggest any quantitative values. The magnetic field of the magnet must be oriented so that the lines of flux of the magnetic field are in the direction of the axis of symmetry of the cylinders. All the other physical components of the device are the same as before, as described above. A drawing would perhaps, and typically, show, but not be limited to, a large "C" shaped magnet with the above described cylinders located within the gap of the magnet. The faces of the cylinders would be parallel to the faces of the gap of the magnet. For patentability purposes, the important basic claim is to have the strong static magnetic field present in this version, which is considered to have an improvement on the operation of the devices of 6-1-00. The principal and practices of Nuclear Magnetic Resonance (called NMR) could then be used to tune the device for proper operation and for maximum output. There is a commonly known list of material elements (in the periodic table of the elements) which are affected by and can be used in conjunction with NMR techniques. This list of materials for the most part is the same as the materials I specified in the inventions of 6-1-00. These are materials which have unpaired protons or neutrons. Therefore, they have a net overall spin value for the nucleus. Materials which have no unpaired protons or neutrons typically have no net spin, and thus will have no net Gravity "B" field. So, really the only change from the earlier devices is the presence of a large static magnetic field, and now the ability to use the well-known formulas of NMR to calculate what frequency and strength the microwave sources must be to either flip the atomic nuclei of the material by 90 degrees four times in succession (to equal 360 degrees total) per iteration cycle, or perhaps it would be possible to rotate the nuclei of the atoms by the full 360 degrees at once per iteration cycle. By iteration cycle, I mean forcing the nuclei to rotate by 360 degrees total per cycle. Most commonly used NMR techniques today are set up to pulse the electromagnetic source field just enough to rotate the nuclei of the target material by 90 degrees, but it is probably possible to do a full 360 degree rotation per pulse, and instead of having separate pulses, just merge these individual microwave source pulses into a smooth, continuous sinusoidal wave. The NMR formulas (not shown) are used to calculate, based upon a "gamma" value (Hz/Tesla) of the material used in the cylinders and based upon the source frequency used, the required strength of the fixed static magnetic field, and the required strength (magnitude) of the magnetic component of the electromagnetic source signals themselves. The fixed field is often referred to in the current literature as B0 ("B zero"), and the electromagnetic wave source as B1 ("B one"). I have calculated that B0 should be, perhaps, an order of magnitude (ten times) larger than the B1 field, but the device will operate anyway if this rule is not followed. I recommend for device operation that the temperature of the cylinders be very low, perhaps under 100 degrees Kelvin. This is because, according to the principles of NMR, low temperature materials will have a larger fraction of atoms aligned with the static magnetic field. Then, when the electromagnetic sources are applied (remember, there are two of them, 90 degrees out of phase as before in the 6-1-00 devices), a larger number of atoms will perform the synchronized sinusoidal reversals of spin direction. Thus a larger effect is produced. One material I would highly recommend for the device is liquid Hydrogen, confined to two shells (not necessarily cylindrical), because of liquid Hydrogen's low temperature, and because of its high "gamma" value.

[PROCEED TO NEXT PAGE].

DATED RECORD OF INVENTIONS, #3: Also on February-06-2001, I invented the following improvement to the device: Rather than have the electromagnetic radiation sources directed at the materials at close proximity, I propose placing each one of the two pieces of material inside the end of its own separate short microwave waveguide, where each one of the two waveguides is connected to its own private source of microwaves (i.e. a source is a microwave oscillator, examples of which are, but are not limited to , a Klystron, a Magnetron, a Twystron, a TWT, etc…). The waveguides should have a rectangular cross-section. A waveguide is basically a hollow metallic (conductive material) tube of constant cross-section. The sources are, as specified before, about 90 degrees out of phase for optimal operation. The device (now, the two parallel waveguides which are separated by a small gap) can be placed, if desired, in the gap of a large magnet as before to use NMR principles of operation. The waveguides should be short and should be terminated by metal sheet at the end (thus a metallic sealed enclosure), and they are more efficient than the external horn source method, as all the microwaves are trapped in the waveguides so they are not wasted to external space, and more importantly the waveguides provide shielding by protecting the user of the device from the microwaves. The optimum shape of the waveguides would be as a resonant cavity, which increases the microwave energy by definition.

There are many textbooks that describe how to excite different TE ("transverse electric") and TM ("transverse magnetic") modes in a waveguide by connecting coaxial source cables at different locations on a waveguide. A lowest order appropriate TM mode should be required for the device. Again, for this version of the gravity wave propulsion device, I recommend that liquid hydrogen be used as the material inside the waveguides.

DATED RECORD OF INVENTIONS, #4: Also on February-06-2001, I invented a helix-shaped version of the device. It would be difficult to tune the device for proper operation, but it could in theory be done. The idea is somewhat similar to commonly known helical antennas, except now the wound helix is not created from a long conductive wire, twisted into a helix shape, but the device is constructed from a very long rectangular waveguide, which is then twisted into a multi turn helix shape. The helix waveguide contains as before the proper type material as specified earlier. The device may employ two separate helix parts, or could possibly operate with just one helix, if properly tuned. In the two helix version, the two helices would share a common central axis (the imaginary axis the device was twisted about in order to construct it). The helices would be separated by a distance along the said imaginary central axis, so they are completely separate and not touching each other.

In either case, the spacing between and diameter of the windings must be calculated so that the gravity "B" fields produced don't cancel each other, and so that the overall gravity wave radiation will be mostly in one (average) direction, as opposed to evenly distributed omnidirectional radiation.
[PROCEED TO NEXT PAGE].

DATED RECORD OF INVENTIONS, #5: **On February-12-2001,** I invented a ring or triangle or square shaped version of the device. They are all actually similar, and the exact shape doesn't matter. Just so it's a closed loop. The device is imagined to consist of a large ring (or loop) of many copies of the individual devices described above. The loop is a closed loop. Rather than having many individual fixed magnets, one large overall (perhaps the same shape as the loop) magnet could be used instead. Also, importantly, each short rectangular waveguide would actually have two of its opposite sides removed, and then all of the short waveguides would be connected together side-by-side to form one large loop-shaped cavity, where the the hydrogen (or other suitable material) would be free to flow around the entire device. Of course, there would have to be two copies of the described loop or ring, one above and parallel to the other one. There would be many coaxial microwave source cables connected to the inner edge of each loop or ring. The cables would be like spokes on a wheel, and would all terminate at a central source (hub of the wheel) of microwave or RF signal. So, there would be two "spoked" wheels (or loops), one directly above and parallel to the other one. The sinusoidal microwave source at the center or hub of the top wheel would be set so it is approximately 90 degrees out of phase with the source at the center or hub of the bottom wheel.
The microwaves, upon being fed into the large ring at many points, would travel only a short distance from the inner edge of the wheel to the outer edge of the wheel.

DATED RECORD OF INVENTIONS, #6: **On February-19-2001,** I invented a phased-array version of the device. It can either be used to produce a variable direction (i.e. steerable) beam of gravity waves, or it can be used as a variable thrust direction (i.e. steerable) propulsion system. In its first version, it consists of a number of parallel waveguides, all with equal spacing between them. They are all arranged along a reference straight line, and they are all perpendicular to the reference line. Each waveguide is connected to, as usual, its own private oscillator source of microwaves. One could perhaps visualize this with the following analogy: Imagine several flagpoles stuck into the ground. They are arranged along the ground in a straight line, left to right. There is equal distance between neighboring poles. Now imagine each pole to actually be a long waveguide, and then imagine all of these waveguides to be, instead of stuck in the ground, stuck to or welded to a structural metal baseplate (which takes the place of the ground). This solidly constructed conglomerate forms the complete overall device (or system, if you prefer to call it that).

In the second version of this device (or system), imagine the flagpoles to be arranged in a circle. Then, as before replace each flagpole with a vertical waveguide, and then replace the ground with a structural baseplate. This overall configuration then forms the second version I invented for this device.

For either version of the design, the direction of the gravity wave beam produced, and thus the direction of thrust (force) upon the device-as-a-whole, can be dynamically controlled in real-time by varying the relative phases of the different microwave oscillator sources. In the two waveguide version of the device discussed in previous paragraphs, we normally set the two oscillator sources so they are 90 degrees apart in phase. For this many-oscillator device described here, there are well-known formulas and equations (developed for ordinary wire antenna arrays) that can be used to compute the required phase value needed for each of the sources, given what current direction one desires to aim the gravity beam that is being produced.
[PROCEED TO NEXT PAGE].

The geometrical arrangements described above are commonly used for phased-arrays of electromagnetic-wave antennas, where each individual antenna is a wire of conductive material. However, what's completely novel about my design and never-before-thought-of is the use of a waveguide instead of each antenna. So we have an array of parallel waveguides instead of an array of parallel conductive wires.

Also regarding the above two phased-array versions of my invention, here is some additional information for setting up the NMR-related parts of the system. Simply have a separate powerful magnet or electromagnet for each waveguide, such that each waveguide is centered in the gap (and thus field) of its corresponding magnet. This would work if the waveguides are spaced rather far apart. In the special case where we are using the first version (all in a line, NOT the circle), AND the parallel waveguides have a small distance between them, we can use only one magnet, so the configuration looks like a stack of closely separated, parallel waveguides where the stack arrangement is centered in the gap of the one large magnet.

DATED RECORD OF INVENTIONS, #7: On June-26-2000, I invented a gyroscope device for testing gravity. Specifically, it was to be used to test for what I call Gravity "C" and Gravity "D". The device consists of a spinning gyroscope hanging from a sensitive linear scale, such that the entire device, or at least the wheel itself of the gyroscope, is enclosed in a vacuum or a vacuum chamber. By "scale", I mean a device that measures either weight or force. The gyroscope wheel and spokes (if any) and axis must be made of material which is unaffected by electric or magnetic fields. Therefore the material cannot be either metallic or magnetic-related. An obvious choice would be glass (silicon dioxide). Other choices are possible. The orientation of the spinning gyroscope axis determines whether we are able to test for Gravity "C" or Gravity "D". I am not disclosing at this time which orientation can test for which of the two types of gravity. The net result of the test is as follows. The scale reading for the weight of the gyroscope when it is not spinning is noted. Then a scale reading for the weight of the gyroscope while it is spinning rapidly is noted. The rotation speed must be very high, perhaps 30,000 RPM, or so. The difference between the two weight readings is proportional to the amount of said gravity type. In this case, the Gravity "C" or "D" is being produced by the Earth. Gravity "C" and "D" in general are generated by the mass of an object (in this specific case, the Earth), and not by the motion of that same object. However, the force effects of "C" or "D" can only be sensed by a second object (in this specific case, a gyroscope), which has relative motion with respect to the generating object.

Note that the device is affected by a Gravity "B" field which is generated by the rotation of the Earth, but this only produces a torque on the gyroscope which does not affect its measured weight by the scale (if the gyroscope is mounted properly so it is free to precess). It is, in fact, possible to measure Gravity "B" of the Earth by measuring the net torque produced on the gyroscope. There is a formula for this, not shown here.
[PROCEED TO NEXT PAGE].

What's not new about these inventions (pre-existing knowledge and physics theories): (this section may not be complete)

Electrons, protons, quarks, spins of particles, electromagnetic fields, microwave devices, NMR devices in general, the term "gravitomagnetic" field, General Relativity formulas, Maxwell-like re-formulation (done by others, but unknown to me at the time I co-developed my own version of a Maxwell-like re-formulation) of General Relativity formulas, waveguides (basically a hollow metallic tube of constant cross-section) and horn antennas, microwave oscillator sources (whose power source can be just a standard AC wall outlet), coaxial cables (these look just like a "cable-TV" type cable) for connecting microwave sources to microwave waveguides, superconducting magnets, knowledge of the existence of gravity waves of a general form (in all textbooks I've seen, they only predict the presence of quadrapole waves, NOT of dipole waves like I derived in my Maxwell-like re-formulation; in some major books they insist that dipole waves could not possibly exist!), conservation of momentum principles in general, **dipolar** antennas for producing **electromagnetic** waves, NMR principles and formulae, materials effective for use with NMR, Helix **wire** antennas for electromagnetic waves, and methods of exciting different electromagnetic wave propagation modes in waveguides (this just amounts to connecting the coaxial cable from the microwave source to the different standard connector ports which are already affixed to the waveguide).
[PROCEED TO NEXT PAGE].

What's new about these inventions (original work done my me or concepts invented by me, with no knowledge at the time of invention of any possible similar work or ideas or patents in existence).
This section will contain items which are outright completely new, or which are at least new, unusual, or never-before-tried or thought of designs or combinations of existing designs:
(this section may not be complete)

NOTE: I am guessing that only the items marked with asterisks ("***") are actually and legally patentable: So it will later be determined which items are in fact patentable.

1) The re-formulation of the common General Relativity formulae into Maxwell-like formulas, similar to Maxwell's well-known formulas for electricity and magnetism, but with electric charge being replaced with mass and mass flow rates, and with the electric field replaced by what I call the Gravity "E" or "A" field and the magnetic field being replaced with what I call the Gravity "B" field (now referred to by others as a "gravitomagnetic" field). Gravity "A" is just the gravity we are all familiar with, being the ordinary gravitational attraction of the Earth. I also discovered what I call Gravity "C" and "D" fields, whose manifested forces are both dependent on the velocity of the object which is sensing the force (as is the case for the "B" field), and both of which are generated solely by the static mass of (Not the movement of) any large nearby gravitating object. Both the "C" and "D" fields are dependent on cross-combinations of the rectangular coordinate velocity components of the object sensing the field. The "B" field forces are dependent only on the simple velocity components of the object sensing the field, and not on any cross-combinations. Importantly, for my theory of how to model the gravity effects contained in subatomic particles, I found that the "C" and "D" fields do not exist, which simplifies the situation down to having only "A" and "B" fields. In fact in many situations, even the "A" field does not exist. I thought these derived formulas were new at the time I derived them, but later found several people had recently done the same type of derivations of formulas (within the last 30 years, but most in just the last 5 years). However, these authors have not yet, to my knowledge, made mention of what I call the "C" and "D" fields. I have only seen the "A" (ordinary static) and "B" or gravitomagnetic field talked about. I have invented, many months ago, designs for devices for measuring the "C" and "D" fields of the Earth, preferably done at the poles, which are based on gyroscopes made of inert (non-metallic and non-magnetic) material. I may present these here, or in a separate patent application. I should mention that the "A", "B", "C", and "D" fields I derived were the result of solving the common General Relativity equation for "weak field solutions" to the general non-linear equation. Most authors on the subject of General Relativity mention that "weak field solutions" are possible, and a few of the authors actually write down the general beginnings of such solutions. Very, very few authors have gone to the extra step of deriving exact Maxwell-like versions of the general weak field solutions, and I've only heard about such writings, but I have not actually seen any one of them. As far as I know, there may even be different versions of such Maxwell-like solutions, and it is possible that my solutions are different from other people's solutions.

2) *** Creating an electromagnetic wave or gravity wave antenna for the specific purposes of producing a <u>net (or overall)</u> **force** on the antenna. Antennas are currently only used for transmitting or receiving radio-related (i.e. electromagnetic wave) signals.

3) Concluding that there may be such a thing as negative mass, since reasons for its existence were compelling and prevalent in connection with the formulas I derived.
<u>[PROCEED TO NEXT PAGE].</u>

4) **Concluding (on 6-01-2000) that electrons and other lowest-order constituent (i.e. non-divisible, as we know it) subatomic particles can be modeled as counter-rotating rings (or spheroids or wavelets, etc.) of positive and "negative" mass, such that the magnitude of the positive mass essentially equals the magnitude of the negative mass. Therefore, there is no net effective mass due to these "masses" present (i.e. they cancel), but there is a net "effective" mass of the particle due to the effective mass of an extremely strong Gravity "B" field present. The dipolar Gravity "B" field results from the rotational flow of both the positive and negative mass components.** There is little or no gravity "A" field generated. Since the generated Gravity "B" field is present, and since it is extremely strong and localized, **I hypothesize that what we commonly measure as the usual mass of the electron (or other subatomic particle) is actually just the effective mass caused by the "B" field.** It is commonly known in today's physics that pure forms of energy such as magnetic fields, electromagnetic fields, gravity waves, and static gravity fields actually possess a mass (or rather, "effective" mass), albeit a very tiny one. They also can possess a momentum. I proposed (on 6-1-2000) that the effective mass we routinely measure for subatomic particles is entirely due to their extremely strong, localized dipolar Gravity "B" field, and there is no "real" or "actual" mass present. **The theory just described was still unknown to any other persons as of the original date of my first writing this disclosure document [Feb. 20, 2001], as far as I know and to the best of my very informed knowledge.**
Calculations show that the Gravity "B" field intensity of typical subatomic particles must be extremely large, due to the fact that it takes an incredible amount of energy field density to create even a miniscule amount of effective mass. The consequences of this are staggering. It should be possible to generate extremely powerful gravitational forces by the proper manipulation of large numbers of atoms in unison. This easily opens the door to such possibilities as personal flying craft capable of ground to Earth orbit travel and beyond, with only reasonable and moderate amounts of energy necessary to power the required gravity wave propulsion engine. Also the intriguing idea occurs that, since such a craft would be enveloped by an artificially generated gravity wave field, the passengers of the craft would probably not experience any subjectively observable forces that are normally due to inertia and acceleration. This is because, in any gravity field, objects just fall and accelerate in the direction of the field, and don't feel any forces as a result of the acceleration. In essence, the craft could accelerate without limit and even make quick right angle turns, and the occupants of the craft wouldn't feel a thing. Finally, an even more intriguing thought occurs. If the craft can produce extremely powerful gravity fields as predicted for just moderate expenditure of input energy (from the microwave sources, in this case), then the craft could accelerate very rapidly to a large percentage of the speed of light in a reasonably short amount of time, without hurting the passengers in any way. Then, from the passenger's viewpoint and as such a physically real experience for that person, and due to the well-known fact that objects travelling near the speed of light experience certain length contraction effects as well as a change in the rate of time itself, the passenger will notice that he or she is able to travel many, many light years in just a short amount of time. Thus, travel to other worlds in the far reaches of the galaxy is now a real possibility.

5) Concluding by energy density calculations that the electron, proton, and neutron must emit an incredibly huge (at least locally) Gravity "B" field. Only a very weak "A" field (which is the normal type of gravity we are all familiar with) is emitted due to the effective mass of the "B" field.

6) *** Finding ways to convert Gravity "B" to ordinary gravity waves.

7) Concluding also that electrons must be many orders of magnitude smaller in size than is currently thought. [PROCEED TO NEXT PAGE].

8) *** Using NMR techniques for creating powerful asymmetrically emitted gravity waves or for creating a net force on an overall device.

9) *** Using two (or more) "targets" in NMR, one target being exposed to an electromagnetic radiation source which is out of phase with the other (90 degrees is the optimum amount) target's source of electromagnetic radiation. Normally, just <u>one</u> target and <u>one</u> electromagnetic radiation source is used in NMR.

10) *** Use of parallel, closely spaced waveguides, to produce gravity waves in a preferred overall direction, as well as to produce a net force on the device.

11) The existence of dipole fields of gravity waves. All graduate-level books about gravitation that I've seen mention quadrapole waves only, and some even appear to say that dipole fields are not possible.

12) *** The helix structured version of the gravity wave propulsion device. Also, the large loop version of the device.

13) *** Identifying 90 degrees as the optimal phase shift angle between the outputs of the two electromagnetic wave sources that are typically part of any gravity wave propulsion device. This assumes equal distance from each source to its respective target material. Just the fact that at least two sources of electromagnetic waves are necessary to build a gravity wave propulsion device should in itself be a separate patentable claim.

14) *** Identifying that typical frequencies of the sources would have to be in, or greater than, the microwave portion of the electromagnetic spectrum. This is to accommodate the fairly close spacing required between the two material pieces (targets), and considering that the net force effect depends upon the speed of light transit time between the two material pieces.

15) *** Using <u>**two**</u> (or more) pieces of, or containers of, material to produce gravity waves or wave-based propulsion (since the net force effect occurs because of the speed of light delay of <u>any</u> types of energy fields (or fields of force) travelling between the pieces). The Gravity "B" field of one piece of material produces an instantaneous force on the other piece, and vice versa. It takes two separate pieces of material, arranged in an opposing way, to produce a time-averaged net force upon an entire device. Note however that the single helix version of the device would only require one piece of material.

16) *** Using <u>**waves**</u> (either electromagnetic or gravitational) for a device's propulsion through empty space, instead of using static fields or slowly changing fields. In fact, although some have proposed using static fields for this, static fields will not work for such propulsion (through empty space, such as far from the Earth). I have seen numerous attempts to specify the use of static fields of various sorts to attempt to move objects which are at or near the surface of the Earth, and some of these may have some possibility of working, although to me, most of these schemes simply won't work either.

17) *** Created a design for the first known <u>high-output</u> gravity wave producing antenna.

18) *** Use of liquid hydrogen or other low temperature elemental materials to produce gravity waves.

19) *** Design of a gravity-based propulsion device that does not require any fields or forces produced by the Earth to "push against" or to interact with to produce lift.
<u>[PROCEED TO NEXT PAGE]</u>.

20) *** Use of conservation of momentum principles in the design of gravity wave-based propulsion, or for electromagnetic wave-based propulsion. This principle, for energy waves, states that waves emitted in one direction will cause a reaction force upon the emitter in the opposite direction.

21) *** Performing NMR-like (sinusoidal synchronized re-directioning of nuclear spins) techniques where both the target and the generated electromagnetic waves (coming from some source) are confined to a waveguide.

22) *** Inventing a phased-array system of several (or more) parallel waveguides for the purpose of generating a dynamically steerable beam of gravitational waves, or for the purpose of creating a propulsion system having a dynamically steerable thrust. By "dynamically steerable", I mean that the direction of the beam or of the force can be changed in real-time by simply altering the relative phases of the several constituent waveguide sources (i.e. the sinusoidal electromagnetic wave source provided for each waveguide).

23) *** Designing or using a gyroscope to measure what I call Gravity "C" and Gravity "D".

[PROCEED TO NEXT PAGE].

Physical Construction of these systems and devices, understanding of:
It is believed that almost anyone who is familiar with catalogs of, or descriptions of, the "off-the-shelf" readily-available components for these specified devices, such as Klystrons, Magnetrons (this is the same tube used in almost all consumer microwave ovens), waveguides, large electromagnets, and so on, could easily geometrically arrange and then anchor (secure) the specified devices into place, in the physically correct configurations as are described herein this document. The selected material could easily be placed in the waveguides, and the microwave sources could be connected easily by supplied coaxial cable to the coaxial connectors on their respective waveguides. The microwave sources need a power source, and the manufacturer specifies whether to connect the device (Klystron or Magnetron, etc.) to a standard AC power source of 110 volts, 220 volts, or 440 volts.

However, it is believed that to tune the devices, it would typically require a person degreed in Electrical Engineering or Engineering Physics or simply Physics to be able to properly "tune" the device for proper operation. By "tune", I mean carefully adjusting distances between waveguides, adjusting the frequency and phase of the microwave sources, and adjusting the intensity of both the microwave sources and / or the large external electromagnet. It is probable that such degreed person may need to briefly study the various widely-available sources of information on just basic NMR theory and equations. Or, such persons could acquire the required tuning formulas and equations from me (not shown in this document). The equations are quite simple, in fact much simpler that most equations that such degreed persons have to normally use and deal with in their line of work.

Once tuned, virtually anyone could operate the device, first given safety instructions and then told how to simply turn on and off the microwave sources (there should be an AC power switch for each one).

Claims (as required for the purpose of Patentability):
Note: this list of claims is still a subject of change and has definitely NOT been finalized at the time of this document. I fully expect that additions and deletions and changes will have to be made before this list of "Claims" is put into the final Patent Application Document.

(This list of claims is to be determined later by the attorneys, but I suspect it will be substantially the same as the list of items marked with asterisks ("****") in the above "what's new" section).

.

(To Be Determined)...

.

Disclaimers: The information contained herein is not final, but is intended to later be re-written as a formal and standard patent application, or firstly as a patent search type document. The information is believed to be complete to the best of my knowledge (except for the "Claims" section, the "What's not new" section, and the "What's new" section) and correct. Drawings may be added later to supplement the document. Still, the possibility of omissions of pertinent information or of incorrect recollection of thoughts or of incorrect statements about device operation and physics does exist. Therefore, I retain the right to make appropriate changes, deletions, and additions to correct for these possibilities in any future versions of either this document or of other documents. [PROCEED TO NEXT PAGE].

Separate Disclaimer regarding any possible related and/or attached drawings for this document or even related drawings which are a part of other documents: The wording of this document stands legally by itself and is complete even if no drawings are attached. Any drawings related to this document are for helpful visualization purposes only, and in cases where a drawing conflicts with the text-based descriptions in this document, the text wording in this document will prevail and will take precedence over any such drawings. If however, a drawing contains some visual information which was overlooked and was accidentally not described in this document, then I reserve the right to later change this document to include the omitted information derived from the drawing.

[PROCEED TO NEXT PAGE].

...n Disclosure Document For Patent Off. page 15 of 15. Date(02/22/01).

...Rights Reserved for the author. DO NOT COPY, PENALTY OF LAW.

SIGNATURE PAGE SIGNATURE PAGE SIGNATURE PAGE

By signing below, I the INVENTOR swear and affirm that all of the information herein this complete document, including dates for inventions, is complete and true and accurate to the best of my knowledge, except for the "Claims" section, the "What's not new" section, and the "What's new" section of this document, where these three exceptioned sections are true and accurate, but not necessarily complete. I also, by signing below, swear and affirm that I believe myself to be the original, first, and sole inventor of the described complete systems and devices and of the described UNIQUE and NOVEL uses and configurations/combinations of existing available subsystems and devices:

Richard Perry Crandall 2-22-2001

Richard Perry Crandall, INVENTOR SS# 453 80 7642 Date Signed:

Location Signed:

Notary Witness information:
By signing and dating below, you the swear and affirm and acknowledge that you were a witness to the act of the above named INVENTOR signing this document as you watched, and you have verified the identity of the above signer to be Richard Perry Crandall by looking at the signer's Colorado driver's license, and by looking at the signer's Social Security Card.

This 22nd day of February, 2001. Cathy R. Dudley State of Colorado County of Arapahoe My commission expires 1-24-04.

CATHY R. DUDLEY NOTARY PUBLIC STATE OF COLORADO

Other Witness Information:
By signing and dating below, I am stating and agreeing that I have read this document and I understand the physical construction of (using the described "off-the-shelf" readily available components such as waveguides and sources) and use of / for the inventions described herein.

FEB 22, 2001 2/22/2001

Witness #1 date Witness #2 date

RAND FANSHIER Tyler Gagnon

Extremely Sensitive Personal Data. Only my AUTHORIZED AGENTS or the U.S. Patent Office may read or reproduce. All other access is strictly forbidden.

U.S. Patent Office may read or reproduce. All other access is strictly forbidden.

Extremely Sensitive Personal Data. Only my AUTHORIZED AGENTS or the U.S. Patent Office may read or reproduce. All other access is strictly forbidden.

[ ]

[THIS PAGE IS INTENTIONALLY BLANK]

Today's date:
April 09, 2001

Audience: NO UNAUTHORIZED ACCESS to this document is allowed. PENALTY OF LAW. The only individuals who may read or possess this document are responsible employees of the U. S. Patent and Trademark Office, and my duly authorized agents. If you are one of my authorized agents (and not a member of the Patent Office), you must have obtained prior permission and authorization from me personally as well as having signed a non-disclosure / non-co-inventor / non-compete / confidentiality document, before you can read this document. I must retain a copy of your signed authorization. The U. S. Patent and Trademark office may read and make copies of this document without any prior permission.

Author and Inventor:
Richard Perry Crandall, SS# 453-80-7642
6229 Zenobia Ct.
Arvada, CO 80003
(303) – 429-3939

Subject:
ADDITIONAL (ADDENDUM) summary and history of my inventions, theory and claims, and significant related activity and dates of invention, for my impending patent application.

This ADDENDUM to my IDD serves as an additional part of my "INVENTION DISCLOSURE DOCUMENT", and I request that it ALSO be retained by the U. S. Patent Office along with my "INVENTION DISCLOSURE DOCUMENT" for however maximum period as it can be kept on file.
[PROCEED TO NEXT PAGE, IF DULY AUTHORIZED].

Possible title for the final patent (may change):

Devices for Gravity Wave Propulsion, Generation of Gravity Waves, and Directional Gravity Wave Antennas, and an Antenna-based Gravity Wave Communication System.

Author and Inventor:
Richard Perry Crandall, SS# 453-80-7642

Dated History of, theory of, and specification of patentable inventions:

DATED RECORD OF INVENTIONS, #8: On April-07-2001, I decided that a new feature may need to be added to my previous inventions. In all those inventions which specify two co-axial, opposed cylinders of special material (where there is a gap between the parallel end faces of the two cylinders), it may be necessary to add two high-gear-ratio electric motors, for the purpose of spinning each cylinder independently at about 30,000 revolutions per minute. Each cylinder would spin about its central axis. For the design variation which uses two parallel waveguides, such that each contains a cylinder of the special material, there would have to be a hole in each waveguide for the respective motor shaft to go through. The motors cannot be inside of the waveguides, as this would cause damage and impair operation.

I can envision several scenarios of operation, but I'm not sure at this time which would be the best. Here is a partial list:

1) The fixed magnetic field (if used) is "turned on" first. The two cylinders are spun up to maximum speed, left that way for some time, and then spun down quickly to a stop. Then, the electromagnetic oscillator sources are turned on, and normal device operation begins.

2) The fixed magnetic field (if used) is "turned on" first. The two cylinders are spun up to maximum speed, and keep spinning at maximum speed. Then, the electromagnetic oscillator sources are turned on, and normal device operation begins.

3) Same as 1) above, except the fixed magnetic field is "turned on" last, when the oscillator sources are turned on.

4) Same as 2) above, except the fixed magnetic field is "turned on" last, when the oscillator sources are turned on.

I am including the two motors in this new variation of the devices, not to align the spins of most of the nuclei in each cylinder, since we have the large fixed magnetic field to do that (although the spinning will have this beneficial effect anyway), but rather to cause each individual atom's nucleus in the cylinders to have a very large self angular momentum (about its own axis of rotation). Therefore, in cases 1) and 3), even when the cylinders are stopped, the individual nuclei in them should each still retain a large rate of spin.

I believe it may be necessary for each nucleus to continue to have a large amount of spin (as long as the overall propulsion device or gravity wave antenna device is in operation), as opposed to the normal spins of nuclei as found in non-moving matter, because it may require a large nuclear spin to cause each nucleus to behave as one composite pair of comparatively large-diameter counter-rotating rings of positive and negative mass, as opposed to having each nucleus behave as a cluster of many smaller particles, each behaving as a separate pair of comparatively small-diameter counter-rotating rings of positive and negative mass.

The larger the effective diameter of the rings, the greater is the force produced. This is apparent from the force formulas found on pages H37 through H39 in the lab notebook. They were based upon the formula for the inductance of a ring of electrical current, but were modified to the model of particles having counter-rotating rings of positive and negative mass flow. So hopefully, each nucleus can be made to act (or perhaps naturally does act) as one large effectively monolithic set of such rings whose diameter is about the same as the diameter of the nucleus.

DATED RECORD OF INVENTIONS, #9: **On April-07-2001,** In the descriptions of all my previous inventions, I overlooked the fact that larger diameter nuclei will create a much larger overall force (again, refer to pages H37 through H39 in the lab notebook). I specified liquid Hydrogen as a possible candidate material, but it really may not work well due to its small nuclear diameter. The devices would probably work much better if they use, for the special material, elemental matter whose nuclei are comparitively very large diameter, and then just use liquid hydrogen or liquid nitrogen to cool the two cylinders of material to very low temperatures so the population of coherent spins is a large number. The material used, as before, must still be such that its nuclei have un-paired neutrons or protons. Also, it would be better to use materials with as high "gamma" value as possible, for the NMR operation to work well.

DATED RECORD OF INVENTIONS, #10: **On April-01-2001,** I thought of the idea of a sort of gravity wave "radio" communication system, analogous to ordinary radio transmitters and receivers. The transmitter would consist of just one waveguide with special material inside. This waveguide is connected to a microwave oscillator source, which provides the "carrier" frequency. This source oscillator can then be AM (amplitude) modulated or FM (frequency) modulated, in order to add information content to the gravity wave carrier being transmitted.

The receiver would consist of one waveguide with special material inside. But instead of having a source microwave oscillator connected to the waveguide, we have connected to the waveguide a tuned electronic circuit for detecting the carrier frequency (specified above), and frequencies in a band close to that carrier frequency. The receiver works based upon the principle of reciprocity, commonly seen in Electrical Engineering literature and textbooks. The atoms of the special material should be affected by an incoming gravity wave carrier, and should therefore have the spin vectors of the nuclei oscillate their orientation at the carrier frequency. This should produce microwaves (electromagnetic waves) which could be detected by the aforementioned tuned circuit. Any instantaneous deviations of the carrier frequency will be detected at the receiver's tuned circuit as information "riding" on the carrier wave. This information can be separated from the carrier sine wave by standard procedures for receiving AM or FM signals.

IDD Addendum- Attach to the IDD For Patent Off. page 4 of 4. Date(04/09/01).

All Rights Reserved for the author. DO NOT COPY, PENALTY OF LAW.

SIGNATURE PAGE SIGNATURE PAGE SIGNATURE PAGE

By signing below, I the INVENTOR swear and affirm that all of the information herein this document is true and accurate, including dates for inventions, to the best of my knowledge. I also, by signing below, swear and affirm that I believe myself to be the original, first, and sole inventor of the described complete systems and devices and of the described UNIQUE and NOVEL uses and configurations/combinations of existing available subsystems and devices:

Richard P. Crandall 4-12-2001

Richard P. Crandall, INVENTOR SS# 453 80 7642 Date Signed: 4-12-01

Signed by Richard P. Crandall on 4-12-01.

Patricia A. Stoltenberg

PATRICIA A STOLTENBERG
NOTARY PUBLIC
STATE OF COLORADO
My Commission Expires 09-18-03

Extremely Sensitive Personal Data. Only my AUTHORIZED AGENTS or the U.S. Patent Office may read or reproduce. All other access is strictly forbidden.

Extremely Sensitive Personal Data. Only my AUTHORIZED AGENTS or the U.S. Patent Office may read or reproduce. All other access is strictly forbidden.

Today's date:
April 4th, 2001

Audience: NO UNAUTHORIZED ACCESS to this document is allowed. PENALTY OF LAW. The only individuals who may read or possess this document are responsible employees of the U. S. Patent and Trademark Office, and my duly authorized agents. If you are one of my authorized agents (and not a member of the Patent Office), you must have obtained prior permission and authorization from me personally as well as having signed a non-disclosure / non-co-inventor / non-compete / confidentiality document, before you can read this document. I must retain a copy of your signed authorization. The U. S. Patent and Trademark office may read and make copies of this document without any prior permission.

Author and Inventor:
Richard Perry Crandall, SS# 453-80-7642
6229 Zenobia Ct.
Arvada, CO 80003
(303) – 429-3939

Subject:
Explanatory Supplemental Information. This document explains the connections between my IDD and my Lab Notebook (not included here).

I respectfully and gratefully request that this helpful "EXPLANATORY SUPPLEMENT" document be retained by the U. S. Patent Office <u>IN ADDITION TO</u> and in the <u>SAME FILE WITH</u> my Invention Disclosure Document, for however maximum period as it can be kept on file.

I understand that this request may not be possible under your Invention Disclosure Document submission regulations. If you cannot file this supplemental document with my IDD document, then this supplemental document may be simply shredded, or otherwise destroyed.

Thank you.

[PROCEED TO NEXT PAGE, IF DULY AUTHORIZED].

Today's Date:
04-04-2001

SUBJECT:
Explanation of Connections Between Invention Disclosure Document (IDD) and Lab Notebook.

[The IDD was notarized on 02-22-01.]
[The Lab Notebook was signed and dated on 09-02-00, and was notarized on 02-28-01.]

The Lab Notebook is NOT contained herein. It is a separate document.

AUTHOR:
Richard P. Crandall, SS# 453-80-7642.

PART 1 of 2: Explanation of the complete content of my Lab Notebook:

The complete original Lab Notebook is effectively 139 pages, and its content has not been (since 9-2-2000) and will not be changed or altered in any way. I have preserved it as-is since when all 139 effective pages of it were signed and dated by me on 9-2-2000. There was no notary to witness this signing, but I have witnesses who can verify the timeframes, and earlier dates claimed in the notebook. I was quite chronically ill around that time of 9-2-2000, and a lot of things went by un-accomplished during that time, including creating a better proof-of-date than just signing the documents myself.

I currently have several printed copies of the 139 page document, as well as have all 139 pages scanned and recorded onto a CD (i.e. a Compact Disk). The CD was created in January of 2001. All of my printed copies were created by using a word processor program to print out the scanned in pages. I had the most critical pages of the ORIGINAL DOCUMENT notarized on 02-28-2001, as is explained in a notarized memo I authored on 02-28-2001. This was because of advice of attorneys, but I realize I should have had the pages notarized many months ago. The pages I had notarized were 1AA1, 1AA2-3, A1, A2, A3, A37A, A37D, G1, G38, G39, G45, H1, H37, H38, H39, and HX5. I also had a separate NMR (Nuclear Magnetic Resonance) calculation page (not part of the original lab notebook) notarized on 02-28-2001 which calculates example values for the B0 (the fixed magnetic field) and B1 (microwave field, generated from oscillator source) fields, based upon given values of material gamma (MHz / T), and upon microwave source oscillator frequency f (sub c). This single calculation page was created on 02-06-2001.

I had two memos notarized on 02-28-2001. **One describes why my Lab Notebook contains no drawings of my inventions.** In essence, I very seldom ever do any drawings (even simple ones), as I keep these mechanical shapes and designs in my head (my professors in college really hated this, and wondered how I could still solve problems and get the right answer without any drawings). **Also, none of my inventions were mentioned by name in the Lab Notebook, except on what I call pages 1AA1, 1AA2, and 1AA3.** Actually these are physically just 2 pages, as one has writing on both sides of the paper. These were the final 3 pages which I created as part of the Lab Notebook, and they were handwritten and signed on 09-02-2000, and were on that date added as the front 3 (physically 2) pages of the Notebook. The Notebook was then complete, and has not changed since then. The other memo (notarized 02-28-2001) describes **why I have no further Lab Notebook pages nor other document pages, except for the sole one-page NMR calculation mentioned above, for my inventions dated 02-06-2001 and onward.**

I will now explain the numbering scheme of my Lab Notebook. The vast majority of the Lab Notebook just contains many, many mathematical formulae, most derived by me and others taken from books (usually, I mention the Author). Often, the formulae have no comments to describe their significance or meaning. There are very few diagrams or wording. There are a total of 139 pages (any cover sheets or back-end blank sheets are not included in the count), when considered as front side only. A few pages in the original document used both front and back, but in all that follows, I will talk about the copies of the document, which were all printed out as front-side only. More precisely stated, the one-sided copies (of the original notebook) contain exactly 139 pages. If a cover page is present, it is not numbered.

No handwritten numbering appears on the first three pages, but I refer to these pages as 1AA1, 1AA2, and 1AA3, because that reflects the file names used for them when I stored each and every individual scanned-in page of the Notebook onto a CD. The next 42 pages are each hand-numbered. I refer to them as A1 through A37D, because that reflects the file names used for them as stored on the CD. However, the hand-numbering does not have the letter "A" in front of the page number. I pre-pended the letter "A" for the purposes of storing the page scans as files. The next 49 pages are hand-numbered (except for G17B and G19B). I refer to them as G0A through G45, due to the file names used. The letter "G" does appear in the hand-numbering scheme. The next 40 pages are all hand-numbered. I refer to them as H1 through H39, which reflects both the hand-numbering and the file names used to store them on CD. The letter "H" does appear in the hand-numbering scheme. The next (and final) 5 pages are all hand-numbered. I refer to them as HX1 through HX5, which reflects both the hand-numbering and the file names used to store them on CD. The letters "HX" do appear in the hand-numbering scheme. Thus, the total number of pages equals 3 + 42 + 49 + 40 + 5, which is equal to 139.

I refer to the pages 1AA1, 1AA2, 1AA3, and A1 through A37D collectively as the "A" series papers. I refer to the pages G0A through G45 collectively as the "G" series papers. I refer to the pages H1 through H39, and HX1 through HX5 collectively as the "H" series papers.

Chronologically, the "A" series papers (EXCEPT for 1AA1, 1AA2, and 1AA3, which are the first three pages, and EXCEPT for A37B, A37C, and A37D) were written first, starting 10-02-1999 and ending ABOUT 11-01-1999.

The "G" series papers were written next, starting ABOUT 11-01-1999 and ending on 05-07-2000. On 06-01-2000, the date of my first invention, I wrote pages A37B, A37C, and A37D, and I also changed the name of page A37 (just 37 on the original) to A37A (just 37A on the original).

The "H" series papers were written next, starting on 07-01-2000, and ending on 09-02-2000. In fact, the chronologically last page I wrote in the "H" series was H39, being done on 09-02-2000. I had written pages HX1 through HX5 before I wrote page H39, but at the time they were un-numbered "scratch" pages for some calculations. So on 09-02-2000 I numbered these un-numbered pages as pages HX1 through HX5 and added them to the very end of the Lab Notebook document.

So on 09-02-2000, upon having just written page H39 (and thus realizing what an incredibly large force could be produced by what I'm calling my first invention (as was invented on 06-01-2000)), I immediately wrote pages 1AA1, 1AA2, and 1AA3, since I then realized I should have been keeping track of important dates all along, and that my invention of 06-01-2000 might be an incredibly significant find, perhaps being able to change the world as we know it. The pages 1AA1 through 1AA3 contain an accounting of dates of significant events, discoveries, inventions, writings, thoughts, and calculations. Upon having added the last three pages (i.e. 1AA1, 1AA2, and 1AA3), the Lab Notebook document was then complete. I then signed and dated all 139 pages of this Lab Notebook document on September 2nd, year 2000. The Lab Notebook document has remained (and will remain) unchanged since then. All pages of this original document were scanned into a computer in the form of files during January, year 2001, and the resulting 139 files were written to non-alterable storage in the form of a write-only Compact Disk (CD).

Now, regarding page numbering, I will be more specific. For convenience, I will always refer to the page numbers as their corresponding file names as taken from the CD. As discussed earlier, the handwritten page numbers differ slightly from these file names. If the entire document is printed on the front side of pages only, there will be 139 pages total. Here is an accounting of all 139 pages, in the order in which they appear in the document, where ranges of pages are used where possible to shorten the description: 1AA1, 1AA2, 1AA3, A1, A2, A3, A4 through A12, A13A, A13B, A13C, A14 through A36, A37A, A37B, A37C, A37D, G0A, G0B, G1, G2, G3 through G17, G17B, G18, G19, G19B, G20 through G37, G38, G39, G40 through G44, G45, H1, H2 through H33, H34A, H34B, H35, H36, H37, H38, H39, HX1, HX2, HX3, HX4, and HX5.

The original purpose of and content of the 3 major sections of the Lab Notebook document was as follows:

The "A" series of pages were written in order to re-develop and expand upon some formulas and designs, original work done by me, that I had developed back in 1984 (the year I graduated with a BSEE degree (Bachelor of Science in Electrical Engineering)). My original papers from back then have long since been lost. What the "A" papers describe (except for the last 4 pages: A37A, A37B, A37C, and A37D, and except for pages 1AA1, 1AA2, and 1AA3) is the following:

1) Page A1: The already known fact that a steady electrical current in one wire will cause a force to be exerted on a parallel wire which also carries a steady electrical current.

2) Page A2: An original idea I developed in 1984 that by applying a periodic square-wave electrical current to each one of the two parallel wires, such that the frequency of each is the same and is a constant, **and such that the waveforms are ninety degrees out-of-phase,** then **there will be a NET average force on the device-as-a-whole.** This situation is different from the device on Page A1. For THAT device, yes there is a force exerted on each of the two wires, but the forces are of equal amount and are of opposite direction. Thus, in THAT case, the net total of the forces on the whole device is ZERO. For THIS case, the device produces a net total average force upon itself because it relies on the fact that any change in electric or magnetic fields will propagate through space at the speed of light, as a maximum velocity. So if we suddenly reverse the direction of the current in one of the wires in a two-wire device, then the force on that wire will reverse direction instantaneously, but the force on the other wire will not change for a small fraction of time, due to the propagation delay of the magnetic field coming from the other wire. So, for that very small amount of time, there will be a net force upon the whole device. The frequency of the square waves is

important, and must be calculated based upon the distance between the parallel wires. It is calculated as the speed of light, "c", divided by 4 times the distance between the wires. The idea can be extended to having two opposed square loops of wire (the plane of each loop is parallel to the plane of the other loop), instead of the two simple parallel wires. The other rules are the same. On pages A3 through A6, I expand the idea to using a sine or cosine wave (i.e. a so-called sinusoidal wave) instead of the square wave. The bottom line is this for this device: This electromagnetic device could in theory be used as a PROPULSION SYSTEM, provided the force produced is strong enough. This is because the device has a net average force acting upon it. However, all calculations I've ever done show that the force achievable in any realistically built design of this type would be very, very feeble indeed. Not really acceptable to achieve propulsion or levitation. Incidentally, as any BSEE degreed person could tell you, this design could also incorporate circular loops instead of the square ones. The device could also be specified as two opposing (multi-turn) coils of wire. Such person would also tell you that since microwave frequencies are involved, the number of turns (loops) in each coil would have to be a small number (1, 2, or maybe 3). The reason that microwave (i.e. greater than about 1 GigaHertz) frequencies must be used for the current sources is that for most reasonable distances of separation between the two coils, the formula mentioned above for calculating the frequency required would usually yield frequencies in the stated region (microwave).

3) Page A7 through A30: This is a lengthy set of calculations that was done in order to help verify the design concepts stated in item "2)" above. The force calculations done in the above item are just a bit too simplified and approximating to be completely trusted. These pages (A7 – A30) do a rigorous (much more accurate) calculation in the far-field of the two parallel wire device of above. The two parallel wire device of above is treated as a radiating antenna of electromagnetic waves. Due to the "conservation of momentum" principle of physics, the rate of momentum being "ejected" from this (antenna) device should equal the net average force produced upon this device. The device ejects momentum because it radiates most of its electromagnetic waves in one overall direction, and electromagnetic waves are known to possess a momentum. To produce a thrust upon any object, there must be an "exhaust", according to physics principles. In this case however, instead of having extremely hot gases or particles as exhaust (such as in a rocket or jet engine), we have powerful electromagnetic waves ejected instead. This is ok since these waves have momentum, just like the got gases or particles. The overall result of these lengthy calculations of rate-of-ejected-momentum yielded some integral formulas which were too complicated to integrate. However, it could easily be seen from them that the resulting force values of these integral formulas were definitely "in the ballpark" of the force values I calculated in pages A1 through A6. Thus, I could then trust even the simple calculations for the force produced by the (antenna) devices as were shown in pages A1 through A6. Result: even simple intuitive formulas for forces produced in opposed-coil devices are valuable and fairly reliable tools for investigative purposes.

4) Page A31 through A36: I calculate the exact near-field equations for the total average force generated by a two-parallel-wire device (which can be extrapolated to describe opposed coil devices, as can be the equations of section "3)"). This was also intended to be used as a check upon the crude calculations done in Pages A1 through A6. Again, the resulting integral formulas were too difficult to fully integrate, but it could be seen that their values were at least in the range of values as were calculated in Pages A1 through A6.

The additional papers in the "A" series are described as follows:

Page A37A was written on ABOUT 11-01-1999 (in fact it was chronologically the last page written in the original "A" series (EXCEPT for pages 1AA1, 1AA2, and 1AA3, and EXCEPT for pages A37B, A37C, and A37D)), and it contained equations which showed a simple (but essentially correct, as it would turn out later (AFTER I had completed the "G" series papers)) correspondence (which I had proposed, on 10-02-99) between static electric fields and static gravitational fields (such as generated by the Earth's mass). Assuming the correspondence was correct, I was able to calculate a value for epsilon for a static gravitational field, based upon the well-known gravitational constant "G". There is already a value of epsilon for electromagnetic fields, which is well-known. Epsilon (for electromagnetism) is called the permittivity constant of free space. Then, using the standard relation between the speed of light and epsilon and a constant called mu, I was able to calculate a value of mu for gravity fields. Mu (for electromagnetism) is called the permeability constant of free space. When I wrote this, I knew that if mu is to have any meaning at all for gravity fields, then there would have to exist a new form of gravity which mainly no one has ever known about the existence of before. I named this the Gravity "B" field, analogous to the well-known magnetic "B" field of electromagnetism, although I never actually used the name I had created in the Lab Notebook (except on page 1AA1, where I stated that "the electron is really a strong Bg field", where this refers to the gravity "B" field, the subscript "g" meaning "for gravity", and also on page 1AA3). In fact, I later discovered Gravity "C" and "D" fields (on page G38, which shows four terms in an equation, where each term corresponds to Gravity "B", Gravity "A", Gravity "C", and Gravity "D", respectively). The existence of these four fields is obvious from the equations on page G38, but these fields were just never called out by name in the Lab Notebook (excepting page 1AA1 and page 1AA3). Anyway, this new Gravity "B" field would have to behave in an analogous manner to the well-known magnetic field of electromagnetism. In many places in the Lab Notebook, I refer to this Gravity "B" field in equations as the letter "B" (or "H") subscripted by the letter "g". An ordinary magnetic field is referred to by a letter "B" (or "H") with no subscript, as is commonly done in electromagnetism. In fact, I commonly use the subscript "g" to distinguish the gravitational equivalent of any electromagnetic quantity. Nowadays, it is apparent that my Gravity "B" field has also been co-developed by others, and it is generally known as the "gravitomagnetic" field. Now to make a long story short, the very existence of my Gravity "B" field as well as its corresponding mu value meant to me that I could do calculations similar to the parallel wire device, but do them based upon mass and gravity-related forces, instead of upon electric charge (or current) and electromagnetic type forces. So, I could do a calculation similar to the one on page A1, but instead of using mu for electromagnetism in the formula, I would use mu for gravity. Rather than having "I" in the formula be electric (charge) current, it would instead be mass flow (such as kilograms per second). The two parallel wires of electric current would instead be replaced by two parallel tubes containing mass (i.e. fluid) flow. What this all means is that if you have two parallel tubes, each containing a constant (very, very high velocity, no doubt) mass flow, then each tube will exert a measurable force on the other tube. In chronologically later pages of the Lab Notebook, I tried to create a gravity wave (instead of electromagnetic wave) "antenna"-based propulsion device, which would consist of the two parallel tubes containing "square wave" or "sinusoidal wave" mass flows (analogous to what was done with electric current flows described in Pages A2 through A6). The idea failed miserably when I calculated the mass flow rates needed for a practical device, as will be seen on page A37D.

Pages A37B, A37C, and A37D were written on 06-01-2000, and contained equations which showed that gravity wave propulsion based upon parallel tubes of sinusoidal mass flow (each 90 degrees out of phase with the other) is in theory possible, but would be VIRTUALLY IMPOSSIBLE to make work in actual practice. It would require somehow forcing ten kilograms of mass to flow at 99.999 percent the speed of light, AND WOULD ALSO REQUIRE reversing the direction of this mass flow at the rate of billions of times per second!! This miserable design failure was what prompted me that same day to think of my first invention, as is described in the Invention Disclosure Document (IDD). **Exactly what conclusions I came to that day, which gave me the idea for my first invention, will be detailed later in this memo.**

Finally, Pages 1AA1, 1AA2, and 1AA3 were written on 09-02-2000. They detail important dates and significant events throughout my original idea and design development period. Some of the interesting things the reader will see in the three pages are as follows: All of my research was prompted by a very large interest in UFO's and UFO propulsion. "IUFOMRC" stands for the International UFO Museum and Research Center. I developed my (still unknown to anyone on Earth to the best of my knowledge) "counter-rotating rings" theory of the electron and (by obvious extension, as per anybody with a physics degree) therefore of quarks and other subatomic and atomic particles on 06-01-2000. Quarks are currently considered to be the lowest order constituent particles which protons and neutrons in atoms are comprised of. Also, on 06-01-2000, I discovered (in my head, of course) the two variations of my first invention, as talked about in the IDD document. Only one of the two variations was actually described in the 1AA3 page. As per the IDD document, one variation was having opposed coils with a (typically cylindrical shaped) <u>special</u> piece of material inside of each coil, and the other variation was having just the material pieces with no coils around them and having a separate microwave frequency electromagnetic oscillator source antenna (such as a horn antenna) aimed at each of the two material pieces. From my Lab Notebook, I quote page 1AA3: "Also thought of an opposed coils device with 90 degrees out of phase and materials inside the coils". Also, from page 1AA3: "Completed numerical calcs of B field force for the opposed coils device (with material = copper or Iron).". I am stating here and now in this memo that both versions had actually occurred to me on 06-01-2000, even though only the first version was recorded on paper. The microwave oscillator source antenna version (with no wire coils) was obvious to me on that date, and would be an immediately obvious variation on the first design to any person degreed in Electrical Engineering, upon such person having first gained knowledge about the first design. I also realized on 06-01-2000 (and this is not recorded in the Lab Notebook) that the materials of my device must be made of elements whose nuclei contain unpaired protons or neutrons, because it was obvious to me then that in nuclei where all particles are paired, there will be no net gravitomagnetic (or Gravity "B" field) since the directed such field of each proton or neutron is exactly cancelled by the oppositely directed such field of another proton or neutron. The critical 90 degree phase feature is mentioned on Page 1AA3. On Page 1AA3 I mention that I completed numerical calculations of the Gravity "B" field where the material in the "cylinders" is copper or iron. Those calculations were for materials whose weight and density and atomic RADIUS were similar to copper or iron (but the material didn't actually have to BE copper or iron), but these scratch pages were actually thrown away, and what was included in the Lab Notebook was a calculation for materials whose nuclei were at least of the RADIUS of the Uranium nucleus (but do not actually have to BE Uranium). This calculation appears on Page H39. I had discovered in pages H34B through H38 that the radius of the nuclei of the materials used in the device is an important factor, and that nuclei with larger radius would produce a larger net overall force on the device.

Also, based upon the entries in my Lab Notebook for 5-7-00 and for 6-7-00, I can **rightfully claim that I am an independent co-developer of the "Maxwellized Equations of Gravity" and of the "gravitomagnetic field" itself.** This is due to the fact that at the time I developed these, I had no knowledge of any similar work or ideas of others whatsoever.

The "G" series of pages were written by me in order to create and examine "weak field" or "linear" solutions to the well-known main Einstein equation of gravity (which is for strong, non-linear fields), and to then see if these resulting solutions are in any ways analogous to the well-known Maxwell equations of electromagnetism. This series of pages contains a successful carrying out of that goal, which was original work done by me with no knowledge at the time that others had done essentially the same thing. Since a successful analogy was achieved finally on page G45, then the "G" series of pages lends complete validation to using formulas such as those found on pages A1, A2, and A37A for the purposes of calculating gravitational and gravitomagnetic forces between mass flow currents, in complete analogy to how those formulas are presently used for calculating electric and magnetic forces between electrical (charge) flow currents. What the "G" papers describe is the following:

1) Page G0A through G0B: These formulas came from a book, but I decided not to use them.

2) Page G1 through G2: These are equations taken from various books on tensor calculus and on General Relativity (authors Synge, Spiegel, Einstein, and Dirac).

3) Page G3 through G8: These are equations taken mainly from a book by Einstein, but re-written slightly for clarity. They are used to compute the famous "K" constant in the main Einstein equation, which can be seen at the top of page G9.

4) Page G9 through G14: All these equations for electromagnetism can be found in a book by Synge ("Relativity: The Special Theory"), but I re-derive these for clarity in a different order from which they are found in Synge's book. It should be noted that all equations in Synge's book are using "CGS" units of measurement. I also heavily modified one of Synge's equations for use on my page G9. These pages contain NO gravity-related formulas.

5) Page G15 through G21: This is where my original work starts. I convert all of the important "Synge" electromagnetic equations from pages G9 through G14 to a similar set of equations which are based upon different units of measurement and NOT upon the "CGS" units of measurement used by Synge. This new special system of units was devised by me in order to explicitly bring out the values for mu and epsilon, which are implied and/or hidden in the Synge formulations.

6) Page G22 through G25: I compute some various relations between "E", "A", and Phi, which are seen to be in agreement with commonly known electromagnetic relations.

7) Page G25 through G29: I calculate the force equations and equations of motion for electric charges moving in an electromagnetic field, starting out with equations derived on previous pages, and also using an equation from Synge shown on page G25. Mu and epsilon are prominent in the equations derived in these pages. The final force law I derive, which is on page G29, is seen to agree with the well-known "Lorentz force law".

8) Page G30: This is where the really most significant part of my original work begins. I discover a relationship between the Einstein "K" constant and the "mu" constant I had created for gravity. This "mu" value for gravity was created by me on ABOUT 11-01-1999, as shown on page A37A. For electromagnetism, "mu" is called the permeability. I define a new gravitational "16-potential" tensor "phi" in terms of an integral of the well-known mass flow current tensor "T", assuming that such can be done (it later turns out it can be done). This is in analogy to defining or calculating the "4-potential" "phi" for electromagnetism in terms of an integral of electrical charge flow current as was shown on page G19. For gravity though, there are 16 components instead of 4.

9) Page G31: Therefore, due to the new "K" – "mu" relation, I can write the weak field solution tensor "gamma", from page G8, in terms of "mu" multiplied by an integral of the well-known

mass flow tensor "T". I therefore note a correspondence between the weak field solution tensor "gamma" and the gravitational "16-potential" tensor "phi". In fact, they are the same except for a constant factor which depends on "mu" for gravity and the speed of light value "c".

10) Page G32: Starting with the well-known geodesic equation of motion for a particle in a gravity field, I calculate the "4-force" representative "Z". The actual "4-force" "X" is just equal to the particle mass "m" multiplied by "Z". This "Z" must therefore be equal to the Christoffel symbol for the gravity field multiplied by "lambda" sub m and multiplied by "lambda" sub n. Here, "lambda" is the well-known "4-velocity" for a particle. Now, since we are seeking a weak-field solution ("gamma"), we can re-write the Christoffel symbol in terms of various partial derivatives of "gamma". The final result of this page is then my formula for calculating "Z" in terms of "gamma" and in terms of "lambda".

11) Page G33: Replacing the tensor "gamma" variables by the equivalent "16-potential" "phi" variables, I arrive at a formula for "Z" in terms of "mu", derivatives of "phi", and "lambda". For brevity, I will in all that follows refer to the "16-potential" "phi" as just the "16-phi", and I will refer to any "4-potential" "phi" as just the "4-phi". Or, I may refer to these two possibilities as the "16-potential" and as the "4-potential", respectively. Using this notation, "Z" is thus written in terms of "mu", derivatives of "16-potential", and "lambda" (the 4-velocity). I then, for notational convenience, replace the derivatives of "16-potential" with a new tensor quantity "psi", where "psi" has three tensor indices, and it thus has 4 x 4 x 4 = 64 components. The final result for this page is then that "Z" is a product of "mu" and "psi" and "lambda" "lambda".

12) Page G34 and G35: I re-write the formula for "Z" by eliminating irrelevant higher order terms. Then, "Z" is written in terms of new variable "Gprime", which is written in terms of "psi".

13) Pages G35 through G37: Neglecting higher order terms, I examine all the different cases for calculating "psi". The "psi" values are generally computed to be the sum of various derivatives of the "16-potential", being different sums for the different possible cases.

14) Page G38: This is perhaps my most important result page in the Lab Notebook, when it comes to understanding all the effects of gravity fields upon moving objects. The "4-force" (and thus the regular force) upon a moving test mass is the sum of four terms. These are due to gravity "B", gravity "A", gravity "C", and gravity "D", respectively, with reference to the equation of page G38, which I could call the "Crandall Gravity Force Formula". The formula shows that there are really 4 types of gravity. This is to be compared, in analogy, to the well-known force upon a moving test charge due to an electric and magnetic field as seen on pages G28 and G29. The electromagnetic force formula (G28 and G29) shows a force due to a static "E" (electrostatic) field and a velocity dependent force due to a "mu" "H" (magnetic) field. Written equivalently, the electromagnetic force upon the test charge particle is due to a static "A" (i.e. "E") field, and a velocity-related "B" (i.e. "mu" times "H") field. The gravity force formula (page G38) looks analogous to the electromagnetic force formula, in as much as the "A" (ordinary static gravity) and "B" (gravitomagnetic) fields are concerned, but there are two additional forces, namely due to gravity "C" and gravity "D". Looking at the simplified version of the force equation of page G38 (i.e. the last equation on that page), we can draw some interesting conclusions. If there is any net mass present at all, the force produced upon an infinitesimally massive test mass due to the ordinary static gravity "A" force generated by the net mass will **far** exceed any forces produced upon the test mass due to the other "B", "C" and "D" fields, which are also generated by the net mass. This is true no matter how fast the small test mass may be moving. Therefore, if there is any net mass present, the gravity "B", "C" and "D" fields cannot be detected, as they are drowned out by the "A" (ordinary gravity)

field by a factor of "c" squared. Also, just as bad, it's in general very hard to detect the "gravitomagnetic", or "B", field, as it is usually drowned out by the "C" and "D" fields when the velocity of the test mass is large. Therefore, to get any physically non-trivial and meaningful results at all, we must have the net mass be zero. However, we can still have mass flow currents, just so long as the sum of all of the mass present (both positive mass, and possibly negative mass) is zero. **So, in order to get non-trivial results, we conclude that there may be such a thing as "negative" mass or matter (not to be confused with well-known "antimatter"), and that the signed algebraic sum of all mass present must be zero. Only then can we generate (via mass flow currents) and detect gravity "B", the elusive "gravitomagnetic" field.** This observation and conclusion later led me to a further conclusion, on 6-1-2000, that subatomic particles, such as electrons and quarks, can be modeled as being comprised of counter-rotating rings of equal positive and negative mass. All that's left, then, is just a very strong, localized gravity "B" field. What we routinely measure as the mass of such a particle is actually just the "effective" mass due to a strong localized "B" field, given by the relation effective mass = field energy / ("c" squared). Therefore, there is no "real" mass present in a particle, just an equivalent "effective" mass! The electromagnetic spin (vector) of a particle must be aligned with the gravity "B" field lines. This conclusion then immediately led me to the idea, also on 6-1-2000, of an opposed coils device with material cylinders inside the coils. The coils are driven by electrical current oscillator sources, in order to sinusoidally flip the nuclei of the material atoms and thus produce an extended overall oscillating gravity "B" field in each cylinder. In analogy with page A2, if the sources are 90 degrees out of phase, there should then be produced an overall gravity force upon the device-as-a-whole. Thus we have achieved a gravity wave propulsion system.

15) Page G39 through G44: I assume for all that follows that the static or net mass is zero, or very, very small. Then, in the force formula, the "A", "C", and "D" terms drop out, leaving only the Gravity "B" term (which is the gravitomagnetic field). The "16-phi" is then seen to reduce down to a "4-phi", since only one column of the "phi" matrix then remains non-zero. We finally see, on page G44, that the force "Z" upon a small test mass "m star" depends on a quantity called "G" and upon the test mass velocity "lambda". Also, "G" is seen to depend upon the derivatives of the "4-phi" or "4-potential". Finally, the "4-potential" is calculated from an integral of "L tilde", where "L tilde" is the modified "4-momentum" vector which completely describes whatever mass current flows are creating all the gravity fields.

16) Page G45 (the final result page): Finally, the assumption on page G30 regarding the ability to define a "16-potential" as an integral of the mass flow tensor "T" is now seen to be justified. This is because on page G45 we can arrive at a final result which shows that the equations of the gravity and gravitomagnetic field are very much analogous to Maxwell's equations for electric and magnetic fields. Compare Maxwell's equations on page G17 to the Maxwell's for gravity waves I derived on page G45. On page G45, "Eg" is the static (ordinary) gravity field, and "Hg" is the gravitomagnetic field, and "rho star" is the mass density. The vector "u" is the velocity of the mass density. Also, the vector force equation on page G45 shows how to detect a gravitomagnetic field using a moving test mass "m star".

The "H" series of pages were written by me in order to recreate all the content of the "G" pages, but using a different "metric", such as that used in the book "Gravitation" by Misner, Thorne, and Wheeler. This goal was not completed. But, more importantly within the "H" series, I perform some calculations of what force would be produced by a device such as that described in the IDD in the section "DATED RECORD OF INVENTIONS, #1". This is an opposed cylinders device with either two microwave antenna oscillator sources, or with opposed coils of wire connected to separate electrical oscillator sources. What the "H" papers describe is the following:

1) Page H1 through H10: Standard formulas from the "G" pages, but re-written using Wheeler's metric.

2) Page H11 through H17: A review of Newtonian (i.e. pre-relativistic) physics. This section also contains a "mathematical digression" which looks at some useful integration formulas.

3) Page H18 through H20: I look at weak field solutions of the well-known geodesic equation of motion, and end with several useful Newtonian correspondence formulae.

4) Page H21 through H32: I write equations describing a linearized (i.e. weak field) theory of gravity, starting out with the main Einstein non-linear equation. I show on page H31 a solution to the linearized equations. On page H32, I finally calculate the value of the well-known Einstein "K" constant.

5) Page H33 through H36: I write an approximation of the product "gamma" "gamma", where "gamma" is a Christoffel symbol of the gravity field. This could be useful in later writings.

6) Page H37 through H38: This is quite important. I start with a known formula for the inductance of a ring of electrical current, and derive a formula for calculating the maximum gravity "B" field of a typical electron, proton, or atomic nucleus. This is completely original work done by me.

7) Page H39: This is the "Grand Finale". I take the formula for the maximum gravity "B" field from the previous two pages and derive a formula for calculating the gravitomagnetic force between two atomic nuclei. I then use this formula to calculate the force between two nuclei which are the size of a uranium nucleus (only size matters here, not numbers of protons and neutrons). I then apply a correction which accounts for the fact that in an opposed cylinders device, there are many, many atomic nuclei present, and so I apply a correction factor which is related to Avogadros' number. The resulting value, which is 4 times 10 to the tenth power newtons, is the maximum force which could be produced from an opposed cylinders gravity wave propulsion device made of material whose nuclei are as large as a uranium atoms' nucleus. This is a very large number!! I did this calculation on 09-02-2000, and it then prompted me to write pages 1AA1 through 1AA3 to start accounting for dates and events. I had also done the calculation for copper or iron as the material, but these scratch pages were misplaced and lost. Note on page 1AA3, the entry for 9-2-00 reads "Completed numerical calcs of B field force for the opposed coils device (with material = copper or iron)."

8) Page HX1 through HX5: These are essentially scratch pages, where I look at torque on loops and rings. On page HX5, I look at a loop geometry, since I was at the time thinking that electrons and other particles could be composed of counter-rotating rings of mass. These pages were written a few days before Page H39 was written, and were numbered and added to the end of the notebook on 09-02-2000.

End of Part 1.

PART 2 of 2: Explanation of how and when I invented the items listed in the Invention Disclosure Document based upon contents of my Lab Notebook, and upon the calculational findings and formulae discovered / derived in that Lab Notebook:

Regarding inventions described in "DATED RECORD OF INVENTIONS, #1" in the IDD, I now explain my train of thought which led up to those inventions (thought of on 06-01-2000).

Chronologically, the "A" series papers (EXCEPT for 1AA1, 1AA2, and 1AA3, which are the first three pages, and EXCEPT for A37B, A37C, and A37D) were written first, starting 10-02-1999 and ending ABOUT 11-01-1999. The "A" series of pages were written in order to re-develop and expand upon some formulas and designs, original work done by me, that I had developed back in 1984 (the year I graduated with a BSEE degree (Bachelor of Science in Electrical Engineering)).

Shown on Page A2 is an original idea I developed in 1984 which states that by applying a periodic square-wave electrical current to each of **two parallel wires**, such that the frequency of each is the same and is a constant, **and such that the waveforms are ninety degrees out-of-phase,** then **there will be a NET average force on the device-as-a-whole.** The device works due to the speed of light propagation delay effect, as was explained earlier. This electromagnetic device could in theory be used as a PROPULSION SYSTEM, provided the force produced is strong enough. This is because the device has a net average force acting upon it. However, all calculations I've ever done show that the force achievable in any realistically built design of this type would be very, very feeble indeed. Not really acceptable to achieve propulsion or levitation. The device could also be specified as **two opposing (multi-turn) coils of wire.** The reason that microwave (i.e. greater than about 1 GigaHertz) frequencies must be used for the current sources is that for most reasonable distances of separation between the two coils, the formula mentioned above for calculating the frequency required would usually yield frequencies in the stated region (microwave).

Page A37A (then A37) was written on ABOUT 11-01-1999, and it contained equations which showed a simple (but essentially correct, as it would turn out later (AFTER I had completed the "G" series papers)) correspondence (which I had proposed, on 10-02-99) between static electric fields and static gravitational fields (such as generated by the Earth's mass).

The "G" series papers were written next, starting ABOUT 11-01-1999 and ending on 05-07-2000. The "G" series of pages were written by me in order to create and examine "weak field" or "linear" solutions to the well-known main Einstein equation of gravity (which is for strong, non-linear fields), and to then **see if these resulting solutions are in any ways analogous to the well-known Maxwell equations of electromagnetism.** This series of pages contains a successful carrying out of that goal, which was original work done by me with no knowledge at the time that others had done essentially the same thing.

On 06-01-2000, I changed the name of page A37 (just 37 on the original) to A37A (just 37A on the original). Pages A37B, A37C, and A37D were then written on 06-01-2000, and contained equations which showed that gravity wave propulsion based upon parallel tubes of sinusoidal mass flow (**each 90 degrees out of phase with the other**) is in theory possible. It would rely on gravitomagnetic fields to be produced by the tubes, in turn causing a net average force to be felt by the device, due to the speed of light propagation delay effect (same principle as for the electromagnetic case described on page A2). But, the device would be VIRTUALLY IMPOSSIBLE to make work in actual practice. It would require somehow forcing ten kilograms of mass to flow at 99.999 percent the speed of light, AND WOULD ALSO REQUIRE reversing the direction of this mass flow at the rate of billions of times per second!! This miserable design failure was what prompted me that same day to think of my first invention, as is described in the Invention Disclosure Document (IDD).

So how did this failure prompt me to think of my first invention? Partly by process of elimination, and partly by a force equation I had developed in the "G" series pages. The only other way I could think of that powerful enough gravitomagnetic fields could be produced would have to be that strong gravitomagnetic fields already exist, and are produced by atoms (electrons and/or nuclei) themselves naturally. We just never detected these fields since nuclei are arranged randomly most of the time, and so the fields cancel out. From Page G38, I saw that the "4-force" (and thus the regular force) upon any moving test <u>mass</u> is the sum of four terms. These are due to gravity "B", gravity "A", gravity "C", and gravity "D", respectively, with reference to the equation of page G38, which I could call the "Crandall Gravity Force Formula". The formula shows that there are really 4 types of gravity. The gravity force formula (page G38) looks analogous to the electromagnetic force formula, in as much as the "A" (ordinary static gravity) and "B" (gravitomagnetic) fields are concerned, but there are two additional forces, namely due to gravity "C" and gravity "D". Looking at the simplified version of the force equation of page G38 (i.e. the last equation on that page), I drew some interesting conclusions. <u>If there is any net mass present, the gravity "B", "C" and "D" fields cannot be detected, as they are drowned out by the "A" (ordinary gravity) field by a factor of "c" squared.</u> Also, just as bad, it's in general very hard to detect the "gravitomagnetic", or "B", field, as it is usually drowned out by the "C" and "D" fields when the velocity of the test mass is large. Therefore, to get any physically non-trivial and meaningful results at all, we must have the <u>net mass</u> be zero. However, we can still have <u>mass flow currents</u>, just so long as the sum of all of the mass present (both positive mass, and possibly negative mass) is zero. **So, in order to get non-trivial results, we conclude that there may be such a thing as "negative" mass or matter (not to be confused with well-known "antimatter"), and that the signed algebraic sum of all mass present must be zero. Only then can we generate (via mass flow currents) and detect gravity "B", the elusive "gravitomagnetic" field.** This observation and conclusion led me to a further conclusion, on 6-1-2000, that **subatomic particles, such as electrons and quarks, can be modeled as being comprised of counter-rotating rings of equal positive and negative mass (for cancellation purposes). All that's left, then, is just a very strong, localized gravity "B" field (due to the rotation).** What we routinely measure as the mass of such a particle is actually just the "effective" mass due to a strong localized "B" field, given by the relation effective mass = field energy / ("c" squared). **Therefore, there is no "real" mass present in a particle, just an equivalent "effective" mass! The electromagnetic spin (vector) of a particle must be aligned with the gravity "B" field lines.** This conclusion then immediately led me to the idea, also on 6-1-2000, of an opposed coils device with a special material cylinder inside of each coil. The coils are driven by separate electrical current oscillator sources, in order to **sinusoidally flip the nuclei of the material atoms** and thus produce an extended overall oscillating gravity "B" (gravitomagnetic) field in each cylinder. In analogy with page A2, due to the speed of light propagation delay effect, if the sources **are 90 degrees out of phase,** there should then be produced an overall gravity force upon the device-as-a-whole. Thus we have achieved a gravity wave propulsion system. This was the basis for my first invention.

So, on 06-01-2000, I discovered (in my head, of course, with no drawings) the two variations of my first invention, as talked about in the IDD document. Only one of the two variations was actually described in the 1AA3 page. As per the IDD document, one variation was having opposed coils with a (typically cylindrical shaped) special piece of material inside of each coil, and the other variation was having just the material pieces with no coils around them and having a separate microwave frequency electromagnetic oscillator source antenna (such as a horn antenna) aimed at each of the two material pieces. From my Lab Notebook, I quote page 1AA3: "Also thought of an opposed coils device with 90 degrees out of phase and materials inside the coils". Also, from page 1AA3: "Completed numerical calcs of B field force for the opposed coils device (with material = copper or Iron).". I am stating here and now in this memo that both versions had actually occurred to me on 06-01-2000, even though only the first version was recorded on paper. The microwave oscillator source antenna version (with no wire coils) was obvious to me on that date, and would be an immediately obvious variation on the first design to any person degreed in Electrical Engineering, upon such person having first gained knowledge about the first design. I also realized on 06-01-2000 (and this is not recorded in the Lab Notebook) that the materials of my device must be made of elements whose nuclei contain unpaired protons or neutrons, because it was obvious to me then that in nuclei where all particles are paired, there will be no net gravitomagnetic (or Gravity "B" field) since the directed such field of each proton or neutron is exactly cancelled by the oppositely directed such field of another proton or neutron. The critical 90 degree phase feature is mentioned on Page 1AA3.

This concludes the description of how and why I thought of the inventions described in **"DATED RECORD OF INVENTIONS, #1" (06-01-2000) in the IDD.**

For all of my inventions described in the rest of this memo, THERE WAS NO LAB NOTEBOOK. I was able to conceive of these later inventions without resort to writing down calculations, with the exception of the one page of NMR calculations.

Regarding inventions described in "DATED RECORD OF INVENTIONS, #2" in the IDD, I now explain my train of thought which led up to those inventions (thought of on 02-06-2001).

After looking at a great deal of material on the Internet regarding Nuclear Magnetic Resonance (abbreviated NMR), I concluded that my devices of 06-01-00 would work much, much better if I used NMR techniques in their design. The only new item needed was a large fixed magnet. NMR is typically used in practice to first align the spins of many of the atomic nuclei in a target material (using a very large fixed magnetic field), and to then cause most of these same nuclei to flip their direction of spin to some other direction (typically 90 degrees). I saw that this kind of process was exactly what I needed to be using in my design, except I needed a flip of 180 or 360 degrees, not 90 degrees. Since there is so much information available on the Internet regarding NMR and NMR formulas, I only needed to write up just one page of simple calculations to fully understand how I could apply NMR to my design. That one page was notarized on 02-28-2001. It calculates example strengths needed for the B0 (the fixed magnetic field) and B1 (microwave field, generated from each of two oscillator sources) fields, based upon given values of material gamma (MHz / T), and upon microwave source oscillator frequency f (sub c). This single calculation page was created on 02-06-2001. As before, we need to use special materials for the cylinders, whose nuclei contain unpaired protons or neutrons. In addition, according to NMR principles, it is better to use very low temperature materials.

This concludes the description of how and why I thought of the inventions described in **"DATED RECORD OF INVENTIONS, #2" (02-06-2001) in the IDD.**

Regarding inventions described in "DATED RECORD OF INVENTIONS, #3" in the IDD, I now explain my train of thought which led up to those inventions (thought of on 02-06-2001).

I did not need to resort to calculations or drawings for this version of the device, as excellent example calculations and drawings for waveguides and different waveguide modes can be found in a book which I am very familiar with, namely "Microwave Devices and Circuits" by Samuel Y. Liao (from publisher Prentice Hall). See pages 95 through 110 in that book. Waveguide theory is fairly old and well-established. Also, the devices mentioned in the IDD such as Klystrons, Magnetrons, etc, are described in detail in the same book. There were really no formulas worth writing down, as these can be looked up in such book, or other similar textbooks. This is what I meant when I said, in the memo notarized on 02-28-01, that "...these later inventions were obvious (to me) and needed no further recordable formulae to be written into a "lab notebook" and of course the physical designs in my head were all I needed to understand the devices and their operations."

This concludes the description of how and why I thought of the inventions described in **"DATED RECORD OF INVENTIONS, #3" (02-06-2001) in the IDD.**

Regarding inventions described in "DATED RECORD OF INVENTIONS, #4" in the IDD, I now explain my train of thought which led up to those inventions (thought of on 02-06-2001).

I did not need to resort to calculations or drawings for this version of the device, as excellent examples of helix antenna theory can be found for example in a book "Antenna Theory and Design" by Stutzman and Thiele (from publisher Wiley). See section 6.1, "Helical Antennas", in that book. Again, there were really no formulas worth writing down. The helix would consist of a twisted waveguide, as opposed to the usual twisted wire conductor. The same formulas and modes of operation apply, just like for the wire helix. The device is to be used in the "axial" mode of radiation. Since the device will emit circularly polarized gravity waves in one particular direction (which is along the axis of the helix), there will be a net force experienced by the device in the opposite direction.

This concludes the description of how and why I thought of the inventions described in **"DATED RECORD OF INVENTIONS, #4" (02-06-2001) in the IDD.**

Regarding inventions described in "DATED RECORD OF INVENTIONS, #5" in the IDD, I now explain my train of thought which led up to those inventions (thought of on 02-12-2001).

I did not need to resort to calculations or drawings for this device, as it is obviously and simply constructed from many of the earlier described devices, all arranged side-by-side in a circle or loop. Again, there were really no formulas worth writing down.

This concludes the description of how and why I thought of the inventions described in **"DATED RECORD OF INVENTIONS, #5" (02-12-2001) in the IDD.**

Regarding inventions described in "DATED RECORD OF INVENTIONS, #6" in the IDD, I now explain my train of thought which led up to those inventions (thought of on 02-19-2001).

I did not need to resort to calculations or drawings for this version of the device, as excellent examples of phased-array antenna theory can be found for example in a book "Antenna Theory and Design" by Stutzman and Thiele (from publisher Wiley). See section 3.7, "Phased Arrays", in that book. Also see chapter 10, "Antenna Synthesis". Again, there were really no formulas worth writing down, since they are all in the book. The formulas in the book for calculating phases of different array elements for electromagnetic array antennas can also be used to calculate the phases of the different waveguide oscillator sources in my gravity wave system of parallel waveguides.

This concludes the description of how and why I thought of the inventions described in **"DATED RECORD OF INVENTIONS, #6" (02-19-2001) in the IDD.**

Regarding inventions described in "DATED RECORD OF INVENTIONS, #7" in the IDD, I now explain my train of thought which led up to those inventions (thought of on 06-26-2000).

This invention was just a simple consequence of looking at the "unabridged" force formula on page G38, and then seeing how the circular motion of the atoms in a gyroscopic spinning wheel would be affected by forces resulting from the various components of gravity, namely "B", "A", "C", and "D", which are generated by the mass and spin of the Earth.

This concludes the description of how and why I thought of the inventions described in **"DATED RECORD OF INVENTIONS, #7" (06-26-2000) in the IDD.**

End of Part 2.

Explanatory Supplement to IDD For Patent Off. pg 17 of 17. Date(04/04/01).

All Rights Reserved for the author. DO NOT COPY, PENALTY OF LAW.

SIGNATURE PAGE SIGNATURE PAGE SIGNATURE PAGE

By signing below, I the INVENTOR swear and affirm that all of the information herein this complete document is complete, true, and accurate to the best of my knowledge:

Richard P. Crandall 4-4-2001

Richard P. Crandall, INVENTOR SS# 453 80 7642 Date Signed:

Notary Witness Information:
By signing and dating below, you swear and affirm and acknowledge that you were a witness to the act of the above named INVENTOR signing this document as you watched, and you have verified the identity of the above signer to be Richard P. Crandall by looking at the signer's Colorado driver's license and any additional required identification as you see fit.

Arlyn Carl Baak, Notary Public 4-4-2001

State of Colorado
County of Arapahoe
Commission Expires 11-12-2004

ARLYN CARL BAAK NOTARY PUBLIC STATE OF COLORADO

Extremely Sensitive Personal Data. Only my AUTHORIZED AGENTS or the U.S. Patent Office may read or reproduce. All other access is strictly forbidden.

Extremely Sensitive Personal Data. Only my AUTHORIZED AGENTS or the U.S. Patent Office may read or reproduce. All other access is strictly forbidden.

[ ]

[THIS PAGE IS INTENTIONALLY BLANK]

Appendix H4

The OLD A, G, & H Papers

[ ]

[THIS PAGE IS INTENTIONALLY BLANK]

≈ 5/84 I calculated that two opposed square loops could be excited with a 90° out of phase electrical sine wave to produce a <u>net force</u> on the whole device.

On 10-2-99 visited the IUFOMRC for the first time. This was the date I first realized that gravity waves must work much the same as electromagnetic waves. Could instead use gravity the same way as the e/m version (of 5/84).

On <u>11-1-99</u> I remember discussing my study of gravity waves at a job interview (MeleaStar). I did not mention the two square loops or two long cylinders containing sinusoidal mass flows. (which I had in mind).

On <u>12-28-99</u> re-visited IUFOMRC.

Glenn Dennis and I talked about physics some, and he told me his son was into physics too.

<u>1-1-00 thru 3-15-00</u> <u>sick</u>, and trying to get other work done.

<u>3-15-00 thru 4-26-00</u> I remember talking to Alex K. and Rocco L. about my mass-flow tubes idea.

On 6-1-00 Finished calculations showing the opposing mass-flow tube device could only produce a <u>very</u> small net force. It was then that I got the Idea that the electron, + other particles, are (<u>must</u> be!) made of counter rotating rings of equal positive and negative mass. How else could UFO's fly! The electron is really a strong Bg-field.

Richard P. Crandall 9-2-00

5-7-00

. e-mail To Director IUFOMRC

(Had already shown that weak-field sol'n's of Gravity are "remarkably similar" to Maxwell's equations.) Had not yet completed the <u>numerical</u> calculations.

Got response from Carol Syska on 5-17-00.

5-18-00

Sent e-mail to Stan Friedman.

"Have you seen any writeups on Gravity Wave Propulsion?"

You may be very interested in what I have got.

"I think I have stumbled upon something very significant, so please keep this confidential."

Richard P. Crandall 9-2-00

[redacted] 6-1-00 Also thought of an opposed coils device with 90° out of phase and materials inside the coils.

6-26-00 to 5 prof's of Relativity at C.U.:
I have devised new ways (gyroscopes) to test gravity.
Did not mention my new secret electron theory.
Had already thought of using permanent magnet as B field

Rich P. Crandall

7-19-00 to Judy Dressler of C.U. source.

6-7-00 Bought "Gravitation" by Misner, Thorne, and Wheeler at Barnes and Noble. Saw mention of a gravito-magnetic force. First time I realized someone else had thought of this B_g field concept. Also found about Ning Li of Nasa on the internet.

7-25-00 Bought more books. One book mentioned electrons could be black holes! This fit well with my idea!

8-6-00 more books.

9-2-00 Completed numerical calcs of B field force for the opposed coils device (with material = copper or Iron).

7-1-00 started on the "H" pages of my notes. Also borrowed lots of books from CU on Q.E.D. .

Rich P. Crandall 9-2-00

Force on a straight wire due to another wire: [1]

(current is D.C.)

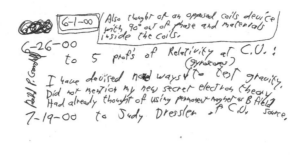

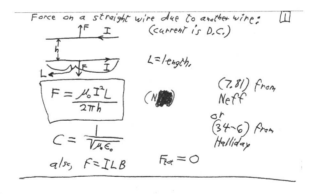

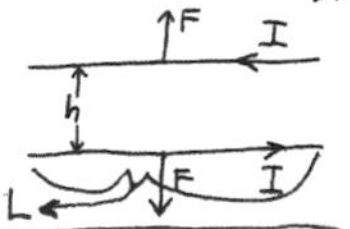

L = length.

$$F = \frac{\mu_0 I^2 L}{2\pi h}$$

(N[redacted])

(7.81) from Neff or (34-6) from Halliday

$$C = \frac{1}{\sqrt{\mu_0 \epsilon_0}}$$

also, $F = ILB$

$F_{tot} = 0$

Rich P. Crandall 9-2-00

Square-wave A.C. case: [2]

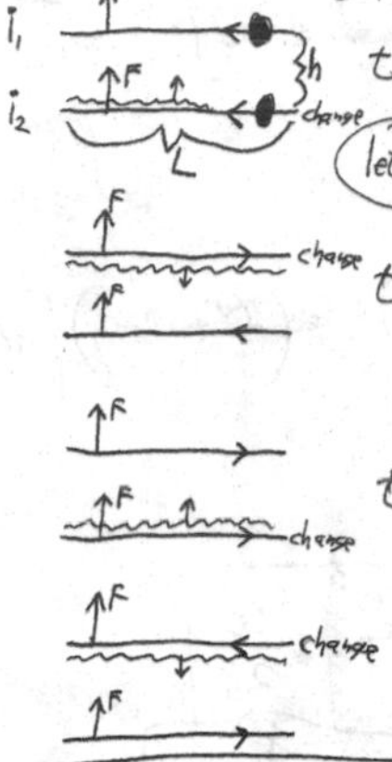

$t = 0^+ \ (= 4t_0^+)$

$\text{let } t_0 = \frac{h}{c}$

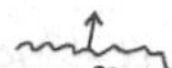

represents new wave travelling at c.

$t = t_0^+$

frequency, f, is

$t = 2t_0^+$

$$f = \frac{1}{4t_0} = \frac{1}{T} \ (Hz)$$

$t = 3t_0^+$

$$\boxed{f = \frac{c}{4h}}$$

$$\omega = 2\pi f$$

$$\boxed{F_{tot,avg} = 2F = \frac{2\mu_0 I^2 L}{2\pi h}}$$

wavelength $\equiv \delta = \frac{c}{f}$

$$\boxed{\delta = 4h} \quad h = \frac{\delta}{4}$$

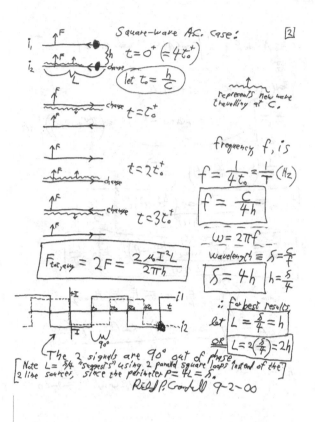

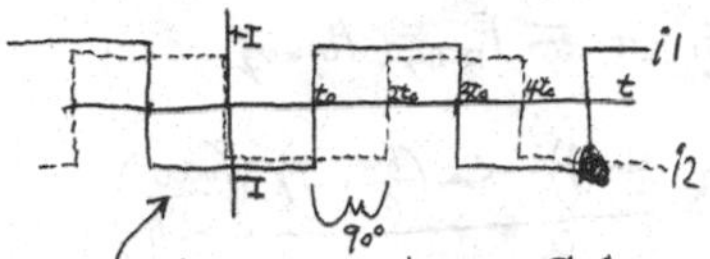

$\therefore$ for best results,

let $\boxed{L = \frac{\delta}{4} = h}$

OR $\boxed{L = 2\left(\frac{\delta}{4}\right) = 2h}$

The 2 signals are 90° out of phase

[Note $L = \delta/4$ "suggests" using 2 parallel square loops instead of the 2 line sources, since the perimeter $P = 4L = \delta$.]

Richard P. Crandall 9-2-00

Sine-wave A.C. Case: [3]

(assuming L is short enough)

$$\dot{I}_{11} e^{j\omega t} = k_1 e^{j\omega t} = k_{1m} e^{j(\omega t + \phi_1)} = I_a e^{j(\omega t + \phi_1)}$$

$$\dot{I}_{12} e^{j\omega t} = I_a e^{j(\omega(t - \frac{h}{c}) + \phi_1)}$$

$$\dot{I}_{22} e^{j\omega t} = k_2 e^{j\omega t} = k_{2m} e^{j(\omega t + \phi_2)} = I_a e^{j(\omega t + \phi_2)}$$

$$\dot{I}_{21} e^{j\omega t} = I_a e^{j(\omega(t - \frac{h}{c}) + \phi_2)}$$

$$F_{12_{inst}} = \frac{\mu_0 \dot{i}_{12_{inst}} \dot{i}_{22_{inst}} L}{2\pi h} = \frac{\mu_0 I_a^2 \cos(\omega(t - \frac{h}{c}) + \phi_1) \cos(\omega t + \phi_2) L}{2\pi h}$$

$$F_{21_{inst}} = \frac{-\mu_0 \dot{i}_{21_{inst}} \dot{i}_{11_{inst}} L}{2\pi h} = \frac{-\mu_0 I_a^2 \cos(\omega(t - \frac{h}{c}) + \phi_2) \cos(\omega t + \phi_1) L}{2\pi h}$$

$F_{tot,inst} = F_{12_{inst}} + F_{21_{inst}}$, and let $\theta = \phi_2 - \phi_1$

Consider ϕ_1 to be constant.

The only variables are ϕ_2 and ω. (and t)

Richard P. Crandall 9-2-00

4

$$\cos A \cos B = \frac{1}{2}\left[\cos(A+B) + \cos(A-B)\right]$$

$$\int_0^{T=\frac{2\pi}{\omega}} \cos\left(\omega\left(t-\frac{h}{c}\right)+\phi_1\right)\cos(\omega t+\phi_2)\,dt$$

$$= \int_0^{2\pi/\omega} \frac{1}{2}\Big[\underbrace{\cos\left(2\omega t+\phi_1+\phi_2-\frac{h\omega}{c}\right)}_{=0} + \cos\left(\frac{h\omega}{c}+\phi_2-\phi_1\right)\Big]dt$$

$$= \frac{1}{2}\left(\cos\left(\frac{h\omega}{c}+\theta\right)\right)\left(\frac{2\pi}{\omega}\right)$$

$$\boxed{F_{12,avg} = \frac{\mu_0 I_a^2 \cos\left(\frac{h\omega}{c}+\theta\right)L}{2h\omega} \cdot \frac{\omega}{2\pi}}$$

$$\boxed{F_{21,avg} = \frac{-\mu_0 I_a^2 \cos\left(\frac{h\omega}{c}-\theta\right)L}{2h\omega} \cdot \frac{\omega}{2\pi}}$$

$$F_{tot,avg} = \int_0^{2\pi/\omega}\left(F_{12,ins} + F_{21,ins}\right)dt = F_{12,avg} + F_{21,avg}$$

$$= \frac{\mu_0 I_a^2 L}{2h}\left(\frac{\cos\left(\frac{h\omega}{c}+\theta\right) - \cos\left(\frac{h\omega}{c}-\theta\right)}{\omega}\right)\left(\frac{\omega}{2\pi}\right)$$

note: $-\sin A \sin B = \frac{1}{2}\left[\cos(A+B) - \cos(A-B)\right]$

Rich'd P. Crandall 9-2-00

5

$$F_{tot,avg} = \overbrace{\frac{-\mu_0 I_a^2 L}{2\pi h}}^{Q}\left(\frac{\sin\left(\frac{h\omega}{c}\right)\sin\theta}{}\right)$$

$$\frac{\partial F}{\partial \omega} = (Q\sin\theta)\left(\cos\left(\frac{h\omega}{c}\right)\left(\frac{h}{c}\right)\right)$$

Set $= 0 \quad \therefore \frac{h\omega}{c} = \frac{\pi}{2} + n\pi$

We get max force when $\frac{h\omega}{c} = \frac{\pi}{2} + n\pi = \pi\left(n+\frac{1}{2}\right)$.

Largest when $n=0$.

$\therefore \; 2\pi f = \frac{c\pi\left(n+\frac{1}{2}\right)}{h}$, $n = 0, 1, 2, \ldots$

or $\boxed{f = \frac{c\left(n+\frac{1}{2}\right)}{2h}, \; n=0,1,2,\ldots \quad \text{if } n=0,\; f = \frac{c}{4h}}$

We also get max force if $\theta = \pm \pi/2$

Rich'd P. Crandall 9-2-00

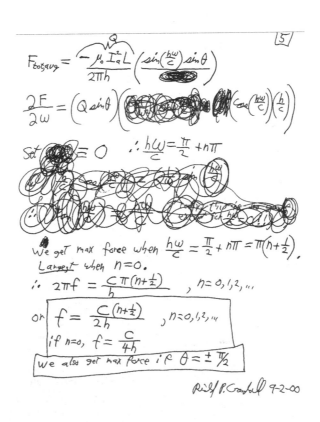

[6]

let $\boxed{f = \frac{c}{4h} \text{ and } \theta = \phi_2 - \phi_1 = \frac{-\pi}{2}}$

$$\therefore \quad \sin\left(\frac{h\omega}{c}\right) = \sin\left(\frac{h2\pi f}{c}\right) = \sin\left(\frac{2\pi}{4}\right) = 1$$

$$\sin\theta = -1$$

$$\therefore \boxed{F_{tot,avg} = \frac{\mu_o I_a^2 L}{2\pi h} = \left(\frac{\mu_o I_a^2 L}{2\pi h}\right)}$$

Reinhold P. Gandell 9-2-00

---

[7]

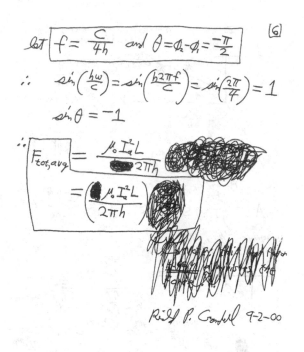

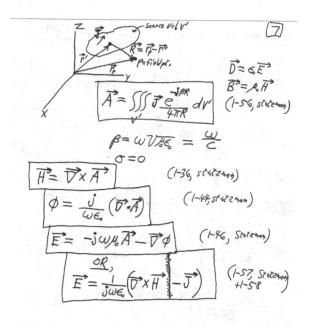

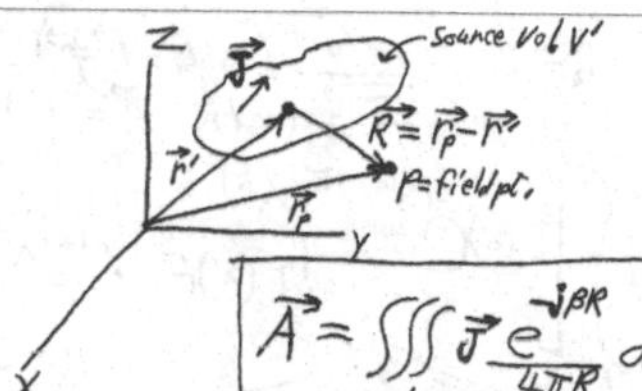

$\vec{D} = \epsilon_o \vec{E}$

$\vec{B} = \mu_o \vec{H}$

$$\boxed{\vec{A} = \iiint_{V'} \vec{J} \frac{e^{-j\beta R}}{4\pi R} dv'}$$ (1-56, Stutzman)

$$\beta = \omega\sqrt{\mu_o\epsilon_o} = \frac{\omega}{c}$$

$$\sigma = 0$$

$$\boxed{\vec{H} = \vec{\nabla} \times \vec{A}}$$ (1-36, Stutzman)

$$\boxed{\phi = \frac{j}{\omega\epsilon_o}(\vec{\nabla} \cdot \vec{A})}$$ (1-44, Stutzman)

$$\boxed{\vec{E} = -j\omega\mu_o\vec{A} - \vec{\nabla}\phi}$$ (1-46, Stutzman)

OR,

$$\boxed{\vec{E} = \frac{1}{j\omega\epsilon_o}\left(\vec{\nabla} \times \vec{H} - \vec{J}\right)}$$ (1-57, Stutzman +1-58)

Reinhold P. Gandell 9-2-00

2 z-directed line current sources: [8]
(both in yz plane)

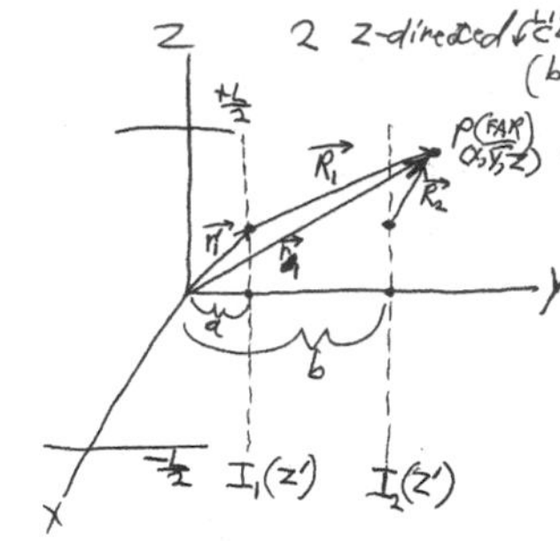

Each source has total length L,

P is not necessarily in yz plane.

Richard P. Goodall 9-2-00

Here, A_y and A_x are zero.

$$A_z = \int_{-L/2}^{+L/2} I_1(z')\frac{e^{-j\beta R_1}}{4\pi R_1}\,dz' + \int_{-L/2}^{+L/2} I_2(z')\frac{e^{-j\beta R_2}}{4\pi R_2}\,dz'$$

in general:

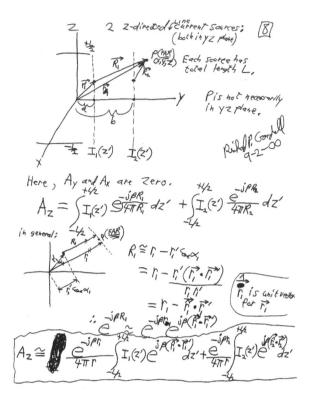

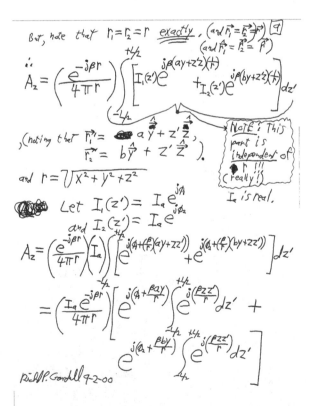

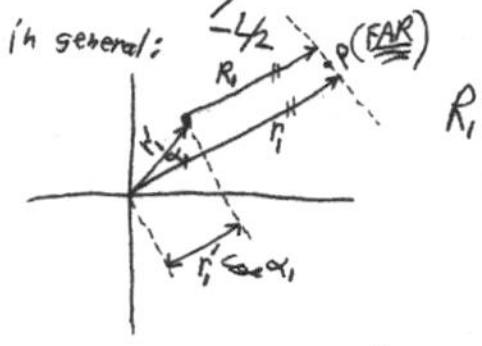

$$R_1 \cong r_1 - r_1'\cos\alpha_1$$
$$= r_1 - r_1'\frac{(\vec{r}_1\cdot\vec{r}_1{}')}{r_1 r_1'}$$
$$= r_1 - \hat{r}_1\cdot\vec{r}_1{}'$$

$\hat{r}_1$ is unit vector for $\vec{r}_1$

$$\therefore\ e^{-j\beta R_1} \cong e^{-j\beta r_1}e^{j\beta(\hat{r}_1\cdot\vec{r}_1{}')}$$

$$A_z \cong \frac{e^{-j\beta r_1}}{4\pi r}\int_{-L/2}^{+L/2} I_1(z')e^{j\beta(\hat{r}_1\cdot\vec{r}_1{}')}\,dz' + \frac{e^{-j\beta r_2}}{4\pi r}\int_{-L/2}^{+L/2} I_2(z')e^{j\beta(\hat{r}_2\cdot\vec{r}_2{}')}\,dz'$$

But, note that $r_1 = r_2 = r$ exactly, (and $\vec{r}_1 = \vec{r}_2 = \vec{r}$) [9]
(and $\hat{r}_1 = \hat{r}_2 = \hat{r}$)

$$\therefore\ A_z = \left(\frac{e^{-j\beta r}}{4\pi r}\right)\int_{-L/2}^{+L/2}\left[I_1(z')e^{j\beta(ay+zz')\left(\frac{1}{r}\right)} + I_2(z')e^{j\beta(by+zz')\left(\frac{1}{r}\right)}\right]dz'$$

NOTE: This part is independent of r !!! (really!)

(noting that $\vec{r}_1{}' = a\hat{y} + z'\hat{z}$,
$\vec{r}_2{}' = b\hat{y} + z'\hat{z}$).

and $r = \sqrt{x^2+y^2+z^2}$

Let $I_1(z') = I_a e^{j\phi_1}$
and $I_2(z') = I_a e^{j\phi_2}$

I_a is real.

$$A_z = \left(\frac{e^{-j\beta r}}{4\pi r}\right)(I_a)\int_{-L/2}^{+L/2}\left[e^{j\left(\phi_1+\left(\frac{\beta}{r}\right)(ay+zz')\right)} + e^{j\left(\phi_2+\left(\frac{\beta}{r}\right)(by+zz')\right)}\right]dz'$$

$$= \left(\frac{I_a e^{-j\beta r}}{4\pi r}\right)\left[e^{j\left(\phi_1+\frac{\beta a y}{r}\right)}\int_{-L/2}^{+L/2} e^{j\left(\frac{\beta z z'}{r}\right)}dz' + e^{j\left(\phi_2+\frac{\beta b y}{r}\right)}\int_{-L/2}^{+L/2} e^{j\left(\frac{\beta z z'}{r}\right)}dz'\right]$$

Richard P. Goodall 9-2-00

[10]

note $\int e^{kz'}dz' = \frac{e^{kz'}}{k} + \text{Const.}$

and $k = \frac{j\beta z}{r}$

$$\int_{-L/2}^{+L/2} e^{j\left(\frac{\beta z z'}{r}\right)}dz' = \frac{e^{k\left(\frac{L}{2}\right)} - e^{-k\left(\frac{L}{2}\right)}}{k}$$

$$= \frac{e^{\frac{j\beta zL}{2r}} - e^{\frac{-j\beta zL}{2r}}}{k}$$

$$= \frac{\left(2j\sin\left(\frac{\beta zL}{2r}\right)\right)r}{j\beta z}$$

$$= \frac{L\sin\left(\frac{\beta zL}{2r}\right)}{\left(\frac{\beta zL}{2r}\right)}$$

Richard P. Gandell 9-2-00

[11]

$A_z =$

$$\left(\frac{I_a e^{-j\beta r}}{4\pi r}\right)\left(\frac{L\sin\left(\frac{\beta zL}{2r}\right)}{\left(\frac{\beta zL}{2r}\right)}\right)\left(e^{j\left(\phi_1+\frac{\beta ay}{r}\right)} + e^{j\left(\phi_2+\frac{\beta by}{r}\right)}\right)$$

Let $a = \frac{-h}{2}$, $b = +\frac{h}{2}$ and $\omega = \frac{\pi c}{2h}$

$\therefore \beta = \frac{\omega}{c} = \frac{\pi}{2h}$

$$\therefore A_z = \left(\frac{I_a e^{-j\left(\frac{\pi r}{2h}\right)} L}{4\pi r}\right)\left(\frac{\sin\left(\frac{\pi zL}{4hr}\right)}{\left(\frac{\pi zL}{4hr}\right)}\right)\Bigg(\cos(\phi_1)\cos\left(\frac{-\beta hy}{2r}\right) -$$

$$\sin(\phi_1)\sin\left(\frac{-\beta hy}{2r}\right) + j\left(\sin(\phi_1)\cos\left(\frac{-\beta hy}{2r}\right) + \cos(\phi_1)\sin\left(\frac{-\beta hy}{2r}\right)\right)$$

$$+ \cos(\phi_2)\cos\left(\frac{\beta hy}{2r}\right) - \sin(\phi_2)\sin\left(\frac{\beta hy}{2r}\right) +$$

$$j\left(\sin(\phi_2)\cos\left(\frac{\beta hy}{2r}\right) + \cos(\phi_2)\sin\left(\frac{\beta hy}{2r}\right)\right)\Bigg)$$

Richard P. Gandell 9-2-00

12

Let $\boxed{\phi_2 = \frac{-\pi}{4},\ \phi_1 = \frac{\pi}{4}}$

then, $\boxed{\phi_2 - \phi_1 = \frac{-\pi}{2}}$

$$\therefore\ \phi_1 + \frac{\beta a y}{r} = \frac{\pi}{4} - \frac{\beta h y}{2r} = \frac{\pi}{4} - \frac{\pi y}{4r} \equiv g$$

$$\phi_2 + \frac{\beta b y}{r} = -\frac{\pi}{4} + \frac{\beta h y}{2r} = -\frac{\pi}{4} + \frac{\pi y}{4r} \equiv -g$$

$$e^{jg} + e^{-jg} = 2\cos g = 2\cos\left(\frac{\pi}{4}\left(1 - \frac{y}{r}\right)\right)$$

$$A_z = \left(\frac{I_a e^{-j\left(\frac{\pi r}{2h}\right)} L}{4\pi r}\right)\left[\left(\frac{\sin\left(\frac{\pi z L}{4hr}\right)}{\left(\frac{\pi z L}{4hr}\right)}\right)\left(2\cos\left(\frac{\pi}{4}\left(1 - \frac{y}{r}\right)\right)\right)\right]$$

NOTE: The part in square brackets is independent of r.

Believe it or not,

$$\therefore\ A_z = \left(f_a(r)\right)\left[f_b(\theta,\phi)\right] = A_z(r,\theta,\phi) \text{ in spherical coordinates,}$$

Richard P. Gordell 9-2-00

Far Fields only: 13A

Per (Stutzman, 1-55) if (i.e. z-directed currents)

$$A_z = \iiint_{V'} J_z \frac{e^{-j\beta R}}{4\pi R}\, dV', \quad (J_x = J_y = 0,\ A_x = A_y = 0)$$

then we can write: (as explained on p. 22),

$$\boxed{\begin{aligned} \vec{E} &= \hat{\theta}\, j\omega\mu_0 \sin\theta A_z \quad (1\text{-}90)\ (E_r = E_\phi = 0) \\ &\text{and} \\ \vec{H} &= \hat{\phi}\, j\beta \sin\theta A_z \quad (1\text{-}88)\ (H_r = H_\theta = 0) \end{aligned}}$$

FAR fields only.

$$\text{So } \frac{E_\theta}{H_\phi} = \frac{\omega\mu_0}{\beta} = \frac{\omega\mu_0}{\omega\sqrt{\mu_0\epsilon_0}} = \eta = \sqrt{\frac{\mu_0}{\epsilon_0}}$$

$$\boxed{H_\phi = \frac{E_\theta}{\eta}}$$

Richard P. Gordell 9-2-00

13B
(proof of 13A)

$$\vec{H} = \vec{\nabla} \times \vec{A}$$

$$= \hat{\vec{r}}\left(\frac{1}{r\sin\theta}\right)\left[\frac{\partial}{\partial\theta}(A_\phi \sin\theta) - \frac{\partial A_\theta}{\partial\phi}\right] +$$

$$\hat{\vec{\theta}}\left(\frac{1}{r}\right)\left[\frac{1}{\sin\theta}\frac{\partial A_r}{\partial\phi} - \frac{\partial}{\partial r}(rA_\phi)\right] +$$

$$\hat{\vec{\phi}}\left(\frac{1}{r}\right)\left[\frac{\partial}{\partial r}(rA_\theta) - \frac{\partial A_r}{\partial\theta}\right]$$

$$\hat{\vec{z}} = \hat{\vec{r}}\cos\theta - \hat{\vec{\theta}}\sin\theta$$

Suppose $\vec{A} = 0\hat{\vec{x}} + 0\hat{\vec{y}} + A_z\hat{\vec{z}}$.

$$\therefore \vec{A} = \hat{\vec{r}}A_z\cos\theta - \hat{\vec{\theta}}A_z\sin\theta + \hat{\vec{\phi}}\,0$$

$$\begin{bmatrix} A_r = A_z\cos\theta \\ A_\theta = -A_z\sin\theta \\ A_\phi = 0 \end{bmatrix}$$

z directed sources only.

$$\vec{H} = \hat{\vec{r}}\left(\frac{1}{r\sin\theta}\right)\left[\sin\theta\frac{\partial A_z}{\partial\phi}\right] +$$

$$\hat{\vec{\theta}}\left(\frac{1}{r}\right)\left[\frac{1}{\tan\theta}\frac{\partial A_z}{\partial\phi}\right] +$$

$$\hat{\vec{\phi}}\left(\frac{1}{r}\right)\left[-A_z\sin\theta - r\sin\theta\frac{\partial A_z}{\partial r} - \frac{\partial A_z}{\partial\theta}\cos\theta + A_z\sin\theta\right]$$

Richard P. Gombell 9-2-00

13C
(proof of 13A)

FAR fields only:

Since, for z-directed sources;

$$A_z \approx \frac{e^{-j\beta r}}{4\pi r} f(\theta,\phi) \quad \text{if } A_z \overset{\text{exactly}}{=} \iiint_{V'} J_z \frac{e^{-j\beta R}}{4\pi R}\, dV',$$

$$\therefore \frac{\partial A_z}{\partial r} = \left(\frac{-j\beta}{4\pi r}e^{-j\beta r} + \frac{-(e^{-j\beta r})}{4\pi r^2}\right) f(\theta,\phi)$$

$$= \left(-j\beta - \frac{1}{r}\right)A_z$$

∴, neglecting all powers greater than or equal to $\left(\frac{1}{r}\right)$,

$$\vec{H} \approx \hat{\vec{\phi}}\left[(-\sin\theta)\left(-j\beta - \cancel{\tfrac{1}{r}}\right)A_z\right]$$

$$= \boxed{\hat{\vec{\phi}}\left(+(\sin\theta)\, j\beta A_z\right)}$$

Richard P. Gombell 9-2-00

14

∴ For this particular problem:

$$\boxed{H_\phi = \frac{j\pi \sin\theta}{2h} A_z}$$

$$\boxed{E_\theta = \sqrt{\frac{\mu_0}{\epsilon_0}}\, H_\phi}$$

$$\vec{H} = H_\phi \hat{\vec{\phi}} = H_0 e^{j\alpha} \hat{\vec{\phi}}$$

(H_0 real (may be negative))

$$\vec{E} = E_\theta \hat{\vec{\theta}} = \sqrt{\frac{\mu_0}{\epsilon_0}}\, H_0 e^{j\alpha} \hat{\vec{\theta}}$$

$$= E_0 e^{j\alpha} \hat{\vec{\theta}}$$

(E_0 real (may be negative))

Richard P. Crandall 9-2-00

---

per Synge: Richard P. Crandall 9-2-00 15

$$T_{rs} = \mu \lambda_r \lambda_s$$

V_0 = proper volume
V = relative volume

$$\lambda_n = \text{4-vel}, \quad \mu = \frac{\sum m}{V_0}$$

$$\lambda_n = \frac{dx_n}{ds} \qquad \mu^* = \frac{\sum m}{V} \qquad m\gamma = m^*$$

$$ds^2 = -dx_n dx_n \qquad \mu^{**} = \frac{\sum m^*}{V} \qquad \gamma = \frac{1}{\sqrt{1-\frac{u^2}{c^2}}}$$

$$\lambda_r \lambda_r = -1$$

$$\lambda_4 = i(1+\lambda_\sigma \lambda_\sigma)^{\frac{1}{2}}$$

m^* = relative mass
m = proper mass

$$u_\sigma = \frac{dx_\sigma}{dt} \qquad x_4 = ict$$

$$ds^2 = -dx^2 - dy^2 - dz^2 + c^2 dt^2$$

For low speeds, $ds \approx c\,dt$

$$\lambda_n \cong \frac{dx_n}{c\,dt}, \quad \boxed{\lambda_\sigma \approx \tfrac{1}{c} u_\sigma, \ \lambda_4 \approx i}$$

$$\mu \approx \mu^* = \frac{\sum \frac{1}{\gamma} m^*}{V} = \frac{\frac{1}{\gamma}\sum m^*}{V} = \frac{\frac{1}{\gamma} m^*}{V} = \frac{m}{V}$$

∴ $\mu \approx D$ = ordinary density

$$\therefore T = \begin{bmatrix} \frac{1}{c^2} D u_x^2 & \frac{1}{c^2} D u_x u_y & \frac{1}{c^2} D u_x u_z & i\frac{1}{c} D u_x \\ & \ddots & & \\ & & & \\ & & & -D \end{bmatrix}$$

per Synge: (16)

Exact values for high speeds:

$$V_0 = V\gamma \qquad \mu = \frac{m}{\gamma V}$$

$$\boxed{\begin{aligned} \delta_\sigma &= \frac{\gamma}{c} u_\sigma \\ \delta_4 &= i\gamma \end{aligned}}$$ (59) from (Synge)

$$(T_{rs})(V) = \frac{m}{\gamma}\delta_r\delta_s = \frac{m^*}{\gamma^2}\delta_r\delta_s$$

relative momentum $p_\sigma = m^* u_\sigma$, relative energy $= m^* c^2$

momentum 4-vector $M_r = m\dfrac{dx_r}{ds} = m\delta_r$

$$M_\sigma = \frac{1}{c} m^* u_\sigma$$

$$M_4 = i m^*$$

$$\therefore (-c^2)(V)(T_{rs}) = \frac{-m^* c^2}{\gamma^2}\delta_r\delta_s \equiv \hat{T}_{rs}$$

$$\therefore \hat{T}_{rs} = \begin{bmatrix} -m^* u_1^2 & -m^* u_1 u_2 & -m^* u_1 u_3 & -i m^* c u_1 \\ -m^* u_2 u_1 & -m^* u_2^2 & -m^* u_2 u_3 & -i m^* c u_2 \\ -m^* u_3 u_1 & -m^* u_3 u_2 & -m^* u_3^2 & -i m^* c u_3 \\ -i m^* c u_1 & -i m^* c u_2 & -i m^* c u_3 & m^* c^2 \end{bmatrix}$$

$\therefore$ momentum density$_\sigma$ $= -ic\,T_{\sigma 4}$

$\therefore$ energy density $= -c^2 T_{44}$

Richard P. Crandall 9-2-00

---

per Neff, instantaneous energy density $= \dfrac{\epsilon_0 E^2}{2} + \dfrac{\mu_0 H^2}{2}$] for time dependent fields. E and H are real amplitudes. [17]

instantaneous real power flux $= \iint_S (\vec{E}\times\vec{H})\cdot d\vec{S}$

---

phasor forms: (averages, not instantaneous)

avg. power flux $= \dfrac{1}{2}\iint_S \mathrm{Re}\{\vec{E}\times\vec{H}^*\}\cdot d\vec{S}$

---

page 4 of (Stutzman)

$$\vec{\nabla}\times\vec{E} = \frac{-\partial\vec{B}}{\partial t} \qquad \vec{D} = \epsilon_0\vec{E}$$

$$\vec{\nabla}\times\vec{H} = \frac{\partial\vec{D}}{\partial t} + \vec{J} \qquad \vec{B} = \mu_0\vec{H}$$

$$\vec{\nabla}\cdot\vec{D} = \rho \qquad c = \frac{1}{\sqrt{\mu_0\epsilon_0}}$$

$$\vec{\nabla}\cdot\vec{B} = 0$$

$\therefore$ in vacuo:

$$\vec{\nabla}\times(\vec{\nabla}\times\vec{H}) = \epsilon_0\frac{\partial}{\partial t}(\vec{\nabla}\times\vec{E}) = -\mu_0\epsilon_0\frac{\partial}{\partial t}\left(\frac{\partial\vec{H}}{\partial t}\right) = -\frac{1}{c^2}\frac{\partial^2\vec{H}}{\partial t^2}$$

$$= \vec{\nabla}(\vec{\nabla}\cdot\vec{H}) - \nabla^2\vec{H}$$

$$= -\nabla^2\vec{H}$$

also, $\boxed{\begin{aligned} \vec{\nabla}\times\vec{E} &= -\mu_0\frac{\partial\vec{H}}{\partial t} \\ \vec{\nabla}\times\vec{H} &= \epsilon_0\frac{\partial\vec{E}}{\partial t} \end{aligned}}$

Richard P. Crandall 9-2-00

[18]

From Synge (p316)

$$\frac{1}{c}\frac{\partial \overset{s}{E}_\sigma}{\partial t} = \epsilon_{\sigma\mu\nu}\frac{\partial \overset{s}{H}_\nu}{\partial x_\mu}, \quad \frac{\partial \overset{s}{E}_\sigma}{\partial x_\sigma} = 0,$$

$$\frac{-1}{c}\frac{\partial \overset{s}{H}_\sigma}{\partial t} = \epsilon_{\sigma\mu\nu}\frac{\partial \overset{s}{E}_\nu}{\partial x_\mu}, \quad \frac{\partial \overset{s}{H}_\sigma}{\partial x_\sigma} = 0$$

$$\therefore \frac{1}{c}\frac{\partial}{\partial t}\left(\epsilon_{\sigma\mu\nu}\frac{\partial \overset{s}{E}_\nu}{\partial x_\mu}\right) = \epsilon_{\sigma\mu\nu}\frac{\partial\left(\epsilon_{\nu\alpha\beta}\frac{\partial \overset{s}{H}_\beta}{\partial x_\alpha}\right)}{\partial x_\mu}$$

$$\therefore -\left(\frac{\partial^2}{\partial x_\psi \partial x_\psi}\right)\overset{s}{H}_\sigma = -\overset{s}{H}_{\sigma,\psi\psi} = \frac{1}{c}\frac{\partial}{\partial t}\left(\frac{-1}{c}\frac{\partial \overset{s}{H}_\sigma}{\partial t}\right)$$

$$\therefore -\overset{s}{H}_{\sigma,\psi\psi} = \frac{-1}{c^2}\frac{\partial^2}{\partial t^2}\left(\overset{s}{H}_\sigma\right)$$

let $\overset{s}{H}_\sigma = k_H H_\sigma$ and $\overset{s}{E}_\sigma = k_E E_\sigma$

$$\therefore -H_{\sigma,\psi\psi} = \frac{-1}{c^2}\frac{\partial^2}{\partial t^2}\left(H_\sigma\right)$$

---

also:

$$\sqrt{\mu_0}\sqrt{\epsilon_0}\, k_E \frac{\partial E_\sigma}{\partial t} = \left(\vec{\nabla}\times k_H \vec{H}\right)_\sigma$$

and $$-\sqrt{\mu_0}\sqrt{\epsilon_0}\, k_H \frac{\partial H_\sigma}{\partial t} = \left(\vec{\nabla}\times k_E \vec{E}\right)_\sigma$$

Richard P. Gandell 9-2-00

or

$$\vec{\nabla}\times\vec{E} = \frac{-\sqrt{\mu_0}\sqrt{\epsilon_0}\, k_H}{k_E}\frac{\partial \vec{H}}{\partial t}$$

$$\vec{\nabla}\times\vec{H} = \frac{\sqrt{\mu_0}\sqrt{\epsilon_0}\, k_E}{k_H}\frac{\partial \vec{E}}{\partial t}$$

---

[19]

So,

$$-\mu_0 = -\frac{\sqrt{\mu_0}\sqrt{\epsilon_0}\, k_H}{k_E}$$

and

$$\epsilon_0 = \frac{\sqrt{\mu_0}\sqrt{\epsilon_0}\, k_E}{k_H}$$

$$\text{, or, } \quad k_E = \frac{\epsilon_0 k_H}{\sqrt{\mu_0}\sqrt{\epsilon_0}}$$

$$\therefore \mu_0 = \left(\sqrt{\mu_0}\sqrt{\epsilon_0}\, \cancel{k_H}\right)\left(\frac{\sqrt{\mu_0}\sqrt{\epsilon_0}}{\epsilon_0 \cancel{k_H}}\right)$$

This is an identity!! It doesn't matter what k_E, k_H are!

So, do energy approach:

per synge, p323, energy density $= -c^2 T_{44} = \frac{1}{2}\left(\overset{s}{E}^2 + \overset{s}{H}^2\right)$

$$= \frac{1}{2}\left((k_E E)^2 + (k_H H)^2\right) = \frac{1}{2}\left(\epsilon_0 E^2 + \mu_0 H^2\right)$$

$$\therefore \quad k_E = \sqrt{\epsilon_0}, \quad k_H = \sqrt{\mu_0}$$

$$\therefore \quad \overset{s}{H}_\sigma = \sqrt{\mu_0}\, H_\sigma, \quad \overset{s}{E}_\sigma = \sqrt{\epsilon_0}\, E_\sigma$$

"s" means synge

Richard P. Gandell 9-2-00

(20)

as a check,

$$\frac{1}{c}\frac{\partial \overset{s}{E}_\sigma}{\partial t} = \epsilon_{\sigma\mu\nu}\frac{\partial \overset{s}{H}_\nu}{\partial X_\mu}$$

$$\downarrow$$

$$\frac{\sqrt{\epsilon_0}}{c}\frac{\partial E_\sigma}{\partial t} = \sqrt{\mu_0}\,\epsilon_{\sigma\mu\nu}\frac{\partial H_\nu}{\partial X_\mu}$$

or

$$\vec{\nabla}\times\vec{H} = \frac{\sqrt{\epsilon_0}}{c\sqrt{\mu_0}}\frac{\partial \vec{E}}{\partial t}$$

$$= \left(\sqrt{\mu_0\epsilon_0}\right)\left(\frac{\sqrt{\epsilon_0}}{\sqrt{\mu_0}}\right)\frac{\partial \vec{E}}{\partial t}$$

$$= \sqrt{\frac{\mu_0\epsilon_0^2}{\mu_0}}\,\frac{\partial \vec{E}}{\partial t}$$

$$= \epsilon_0\frac{\partial \vec{E}}{\partial t}$$ ✓✓ ok

Rich'd P. Goodall 9-2-00

(21)

another check:

$$-\frac{1}{c}\frac{\partial \overset{s}{H}_\sigma}{\partial t} = \epsilon_{\sigma\mu\nu}\frac{\partial \overset{s}{E}_\nu}{\partial X_\mu}$$

$$\downarrow$$

$$-\frac{\sqrt{\mu_0}}{c}\frac{\partial H_\sigma}{\partial t} = \sqrt{\epsilon_0}\,\epsilon_{\sigma\mu\nu}\frac{\partial E_\nu}{\partial X_\mu}$$

or

$$\vec{\nabla}\times\vec{E} = \frac{-\sqrt{\mu_0}}{c\sqrt{\epsilon_0}}\frac{\partial \vec{H}}{\partial t}$$

$$= -\left(\sqrt{\mu_0\epsilon_0}\right)\left(\frac{\sqrt{\mu_0}}{\sqrt{\epsilon_0}}\right)\frac{\partial \vec{H}}{\partial t}$$

$$= -\sqrt{\frac{\mu_0^2\epsilon_0}{\epsilon_0}}\,\frac{\partial \vec{H}}{\partial t}$$

$$= -\mu_0\frac{\partial \vec{H}}{\partial t}$$ ✓✓ ok

Rich'd P. Goodall 9-2-00

∴ per Synge for e/m fields (p.323) [22]

$$T_{11} = \frac{1}{2}\left(\frac{1}{c^2}\right)\left(\epsilon_0\left(-E_1^2+E_2^2+E_3^2\right)+\mu_0\left(-H_1^2+H_2^2+H_3^2\right)\right)$$

$$T_{22} = \frac{1}{2}\left(\frac{1}{c^2}\right)\left(\epsilon_0\left(E_1^2-E_2^2+E_3^2\right)+\mu_0\left(H_1^2-H_2^2+H_3^2\right)\right)$$

$$T_{33} = \frac{1}{2}\left(\frac{1}{c^2}\right)\left(\epsilon_0\left(E_1^2+E_2^2-E_3^2\right)+\mu_0\left(H_1^2+H_2^2-H_3^2\right)\right)$$

$$T_{44} = -\frac{1}{2}\left(\frac{1}{c^2}\right)\left(\epsilon_0 E^2+\mu_0 H^2\right)$$

$$T_{23} = -\left(\frac{1}{c^2}\right)\left(\epsilon_0 E_2 E_3 + \mu_0 H_2 H_3\right)$$

$$T_{31} = -\left(\frac{1}{c^2}\right)\left(\epsilon_0 E_3 E_1 + \mu_0 H_3 H_1\right)$$

$$T_{12} = -\left(\frac{1}{c^2}\right)\left(\epsilon_0 E_1 E_2 + \mu_0 H_1 H_2\right)$$

$$T_{14} = \left(\frac{i}{c^2}\right)\left(\frac{1}{c}E_2 H_3 - \frac{1}{c}E_3 H_2\right)$$

$$T_{24} = \left(\frac{i}{c^2}\right)\left(\frac{1}{c}E_3 H_1 - \frac{1}{c}E_1 H_3\right)$$

$$T_{34} = \left(\frac{i}{c^2}\right)\left(\frac{1}{c}E_1 H_2 - \frac{1}{c}E_2 H_1\right)$$

momentum density$_\alpha$	$= -ic\,T_{\alpha 4}$
energy density	$= -c^2 T_{44}$

Just like the mechanical counter part.

Richard P. Goodall 9-2-00

---

[23]

Now, switching to phasor form, we get the time-averaged values:

$$T_{14} = \mathrm{Re}\left[\left(\frac{i}{2c^3}\right)\left(E_2 H_3^* - E_3 H_2^*\right)\right]$$

$$T_{24} = \mathrm{Re}\left[\left(\frac{i}{2c^3}\right)\left(E_3 H_1^* - E_1 H_3^*\right)\right]$$

$$T_{34} = \mathrm{Re}\left[\left(\frac{i}{2c^3}\right)\left(E_1 H_2^* - E_2 H_1^*\right)\right]$$

So

$$(\text{momentum density})_1 = \mathrm{Re}\left[\frac{1}{2c^2}\left(E_2 H_3^* - E_3 H_2^*\right)\right]$$

$$(\text{momentum density})_2 = \mathrm{Re}\left[\frac{1}{2c^2}\left(E_3 H_1^* - E_1 H_3^*\right)\right]$$

$$(\text{momentum density})_3 = \mathrm{Re}\left[\frac{1}{2c^2}\left(E_1 H_2^* - E_2 H_1^*\right)\right]$$

Since we are using z-directed sources, we can, since E and H components are in phase, delete the "Re".

Richard P. Goodall 9-2-00

[24]

Ricd P. Gardell 9-2-00

note that:

$$\vec{E} \times \vec{H}^* = \begin{vmatrix} \vec{i} & \vec{j} & \vec{k} \\ E_1 & E_2 & E_3 \\ H_1^* & H_2^* & H_3^* \end{vmatrix}$$

$$= \vec{i}\left(E_2 H_3^* - E_3 H_2^*\right) + \vec{j}\left(E_3 H_1^* - E_1 H_3^*\right) + \vec{k}\left(E_1 H_2^* - E_2 H_1^*\right)$$

$\therefore$

momentum Density (average) $\equiv$ $\boxed{\vec{p} = \frac{1}{2c^2}\vec{E} \times \vec{H}^*}$

For <u>all</u> far field problems, $\vec{E} \times \vec{H}^*$ will <u>always</u> point radially out from the origin.

$\therefore \vec{p} = p\hat{\vec{r}}$

(p: must be real, (and positive))

For <u>z-directed</u> sources, at least, $\vec{E}$ will be <u>perpendicular</u> to $\vec{H}^*$, so $\vec{E} \times \vec{H}^* = E_0 e^{j\alpha}\hat{\vec{\theta}} \times H_0 e^{-j\alpha}\hat{\vec{\phi}}$

$= E_0 H_0 \hat{\vec{r}}$

(E_0, H_0: real)

$\therefore$ For our <u>particular</u> problem: $\boxed{\vec{p} = \frac{E_0 H_0}{2c^2}\hat{\vec{r}}}$

---

[25]

Also, for our particular problem (cont'd from [14]):

$$E_0 = \left(\sqrt{\frac{\mu_0}{\epsilon_0}}\right) H_0 \quad \text{, per [14].}$$

$\therefore$

$$\boxed{\vec{p} = \frac{\left(\sqrt{\frac{\mu_0}{\epsilon_0}}\right) H_0^2}{2c^2}\hat{\vec{r}}}$$

H_0 may be negative, but H_0^2 is always positive.

$$H_0 e^{j\alpha} = H_\phi = \left(\frac{\pi \sin\theta}{2h}\right)\left(e^{j\pi/2}\right)\left[\left(e^{-j\left(\frac{\pi r}{2h}\right)}\right) \cdot \left(\frac{I_a L}{4\pi r}\right)\left[\left(\frac{\sin\left(\frac{\pi z L}{4hr}\right)}{\left(\frac{\pi z L}{4hr}\right)}\right)\left(2\cos\left(\frac{\pi}{4}\left(1-\frac{y}{r}\right)\right)\right)\right]\right]$$

(bracketed term: actually independent of r!!)

Ricd P. Gardell 9-2-00

26

$$\text{let } U_1 = U_1(r) = \frac{I_a^2 L^2}{16 \hbar^2 r^2}$$

$$\text{let } U_2 = U_2(\phi,\theta) = \left(\sin^2\theta\right)\left(\frac{\sin^2\left(\frac{\pi z L}{4\hbar r}\right)}{\left(\frac{\pi z L}{4\hbar r}\right)^2}\right)\left(\cos^2\left(\frac{\pi}{4}\left(1-\frac{y}{r}\right)\right)\right)$$

$$\therefore H_0^2 = U_1 U_2$$

$$\text{let } W = \frac{\sqrt{\frac{\mu_0}{\epsilon_0}}}{2C^2} \quad \text{(a constant)}$$

$$\therefore \vec{p} = W U_1 U_2 \hat{r}$$

Momentum Density, in Far Field.

note $y = r\sin\theta\sin\phi$ $\qquad \frac{y}{r} = \sin\theta\sin\phi$

$z = r\cos\theta$ $\qquad \frac{z}{r} = \cos\theta$

Richard P. Crandall 9-2-00

27

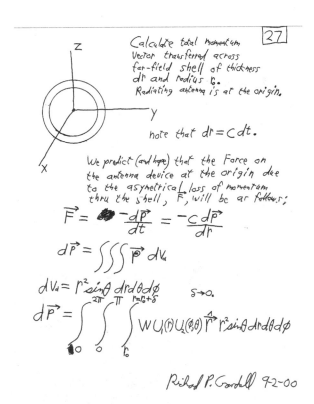

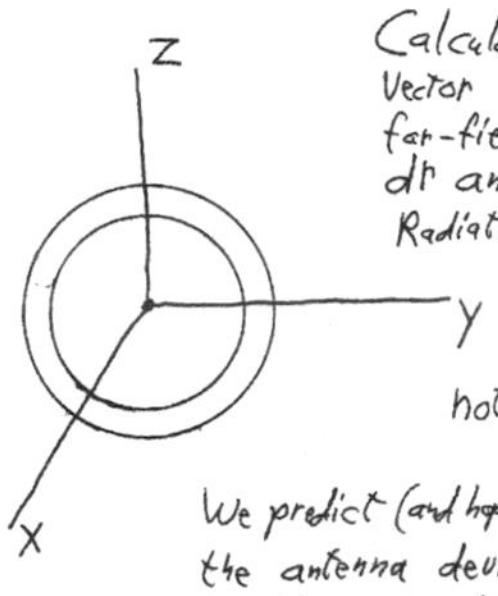

Calculate total momentum vector transferred across far-field shell of thickness dr and radius r_0. Radiating antenna is at the origin.

note that $dr = c\,dt$.

We predict (and hope) that the Force on the antenna device at the origin due to the asymetrical loss of momentum thru the shell, $\vec{F}$, will be as follows:

$$\vec{F} = -\frac{d\vec{P}}{dt} = -c\frac{d\vec{P}}{dr}$$

$$d\vec{P} = \iiint \vec{p}\, dV_d$$

$$dV_d = r^2 \sin\theta\, dr\, d\theta\, d\phi \qquad \delta \to 0.$$

$$d\vec{P} = \int_0^{2\pi}\int_0^{\pi}\int_{r_0}^{r=r_0+\delta} W U_1(r) U_2(\phi,\theta)\, \hat{r}\, r^2 \sin\theta\, dr\, d\theta\, d\phi$$

Richard P. Crandall 9-2-00

28

$\therefore$

$$\vec{F} = -WC \int_0^{2\pi}\int_0^{\pi} U_1(r_0)U_2(\phi,\theta)\,\hat{\vec{r}}\,r_0^2 \sin\theta \,d\theta\, d\phi$$

$$\hat{\vec{r}} = \hat{\vec{x}}\sin\theta\cos\phi + \hat{\vec{y}}\sin\theta\sin\phi + \hat{\vec{z}}\cos\theta$$

$$\vec{F} = -WC r_0^2 U_1(r_0) \int_0^{\pi}(\sin\theta)\int_0^{2\pi} U_2(\phi,\theta)\,\hat{\vec{r}}\, d\phi\, d\theta$$

$$F_x = \frac{-WCI_a^2L^2}{16h^2}\int_0^{\pi}(\sin\theta)\int_0^{2\pi} U_2 \sin\theta\cos\phi \, d\phi\, d\theta$$

$$F_y = \frac{-WCI_a^2L^2}{16h^2}\int_0^{\pi}(\sin\theta)\int_0^{2\pi} U_2 \sin\theta\sin\phi \, d\phi\, d\theta$$

$$F_z = \frac{-WCI_a^2L^2}{16h^2}\int_0^{\pi}(\sin\theta)(\cos\theta)\int_0^{2\pi} U_2 \, d\phi\, d\theta$$

Rich P. Crandall 9-2-00

29

let $U_3 = U_3(\theta) = \dfrac{\sin^2\theta \sin^2\left(\frac{\pi L\cos\theta}{4h}\right)}{\left(\frac{\pi L\cos\theta}{4h}\right)^2}$

let $U_4 = U_4(\phi,\theta) = \cos^2\left(\frac{\pi}{4}\left(1-\sin\theta\sin\phi\right)\right)$

$\therefore\ U_2 = U_3(\theta)U_4(\phi,\theta)$

$\eta = \sqrt{\frac{\mu_0}{\epsilon_0}}$

define $\hat{W} = \dfrac{-WCI_a^2L^2}{16h^2}$

$$\therefore\ \hat{W} = \frac{-\left(\frac{\eta}{2C}\right)I_a^2L^2}{16h^2} = \frac{-\eta I_a^2L^2}{32h^2C}$$

$$\hat{W} = \frac{-\eta I_a^2L^2}{32h^2C}$$

$$\frac{\eta}{C} = \frac{\sqrt{\frac{\mu_0}{\epsilon_0}}}{\left(\frac{1}{\sqrt{\mu_0\epsilon_0}}\right)} = \sqrt{\frac{\mu_0}{\epsilon_0}(\mu_0\epsilon_0)} = \mu_0$$

$$\hat{W} = \frac{-\mu_0 I_a^2L^2}{32h^2}$$

Rich P. Crandall 9-2-00

30

$$F_x = \hat{W} \int_0^{\pi} (\sin^2\theta) U_3 \int_0^{2\pi} U_4 \cos\phi \, d\phi \, d\theta \quad \leftarrow \text{(zero)}$$

$$F_y = \hat{W} \int_0^{\pi} (\sin^2\theta) U_3 \int_0^{2\pi} U_4 \sin\phi \, d\phi \, d\theta \quad \leftarrow \text{(non-zero)}$$

$$F_z = \hat{W} \int_0^{\pi} (\sin\theta)(\cos\theta) U_3 \int_0^{2\pi} U_4 \, d\phi \, d\theta \quad \leftarrow \text{(zero)}$$

$F_x = 0$ due to symmetry of $U_4 \cos\phi$ about $\phi = \frac{\pi}{2}$.

$F_z = 0$ due to symmetry of $(\cos\theta)$ about $\theta = \frac{\pi}{2}$.

let $G = \int_0^{2\pi} (\sin\phi) \cos^2\left(\frac{\pi}{4}(1 - k\sin\phi)\right) d\phi$

Where $k = \sin\theta$, $0 \le k \le 1$
(k doesn't include ϕ).

$$\therefore \; F_y = \hat{W} \int_0^{\pi} (\sin^2\theta) U_3 \, G \, d\theta$$

note: G is a function of θ !!

Richard P. Crandall 9-2-00

31

Near field example:

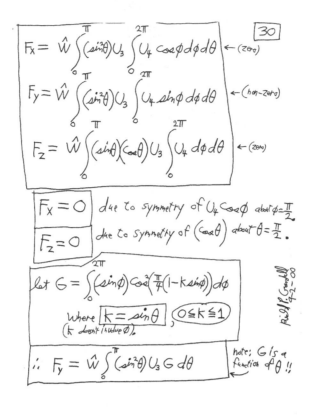

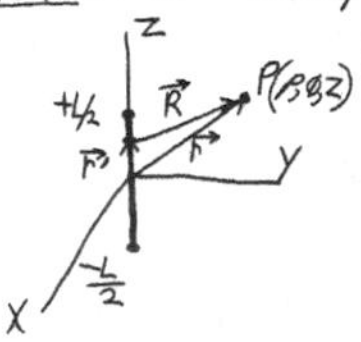

Uniform (sinusoidal) current,
Amplitude $= I_a$.

Use $L \to \infty$.

A_y and A_x are zero.

No approximations will be used, as we seek an exact solution in the near field.

$$A_z = \int_{-L/2}^{+L/2} I(z') \frac{e^{-j\beta R}}{4\pi R} dz' = \frac{I_a}{4\pi} \int_{-L/2}^{+L/2} \frac{e^{-j\beta R}}{R} dz'$$

$$\vec{R} = \vec{r} - \vec{r}'$$

Cylindrical coords:

$$\vec{r} = \rho\hat{\rho} + z\hat{z}$$

$$\vec{r}' = z'\hat{z}$$

$$\vec{R} = \rho\hat{\rho} + (z - z')\hat{z}$$

$$R = \sqrt{\rho^2 + (z - z')^2}$$

Richard P. Crandall 9-2-00

[32] NEAR Field example [32]

$$A_z = \frac{I_a}{4\pi}\int_{-L/2}^{+L/2} \frac{e^{-j\beta\sqrt{\rho^2+(z-z')^2}}}{\sqrt{\rho^2+(z-z')^2}}\,dz'$$

To simplify, let $Z=0$. (ok, since $L\to\infty$)

$$\therefore \boxed{A_z = \frac{I_a}{4\pi}\int_{-L/2}^{+L/2} \frac{e^{-j\beta\sqrt{\rho^2+z'^2}}}{\sqrt{\rho^2+z'^2}}\,dz'}$$

$R=\sqrt{\rho^2+z'^2}$

let $X=\sqrt{\rho^2+z'^2}$

$dX = (\tfrac{1}{2})\,2z'\,dz'$

$Z'=\sqrt{X^2-\rho^2}$

$$\therefore \frac{dz'}{\sqrt{\rho^2+z'^2}} = \frac{dz'}{X} = \frac{dX}{\sqrt{X^2-\rho^2}}$$

$$\therefore A_z = \frac{I_a}{4\pi}\int \frac{e^{-j\beta X}}{\sqrt{X^2-\rho^2}}\,dX \quad \text{NoGo}$$

Richard P. Goodell, 4-2-00

Near field ex. cont'd; [33]

$$\vec{\nabla}\times\vec{A} = \vec{H} =$$

$$\hat{\vec{\rho}}\left(\frac{1}{\rho}\frac{\partial A_z}{\partial\phi}-\frac{\partial A_\phi}{\partial z}\right)+\hat{\vec{\phi}}\left(\frac{\partial A_\rho}{\partial z}-\frac{\partial A_z}{\partial\rho}\right)+\hat{\vec{z}}\left(\frac{1}{\rho}\right)\left[\frac{\partial}{\partial\rho}(\rho A_\phi)-\frac{\partial A_\rho}{\partial\phi}\right]$$

$$=\hat{\vec{\rho}}\left(\frac{1}{\rho}\frac{\partial A_z}{\partial\phi}\right)+\hat{\vec{\phi}}\left(-\frac{\partial A_z}{\partial\rho}\right)$$

$$= -\left(\frac{\partial A_z}{\partial\rho}\right)\hat{\vec{\phi}}$$

, since $A_z = A_z(\rho)$, not $A_z(\rho,\phi,z)$

$$H_\phi = \frac{-I_a}{4\pi}\int_{-\infty}^{+\infty}\frac{\partial}{\partial\rho}\left(\frac{e^{-j\beta\sqrt{\rho^2+u^2}}}{\sqrt{\rho^2+u^2}}\right)du$$

$\bar{B} = \mu_0 H_\phi$ consider $\bar{B}$ a phasor

$\boxed{\rho = h}$

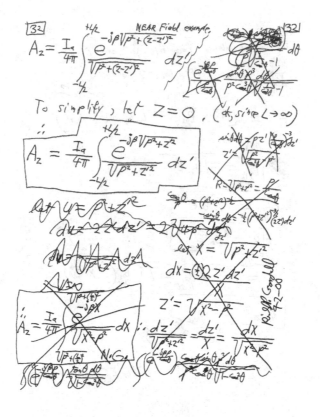

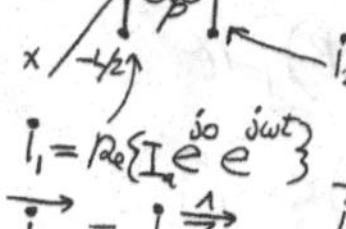

$i_2 = Re\{I_a e^{j\theta} e^{j\omega t}\}$

$i_1 = Re\{I_a e^{j0} e^{j\omega t}\}$

$\vec{i_1} = i_1\hat{\vec{z}}$ $\quad \vec{i_2} = i_2\hat{\vec{z}}$

Richard P. Goodell 4-2-00

Near Field

34

$\vec{F}_{12_{inst}}$ = instantaneous force felt by 2 due to i_1

$\vec{B}_{12_{inst}}$ = instantaneous field at 2 due to i_1, $= -\hat{y}$

$$\vec{F}_{12_{inst}} = L\vec{i}_2 \times \vec{B}_{12_{inst}} = L i_2 B_{12_{inst}} (\hat{z} \times \hat{\phi})$$

$$\boxed{F_{y12_{inst}} = -L i_2 B_{12_{inst}}}$$

$$B_{12_{inst}} = Re\{\bar{B} e^{j\omega t}\}$$

let $\boxed{\bar{B} = B_m e^{j\alpha}}$

$+\hat{y}$

$$\vec{F}_{21_{inst}} = L\vec{i}_1 \times \vec{B}_{21_{inst}} = L i_1 B_{21_{inst}} (\hat{z} \times \hat{\phi})$$

$$\boxed{F_{y21_{inst}} = L i_1 B_{21_{inst}}}$$

$$B_{21_{inst}} = Re\{\bar{B} e^{j\theta} e^{j\omega t}\}$$

Rich P. Gordell 9-2-00

35

$$B_{12_{inst}} = B_m \cos(\omega t + \alpha)$$

$$B_{21_{inst}} = B_m \cos(\omega t + \alpha + \theta)$$

$$\boxed{\begin{aligned} F_{y12_{inst}} &= -L I_a \cos(\omega t + \theta) B_m \cos(\omega t + \alpha) \\ F_{y21_{inst}} &= L I_a \cos(\omega t) B_m \cos(\omega t + \alpha + \theta) \end{aligned}}$$

Looking at the H_ϕ formula, my guess is that $\alpha \simeq \pm\pi/2$, but we really don't know. (since $\beta r = \beta h = \frac{\omega h}{c} = \frac{2\pi h}{c} f = \frac{2\pi h}{c}\left(\frac{c}{4h}\right) = \frac{\pi}{2}$).

This strongly suggests we let $\theta = \pm\alpha$, but we will not do this (yet).

$$\int_0^{T=\frac{2\pi}{\omega}} \cos(\omega t + A)\cos(\omega t + B)\,dt$$

$$= \frac{1}{2}\int_0^{2\pi/\omega} \left[\cos(2\omega t + A + B) + \cos(A - B)\right] dt$$

($\cos(2\omega t + A + B)$ term $= 0$)

$$= \frac{1}{2}\left(\cos(A-B)\right)\left(\frac{2\pi}{\omega}\right) = \left(\frac{\pi}{\omega}\right)\cos(A-B)$$

Rich P. Gordell 9-2-00

36

$$F_{y12,avg} = -I_a L B_m\left(\frac{\pi}{\omega}\right)\cos(\theta-\alpha)\cdot\frac{\omega}{2\pi}$$

$$F_{y21,avg} = I_a L B_m\left(\frac{\pi}{\omega}\right)\cos(\theta+\alpha)\cdot\frac{\omega}{2\pi}$$

$$F_{ytot,avg} = \frac{I_a L B_m}{2}\left(\cos(\theta+\alpha) - \cos(\theta-\alpha)\right)$$

$\therefore$

$$F_{ytot,avg} = -I_a L B_m \sin\theta \sin\alpha$$

where

$$B_m e^{j\alpha} = -\frac{\mu_0 I_a}{4\pi}\int_{-\infty}^{+\infty}\frac{\partial}{\partial h}\left(\frac{e^{-j\left(\frac{\pi}{2h_0}\sqrt{h^2+u^2}\right)}}{\sqrt{h^2+u^2}}\right)du$$

Note: consider h_0 a constant, which is later set equal to h.

let $h_0 = \frac{c\pi}{2\omega}$ therefore, $\beta = \frac{\omega}{c}$

Richard P. Crandall 9-2-00

37a

$$F = \frac{(G) M_1 M_2}{r^2} \qquad G = 6.67\times10^{-11}\ \frac{N\cdot m^2}{kg^2}$$

$$= \left(\frac{1}{4\pi\epsilon_g}\right)\frac{M_1 M_2}{r^2} \qquad \epsilon_g = \frac{1}{4\pi G} = 1.193\times10^{9}\ \frac{kg^2}{N\cdot m^2}$$

$$\mu_g = \frac{1}{c^2\epsilon_g} \qquad \left(\frac{s^2}{m^2}\right)\left(\frac{N\,m^2}{kg^2}\right) \text{ or } \frac{N\cdot s^2}{kg^2} \text{ or } \frac{N}{A_g^2}$$

$$F_{mg} = \frac{\mu_g I_g^2 L}{2\pi h}\ (N) \qquad \therefore \mu_g = 9.37\times10^{-27} \qquad 1A_g = 1\frac{kg}{s}$$

I_g is units of A_g

$$F = \left(\frac{1}{4\pi\epsilon_0}\right)\frac{Q_1 Q_2}{r^2} \qquad \epsilon_0 = 8.85\times10^{-12}\ F/m$$

$$\frac{\mu_g}{\mu_0} = \frac{\epsilon_0}{\epsilon_g} = 4\pi G\epsilon_0 \qquad \text{or } \frac{C^2}{N\cdot m^2}$$

$$c = \frac{1}{\sqrt{\mu_0\epsilon_0}} \approx 2.99\times10^{8}\ m/s$$

$$\mu_0 = \frac{1}{c^2\epsilon_0} \qquad \left(\frac{s^2}{m^2}\right)\left(\frac{N\,m^2}{C^2}\right) \text{ or } \frac{N\cdot s^2}{C^2} \text{ or } \frac{N}{A^2}$$

note $1A = \frac{1C}{s}$

$$F_m = \frac{\mu_0 I^2 L}{2\pi h}\ (N)$$

I is units of A

$$\mu_0 = 1.26\times10^{-6}$$

Richard P. Crandall 9-2-00

$$I_g = D^* \cdot Area \cdot u \qquad \left(\frac{kg}{m^3}\right)(m^2)\left(\frac{m}{s}\right) \quad \boxed{37b} \ or \ \frac{kg}{s}$$

$$\boxed{D = \frac{M_0}{Vol}}$$

$$D^* = \frac{M^*_{\bullet}}{Vol} = \frac{M_0 \gamma}{Vol} = D\gamma$$

$$\gamma = \frac{1}{\sqrt{1-\frac{u^2}{c^2}}}$$

$$\boxed{I_g = D \cdot Area \cdot u\gamma}$$

$$u = \frac{L_0}{t_0}$$

$$V_{ol} = L_0 A_{rea}$$

$$I_g = \frac{M_0}{Vol} \cdot Area \cdot \frac{L_0}{t_0} \gamma$$

$$\gamma u = \frac{u}{\sqrt{1-\frac{u^2}{c^2}}} = \sqrt{\frac{u^2}{1-u^2/c^2}}$$

$$= \frac{M_0 \cdot Area \cdot L_0 \cdot \gamma}{L_0 \cdot Area \cdot t_0} = \frac{M_0 \gamma}{t_0} = \frac{M^*}{t_0}$$

$$\rightarrow I_g = M_0 \left(\frac{1}{t_0}\right) \cdot \gamma \cdot u = \frac{M_0 (\gamma u)}{L_0}$$

$$= \left(\frac{M_0}{L_0}\right)\left(\sqrt{\frac{1}{\frac{1}{u^2}-\frac{1}{c^2}}}\right) = \frac{\left(\frac{M_0}{L_0}\right)}{\sqrt{\frac{1}{u^2}-\frac{1}{c^2}}}$$

Richard P. Goodell 9-2-00

$$\boxed{37c}$$

$$\sqrt{\frac{1}{u^2}-\frac{1}{c^2}} = \frac{M_0}{L_0 I_g}$$

$$\sqrt{\left(\frac{1}{c^2}\right)\left(\frac{1}{\left(\frac{u^2}{c^2}\right)}-1\right)} = \frac{M_0}{L_0 I_g}$$

$$\frac{1}{\left(\frac{u^2}{c^2}\right)}-1 = \frac{c^2 M_0^2}{L_0^2 I_g^2}$$

$$\frac{u^2}{c^2} = \frac{1}{\frac{c^2 M_0^2}{L_0^2 I_g^2}+1}$$

$$\boxed{\frac{u}{c} = \frac{1}{\sqrt{\frac{c^2 M_0^2}{L_0^2 I_g^2}+1}}}$$

Richard P. Goodell 9-2-00

we need $I_g = 10^{13}$ (to get $F_{ng} \approx 1$) [37d]

let $L_o = 1$

try $M_o = .1$ kg

$\therefore \frac{u}{c} = .99999$

---

try $M_o = 10$ kg

Conclusion: $\frac{u}{c} = .99999...$

Can't be done!!

(produce strong enough gravity wave force)

Richard P. Gardell 9-2-00

← per einstien and Synge (and Dirac)

$$R_{\mu\nu} - \frac{1}{2} g_{\mu\nu} R = -\underline{k}\, T_{\mu\nu} \quad (1)$$

per Synge ($-ds^2 = d\phi = g_{\mu\nu}dx\,dx$)

assume, for weak fields, $g_{\mu\nu} = \delta_{\mu\nu} + h_{\mu\nu}$ (2)

($h_{\mu\nu}$ is a first order quantity)

DO NOT USE

$$h_{\mu}^{\cdot\delta} = \delta^{\delta\alpha} h_{\mu\alpha} \quad (3)$$

$$h = h_{\alpha}^{\cdot\alpha} = \delta^{\sigma\delta} h_{\sigma\delta} \quad (4)$$

$$-R_{\mu\nu} = \Gamma^{\beta}_{\mu\nu,\beta} - \Gamma^{\beta}_{\mu\beta,\nu} + \Gamma^{\beta}_{\mu\nu}\Gamma^{\alpha}_{\beta\alpha} - \Gamma^{\alpha}_{\mu\beta}\Gamma^{\beta}_{\nu\alpha} \quad (5)$$

∴ We can write $R_{\mu\nu}$ to first order as:

$$-R_{\mu\nu} = -\frac{1}{2}\delta^{\sigma\delta} h_{\mu\nu,\sigma\delta} - \frac{1}{2}\left(h_{,\mu\nu} - h_{\mu\ \ ,\nu\beta}^{\cdot\beta} - h^{\beta}_{\cdot\nu,\mu\beta}\right) \quad (6)$$

By rearrangement,

$$h_{,\mu\nu} - h_{\mu\ \ ,\nu\beta}^{\cdot\beta} - h^{\beta}_{\cdot\nu,\mu\beta} = \left(\frac{1}{2}\delta_{\mu}^{\cdot\beta} h - h_{\mu}^{\cdot\beta}\right)_{,\beta\nu} + \left(\frac{1}{2}\delta_{\nu}^{\cdot\beta} h - h_{\nu}^{\cdot\beta}\right)_{,\beta\mu} \quad (7)$$

Equation (7) can be made to vanish by choosing the coordinate conditions such that:

$$\left(h_{\mu}^{\cdot\beta} - \frac{1}{2}\delta_{\mu}^{\cdot\beta}\right)_{,\beta} = 0 \quad (8)$$

∴

$$-\frac{1}{2}\delta^{\sigma\delta} h_{\mu\nu,\sigma\delta} + g_{\mu\nu}\left[\frac{1}{4}\delta^{\sigma\delta} h_{,\sigma\delta}\right] = \underline{k}\, T_{\mu\nu} \quad (9)$$

Make sure the order of the $T_{\mu\nu}$ components match left side.

G0a Rild P. Gandall 9-2-00

define the following: DO NOT USE

$$\phi_{\mu}^{\cdot\nu} = h_{\mu}^{\cdot\nu} - \frac{1}{2}\delta_{\mu}^{\cdot\nu} h \quad (10)$$

∴ Equation (8) may be written as:

$$\boxed{\phi_{\mu\ ,\nu}^{\cdot\nu} = 0} \quad (11)$$ "Coordinate Conditions"

definition: $\boxed{\Box \cdot = \cdot_{,\sigma\sigma}}$

By raising ν in (9) and using (10):

$$\boxed{\Box\, \phi_{\mu}^{\cdot\nu} = -2\underline{k}\, T_{\mu}^{\cdot\nu}} \quad (12)$$

$$\boxed{\phi_{\mu}^{\cdot\nu}(r,t) = \frac{\underline{k}}{2\pi} \iiint_{V'} \frac{(T_{\mu}^{\cdot\nu})\big|_{\text{retarded}}}{R}\, dV' \qquad R = |\vec{r} - \vec{r}'|} \quad (13)$$

Side note, Yukawa potential for forces transmitted by "particles" with rest mass m is

$$\boxed{\phi \sim \frac{e^{-mcr/\hbar}}{r}}.$$

(Has infinite range only if $m \to 0$.)

GOb   Richard P. Gandell 9-2-00

G1

note $\{\mu\tau, \rho\} = \left\{ {\rho \atop \mu\tau} \right\}$ etc.,

also: $\Gamma^{\tau}_{\mu\nu} = +\{\mu\nu, \tau\}$ → per synge + others

$$\therefore B^{\rho}_{\mu\sigma\tau} = \frac{\partial}{\partial X_\sigma}\left\{ {\rho \atop \mu\tau} \right\} - \frac{\partial}{\partial X_\tau}\left\{ {\rho \atop \mu\sigma} \right\} + \left\{ {\alpha \atop \mu\tau} \right\}\left\{ {\rho \atop \alpha\sigma} \right\} - \left\{ {\alpha \atop \mu\sigma} \right\}\left\{ {\rho \atop \alpha\tau} \right\}$$

$= R^{\rho}_{\cdot\mu\sigma\tau}$

→ per einstein, and dirac, synge, spiegel

$$\therefore R^{\beta}_{\mu\nu\tau} = -\Gamma^{\beta}_{\mu\nu,\tau} + \Gamma^{\beta}_{\mu\tau,\nu} + \Gamma^{\alpha}_{\mu\tau}\Gamma^{\beta}_{\alpha\nu} - \Gamma^{\alpha}_{\mu\nu}\Gamma^{\beta}_{\alpha\tau}$$

← per dirac, synge, spiegel, NOT Einstein

per spiegel and synge:

$$[pq, r] = \frac{1}{2}\left(\frac{\partial g_{pr}}{\partial x^q} + \frac{\partial g_{qr}}{\partial x^p} - \frac{\partial g_{pq}}{\partial x^r}\right)$$

$$\left\{ {s \atop pq} \right\} = g^{sr}[pq, r]$$

per synge and spiegel:

$$R^{s}_{\cdot rmn} = \frac{\partial}{\partial x^m}\left\{ {s \atop rn} \right\} - \frac{\partial}{\partial x^n}\left\{ {s \atop rm} \right\} + \left\{ {p \atop rn} \right\}\left\{ {s \atop pm} \right\} - \left\{ {p \atop rm} \right\}\left\{ {s \atop pn} \right\}$$

G1 Richard P. Gandall 9-2-00

Per Einstein: and Synge and Dirac

$$R_{\mu\nu} - \frac{1}{2} g_{\mu\nu} R = -kT_{\mu\nu} \quad (96)$$

Main Equation for Gravity.

62

We multiply (96) $g^{\mu\nu}$, summed over by μ and ν, and remembering that $g_{\mu\nu} g^{\mu\nu} = 4$, we get:

$$R = k g^{\mu\nu} T_{\mu\nu} = kT$$

Putting this value of R into (96), we get:

$$R_{\mu\nu} = -k\left(T_{\mu\nu} - \frac{1}{2} g_{\mu\nu} T\right) = -kT^{*}_{\mu\nu} \quad (96a)$$

Per einstein: (equation of motion)

$$\frac{d^2 x^r}{ds^2} = -\left\{ {r \atop \mu\nu} \right\} \frac{dx^\mu}{ds} \frac{dx^\nu}{ds}$$

what does Synge Say??

Same thing exactly.

62

Reginald P. Crandall 9-2-00

1st order approx;

$$g_{\mu\nu} = \delta_{\mu\nu} + \gamma_{\mu\nu}$$

$$-ds^2 = dX_\mu dX_\mu = dt^2(u^2 - c^2)$$

$$\therefore -ds^2 \simeq -c^2 dt^2$$

$$\text{or } ds = c\,dt$$

$$-\left\{ {\mu \atop \alpha\beta} \right\} = -\delta_{\mu\sigma}[\alpha\beta,\sigma] = -[\alpha\beta,\mu] = \frac{1}{2}\left(\frac{\partial\gamma_{\alpha\beta}}{\partial X_\mu} - \frac{\partial\gamma_{\alpha\mu}}{\partial X_\beta} - \frac{\partial\gamma_{\beta\mu}}{\partial X_\alpha}\right)$$

Also, we neglect derivatives of $\gamma_{\mu\nu}$ w respect to X_4.

$\therefore$ For $\mu = 1,2,3$ we get;

$$\frac{d^2X_r}{ds^2} + \left\{ {r \atop mn} \right\} \frac{dX_m}{ds}\frac{dX_n}{ds} = 0$$

(per synge; and einstein;)

$$\frac{d^2X_\mu}{ds^2} = -\left\{ {\mu \atop \alpha\beta} \right\} \frac{dX_\alpha}{ds}\frac{dX_\beta}{ds}$$

$$\cancel{\left(\frac{1}{c^2}\right)}\frac{d^2X_\mu}{dt^2} = \frac{1}{2}\left(\frac{\partial\gamma_{\alpha\beta}}{\partial X_\mu} - \frac{\partial\gamma_{\alpha\mu}}{\partial X_\beta} - \frac{\partial\gamma_{\beta\mu}}{\partial X_\alpha}\right)\cancel{\left(\frac{1}{c^2}\right)}\frac{dX_\alpha}{dt}\frac{dX_\beta}{dt}$$

Richard P. Crandall 9-2-00

64

$$\therefore \quad = O(\gamma)\left(\sum_{\alpha\beta}^{1..3} u_\alpha u_\beta\right) + O(\gamma)\left(\sum_{\alpha}^{1..3}(u_\alpha)(ic)\right) + O(\gamma)\left(-c^2\right)$$

$\therefore$ Dropping powers of c less than 2, we get:

$\mu = 1,2,3$

$$\frac{d^2 X_\mu}{dt^2} = \frac{1}{2}\left(\frac{\partial \gamma_{44}}{\partial X_\mu} - \frac{\partial \gamma_{4\mu}}{\partial X_4} - \frac{\partial \gamma_{4\mu}}{\partial X_4}\right)\left(-c^2\right)$$

neglect, since this equals $-2\left(\frac{1}{ic}\right)\frac{\partial \gamma_{4\mu}}{\partial t} \approx \left(\frac{1}{c}\right)\frac{\partial \gamma_{4\mu}}{\partial t}$

$$\iint(-\vec{a})\cdot d\vec{s} = \iint \frac{GM}{r^2}(\vec{u})\cdot d\vec{s} = \frac{GM}{R^2}\iint \vec{u}\cdot d\vec{s} = \frac{GM}{R^2}(\text{[redacted]}\,4\pi R^2) = 4\pi GM$$

$\therefore$

$$\frac{d^2 X_\mu}{dt^2} = \left(\frac{-c^2}{2}\right)\frac{\partial \gamma_{44}}{\partial X_\mu} = -\frac{\partial\left(\frac{c^2}{2}\gamma_{44}\right)}{\partial X_\mu}$$

$$\therefore \boxed{a_\mu = \frac{-\partial \phi}{\partial X_\mu}, \quad \text{where } \phi = \frac{c^2}{2}\gamma_{44}}$$

or $\boxed{\vec{a} = -\vec{\nabla}\phi}$

From Newton's Laws: $\frac{\vec{F}}{m} = \vec{a} = -\vec{\nabla}\phi = \frac{-GM}{r^2}\vec{u}$ $\therefore \phi = \frac{-GM}{r}$

and [redacted] $\iint -\vec{a}\cdot d\vec{s} = \iiint \vec{\nabla}\cdot(-\vec{a})\,dV = \iiint$ [redacted] $dV =$ [redacted] $4\pi GM$

$\therefore \vec{\nabla}\cdot(\vec{\nabla}\phi) = 4\pi G\rho$, or $\boxed{\nabla^2\phi = \text{[redacted]}\,4\pi G\rho}$

$$\therefore \boxed{\frac{c^2}{2}\nabla^2\gamma_{44} = \text{[redacted]}\,4\pi G\rho}$$

Richard P. Gandall
9-2-00

$$\boxed{R_{\mu\nu} - \frac{1}{2} g_{\mu\nu} R = -k T_{\mu\nu}}$$

Main equation, repeated.

$$\boxed{T = g^{\mu\nu} T_{\mu\nu}}$$

and, as shown before:

$$R_{\mu\nu} = -k\left(T_{\mu\nu} - \frac{1}{2} g_{\mu\nu} T\right) = -k T^{*}_{\mu\nu}$$

$$\left(\text{again, } g_{\mu\nu} = \delta_{\mu\nu} + \gamma_{\mu\nu}\right).$$

Now, the left hand side equals:

$$R_{\mu\nu} = R^{\tau}_{\mu\nu\tau} = -\frac{\partial \Gamma^{\tau}_{\mu\nu}}{\partial X_\tau} + \frac{\partial \Gamma^{\tau}_{\mu\tau}}{\partial X_\nu} + \Gamma^{\beta}_{\mu\tau}\Gamma^{\tau}_{\beta\nu} - \Gamma^{\beta}_{\mu\nu}\Gamma^{\tau}_{\beta\tau}$$

$$= -\frac{\partial \Gamma^{\alpha}_{\mu\nu}}{\partial X_\alpha} + \Gamma^{\alpha}_{\mu\beta}\Gamma^{\beta}_{\nu\alpha} + \frac{\partial \Gamma^{\alpha}_{\mu\alpha}}{\partial X_\nu} - \Gamma^{\alpha}_{\mu\nu}\Gamma^{\beta}_{\alpha\beta}$$

order $= \gamma$; order $= \gamma^2$, so eliminate.

$$\Gamma^{m}_{ab} = \delta_{mr}[ab,r] - \tilde{\gamma}_{mr}[ab,r] = [ab,m]$$

these are $\sum \frac{\partial \gamma_{de}}{\partial X_f}$

$$\therefore \Gamma^{m}_{ab} = \frac{1}{2}\left(\frac{\partial \gamma_{am}}{\partial X_b} + \frac{\partial \gamma_{bm}}{\partial X_a} - \frac{\partial \gamma_{ab}}{\partial X_m}\right)$$

Richard P. Gardell 9-2-00

In $R_{\mu\nu}$,

66

The Γ terms are of order γ, and the $\Gamma\Gamma$ terms are of order γ^2, so we eliminate the $\Gamma\Gamma$ terms in $R_{\mu\nu}$, leaving:

$$R_{\mu\nu} = -\frac{\partial \Gamma^{\alpha}_{\mu\nu}}{\partial X_\alpha} + \frac{\partial \Gamma^{\alpha}_{\mu\alpha}}{\partial X_\nu}$$

(This is of order γ.)

$$R_{\mu\nu} = -\frac{1}{2}\left(\frac{\partial^2 \gamma_{\mu\alpha}}{\partial X_\alpha \partial X_\nu} + \frac{\partial^2 \gamma_{\nu\alpha}}{\partial X_\alpha \partial X_\mu} - \frac{\partial^2 \gamma_{\mu\nu}}{\partial X_\alpha \partial X_\alpha}\right) +$$

$$\frac{1}{2}\left(\frac{\partial^2 \gamma_{\mu\alpha}}{\partial X_\nu \partial X_\alpha} + \frac{\partial^2 \gamma_{\alpha\alpha}}{\partial X_\nu \partial X_\mu} - \frac{\partial^2 \gamma_{\mu\alpha}}{\partial X_\nu \partial X_\alpha}\right)$$

$$R_{\mu\nu} = +\frac{1}{2}\left(\frac{\partial^2 \gamma_{\mu\nu}}{\partial X_\alpha \partial X_\alpha} + \frac{\partial^2 \gamma_{\alpha\alpha}}{\partial X_\mu \partial X_\nu} - \frac{\partial^2 \gamma_{\mu\alpha}}{\partial X_\nu \partial X_\alpha} - \frac{\partial^2 \gamma_{\nu\alpha}}{\partial X_\mu \partial X_\alpha}\right)$$

Let $\gamma'_{\mu\nu} = \gamma_{\mu\nu} - \frac{1}{2}\gamma_{\sigma\sigma}\delta_{\mu\nu}$

Richard P. Gandall 9-2-00

Now note the following:

$$\frac{\partial}{\partial X_\mu}\left(\frac{\partial \gamma'_{\nu\alpha}}{\partial X_\alpha}\right) = \frac{\partial \gamma_{\nu\alpha}}{\partial X_\mu \partial X_\alpha} - \frac{1}{2}\delta_{\nu\alpha}\frac{\partial \gamma_{\sigma\sigma}}{\partial X_\mu \partial X_\alpha}$$

$$\frac{\partial}{\partial X_\nu}\left(\frac{\partial \gamma'_{\mu\alpha}}{\partial X_\alpha}\right) = \frac{\partial \gamma_{\mu\alpha}}{\partial X_\nu \partial X_\alpha} - \frac{1}{2}\delta_{\mu\alpha}\frac{\partial \gamma_{\sigma\sigma}}{\partial X_\nu \partial X_\alpha}$$

Then, adding these two:

$$-\frac{1}{2}\frac{\partial}{\partial X_\nu}\left(\frac{\partial \gamma'_{\mu\alpha}}{\partial X_\alpha}\right) - \frac{1}{2}\frac{\partial}{\partial X_\mu}\left(\frac{\partial \gamma'_{\nu\alpha}}{\partial X_\alpha}\right) =$$

$$-\frac{1}{2}\left(\frac{\partial \gamma_{\mu\alpha}}{\partial X_\nu \partial X_\alpha} + \frac{\partial \gamma_{\nu\alpha}}{\partial X_\mu \partial X_\alpha} - \frac{\partial^2 \gamma_{\alpha\alpha}}{\partial X_\mu \partial X_\nu}\right)$$

$\therefore$

$$\boxed{R_{\mu\nu} = +\frac{1}{2}\frac{\partial^2 \gamma_{\mu\nu}}{\partial X_\alpha \partial X_\alpha} - \frac{1}{2}\frac{\partial}{\partial X_\nu}\left(\frac{\partial \gamma'_{\mu\alpha}}{\partial X_\alpha}\right) - \frac{1}{2}\frac{\partial}{\partial X_\mu}\left(\frac{\partial \gamma'_{\nu\alpha}}{\partial X_\alpha}\right)}$$

Now, choose special coordinates such that:

$$\boxed{\frac{\partial \gamma'_{\mu\nu}}{\partial X_\nu} = \gamma'_{\mu\nu,\nu} \equiv 0} \leftarrow \text{"Coordinate conditions"}$$

$$\therefore \boxed{\frac{\partial \gamma_{\mu\nu}}{\partial X_\nu} - \frac{1}{2}\frac{\partial \gamma_{\sigma\sigma}}{\partial X_\mu} = 0} \leftarrow$$

Richard P. Crandall
9-2-00

Then, $R_{\mu\nu} = -kT^*_{\mu\nu}$ takes the form; [G8]

$$\therefore R_{\mu\nu} = +\frac{1}{2}\frac{\partial^2 \gamma_{\mu\nu}}{\partial X_\alpha^2}$$

per Synge:
$T_{rs} = \mu \frac{dx_r}{ds}\frac{dx_s}{ds}$
for SPECIAL Relativity

$$\therefore \frac{\partial^2 \gamma_{\mu\nu}}{\partial X_\alpha^2} = -2kT^*_{\mu\nu}$$

$$\boxed{\Box \gamma_{\mu\nu} = \gamma_{\mu\nu,\alpha\alpha} = -2kT^*_{\mu\nu}} = \nabla^2\gamma_{\mu\nu} - \frac{1}{c^2}\frac{\partial^2\gamma_{\mu\nu}}{\partial t^2}$$

$$\therefore \boxed{\gamma_{\mu\nu}(r,t) = \frac{k}{2\pi}\iiint_{V'} \frac{(T^*_{\mu\nu})\big|_{retarded}}{R}\, dV' \qquad R = |\vec{r} - \vec{r}'|}$$ ✓✓

Richard P. Crandall 9-2-08

Note that $\nabla^2 \gamma_{44} = -2kT^*_{44}$ [assume, here, $\frac{\partial}{\partial t} = 0$]

per Synge $T_{44} = -\rho$

$$\text{or, } \frac{c^2}{2}\nabla^2\gamma_{44} = -kc^2 T^*_{44} = 4\pi G\rho$$

$$T^*_{44} = T_{44} - \frac{1}{2}g_{44}\left(g^{\mu\nu}T_{\mu\nu}\right) \approx T_{44} - \frac{1}{2}g_{44}\,g^{44}T_{44}$$ ($g_{44}g^{44} = 1$)

$$\therefore T^*_{44} = \frac{1}{2}T_{44}, \quad \therefore -\frac{kc^2}{2}T_{44} = 4\pi G\rho$$

$$\therefore k = \frac{8\pi G\rho}{-c^2 T_{44}} = \frac{8\pi G\rho}{+c^2\rho} = \boxed{\frac{8\pi G}{c^2}}$$ ✓✓

Therefore,

$$R_{\mu\nu} - \frac{1}{2} g_{\mu\nu} R = -\left(\frac{8\pi G}{c^2}\right) T_{\mu\nu}$$ ✓✓

"s" means "synge" G is the gravitational const.

For e-m fields (per synge): [single charge q]

$$\overset{s}{\phi}_r = \frac{\overset{s}{q}}{4\pi}\left(\frac{\lambda'_r}{W'}\right) \qquad \lambda'_r = \frac{dX'_r}{dS}$$

$$W' = \lambda'_n (X'_n - X_n)$$

[$\overset{s}{\phi}_r$ is the field at point X_n due to a charge q at X'_n.]

But in general, we have:

$$\overset{s}{\phi}_r = \overset{s}{\phi}_r(X, Y, Z, t) \qquad r = 1, 2, 3, 4$$

[charge and current distribution]

$$\overset{s}{\phi}_r = \frac{1}{4\pi} \iiint_{V'} \frac{\overset{s}{J}_r(X', Y', Z', t' - R/c)}{R} dV'$$

$X_1 = X$
$X_2 = Y$
$X_3 = Z$
$X_4 = ict$

[Consider this to be the definition of $\overset{s}{\phi}_r$.]

$$R^2 = (X_\sigma - X'_\sigma)(X_\sigma - X'_\sigma)$$

$$\sigma = 1, 2, 3.$$

This is from Synge, p.407 (104), modified HEAVILY to make it easier to use!!

from Synge p.403 we have

$$\overset{s}{J}_r = \overset{s}{\rho}_c \lambda_r, \quad \lambda_\sigma = \frac{\gamma}{c} u_\sigma, \quad \lambda_4 = i\gamma$$

[$\overset{s}{\rho}_c$ = "proper" chg. density]

G10

$$\therefore \quad \overset{s}{J}_\sigma = \frac{\overset{s}{\rho}_c \gamma}{c} u_\sigma = \frac{(\overset{s}{e})(\#\text{charges})}{V_0}\left(\frac{u_\sigma}{c}\right)\gamma$$

$$= \frac{(\overset{s}{e})(\#\text{charges})}{V}\left(\frac{u_\sigma}{c}\right)$$

$$= \frac{\overset{s*}{\rho}_c}{c} u_\sigma$$

define $\overset{\wedge s}{J}_\sigma$ = relative 3-current density. (the one we mortals normally use!)

$$\therefore \quad \overset{\wedge s}{J}_\sigma = \overset{s*}{\rho}_c u_\sigma = c\,\overset{s}{J}_\sigma$$

$\overset{s*}{\rho}_c$ = rel. chg. density (per Synge??)

$$\overset{s}{J}_4 = \overset{s}{\rho}_c i \gamma = i \overset{s*}{\rho}_c$$

So, $\overset{s*}{\rho}_c = \overset{s}{J}_4 / i$

The necessary conservation equation is:

$$\overset{s}{J}_{r,r} = 0$$

written out, this is:

$$\frac{\partial}{\partial x_1}\overset{s}{J}_1 + \frac{\partial}{\partial x_2}\overset{s}{J}_2 + \frac{\partial}{\partial x_3}\overset{s}{J}_3 + \frac{1}{ic}\frac{\partial (i\overset{s*}{\rho}_c)}{\partial t} = 0$$

or $\frac{1}{c}\frac{\partial}{\partial x_1}\overset{\wedge s}{J}_1 + \text{''} + \text{''} + \frac{1}{c}\frac{\partial \overset{s*}{\rho}_c}{\partial t} = 0$

or

$$\frac{\partial \overset{\wedge s}{J}_1}{\partial x_1} + \frac{\partial \overset{\wedge s}{J}_2}{\partial x_2} + \frac{\partial \overset{\wedge s}{J}_3}{\partial x_3} = -\frac{\partial \overset{s*}{\rho}_c}{\partial t}$$

Rich P. Gondall 9-2-00

Comparing the $\overset{s}{\Phi}_r$ formula (integral) on p69 with a typical $\Box\psi$ solution on page 68, we see that: (in general)

$$\iiint_{V'} \frac{(-\Omega)|_{ret}}{R}\, dV' = \frac{2\pi}{k}(\psi)$$

and $\Box\psi = -2k\Omega$

$$\therefore \frac{1}{4\pi}\iiint_{V'} \frac{(\Omega)|_{ret}}{R}\, dV' = \frac{1}{2k}(\psi) = \tilde{\psi}$$

and $\Box\tilde{\psi} = \frac{1}{2k}\Box\psi = \frac{-2k\Omega}{2k} = -\Omega$

$$\therefore \boxed{\Box\overset{s}{\Phi}_r = -\overset{s}{J}_r}$$

and, since $\overset{s}{J}_{r,r} = 0$, we get:

$$\therefore \boxed{\overset{s}{\Phi}_{r,r} = 0}$$

these are identical to (86) on p.404 of <u>Synge</u>.

Richard P. Gardell 9-2-00

612

Now, define $\overset{s}{F}_{rs} = \overset{s}{\phi}_{s,r} - \overset{s}{\phi}_{r,s}$ $\qquad$ $\overset{s}{F}_{rs} = -\overset{s}{F}_{sr}$

So, there are only 6 non-zero independent components.

$$\therefore \overset{s}{F}_{rs,s} = \left(\overset{s}{\phi}_{s,s}\right)_{,r} - \Box \overset{s}{\phi}_r = \overset{s}{J}_r$$

$$\overset{s}{F}_{mn,r} = \overset{s}{\phi}_{n,mr} - \overset{s}{\phi}_{m,nr}$$

$$\overset{s}{F}_{nr,m} = \overset{s}{\phi}_{r,nm} - \overset{s}{\phi}_{n,rm}$$

$$\overset{s}{F}_{rm,n} = \overset{s}{\phi}_{m,rn} - \overset{s}{\phi}_{r,mn}$$

$$\therefore \quad \overset{s}{F}_{rs,s} = \overset{s}{J}_r$$

$$\overset{s}{F}_{mn,r} + \overset{s}{F}_{nr,m} + \overset{s}{F}_{rm,n} = 0$$

Now, define: $\overset{s}{F}{}^{*}_{rs} = \frac{1}{2} i \epsilon_{rsmn} \overset{s}{F}_{mn}$ $\qquad \therefore \overset{s}{F}{}^{*}_{rs} = -\overset{s}{F}{}^{*}_{sr}$

$\therefore \overset{s}{F}_{rs} = -\frac{1}{2} i \epsilon_{rsmn} \overset{s}{F}{}^{*}_{mn}$ (not proved, but could be)

Now, define: $\overset{s}{E}_p = i \overset{s}{F}_{p4}$ and $\overset{s}{H}_p = -i \overset{s}{F}{}^{*}_{p4}$

$p = 1, 2, 3$

Richard P. Gandell 9-2-00

G13

$\therefore \quad \overset{s}{E}_\rho = i\,\overset{s}{F}_{\rho 4} \,, \quad \overset{s}{H}_\rho = \frac{1}{2}\epsilon_{\rho\mu\nu}\overset{s}{F}_{\mu\nu} \,,$

$\overset{s}{E}_\rho = \frac{1}{2}\epsilon_{\rho\mu\nu}\overset{s}{F}{}^{*}_{\mu\nu} \,, \quad \overset{s}{H}_\rho = -i\,\overset{s}{F}{}^{*}_{\rho 4}$

or, explicitly:

$$\overset{s}{E}_1 = i\,\overset{s}{F}_{14} \,, \quad \overset{s}{E}_2 = i\,\overset{s}{F}_{24} \,, \quad \overset{s}{E}_3 = i\,\overset{s}{F}_{34}$$

$$\overset{s}{H}_1 = \overset{s}{F}_{23} \,, \quad \overset{s}{H}_2 = \overset{s}{F}_{31} \,, \quad \overset{s}{H}_3 = \overset{s}{F}_{12}$$

$$\overset{s}{E}_1 = \overset{s}{F}{}^{*}_{23} \,, \quad \overset{s}{E}_2 = \overset{s}{F}{}^{*}_{31} \,, \quad \overset{s}{E}_3 = \overset{s}{F}{}^{*}_{12}$$

$$\overset{s}{H}_1 = -i\,\overset{s}{F}{}^{*}_{14} \,, \quad \overset{s}{H}_2 = -i\,\overset{s}{F}{}^{*}_{24} \,, \quad \overset{s}{H}_3 = -i\,\overset{s}{F}{}^{*}_{34}$$

Inversely, :

$$\overset{s}{F}_{\rho 4} = -i\,\overset{s}{E}_\rho \,, \quad \overset{s}{F}_{\rho\sigma} = \epsilon_{\rho\sigma\mu}\overset{s}{H}_\mu$$

$$\overset{s}{F}{}^{*}_{\rho 4} = i\,\overset{s}{H}_\rho \,, \quad \overset{s}{F}{}^{*}_{\rho\sigma} = \epsilon_{\rho\sigma\mu}\overset{s}{E}_\mu$$

Then, $\overset{s}{F}{}^{*}_{rs,s} = \overset{s}{F}{}^{*}_{21,1} + \overset{s}{F}{}^{*}_{22,2} + \overset{s}{F}{}^{*}_{23,3} + \overset{s}{F}{}^{*}_{24,4}$ (example, where $r=2$)

$= -i\,\overset{s}{F}_{34,1} + 0 + i\,\overset{s}{F}_{14,3} - i\,\overset{s}{F}_{13,4} = +i\,\overset{s}{F}_{43,1} + i\,\overset{s}{F}_{14,3} + i\,\overset{s}{F}_{31,4} = 0$

In general, $\boxed{\overset{s}{F}{}^{*}_{rs,s} = 0}$

Richard P. Crandall 9-2-00

G14

$$\therefore \quad \boxed{\frac{\partial \overset{s}{F}_{rs}}{\partial X_s} = \overset{s}{J}_r \;, \quad \frac{\partial \overset{s*}{F}_{rs}}{\partial X_s} = 0}$$

Maxwell's equations in Tensor form (for Special Relativity)

$$\frac{\partial \overset{s}{F}_{\rho s}}{\partial X_s} = \frac{\partial \overset{s}{F}_{\rho\sigma}}{\partial X_\sigma} + \frac{\partial \overset{s}{F}_{\rho 4}}{\partial X_4} = \epsilon_{\rho\sigma\mu}\frac{\partial \overset{s}{H}_\mu}{\partial X_\sigma} + \frac{(-i)\partial \overset{s}{E}_\rho}{(ic)\partial t} = \frac{1}{c}\overset{1s}{J}_\rho$$

$$\frac{\partial \overset{s}{F}_{4s}}{\partial X_s} = \frac{\partial \overset{s}{F}_{4\sigma}}{\partial X_\sigma} + \frac{\partial \overset{s}{F}_{44}}{\partial X_4} = i\frac{\partial \overset{s}{E}_\sigma}{\partial X_\sigma} + 0 = i\overset{s*}{\rho}_c$$

$$\frac{\partial \overset{s*}{F}_{\rho s}}{\partial X_s} = \frac{\partial \overset{s*}{F}_{\rho\sigma}}{\partial X_\sigma} + \frac{\partial \overset{s*}{F}_{\rho 4}}{\partial X_4} = \epsilon_{\rho\sigma\mu}\frac{\partial \overset{s}{E}_\mu}{\partial X_\sigma} + \frac{(i)\partial \overset{s}{H}_\rho}{(ic)\partial t} = 0$$

$$\frac{\partial \overset{s*}{F}_{4s}}{\partial X_s} = \frac{\partial \overset{s*}{F}_{4\sigma}}{\partial X_\sigma} + \frac{\partial \overset{s*}{F}_{44}}{\partial X_4} = \frac{(-i)\partial \overset{s}{H}_\sigma}{\partial X_\sigma} + 0 = 0$$

$$\therefore \quad \epsilon_{\rho\sigma\mu}\frac{\partial \overset{s}{H}_\mu}{\partial X_\sigma} = \frac{1}{c}\frac{\partial \overset{s}{E}_\rho}{\partial t} + \frac{1}{c}\overset{1s}{J}_\rho \;, \quad \frac{\partial \overset{s}{E}_\sigma}{\partial X_\sigma} = \overset{s*}{\rho}_c$$

$$\epsilon_{\rho\sigma\mu}\frac{\partial \overset{s}{E}_\mu}{\partial X_\sigma} = \frac{-1}{c}\frac{\partial \overset{s}{H}_\rho}{\partial t} \;, \quad \frac{\partial \overset{s}{H}_\sigma}{\partial X_\sigma} = 0$$

Maxwell's Equations in pseudo-Vector form.

Richard P. Gandall 9-2-00

615

Now, let
$$\overset{s}{H}_\sigma = \sqrt{\mu_0}\, H_\sigma$$
$$\overset{s}{E}_\sigma = \sqrt{\epsilon_0}\, E_\sigma$$

and note that $c = \frac{1}{\sqrt{\mu_0 \epsilon_0}}$

$$\eta = \sqrt{\frac{\mu_0}{\epsilon_0}}$$

$$\sqrt{\mu_0}\,\epsilon_{\rho\sigma\mu}\frac{\partial H_\mu}{\partial X_\sigma} = \frac{\sqrt{\epsilon_0}}{c}\frac{\partial E_\rho}{\partial t} + \frac{1}{c}\overset{\wedge s}{J}_\rho$$

$$\epsilon_{\rho\sigma\mu}\frac{\partial H_\mu}{\partial X_\sigma} = \frac{\sqrt{\epsilon_0}\sqrt{\mu_0\epsilon_0}}{\sqrt{\mu_0}}\frac{\partial E_\rho}{\partial t} + \frac{\sqrt{\mu_0\epsilon_0}}{\sqrt{\mu_0}}\overset{\wedge s}{J}_\rho$$

$$= \frac{\partial(\epsilon_0 E_\rho)}{\partial t} + \sqrt{\epsilon_0}\,\overset{\wedge s}{J}_\rho$$

define: $\hat{J}_\rho = \sqrt{\epsilon_0}\,\overset{\wedge s}{J}_\rho \quad \rho = 1,2,3$ only

$$\sqrt{\epsilon_0}\,\epsilon_{\rho\sigma\mu}\frac{\partial E_\mu}{\partial X_\sigma} = -\frac{\sqrt{\mu_0}}{c}\frac{\partial H_\rho}{\partial t}$$

$$\epsilon_{\rho\sigma\mu}\frac{\partial E_\mu}{\partial X_\sigma} = -\frac{\sqrt{\mu_0}\sqrt{\mu_0\epsilon_0}}{\sqrt{\epsilon_0}}\frac{\partial H_\rho}{\partial t}$$

$$= -\frac{\partial(\mu_0 H_\rho)}{\partial t}$$

Richard P. Gondall 9-2-00

$$\sqrt{\epsilon_0}\,\frac{\partial E_\sigma}{\partial X_\sigma} = \overset{s}{\rho}{}_c^*$$

Define: $\boxed{\rho_c^* = \sqrt{\epsilon_0}\,\overset{s}{\rho}{}_c^*}$

$$\therefore\ \sqrt{\epsilon_0}\,\frac{\partial E_\sigma}{\partial X_\sigma} = \frac{1}{\sqrt{\epsilon_0}}\,\rho_c^*$$

$$\therefore\ \frac{\partial(\epsilon_0 E_\sigma)}{\partial X_\sigma} = \rho_c^*$$

and, $$\frac{\partial(\mu_0 H_\sigma)}{\partial X_\sigma} = 0$$

In summary;

$$\epsilon_{\rho\sigma\mu}\frac{\partial H_\mu}{\partial X_\sigma} = \frac{\partial(\epsilon_0 E_\rho)}{\partial t} + \hat{J}_\rho \quad , \quad \frac{\partial(\epsilon_0 E_\sigma)}{\partial X_\sigma} = \rho_c^*$$

$$\epsilon_{\rho\sigma\mu}\frac{\partial E_\mu}{\partial X_\sigma} = \frac{-\partial(\mu_0 H_\rho)}{\partial t} \quad , \quad \frac{\partial(\mu_0 H_\sigma)}{\partial X_\sigma} = 0$$

Maxwell's equations, MKS units, pseudo-vector form

Richard P. Gandell 9-2-00

G17

∴ if $\vec{E} = E_\sigma \hat{\vec{u}}_\sigma$

$\vec{H} = H_\sigma \hat{\vec{u}}_\sigma$

$\hat{\vec{J}} = \hat{J}_\rho \hat{\vec{u}}_\rho$

Then

$$\vec{\nabla}\times\vec{H} = \frac{\partial(\epsilon_0\vec{E})}{\partial t} + \hat{\vec{J}}, \quad \vec{\nabla}\cdot(\epsilon_0\vec{E}) = \rho^*$$

$$\vec{\nabla}\times\vec{E} = -\frac{\partial(\mu_0\vec{H})}{\partial t}, \quad \vec{\nabla}\cdot(\mu_0\vec{H}) = 0$$

Maxwell's vector equations, MKS units.

Now, the $\overset{s}{F}_{rs}$ and $\overset{s}{F}{}^*_{rs}$ must be changed to F_{rs} and F^*_{rs}.

define: $iF_{\rho 4} = E_\rho, \quad -iF^*_{\rho 4} = H_\rho$

$\rho = 1, 2, 3$

and, of course:

where: $F^*_{rs} = \frac{1}{2} i \epsilon_{rsmn} F_{mn}$

and $F_{rs} = -\frac{1}{2} i \epsilon_{rsmn} F^*_{mn}$

Richard P. Crandall 9-2-00

$$X_r = \frac{\overset{s}{q}}{c^2} \overset{s}{F}_{rs} \delta_s$$

gives

$$\frac{d}{dt}(m\gamma u_p) = \overset{s}{q}\left(\overset{s}{E}_p + \frac{1}{c}\epsilon_{p\mu\nu} u_\mu \overset{s}{H}_\nu\right)$$

$$\frac{d}{dt}(m\gamma c^2) = \overset{s}{q}\overset{s}{E}_p u_p$$

Now, you'd think we would alter the $\overset{s}{F}_{rs}$ to make a F_{rs}, but we can't, since the new F_{rs} would not be symmetric (skew symmetric) since the units are now un-balanced (see p. G14). Also, the relation

$$\overset{s}{F}{}^{*}_{rs} = \frac{1}{2} i \, \epsilon_{rsmn} \overset{s}{F}_{mn}$$ would not work

anymore.

Therefore, we define:

$$F_{rs} = \overset{s}{F}_{rs}, \quad \text{and} \therefore \; F^{*}_{rs} = \overset{s}{F}{}^{*}_{rs}$$

However, we can now say that:

$$J_r = \sqrt{\epsilon_0}\, \overset{s}{J}_r \quad \text{or} \quad \overset{s}{J}_r = \frac{1}{\sqrt{\epsilon_0}} J_r$$

(The 4-current.) $r = 1, 2, 3, 4$

$$\therefore \quad \frac{\partial F_{rs}}{\partial X_s} = \frac{1}{\sqrt{\epsilon_0}} J_r \; , \quad \frac{\partial F^{*}_{rs}}{\partial X_s} = 0$$

$$\therefore \quad F_{rs,s} = \frac{1}{\sqrt{\epsilon_0}} J_r \quad \text{and} \quad F_{mn,r} + F_{nr,m} + F_{rm,n} = 0$$

Richard P. Crandall 9-2-00

Richard D. Gandell
9-2-00

619

$$F_{rs} = \sqrt{\mu_0}\left(\phi_{s,r} - \phi_{r,s}\right)$$

New definition of ϕ_r.

$$\therefore \quad \overset{s}{\phi}_r = \sqrt{\mu_0}\,\phi_r$$

$$\phi_r = \frac{1}{4\pi\sqrt{\mu_0}} \iiint_{V'} \frac{\left(\frac{1}{\sqrt{\epsilon_0}}\right) J_r(x', y', z', t' - R/c)}{R}\, dV'$$

$$\phi_r = \frac{c}{4\pi} \iiint_{V'} \frac{J_r(x', y', z', t' - R/c)}{R}\, dV'$$

$$R^2 = (x_\sigma - x'_\sigma)(x_\sigma - x'_\sigma) \qquad \sigma = 1, 2, 3$$

$r = 1, 2, 3, 4 \qquad x_1 = x,\ x_2 = y,\ x_3 = z,\ x_4 = ict$

$\phi_r = \phi_r(x, y, z, t)$

$$\hat{J}_\sigma = \rho_c^* u_\sigma = c J_\sigma$$

$$J_4 = \rho_c i \gamma = i\rho_c^*$$

$$\rho_c^* = J_4 / i$$

4-current vs. ordinary 3-current. ρ_c = "proper" charge density. ρ_c^* = "relative" charge density. $\hat{J}_\sigma$ = the relative 3-current density. $\sigma = 1, 2, 3$

$$\therefore \quad J_{r,r} = 0$$ ← conservation equation.

$$x\frac{1}{\sqrt{\epsilon}} = \frac{1}{\epsilon} \qquad \sqrt{\mu\epsilon} = \frac{1}{c}$$

$$x = \frac{\sqrt{\epsilon}}{\epsilon} = \frac{1}{\epsilon c \sqrt{\mu}}$$

$$\mu_g = \frac{1}{\epsilon_g c^2}$$

$$\overset{s}{\Phi}_r = \frac{1}{4\pi\sqrt{\epsilon_0}} \iiint_{V'} \frac{J_r}{R}\, dv'$$

$$\gamma = \frac{K}{2\pi} I = \frac{K}{2\pi}\left(\frac{4\pi}{c}\Phi_r\right) = \frac{2K}{c}\Phi_r \qquad \frac{X\sqrt{\mu_0}}{\sqrt{\epsilon_0}} = \frac{1}{\epsilon_0}$$

$$= \frac{16\pi G}{c^3}\Phi_r = \frac{16\pi}{c^3}\left(\frac{1}{4\pi\epsilon_g}\right)\Phi_r = \frac{4}{c^3\epsilon_g}\Phi_r = \left(\frac{4}{c}\right)(\mu_g)\Phi_r$$

$$X = \frac{\epsilon_0\sqrt{\epsilon_0}}{\sqrt{\mu_0}} = \frac{\epsilon_0}{c\sqrt{\mu_0}\sqrt{\mu_0}} = \frac{\epsilon_0}{c\mu_0}$$

$$\overset{s}{F}_{r_s} = \vec{\nabla} \times \overset{s}{\Phi}_r$$

$$\sqrt{\mu_0} H_r = \sqrt{\mu_0}\,\vec{\nabla} \times \left(\frac{\overset{s}{\Phi}_r}{\sqrt{\mu_0}}\right)$$

$$H_r = \vec{\nabla} \times \Phi_p$$

$$\Phi_r = \frac{\overset{s}{\Phi}_r}{\sqrt{\mu_0}} = \frac{c}{4\pi}\iiint_{V'} \frac{J_r}{R}\, dv' \text{ or } \frac{4\pi}{c} = \frac{c}{4\pi} I$$

$$\epsilon_g = \frac{1}{4\pi G} \qquad G = \frac{1}{4\pi\epsilon_g}$$

$$\frac{\sqrt{\epsilon}}{\mu\epsilon} = \sqrt{\frac{\epsilon}{\mu^2\epsilon^2}} = \frac{1}{\mu\sqrt{\epsilon}}$$

$$\Phi_p = \frac{1}{4\pi}\iiint \frac{\hat{J}_p}{R}\, dv' \quad \text{VV}$$

$$\Phi_4 = \frac{c}{4\pi}\iiint \frac{\rho}{R}\, dv' =$$

$$\sqrt{\epsilon_0} E_\sigma = \left[\frac{1}{4\pi\sqrt{\mu_0}\sqrt{\epsilon_0}}\iiint \frac{\rho}{R}\, dv'\right]\sqrt{\mu_0}$$

$$E_\sigma = \frac{1}{4\pi\sqrt{\mu_0}\,\epsilon_0}\iiint \frac{\rho}{R}\, dv' \quad \text{VV}$$

G20

∴

$$\Box \phi_r = \frac{1}{\sqrt{\mu_0}}\left(-\frac{1}{\sqrt{\epsilon_0}} J_r\right)$$

$$\boxed{\Box \phi_r = -c\, J_r}$$

$$\boxed{\phi_{r,r} = 0}$$

Components of ϕ_r:

$$\phi_\sigma = \frac{1}{4\pi}\iiint_{V'} \frac{\hat{J}_\sigma(x', y', z', t' - R/c)}{R}\, dV' \quad , \sigma = 1,2,3$$

$$\phi_4 = \frac{i}{\sqrt{\mu_0 \epsilon_0}\, 4\pi}\iiint_{V'} \frac{\rho_e^*(x', y', z', t' - R/c)}{R}\, dV'$$

Components of F_{rs} =

$$\begin{bmatrix} 0 & \sqrt{\mu_0} H_3 & -\sqrt{\mu_0} H_2 & \frac{1}{i}\sqrt{\epsilon_0} E_1 \\ -\sqrt{\mu_0} H_3 & 0 & \sqrt{\mu_0} H_1 & \frac{1}{i}\sqrt{\epsilon_0} E_2 \\ \sqrt{\mu_0} H_2 & -\sqrt{\mu_0} H_1 & 0 & \frac{1}{i}\sqrt{\epsilon_0} E_3 \\ -\frac{1}{i}\sqrt{\epsilon_0} E_1 & -\frac{1}{i}\sqrt{\epsilon_0} E_2 & -\frac{1}{i}\sqrt{\epsilon_0} E_3 & 0 \end{bmatrix}$$

Richard P. Gandell
9-2-00

Components of F^*_{rs} =

G21

$$\begin{bmatrix} 0 & \sqrt{\epsilon_0}E_3 & -\sqrt{\epsilon_0}E_2 & i\sqrt{\mu_0}H_1 \\ -\sqrt{\epsilon_0}E_3 & 0 & \sqrt{\epsilon_0}E_1 & i\sqrt{\mu_0}H_2 \\ \sqrt{\epsilon_0}E_2 & -\sqrt{\epsilon_0}E_1 & 0 & i\sqrt{\mu_0}H_3 \\ -i\sqrt{\mu_0}H_1 & -i\sqrt{\mu_0}H_2 & -i\sqrt{\mu_0}H_3 & 0 \end{bmatrix}$$

Richard J. Gronbeck
1-2-00

define $\vec{A} = A_\sigma \hat{\vec{u}}_\sigma = \phi_\sigma \hat{\vec{u}}_\sigma$ $\quad \sigma = 1,2,3$

and, $\phi = \left(\frac{1}{i}\right)\sqrt{\frac{\mu_0}{\epsilon_0}}\,\phi_4 = \frac{\eta}{i}\phi_4$ $\quad$ Compare to Neff, (9.68, 9.69), VV

$$\vec{\nabla}\times\vec{A} = \begin{vmatrix} \hat{\vec{u}}_1 & \hat{\vec{u}}_2 & \hat{\vec{u}}_3 \\ \frac{\partial}{\partial x_1} & \frac{\partial}{\partial x_2} & \frac{\partial}{\partial x_3} \\ \phi_1 & \phi_2 & \phi_3 \end{vmatrix} = (\phi_{3,2}-\phi_{2,3})\hat{\vec{u}}_1 + (\phi_{1,3}-\phi_{3,1})\hat{\vec{u}}_2 + (\phi_{2,1}-\phi_{1,2})\hat{\vec{u}}_3$$

$$F_{23} = \sqrt{\mu_0}\,H_1 = \sqrt{\mu_0}(\phi_{3,2}-\phi_{2,3}) = (\vec{\nabla}\times\vec{A})\cdot\hat{\vec{u}}_1\sqrt{\mu_0}$$

$$F_{31} = \sqrt{\mu_0}\,H_2 = \sqrt{\mu_0}(\phi_{1,3}-\phi_{3,1}) = (\vec{\nabla}\times\vec{A})\cdot\hat{\vec{u}}_2\sqrt{\mu_0}$$

$$F_{12} = \sqrt{\mu_0}\,H_3 = \sqrt{\mu_0}(\phi_{2,1}-\phi_{1,2}) = (\vec{\nabla}\times\vec{A})\cdot\hat{\vec{u}}_3\sqrt{\mu_0}$$

$\therefore$ $\boxed{\vec{H} = \vec{\nabla}\times\vec{A}}$ since $\vec{H} = H_1\hat{\vec{u}}_1 + H_2\hat{\vec{u}}_2 + H_3\hat{\vec{u}}_3$

$H_1 = \vec{H}\cdot\hat{\vec{u}}_1$, $H_2 = \vec{H}\cdot\hat{\vec{u}}_2$, $H_3 = \vec{H}\cdot\hat{\vec{u}}_3$

Also, **G22**

$$F_{14} = \frac{1}{i}\sqrt{\epsilon_0}\, E_1 = \sqrt{\mu_0}\left(\phi_{4,1} - \phi_{1,4}\right) = \sqrt{\mu_0}\left(\frac{\partial \phi_4}{\partial x_1} - \frac{1}{ic}\frac{\partial A_1}{\partial t}\right)$$

$$= \sqrt{\mu_0}\left(\frac{\partial \phi_4}{\partial x_1} - \frac{\sqrt{\mu_0 \epsilon_0}}{i}\frac{\partial A_1}{\partial t}\right)$$

$$\therefore E_1 = \frac{i\sqrt{\mu_0}}{\sqrt{\epsilon_0}}\frac{\partial \phi_4}{\partial x_1} - \mu_0\frac{\partial A_1}{\partial t}$$

$$= -\left(\frac{1}{i}\right)\sqrt{\frac{\mu_0}{\epsilon_0}}\frac{\partial \phi_4}{\partial x_1} - \mu_0\frac{\partial A_1}{\partial t}$$

$$\therefore E_1 = -\frac{\partial \phi}{\partial x_1} - \mu_0\frac{\partial A_1}{\partial t}$$

$$F_{24} = \frac{1}{i}\sqrt{\epsilon_0}\, E_2 = \sqrt{\mu_0}\left(\phi_{4,2} - \phi_{2,4}\right) = \sqrt{\mu_0}\left(\frac{\partial \phi_4}{\partial x_2} - \frac{1}{ic}\frac{\partial A_2}{\partial t}\right)$$

$$\therefore E_2 = -\frac{\partial \phi}{\partial x_2} - \mu_0\frac{\partial A_2}{\partial t}$$

$$\therefore E_3 = -\frac{\partial \phi}{\partial x_3} - \mu_0\frac{\partial A_3}{\partial t}$$

$\therefore$

$E_\sigma = -\dfrac{\partial \phi}{\partial x_\sigma} - \mu_0\dfrac{\partial A_\sigma}{\partial t}$, or, $\sigma = 1,2,3$	$\vec{E} = -\vec{\nabla}\phi - \mu_0\dfrac{\partial \vec{A}}{\partial t}$

Richard P. Crandall 9-2-00

G23

Now, From G17,

$$\vec{\nabla} \times (\mu_0 \vec{H}) = \mu_0 \epsilon_0 \frac{\partial \vec{E}}{\partial t} + \mu_0 \vec{\hat{J}}$$

$$\vec{\nabla} \times (\vec{\nabla} \times (\mu_0 \vec{A})) - \mu_0 \epsilon_0 \frac{\partial}{\partial t}\left(-\vec{\nabla}\phi - \frac{\partial(\mu_0 \vec{A})}{\partial t}\right) = \mu_0 \vec{\hat{J}}$$

$$\vec{\nabla} \times (\vec{\nabla} \times (\mu_0 \vec{A})) + \mu_0 \epsilon_0 \frac{\partial^2(\mu_0 \vec{A})}{\partial t^2} + \mu_0 \epsilon_0 \vec{\nabla}\left(\frac{\partial \phi}{\partial t}\right) = \mu_0 \vec{\hat{J}}$$

Now, $\vec{\nabla} \times (\vec{\nabla} \times \vec{L}) = \vec{\nabla}(\vec{\nabla} \cdot \vec{L}) - \nabla^2 \vec{L}$

$$\therefore \nabla^2(\mu_0 \vec{A}) - \vec{\nabla}\left(\vec{\nabla} \cdot (\mu_0 \vec{A}) + \mu_0 \epsilon_0 \frac{\partial \phi}{\partial t}\right) - \mu_0 \epsilon_0 \frac{\partial^2(\mu_0 \vec{A})}{\partial t^2} = -\mu_0 \vec{\hat{J}}$$

Now, recall that $\square \phi_r = -c J_r$ (page G20)

or [redacted] $\phi_{r,ss} = -c J_r$

So $\nabla^2 \phi_r - \frac{1}{c^2} \frac{\partial^2 \phi_r}{\partial t^2} = -c J_r$

$$\therefore \nabla^2 \phi_\sigma - \mu_0 \epsilon_0 \frac{\partial^2 \phi_\sigma}{\partial t^2} = -\hat{J}_\sigma \qquad \sigma = 1,2,3$$

and

$$\nabla^2\left(\mathbf{i}\sqrt{\tfrac{\epsilon_0}{\mu_0}}\,\phi\right) - \mu_0 \epsilon_0 \frac{\partial^2}{\partial t^2}\left(\mathbf{i}\sqrt{\tfrac{\epsilon_0}{\mu_0}}\,\phi\right) = \frac{-1}{\sqrt{\mu_0 \epsilon_0}}\, \mathbf{i}\rho^*$$

Richard P. Gandell 9-2-00

$\therefore$ [G24]

$$\nabla^2\phi - \mu_0\epsilon_0\frac{\partial^2\phi}{\partial t^2} = \left(\frac{1}{i}\sqrt{\frac{\mu_0}{\epsilon_0}}\right)\left(\frac{-i}{\sqrt{\mu_0\epsilon_0}}\right)\rho^*$$

$$\boxed{\nabla^2\phi - \mu_0\epsilon_0\frac{\partial^2\phi}{\partial t^2} = -\frac{\rho^*}{\epsilon_0}}$$

and

$$\boxed{\nabla^2(\mu_0\vec{A}) - \mu_0\epsilon_0\frac{\partial^2(\mu_0\vec{A})}{\partial t^2} = -\mu_0\hat{\vec{J}}}$$

these are equivalent to $\Box\phi_r = -cJ_r$

$\therefore$, referring to previous page (G23), the circled item, we see that: (we must have) (and setting integration constants to zero)

$$\boxed{\vec{\nabla}\cdot(\mu_0\vec{A}) = -\mu_0\epsilon_0\frac{\partial\phi}{\partial t}}$$

This relation can also be proved (more easily!!) from the second equation back on page G20, namely $\phi_{r,r} = 0$, as follows:

Richard P. Crandall
9-2-00

G25

$$\frac{\partial A_1}{\partial X_1}+\frac{\partial A_2}{\partial X_2}+\frac{\partial A_3}{\partial X_3}+\frac{1}{ic}\frac{\partial}{\partial t}\left(i\sqrt{\frac{\epsilon_0}{\mu_0}}\,\phi\right)=0$$

$$\therefore \mu_0\vec{\nabla}\cdot\vec{A}+\mu_0\frac{\partial}{\partial t}\left(\sqrt{\mu_0\epsilon_0}\,\frac{\sqrt{\epsilon_0}}{\sqrt{\mu_0}}\,\phi\right)=0$$

$$\therefore \boxed{\vec{\nabla}\cdot(\mu_0\vec{A})=-\mu_0\epsilon_0\frac{\partial\phi}{\partial t}}$$

Finally, for e/m fields, we have the force equation: (from Synge:)(modified) $\lambda_r=\frac{dx_r}{ds}$

Four-Force: $$\boxed{X_r=\frac{q}{\sqrt{\epsilon_0}\,c^2}F_{rs}\lambda_s=\mu_0 q\sqrt{\epsilon_0}\,F_{rs}\lambda_s}$$

q is the charge of the affected particle.

$$\gamma=\frac{1}{\sqrt{1-u^2/c^2}}$$

$$\left(\text{since } \sqrt{\epsilon_0}\,c^2=\frac{\sqrt{\epsilon_0}}{\mu_0\epsilon_0}=\frac{1}{\mu_0\sqrt{\epsilon_0}}\right)$$

$M_p=\frac{1}{c}m\gamma u_p$

$M_4=im\gamma$

$M_r=m\frac{dx_r}{ds}$

From Synge:
The relative 3-force $$\boxed{P_\sigma=\frac{c^2X_\sigma}{\gamma}}$$

$$X_r=\frac{dM_r}{ds}=\frac{d}{ds}(m\lambda_r)=m\frac{d\lambda_r}{ds}$$

ONLY if $X_r\lambda_r=0$!!

Richard P. Gandell 9-2-00

G26

$$ds^2 = -dx_n dx_n$$

$$\delta_n \delta_n = -1$$

$$\delta_4 = i\sqrt{1 + \delta_\sigma \delta_\sigma}$$

$$ds = \sqrt{c^2 dt^2 - dx_\sigma dx_\sigma} = c\,dt\left(1 - u^2/c^2\right)^{\frac{1}{2}}$$

$$\boxed{c\,dt = \gamma\, ds}$$

$$\boxed{\begin{aligned} \delta_\sigma &= \frac{\gamma}{c} u_\sigma \\ \delta_4 &= i\gamma \end{aligned}}$$

$$\frac{d\delta_n}{ds} = \frac{d^2 x_n}{ds^2}$$

$$\frac{d\delta_\sigma}{ds} = c^{-2}\gamma \frac{d(\gamma u_\sigma)}{dt}$$

$$\frac{d\delta_4}{ds} = i c^{-1} \gamma \frac{d\gamma}{dt}$$

$$\delta_r \frac{d\delta_r}{ds} = 0$$

$$\boxed{\frac{dm}{ds} = -X_r \delta_r}$$

$$\boxed{\begin{aligned} \frac{d(m\gamma u_\sigma)}{dt} &= \frac{c^2 X_\sigma}{\gamma} \\ \frac{d(m\gamma c^2)}{dt} &= \frac{c^3 X_4}{i\gamma} \end{aligned}} \begin{aligned} &= P_\sigma \\ &= P_\sigma u_\sigma \end{aligned}$$

Richard P. Gandell 9-2-00

∴ equations of motion are:

$$\frac{d}{ds}\left(m\frac{dX_r}{ds}\right) = \mu_0 q\sqrt{\epsilon_0}\, F_{rs}\delta_s = \frac{q}{\sqrt{\epsilon_0}\,c^2}F_{rs}\delta_s$$

(see p. G20)

Since we can see that $X_r\delta_r = \mu_0 q\sqrt{\epsilon_0}\, F_{rs}\delta_s\delta_r = 0$,

∴ $$m\frac{d\delta_r}{ds} = ~~q\mu_0\sqrt{\epsilon_0}\,F_{rs}\delta_s~~ = \frac{q}{\sqrt{\epsilon_0}\,c^2}F_{rs}\delta_s$$

note that:
$\delta_0 = \sqrt{\mu_0\epsilon_0}\,\gamma u_0$
$\delta_4 = i\gamma$

~~$Y_r = \frac{m}{q}\frac{d^2X_r}{ds^2} = \mu_0\sqrt{\epsilon_0}\,F_{rs}\delta_s$~~

~~$Y_1 = \mu_0\sqrt{\epsilon_0}\,(0 + \mu_0\sqrt{\epsilon_0}\,H_3\gamma u_2 - \mu_0\sqrt{\epsilon_0}\,H_2\gamma u_3 + \sqrt{\epsilon_0}\,E_1\gamma)$~~

$$Y_r c^2 = \frac{mc^2}{q}\frac{d^2X_r}{ds^2} = \frac{1}{\sqrt{\epsilon_0}}F_{rs}\delta_s$$ definition of Y_r.

$$Y_1c^2 = \frac{1}{\sqrt{\epsilon_0}}\left(0 + \mu_0\sqrt{\epsilon_0}\,H_3\gamma u_2 - \mu_0\sqrt{\epsilon_0}\,H_2\gamma u_3 + \sqrt{\epsilon_0}\,E_1\gamma\right)$$

$$Y_2c^2 = \frac{1}{\sqrt{\epsilon_0}}\left(-\mu_0\sqrt{\epsilon_0}\,H_3\gamma u_1 + 0 + \mu_0\sqrt{\epsilon_0}\,H_1\gamma u_3 + \sqrt{\epsilon_0}\,E_2\gamma\right)$$

$$Y_3c^2 = \frac{1}{\sqrt{\epsilon_0}}\left(\mu_0\sqrt{\epsilon_0}\,H_2\gamma u_1 - \mu_0\sqrt{\epsilon_0}\,H_1\gamma u_2 + 0 + \sqrt{\epsilon_0}\,E_3\gamma\right)$$

$$Y_4c^3 = \frac{c}{\sqrt{\epsilon_0}}\left(i\sqrt{\epsilon_0}\,E_1\frac{\gamma}{c}u_1 + i\sqrt{\epsilon_0}\,E_2\frac{\gamma}{c}u_2 + i\sqrt{\epsilon_0}\,E_3\frac{\gamma}{c}u_3 + 0\right)$$

Rich P. Goodell 9-2-00

$$X_r = Y_r q$$

G28

$\therefore$

$$Y_1 c^2 = \frac{c^2 X_1}{q} = \gamma\left(E_1 + \mu_0\left(u_2 H_3 - u_3 H_2\right)\right)$$

$$Y_2 c^2 = \frac{c^2 X_2}{q} = \gamma\left(E_2 + \mu_0\left(u_3 H_1 - u_1 H_3\right)\right)$$

$$Y_3 c^2 = \frac{c^2 X_3}{q} = \gamma\left(E_3 + \mu_0\left(u_1 H_2 - u_2 H_1\right)\right)$$

$$Y_4 c^3 = \frac{c^3 X_4}{q} = \gamma\left(i\right)\left(u_1 E_1 + u_2 E_2 + u_3 E_3\right)$$

$\therefore$

$$\frac{d(m\gamma u_1)}{dt} = \frac{c^2 X_1}{\gamma} = P_1 = q\left(E_1 + \mu_0\left(u_2 H_3 - u_3 H_2\right)\right)$$

$$\frac{d(m\gamma u_2)}{dt} = \frac{c^2 X_2}{\gamma} = P_2 = q\left(E_2 + \mu_0\left(u_3 H_1 - u_1 H_3\right)\right)$$

$$\frac{d(m\gamma u_3)}{dt} = \frac{c^2 X_3}{\gamma} = P_3 = q\left(E_3 + \mu_0\left(u_1 H_2 - u_2 H_1\right)\right)$$

$$\frac{d(m\gamma c^2)}{dt} = \frac{c^3 X_4}{i\gamma} = P_\sigma u_\sigma = q E_\sigma u_\sigma$$

$\sigma = 1, 2, 3$

Rich P. Gardell 9-2-00

note that:

$$\vec{u} \times \vec{H} = \begin{vmatrix} \hat{\vec{X}}_1 & \hat{\vec{X}}_2 & \hat{\vec{X}}_3 \\ u_1 & u_2 & u_3 \\ H_1 & H_2 & H_3 \end{vmatrix}$$

$$= (u_2 H_3 - u_3 H_2)\,\hat{\vec{X}}_1 + (u_3 H_1 - u_1 H_3)\,\hat{\vec{X}}_2 + (u_1 H_2 - u_2 H_1)\,\hat{\vec{X}}_3$$

$\therefore$

$$\boxed{\vec{p} = q\vec{E} + q\vec{u} \times (\mu_0 \vec{H})}$$ Lorentz force law.

$$\boxed{\vec{p} \cdot \vec{u} = (q\vec{E}) \cdot \vec{u}}$$

This concludes all the electromagnetic formulas.

Now, on to more gravity formulas, continuing where we left off on page G8 and G9....

Richard P. Gombell 9-2-00

G30

The static gravity force law gives:

$$\frac{G m_1 m_2}{r^2}$$

∴ By analogy with the electric force law, $\left(\frac{1}{4\pi\epsilon_0}\frac{Q_1 Q_2}{r^2}\right)$ we intuitively define ϵ_g, the epsilon for gravity such that:

$$\boxed{G = \frac{1}{4\pi\epsilon_g}} \quad (\text{defines } \epsilon_g).$$

And then we define μ_g, the mu for gravity such that:

$$\boxed{c = \frac{1}{\sqrt{\mu_g \epsilon_g}}} \quad (\text{defines } \mu_g).$$

Note that $\frac{k}{2\pi}\left(\frac{4\pi}{c}\right) = \frac{2k}{c} = \frac{16\pi G}{c^3} = \frac{16\pi}{c^3}\left(\frac{1}{4\pi\epsilon_g}\right)$

$$= \frac{4}{c^3\epsilon_g} = \frac{4\mu_g}{c}$$

Now Define:

$$\Phi_{rs} = \frac{c}{4\pi}\iiint_{V'} \frac{T_{rs}^{*}(x',y',z',t'-R/c)}{R}\,dV'$$

(compare to page G19).

Richd P. Crandall 9-2-00

G31

∴ (see p. G8)

$$\gamma_{rs} = \frac{4\mu_g}{c}\Phi_{rs}$$

∴

$$\gamma_{rs} = \frac{\mu_g}{\pi}\iiint_{V'} \frac{T^*_{rs}(x', y', z', t' - R/c)}{R}\, dV'$$

$${}_{\mu_g}T^*_{rs} = {}_{\mu_g}T_{rs} - \frac{1}{2} g_{rs} T_{\mu_g}, \quad {}_{\mu_g}T = g^{rs} T_{rs\,\mu_g} \text{ (exactly).}$$

$$g_{rs} = \delta_{rs} + \gamma_{rs}$$

$$g^{rs} = \delta_{rs} - \tilde{\gamma}_{rs}$$

diagonal elements of $\tilde{\gamma}_{rs}$ equal γ_{rs}. off-diagonal elements do not, but they are small as well.

From page G8, we note that kT^*_{rs} is of order γ_{rs}, and therefore $\mu_g T^*_{rs}$ is of order γ_{rs}.

∴ ${}_{\mu_g}T$ is of order γ.

∴ we can say $T = \delta_{rs} T_{rs}$, by neglecting higher powers of γ.

∴ $T = T_{mm}$

Since T_{11}, T_{22}, T_{33} are all of the order $\frac{T_{44}}{c^2}$, we will say:

$$T = T_{44} \qquad \therefore \quad T^*_{rs} = T_{rs} - \frac{1}{2}\delta_{rs} T_{44}$$

Richard P. Crandall
9-2-00

Richd P. Crandall 9-2-00

G32

Now, to the equations of motion: (using X_m notation (not X^m) for convenience)

$$\frac{d^2 X_r}{ds^2} + \left\{ {r \atop m\,n} \right\} \frac{dX_m}{ds}\frac{dX_n}{ds} = 0$$

(per synge and einstein)

definition of Z_r, (it's similar to Y_r.)

$$\therefore \frac{d^2 X_r}{ds^2} = \frac{d\lambda_r}{ds} = \frac{X_r}{m} = Z_r = -\left\{ {r \atop m\,n} \right\} \lambda_m \lambda_n$$

Since $\lambda_k = \frac{dX_k}{ds}$,

The quantities $Z_r c^2$ should prove to be interesting just like the $Y_r c^2$ values were in the previous electromagnetic force problem.

order $= \gamma$; order γ^2, so eliminate!

$$\left\{ {r \atop m\,n} \right\} = g^{rs}[mn,s] = \delta_{rs}[mn,s] - \tilde{\gamma}_{rs}[mn,s]$$

$$= [mn,r] = \frac{1}{2}\left(\frac{\partial \gamma_{mr}}{\partial X_n} + \frac{\partial \gamma_{nr}}{\partial X_m} - \frac{\partial \gamma_{mn}}{\partial X_r}\right)$$

$$\therefore \boxed{Z_r c^2 = -\frac{c^2}{2}\left(\frac{\partial \gamma_{mr}}{\partial X_n} + \frac{\partial \gamma_{nr}}{\partial X_m} - \frac{\partial \gamma_{mn}}{\partial X_r}\right)\lambda_m \lambda_n}$$

G33

$$Z_r c^2 = -2\mu_g c \underbrace{\left(\frac{\partial \Phi_{mr}}{\partial X_n} + \frac{\partial \Phi_{nr}}{\partial X_m} - \frac{\partial \Phi_{mn}}{\partial X_r}\right)}_{\to\ \Psi_{rmn}} \delta_m \delta_n$$

order of δ_m [redacted]: $\begin{bmatrix} 1/c \\ 1/c \\ 1/c \\ 1 \end{bmatrix}$

$$Z_r c^2 = -2\mu_g c\, \Psi_{rms}\, \delta_m \delta_s$$

$$\Psi_{rms} = \frac{\partial \Phi_{rm}}{\partial X_s} + \frac{\partial \Phi_{rs}}{\partial X_m} - \frac{\partial \Phi_{sm}}{\partial X_r}$$

Richard P. Gandall 9-2-00

G34

The order of Ψ_{rms} is such that it always contains summed terms of order $1, \frac{1}{c}, \frac{1}{c^2}$, or $\frac{1}{c^3}$.

The order of δ_k is $\begin{bmatrix} 1/c \\ 1/c \\ 1/c \\ 1 \end{bmatrix}$.

The terms of $\Psi\delta\delta$ are summed terms ranging in order from 1 to $\frac{1}{c^5}$.

[We only want to retain the terms of $\Psi\delta\delta$ that are of order 1 and order $\frac{1}{c}$, and order $\frac{1}{c^2}$.]

$\therefore$ We want to ignore terms $<= \frac{1}{c^3}$.

$\therefore$ we write (since $\Psi_{rms} = \Psi_{rsm}$ and $\delta_m \delta_s = \delta_s \delta_m$):

$$Z_r c^2 = -2\mu_g c\left[2\left(\Psi_{r41}\delta_4\delta_1 + \Psi_{r42}\delta_4\delta_2 + \Psi_{r43}\delta_4\delta_3\right) + \Psi_{r44}\delta_4\delta_4\right]$$

+3 diagonal terms.

$$= -4\mu_g c\,\delta_4\left[\Psi'_{r4s}\delta_s\right] - 2\mu_g c\left[\Psi_{r11}\delta_1^2 + \Psi_{r22}\delta_2^2 + \Psi_{r33}\delta_3^2\right].$$

(diagonal terms)

$$\boxed{\Psi'_{rms} = \Psi_{rms} - \frac{1}{2}\delta_{4m}\delta_{4s}\Psi_{r44}}$$

Richard P. Crandall 4-2-00

G35

$$\therefore \quad \psi'_{r4s} = \psi_{r4s} - \frac{1}{2}\delta_{4s}\psi_{r44}$$

define: $$G'_{rs} = \psi'_{r4s} = \frac{\partial \Phi_{r4}}{\partial X_s} + \frac{\partial \Phi_{rs}}{\partial X_4} - \frac{\partial \Phi_{s4}}{\partial X_r} - \frac{1}{2}\delta_{4s}\left[\frac{\partial \Phi_{r4}}{\partial X_4} + \frac{\partial \Phi_{r4}}{\partial X_4} - \frac{\partial \Phi_{44}}{\partial X_r}\right]$$

$$\therefore \quad Z_r c^2 = -\left(4\mu_g c \lambda_4\right) G'_{rs}\lambda_s - 2\mu_g c\left[\psi_{r11}\lambda_1^2 + \psi_{r22}\lambda_2^2 + \psi_{r33}\lambda_3^2\right].$$

(Compare this to the formula on page G27!!)

There are 4 cases of r and s; (keep $G'\lambda$ order $>= \frac{1}{c^2}$)

Case $r \neq 4$, $s \neq 4$:

$$G'_{rs} = \frac{\partial \Phi_{r4}}{\partial X_s} - \frac{\partial \Phi_{s4}}{\partial X_r} - \delta_{rs}\frac{\partial \Phi_{44}}{\partial X_4}$$

Case $r \neq 4$, $s = 4$;

$$G'_{rs} = \frac{\partial \Phi_{r4}}{\partial X_4} - \frac{1}{2}\frac{\partial \Phi_{44}}{\partial X_r}$$

Richard P. Goodell 9-2-00

G36

Case $r=4,\ s\neq 4$:

$$G'_{rs} = \frac{\partial \Phi_{44}}{\partial X_s}$$

Case $r=4,\ s=4$:

$$G'_{rs} = \frac{1}{2}\frac{\partial \Phi_{44}}{\partial X_4}$$

Also, ...

Note: $\Phi_{11} = \Phi_{22} = \Phi_{33} = -\Phi_{44}$

CASE $r=1$:

$$\Psi_{r11} = \frac{\partial \Phi_{11}}{\partial X_1},\quad \Psi_{r22} = \frac{-\partial \Phi_{22}}{\partial X_1},\quad \Psi_{r33} = \frac{-\partial \Phi_{33}}{\partial X_1}$$

CASE $r=2$:

$$\Psi_{r11} = -\frac{\partial \Phi_{11}}{\partial X_2},\quad \Psi_{r22} = \frac{\partial \Phi_{22}}{\partial X_2},\quad \Psi_{r33} = \frac{-\partial \Phi_{33}}{\partial X_2}$$

CASE $r=3$:

$$\Psi_{r11} = -\frac{\partial \Phi_{11}}{\partial X_3},\quad \Psi_{r22} = \frac{-\partial \Phi_{22}}{\partial X_3},\quad \Psi_{r33} = \frac{\partial \Phi_{33}}{\partial X_3}$$

CASE $r=4$:

$$\Psi_{r11} = 0,\quad \Psi_{r22} = 0,\quad \Psi_{r33} = 0$$

Richard P. Crandall 9-2-00

G37

Actually, I missed some cross terms, which also need to be added to $Z_r c^2$,

$$Z_r c^2 \text{ "+=" } -2\mu_g c\left[2\Psi_{r12}\,\xi_1\xi_2 + 2\Psi_{r13}\,\xi_1\xi_3 + 2\Psi_{r23}\,\xi_2\xi_3\right]$$

CASE r=1:

$$\Psi_{r12} = \frac{\partial \Phi_{11}}{\partial X_2}, \quad \Psi_{r13} = \frac{\partial \Phi_{11}}{\partial X_3}, \quad \Psi_{r23} = 0$$

CASE r=2:

$$\Psi_{r12} = \frac{\partial \Phi_{22}}{\partial X_1}, \quad \Psi_{r13} = 0, \quad \Psi_{r23} = \frac{\partial \Phi_{22}}{\partial X_3}$$

CASE r=3:

$$\Psi_{r12} = 0, \quad \Psi_{r13} = \frac{\partial \Phi_{33}}{\partial X_1}, \quad \Psi_{r23} = \frac{\partial \Phi_{33}}{\partial X_2}$$

CASE r=4:

$$\Psi_{r12} = 0, \quad \Psi_{r13} = 0, \quad \Psi_{r23} = 0$$

Richard P. Crandall 9-2-00

Richard P. Gordall 9-2-00

G38

$$Z_r c^2 = -4\mu_g c\left[i\gamma\, G'_{r\sigma}\,\gamma \frac{1}{c} u_\sigma\right] +$$

order $= \frac{1}{c}$

$$+4\mu_g c\left[\gamma\, G'_{r4}\,\gamma\right] +$$

order $= \frac{1}{c}, c$

$$-2\mu_g c\left[\Psi_{r11}\gamma^2\left(\frac{1}{c^2}\right)u_1^2 + \Psi_{r22}\gamma^2\left(\frac{1}{c^2}\right)u_2^2 + \Psi_{r33}\gamma^2\left(\frac{1}{c^2}\right)u_3^2\right]$$

order $= \frac{1}{c}$

$$-4\mu_g c\left[\Psi_{r12}\gamma^2\left(\frac{1}{c^2}\right)u_1u_2 + \Psi_{r13}\gamma^2\left(\frac{1}{c^2}\right)u_1u_3 + \Psi_{r23}\gamma^2\left(\frac{1}{c^2}\right)u_2u_3\right].$$

→ order $= \frac{1}{c}$

large static mass term.

$$\therefore Z_r c^2 \sim (\mu_g)\left[\left(c + \frac{1}{c}\right)u^0 + \left(\frac{1}{c}\right)u^1 + \left(\frac{1}{c}\right)u^2\right](c)$$

$\mu_g \approx 10^{-26}$

$\left(c + \frac{1}{c}\right)$: could be zero if there is negative mass present which equals any positive mass!

(c): "forgotten" c factor. See p. G30, the Φ_{rs} integral.

∴ The weird 2nd order terms in the velocity (i.e. $u_a u_b$ terms) actually <u>dominate</u> the velocity dependency of the force. (They contribute <u>much</u> more than the linear terms in u).

However the situation can be reversed if we have negative mass exactly equalling positive mass. Since all u^2 terms depend on derivatives of Φ_{44}, and since Φ_{44} would be <u>zero</u>, the u^2 terms then vanish completely! Also, the large static mass term vanishes.

G39

It's also very possible that the u^2 terms wouldn't contribute anything in the "far-field" from a gravity wave source. The large static term would disappear also.

FOR ALL THAT FOLLOWS, I will assume that T_{44}, and thus Φ_{44} equal ZERO, due to either far-field conditions, or due to having very large mass currents, but with zero net mass (such as having negative mass currents to balance out the positive mass). (OR, maybe we happen to have a situation where all u^2 terms just cancel out.)

$$\therefore \quad Z_r c^2 = -4\mu_g c i \gamma G'_{rs} \eta_s$$

Actually, I will leave in terms such as $\dfrac{\partial \Phi_{44}}{\partial X_\sigma}$, but will leave out $\dfrac{\partial \Phi_{44}}{\partial X_4}$, just in case there is a small static field.

$\therefore$

Case $r \neq 4,\ s \neq 4$:

$$G'_{rs} = \frac{\partial \Phi_{r4}}{\partial X_s} - \frac{\partial \Phi_{s4}}{\partial X_r}$$

Case $r = 4,\ s \neq 4$:

$$G'_{rs} = \frac{\partial \Phi_{44}}{\partial X_s}$$

Case $r \neq 4,\ s = 4$:

$$G'_{rs} = \frac{\partial \Phi_{r4}}{\partial X_4} - \frac{1}{2}\frac{\partial \Phi_{44}}{\partial X_r}$$

Case $r = 4,\ s = 4$:

$$G'_{rs} = 0$$

Richard P. Crandall 9-2-00

[illegible]

G40

From Synge, after dividing by relative volume V,

$$M_r = \rho_m \frac{dx_r}{ds} = \rho_m \lambda_r$$

$$M_\sigma = \frac{1}{c}\rho_m \gamma u_\sigma = \frac{1}{c}\rho_m^* u_\sigma$$

$$M_4 = i\rho_m \gamma = i\rho_m^*$$

$$\hat{M}_\sigma = \rho_m^* u_\sigma = c M_\sigma$$

$$\rho_m^* = M_4 / i$$

$$T_{r4}^* = \left[\frac{i}{c}\rho_m^* u_1, \frac{i}{c}\rho_m^* u_2, \frac{i}{c}\rho_m^* u_3, -\frac{1}{2}\rho_m^*\right]$$

$$M_r = \left[\frac{1}{c}\rho_m^* u_1, \frac{1}{c}\rho_m^* u_2, \frac{1}{c}\rho_m^* u_3, i\rho_m^*\right]$$

Define $G''_{rs} = \frac{+1}{i} G'_{rs}$

Define $\Phi'_{r} = \frac{-1}{i}\Phi_{r4} = -\left(\frac{c}{4\pi}\right)\left(\frac{1}{i}\right)\iiint T_{r4}^*$

$$Z_r c^2 = 4\mu_g c \gamma G''_{rs} \lambda_s$$

Richard P. Crandall 9-2-00

G41

Case $r \neq 4, s \neq 4$:

$$G''_{rs} = \frac{-1}{1}\left(\frac{\partial \Phi_{s4}}{\partial X_r} - \frac{\partial \Phi_{r4}}{\partial X_s}\right),$$

Case $r = 4, s \neq 4$:

$$G''_{rs} = \frac{-1}{1}\left(-\frac{\partial \Phi_{44}}{\partial X_s}\right)$$

Case $r \neq 4, s = 4$:

$$G''_{rs} = \frac{-1}{1}\left(\frac{1}{2}\frac{\partial \Phi_{44}}{\partial X_r} - \frac{\partial \Phi_{r4}}{\partial X_4}\right),$$

Case $r = 4, s = 4$:

$$G''_{rs} = 0$$

$\therefore$

Case $r \neq 4, s \neq 4$:

$$G''_{rs} = \underbrace{\frac{\partial \Phi'_s}{\partial X_r}}_{\text{order } 1/c} - \underbrace{\frac{\partial \Phi'_r}{\partial X_s}}_{\text{order } 1/c},$$

Case $r = 4, s \neq 4$:

$$G''_{rs} = \underbrace{(2)\left(\frac{\partial \Phi'_s}{\partial X_4}\right)}_{\text{order } 1/c^2 \text{, so not a significant error by keeping this term.}} - \underbrace{\frac{\partial \Phi'_4}{\partial X_s}}_{\text{order } 1}$$

Case $r \neq 4, s = 4$:

$$G''_{rs} = \frac{1}{2}\underbrace{\frac{\partial \Phi'_4}{\partial X_r}}_{\text{order } 1} - \underbrace{\frac{\partial \Phi'_r}{\partial X_4}}_{\text{order } 1/c^2},$$

Case $r = 4, s = 4$:

$$G''_{rs} = \underbrace{\frac{\partial \Phi'_4}{\partial X_4}}_{\text{order } 1/c} - \underbrace{\frac{\partial \Phi'_4}{\partial X_4}}_{\text{order } 1/c} = 0$$

Richard P. Crandall 9-2-00

$$\therefore \Phi'_r = -\frac{c}{4\pi}\iiint\left[\frac{1}{c}\rho_m^* u_1, \frac{1}{c}\rho_m^* u_2, \frac{1}{c}\rho_m^* u_3, \frac{1}{2} i \rho_m^*\right]$$

Define $\Phi''_k = -\frac{c}{4\pi}\iiint L_k$

where, as defined: $L_k = \left[\frac{1}{c}\rho_m^* u_1, \frac{1}{c}\rho_m^* u_2, \frac{1}{c}\rho_m^* u_3, \frac{1}{4} i \rho_m^*\right]$

(remember, per Synge,: $M_r = \rho_m \lambda_r = \left[\frac{1}{c}\rho_m^* u_1, \frac{1}{c}\rho_m^* u_2, \frac{1}{c}\rho_m^* u_3, i \rho_m^*\right]$)

case $r \neq 4$, $s \neq 4$:

$$G''_{rs} = \frac{\partial \Phi''_s}{\partial X_r} - \frac{\partial \Phi''_r}{\partial X_s},$$

case $r = 4$, $s \neq 4$:

$$G''_{rs} = 2\frac{\partial \Phi''_s}{\partial X_4} - 2\frac{\partial \Phi''_4}{\partial X_s}$$

case $r \neq 4$, $s = 4$:

$$G''_{rs} = \frac{\partial \Phi''_4}{\partial X_r} - \frac{\partial \Phi''_r}{\partial X_4},$$

case $r = 4$, $s = 4$:

$$G''_{rs} = \left(\frac{\partial \Phi''_4}{\partial X_4} - \frac{\partial \Phi''_4}{\partial X_4}\right)2 = 0$$

$\therefore$ define $G'''_{rs} = \frac{\partial \Phi''_s}{\partial X_r} - \frac{\partial \Phi''_r}{\partial X_s}$

Richard P. Crandall 9-2-00

$$\therefore \boxed{G''_{rs} = G'''_{rs} + \delta_{4r}\left(G'''_{4s}\right)}$$

$$Z_r c^2 = 4\mu_g c \gamma G''_{rs} \Lambda_s$$

$$= 4\mu_g c \gamma \left(G'''_{rs} + \delta_{4r} G'''_{4s}\right)\Lambda_s$$

$$\Phi''_r = \frac{-c}{4\pi}\iiint_{V'} \frac{L_r(x', y', z', t' - R/c)}{R}\, dV'$$

$$R^2 = (x_\sigma - x'_\sigma)(x_\sigma - x'_\sigma) \quad \sigma = 1,2,3$$

$$x_1 = x,\ x_2 = y,\ x_3 = z,\ x_4 = ict$$

$r = 1,2,3,4$

$$\Phi''_r = \Phi''_r(x, y, z, t)$$

note the "–" sign, as compared to formula on page G19

$$\boxed{G'''_{rs} = \Phi''_{s,r} - \Phi''_{r,s}}$$ (compare to p. G19)

Define: $\boxed{\overline{G}_{rs} = \sqrt{\mu_g}\left(\Phi''_{s,r} - \Phi''_{r,s}\right)} = \sqrt{\mu_g}\, G'''_{rs}$

$$\therefore Z_r c^2 = \frac{4\mu_g \gamma}{\sqrt{\mu_g}\sqrt{\mu_g}\sqrt{\epsilon_g}}\left(\overline{G}_{rs} + \delta_{4r}\overline{G}_{4s}\right)\Lambda_s$$

Richard P. Gandall 9-2-00

G44

$$Z_r c^2 = \frac{4}{\sqrt{\epsilon_g}}\left(\overline{G}_{rs} + \delta_{4r}\overline{G}_{4s}\right)\left(\gamma \Lambda_s\right)$$

note the "4" and the "$\frac{1}{4}$" factors!

$$Z_r = \frac{X_r}{m}$$

$$L_r = \left[\frac{1}{c}\rho_m^* u_1, \frac{1}{c}\rho_m^* u_2, \frac{1}{c}\rho_m^* u_3, \frac{1}{4} i \rho_m^*\right]$$

Now, DEFINE:

$$\tilde{L}_r = \left[\frac{-4}{c}\rho_m^* u_1, \frac{-4}{c}\rho_m^* u_2, \frac{-4}{c}\rho_m^* u_3, -i\rho_m^*\right]$$

the MODIFIED Momentum Density 4-Vector.

(Recall the standard Momentum Density 4-vector is:

$$M_r = \rho_m \Lambda_r = \left[\frac{1}{c}\rho_m^* u_1, \frac{1}{c}\rho_m^* u_2, \frac{1}{c}\rho_m^* u_3, i\rho_m^*\right]$$
)

Also Define:

$$\Phi_r = \frac{c}{4\pi}\iiint_{V'} \frac{\tilde{L}_r(x', y', z', t' - R/c)}{R}\, dV'$$

compare to formula on page G19.

Define:

$$G_{rs} = \sqrt{\mu_g}\left(\Phi_{s,r} - \Phi_{r,s}\right)$$

Richard P. Crandall
9-2-00

G45

$$\therefore \quad \frac{Z_r c^2}{\gamma} = \boxed{\frac{X_r c^2}{m^*} = \frac{1}{\sqrt{\epsilon_g}}\left(G_{rs} + \delta_{4r} G_{4s}\right)\Lambda_s}$$

Compare to page G25.

Force law. (equation of motion)

Define $\vec{A}_g = A_{g\sigma}\hat{\vec{u}}_\sigma = \Phi_\sigma \hat{\vec{u}}_\sigma$

and $\Phi = \left(\frac{1}{i}\right)\sqrt{\frac{\mu_g}{\epsilon_g}}\,\Phi_4 = \frac{\eta_g}{i}\Phi_4$

$$\therefore \quad \vec{H}_g = \vec{\nabla} \times \vec{A}_g$$

and $\vec{E}_g = -\vec{\nabla}\Phi - \mu_g \frac{\partial \vec{A}_g}{\partial t}$

actual force on a mass m^*

$$= \vec{P} = m^* \vec{E}_g + m^* \vec{U} \times (\mu_g \vec{H}_g)$$

Maxwell's For Gravity Waves

$$\vec{\nabla} \times \vec{H}_g = \frac{\partial(\epsilon_g \vec{E}_g)}{\partial t} - 4(\rho_m^* \vec{U}), \quad \vec{\nabla}\cdot(\epsilon_g \vec{E}_g) = -\rho_m^*$$

$$\vec{\nabla} \times \vec{E}_g = -\frac{\partial(\mu_g \vec{H}_g)}{\partial t}, \quad \vec{\nabla}\cdot(\mu_g \vec{H}_g) = 0$$

Richard J. Cardell 4-2-00

H1

Flat space only:

$$\left(\overset{S}{ds}\right) = \left(\overset{W'}{d\tau}\right) = d\tau' = \sqrt{c^2dt^2 - dx^2 - dy^2 - dz^2}$$

$\overset{S}{ds}$ and $\overset{W'}{d\tau}$ are in meters.

"W" = wheeler where "c" is removed and no longer appears.

"W'" = wheeler with c present, but $c \equiv 1$.

"S" = Synge.

Flat space:

$$\left(\overset{S}{ds^2}\right) = \left(\overset{W'}{d\tau^2}\right) = c^2dt^2 - \left((dX_\sigma)(dX^\sigma)\right) \qquad \sigma = 1,2,3$$

Define $X^0 = ct$ per Wheeler ($X^4 = ict$ per Synge.)

For flat space, define η_{rs}:

$$\eta_{rs} = \underset{FLAT}{g_{rs}} = \begin{bmatrix} -1 & 0 & 0 & 0 \\ 0 & 1 & 0 & 0 \\ 0 & 0 & 1 & 0 \\ 0 & 0 & 0 & 1 \end{bmatrix}$$

η_{rs} is a constant.

$$\overset{W}{d\tau} = \sqrt{dt^2 - dx^2 - dy^2 - dz^2}$$

$$\therefore -\underset{Flat}{\overset{W'}{d\tau^2}} = -\underset{Flat}{\overset{W}{d\tau^2}} = dX_\sigma dX^\sigma + dX_0 dX^0$$
$$= dX_m dX^m \qquad m = 0,1,2,3$$
$$= \underset{FLAT}{g_{rs}} dX^r dX^s = \eta_{rs} dX^r dX^s$$

In general, (non-flat space):

$$d\tau^2 = -dX_s dX^s = -g_{rs} dX^r dX^s = \overset{W'}{d\tau^2} = d\tau'^2$$

(I forthwith drop the "W'" notation for $d\tau$.)

Richard P. Crandall 9-2-00

H2

In curved space, the "length" "squared" of any vector A_m is calculated by $A_s A^s = g_{rs}A^r A^s$.

This squared length is negative for timelike vectors and positive for spacelike vectors.

$$-d\tau'^2 = dt^2(u^2 - c^2) \qquad \text{where } u = \text{ordinary 3-velocity.}$$

(note, "W" was dropped, keeping the "'".)

$$\therefore d\tau' = \sqrt{dt^2(c^2 - u^2)} = c\,dt\left(1 - \frac{u^2}{c^2}\right)^{\frac{1}{2}}$$

define $\gamma = \dfrac{1}{\sqrt{1 - \frac{u^2}{c^2}}}$, $\therefore \overset{W}{\gamma} = \dfrac{1}{\sqrt{1 - u^2}}$

recall $\overset{S}{\gamma} = \dfrac{1}{\sqrt{1 - \frac{u^2}{c^2}}} = \gamma$

$$\therefore c\,dt = \gamma\, d\tau' \quad \text{and} \quad dt = \overset{W}{\gamma}\, d\tau$$

Richard P. Crandall 9-2-00

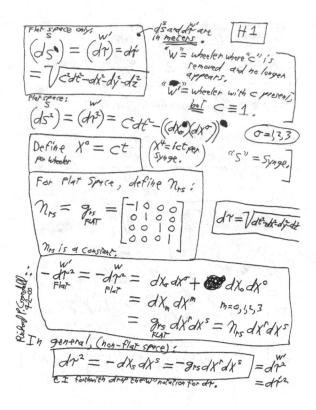

H3

DEFINITION of 4-velocity vector:
per Wheeler, p. 50, <u>modified</u> notation:

$$\boxed{\acute{\delta}^{j} = \frac{dX^{j}}{d\acute{\tau}} \quad \text{and} \quad \overset{w}{\delta}^{j} = \frac{dX^{j}}{d\tau}} \qquad j = 0,1,2,3$$

$$u^{\sigma} = \frac{dX^{\sigma}}{dt} = \text{ordinary 3-velocity.}$$

Richd P. Crandall 9-2-00

$$\acute{\delta}_{j}\acute{\delta}^{j} = \frac{dX_{j}dX^{j}}{d\acute{\tau}^{2}} = \frac{-d\acute{\tau}^{2}}{d\acute{\tau}^{2}} = -1$$

$$\therefore \boxed{\acute{\delta}_{j}\acute{\delta}^{j} = -1 \quad \text{and} \quad \overset{w}{\delta}_{j}\overset{w}{\delta}^{j} = -1} \qquad j = 0,1,2,3$$

$$g_{kj}\acute{\delta}^{k}\acute{\delta}^{j} = -1 \qquad \alpha, \beta = 1,2,3$$

$$g_{\alpha\beta}\acute{\delta}^{\alpha}\acute{\delta}^{\beta} + g_{0\beta}\acute{\delta}^{0}\acute{\delta}^{\beta} + g_{\alpha 0}\acute{\delta}^{\alpha}\acute{\delta}^{0} + g_{00}\acute{\delta}^{0}\acute{\delta}^{0}$$

$$= -1 = g_{\alpha\beta}\acute{\delta}^{\alpha}\acute{\delta}^{\beta} + \left(2 g_{0\beta}\acute{\delta}^{\beta}\right)\left(\acute{\delta}^{0}\right) + g_{00}\left(\acute{\delta}^{0}\right)^{2}$$

This <u>could</u> be solved for $\acute{\delta}^{0}$ if desired!!
However, $\acute{\delta}_{0}\acute{\delta}^{0} + \acute{\delta}_{\sigma}\acute{\delta}^{\sigma} = -1 \qquad \sigma = 1,2,3$

So, $\boxed{\acute{\delta}_{0}\acute{\delta}^{0} = (-1)\left(1 + \acute{\delta}_{\sigma}\acute{\delta}^{\sigma}\right)}$ (Dependence of $\acute{\delta}^{0}$ on $\acute{\delta}^{1}, \acute{\delta}^{2}, \acute{\delta}^{3}$.)

H4

$$\acute{\delta}^{\sigma} = \frac{dX^{\sigma}}{d\acute{\tau}} = \frac{dX^{\sigma}}{\left(\frac{c\,dt}{\acute{\gamma}}\right)} = \frac{\acute{\gamma}}{c}\frac{dX^{\sigma}}{dt} = \frac{\acute{\gamma}}{c}u^{\sigma}$$

$$\acute{\delta}^{0} = \frac{dX^{0}}{d\acute{\tau}} = \frac{\acute{\gamma}\,dX^{0}}{c\,dt} = \frac{\acute{\gamma}\,c\,dt}{c\,dt} = \acute{\gamma}$$

So,

$$\boxed{\begin{array}{lcl} \acute{\delta}^{\sigma} = \frac{\acute{\gamma}}{c}u^{\sigma} = \frac{dX^{\sigma}}{d\acute{\tau}} & \underline{\text{and}} & \overset{w}{\delta}^{\sigma} = \overset{w}{\gamma}u^{\sigma} = \frac{dX^{\sigma}}{d\tau} \\ \acute{\delta}^{0} = \acute{\gamma} = c\frac{dt}{d\acute{\tau}} & & \overset{w}{\delta}^{0} = \overset{w}{\gamma} = \frac{dt}{d\tau} \\ \sigma = 1,2,3 & & \end{array}}$$

For a particle of rest mass m, define:

relative 3-momentum as $\acute{p}^{\sigma} = \acute{m}^{*}u^{\sigma} = m\acute{\gamma}u^{\sigma}$
and $\overset{w}{p}^{\sigma} = \overset{w}{m}^{*}u^{\sigma} = m\overset{w}{\gamma}u^{\sigma}$,
(little letter p)
and
relative total energy as $\acute{e} = \acute{m}^{*}c^{2} = m\acute{\gamma}c^{2}$
and $\overset{w}{e} = \overset{w}{m}^{*} = m\overset{w}{\gamma}$

Richd P. Crandall 9-2-00

H5

For a particle, define the (rest mass = m) 4-momentum/energy vector as:

$$\acute{P}^r = m\acute{\delta}^r = m\frac{dx^r}{d\acute{\tau}} \;;\quad \overset{w}{P}^r = m\overset{w}{\delta}^r = m\frac{dx^r}{d\overset{w}{\tau}}$$

$\therefore$

$$\acute{P}^\sigma = \frac{m\acute{\gamma}}{c}u^\sigma = \frac{1}{c}\acute{m}^* u^\sigma = \frac{1}{c}\acute{p}^\sigma \;;\quad \overset{w}{P}^\sigma = m\overset{w}{\gamma}u^\sigma = \overset{w}{m}^* u^\sigma = \overset{w}{p}^\sigma$$

$$\acute{P}^0 = m\acute{\gamma} = \acute{m}^* = \frac{\acute{e}}{c^2} \;;\quad \overset{w}{P}^0 = m\overset{w}{\gamma} = \overset{w}{m}^* = \overset{w}{e}$$

For a fluid, of rest-frame density $\rho_{mrest} = \frac{\sum_k m_k}{V_0}$, where m_k is the rest mass of the k'th particle and V_0 is the volume in the rest frame we see the following: Let V = the same volume as seen from an arbitrary moving frame of reference. Then, $V_0 = V\acute{\gamma}$. Now, Define $\rho_m = \frac{\sum_k m_k}{V}$. $\therefore$ $\boxed{\rho_m = \acute{\gamma}\rho_{mrest}}$. Also, $\boxed{\rho_{mrest} = \frac{\rho_m}{\acute{\gamma}}}$

Richard P. Gondell 9-2-00

H6

$\therefore$ For the fluid,

Define the 4-momentum/energy density vector as:

$$\acute{M}^r = \rho_m\acute{\delta}^r = \rho_m\frac{dx^r}{d\acute{\tau}} \;;\quad \overset{w}{M}^r = \rho_m\overset{w}{\delta}^r = \rho_m\frac{dx^r}{d\overset{w}{\tau}}$$

$\therefore$

$$\acute{M}^\sigma = \frac{\rho_m\acute{\gamma}}{c}u^\sigma = \frac{1}{c}\acute{\rho}_m^* u^\sigma \;;\quad \overset{w}{M}^\sigma = \rho_m\overset{w}{\gamma}u^\sigma = \overset{w}{\rho}_m^* u^\sigma$$

$$\acute{M}^0 = \rho_m\acute{\gamma} = \acute{\rho}_m^* \;;\quad \overset{w}{M}^0 = \rho_m\overset{w}{\gamma} = \overset{w}{\rho}_m^*$$

Now, the Definition of the Energy Tensor of the fluid:

$$\boxed{\acute{T}^{rs} = \rho_{mrest}\,\acute{\delta}^r\acute{\delta}^s} = \frac{\rho_m}{\acute{\gamma}}\acute{\delta}^r\acute{\delta}^s = \left(\frac{\acute{\rho}_m^*}{\acute{\gamma}^2}\right)\acute{\delta}^r\acute{\delta}^s$$

$\therefore$

$$\acute{T}^{rs} = \begin{bmatrix} \acute{\rho}_m^* & (\acute{\rho}_m^*/c)u^1 & (\acute{\rho}_m^*/c)u^2 & (\acute{\rho}_m^*/c)u^3 \\ (\acute{\rho}_m^*/c)u^1 & (\acute{\rho}_m^*/c^2)u^1u^1 & (\acute{\rho}_m^*/c^2)u^1u^2 & (\acute{\rho}_m^*/c^2)u^1u^3 \\ (\acute{\rho}_m^*/c)u^2 & (\acute{\rho}_m^*/c^2)u^2u^1 & (\acute{\rho}_m^*/c^2)u^2u^2 & (\acute{\rho}_m^*/c^2)u^2u^3 \\ (\acute{\rho}_m^*/c)u^3 & (\acute{\rho}_m^*/c^2)u^3u^1 & (\acute{\rho}_m^*/c^2)u^3u^2 & (\acute{\rho}_m^*/c^2)u^3u^3 \end{bmatrix}$$

The corresponding $\overset{w}{T}^{rs}$ is obvious. (remove c everywhere)

Richard P. Gondell 9-2-00

H7

Note that $\boxed{\acute{T}^{0s} = \acute{M}^{s}}$.

$$\begin{pmatrix}\text{momentum}\\ \text{density}\end{pmatrix}'^{\sigma} = c\,\acute{T}^{0\sigma}$$

$$\begin{pmatrix}\text{energy}\\ \text{density}\end{pmatrix}' = c^2\,\acute{T}^{00}$$

---

per Wheeler, p210:

$\alpha, \beta, \gamma, \mu = 0, 1, 2, 3.$

$$[\beta\gamma, \mu]' = \Gamma'_{\mu\beta\gamma} = \frac{1}{2}\left(\frac{\partial g_{\mu\beta}}{\partial x^{\gamma}} + \frac{\partial g_{\mu\gamma}}{\partial x^{\beta}} - \frac{\partial g_{\beta\gamma}}{\partial x^{\mu}}\right)$$

$$\left\{\begin{matrix}\alpha\\ \beta\,\gamma\end{matrix}\right\}' = \acute{\Gamma}^{\alpha}_{\cdot\beta\gamma} = g^{\alpha\mu}\acute{\Gamma}_{\mu\beta\gamma} = g^{\alpha\mu}[\beta\gamma, \mu]'$$

These agree exactly (symbolically) with their "Synge" counterparts on page G1.

Don't forget these $g_{\alpha\beta}$'s are quite different from "Synge" since he uses $x_4 = ict$ (or $x^4 = ict$) and Wheeler uses $x^0 = ct$.

(Synge uses $\underset{\text{flat}}{g_{rs}} = \begin{bmatrix}1&0&0&0\\0&1&0&0\\0&0&1&0\\0&0&0&1\end{bmatrix}$.)

Richard P. Goodell 4-2-00

---

per Wheeler, p. 219:

$\alpha, \beta, \gamma, \delta, \mu = 0, 1, 2, 3$

H8

$$\acute{R}^{\alpha}_{\cdot\beta\gamma\delta} = \frac{\partial\acute{\Gamma}^{\alpha}_{\cdot\beta\delta}}{\partial x^{\gamma}} - \frac{\partial\acute{\Gamma}^{\alpha}_{\cdot\beta\gamma}}{\partial x^{\delta}} + \acute{\Gamma}^{\alpha}_{\cdot\mu\gamma}\acute{\Gamma}^{\mu}_{\cdot\beta\delta} - \acute{\Gamma}^{\alpha}_{\cdot\mu\delta}\acute{\Gamma}^{\mu}_{\cdot\beta\gamma}$$

This agrees exactly with it's "Synge" counterpart(s) on page G1. $\boxed{\acute{R}^{\alpha}_{\cdot\beta\gamma\delta} = \overset{s}{R}{}^{\alpha}_{\cdot\beta\gamma\delta}}$ (Symbolic only)

Now comes the trouble...:

per Wheeler, p. 222:

all indices range from 0, 1, 2, 3

$$\acute{R}_{\mu\nu} = \acute{R}^{\alpha}_{\cdot\mu\alpha\nu} = \frac{\partial\acute{\Gamma}^{\alpha}_{\cdot\mu\nu}}{\partial x^{\alpha}} - \frac{\partial\acute{\Gamma}^{\alpha}_{\cdot\mu\alpha}}{\partial x^{\nu}} + \acute{\Gamma}^{\alpha}_{\cdot\beta\alpha}\acute{\Gamma}^{\beta}_{\cdot\mu\nu} - \acute{\Gamma}^{\alpha}_{\cdot\beta\nu}\acute{\Gamma}^{\beta}_{\cdot\mu\alpha}$$

$$\text{and } \acute{R} = \acute{R}^{\mu}_{\cdot\mu} = g^{\epsilon\beta}\acute{R}_{\beta\mu}\Big|_{\epsilon=\mu} = \acute{R}^{\epsilon}_{\cdot\mu}\Big|_{\epsilon=\mu}$$

<u>But</u>, $\overset{s}{R}_{\mu\nu} = \overset{s}{R}{}^{\alpha}_{\cdot\mu\nu\alpha} = \overset{s}{R}{}^{\tau}_{\cdot\mu\nu\tau} = -\acute{R}_{\mu\nu}$

$$\therefore\quad \boxed{\acute{R}_{\mu\nu} = -\overset{s}{R}_{\mu\nu} \quad\text{and}\quad R' = -\overset{s}{R}}$$

(Symbolic only) — since we are comparing apples and oranges, since ′indices = 0, 1, 2, 3, while s indices = 1, 2, 3, 4, and Synge has $x_4 = ict$ where Wheeler has $x^0 = ct$.

The g's are quite different!

Richard P. Goodell 4-2-00

H9

Now, we look at the components of

$\acute{T}^{rs}$, $\acute{T}^{r}_{\cdot s}$, and $\acute{T}_{rs}$. (in *any* curved space)

$$\acute{T}^{r}_{\cdot s} = g^{cr}\acute{T}_{cs} \quad \text{and} \quad \acute{T}^{rs} = g^{Bs}\acute{T}^{r}_{\cdot B} .$$

Also, (ok, sr, by symmetry,)

$$[\,]_{dr} = g_{ds}\acute{T}^{rs} = g_{ds}\, g^{Bs}\acute{T}^{r}_{\cdot B} = \delta^{B}_{d}\acute{T}^{r}_{\cdot B} = \acute{T}^{r}_{\cdot d} = [\,]_{dr} \quad \text{reversed!}$$

$$[\,]_{es} = g_{er}\acute{T}^{r}_{\cdot s} = g_{er}\, g^{cr}\acute{T}_{cs} = \delta^{c}_{e}\acute{T}_{cs} = \acute{T}_{es} = [\,]_{es} \quad \text{not reversed!}$$

In FLAT SPACE, where $g_{rs} = \begin{bmatrix} -1 & 0 & 0 & 0 \\ 0 & 1 & 0 & 0 \\ 0 & 0 & 1 & 0 \\ 0 & 0 & 0 & 1 \end{bmatrix}$;

$$\therefore g^{rs} = \begin{bmatrix} -1 & 0 & 0 & 0 \\ 0 & 1 & 0 & 0 \\ 0 & 0 & 1 & 0 \\ 0 & 0 & 0 & 1 \end{bmatrix} = g_{rs}$$

$$\acute{T}^{rs} = \underset{\text{symmetric}}{\begin{bmatrix} + & + & + & + \\ + & + & + & + \\ + & + & + & + \\ + & + & + & + \end{bmatrix}}, \quad \acute{T}^{r}_{\cdot d} = \underset{\text{not symmetric}}{\begin{bmatrix} - & + & + & + \\ - & + & + & + \\ - & + & + & + \\ - & + & + & + \end{bmatrix}}$$

same values, diff signs

$$\therefore \acute{T}_{es} = \underset{\text{symmetric}}{\begin{bmatrix} + & - & - & - \\ - & + & + & + \\ - & + & + & + \\ - & + & + & + \end{bmatrix}}$$

Same values, diff signs

Richard J. Crandall 9-2-00

H10

For *any curved space*,

$$\acute{T}_{es} = g_{er}\acute{T}^{r}_{\cdot s}$$

$$\acute{T}^{r}_{\cdot s} = g_{sc}\acute{T}^{rc}$$

$$\boxed{\acute{T}_{es} = g_{er}\, g_{sc}\acute{T}^{rc}} \quad \text{for later use.}$$

$$\acute{T}^{rs} = g^{bs}\acute{T}^{r}_{\cdot b}$$

$$\acute{T}^{r}_{\cdot b} = g^{cr}\acute{T}_{cb} \rightarrow \boxed{= g^{sb}g^{rc}\acute{T}_{cb}}$$

$$\boxed{\acute{T}^{rs} = g^{bs}g^{cr}\acute{T}_{cb}} \quad \text{for later use.}$$

Richard J. Crandall 9-2-00

From Wheeler, p321, with $\int = \acute{\tau}$:

$$\boxed{\frac{d^2X^{\beta}}{d\acute{\tau}^2} + \acute{\Gamma}^{\beta}_{\cdot\mu\nu}\frac{dX^{\mu}}{d\acute{\tau}}\frac{dX^{\nu}}{d\acute{\tau}} = 0}$$

all indices = 0,1,2,3

This agrees exactly with (symbolically only) its "Synge" counterpart on p. G2 and G32.
(Remember the g_{mn} values are quite different as are the indices.)

$$\therefore \boxed{\frac{d^2X^{\beta}}{d\acute{\tau}^2} = \frac{d}{d\acute{\tau}}\left(\acute{\delta}^{\beta}\right) = -\acute{\Gamma}^{\beta}_{\cdot\mu\nu}\acute{\delta}^{\mu}\acute{\delta}^{\nu}}$$

equation of motion of particle

Rich P. Gandell 9-2-00

H11

Quick review of Newtonian formulae:

$$\frac{\vec{F}}{m} = \boxed{\vec{a} = \sum_k -\frac{GM_k}{R_k^2}\frac{\vec{R_k}}{R_k}} = \sum_k \vec{\nabla}\left(-\frac{GM_k}{R_k}\right) = \text{acceleration of test mass } m \text{ due to all } M_k \text{ masses.}$$

m = test mass.

where $\vec{R_k} = \vec{r} - \vec{r_k}'$. $\vec{r}$ is the position of m (the test mass).
$\vec{r_k}'$ is the position of M_k (a source mass).

The $\vec{\nabla}$ operater operates on $\vec{r}$, not $\vec{r_k}'$.

Also, $R_k = |\vec{r} - \vec{r_k}'|$.

[The above equation is proved by recalling the standard formula: $\vec{\nabla}\frac{1}{R} = -\vec{\nabla}'\frac{1}{R} = \frac{-\vec{R}}{R^3} = -\frac{1}{R^2}\left(\frac{\vec{R}}{R}\right)$.

Define ϕ, the gravitational potential as

$$\boxed{\phi = \sum_k \frac{-GM_k}{R_k}} \quad \therefore \quad \boxed{\vec{a} = -\vec{\nabla}\phi}.$$

If we consider the path of a test mass m as it moves from point A to point B, we see that:

$$\Delta\phi = [\phi]_A^B = \phi_B - \phi_A = \int_A^B d\phi = \int_A^B \vec{\nabla}\phi \cdot d\vec{r}$$

$$= -\int_A^B \vec{a} \cdot d\vec{r}.$$

$$\therefore \boxed{-m\Delta\phi = \int_A^B \vec{F} \cdot d\vec{r}} = \int_A^B \vec{F} \cdot \vec{V}\,dt = m\int_A^B \left(\left(\frac{d\vec{V}}{dt}\right) \cdot \vec{V}\right)dt$$

$$= m\int_A^B \frac{1}{2}\frac{d(\vec{V}\cdot\vec{V})}{dt}dt = \frac{1}{2}m\int_A^B d(V^2) = \int_A^B d\left(\frac{1}{2}mV^2\right)$$

H12

Now define kinetic energy as: (for a test mass m)

$$K.E. = \frac{1}{2}mV^2.$$

Define potential energy as (for a test mass m)

$$P.E. = m\phi.$$

$$\therefore \quad -m(\phi_B - \phi_A) = -m\Delta\phi = K.E._B - K.E._A = \Delta K.E.$$

$$\therefore \quad -(P.E._B - P.E._A) = -\Delta P.E. = \Delta K.E.$$

Define "work done" on m as:

$$W = \int_A^B \vec{F} \cdot d\vec{r}$$

$$\therefore \quad \boxed{W = \int \vec{F} \cdot d\vec{r} = \Delta K.E. = -\Delta P.E.}$$

$$\therefore \quad \boxed{K.E. + P.E. = \text{a constant}}$$

By Gauss's divergence theorem;

$$\iiint_{\substack{\text{total volume}\\\text{(encloses all } M_k\text{'s)}}} \vec{\nabla}\cdot(-\vec{a})\,dv = \iint_{\substack{\text{total surface}\\\text{(encloses all } M_k\text{'s)}}} \left((-\vec{a})\cdot d\vec{s}\right) = \iint_{\substack{\text{total}\\\text{surface}}} \left(\left(\sum_k -\vec{a_k}\right)\cdot d\vec{s}\right)$$

$$= \sum_k \left(\iint_{\substack{\text{total}\\\text{surface}}} \left((-\vec{a_k})\cdot d\vec{s}\right)\right)$$

Rich P. Gandell 9-2-00

H13

A quick mathematical digression is in order:

In spherical coordinates, we have the standard formula:

$$\nabla^2\alpha = \frac{1}{r^2}\frac{\partial}{\partial r}\left(r^2\frac{\partial\alpha}{\partial r}\right) + \text{terms depending on } \frac{\partial\alpha}{\partial\theta} \text{ and } \frac{\partial\alpha}{\partial\phi}.$$

Let $\alpha = \begin{cases} \frac{1}{r}, & r>\epsilon \\ \frac{1}{\epsilon}, & r\le\epsilon \end{cases}$ ϵ is a small constant.

Let S_v be a spherical volume centered on the origin, whose radius is R ($>\epsilon$).

$$\iiint_{S_v}(\nabla^2\alpha)\,dx\,dy\,dz = \int_0^R(\nabla^2\alpha)(4\pi r^2 dr)$$

$$= 4\pi\int_0^R\left[\frac{d}{dr}\left(r^2\frac{d\alpha}{dr}\right)\right]dr = 4\pi\int_{r=0}^{r=R}dU$$

where definition of U is: $U = r^2\frac{d\alpha}{dr}$.

Now $\frac{d\alpha}{dr} = \begin{cases} -\frac{1}{r^2}, & r>\epsilon \\ 0, & r\le\epsilon \end{cases}$, $\therefore$ $U = \begin{cases} -1, & r>\epsilon \\ 0, & r\le\epsilon \end{cases}$

Rich'd P. Crandall 9-2-00

H14

$$\therefore 4\pi\int_{r=0}^{r=R}du = 4\pi\int_{U=0}^{U=-1}du = -4\pi$$

Note that if $r>\epsilon$,

$$\nabla^2\alpha = \frac{1}{r^2}\left(\frac{\partial}{\partial r}\left(r^2\frac{\partial\left(\frac{1}{r}\right)}{\partial r}\right)\right) = \frac{1}{r^2}\left(\frac{\partial}{\partial r}\left(r^2\left[\frac{-1}{r^2}\right]\right)\right)$$

$$= \frac{1}{r^2}\left(\frac{\partial}{\partial r}(-1)\right) = 0$$

$\therefore$ $\nabla^2\alpha = 0$ if $r>\epsilon$

also, note that $\nabla^2\alpha = 0$ if $r<\epsilon$

$\therefore$ ($\nabla^2\alpha \to -\infty$ at $r=\epsilon$)

$\therefore$ For the spherical surface,

$$\iiint_{\text{Spherical Volume}}\nabla^2(\alpha)\,dV = -4\pi$$

But, since $\nabla^2\alpha = 0$ as long as we are outside ϵ, we can integrate over any shaped volume as long as its defining boundary surface stays outside ϵ.

$$\therefore \iiint_{\text{arbitrary Volume}}(\nabla^2\alpha)\,dV = \iiint_{\text{arb. Volume}}\nabla\cdot(\nabla\alpha)\,dV = \iint_{\text{arbitrary closed surface}}(\nabla\alpha)\cdot d\vec{s}$$

per gauss's theorem

$$= -4\pi$$

Rich'd P. Crandall 9-2-00

H15

Suppose α is shifted to a point $\vec{P} = P_x\vec{i} + P_y\vec{j} + P_z\vec{k}$.

There are 2 cases to consider:

Case A: Case B:

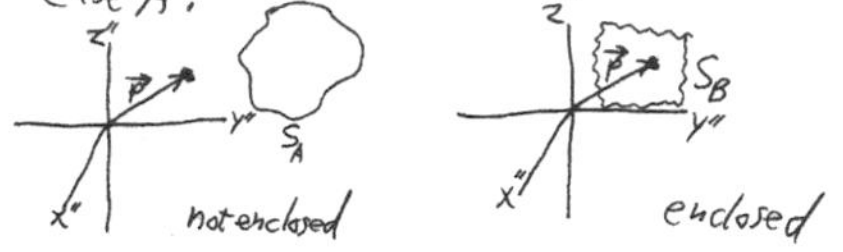

Now, shift coordinates so that α is back at the origin.:

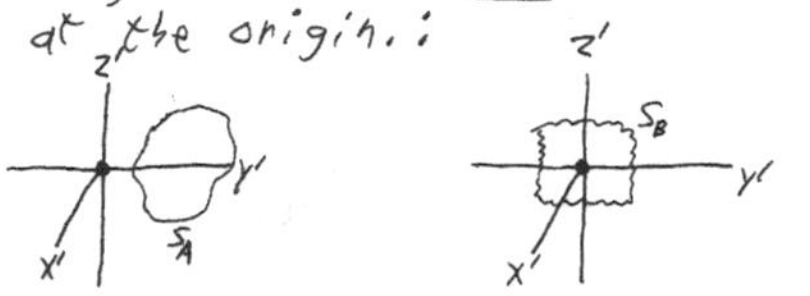

∴ The field values (α) at each point on the surfaces and in the volumes remain the same regardless of if we are using (′) or (″) coordinates. Therefore, the shift does not affect the value of the integral.

∴

	case A	Case B	
(″)	0	-4π	$= \iint_S \vec{\nabla}\left(\frac{1}{R}\right) \cdot d\vec{s}$
(′)	0	-4π	

Letting $\epsilon \to 0$, this ends the "Mathematical Digression."

Richard P. Gondell 4-2-00

H16

Now resuming from page H12...:

$$\iiint_{\substack{\text{total Volume} \\ \text{(encloses all } M_k\text{'s)}}} \vec{\nabla} \cdot (-\vec{a})\, dV = -\sum_k \left(\iint_{\substack{\text{total} \\ \text{Surface}}} \vec{a}_k \cdot d\vec{s} \right)$$

$$= -\sum_k \left(\iint_S \vec{\nabla}\left(\frac{GM_k}{R_k}\right) \cdot d\vec{s} \right)$$

$$= -\sum_k \left(GM_k \left[-4\pi \right] \right) \quad \text{(per page H15)}$$

$$= 4\pi G \sum_k M_k = 4\pi G M$$

where M = the total source mass enclosed by the surface. Even if mass exists outside of the surface, it will not contribute to the surface integral or the volume integral.

$$\therefore\ 4\pi G M_{encl} = \iiint_{\text{any Volume}} \vec{\nabla} \cdot (\vec{\nabla}\phi)\, dV \quad (\text{since } \vec{a} = -\vec{\nabla}\phi)$$

$$= \iiint_{\text{any Volume}} (\nabla^2 \phi)\, dV$$

Richard P. Gondell 4-2-00

H17

Going to the continuum limit for dense M_k's:

$$\underset{\text{any volume}}{\iiint} 4\pi G\rho \, dV = \underset{\text{same volume}}{\iiint} \nabla^2\phi \, dV$$

Since the volume is arbitrary, we get

$$\boxed{\boxed{\nabla^2\phi = 4\pi G\rho}}$$

This concludes the review of Newtonian physics.

Also,

$$\boxed{\begin{gathered}\vec{\nabla}\cdot\vec{a} = -4\pi G\rho \\ \underset{\substack{\text{any closed}\\ \text{surface}}}{\oiint} \vec{a}\cdot d\vec{s} = -4\pi G M_{enclosed}\end{gathered}}$$

$$G = \frac{1}{4\pi\epsilon_g} \qquad \boxed{\phi = \underset{\substack{\text{All}\\ \text{space}}}{\iiint} \frac{-G\rho}{|\vec{r}-\vec{r}'|} dV'}$$

$$\vec{a} = -\vec{\nabla}\phi = \underset{\substack{\text{All}\\ \text{space}}}{\iiint} \frac{-G\rho(\vec{r}-\vec{r}')}{|\vec{r}-\vec{r}'|^3} dV' \qquad \vec{F} = m\vec{a}$$

Richard P. Crandall
9-2-00

H18

Geodesic equation of motion. (Newtonian Limit)
(see page H10).

For a first order approx, let (ie, weak fields)

$$\boxed{g_{\mu\nu} = \eta_{\mu\nu} + h_{\mu\nu}} \qquad |h_{\mu\nu}| << 1.$$

Also, since $c\,dt = \gamma\, d\tau$, we have approximately $d\tau = c\,dt$. (for non-relativistic motions)

In general,

$$\frac{d^2X^\mu}{d\tau^2} = -\Gamma^{\mu}_{\cdot\alpha\beta}\frac{dX^\alpha}{d\tau}\frac{dX^\beta}{d\tau}.$$

$$\Gamma^{\mu}_{\cdot\alpha\beta} = g^{\mu\sigma}\Gamma_{\sigma\alpha\beta} = \frac{1}{2}g^{\mu\sigma}\left(g_{\sigma\alpha,\beta} + g_{\sigma\beta,\alpha} - g_{\alpha\beta,\sigma}\right)$$

$$\cancel{\left(\frac{1}{c^2}\right)}\frac{d^2X^\mu}{dt^2} = \frac{1}{2}g^{\mu\sigma}\left(g_{\alpha\beta,\sigma} - g_{\sigma\alpha,\beta} - g_{\sigma\beta,\alpha}\right)\cancel{\left(\frac{1}{c^2}\right)}\frac{dX^\alpha}{dt}\frac{dX^\beta}{dt}$$

$\mu = 1, 2, 3.$

$g^{\mu\sigma} \overset{\text{(exactly)}}{=} \eta^{\mu\sigma} - \tilde{h}^{\mu\sigma}$, where $\tilde{h}^{nn}$ (not summed) $\approx h_{nn}$ (not summed).
$\tilde{h}^{\alpha\beta}$ $(\alpha\neq\beta)$ are of order $h_{\alpha\beta}$.
Note, that to first order in h: $(1-h)(1+h) = 1-h^2 \simeq 1$.
$\therefore$ $\boxed{\tilde{h}^{\mu\sigma} \text{ is all of order } h_{\mu\sigma}}$.

Richard P. Crandall 9-2-00

Richard P. Crandall 4-2-00

[H19]

$$\frac{d^2x^\mu}{dt^2} = \left(\frac{1}{2}\left(\eta^{\mu\sigma} - h^{\mu\sigma}\right)\left(h_{\alpha\beta,\sigma} - h_{\sigma\alpha,\beta} - h_{\sigma\beta,\alpha}\right)\right)\frac{dx^\alpha}{dt}\frac{dx^\beta}{dt}$$

$\mu = 1,2,3.$ $\alpha, \beta, \sigma = 0,1,2,3.$

∴, to first order in $h_{\alpha\beta}$:

$$\frac{d^2x^\mu}{dt^2} = \left(\frac{1}{2}\eta^{\mu\sigma}\left(h_{\alpha\beta,\sigma} - h_{\sigma\alpha,\beta} - h_{\sigma\beta,\alpha}\right)\right)\frac{dx^\alpha}{dt}\frac{dx^\beta}{dt}$$

$$= O(h)\left(\sum_{\alpha,\beta}^{1\ldots3} u_\alpha u_\beta\right) + O(h)\left(2\sum_{\alpha}^{1\ldots3} u_\alpha(c)\right) + O(h)\left(c^2\right)$$

Dropping powers of c less than 2, we get:

$$a^\mu = \frac{1}{2}\eta^{\mu\sigma}\left(h_{00,\sigma} - h_{\sigma 0,0} - h_{\sigma 0,0}\right)c^2$$

$$= \frac{1}{2}c^2\left(h_{00,\mu} - h_{\mu 0,0} - h_{\mu 0,0}\right) \quad (\mu = 1,2,3)$$

→ neglect, since this equals $-2\left(\frac{1}{c}\right)\frac{\partial h_{\mu 0}}{\partial t}$

$$\therefore a^\mu = \frac{-\partial\left(\frac{-c^2}{2}h_{00}\right)}{\partial x^\mu} = \frac{-\partial\phi}{\partial x^\mu}, \text{ per Newtonian Theory.}$$

since ($\vec{a} = -\vec{\nabla}\phi$)

Newtonian Limit correspondence: [H20]

$$\therefore \phi = \frac{-c^2}{2}h_{00}$$

$$h_{00} = \frac{-2\phi}{c^2}$$

$$\boxed{g_{00} = -1 - \frac{2\phi}{c^2}}$$

$$\nabla^2\phi = 4\pi G\rho$$

$$\boxed{\frac{-c^2}{2}\nabla^2 h_{00} = 4\pi G\rho}$$

$$\frac{-2\phi}{c^2} = g_{00} + 1, \quad \phi = \frac{-c^2}{2}g_{00} - \frac{c^2}{2}$$

$$\boxed{-\frac{c^2}{2}\nabla^2 g_{00} = 4\pi G\rho}$$

Richard P. Crandall 4-2-00

Linearized theory of gravity: [H21]

(and also a way to determine the "k" constant of the general, nonlinear theory).

(The main Einstein non-linear theory equation reads: $\boxed{R'_{\mu\nu} - \frac{1}{2} g_{\mu\nu} R' = k T'_{\mu\nu}}$, per Wheeler.)

[Note this does not agree with the "Synge" equation, which has $-k$ in the equation instead of $+k$ above.]

Define $\boxed{T' = T'^{\,r}_{\bullet r}}$

$\therefore T' = g^{cr} T'_{cr} = g_{rc} T'^{\,rc}$;

Define the "—" bar operation, which acts on mixed tensors as: (δ^r_s is the kronecker delta tensor) $= g_{cs} g^{cr}$

$$\boxed{\overline{A}^r_{\bullet s} = A^r_{\bullet s} - \frac{1}{2}\delta^r_s A \quad , \text{where } A = A^b_{\bullet b}}$$

$$\therefore \overline{\overline{A}}^m_{\bullet n} = A^m_{\bullet n} - \frac{1}{2}\delta^m_n A - \frac{1}{2}\delta^m_n\left(A^b_{\bullet b} - \frac{1}{2}\delta^b_b A\right)$$
$$= A^m_{\bullet n} - \frac{1}{2}\delta^m_n A - \frac{1}{2}\delta^m_n(-A)$$

$$\therefore \boxed{\overline{\overline{A}}^m_{\bullet n} = A^m_{\bullet n}}$$

Now, as before, for a first order (weak field) approx., we can write: $\boxed{g_{\mu\nu} = \eta_{\mu\nu} + h_{\mu\nu}}$ $\quad |h_{\mu\nu}| \ll 1$.

This will constitute a Linear Theory.

Rich'd P. Goodall 9-2-00

[H22]

Define $h = h^{\beta}_{\bullet\beta} = g^{\alpha\beta} h_{\alpha\beta} = g^{\alpha\nu} h_{\alpha\beta}\Big|_{\nu=\beta} = h^{\nu}_{\bullet\beta}\Big|_{\nu=\beta}$

As before (see p H18), we have, in general:

$$\left[\; \Gamma'^{\mu}_{\bullet\alpha\beta} = g^{\mu\nu}\Gamma'_{\nu\alpha\beta} = \frac{1}{2} g^{\mu\nu}\left(g_{\nu\alpha,\beta} + g_{\nu\beta,\alpha} - g_{\alpha\beta,\nu}\right) \quad \text{exactly.} \;\right]$$

note $\Gamma'^{\mu}_{\bullet\alpha\beta} = \frac{1}{2} g^{\mu\nu}\left(h_{\nu\alpha,\beta} + h_{\nu\beta,\alpha} - h_{\alpha\beta,\nu}\right)$ exactly

$$= \frac{1}{2}\left(\eta^{\mu\nu} - \tilde{h}^{\mu\nu}\right)\left(h_{\nu\alpha,\beta} + h_{\nu\beta,\alpha} - h_{\alpha\beta,\nu}\right)$$

$$\approx \frac{1}{2}\eta^{\mu\nu}\left(h_{\nu\alpha,\beta} + h_{\nu\beta,\alpha} - h_{\alpha\beta,\nu}\right)$$

(to the first order in h).

$\therefore \boxed{\Gamma'^{\mu}_{\bullet\alpha\beta} \text{ are of order } h}$

$\therefore \boxed{\Gamma'\Gamma' \text{ is of order } h^2}$

Rich'd P. Goodall 9-2-00

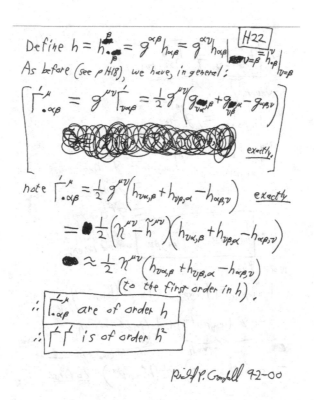

H23

[Therefore, to first order in h: (eliminate $\Gamma\Gamma$ terms)

$$R'_{\mu\nu} = \Gamma'^{\alpha}_{\cdot\mu\nu,\alpha} - \Gamma'^{\alpha}_{\cdot\mu\alpha,\nu}$$

(see p. H8)

Now, $$\Gamma'^{\mu}_{\cdot\alpha\beta} = \frac{1}{2} g^{\mu\nu}\left(h_{\nu\alpha,\beta} + h_{\nu\beta,\alpha} - h_{\alpha\beta,\nu}\right) \quad \text{exactly}$$

$$= \frac{1}{2}\left(h^{\mu}_{\cdot\alpha,\beta} + h^{\mu}_{\cdot\beta,\alpha} - h_{\alpha\beta,\cdot}{}^{\mu}\right) \quad \text{(by definition) exactly.}$$

So, to first order: (in h)

$$R'_{\mu\nu} = \frac{1}{2}\left(h^{\alpha}_{\cdot\mu,\nu\alpha} + h^{\alpha}_{\cdot\nu,\mu\alpha} - h_{\mu\nu,\cdot\alpha}{}^{\alpha}\right)$$

$$- \frac{1}{2}\left(h^{\alpha}_{\cdot\mu,\alpha\nu} + h^{\alpha}_{\cdot\alpha,\mu\nu} - h_{\mu\alpha,\cdot\nu}{}^{\alpha}\right)$$

(see p. H25 for a justification of the correctness of $(h^{\mu}_{\cdot\alpha,\beta})_{,\nu} = h^{\mu}_{\cdot\alpha,\beta\nu}$)

note that $g^{\beta\epsilon} h_{\mu\alpha,\beta\nu} = h_{\mu\alpha,\cdot\nu}{}^{\epsilon} = h_{\mu\alpha,\nu\cdot}{}^{\epsilon}$ (due to non-order of partial derivates.)

and $g^{\alpha\delta} h_{\mu\alpha,\beta\nu} = h_{\mu\cdot,\beta\nu}{}^{\delta} = h^{\delta}_{\cdot\mu,\beta\nu}$ (due to symmetry of $h_{\mu\alpha}$.)

$\therefore$ $h_{\mu\delta,\cdot\nu}{}^{\beta} = g^{\gamma\beta} h_{\mu\delta,\gamma\nu}$

Rich P. Goodall 92-00

H24

$$\therefore \boxed{h_{\mu\delta,\cdot\nu}{}^{\beta} \neq h_{\mu\cdot,\beta\nu}{}^{\delta}}$$

However,

$$h_{\mu\beta,\cdot\nu}{}^{\beta} = g^{\gamma\beta} h_{\mu\beta,\gamma\nu}$$

and,

$$h_{\mu\cdot,\beta\nu}{}^{\beta} = g^{\alpha\beta} h_{\mu\alpha,\beta\nu} = g^{sc} h_{\mu s,c\nu} = g^{cs} h_{\mu s,c\nu} = g^{\gamma\beta} h_{\mu\beta,\gamma\nu}$$

$$\therefore \boxed{h_{\mu\beta,\cdot\nu}{}^{\beta} = h_{\mu\cdot,\beta\nu}{}^{\beta}}$$

useful formula to be used later.

Note that

$$\Gamma'^{\mu}_{\cdot\alpha\beta,\nu} = \left(\Gamma'^{\mu}_{\cdot\alpha\beta}\right)_{,\nu} = \left(\frac{1}{2}\left(g^{\mu\epsilon}\left(h_{\epsilon\alpha,\beta} + \cdots\right)\right)\right)_{,\nu} \quad \text{exactly}$$

$$= \underbrace{\frac{1}{2} g^{\mu\epsilon}\left(h_{\epsilon\alpha,\beta\nu} + \cdots\right)}_{\text{order } h} + \underbrace{\frac{1}{2}\left(\left(\eta^{\mu\epsilon} - h^{\mu\epsilon}\right)_{,\nu}\right)\left(h_{\epsilon\alpha,\beta} + \cdots\right)}_{\text{order } h^2} \quad \text{exactly}$$

$$= \frac{1}{2} g^{\mu\epsilon}\left(h_{\epsilon\alpha,\beta\nu} + \cdots\right) = \frac{1}{2}\left(h^{\mu}_{\cdot\alpha,\beta\nu} + \cdots\right) \quad \text{to first order.}$$

Rich P. Goodall 92-00

Richard P. Crandall 9-2-00

H25

Done a different way, we could also say:

$$\overset{1}{\Gamma}{}^{\mu}_{\bullet\alpha\beta,\nu} = \left(\overset{1}{\Gamma}{}^{\mu}_{\bullet\alpha\beta}\right)_{,\nu} = \left(\frac{1}{2}\left(g^{\mu\epsilon}\left(h_{\epsilon\alpha,\beta}+\cdots\right)\right)\right)_{,\nu}$$ exactly

(where $g^{\mu\epsilon}(h_{\epsilon\alpha,\beta}+\cdots) = h^{\mu}_{\bullet\alpha,\beta}{}^{+\cdots}$ by definition)

$$= \left(\frac{1}{2}\left(\eta^{\mu\epsilon}-\tilde{h}^{\mu\epsilon}\right)\left(h_{\epsilon\alpha,\beta}+\cdots\right)\right)_{,\nu}$$ exactly

$$= \left(\frac{1}{2}\left(\underbrace{\eta^{\mu\epsilon}\left(h_{\epsilon\alpha,\beta}+\cdots\right)}_{\text{order } h}-\underbrace{\tilde{h}^{\mu\epsilon}\left(h_{\epsilon\alpha,\beta}+\cdots\right)}_{\text{order } h^2}\right)\right)_{,\nu}$$ exactly

$$\approx \left(\frac{1}{2}\left(\eta^{\mu\epsilon}\left(h_{\epsilon\alpha,\beta}+\cdots\right)\right)\right)_{,\nu}$$ to first order in h

[scribbled out]

$$= \frac{1}{2}\eta^{\mu\epsilon}\left(h_{\epsilon\alpha,\beta\nu}+\cdots\right)$$ "

$$= \frac{1}{2}\left(g^{\mu\epsilon}+\tilde{h}^{\mu\epsilon}\right)\left(h_{\epsilon\alpha,\beta\nu}+\cdots\right)$$ "

$$= \underbrace{\frac{1}{2}g^{\mu\epsilon}\left(h_{\epsilon\alpha,\beta\nu}+\cdots\right)}_{\text{order } h} + \underbrace{\frac{1}{2}\tilde{h}^{\mu\epsilon}\left(h_{\epsilon\alpha,\beta\nu}+\cdots\right)}_{\text{order } h^2}$$ "

$$\approx \frac{1}{2}g^{\mu\epsilon}\left(h_{\epsilon\alpha,\beta\nu}+\cdots\right) = \frac{1}{2}\left(h^{\mu}_{\bullet\alpha,\beta\nu}{}^{+\cdots}\right)$$ To first order in h.

H26

Therefore we have justified the correctness to first order in h of expressions such as

$$\boxed{\left(h^{\mu}_{\bullet\alpha,\beta}\right)_{,\nu} = h^{\mu}_{\bullet\alpha,\beta\nu}}$$

Richard P. Crandall 9-2-00

Note that, due to the above, we can say that $h^{\alpha}_{\bullet\alpha,\mu\nu} = \left(h^{\alpha}_{\bullet\alpha}\right)_{,\mu\nu} = h_{,\mu\nu}$ to first order in h.

$\therefore$

$$\boxed{2\overset{1}{R}_{\mu\nu} = -h_{\mu\nu,\alpha\bullet}{}^{\bullet\alpha} - h_{,\mu\nu} + h_{\mu\alpha,\bullet\nu}{}^{\alpha} + \underbrace{h_{\nu\alpha,\bullet\mu}{}^{\alpha}}}$$
To first order in h.

$= h^{\alpha}_{\nu\bullet,\alpha\mu}$

$= h^{\alpha}_{\bullet\nu,\mu\alpha}$

Like on p H23.

Now, we multiply both sides by $g^{\epsilon\mu}$ to get:

$$\boxed{2\overset{1}{R}{}^{\epsilon}_{\bullet\nu} = -h^{\epsilon}_{\bullet\nu,\alpha\bullet}{}^{\alpha} - h_{,\bullet\nu}^{\epsilon} + h^{\epsilon}_{\bullet\alpha,\bullet\nu}{}^{\alpha} + h_{\nu\alpha,\bullet\bullet}{}^{\alpha\epsilon}}$$ (not used)

Multiply the "main equation" on p H21 by $g^{\beta\mu}$ to get:

$$\overset{1}{R}{}^{\beta}_{\bullet m} - \frac{1}{2}g^{\beta\mu}g_{\mu m}\overset{1}{R} = k\overset{1}{T}{}^{\beta}_{\bullet m}$$ exactly (even non-linear)

$$\therefore \boxed{\overset{1}{R}{}^{\beta}_{\bullet m} - \frac{1}{2}\delta^{\beta}_{m}\overset{1}{R} = \overset{1}{\bar{R}}{}^{\beta}_{\bullet m} = k\overset{1}{T}{}^{\beta}_{\bullet m} \text{ exactly}} \quad \therefore \boxed{\overset{1}{\bar{R}}{}^{\beta}_{\bullet m} = k\overset{1}{\bar{T}}{}^{\beta}_{\bullet m}}$$

H27

Again, to first order in h (after minor manipulation):

$$2R'_{\mu\nu} = -h_{\mu\nu,\alpha}{}^{\alpha} + \left(-h_{,\mu\nu} + h^{\alpha}_{\cdot\mu,\alpha\nu} + h^{\alpha}_{\cdot\nu,\alpha\mu}\right)$$

Now, using the bar "—" operator, we see that $\bar{h}^{r}_{\cdot s} = h^{r}_{\cdot s} - \frac{1}{2}\delta^{r}_{s} h$ by definition.

$$\therefore \bar{h}^{r}_{\cdot s,\alpha} = h^{r}_{\cdot s,\alpha} - \frac{1}{2}\delta^{r}_{s} h_{,\alpha}$$

$$\therefore \bar{h}^{\alpha}_{\cdot s,\alpha} = h^{\alpha}_{\cdot s,\alpha} - \frac{1}{2}\delta^{\alpha}_{s} h_{,\alpha} \quad \left(\frac{1}{2}\delta^{\alpha}_{s} h_{,\alpha} = \frac{1}{2}h_{,s}\right)$$

Therefore:

$$\bar{h}^{\alpha}_{\cdot\mu,\alpha\nu} = h^{\alpha}_{\cdot\mu,\alpha\nu} - \frac{1}{2}\delta^{\alpha}_{\mu} h_{,\alpha\nu} = h^{\alpha}_{\cdot\mu,\alpha\nu} - \frac{1}{2}h_{,\mu\nu}$$

and

$$\bar{h}^{\alpha}_{\cdot\nu,\alpha\mu} = h^{\alpha}_{\cdot\nu,\alpha\mu} - \frac{1}{2}\delta^{\alpha}_{\nu} h_{,\alpha\mu} = h^{\alpha}_{\cdot\nu,\alpha\mu} - \frac{1}{2}h_{,\mu\nu}$$

$$\therefore \quad 2R'_{\mu\nu} = -h_{\mu\nu,\alpha}{}^{\alpha} + \bar{h}^{\alpha}_{\cdot\mu,\alpha\nu} + \bar{h}^{\alpha}_{\cdot\nu,\alpha\mu}$$

To first order.

also,

$$2R'^{\beta}_{\cdot\nu} = -h^{\beta}_{\cdot\nu,\alpha}{}^{\alpha} + \bar{h}^{\alpha\beta}_{\cdot\cdot,\alpha\nu} + \bar{h}^{\alpha}_{\cdot\nu,\alpha}{}^{\beta}$$

to first order

Richard P. Gondell 9-2-00

H28

Rewriting the main equation in its form as on p. H26:

$$R'^{s}_{\cdot m} - \frac{1}{2}\delta^{s}_{m} R' = \bar{R}'^{s}_{\cdot m} = k\bar{T}'^{s}_{\cdot m} \quad \text{exactly.}$$

Also, recall that $R' = R'^{\beta}_{\cdot\beta}$.

So we have

$$2R' = 2R'^{\beta}_{\cdot\beta} = -h^{\beta}_{\cdot\beta,\alpha}{}^{\alpha} + \bar{h}^{\alpha\beta}_{\cdot\cdot,\alpha\beta} + \bar{h}^{\alpha}_{\cdot\beta,\alpha}{}^{\beta}$$

$$= -h_{,\alpha}{}^{\alpha} + 2\bar{h}^{\alpha}_{\cdot\beta,}{}^{\beta}{}_{\cdot\alpha}$$

$\therefore$

$$2k\bar{T}^{s}_{\cdot m} = 2R'^{s}_{\cdot m} - \delta^{s}_{m} R'$$

$$= -h^{s}_{\cdot m,\alpha}{}^{\alpha} + \bar{h}^{s}_{\cdot\alpha,}{}^{\alpha}{}_{\cdot m} + \bar{h}^{\alpha}_{\cdot m,}{}^{s}{}_{\cdot\alpha} \quad (\text{see H27})$$

$$- \delta^{s}_{m}\left(-\frac{1}{2}h_{,\alpha}{}^{\alpha} + \bar{h}^{\alpha}_{\cdot\beta,}{}^{\beta}{}_{\cdot\alpha}\right)$$

$$= -h^{s}_{\cdot m,\alpha}{}^{\alpha} + \frac{1}{2}\delta^{s}_{m} h_{,\alpha}{}^{\alpha} - \delta^{s}_{m}\bar{h}^{\alpha}_{\cdot\beta,}{}^{\beta}{}_{\cdot\alpha} + \bar{h}^{s}_{\cdot\alpha,}{}^{\alpha}{}_{\cdot m} + \bar{h}^{\alpha}_{\cdot m,}{}^{s}{}_{\cdot\alpha}$$

$$= -\bar{h}^{s}_{\cdot m,\alpha}{}^{\alpha} - \delta^{s}_{m}\bar{h}^{\alpha}_{\cdot\beta,}{}^{\beta}{}_{\cdot\alpha} + \bar{h}^{s}_{\cdot\alpha,}{}^{\alpha}{}_{\cdot m} + \bar{h}^{\alpha}_{\cdot m,}{}^{s}{}_{\cdot\alpha}$$

to first order in h.

Richard P. Gondell 9-2-00

H29

By lowering s, we get:
(and renaming $s \rightarrow \mu$ and $m \rightarrow \nu$):

$$-\bar{h}_{\mu\nu,\alpha\cdot}{}^{\alpha} - [redacted]\, g_{\mu\nu}\bar{h}^{\alpha}{}_{\cdot\beta,\cdot\alpha}{}^{\beta} + \bar{h}_{\mu\alpha,\cdot\nu}{}^{\alpha} + \bar{h}^{\alpha}{}_{\cdot\nu,\mu\alpha} = 2k\acute{T}_{\mu\nu}$$

Therefore, to first order in h (or $\bar{h}$):

$$-\bar{h}_{\mu\nu,\alpha\cdot}{}^{\alpha} - \eta_{\mu\nu}\bar{h}_{\alpha\beta,\cdot\cdot}{}^{\alpha\beta} + \bar{h}_{\mu\alpha,\cdot\nu}{}^{\alpha} + \bar{h}_{\nu\alpha,\cdot\mu}{}^{\alpha} = 2k\acute{T}_{\mu\nu}$$

See Wheeler, p. 437 The linearized field equ's.

We now impose the following gauge conditions (or coordinate conditions):

$$\bar{h}^{\mu\alpha}{}_{\cdot\cdot,\alpha} = 0 = \bar{h}^{\mu}{}_{\cdot\alpha,\cdot}{}^{\alpha} = \bar{h}_{\mu\alpha,\cdot}{}^{\alpha} = \bar{h}^{\alpha}{}_{\cdot\mu,\alpha}$$

$\therefore$ $$-\bar{h}_{\mu\nu,\alpha\cdot}{}^{\alpha} = 2k\acute{T}_{\mu\nu}$$ in this gauge. (to first order)

Also, $$-\bar{h}^{s}{}_{\cdot m,\alpha\cdot}{}^{\alpha} = 2k\acute{T}^{s}{}_{\cdot m} = 2\bar{\acute{R}}^{s}{}_{\cdot m}$$

From p. 27H, we see that in this gauge, $\acute{R}^{\beta}{}_{\cdot\nu}$ is:

$$2\acute{R}^{\beta}{}_{\cdot\nu} = -h^{\beta}{}_{\cdot\nu,\alpha\cdot}{}^{\alpha}$$ Note this is h, not $\bar{h}$.

$\therefore$ $$2k\bar{\acute{T}}^{\mu}{}_{\cdot\nu} = 2\acute{R}^{\mu}{}_{\cdot\nu} = -h^{\mu}{}_{\cdot\nu,\alpha\cdot}{}^{\alpha}$$ and $$-h_{\mu\nu,\alpha\cdot}{}^{\alpha} = 2k\bar{\acute{T}}_{\mu\nu}$$

(See p. H26)

Richard P. Crandall 9-2-00

H30

It still remains to find the "k" constant in the general non-linear theory $\acute{R}_{\mu\nu} - \frac{1}{2}g_{\mu\nu}\acute{R} = k\acute{T}_{\mu\nu}$, and also to write a solution to the Linearized theory.

The linearized equation to be solved is:

$$h_{\mu\nu,\alpha\cdot}{}^{\alpha} = -2k\bar{\acute{T}}_{\mu\nu}$$

$$\bar{\acute{T}}^{\epsilon}{}_{\cdot\nu} = \acute{T}^{\epsilon}{}_{\cdot\nu} - \frac{1}{2}\delta^{\epsilon}_{\nu}\acute{T}$$

$$\therefore \bar{\acute{T}}_{\mu\nu} = \acute{T}_{\mu\nu} - \frac{1}{2}g_{\mu\nu}\acute{T}$$

$$h_{\mu\nu,\alpha\cdot}{}^{\alpha} = \left[h_{\mu\nu,\alpha\cdot}{}^{\beta}\right]_{\beta=\alpha} = \left[g^{\epsilon\beta}h_{\mu\nu,\alpha\epsilon}\right]_{\beta=\alpha}$$

keeping no terms smaller than first order in h, we get:

$$h_{\mu\nu,\alpha\cdot}{}^{\alpha} \equiv \Box h_{\mu\nu} = \eta^{\alpha\epsilon}h_{\mu\nu,\alpha\epsilon} = \frac{\partial^2 h_{\mu\nu}}{\partial x^2} + \frac{\partial^2 h_{\mu\nu}}{\partial y^2} + \frac{\partial^2 h_{\mu\nu}}{\partial z^2} - \frac{\partial^2 h_{\mu\nu}}{\partial (ct)^2}$$

$$= \nabla^2 h_{\mu\nu} - \frac{1}{c^2}\frac{\partial^2 h_{\mu\nu}}{\partial t^2} = -2k\bar{\acute{T}}_{\mu\nu}$$

Richard P. Crandall 9-2-00

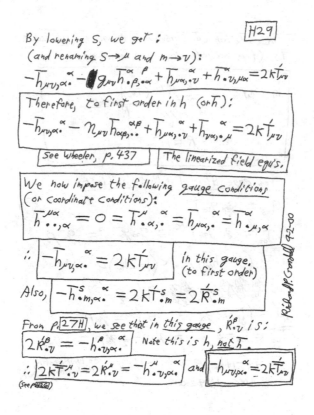

H31

The solution to this wave equation is (from electrodynamics-type formulae):

$$h_{\mu\nu}(x,y,z,t) = \frac{k}{2\pi}\iiint_{V'} \frac{(\bar{T}'_{\mu\nu})\big|_{retarded}}{R}\, dV'$$

where $R = |\vec{r} - \vec{r}'|$

$\mu, \nu = 0, 1, 2, 3.$

From page H10, For Newtonian Limit:

$$T'_{es} = g_{er}\, g_{sc}\, T'^{rc}$$

$$\therefore T'_{00} = g_{0r}\, g_{0c}\, T'^{rc} = g_{00}\, g_{00}\, T'^{00} = g_{00}^2\, \rho$$ (see p. H6)

$$= (-1 + h_{00})^2 \rho \approx (1 - 2h_{00})\rho \quad \text{to 1st order}$$

Also, $T'^{r}_{\cdot s} = g_{sc}\, T'^{rc}$

$$T' = T'^{r}_{\cdot r} = g_{rc}\, T'^{rc} = g_{00}\, T'^{00} = g_{00}\, \rho$$

$$\therefore \bar{T}'_{00} = T'_{00} - \tfrac{1}{2} g_{00}\, T' = g_{00}^2 \rho - \tfrac{1}{2} g_{00}^2 \rho = \tfrac{1}{2} g_{00}^2 \rho$$

$$= \tfrac{1}{2}(-1 + h_{00})^2 \rho \approx \tfrac{1}{2}(1 - 2h_{00})\rho$$, to first order, in the Newtonian Limit.

Richd P. Crandall 9-2-00

H32

So $\bar{T}'_{00} = \tfrac{1}{2}(1 - 2h_{00})\rho$ to 1st order in h.

$$= \tfrac{1}{2}\left(1 + \frac{4\phi}{c^2}\right)\rho$$

From page H30, assuming static conditions over time:

$$\nabla^2 h_{\mu\nu} = -2k\, \bar{T}'_{\mu\nu}$$

$$\therefore \nabla^2 h_{00} = -2k\, \bar{T}'_{00} = -k(1 - 2h_{00})\rho$$

But, from p. H20:

$$\nabla^2 h_{00} = \left(\frac{-2}{c^2}\right)(4\pi G \rho)$$

$$= \frac{-8\pi G \rho}{c^2} = -k(1 - 2h_{00})\rho$$

$$\therefore k = \left(\frac{1}{(1 - 2h_{00})}\right)\frac{8\pi G}{c^2} = \left(\frac{1}{(1 + \frac{4\phi}{c^2})}\right)\frac{8\pi G}{c^2}$$

Since k is a constant, we have, in the limit:

$$k = \frac{8\pi G}{c^2}$$

$$\therefore R'_{\mu\nu} - \tfrac{1}{2} g_{\mu\nu} R' = \frac{8\pi G}{c^2}\, \bar{T}'_{\mu\nu}$$

(The General theory).

Richard P. Crandall 9-2-00

H33

$R'_{bm} = K\bar{\bar{T}}'_{bm}$ per p. H26.

$\bar{\bar{T}}'_{\mu\nu} = T'_{\mu\nu} - \frac{1}{2} g_{\mu\nu} T'$ per p. H30.

$$R'_{\mu\nu} = \frac{\partial \Gamma'^{\alpha}_{\cdot\mu\nu}}{\partial X^{\alpha}} - \frac{\partial \Gamma'^{\alpha}_{\cdot\mu\alpha}}{\partial x^{\nu}} + \Gamma'^{\alpha}_{\cdot\beta\alpha}\Gamma'^{\beta}_{\cdot\mu\nu} - \Gamma'^{\alpha}_{\cdot\beta\nu}\Gamma'^{\beta}_{\cdot\mu\alpha} \quad (48)$$

From p. H29,

$\bar{h}^{\alpha}_{\cdot\mu,\alpha} = 0$ (gauge condition)

$\therefore \bar{h}^{\alpha}_{\cdot s,\beta}\Big|_{\beta=\alpha} = \left(h^{\alpha}_{\cdot s,\beta} - \frac{1}{2}\delta^{\alpha}_{s} h_{,\beta}\right)\Big|_{\beta=\alpha} = 0$ (from p. H27)

~~$g = g^{\beta}_{\cdot\beta} = \eta^{\beta}_{\cdot\beta} + h^{\beta}_{\cdot\beta}$~~

~~$\therefore \bar{h}_{ss,\beta} = h_{ss,\beta} - \frac{1}{2} g_{\alpha s}\delta^{\alpha}_{s} h_{,\beta}$~~ ~~$g^{\beta}_{\cdot\beta} = \delta^{\beta}_{\beta} = 4$, $h_{\alpha} = g_{\alpha}$~~

~~$\therefore g^{\alpha s}\bar{h}_{ss,\beta}\Big|_{\beta=\alpha} = \left(g^{\alpha s} h_{ss,\beta} - \frac{1}{2} g^{\alpha s} g_{\alpha s} h_{,\beta}\right)\Big|_{\beta=\alpha} = 0$~~

~~$\therefore g^{\alpha s}\left(g_{ss,\alpha} - \frac{1}{2}\delta_{\alpha s} g_{,\alpha}\right) = 0$~~ ($g=4$) ($=0$)

~~$\therefore g^{\alpha s}(g_{ss,\alpha}) = 0$ (gauge condition)~~

Rich P. Gardell
4-2-00

H34A

$$\bar{h}^{\alpha\beta}{}^{\alpha}_{\cdot\beta,\alpha'}{}^{\beta'}_{\cdot} = h^{\alpha}_{\cdot\beta,\alpha'}{}^{\beta'}_{\cdot} - \frac{1}{2}\delta^{\alpha}_{\beta}\left(h_{,\alpha'}{}^{\beta'}_{\cdot}\right)$$

(ok) ~~$\bar{h}^{\alpha'\beta'}_{,\alpha'}{}^{\beta'} = h_{,\alpha'} - \frac{1}{2}\delta^{\beta'}_{\alpha'}(h_{,z}{}^{z})$~~

~~$\bar{h}^{\alpha'\beta'}_{,\beta'}{}^{\alpha'} = h_{,\beta'}{}^{\alpha'} - \frac{1}{2}\delta^{\alpha'}_{\beta'}\left(h_{,z}{}^{z}\right)$~~

$$\bar{h}^{\alpha\beta}{}^{\alpha}_{\cdot\beta,\alpha'}{}^{\beta'}_{\cdot} = h^{\alpha}_{\cdot\beta,\alpha'}{}^{\beta'}_{\cdot} - \frac{1}{2}\delta^{\alpha}_{\beta}\left(\bar{h}^{\alpha'\beta'}_{,\alpha'}{}^{\beta'}_{\cdot} + \frac{1}{2}\delta^{\beta'}_{\alpha'}\left(h_{,z}{}^{z}\right)\right)$$

$$\bar{h}_{\mu\nu} = h_{\mu\nu} - \frac{1}{2} g_{\mu\nu} h$$

$$\bar{h}_{\mu\nu,}{}^{\mu\nu}_{\cdot\cdot} = h_{\mu\nu,}{}^{\mu\nu}_{\cdot\cdot} - \frac{1}{2} g_{\mu\nu} h_{,}{}^{\mu\nu}_{\cdot\cdot}$$

$$\overline{h_{,\mu\nu}} = h_{,\mu\nu} - \frac{1}{2} g_{\mu\nu}\left(\eta^{rs} h_{,rs}\right)$$

$$h_{\mu\nu,}{}^{rs}_{\cdot\cdot} = \bar{h}_{\mu\nu,}{}^{rs}_{\cdot\cdot} - \frac{1}{2} g_{\mu\nu}\left(\eta^{ab}\bar{h}_{ab,}{}^{rs}_{\cdot\cdot}\right)$$

$$h_{\mu\nu,}{}^{\mu s} = \bar{h}_{\mu\nu,}{}^{\mu s}_{\cdot\cdot} - \frac{1}{2}\bar{h}_{,\nu}{}^{s}_{\cdot}$$

$$g_{sk} h_{\mu\nu,}{}^{\mu s} = g_{sk}\bar{h}_{\mu\nu,}{}^{\mu s}_{\cdot\cdot} - \frac{1}{2}\bar{h}_{,\nu k}$$

Rich P. Gardell
4-2-00

$\frac{1}{2}\mu H^2$ $\iint \vec{J}\cdot d\vec{s} = \oint \vec{H}\cdot d\vec{l}$ [H34B]

$\nabla \times \vec{H} = \frac{\partial(\epsilon \vec{E})}{\partial t} + \vec{J}$

$\iint \nabla \times \vec{H}\, d\vec{s} = \iint \vec{J}\, ds$

$\oint \vec{H}\cdot d\vec{l} = \iint \vec{J}\cdot d\vec{s}$

~~$W = \int I E\, dt = \int I L \frac{dI}{dt} dt$~~

~~$= \frac{1}{2} L I^2 = \frac{1}{2}$~~

Solenoid:

$N I_0 = H L$

$\text{energy} = (LS)\frac{1}{2}\mu\left(\frac{N I_0}{L}\right)^2$

$= \frac{1}{2}\mu S \frac{N^2 I_0^2}{L}$

$= \frac{1}{2}\mu \frac{m^2}{SL}$

$$\frac{\Delta E}{Vol} = \frac{1}{2}\mu (H+H_0)^2 - \frac{1}{2}\mu (H-H_0)^2$$

$$= \frac{1}{2}\mu\left[H^2 + 2HH_0 + H_0^2 - H^2 + 2HH_0 - H_0^2\right]$$

$$= \frac{1}{2}\mu\left[4HH_0\right] = 2\mu H H_0 = \frac{2NI_0}{L}\mu H_0$$

$$(\Delta E)[\text{scribbled out}] = 2(NI_0)(S)\mu H_0 = 2m\mu H_0 = 2mB_0 \quad \therefore (m=H)$$

$$\frac{\Delta E}{Vol} = 2\mu H H_0 = 2m\mu_0 H_0 \quad \therefore \frac{\Delta E}{SL} = 2m\mu H_0$$

Richard P. Crandall 9-2-00

[H35]

$$\Gamma'^{a}_{\cdot ba}\Gamma'^{b}_{\cdot rs} \cong \left[\frac{1}{2}\eta^{a\sigma}\left(g_{\sigma b,a} + g_{\sigma a,b} - g_{ba,\sigma}\right)\right]\left[\frac{1}{2}\eta^{b\gamma}\left(g_{\gamma r,s} + g_{\gamma s,r} - g_{rs,\gamma}\right)\right]$$

$$\Gamma'^{a}_{\cdot ba}\Gamma'^{b}_{\cdot 00} \approx \frac{1}{4}\eta^{a\sigma}\eta^{b\gamma}\left(g_{\sigma b,a} + g_{\sigma a,b} - g_{ba,\sigma}\right)\cdot \left(2\underbrace{g_{\gamma 0,0}}_{\text{since }0}{}^{=0} - g_{00,\gamma}{}^{=0}\right) \doteq 0$$

$$\Gamma'^{a}_{\cdot bs}\Gamma'^{b}_{\cdot ra} \approx \left[\frac{1}{2}\eta^{a\sigma}\left(g_{\sigma b,s} + g_{\sigma s,b} - g_{bs,\sigma}\right)\right]\left[\frac{1}{2}\eta^{b\gamma}\left(g_{\gamma r,a} + g_{\gamma a,r} - g_{ra,\gamma}\right)\right]$$

$$\Gamma'^{a}_{\cdot b0}\Gamma'^{b}_{\cdot 0a} \approx \frac{1}{4}\eta^{a\sigma}\eta^{b\gamma}\left(g_{\sigma b,0}{}^{=0} + g_{\sigma 0,b} - g_{b0,\sigma}\right)\cdot\left(g_{\gamma 0,a} + g_{\gamma a,0}{}^{=0} - g_{0a,\gamma}\right)$$

Richard P. Crandall 9-2-00

note if $\sigma = 0$ or $\gamma = 0$ we get zero, [H36]
since $g_{00,b} = 0$ and $g_{b0,0} = 0$
and $g_{00,a} = 0$, and $g_{0a,0} = 0$,

$\therefore$ we can replace $\eta^{a\sigma}$ with $\delta^{a\sigma}$ and $\eta^{b\gamma}$ with $\delta^{b\gamma}$.

$$\therefore \Gamma^a_{\cdot b0} \Gamma^b_{\cdot 0a} \approx \frac{1}{4}\left(g_{a0,b} - g_{b0,a}\right)\left(g_{b0,a} - g_{0a,b}\right)$$

$$\therefore \left[-\Gamma^\alpha_{\cdot \beta\nu} \Gamma^\beta_{\cdot \mu\alpha}\right]_{\substack{\mu=0\\ \nu=0}} \approx \frac{1}{4}\left(g_{a0,b} - g_{b0,a}\right)\left(g_{a0,b} - g_{b0,a}\right)$$

Richd P. Crandall 9-2-00

[H37]

2:45 wrench

$$L = \mu_0 a\left(\ln \frac{8a}{b} - 2\right)$$

$$E = m_e c^2 = \frac{1}{2} L I^2 = \frac{1}{2}\mu_0 a J^2 \pi^2 b^4 \left(\ln \frac{8a}{b} - 2\right)$$

$$I = J\left(\pi b^2\right)$$

$$E = \frac{\frac{1}{2}\left(I^2 a^4\right)}{a^4} \mu_0 a \left(\ln \frac{8a}{b} - 2\right)$$

$$B = \frac{\mu_0 (Is)}{4\pi r^3} \qquad Is = I\pi a^2$$

$$(Is)^2 = \pi^2\left(I^2 a^4\right)$$

$$m_e c^2 = \frac{\frac{1}{2}\pi^2\left(I^2 a^4\right)\mu_0}{\pi^2 a^3}\left(\ln \frac{8a}{b} - 2\right)$$

Richd P. Crandall 9-2-00

H38

$$\pi^2 (I^2 a^4) = \left(\frac{4\pi r^3 B}{\mu_0}\right)^2$$

$$m_e c^2 = \frac{\frac{1}{2}\mu_0}{\pi^2 a^3}\left(\frac{4\pi r^3 B}{\mu_0}\right)^2 \ln\left(\frac{8a}{b} - 2\right)$$

$$= \frac{\mu_0 16 \pi^2 r^6 B^2}{2\pi^2 a^3 \mu_0^2} \ln(\;)$$

$$= \frac{8 r^6 B^2}{a^3 \mu_0} \ln(\;)$$

$$B^2 = \frac{\mu_g a^3 m_e c^2}{8 r^6 \ln\left(\frac{8a}{b} - 2\right)}$$

let $b = a/4$, let $r = 1$ meter

$$B_{max}^2 = \frac{\mu_g a_{max}^3 m_e c^2}{8 \ln(30)}$$

$\mu_g = 9 \times 10^{-27}$

$a_{max} = 1 \times 10^{-15}$

$m_e = 9 \times 10^{-31}$

$c^2 = 9 \times 10^{16}$

Richard P. Crandall 9-2-00 ; $B_{max} = 8 \times 10^{-43}$

H39

for a uranium nucleus:

$a_{max} = 5 \times 10^{-14}$ m

$m_n = 4 \times 10^{-25}$ kg

$B_{max} = 2 \times 10^{-37} \times 10^{+30}$ correction

$\approx 2 \times 10^{-7}$

$$B = \frac{\mu_0 I}{2\pi h}, \quad I = \frac{2\pi h B}{\mu_0}$$

$$F = ILB = \frac{2\pi h L B^2}{\mu_0}$$

$$= 4 \times 10^{10} \;!!$$

Richard P. Crandall 9-2-00

Rich P. Cordell 4-2-00 HX1

$$\int \vec{F}\cdot d\vec{l} = -2\int_0^{\pi/2} I L B_o\, \vec{a}_z \cdot (\vec{a}_\theta)\left(\frac{L}{2}\right) d\theta$$

$$\vec{a}_\theta = \vec{a}_x \cos\theta\cos\phi + \vec{a}_y \cos\theta\sin\phi - \vec{a}_z \sin\theta$$

note $\phi = -\frac{\pi}{2}$

$$\vec{a}_\theta = -\vec{a}_y \cos\theta - \vec{a}_z \sin\theta$$

$$\int \vec{F}\cdot d\vec{l} = -IL^2B_o \int_0^{\pi/2} \vec{a}_z\cdot\vec{a}_\theta\, d\theta$$

$$= -IL^2B_o \int_0^{\pi/2} (-\sin\theta)\, d\theta$$

$$= -IL^2B_o \left[\cos\theta\right]_0^{\pi/2} = -IL^2B_o(0-1)$$

$\vec{m} = I\vec{S}$

$$= IL^2B_o = ISB_o$$

$$= |\vec{m}||\vec{B}|$$

$\therefore \Delta U = -|\vec{m}||\vec{B}|$

$\therefore |U| = |\vec{m}||\vec{B}|$ or $|U| = \vec{m}\times\vec{B} = \vec{T}$

HX2

$$\int \vec{F}\cdot d\vec{l} = \int \vec{F}\cdot\vec{v}\, dt$$

$$= \int\left(m\frac{d\vec{v}}{dt}\right)\cdot\vec{v}\, dt = m\int \frac{1}{2}\frac{d(\vec{v}\cdot\vec{v})}{dt}\, dt$$

$$= \int \frac{1}{2}m\, d(v^2) = \int d\left(\frac{1}{2}mv^2\right) = \frac{1}{2}mv^2 = \Delta KE = \Delta U$$

Torque on a Square Loop L:

$$\vec{F}_u = ILB_o\vec{a}_z$$

$$\vec{F}_l = -ILB_o\vec{a}_z$$

$$\int \vec{F}_u\cdot d\vec{l} = \int_{\theta=\frac{\pi}{2}}^{\theta=0} ILB_o\vec{a}_z\cdot(-\hat{a}_\theta)\left(+\frac{L}{2}\right)(-d\theta)$$

$$\int \vec{F}_l\cdot d\vec{l} = \int_{\theta=\pi/2}^{\pi} -ILB_o\vec{a}_z\cdot(\hat{a}_\theta)\left(+\frac{L}{2}\right)(d\theta)$$

Rich P. Cordell 4-2-00

HX3

$$\vec{r} \times q\vec{v} = rqv$$

$$I = \frac{1}{T}(q) = \frac{\text{rate}}{\text{dist}}(q)$$
$$= \frac{v}{2\pi r}(q)$$

$\therefore$ $qv = 2\pi r I$

$$\therefore \vec{r} \times q\vec{v} = r 2\pi r I$$
$$= 2(\pi r^2) I$$
$$= 2 S I = 2|\vec{m}|$$

$$\therefore \frac{-q}{2mc}(\vec{r} \times \vec{p}) \cdot \vec{B} = \frac{-1}{2c}(\vec{r} \times q\vec{v}) \cdot \vec{B}$$
$$= \frac{-1}{2c}(2|\vec{m}|B_0) = -\frac{1}{c}|\vec{m}||\vec{B}| = \frac{-U}{c}$$

Richard P. Crandall 4-2-00

HX4

$$|\vec{m}| = |I\vec{S}| = \left|\frac{1}{2}\left(\frac{q}{m}\right)(\vec{r} \times m\vec{v})\right|$$
$$= \left|\frac{q}{2mc}(\vec{r} \times \vec{p})\right|$$
$$= \left|\frac{q}{2mc}\vec{L}\right|$$

halve the chg.

OR

double the mass

to make $\vec{m}_{spin}$ correct

$$|m| = \left|\frac{1}{2}\vec{r} \times q(r\vec{\omega})\right| = \left|\frac{1}{2}r^2 q\omega\right|$$

$\therefore$ change chg. r to $\frac{1}{\sqrt{2}}r$.
(If ω's are same)

Richard P. Crandall 4-2-00

HX5

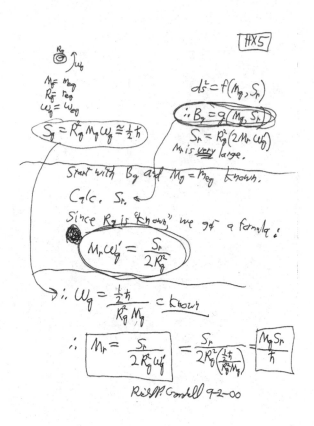

$M_g = m_{eg}$
$R_g = r_{eg}$
$\omega_g = \omega_{eg}$

$S_g = R_g^2 M_g \omega_g \cong \frac{1}{2}\hbar$

$ds^2 = f(M_g, S_r)$

$\therefore B_g = g(M_g, S_r)$

$S_r = R_g^2 (2 M_r \omega_g)$

M_r is very large.

Start with B_g and $M_g = m_{eg}$ known.

Calc. S_r

Since R_g is "known," we get a formula:

$$M_r \omega_g' = \frac{S_r}{2R_g^2}$$

$$\therefore \omega_g = \frac{\frac{1}{2}\hbar}{R_g^2 M_g} \leftarrow \text{known}$$

$$\therefore M_r = \frac{S_r}{2R_g^2\omega_g'} = \frac{S_r}{2R_g^2\left(\frac{\frac{1}{2}\hbar}{R_g^2 M_g}\right)} = \frac{M_g S_r}{\hbar}$$

Rich P. Gardell 9-2-00

Printed in the United States
By Bookmasters